# HAZARDOUS AND INDUSTRIAL WASTES

## HOW TO ORDER THIS BOOK

# HAZARDOUS AND INDUSTRIAL WASTES

**Proceedings of the
Twenty-Seventh Mid-Atlantic
Industrial Waste Conference**

EDITED BY:
## Arup K. Sengupta
LEHIGH UNIVERSITY

TECHNOMIC
PUBLISHING CO., INC.

LANCASTER · BASEL

# Hazardous and Industrial Wastes

a **TECHNOMIC** publication

*Published in the Western Hemisphere by*
Technomic Publishing Company, Inc.
851 New Holland Avenue, Box 3535
Lancaster, Pennsylvania 17604 U.S.A.

*Distributed in the Rest of the World by*
Technomic Publishing AG
Missionsstrasse 44
CH-4055 Basel, Switzerland

Printed in the United States of America
10 9 8 7 6 5 4 3 2 1

Main entry under title:
   Proceedings of the Twenty-Seventh Mid-Atlantic Industrial Waste Conference—
   Hazardous and Industrial Wastes

A Technomic Publishing Company book
Bibliography: p.
Includes index p. 929

ISSN No. 1044-0631
ISBN No. 1-56676-357-6

HAZARDOUS AND
INDUSTRIAL WASTES

Proceedings of the Twenty-Seventh
Mid-Atlantic Industrial Waste Conference

July 9–12, 1995, Lehigh University, Bethlehem, PA

The Mid-Atlantic Industrial Waste Conference
is organized by:

Bucknell University
University of Cincinnati
University of Delaware
Drexel University
Howard University
The Johns Hopkins University
Lehigh University
University of Maryland
The Pennsylvania State University
University of Pittsburgh
Syracuse University
State University of New York at Buffalo
Villanova University
Virginia Polytechnic Institute and State University
West Virginia University

and sponsored by:

Advanced Environmental Recycling Corporation
Air Products and Chemicals, Inc. (Corporate)
American Institute of Chemical Engineers
   (AIChE Lehigh Valley Chapter)
American Society of Civil Engineers
   (ASCE Lehigh Valley Chapter)
Base Engineering, Inc.
Bethlehem Steel Corporation
Dames and Moore, Inc.
Envirogen, Inc.

Environmental Resources Management (ERM), Inc.
Horsehead Resource Development Co. (Corporate)
Iacocca Institute, Lehigh University
Langan Engr. & Envir., Inc.
New Jersey Department of Environmental Protection
Newhart Food Products, Inc.
Pennsylvania Department of Environmental Resources
Pennsylvania Water Environment Association
Technomic Publishing Co., Inc.
W.L. Gore and Associates, Inc.
Water Environment Federation
Wright Environmental, Inc.

# CONTENTS

*Preface* . . . . . . . . . . . . . . . . . . . . . . . . . . . . . . . . . . . . . . . . . . . . . . . . xvii

*Acknowledgements* . . . . . . . . . . . . . . . . . . . . . . . . . . . . . . . . . . . . . . . ix

*Program Committee* . . . . . . . . . . . . . . . . . . . . . . . . . . . . . . . . . . . . xxi

CATEGORY I:
*Waste Minimization and Reuse*

**An Integrated Approach to Waste Minimization/
Pollution Prevention** . . . . . . . . . . . . . . . . . . . . . . . . . . . . . . . . . . . . 3
   P. J. Usinowicz

**Utilizing Paper Mill Sludges as Landfill Cover
Material: A Means of Waste Minimization and Reuse** . . . . . . . . 11
   H. K. Moo-Young, Jr. and T. F. Zimmie

**Environmental Degradation of Air Products' Vinex,
Airflex and Airvol Polymers through Composting** . . . . . . . . . . . 23
   M. A. Larson and N. M. O'Brien

**Selective Recovery of Alum from Clarifier Sludge
Using Composite Ion-Exchange Membranes** . . . . . . . . . . . . . . . . 33
   P. Li and A. K. Sengupta

**Proler Syngas Process for Gasification of
Waste Organic Materials** . . . . . . . . . . . . . . . . . . . . . . . . . . . . . . . 43
   N. Bishop

**Waste Management Practices at an Integrated
Steel Mill** . . . . . . . . . . . . . . . . . . . . . . . . . . . . . . . . . . . . . . . . . . . . . 53
   J. D. Lynn, J. R. Koch and R. D. Lane

**Waste Minimization in Electropolishing:
Process Control** . . . . . . . . . . . . . . . . . . . . . . . . . . . . . . . . . . . . . . . 62
   A. P. Davis, C. Bernstein and P. M. Gietka

**Application of Pyrolized Carbon Black from
Scrap Tires in Asphalt Pavement Design
and Construction** . . . . . . . . . . . . . . . . . . . . . . . . . . . . . . . . . . . . . . 72
   T. Park, B. J. Coree and C. W. Lovell

CATEGORY II:
*Technology, Regulations and Ethics*

Technology Transfer: How It Can Best Be Accomplished ......85
J. H. Saling

Environmental Considerations of Construction Projects.......91
H. S. Laird and P. R. Jacobson

Regulatory and Legal Issues in Industrial and
Hazardous Waste Management ...........................100
A. A. Siess, Jr.

Evolution of Environmental Responsibility
in Civil Engineering.....................................109
C. W. Lovell

Risk Assessment and Communication–
Tools for Site Closure ...................................118
M. Myrick

Remedial Action Suitability for the Cornhusker
Army Ammunition Plant Site ...........................129
S. Nonavinakere and P. Rappa III

A Model for the Permitting of Reactive Chemical
Emergency Projects: Closing the Gap between
Regulation, Safety and Common Sense ...................135
M. F. Richter and R. Hartnett

Comparative Analysis of Risk-Based Cleanup Levels
and Associated Remediation Costs Using Linearized
Multistage Model (Cancer Slope Factor) vs.
Threshold Approach (Reference Dose) for
Three Chlorinated Alkenes ............................144
L. J. Lawton and J. P. Mihalich


CATEGORY III:
*Air Pollution: Modeling and Control*

Simulation of VOC Emissions from Treatment of
an Industrial Wastewater in a Biological
Treatment System ......................................155
R. K. Agarwal and M. A. Larson

Quantifying and Modeling Vapor Phase Diffusion of
Hazardous Substances in Surface Soils–
Impact on Air Emission ...............................165
G. P. Partridge, A. Dasgupta and J. P. Kline

Evaluation of the Effect of Mobile Source
Emissions on Ozone Formation in Memphis, Tennessee ......175
N. Bandyopadhyay, W. T. Davis and T. L. Miller

CATEGORY IV:
*Biotreatment and Bioremediation*

**Anaerobic Biodegradation of Nitroglycerin
by Digester Sludge** ........................................187
   *C. Christodoulatos, N. Pal and S. Bhaumik*

**Bioremediation: A Systematic, Tiered Approach** ............194
   *J. V. Forsyth III, R. Bleam and N. Wrubel*

**Investigation of the Biotransformation of TNT
by a *Rhodococcus sp.*** .....................................203
   *J. P. Tharakan, G. Welsh and J. H. Johnson, Jr.*

**Biological Reductive Dechlorination of
Chlorinated Ethylenes: Implications for
Natural Attenuation and Biostimulation** ....................213
   *T. D. DiStefano*

**Biotreatment of Chromium (VI) Effluents** .................223
   *T. Tavares, C. Martins and P. Neto*

**Desorption and Biodegradation Behavior of
Naphthalene Sorbed to Soil Colloids** .......................233
   *A. Bandyopadhyay and K. G. Robinson*

**A Ceramic Membrane Bioreactor for Aerobic
Treatment of Very High COD and BOD Dairy Effluents** ......241
   *K. L. Smith, J. A. Scott and J. A. Howell*

**Activated Carbons as Biofilm Supports for
Decontamination of Streams Polluted with
Heavy Metals and Organic Residues** .......................248
   *J. A. Scott and A. M. Karanjkar*

**Biologically Catalyzed Reduction of Nitrous Oxide
to Nitrogen Gas Using Waste Water Treatment Systems** ......255
   *D. Endy*

**Removal of Volatile Organic Compounds
(VOCs) Using Biofilters** ...................................264
   *P. E. Carriere, S. D. Mohaghegh
   and B. S. Madabhushi*

**Design of Activated Sludge Settling Tanks for
Operating with a Bulking Sludge** ..........................274
   *Y. Kim and W. O. Pipes*

**Enhanced Degradation of Dinitrotoluene by
*Phanerochaete Chrysosporium* in Packed-Bed Bioreactor** .....284
   *N. Pal, C. Christodoulatos and S. Kodali*

**Enhanced Biodegradation of Creosote
Contaminated Soil Using a Nonionic Surfactant** ............294
   *P. P. E. Carriere and F. A. Mesania*

A Sequential Checklist for the Assessment of
Natural Attenuation of Dissolved Petroleum
Contaminant Plumes from Leaking
Underground Storage Tanks ............................ 304
    *N. De Rose*

Performance Optimization of Biological Waste
Treatment by Flotation Clarification at a
Chemical Manufacturing Facility ......................... 314
    *B. J. Kerecz and D. R. Miller*

Treatment of a Dilute Waste Oil Emulsion from
Aluminum Rolling Mill Operations by a
Biological Aerated Filter ............................... 322
    *P. E. Carriere, B. E. Reed and J. Brooks*


CATEGORY V:
*Advanced Oxidation Process*

Update on Wet Oxidation Technologies for
Wastewater and Sludge Management ..................... 333
    *D. L. Bowers, F. J. Romano and J. E. Sawicki*

Study on the Photocatalytic Degradation of
Monochlorophenol Pollutants by Titanium Dioxide
in Aqueous Solution .................................... 342
    *K.-H. Wang, K.-S. Huang and Y.-H. Hsieh*

Investigation of $H_2O_2$ Catalysis with Iron Oxide
for Removal of Synthetic Organic Compounds .............. 352
    *S.-S. Lin and M. D. Gurol*

$TiO_2$-Mediated Photodegradation of 3-Chlorophenol
in the Presence of Hydrogen Peroxide .................... 361
    *C. Dong, C.-W. Chen and S.-S. Wu*

Oxidation of Chlorobenzene by Ozone and
Heterogeneous Catalytic Ozonation ...................... 371
    *N. N. Bhat and M. D. Gurol*

Investigation and Study on the Photocatalytic
Degradation of Polychlorophenols in Aqueous Solution ...... 383
    *Y.-H. Hsieh and T.-S. Chia*


CATEGORY VI:
*Physical-Chemical Treatment*

Sludge Thickening, Dewatering and Drying:
The Removal of Water by Mechanical
and Thermal Processes ................................. 395
    *P. A. Vesilind*

Research Laboratory Wastewater Treatment Process . . . . . . . . 405
W. A. Jancuk and J. R. Fisher

Treatment of a Dilute Waste Oil Emulsion by
Chemical Addition (CA)-Dissolved Air Flotation (DAF) . . . . . . . 414
B. E. Reed, P. Carriere, X. Zhu and T. Lorkowski

Treatment of Aluminum Manufacturer Coolant
Waste Using Pilot-Scale Ultrafiltration . . . . . . . . . . . . . . . . . . 424
B. E. Reed, P. Carriere and C. Dunn

Mechanisms That Govern the Successful
Application of Sparging Technologies . . . . . . . . . . . . . . . . . . . . 433
S. H. Abrams, M. Marley and E. X. Droste

Construction of Low Permeability Soil-Bentonite
Barrier Caps and Liners for Landfills . . . . . . . . . . . . . . . . . . . . 443
T. Webber and M. Williams

The Use of Soil Washing Processes for the
Reclamation and Reuse of Foundry Waste Sands . . . . . . . . . . . 450
W. M. Kocher

Leachability of Organic Compounds from Sodium
Silicate Grouts Containing Organic Reagents . . . . . . . . . . . . . . 460
J. M. Malone, M. A. Barlaz and R. H. Borden

Remediation and Reuse of Chromium Contaminated
Soils through Cold Top Ex-Situ Vitrification . . . . . . . . . . . . . . 470
J. Meegoda, B. Librizzi, G. F. McKenna,
    W. Kamolpornwijit, D. Cohen, D. A. Vaccari,
    S. Ezeldin, L. Walden, B. A. Noval,
    R. T. Mueller and S. Santora

Removal of Cyanides by Complexation with
Ferrous Compounds . . . . . . . . . . . . . . . . . . . . . . . . . . . . . . . . . 480
C. P. Varuntanya and W. Zabban

Hydrolysis of Fluoroborate in an
Electroplating Wastewater . . . . . . . . . . . . . . . . . . . . . . . . . . . . 490
D. M. Watkins and P. J. Usinowicz

CATEGORY VII:
Adsorption and Separation Processes

A New Dimension in Precoat Technology . . . . . . . . . . . . . . . . . 503
J. R. Smith

Novel Carbon and Ceramic Rigid Microfiltration
and Ultrafiltration Membranes . . . . . . . . . . . . . . . . . . . . . . . . . 511
J. Jablonsky

Vapor Phase Adsorption-Desorption of
1,1,1-Trichloroethane on Dry Soil . . . . . . . . . . . . . . . . . . . . . . . . 518
  *N. H. Dural and C.-H. Chen*

Competitive Adsorption of Chlorinated Aliphatic
Hydrocarbons from Aqueous Mixtures onto Soil . . . . . . . . . . . 528
  *N. H. Dural and D. Peng*

Neural Networks: Can They Predict the Performance
of Adsorption Columns? . . . . . . . . . . . . . . . . . . . . . . . . . . . . . . . 538
  *Y. M. Najjar and I. A. Basheer*

CATEGORY VIII:
*Heavy Metals Removal*

Remediation of Metal/Organic Contaminated Soils
by Combined Acid Extraction and Surfactant Washing . . . . . . . 551
  *J. E. Van Benschoten, M. E. Ryan, C. Huang,*
   *C. Huang, T. C. Healy and P. J. Brandl*

Solid-Phase Heavy-Metal Separation under
Unfavorable Background Conditions by
Composite Membranes . . . . . . . . . . . . . . . . . . . . . . . . . . . . . . . . 561
  *S. Sengupta and A. K. Sengupta*

The Removal of Hexavalent Chromium from Water
by Ferrous Sulfate . . . . . . . . . . . . . . . . . . . . . . . . . . . . . . . . . . . 568
  *C.-J. J. Lin and P. A. Vesilind*

Selective Removal of Metal-Ligand Complexes
using Synthetic Sorbents . . . . . . . . . . . . . . . . . . . . . . . . . . . . . . 578
  *A. Kney and A. Sengupta*

Fly Ash Enhanced Metal Removal Process . . . . . . . . . . . . . . . . 588
  *S. Nonavinakere and B. E. Reed*

Adsorption of Zinc on Magnetite Pellets . . . . . . . . . . . . . . . . . . 600
  *D. A. Cargnel and C. A. Cole*

The Use of Residual Materials for the Adsorption
of Lead in Waste Disposal Facility Subgrades . . . . . . . . . . . . . 610
  *T. M. Laumakis, J. P. Martin,*
   *S. Pamukcu and S.-C. J. Cheng*

Treatment of Organic-Heavy Metal Wastewaters
Using Granular Activated Carbon (GAC) Columns . . . . . . . . . . 622
  *B. E. Reed, M. Jamil, P. Carriere and B. Thomas*

*REDOX* Manipulation to Enhance Chelant
Extraction of Heavy Metals from Contaminated Soils . . . . . . . . 632
  *R. W. Peters, W. Li, G. Miller, M. D. Brewster,*
   *T. L. Patton and L. E. Martino*

Pollution Potential of Murgul Copper Plant Solid Wastes . . . . . 644
R. Gül, A. Yartaşi, A. Ekmekyapar and A. Sert

CATEGORY IX:
Monitoring and Assessment

AOX in Sewer Slime-Identification of Industrial
Wastewater Discharges into Public Sewers . . . . . . . . . . . . . . . . 655
E. Antusch, C. Ripp and H. H. Hahn
Field Assessment Screening Team (FAST)
Technology Process and Economics . . . . . . . . . . . . . . . . . . . . . . 664
M. D. Nickelson and D. D. Long
Program Development to Identify and Characterize
Potential Emergency Situations at a Petroleum
Refinery and Determination of Industrial Hygiene
Emergency Responses . . . . . . . . . . . . . . . . . . . . . . . . . . . . . . . 669
J. J. Oransky, S. N. Delp, E. A. Deppen
and D. Barrett

CATEGORY X:
Groundwater: Modeling and Treatment

Review of Limitations of Pump-and-Treat Simulation
Models for Groundwater Remediation . . . . . . . . . . . . . . . . . . . . 681
G. P. Lennon
Confining Units as Barriers to Regional Ground-Water
Contamination: Hydrogeologic Maps as Planning Tools . . . . . . 689
A. A. Pucci, Jr.
Investigation and Monitoring of a Petroleum
Hydrocarbon Plume in the Blue Ridge Physiographic
Province of Southwestern Virginia . . . . . . . . . . . . . . . . . . . . . . . 700
A. C. Risner
Enhanced Slurry Walls as Treatment Zones for
Inorganic Contaminants . . . . . . . . . . . . . . . . . . . . . . . . . . . . . . 712
J. C. Evans, T. L. Adams and K. A. Dudiak
Use of Field Screening to Delineate a Low-Level
Groundwater Plume of Ethylene Dibromide . . . . . . . . . . . . . . . 722
M. J. Gunderson, N. M. Breton,
E. L. Pesce and R. W. Hammons

CATEGORY XI:
Soil Treatment and Soil Characterization

Ultrasound Enhanced Soil Washing . . . . . . . . . . . . . . . . . . . . . . 733
J. Meegoda, W. Ho, M. Bhattacharjee, C. F. Wei,
M. Cohen, R. S. Magee and R. M. Frederick

**Effect of Thermal Gradient on Soil-Water System** ...........743
M. Pervizpour and S. Pamukcu

**Comparison of Extraction Methodologies for
Desorption of Pyrene** .....................................754
D. Raghavan and J. H. Johnson, Jr.

**Volatilization and Biodegradation of Hazardous Waste
Utilizing Soil Agitation in a Covered Treatment Unit** ........763
J. W. Eplin, S. A. Recker and M. K. Myrick

**Operation of a Soil Vapor Extraction and Air
Sparging System at a Former Gasoline Service Station**.......773
G. J. Gromicko, R. C. Klingensmith
and D. K. Simpson

**The Use of Digital Signal Processing and Neural
Networks for Characterization of Composites of
Residual Materials** ......................................782
S. Romero and S. Pamukcu


CATEGORY XII:
*Electrokinetic Processes*

**Electrokinetic Decontamination of Millpond Sludge** .........795
L. I. Khan and M. Rahman

**Electrokinetic (EK) Remediation of a Fine
Sandy Loam: The Effect of Voltage and
Reservoir Conditioning** .................................804
J. Ramsey and B. Reed

**Transport of Hexavalent Chromium in Porous Media:
Effect of an Applied Electrical Field** .....................814
K. R. McIntosh and C. P. Huang

**Utilization of Solubilizing and Stabilizing
Agents in Electrokinetic Processing of Soils** ...............824
A. Weeks and S. Pamukcu

***In-Situ* Removal of Phenols from Contaminated Soil
by Electro-Osmosis Process**............................835
L. R. Takiyama and C. P. Huang


CATEGORY XIII:
In-Situ *Treatment*

***In-Situ* Bioremediation of Chlorinated Solvents** .............849
W. A. Sack, P. E. Carriere, C. S. Whiteman,
M. P. Davis, S. Raman, J. E. Cuddeback
and A. K. Shiemke

Industrial Soil Vapor Extraction Enhanced by
Hot Air Injection . . . . . . . . . . . . . . . . . . . . . . . . . . . . . . . . . . . 859
    W. H. MacNair, Jr. and K. V. Littlefield

The Design of an *In-Situ* Sparging Trench . . . . . . . . . . . . . . . . 869
    M. C. Marley, S. H. Abrams and E. X. Droste

*In-Situ* Vacuum Extraction/Bioventing of a
Hazardous Waste Landfill . . . . . . . . . . . . . . . . . . . . . . . . . . . . 880
    D. M. Heuckeroth, M. F. Eberle
      and M. J. Rykaczewski

CATEGORY XIV:
*Case Studies and Economic Analysis*

Case Study: Beneficial Use of a Mineral Waste . . . . . . . . . . . . 893
    J. C. Carlton and T. W. Christopher

A Unique Approach to Environmental Evaluation
and Cleanup Activities under ECRA . . . . . . . . . . . . . . . . . . . . . 900
    H. J. Archer and J. Case

Application of a Passive Soil Vapor Survey at
a Former Manufactured Gas Plant . . . . . . . . . . . . . . . . . . . . . . 909
    M. J. Wrigley

Investigation and Remediation of a Chlorobenzene
Spill—A Case Study . . . . . . . . . . . . . . . . . . . . . . . . . . . . . . . . . 918
    E. J. Donovan, Jr. and L. A. Sparrow

*Index* . . . . . . . . . . . . . . . . . . . . . . . . . . . . . . . . . . . . . . . . . . . . 929

*Indices for 22nd, 23rd, 24th, 25th and 26th*
*Industrial Waste Conferences* . . . . . . . . . . . . . . . . . . . . . . . . . 935

# PREFACE

The Mid-Atlantic Industrial and Hazardous Waste Conference is an annual gathering of consultants, industrial environmental managers, regulators and academicians. The purpose is to exchange ideas and information and also to initiate healthy debates on environmental issues facing the industries and the society as a whole. The central theme of the 27th Conference is "Enhancing Industrial Growth and Protecting Our Environment: A Partnership between Industries, Academia and the Government." Approximately ninety technical papers in areas as diverse as Waste Minimization and Reuse, Regulations, and Ethical Issues, Biological Treatment, Soil Treatment and Characterization, Heavy Metals Removal, etc., are to be presented during the three days of this conference. In addition, the conference will be highlighted by technical exhibition and students' posters presentation.

This book is a compilation of papers presented by conference participants. These papers are organized into categories consistent with the sessions of the three-day conference.

# ACKNOWLEDGEMENTS

The actual duration of the conference is truly less than three days but its preparation started almost a year ago. Needless to say, organizing such an event demands the help, time and effort of many individuals, and the support of private and professional organizations. The editor wishes to thank members of the organizing committee for their hard work and the sponsoring organizations for their support. The editor is also thankful to C. P. Huang of the University of Delaware and Mike LaGrega of Bucknell University for their encouragement and moral support from time to time.

Of all the volunteers and committee members, there remain a few who were always prepared to go the extra mile for smooth completion of various activities at different stages of the conference. Allen O'Dell, Bill MacNair, Cheryl Hendricks and Diana Walsh deserve special recognition in this regard and the editor is immensely thankful for their unselfish efforts during the last six months.

Last but not least, the department of Civil and Environmental Engineering of Lehigh University is the host of this conference. Sincere thanks are due to Le-Wu Lu, department chair, for allocating secretarial and administrative help, whenever necessary, and enthusiastic support in hosting this conference.

ARUP K. SENGUPTA

# ORGANIZING COMMITTEE & STEERING COMMITTEE

## Organizing Committee

**Arup K. Sengupta,** Professor, Chairman, *Lehigh University*

Gerard P. Lennon, Vice-Chair, *Lehigh University*
Jim J. Jablonsky, *J&M Associates*
Irwin J. Kugelman, *Lehigh University*
William H. MacNair, *Air Products and Chemicals, Inc.*
Julie K. O'Brien, *Air Products and Chemicals, Inc.*
Allen R. O'Dell, *O'Dell Engineering Co.*
Louis Pacchioli, *AT&T Microelectronics*
Sibel A. Pamukcu, *Lehigh University*
Alfred A. Siess, Jr., Principal, *CE Research Group*
Paul J. Usinowicz, *ERM, Inc.*

Cheryl Hendricks, Conference Coordinator
Diana Walsh, Conference Coordinator

## Steering Committee

Herbert E. Allen, *University of Delaware*
William P. Ball, *The Johns Hopkins University*
Paul L. Bishop, *University of Cincinnati*
Gregory D. Boardman, *Virginia Polytechnic Institute and State University*
Judith Carberry, *University of Delaware*
G. Lee Christensen, *Villanova University*
Charles A. Cole, *The Pennsylvania State University at Harrisburg*
Anthony G. Collins, *Clarkson University*

Allen P. Davis, *University of Maryland*
Brian A. Dempsey, *The Pennsylvania State University*
Steven K. Dentel, *University of Delaware*
Jeff Evans, *Bucknell University*
Mirat D. Gurol, *Drexel University*
Oliver J. Hao, *University of Maryland*
Chin-Pao Huang, *University of Delaware*
James H. Johnson, Jr., *Howard University*
Michael D. LaGrega, *Bucknell University*
Raymond D. Letterman, *Syracuse University*
David A. Long, *The Pennsylvania State University*
Joe Martin, *Drexel University*
Ronald D. Neufeld, *University of Pittsburgh*
Brian E. Reed, *West Virginia University*
Raymond W. Regan, Sr., *The Pennsylvania State University*
William A. Sack, *West Virginia University*
P. Aarne Vesilind, *Duke University*
Radisav D. Vidic, *University of Pittsburgh*

# CATEGORY I:
## *Waste Minimization and Reuse*

# AN INTEGRATED APPROACH TO WASTE MINIMIZATION/ POLLUTION PREVENTION

PAUL J. USINOWICZ, PH.D., P.E.
Environmental Resources Management, Inc.
855 Springdale Drive
Exton, Pennsylvania 19341

## INTRODUCTION

Pollution prevention and waste minimization can be the most successful and cost-effect approaches to waste management. Businesses report reduced materials cost, increased profits, improved environmental awareness among employees, and enhanced public image. Every organization has different requirements for pollution prevention based on this perception of the impact of regulations, costs, future liabilities, public image, and competitive position. A "continuous environment improvement" approach for pollution prevention is a preferred approach because it allows the rational and incremental evaluations, assessments, and implementation steps for preventing or reducing pollution. Elimination or reduction of pollutants at the source is preferable. When this is not possible, the focus shifts to the recycling, reuse, or reclamation of wastes, or proper treatment and disposal of wastes to the land, air, or water. Properly implemented, pollution prevention gives industry the opportunity to have the best of both worlds-increased profitability and reduced environmental liabilities.

## STEPS FOR POLLUTION PREVENTION PROGRAMS

Pollution prevention programs are successful when a complete program is established and implemented. The steps for a successful pollution prevention program include the following:

- Establish corporate commitment.

3

- Develop program guidelines.
- Perform facility assessment.
- Implement action plans.
- Recognize, reward, and publicize results.
- Institute a tracking system.
- Audit for continuous progress.

Figure 1 illustrates elements of a pollution prevention/waste minimization program. By incorporating these elements into a continuous improvement strategy and implementing the plan, a complete program which includes measurement towards goals can be developed an implemented. These steps must incorporate the organizations' goals and perceptions on pollution prevention. Important factors to consider in the development of pollution prevention programs include:

- Expanding statutory and regulatory requirements.
- Strict emissions and discharge requirements/permit limits.
- Regulatory restrictions and market pressure on solids/hazardous waste disposal.
- Increased capital investments for pollution abatement.
- Increased equipment operation and maintenance costs.
- Trade group and international standards of practice.
- Product stewardship, ecolabeling, and packaging initiatives.

## Figure 1. Management System for Pollution Prevention

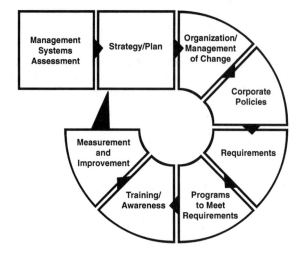

## Figure 2. Driving Forces for Pollution Prevention

- *Economic benefits*
- *Operations flexibiltiy*
- *Liability reduction*
- *Regulatory compliance*
- *Improved public relations*

The practical driving forces for pollution prevention and waste minimization are shown in Figure 2. Economic benefits must accrue as incentives to these programs. These economic benefits can be increased income due to improved yields, or cost savings due to reduction of material inputs or reduced treatment and disposal costs. Operations flexibility can be increased for both the manufacturing steps and treatment steps. Liability reduction related to waste handling and disposal has great appeal. Regulatory compliance is typically the major driver. Improved public relations provides additional incentives.

## POLLUTION PREVENTION/WASTE MINIMIZATION APPROACH

There is a hierarchy for waste minimization/pollution prevention which generally flows from source reduction to the disposal step, as illustrated in Figure 3. These steps include source reduction, reuse, recycle, treatment, and disposal. Typically, the highest benefit is achieved by addressing the source reduction opportunities and then sequentially looking at the reuse, recycle, treatment, and disposal options. In practice, the focus of attention is typically reversed when evaluating a waste minimization program in that the disposal is the driving force and reducing the disposal cost is the initial focal point. However, attention to the source is the recommended initial focus point.

## Figure 3. Waste Minimization Heirarchy

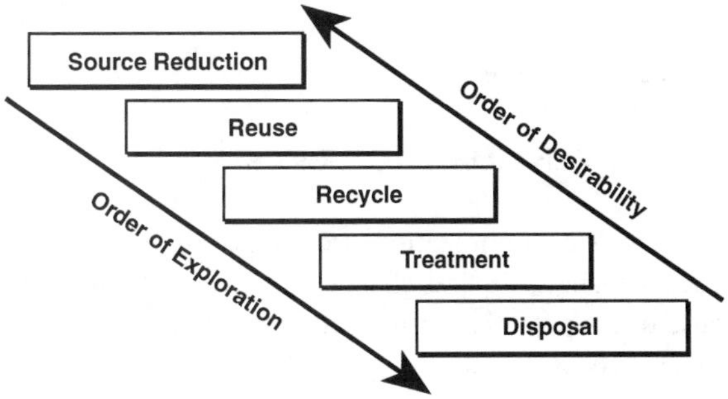

True waste minimization incorporates an integrated analysis, as illustrated in Figure 4. The integrated analysis takes into consideration inter-media effects and cross-media transfer. In essence, cross-media transfer is to be avoided, since such transfers represent future liabilities or additional total pollution control system costs.

To understand the opportunities for pollution prevention and waste minimization, a unit environmental profile approach is recommended. This essentially provides a mass balance on the total production unit or facility, and addresses not only the raw materials and products but also the emissions to the different disposal media, including air, land, and water. The unit environmental profile allows the development of procedures, training requirements, and identification of key personnel to implement the pollution prevention program. The unit environmental profile allows the identification of the major impact areas for a particular production unit or production facility. In this way, focus can be on the areas which have the greatest impact both from an economic and waste reduction standpoint. Since limited resources are available for manufacturing projects, the intelligent application of the resources to achieve the highest impact is necessary. Pollution prevention and waste minimization programs often can be seen as process improvements and yield improvement programs, which are justified on a cost basis. This is a normal approach in industry and should be understood as a fundamental underlying requirement in business.

## Figure 4. Integrated Analysis Approach

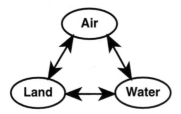

- Media Transfer
- Mass Balance
- Unit Environmental Profile

## TYPICAL POLLUTION PREVENTION PROGRAMS

Pollution prevention and waste minimization programs fall into several categories. The first is process change. Process change can have the highest risk and is often resisted because of the potential for impacting the final product in a negative fashion. However, process change may result in less waste being produced or wastes being less hazardous. Another approach to waste minimization is to incorporate procedural changes which reduce the amounts of resources required in the manufacturing process. These procedures can minimize the amounts of chemicals, energy, or process utilities required in the manufacturing process. Procedures need to be user-friendly and easily understood by operators. The associated training on the procedure and continuous monitoring of procedure adherence is necessary for this to be successful. Operators should be included in the development of the procedures because they have first hand practical knowledge of the processes and how the procedures will impact the process.

A third approach for pollution prevention and waste minimization is to provide focused treatment. In focused treatment, destruction technologies are recommended. Destruction technologies mean that the original species are reacted to a new form which is innocuous. For conservative substances such as metals and many inorganics, destruction technologies do not apply, although conversion processes may decrease their hazards. For organics, destruction to carbon dioxide and water may be achieved. This eliminates the original compounds themselves and the future liabilities of those compounds.

## CASE STUDIES

Four different case studies are summarized. The first case study is a process change. The next case study deals with the procedures change and this deals with brewery water and chemical use in their cleaning processes. The last two case studies deal with focused treatment to provide waste minimization in that residuals are kept out of the environment, or destruction technologies used such that the end products do not create environmental issues.

## PROCESS CHANGE WASTE MINIMIZATION CASE STUDIES

### CYCLIC AMINES PRODUCTION

The process involved was a batch hydrogenation process in which an aromatic was hydrogenated to create a saturated cyclic amine compound in batch reaction processing. The product was then separated in a batch column distillation to give product and water cuts. The problem that developed was that benzene was found in the water cut at concentrations which would make the water cut a hazardous waste by USEPA RCRA standards. Benzene in the water cut was greater than 0.50 mg/l. At that benzene concentration, the wastewater was classified as hazardous, and that meant that it was necessary to remove the benzene. To treat the benzene as part of the mixture of organics or to selectively remove the benzene would be very costly. The disposal operation was to a deep well and deep well discharge of hazardous waste was not permitted in the state in question.

Without removal of the benzene, and having to treat this as a hazardous waste, limited disposal options.

The source of the benzene was identified as the batch distillation processing step. Benzene was formed in the batch distillation, and the production of the benzene was a function of the retention time in the batch distillation column and of the temperature at which the batch distillation was run. Investigation into the column operations showed that the operations could be changed, resulting in satisfactory conditions which maintained product integrity and minimized benzene production. The benzene formed was less than the 0.5 mg/l limit, and therefore the water was no longer a TC (toxicity characteristic) hazardous waste.


## PROCEDURES CHANGE WASTE MINIMIZATION CASE STUDIES

### *BREWERY WATERS/CHEMICAL USE*

A major brewery has many clean-in-place (CIP) processes in the brewing and packaging operations. The cleaning chemicals are caustic (sodium hydroxide) and acid (in this case, phosphoric acid). The brewery had experienced violations of pretreatment discharge requirements for pH, and also faced higher costs for water use surcharges for both water supply and for wastewater treatment. The water supply issue was further complicated in that the allocation to the brewery was being exceeded, and would require the brewery to purchase additional water allocation. The additional water allocation represented a single up-front charge for the right to use additional water.

The solution to the problem of high water use and chemical use focused on the procedures in the brewing and packing operations. By working with the plant operators and defining potential high water use and high chemical use operations, the actual operations were viewed and reviewed to define opportunities for minimizing water use and/or chemical use. Sixteen new brewing and fourteen new packaging operations procedures were developed. These procedures were a compilation of the steps needed to assure proper CIP operations and water and chemical management. The procedures were developed in a user-friendly format, which basically provided a step-by-step flow chart of the critical steps in the operations. The idea was to have a "game plan" versus "play book" operational procedure. The game plan is a focused version of the play book. (By this analogy, we are comparing the play book approach which a sports team has that encompasses all the plays that they can execute. On a given game day, those plays are condensed into a game plan which is a smaller subset of the larger play book.) The flow sheets and the description of the critical steps in the processing identified the steps to be taken, their impacts on water and/or chemical use, and the importance of each step. The results of the user-friendly game-plan approach resulted in decreased chemical usage and water requirements for the processing with a cost savings and/or cost avoidance for the first year of $1.0 million.

8

# FOCUSED TREATMENT WASTE MINIMIZATION CASE STUDIES

## *PHARMACEUTICAL FERMENTATION BROTH*

A pharmaceutical manufacturer had a high strength organic waste from fermentation processing to produce an antibiotic. The high strength organic waste was basically the spent fermentation broth. The existing biological wastewater treatment plant was at capacity, and sludge disposal from the activated sludge process represented a major cost. The problem was that increased production and new products would result in more wastewater treatment capacity being required. The conventional solution, i.e., duplicating what currently is being done by treatment in the activated sludge biological system, meant a doubling of the treatment plant size.

The solution developed was to pretreat the high strength spent fermentation broth prior to the main treatment plant. An innovative biological treatment system incorporating thermophilic aerobic digestion resulted in very high destruction of the fermentation broth organics and a minimization of net sludge production. The focused treatment resulted in capital costs which were less than half of the conventional upgrade. In addition, because the high strength organic stream was treated in a pretreatment process with low net sludge production, sludge disposal decreased significantly.

## *NITRATIONS PROCESS PRETREATMENT*

A continuous nitrations process chemical manufacturer produced nitrations process wastewaters which were sent to a publicly owned treatment works (POTW). Within the nitrations process, the organics were washed to remove organic contaminants. One wash stream contained organics which passed through the POTW. The biological and physical processes of the POTW did not remove these organics. The chemical manufacturer was concerned with the fate in the environment and the effects on the environment of the organics which passed through the POTW. The organics were not regulated by priority pollutant or other hazardous waste lists.

The solution to the problem was to provide a partial oxidation of the organics in the washwater using Fenton's reagent. Doses used were less than stoichiometric requirements for complete oxidation. The oxidation was meant to break down the organics from complex to simple forms, and to make the organics more biodegradable. This is precisely what happened. In addition, Fenton's reagent was reactively selective to the compounds of interest. Other chemical oxidants tested were not as selective. Therefore, economies of chemical use were realized with Fenton's reagent. The POTW biological treatment was capable of destroying the new organics in the activated sludge system. This resulted in no recalcitrant residual organics from this process in the final discharge from the POTW.

## SUMMARY

Pollution prevention and waste minimization can be achieved in multiple ways. The benefits of cost-effectiveness in handling and managing wastes and associated reduction in costs and an increase in profits are major incentives. The ancillary factors of improved environmental awareness among employees and enhanced public images for companies are important benefits as well. These benefits can be achieved through an integrated approach to the pollution prevention and waste minimization programs and by approaching these programs through a continuous improvement process.

# UTILIZING PAPER MILL SLUDGES AS LANDFILL COVER MATERIAL: A MEANS OF WASTE MINIMIZATION AND REUSE

HORACE K. MOO-YOUNG JR.
Department of Civil Engineering
Rensselaer Polytechnic Institute
Troy, NY 12180

THOMAS F. ZIMMIE
Department of Civil Engineering
Rensselaer Polytechnic Institute
Troy, NY 12180

## ABSTRACT

This investigation strives to find a beneficial use for waste water paper mill sludges and to create a program for waste minimization and reuse by using paper mill sludges as the impermeable barrier in landfill covers. This study investigates the geotechnical properties of seven paper mill sludges. Paper mill sludges have a high water content and a high degree of compressibility and behave like a highly organic soil. Consolidation tests reveal a large reduction in void ratio and high strain values that are expected due to the high compressibility. Triaxial shear strength tests conducted on remolded and undisturbed samples show variations in the strength parameters resulting from the differences in sludge composition (i.e., water content and organic content). Laboratory permeability tests conducted on insitu specimens meet the regulatory requirement for the permeability of a landfill cover. Test plots were constructed to simulate a typical landfill cover with paper sludge and clay as the impermeable barrier and were monitored for infiltration rates for five years. Long term permeability values estimated from the leachate generation rates of the test plots indicate that paper sludge provides an acceptable hydraulic barrier.

## INTRODUCTION

The elevating cost of waste disposal may be reduced by the use of unconventional material in the construction of landfills. The high price of disposal has sparked interest in the development of alternative uses for waste sludges (paper mill sludges and water treatment plant sludges). Paper mill sludges in spite of high water contents and low solids contents in comparison to clays, can be compacted to low permeabilities and can substitute for clays in landfill covers. Since 1975, paper mill sludges have been used to cap landfills in Wisconsin [1, 2, 3, 4, 5].

Seven sludges were used in this study. Sludge A is a waste water treatment plant sludge from a deinking recycling paper mill. Sludge B is a blended sludge from a wastewater treatment plant which receives its effluent from a recycling paper mill and the neighboring community. Sludge C is a blended sludge from an integrated paper mill and is comprised of kaolin clay, wood pulp and organics. Sludge C was mined from a sludge monofill which was in operation since 1973. Samples were collected from different sections of the monofill to represent different sludge ages: one week (C1), 2-4 years (C2), and 10-14 years (C3). Sludge D is a primary waste water treatment plant sludge from a recycling paper mill. Sludge E is a primary wastewater treatment plant sludge from a non-integrated paper mill that uses titanium oxide as the primary filler.

## GEOTECHNICAL CHARACTERISTICS

The geotechnical classification of paper mill sludges is not like that of typical clays used in landfill cover systems. For example, Atterberg Limit tests are very difficult to perform on paper sludges and the results may not be meaningful in terms of classical geotechnical classification [6]. Organic content, specific gravity, natural water content, and permeability are the major physical properties of sludges.

The ranges of natural water contents, organic content, specific gravity, and permeability are summarized in Table 1. Water content was determined according to American Society for Testing and Materials (ASTM) procedure D2974. The organic content of the paper sludge was determined according to ASTM procedure D2974, method C for geotechnical classification purposes. Specific gravity tests were performed on the sludges according to ASTM procedure D854. Permeability tests were conducted on remolded specimens of the various sludges using ASTM procedure 5084. Paper sludge specimens were remolded at various water contents in the range of the initial moisture content. Average initial permeability values were measured at a low confining stress of 34.5 kPa.

## COMPACTION TESTS

Proctor tests were performed following the ASTM procedure D698-78. Because of the high water content, tests were conducted from the wet side rather than from the dry side as recommended by ASTM. Furthermore, when

## TABLE 1 SUMMARY OF WATER CONTENT, ORGANIC CONTENT, SPECIFIC GRAVITY, AND AVERAGE INITIAL PERMEABILITY

| SLUDGE | WATER CONTENT (%) | ORGANIC CONTENT (%) | SPECIFIC GRAVITY | AVERAGE INITIAL PERMEABILITY (cm/sec) |
|--------|-------------------|---------------------|------------------|---------------------------------------|
| A  | 150-230 | 45-50 | 1.88-1.96 | $1.0 \times 10^{-7}$ |
| B  | 236-250 | 50-60 | 1.83-1.85 | $1.0 \times 10^{-7}$ |
| C1 | 255-268 | 50-60 | 1.80-1.84 | $1 \times 10^{-6}$ |
| C2 | 183-198 | 45-50 | 1.90-1.93 | $3 \times 10^{-7}$ |
| C3 | 222-230 | 40-45 | 1.96-1.97 | $1 \times 10^{-7}$ |
| D  | 150-185 | 42-46 | 1.93-1.95 | $1 \times 10^{-6}$ |
| E  | 150-200 | 40-45 | 1.86-1.88 | $5 \times 10^{-6}$ |

water was added to dry sludge, large clods formed, the clods were difficult to break apart, and the sludge lost its initial plasticity. During the drying process, the sludge was passed through the number 4 sieve and placed in a pan to air dry. Many trials were conducted to reach the optimum moisture content and density.

Figure 1 shows the Proctor curve, optimum moisture content, and dry density for the various sludges. The Proctor curves are skewed with only a small range of water contents on the dry of optimum side of the curves and with a wide range of moisture contents on the wet of optimum portion of the curve. At higher water contents, the dry density obtained from the Proctor curve for the various sludges is similar. At the optimum density and moisture content, the sludge is dry, stiff, and unworkable. A very high water content is desirable, if the sludge is to be used as a landfill capping material [7]. These test results compare favorably to research conducted on water treatment plant sludges (8, 9, 10).

## CONSOLIDATION TESTS

One dimensional consolidation tests were conducted on all sludge samples following ASTM procedure D2435. Figure 2 displays the plot of strain versus the logarithm of pressure from a typical consolidation test run on paper sludge. Test results show that paper sludge is highly compressible. At higher consolidation pressures, high strain values were measured. Low strains were encountered during the first increment. Large reductions in water content and

# FIGURE 1 PROCTOR CURVE FOR VARIOUS SLUDGES

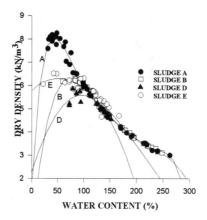

WATER CONTENT (%)

void ratio resulted from application of higher applied stresses. These results compare favorably to consolidation test conducted on water treatment sludges [8, 10, 11].

## SHEAR STRENGTH TESTS

The shear strength behavior of paper sludge A, B, C3, and D was determined using consolidated undrained triaxial compression tests with pore pressure measurements following ASTM procedure D4767. Two sets of tests were conducted on sludge A. Table 2 summarizes the results which compare favorably to other researchers (10, 11).

During the consolidation phase of the triaxial tests, a large reduction in void ratio resulted due to the high compressibility of the sludge. This behavior is consistent with that observed during the consolidation testing reported previously. Moreover, the values of $A_f$ indicate that the sludges behaved in a similar manner to a normally consolidated clay.

Failure is difficult to determine from the stress-strain curves, which are typical of soft compressible material in that they exhibit no sharp yield point. Failure has to be arbitrarily selected at some reasonable strain. For the purpose of this study, failure is defined at 10% strain. Obviously, if failure is defined at a different strain, the strength parameters would change. The variation in shear strength for the various sludges may be attributed to the wide range of water contents, to the variations in sludge production, and to the high organic content. Moreover, differences in the amount of fibers in the sludge matrix may alter the amount of cohesion measured in the paper sludge.

14

FIGURE 2 TYPICAL CONSOLIDATION TEST RESULTS

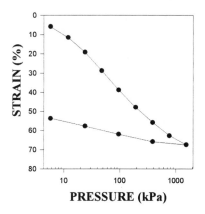

**PRESSURE (kPa)**


**LABORATORY TESTS CONDUCTED ON UNDISTURBED SAMPLES**

Laboratory permeability tests were conducted on undisturbed sludge A samples taken from the Hubbardston Landfill on five occasions: July 1991, October 1991, April 1992, January 1993, and July 1993. Permeability tests were performed following the procedures of ASTM D5084 for measuring the hydraulic conductivity of saturated porous material using a flexible wall permeameter with backpressure. Samples were tested at a low confining stress of 34.5 KPa to simulate the worst case, that is the highest permeability.

The best sampling procedure was discovered through trial and error using Shelby tubes. Slow static pressure (pushing the Shelby tube into the sludge layer with a constant vertical force) compressed the sludge during the sampling process and led to low recovery rates. A dynamic sampling process, like striking the Shelby tube with a hammer, resulted in high rates of recovery and minimal disturbance. Apparently, due to the fibers and tissues in the sludge matrix, a sharp blow was needed to cut through the sludge. The normal field procedure was to place the Shelby tube on the sludge, place a wood block on top of the Shelby tube, and strike the block with a hammer. This procedure resulted in the highest rates of recovery and the least disturbance [12].

In general, the samples met the $1 \times 10^{-7}$ cm/sec regulatory requirement for a low permeability landfill cover system. Table 3 summarizes the permeabilities of the samples. The water contents of the specimens taken from the landfill after construction varied from 150% to 220%. In general all specimens taken from various sections of the landfill immediately after construction either met the regulatory requirement or were very close.

## TABLE 2 SUMMARY OF SHEAR STRENGTH TESTS

| Sludge | Effective Angle of Internal Friction ($\phi'$) | Cohesion (KPa) | $A_f$ |
|--------|------------|----------|------|
| A (Test 1) | 37 | 2.8 | 0.72 |
| A (Test 2) | 25 | 9.0 | 0.74 |
| B | 37 | 5.5 | 0.9 |
| C3 | 32 | 9.0 | 0.7 |
| D | 40 | 5.5 | 0.73 |

Sample 3, taken after 9 months, was dewatered and consolidated under an eighteen inch overburden. It was markedly stiffer and denser than samples obtained shortly after construction. The permeability for the sample meets the regulatory requirements of $1 \times 10^{-7}$ cm/sec.

Sample 5 obtained in January 1993 was taken from the same section of the landfill as sample 3, eighteen months after placement. Permeability tests yielded an average permeability of $3.4 \times 10^{-8}$ cm/sec at a water content of 107 %, which easily meets the $1 \times 10^{-7}$ cm/sec standard for landfill cover design. After 18 months of consolidation the sludge layer met the regulatory requirements. The sludge layer performs as an adequate hydraulic barrier at a water content of 107% and a void ratio of 2.1. Sample 6 was taken two years after placement from the same section of the landfill as samples 3 and 5. Sample 6 meets the permeability requirement. Thus, time, dewatering, consolidation, and organic decomposition have reduced the permeability of sludge A.

## TEST CELL SETUP

In 1989, Erving Paper mill along conducted a study to establish the long term hydraulic conductivity characteristics of paper sludge to obtain approval from the Massachusetts Department of Environmental Protection to use sludge A as a landfill capping material [3]. Six test plots simulating typical landfill final capping design were constructed of primary sludge (test plot 2 and 3), clay (test plot 1), and blended sludge (test plot 4, 5, and 6). Test plots are 7.62 meters by 7.62 meters in area. Fine grained sandy soil is used to prepare a smooth base with a 6 percent bottom slope and containment berms. A protective geotextile filter fabric covers the base of each test plot. The liner of the test plot consists of a 6 ml agricultural plastic. The leachate collection system consists of PVC piping and two plastic drums in series. The PVC pipe is secured to the liner with gaskets

## TABLE 3 SUMMARY OF LABORATORY PERMEABILITY TESTS ON INSITU SAMPLES

| SAMPLE | PERMEABILITY (cm/sec) | WATER CONTENT (%) |
|---|---|---|
| | JULY 1991 | |
| 1 | $1.06 \times 10^{-7}$ | 190 |
| | October 1991 | |
| 2 | $4.0 \times 10^{-8}$ | 185 |
| | April 1992 | |
| 3 | $4.47 \times 10^{-8}$ | 106 |
| 4 | $4.2 \times 10^{-7}$ | 220 |
| | JANUARY 1993 | |
| 5 | $3.4 \times 10^{-8}$ | 107 |
| | JULY 1993 | |
| 6 | $3.8 \times 10^{-8}$ | 91.5 |

and clamps. A geotextile filtered solids from entering and possibly clogging the leachate collection system.

A sand drainage layer (15.24 cm in thick) was placed over the test plots liner system for the collection of leachate infiltrating through the overlying cap. A low permeability cap, either clay or paper sludge, was placed above the drainage layer. A 15.24 cm sand layer was placed above the low permeability to facilitate the lateral flow of rain water. A 30.5 cm layer of top soil was placed on top of the sand layer for vegetative support. Test plot 1 was constructed with a 45.72 cm thick clay barrier, test plot 2 and 3 were constructed with a 45.72 and 91.44 cm thick primary sludge barrier, and test plots 4, 5, and 6 were constructed of 45.72, 91.44, and 91.44 cm thick blended sludge barriers.

The collection drums were emptied periodically to determine the amount of leachate generated. The estimated field permeability or percolation rate was estimated based on the rearrangement of Darcy's law where $k = Q/iA$, where $k$ is the field permeability, $Q$ is the flow rate, $A$ is the area of the test plot, and $i$ is the barrier thickness divided by the hydraulic head. Assuming that the low permeability layer is fully saturated and that the hydraulic head in the sand drainage layer is negligible, the hydraulic gradient is assumed to be 1.

## LEACHATE GENERATION

Figure 3 illustrates the cumulative leachate production of each test plot over five years. The breaks in the data represent the winter month when snow covered the test plots and frozen ground conditions occurred. Analysis of the clay test plot data reveals that during the fall of 1989 very little moisture percolated through the clay test plot. After the first winter, the clay control plot generated greater quantities of leachate. The highest cumulative leachate production occurs for the 45.72 cm and 91.44 primary paper sludge test plots (test plots 2 and 3). The lowest cumulative leachate production occurs for the two 91.44 cm blended paper sludge test plots (test plots 5 and 6). The slope of the cumulative leachate production for test plots 2 and 3 are 3.3 and 2.3, respectively. The slope of the cumulative leachate production for test plots 5 and 6 are 1.53 and 1.6 respectively. After five years, the leachate production for test plots 5 was approximately 20 percent less than the clay control plot, and for test plot 6 the leachate production was approximately 16 percent less than the clay control plot.

## FIELD PERMEABILITY

Figure 4 shows the field permeability or percolation rate for the test plot 1, the clay control, over the five year period. The initial permeability of test plot 1 was $3.2 \times 10^{-7}$ cm/sec. After the first frost in November of 1989, there was a noticeable increase in permeability. Moreover, it was noticed that following the winter of 1989-1990, the calculated field permeability of test plot 1 increased by an order of magnitude and appeared to vary with the amount of precipitation [3]. From the calculated field permeability, after each winter, there is an increase in permeability. The increase in hydraulic conductivity may be caused by ground freezing conditions. The average permeability for the test plot 1 over the five year period is $2.27 \times 10^{-7}$ cm/sec with a standard deviation of $4.8 \times 10^{-7}$. The minimum and maximum permeability is $2.26 \times 10^{-9}$ and $4.11 \times 10^{-6}$ cm/sec, respectively, and the range of the data is $4.11 \times 10^{-6}$ cm/sec.

Figure 4 also shows the calculated field permeability for test plot 2 ( 45.72 cm primary sludge test plot). Initially, the laboratory measured permeability was $2.1 \times 10^{-6}$ cm/sec which is about an order of magnitude higher than the clay control. Consolidation and dewatering of the sludge contributed to the high initial permeability. Test plot 2 fluctuates in a similar pattern to the clay control ( an increase in permeability occurs after each winter). This pattern is illustrated by the similarity in the variances of $2.31 \times 10^{-13}$ ( test plot 1) and $7.96 \times 10^{-13}$ ( test plot 2). For test plot 2, the range in data is narrower than the clay control. Test plot 2 has an average permeability of $3.53 \times 10^{-7}$ cm/sec with a standard deviation, minimum hydraulic conductivity, and maximum hydraulic conductivity of $2.82 \times 10^{-7}$, $1.75 \times 10^{-8}$, and $1.54 \times 10^{-6}$ cm/sec, respectively. The average permeability for the 45.72 cm primary sludge test plot is slightly higher than the

18

FIGURE 3  CUMULATIVE LEACHATE PRODUCTION FOR TEST PLOTS

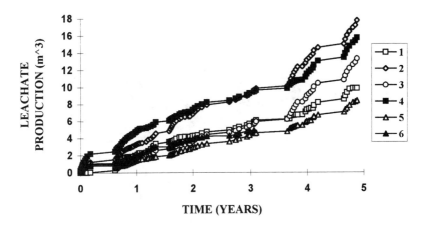

clay control.

Figure 4 illustrates the calculated field permeability for test plot 3 (91.44 cm primary sludge test plot). The initial permeability of test plot 3 is $1.9 \times 10^{-6}$ cm/sec which is approximately one order of magnitude higher than the clay control. Test plot 3 has an average permeability of $3.47 \times 10^{-7}$ cm/sec with a standard deviation of $3.04 \times 10^{-7}$ cm/sec. The maximum and minimum permeability are $1.48 \times 10^{-6}$ and $1.76 \times 10^{-8}$ cm/sec, respectively. Test plot 3 with a variance of $1.03 \times 10^{-13}$ fluctuates in a similar pattern to the clay control with a dramatic increase in permeability after each winter. Moreover, the range in data for test plot 3 is narrower than the clay control.

Figure 4 depicts the field permeability of test plot 4 ( 45.72 cm blended sludge test plot). The initial percolation rate for test plot 4 is $1.4 \times 10^{-6}$ cm/sec which is about one order of magnitude higher than the clay control. The average permeability for test plot 4 is $2.66 \times 10^{-7}$ cm/sec with a standard deviation of $3.21 \times 10^{-7}$ cm/sec. The minimum and maximum permeability are $7.85 \times 10^{-9}$ cm/sec and $1.92 \times 10^{-6}$ cm/sec, respectively . With a similar variance ($1.03 \times 10^{-13}$), test plot 3 varies in permeability in a similar fashion to the clay control with a sharp increase in hydraulic conductivity after each winter.

Figure 4 also display the field permeability of test plot 5 and 6 which are the 91.44 cm blended sludge test plots. Both test plots are characterized by a high initial permeability of $1.4 \times 10^{-6}$ cm/sec and $1.6 \times 10^{-6}$ cm/sec, respectively, which is approximately one order of magnitude higher than the clay test plot. Consolidation and dewatering of the paper sludge contributed to the higher initial permeability. Test plots 5 has an average permeability over the five years of $1.65 \times 10^{-7}$ cm/sec with a standard deviation of $1.82 \times 10^{-7}$ cm/sec, and test plot 6 has an average permeability of $1.11 \times 10^{-7}$ cm/sec with a standard deviation of $1.10 \times 10^{-7}$ cm/sec . The variances of test plots 5 and 6 are approximately one order of magnitude lower than the clay control which indicates that the insitu permeability

FIGURE 4  FIELD PERMEABILITY FOR THE VARIOUS TEST PLOTS

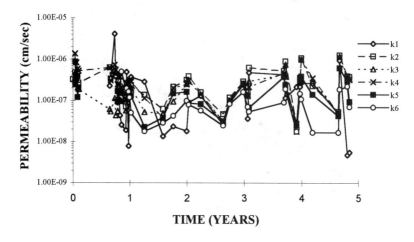

TIME (YEARS)

for the blended paper sludge does not fluctuate as much as the clay test plot. Moreover, the range in data for test plots 5 and 6 show a narrower distribution in permeability in comparison to the clay control. The average permeability for the blended sludge is approximately 40 percent less than the clay control and meets the regulatory requirement for permeability of $1 \times 10^{-7}$ cm/sec. From the data, it can be inferred that with time, consolidation and organic decomposition that the 91.4 cm deep blended sludge provides a better hydraulic barrier than the clay control.

Shelby tube specimens were taken from test plots 1 (clay control), 5 (blended sludge), and 6 (blended sludge) in July 1993 (approximately 4 years after construction). The specimen taken from the clay test plot was dry and stiff with a water content of 15%. The clay test plot specimen had a laboratory permeability of $1.14 \times 10^{-7}$ cm/sec after four years insitu. After four years insitu, test plots 5 and 6 specimens had a water content of 125% and 120% and had permeabilities of $1.1 \times 10^{-8}$ and $1.8 \times 10^{-8}$ cm/sec. Initially, the blended sludge test plots had an organic content of 50%. After four years, the organics of test plots 5 and 6 decreased to 31 and 30 percent, respectively. Thus, from the laboratory permeability testing of insitu samples, it can be inferred that the blended paper sludge provides a better hydraulic barrier than the clay control. The clay control was dry and stiff which increase its susceptibility to cracking if differential settlement were to occur. With dewatering, consolidation, and organic decomposition, blended paper sludge A provided a better hydraulic barrier than the clay used in this study.

## CONCLUSION

The following is concluded from the long term infiltration data on sludge:

1. Paper sludges are characterized by a high water content and organic content and a low specific gravity in comparison to typical clays.
2. Compaction test were conducted from the wet side, since paper sludge loses its plasticity upon drying and rewetting. At the optimum moisture content, paper sludge is unworkable. A high water content is desirable when designing a paper sludge landfill cover.
3. Paper mill sludges are characterized by high compressibility. Consolidation of paper mill sludges results in a large reduction in void ratio and water content.
4. The substitution of paper mill sludge for clay can reduce the cost of disposal and is an excellent alternative in areas that do not have a local source of clay.
5. The initial permeability values from a landfill that uses paper sludge as the hydraulic barrier either satisfy the regulatory requirements or are very close to the requirement. With time, dewatering and consolidation improve the permeability of a paper sludge hydraulic barrier.
6. Initially, the leachate production from the blended sludge test plot was greater than the leachate production from the clay test plot.
7. As time increased, the leachate produced from the blended sludge test plot decreased.
8. After five years, the estimated permeability of the blended paper sludge was lower than the clay control.
9. With time, dewatering, and organic decomposition, the blended paper sludge provided a better hydraulic barrier than clay.

## ACKNOWLEDGMENTS

Support for this research was received from the following organizations: Erving Paper Company, Erving, MA; International Paper Company, Corinth, NY; Marcal Paper Mill Inc., Elmwood Park, NJ; Mead Specialty Paper Division, Lee, MA; Clough Harbour and Assoc., Albany, NY; and the Army Research Office, Research Triangle Park, NC. Their generous support and cooperation is greatly appreciated. However, the opinions expressed herein are solely those of the authors.

## REFERENCES

1. Stoffel, C.M., and R.K. Ham. 1979. "Testing of High Ash Paper Mill Sludge for Use in Sanitary Landfill Construction," Prepared for the City of Eau Claire, WI, by Owen Ayers and Associates, Inc. Eau Claire, WI.

2.  Pepin, R.G. 1984. "The Use of Paper Mill Sludge as a Landfill Cap". *Proceedings of the 1983 NCASI Northeast Regional Meeting*, NCASI, New York, NY.
3.  Aloisi, W. and D.S. Atkinson. 1990. "Evaluation of Paper Mill Sludge for Landfill Capping Material". Prepared for Town of Erving, MA by Tighe and Bond Consulting Engineering, Westfield, MA.
4.  Swann, C.E. 1991. "Study Indicates Sludge Could be Effective Landfill Cover Material," *American PaperMaker*, 34-36.
5.  Zimmie, T.F., and H.K. Moo-Young, 1995. "Hydraulic Conductivity of Paper Sludges Used For Landfill Covers". *GeoEnvironmental 2000*, Eds. Yalcin B. Acar and David E. Daniel, ASCE Geotechnical Special Publication No. 46, New Orleans, LA. 2: pp. 932-946.
6.  LaPlante, K. 1993. "Geotechnical Investigation of Several Paper Mill Sludges for Use in Landfill Covers". Master of Science Thesis. Rensselaer Polytechnic Institute, Troy, NY.
7.  Zimmie T.F., H.K. Moo-Young, and K. LaPlante. 1993. "The Use of Waste Paper Sludge for Landfill Cover Material". *Proceedings from the Green '93-Waste Disposal by Landfill Symposium*. Bolton Institute, Bolton, UK.
8.  Raghu, D., H.N. Hsieh, T. Neilan, and C.T. Yih. 1987. "Water Treatment Plant Sludge as Landfill Liner," *Geotechnical Practice for Waste Disposal 87*. Geotechnical Special Publication No. 13, ASCE, pp. 744-757.
9.  Environmental Engineering & Technology Inc. 1989. "Water Plant Sludge Disposal in Landfills," *Quarterly Report 1*.
10. Wang, M.C., J.Q. He, and M. Joa. 1991. "Stabilization of Water Plant Sludge for Possible Utilization as Embankment Material". Report, Dept. Civil Engineering, The Pennsylvania State Univ., PA.
11. Alvi, P.M. and Lewis, K.H. 1987. "Geotechnical Properties of Industrial Sludges". *Environmental Geotechnology*. H.Y. Fang (Eds). Envo PC: Bethlehem, PA. pp. 57-76.
12. Moo-Young, H.K. 1992. "Evaluation of the Geotechnical Properties of a Paper Mill Sludge for Use in Landfill Covers". Master of Science Thesis, Rensselaer Polytechnic Institute, Troy, NY.

# ENVIRONMENTAL DEGRADATION OF AIR PRODUCTS' VINEX, AIRFLEX AND AIRVOL POLYMERS THROUGH COMPOSTING

MARTIN A. LARSON
Senior Research Technician
Air Products and Chemicals, Inc.
7201 Hamilton Boulevard
Allentown, PA 18195-1501

NEIL M. O'BRIEN
Senior Research Technician
Air Products and Chemicals, Inc.
7201 Hamilton Boulevard
Allentown, PA 18195-1501

## INTRODUCTION

With the growing public concern for environmental awareness, it has become increasingly important to market products which are considered environmentally friendly, i.e., non-toxic, recyclable or compostable. This concern, in part, represents a need to reduce waste streams currently entering landfills. Landfills are essentially engineered as sealed "cells" where entombed waste exhibits very little, if any, degradation. Primarily, degradation here occurs anaerobically--a very slow process by organisms that do not use oxygen in the metabolism of food.

A much more favorable alternative is composting--an aerobic process in which organic materials are decomposed by a variety of microorganisms (primarily bacteria and fungi) to a final, stable, organic end product which is marketable as compost. Most compost facilities now in operation deal mainly with grass, leaves, yard waste and wastewater treatment plant sludges or other specific industrial waste streams. However, there are a growing number of experimental compost operations that are attempting to integrate a larger variety of waste streams, including plastics, into large scale compost operations. Potentially, a larger volume of what is now dumped into landfills could be composted into a useful by-product by using the proper composting conditions. Additionally, an even larger percentage of the waste streams could be composted if industry produced more products which were engineered to be

23

compostable. Compostable products will be favored by manufacturers as the consumer demand for these "green" products increases.

Most plastics that are marketed as biodegradable are merely long polymer chains interspersed with starch. In nature when the plastic is exposed to moisture and organisms, the starch link is biodegraded and the remaining polymer chains are dispersed in the environment. A truly biodegradable plastic however, would be one that is completely broken down into carbon dioxide and water in the environment by naturally occurring microorganisms.

## BACKGROUND

Numerous laboratory tests have been used to measure the biodegradability of various synthetic products [1]. These tests have proven to be invaluable in assessing whether or not such materials can be broken down by microorganisms to yield carbon dioxide and water. Under these controlled laboratory conditions definitive statements can be made as to the degree of biodegradation that occurs and the relative degradation that occurs. From this data predictions can be made about the potential of a material to biodegrade in a variety of real world environments. From a practical point of view it would be valuable to assess the biodegradation of materials in situ. Biodegradation studies of materials in compost piles, landfills or wastewater treatment facilities would yield important information about their survivability under real world conditions. Information derived from such tests would be useful for manufacturers, waste handlers and consumers of these products.

Polyvinyl alcohol or PVOH is a non-toxic, water soluble, biodegradable thermoplastic that has drawn considerable attention lately due to its potential for a large number of biodegradable applications. PVOH is presently used in textile sizing agents, paper coatings, adhesives, disposable films and bottles. It is also blended with starch and sold as a biodegradable film. Various other applications are currently being explored.

Studies with samples in activated sludge systems have shown PVOH to be completely biodegradable [2]. Parameters measured were oxygen uptake, total organic carbon reduction and residual PVOH.

## MECHANISM OF DEGRADATION

Degradation of PVOH can be thought of as occurring at two levels. Macrodegradation occurs at the first level where agents responsible for changes in appearance and physical properties of the polymer prepare the way for the second level of degradation. This second level of degradation or microdegradation occurs at the cellular or molecular level where the enzymes responsible for the degradation catalyze the reactions that cut the carbon backbone into smaller pieces for their eventual fate. The eventual fate of PVOH, like any organic carbon chain utilized as a food source for aerobic

24

organisms, is carbon dioxide and water. This process is known as mineralization, where organic materials are completely degraded to inorganic forms.

The physical and biological agents responsible for macrodegradation are heat (thermal degradation), humidity (hydrolytic degradation), light (photo degradation), microorganisms (biodegradation) and atmospheric (oxidative degradation). These agents can cause changes in the appearance of the polymer such as discoloration, embrittlement, stickiness, solubility and dulling of surface [3]. These changes deteriorate the tensile strength and elasticity and promote the breaking of the polymer into smaller pieces or result in a phase change from solid to liquid. This in turn results in a greatly increased surface area and may reduce the average molecular weight of the material.

All heterotrophic microorganisms obtain energy through the oxidation of chemical compounds. The breakdown or catabolism of these compounds yields energy derived from the chemical bonds that is used in cell maintenance and growth. Large organic compounds and polymeric materials are broken down to smaller carbon units through the action of extracellular and intracellular enzymes. The smaller carbon units produced are used to build structures of the cell and cell wall in a process known as anabolism. Final chemical oxidation of cell constituents converts the organic components to inorganic forms such as $CO_2$, $H_2O$, $NO_x$ and $SO_x$. This mineralization is considered the final step in the decomposition process but may represent a starting point if and when the materials enter another chemical cycle.

PVOH is a polymer consisting of repeating two carbon units with alternating hydroxyl groups and acetate group attached at every other carbon. The ratio of hydroxyl units to acetate groups varies depending on the degree of hydrolysis of the molecule. Additionally, the hydroxyl groups may be substituted with a different organic group at various ratios. Manipulations of length of carbon chain as well as percent hydrolysis, side chain components and other processing techniques can be used to vary polymer properties for a particular application. Regardless of the wide degree of variability there exits a common thread--a long carbon backbone capable of undergoing oxidation and subsequent hydrolysis. Many of these reactions can be mediated by biological enzymes.

## EXPERIMENTAL

A windrow mulching operation operated by Lehigh County's Office of Solid Waste Management was the site for the study. Windrows consisting of approximately two thirds mixed deciduous tree leaves and one third grass and yard waste was mixed and maintained by the county. Important conditions optimized include organic content, humidity, porosity, internal temperature, pH and oxygen. Oxygen, a particularly important parameter, was maintained at levels to provide an aerobic environment by the convective flow of air up through the windrows in a "chimney effect." Conditions were controlled through periodic turning of the windrows to maintain the optimum conditions for growth and proliferation of the biological community of fungi, bacteria, protozoa and the shredders and secondary consumers that are responsible for the entire biodegradation process.

Duplicate test samples were placed into penetrable fiberglass window screen mesh sacks measuring 250 mm by 350 mm with a pore size of 1 mm by 1 mm. The sacks were labelled with non-biodegradable plastic tags. Colored flagging ribbon was attached to each bag to provide quick identification and to simplify retrieval. The screen sacks with sample were hand placed in the center of the windrow into an area of maximum temperature and presumed highest degradatory environment (Figure I).

Samples tested are listed in Table I. Controls used for the study were low density polyethylene (LDPE) for a negative or non-degrading material and cellulose filter paper for a positive.

Samples were retrieved periodically from the windrow before the overturning of the pile. The samples were, therefore, not subjected to mechanical agitation. Degradation was measured by weight loss. Upon retrieval, samples were taken back to the laboratory were they were dried, weighed and photographed. All samples were taken back to the windrow within 24 hours of retrieval. It was not believed that the drying of the samples had a significant effect on the degradation since the high humidity within the windrow pile rehydrated the samples quickly allowing for reactivation and recolonization of the organisms on the sample. Additional measurements taken at the time of retrieval and burial were windrow temperature and dissolved oxygen and ambient air temperature.

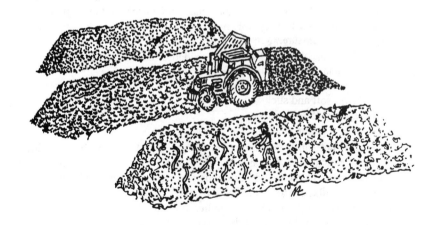

**FIGURE I**
**TEST SAMPLES PLACED INTO WINDROWS**
**WITH COLORED FLAGGING RIBBON**

## RESULTS AND DISCUSSION

The Vinex 2144 samples showed 100% weight loss during the course of the study (Table I). The 1 mil thick film was over 80% gone by day 7 (Graph II) of the study and the 2 mil film was completely gone by day 28 (Graph I). This confirmed earlier respirometry studies with this product that showed complete degradation based on oxygen consumption and total organic carbon reduction [4]. This product also has been shown to be 100% water soluble [5].

Vinex 1003 was tested as a 1 mil thick film, a 2 mil thick annealed and unannealed samples and as a 1 mil coating on 50 lb. bleached greeting card (paper) stock. The 1 mil thick film showed a 60% weight loss by day 91 with a final weight loss on day 190 of only 62%. Visual observation showed an estimated 85% material loss at day 91 with large and small holes throughout, dark brown staining and the sample was friable. The Vinex 1003 paper coating showed 90% weight loss by day 65 and 100% by day 91. Despite gradual weight loss data, at the completion of the study the 2 mil annealed and unannealed samples showed only 15% and 19% weight loss, respectively. Once again, weight loss for these samples was predicted by water solubility and respirometry.

Vinex 2025 was tested as a 1 mil coating on 50 lb. bleached greeting card (paper) stock only. This showed greater than 50% weight reduction by day 28 and 100% weight loss by day 91.

Vinex 2019 as a 2 mil thick film showed complete weight loss on day 17 with 98% of that loss occurring by day 7. This product also showed complete degradation through respirometric oxygen consumption and 100% solubility in water.

A 2 mil thick film of Vinex 5030 showed 100% degradation by day 65 by weight loss but visual observation showed some residual pieces. They were cracked, stained brown and covered with a white fungus. This product showed 100% solubility in water and 100% biodegradability through respirometry.

Weight loss measurements for Vinex EXP 2 mil film erroneously showed 94% weight loss by day 91 and 100% weight loss by day 141. Visual observations and final weight confirmation showed only a 34% weight loss at the end of the study. Observations noted on day 7 that the test film was opaque, stained brown, soft and sticky. This product also showed minimum solubility in water (10%) and 45% biodegradation based on respirometric analysis.

Non-woven paper web treated with binders of Airflex, Airvol and PVOH and polyvinyl acetate blends were tested to compare the effects of these various binders on the degradation of web. The prepared samples were largely made of commercial toilet paper with binder sprayed on to comprise approximately 20% of the total weight. Also tested were two paper towels and two baby wipes and unbound airlaid pulp. The emulsion binders and webs tested are listed in Table I. All paper webs tested showed an initial weight gain with steady weight losses over the course of the study (Graphs III and IV). Initial weight gains were attributed to biomass adhesion as well as associated water weight gain. All toilet paper webs showed 100% degradation with the exception of the

27

12374-83-11 (proprietary blend) binder which showed 95% degradation (Table I). The baby wipes showed the lowest of 59% and 40%. The web in these samples is more dense than the other webs tested and offer a more resistant material that tends to hold up longer in the compost environment.


## CONCLUSION

This study marks the first time that a full scale composting study has been used at Air Products to evaluate and demonstrate the degradation of Vinex thermoplastic PVOH resins and Airflex and Airvol emulsion binders. Results here have served to differentiate the Vinex product line for research and development, product application and marketing.

Vinex 2019, Vinex 2144, Vinex 5030, Vinex 2025 (paper coating) and Vinex 1003 (paper coating) were completely degraded after 190 days in the compost environment. Although they showed less than complete degradation (15-62%) by the end of the study, other Vinex products did show notable physical changes including cracks, holes and staining. It was felt that a longer period in the compost environment would lead to a more complete degradation. The degree of degradation was related to solubility in water and correlated biodegradation results from previous respirometric testing. The non-woven binders, Airvol and Airflex, did not affect the degradation of the paper webs. The toilet paper and paper towel webs treated with Airflex 109, Airflex 124, Airflex 245, Airvol 350 and a polyvinyl acetate/PVOH blend showed the same results of 100% degradation as with commercially available product and untreated airlaid. Other products such as baby wipes seem more influenced by type of web than by the binder.

Further product development and composting work is needed to test product blends that optimize biodegradation while maintaining properties needed in specific applications. Additional research on the toxicity of polymer breakdown by-products in a compost environment should be performed to understand any potential carryover that may be imparted on the food chain once the compost is used as soil fertilizer.


## ACKNOWLEDGMENTS

We would like to thank the Lehigh County Office of Solid Waste Management for their generous cooperation with use of their site, coordinating windrow turning and sample retrieval and for providing needed information on the basics of windrow composting. In particular, we would like to thank Galen Freed, Jack Conklin and Julia Stamm. We would also like to give thanks for the cooperation and the sharing of information with Polymers Technology at Air Products including Dr. Bob Axelrod, Dr. Joel Goldstein and Scott Johnson. We would also like to thank John Phillips for his constant support and early initiatives in spearheading composting at Air Products. Without all of these individual's support, the project would not have been possible.

# REFERENCES

1. J. H. Phillips, "Polymer Biodegradation," internal presentation, Air Products and Chemicals, Inc., Allentown, PA, 1992.

2. J. P. Casey and D. G. Manly, "Polyvinyl Alcohol Biodegradation By Oxygen Activated Sludge," Proc. Int. Biodegradation Symp., 1975, pp. 819-833

3. I. C. McNeill, "Fundamental Aspects of Polymer Degradation," Spec. Publ. - R. Soc. Chem., 1992.

4. K. J. Leinbach, J. H. Phillips, M. A. Larson, "Vinex Biodegradation Respirometry Study," internal memorandum, Air Products and Chemicals, Inc., Allentown, PA, 1991.

5. M. A. Larson, "Vinex Solubility Study," internal memorandum, Air Products and Chemicals, Inc., Allentown, PA, 1991.

# TABLE I

## COMPOST STUDY RESULTS

| Polymer | Sample | % Degradation |
|---|---|---|
| Vinex 1003 (annealed) | 2 mil film | 15 |
| Vinex 1003 (unannealed) | 2 mil film | 19 |
| Vinex 2019 | 2 mil film | 100 |
| Vinex 2144 | 2 mil film | 100 |
| Vinex 5030 | 2 mil film | 90 |
| Vinex EXP | 2 mil film | 34 |
| Vinex 2144 | 1 mil film | 100 |
| Vinex 1003 | 1 mil film | 62 |
| Vinex 2025 | Paper coating | 100 |
| Vinex 1003 | Paper coating | 100 |
| Airflex 109 | Toilet paper | 100 |
| Airflex 124 | Toilet paper | 100 |
| Polyvinyl acetate/PVOH blend | Toilet paper | 100 |
| Airflex 245 | Toilet paper | 100 |
| 12374-82-11* | Toilet paper | 95 |
| Airvol 350 | Toilet paper | 100 |
| Airflex 105 | Paper towel | 100 |
| PVOH | Paper towel | 100 |
| Styrene butadiene | Baby wipe | 59 |
| Airflex 108 | Baby wipe | 40 |
| None | Airlaid | 100 |
| Controls | | |
| LDPE | | 0 |
| Cellulose | | 100 |

* proprietary blend

Note: Polymers are primarily PVOH blends with co-polymers and other additives that vary in percent hydrolysis and water solubility. They are developed for particular physical and chemical characteristics for specific market applications.

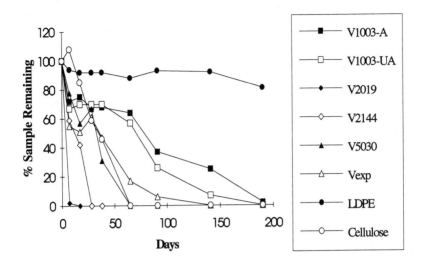

**GRAPH I**

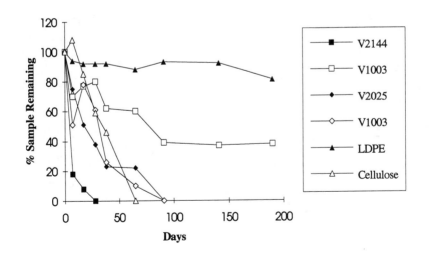

**GRAPH II**

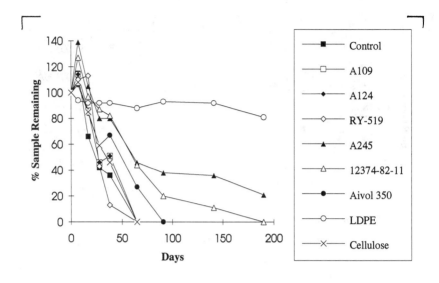

**GRAPH III**

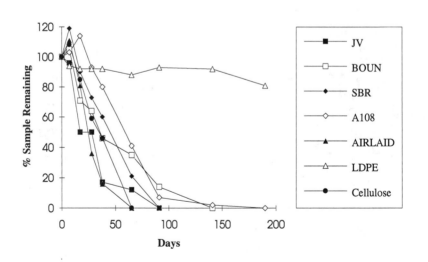

**GRAPH IV**

# SELECTIVE RECOVERY OF ALUM FROM CLARIFIER SLUDGE USING COMPOSITE ION-EXCHANGE MEMBRANES

PING LI,   ARUP K. SENGUPTA
Department of Civil and Environmental Engineering
Lehigh University
Bethlehem, PA 18015

## INTRODUCTION

Alum is widely used as a coagulant in water treatment plants and ends up in clarifier sludge primarily as insoluble aluminum hydroxide. The clarifier sludge is generally discharged into water bodies or disposed of at landfill sites, but the discharge and the disposal of the clarifier sludge are subject to more and more stringent regulations. Alum recovery from clarifier sludge produced by water treatment plants has received considerable attention because of the following reasons: (1) Recovered alum can be reused in water treatment plants so that the operation cost of water treatment can be greatly reduced. (2) Since aluminum hydroxide constitutes almost 30%-50% of the total solid content in clarifier sludge, alum recovery will greatly reduce the volume of clarifier sludge. (3) The toxicity of free and complexed aluminum toward aquatic life has concerned many researchers. (Sengupta and Shi, 1992)

In recent years, alum recovery from clarifier sludge has been studied by many researchers. The most widely used method for alum recovery is acidic extraction using sulfuric acid. The acidic extraction process is based on the simple concept that insoluble aluminum hydroxide in sludge is dissolved into the liquid phase when sulfuric acid is added to sludge. The reaction between aluminum hydroxide and sulfuric acid can be written as follows:

$$2Al(OH)_3 \cdot 3H_2O + 3H_2SO_4 = Al_2(SO_4)_3 + 9H_2O \qquad (1)$$

The dissolved aluminum in the liquid phase can be reused as a coagulant, which provides benefits similar to those of commercial alum. Furthermore, alum recovery can significantly reduce sludge mass and volume (AWWA, 1991). This acidic extraction process for alum recovery has been tested at laboratory scale and pilot scale studies (AWWA,1991). A full scale process has been built in the water treatment plant of Durham, N.C. (Cornwell and Bailey, 1994).

Although the acidic extraction process shows promising results, the following shortcomings may affect widespread application of this process: (1) The process is nonselective. It recovers all other substances along with aluminum under highly acidic conditions. Humic substances, which are removed from raw water by coagulation and exist in clarifier sludge, are dissolved into recovered alum in the acidic extraction process. (2) Although the concentration of heavy metals in clarifier sludge is very low, the accumulation of heavy metals

33

in clarifier sludge may warrant concern if recovered alum is used. (3) Aluminum cannot be concentrated by the acid extraction process. (Sengupta and Shi, 1992)

Because the acidic extraction process is nonselective, it is imperative to develop a selective process. The objectives of this study are to evaluate the feasibility of the two-step process for alum recovery, to understand the characteristics of the membranes: the strong acid membrane and the chelating membrane, and to evaluate the feasibility of recovered alum as a coagulant.

## THE TWO-STEP ALUM RECOVERY PROCESS

In the two-step process, a new class of composite ion-exchange membrane is used. In the composite ion-exchange membrane, ion-exchanger beads are physically enmeshed into a sheet of highly porous polytetrafluoroethylene (PTFE), which facilitates ion-exchanger beads' easy insertion in and withdrawal from sludge. The scanning electron micrograph of the composite membrane is shown in Figure 1.

The conceptual two-step process is shown in Figure 2. The process includes: step 1 — the selective sorption of aluminum on the composite membrane; step 2 — the regeneration of the composite membrane.

In the first step, the pH of sludge is adjusted to 3.5-4.0, at which aluminum hydroxide in sludge is partially dissolved and the concentration of aluminum reaches an appropriate level. The dissolution of aluminum hydroxide in sludge can be expressed as follows:

$$Al(OH)_3(s) + 3H^+ = Al^{3+} + 3H_2O \qquad (2)$$

When the membrane is suspended in the supernatant of sludge, the dissolved aluminum ions in the liquid phase are selectively adsorbed onto the membrane and the hydrogen ions on the membrane are released into the aqueous phase. This ion-exchange process can be expressed as follows:

$$3\overline{R-H} + Al^{3+} = \overline{R_3-Al} + 3H^+ \qquad (3)$$

The release of hydrogen ions from the membrane favors the forward reaction in (2), resulting in aluminum dissolution.

In the second step, the membrane, with sufficiently adsorbed aluminum ions after the first step, is introduced into regenerant. The regenerant is 5-10% sulfuric acid. During regeneration, the aluminum ions on the membrane are replaced by the hydrogen ions in the regenerant; consequently, aluminum ions are concentrated in the regenerant. The process for the desorption of aluminum ions from the membrane can be written as:

$$2\overline{R_3-Al} + 3H_2SO_4 = 6\overline{R-H} + Al_2(SO_4)_3 \qquad (4)$$

In the two-step process, the membrane rolls between the sludge tank and the regeneration tank. Finally, alum can be recovered in the regeneration tank.

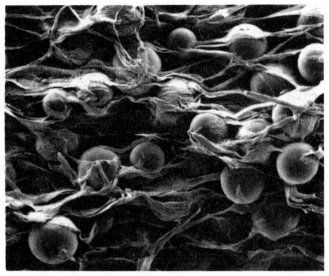

Figure 1    Scanning electron micrograph of composite membrane (300×)

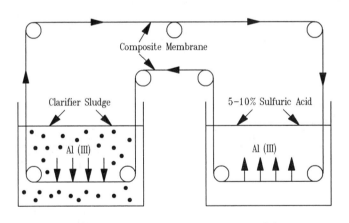

Composite Membrane

Clarifier Sludge

5–10% Sulfuric Acid

Al (III)

Al (III)

Tank 1: Al(III) Sorption

Tank 2: Al(III) Desorption

Figure 2    A conceptual two-step alum recovery process

## MATERIALS AND METHODS

**Composite Ion-Exchange Membrane**   Two kinds of composite membrane were used: the strong acid membrane and the chelating membrane. The membrane used in this study was purchased from Bio-Rad Inc., CA. The commercial name for the strong acid membrane is *AG 50W-X8*, and the commercial name for the chelating membrane is *Chelex 100*. Table I lists the properties and Figure 3 shows the functional groups for these two membranes.

**Clarifier Sludge**   The sludge used in this study was obtained from the Allentown Water Treatment Plant (AWTP), PA. Alum is used to remove the turbidity in the surface water which ranges from 2 NTU to 680 NTU, and alum dosage ranges from 10 to 50 mg/L. The sludge is drained intermittently about two times per week from the storage zone of the settlement tank. The sludge sample obtained from the water treatment plant was immediately used in the experimental simulation of the two-step process or stored in a refrigerator for subsequent use. Table II shows the primary constituent of the AWTP sludge.

Table I   Properties of Composite Membrane

| | |
|---|---|
| Composition | 90% resin; 10% PTFE (by mass) |
| Pore Size | 0.4 μm |
| Nominal Capacity | 0.03 meq/cm$^2$ for Chelating Membrane |
| | 0.15 meq/cm$^2$ for Strong Acid Membrane |
| Thickness of Membrane | 0.2-0.5 mm |
| Resin Matrix | Styrene divinylbenzene |
| Functional Groups | Iminodiacetate for Chelating Membrane |
| | Sulfonic acid for Strong Acid Membrane |

Table II  Composition of the Sludge Obtained from the Allentown Water Treatment Plant

| ELEMENTS | CONCENTRATION (mg/L) |
|---|---|
| Al | 2400.0 |
| Fe | 242.0 |
| Mn | 20.0 |
| Zn | 5.2 |
| Cu | 1.2 |
| Pb | 2.4 |
| Cr | 1.2 |
| DOC | 247.0 |

Note: Concentrations were measured for acidized sludge solution in which pH was less than 1.0.

**Simulation of the Two-step Alum Recovery Process — Cyclic Tests**
The experimental simulation of the two-step process was conducted in one sludge tank and one regeneration tank. Both the sludge tank and the regeneration tank were made from plexiglass. The sludge tank was cylindrical with a diameter of 9.8 inches and a height of 14.0 inches. The regeneration tank is rectangular with a length of 6.5 inches, a width of 4.0 inches, and a height of 12.0 inches. Mixing in the sludge tank was provided by a mixer with a power of 1/14 hp @2000 rpm. The mixing in the regeneration tank was achieved by aerating the contents of the tank with nitrogen gas. 192 square inches of either the strong acid membrane or the chelating membrane was used for the simulation of two-step alum recovery process.

The first step, i.e., the sorption of Al(III) on the membrane was carried out in the sludge tank; the second step, i.e., the desorption of Al(III) from the membrane was carried out in the regeneration tank. Figure 4 shows this procedure.

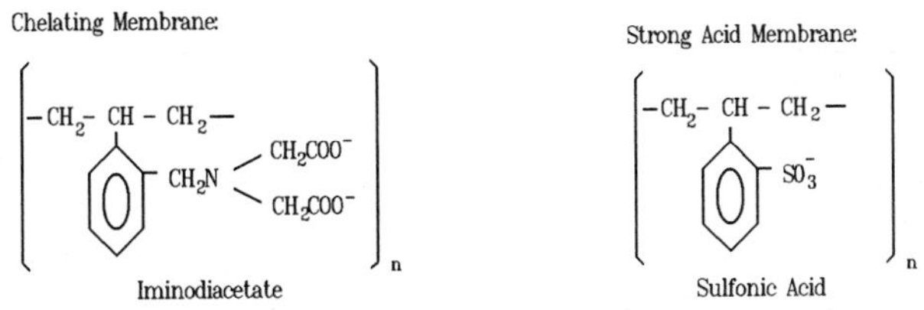

Chelating Membrane:

$$\left[ -CH_2-\ CH\ -\ CH_2- \right]_n$$

Iminodiacetate

Strong Acid Membrane:

$$\left[ -CH_2-\ CH\ -\ CH_2- \right]_n$$

Sulfonic Acid

Figure 3    Functional groups for two kinds of membrane used in this study

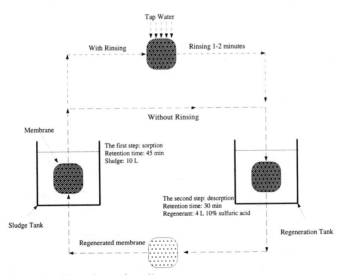

Figure 4    Procedure of cyclic tests

**Effectiveness of Recovered Alum**   Jar-tests were conducted using a conventional jar-test apparatus.   In the jar-tests, the recovered alum obtained from the experimental simulation of the two-step process was used as a coagulant.   For the purpose of comparison, the commercial alum obtained from the Allentown Water Treatment Plant (AWTP) and the mixed alum (50% the recovered alum and 50% the commercial alum) were also employed as a coagulant in the jar tests.   The water from the Little Lehigh River, which is the water source of the AWTP, was used in the jar tests.

**Analytical Methods**   All standards and dilutions in this study were prepared using analytical-grade chemicals and deionized water.   All samples were filtered through a 0.45 µm membrane filter before they were analyzed. Aluminum analysis was performed by using a spectrophotometer (Perkin-Elmer, Model Lambda 2).   The analysis of aluminum employed the eriochrome cyanine R method described in *Standard Methods* (18th edition, 1992).   DOC analysis was conducted by using a TOC analyzer (Dohrmann DC-190).   All metals, other than aluminum, were analyzed by using an atomic absorption spectrophotometer (Perkin-Elmer, Model 2380).   The measurement of pH was achieved by using a pH meter (Fisher Scientific).   Turbidity was measured using a turbidimeter (Hach, Model 2100).

RESULTS AND DISCUSSIONS

**Cyclic Test with the Chelating Membrane**   Figure 5 displays the concentration of Al(III) in the regenerant and pH in the sludge versus the cycle number.   The figure also shows the pH change in the sludge tank.   The concentration of Al(III) in the regeneration tank increased with the cycle number; the concentration of Al(III) in the regeneration tank after 30 cycles was 1800 mg/L.   The pH in the sludge tank decreased as the cycle number increased; after 30 cycles, the pH in the sludge tank was about 1.8.   At this low pH, the increment of aluminum in the regenerant was very small.   After 30 cycles, 10 more cycles were carried out under controlled pH of 3.2.   After pH was raised to about 3.2, the concentration of aluminum in the regenerant continued to increase up to 2400 mg/L.   It is obvious that the sludge pH has a significant impact on the effectiveness of alum recovery with the chelating membrane. Figure 6 presents the composition of recovered alum after 30 cycles in a pie-chart.

The experimental results showed that the chelating membrane had very low recovery efficiency under the condition of low pH.   The shortcomings of the chelating membrane impede the application of the chelating membrane in the two-step process; therefore, it is necessary to use another kind of membrane in the two-step process in order to assure process effectiveness.

**Cyclic Test with the Strong Acid Membrane**   Figure 7 shows the concentration of Al(III) in the regenerant and pH in the sludge versus the cycle number.   The concentration of aluminum(III) in the regeneration tank after 30 cycles was about 4000 mg/L.   Similar to the result of the test with the chelating membrane, pH in the sludge dropped with an increase in the number of cycles. Nevertheless, incremental aluminum recovery in regenerant was not reduced significantly. Figure 8 depicts aluminum concentrations in the regeneration tank for both tests using the strong acid membrane and the chelating membrane. The concentration of aluminum(III) in the regeneration tank increased faster in the test with the strong acid membrane than in the test with the chelating membrane. Figure 9 presents the composition of recovered alum after 30 cycles in a pie-chart.

The cyclic test with the strong acid membrane indicated that the recovery efficiency of the strong acid membrane was not significantly affected by the very low pH and the strong acid membrane had higher recovery effeciency than the chelating membrane.

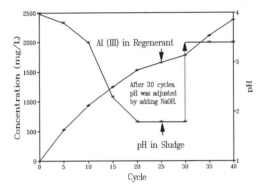

Figure 5    Concentration of Al(III) versus the cycle number for the test with the chelating membrane

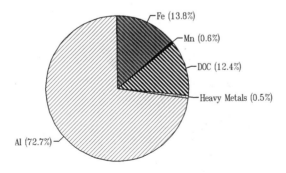

Figure 6    Composition of recovered alum for the test with the chelating membrane after 30 cycles

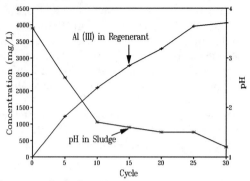

Figure 7   Concentration of Al(III) versus the cycle number for the test
with the strong acid membrane

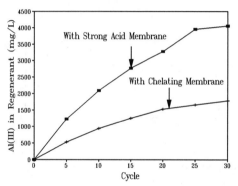

Figure 8   Aluminum concentration versus the cycle number, for the tests
with the strong acid membrane and the chelating membrane

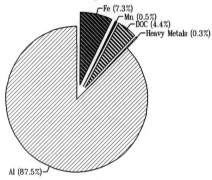

Figure 9   Composition of recovered alum for the test with the strong acid
membrane after 30 cycles

**Effectiveness of Recovered Alum** Figure 10 shows turbidity versus dosage. The turbidity in the jar test using recovered alum as a coagulant was lower than that in the jar test using commercial alum or mixed alum as a coagulant. This indicated that the use of recovered alum resulted in better coagulation. Figure 11 shows the absorbance of UV at 254 nm versus dosage. The UV absorbance is usually used as a surrogate parameter for humic and fulvic acids (Tan and Sudak, 1992). The jar tests using the three kinds of coagulants had similar values of the absorbance. The results indicated that the three kinds of coagulants had the similar effectiveness of removing organic matter.

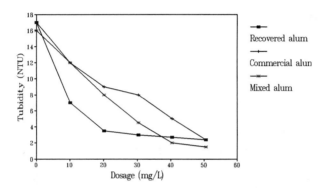

Figure 10   Turbidity versus dosage in jar-tests

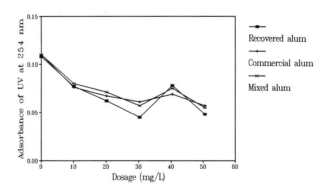

Figure 11   Absorbance of UV versus dosage in jar-tests

## CONCLUSIONS

From this study, conclusions can be summarized as follows:

(1) The two-step process can selectively recover alum from clarifier sludge. Aluminum can be concentrated in recovered alum, while the concentrations of organic matter and heavy metals are very low in recovered alum.

(2) The pH of sludge has an insignificant effect on the effectiveness of the two-step process using the strong acid membrane. On the contrary, the pH of sludge has a significant effect on the effectiveness of the two-step process using the chelating membrane.

(3) The use of the recovered alum as a coagulant is as effective as the commercial alum in removing turbidity from surface water. Also, the removal of natural organic matter (NOM) using the recovered alum is as good as using the commercial alum.

## REFFERENCES

1. APHA, AWWA, and WPCF, 1992. *Standard Methods for the Examination of Water and wastewater*, 18th Edition. APHA, AWWA, and WPCF, Washington,D.C.
2. American Water Works Association Research Foundation, 1991. *Coagulant Recovery: A Critical Assessment*, American Water Works Association Research Foundation, Denver, CO
3. Cornwell, D.A, and Bailey T., 1994. "The Benefits and Performance of a Full-Scale Alum Recovery Facility." Environmental Engineering & Technology, Inc.
4. Sengupta, A.K., and Shi, B., 1992. "Selective Alum Recovery from Clarifier Sludge" *Jour. AWWA* 84:96
5. Tan, L. and Sudak, R.G., 1992. "Removing Color from a Groundwater Source", *Jour. AWWA* 84:85

# PROLER SYNGAS PROCESS FOR GASIFICATION OF WASTE ORGANIC MATERIALS

NORMAN BISHOP
Proler International Corp.
15630 Jacintoport Blvd.
Houston, Texas 77015

## INTRODUCTION

Proler International Corp. of Houston, Texas, has been in the recycling business since its inception in 1925 and is listed on the New York Stock Exchange as NYSE:PS. Proler International Corp. is an environmental services, technology and industrial energy company primarily involved in the recovery, processing and recycling of metals; formulation and sales of speciality chemicals; and gasification of organic waste materials. From 17 operating locations, owned either directly or through joint ventures, Proler provides high-quality raw materials, recycling and energy services to industrial users.

Proler is a developer of resource recovery technologies and patented the first automobile shredding machine in 1960. Today Proler and its joint venture partners are the world's largest exporters of steel scrap, and in 1991 was presented the "E" Award for excellence in exporting by President Bush.

Proler is a leader in detinning technologies for the recovery of tin from waste cans and tin plate scrap and in producing precipitation iron for the copper industry. A new facility in Arizona recovers both tin and copper from waste chemical solutions and electronic circuit board reject and trim materials. The chemicals are reformulated and returned to the suppliers.

In early 1990 Proler teamed with HYLSA, a major steel producer in Mexico to develop an environmentally acceptable process for recovery of remaining elemental resources contained in automobile shredder residue. This program led to the development of the Proler Syngas Process and the filing of U.S. and foreign patents in mid-1992. Allowance of all claims was received from the U.S. Patent Office in late 1994.

43

## AUTOMOBILE SHREDDER RESIDUE

Today over 200 automobile shredders operate in the USA producing between twelve and fifteen million tons of shredded steel and two to three million tons organic and inorganic waste material commonly known as automobile shredder residue (ASR).

ASR is a low density waste and in the United States is generally considered to be nonhazardous; however, in some states and other countries it is classified as a "special waste" and requires special disposal. ASR may contain trace amounts of potentially toxic metals including lead, cadmium, chromium, barium, and lesser amounts of other soluble metals which are subject to rainwater leaching from landfill.

Numerous studies have characterized ASR. A recent US DOE[1] study characterized ASR obtained from 12 shredder locations and the following average composition of ASR after dirt, stone and glass fines removal was determined.

| Component | Weight Percent |
|---|---|
| Fiber | 42.0 % |
| Plastics | 19.3 % |
| Metals | 8.1 % |
| Paper | 6.4 % |
| Tar | 5.8 % |
| Elastomers | 5.3 % |
| Glass | 3.5% |
| Fabric | 3.1 % |
| Wood Splinters | 2.2 % |
| Foam | 2.2 % |
| Wiring | 2.1 % |

## DEVELOPMENT OF THE PROLER SYNGAS PROCESS

In order to develop an "environmentally acceptable" gasification process, certain guidelines were established.

(1) The process should not have a "smokestack."
(2) The process should produce a syngas essentially free of oil/tar contamination.
(3) The process should have zero or near zero solid residue waste for land based disposal.
(4) The process should have zero or near zero waste water discharge.
(5) The process should be designed to eliminate fugitive process dust/gas emissions into the work area.

(6) The process should produce a clean syngas which can be safely pipelined to the consumer.

Gasification research began in Monterrey at HYLSA's R&D facility in early 1990. Work began by using methods of indirect heating "starved air" pyrolysis in order to study the associated problems, then quickly progressed through four separate phases of development. Based on encouraging results achieved in the fourth phase of development, a decision was made by Proler in late 1990 to build a demonstration plant rated at two tons ASR feed per hour in Houston, Texas.

The demonstration plant was completed in May, 1991, and began operating in early June of that year. The direct gasification process proved to be successful from the beginning, and the plant has continued to operate and make improvements. Most of the improvements have involved feeding equipment, residue handling/cooling equipment, and gas cleaning equipment. The primary reforming reactor has operated at up to 25% above design capacity and has not needed revisions.

## PROCESS DESCRIPTION

### Equipment Description
The gasification plant consists of three primary sections:

(1) <u>Raw Material Feeding</u>:  Feedstocks, listed under **"Candidate Feedstocks"**, are metered using a feed hopper set on load cells and then transferred by belt conveyor to a multi-ram stuffing feeder. The stuffing feeder also serves as an atmospheric seal. Some types of raw materials may require feeder modifications.

(2) <u>Primary Reforming Reactor</u>: The reactor is a specially designed rotary vessel, which is heated by a special oxygen/fuel lance. The temperature of gasification is readily controlled for any type of feedstock. The input energy needed to heat, vaporize, dissociate, and reform various types of organic vapors to produce the desired synthesis gas quality is provided by one or more lances.

(3) <u>Product Gas Management</u>: The product gas may be cleaned by either wet or dry gas scrubbing depending on the end use for the solids recovered from the flue gas. A secondary reactor may also be utilized depending on the end use of the syngas. The secondary reactor assures that entrained fractions of tar forming substances and carbon (soot) particles are reformed to $CO + H_2$.

The dry gas scrubbing system employs a Proler designed baghouse in which the fly residue is recovered in a dry state and pneumatically conveyed to the vitrification furnace where it is used as a partial source of fuel and melt stock.

The gasification process, utilizing a dry scrubbing system, is illustrated schematically in Figure 1.

## Description of Syngas

The syngas composition remains relatively consistent for various feedstocks; however, the quantity of syngas generated depends on the richness of the feedstocks. Because the process produces carbon monoxide and hydrogen rich syngas (70% to 80%), the HHV (higher heating value) of the gas remains fairly constant depending on whether or not the secondary reactor is used. For example, when the secondary reactor is operated, methane, ethane, ethylene, acetylene, and remaining fractions of more complex hydrocarbon gases are reformed to $CO + H_2$ which have a lower HHV of only 323 and 325 Btu/scf (12.03 and 12.11 MJ/m³) respectively. Therefore, the HHV of syngas produced when the secondary reactor is operating will be near 280 Btu/scf (11.18 MJ/m³).

When the gasification process operates without using the secondary reactor, the syngas will contain between 350 and 400 Btu/scf (13.04 and 14.90 MJ/m³). The approximate average composition (dry) of product syngas produced from ASR with and without the secondary reactor is shown in Table 1.

Table I
Typical Composition of Proler Syngas
(Produced from ASR)

| Secondary Reactor | With | Without |
|---|---|---|
| Component | Volume% | Volume% |
| $H_2$ | 40.00 | 36.00 |
| CO | 36.00 | 33.00 |
| $CH_4$ | 2.80 | 7.50 |
| ethylene | 0.10 | 3.80 |
| ethane | 0.10 | 0.70 |
| acetylene | 0.10 | 0.60 |
| nitrogen | 3.00 | 3.00 |
| $CO_2$ | 17.90 | 15.40 |
| **HHV** | | |
| Btu/scf | 280.00 | 383.00 |
| MJ/m³ | 10.43 | 14.27 |
| **Density** | | |
| lbs/scf | 0.0528 | 0.0531 |
| kg/m³ | 0.8452 | 0.85 |

# Proler SynGas Process
## with Dry Gas Cleaning

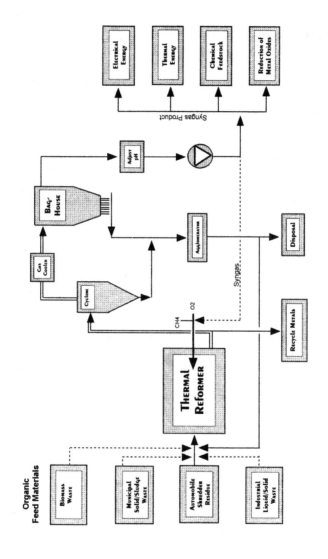

*Figure 1*

### Environmental Considerations

The Proler Syngas Process is designed for minimum releases to the environment. The process does not employ a "smokestack," and all components of the thermal energy which drives the reforming process become part of the syngas product. Fugitive dusts from handling and conveying the raw feed material are collected by a small baghouse and recycled to the primary reforming reactor.

Post reformer process gases pass through several stages of cooling and cleaning before passing through a special fabric filter, which removes the remaining organic and inorganic particulate matter. Acid removal from the syngas can be accomplished either by wet packed tower or by injection of lime dust ahead of the filter house.

Residue from the gasification process is vitrified to form glass, either as "frit" or as a cast product. The glass product has consistently passed the Toxicity Characteristic Leach Procedure tests as regulated by the U.S. Environmental Protection Agency.

Independent environmental and engineering consultants have conducted intensive testing programs to determine if the syngas poses either an environmental or operational problem for the down stream consumer and has been found to pose no significant corrosion, abrasion, or combustion problems for turbines, engines, boilers, or other flame fired heaters.

### Candidate Feedstocks

Materials successfully tested include ASR, MSW, recycled cardboard residue, and combinations of each of these materials with up to 25% chipped automobile and truck tires. Candidate feedstocks not yet tested but believed to be fully acceptable in the demonstration plant include paper mill sludges, agriculture and food industry wastes (e.g., rice hulls, fruit pits, packing plant waste, etc.), wood chips, and industrial packaging waste. Rice hulls, municipal sewage sludge, waste fiberglass, and electronic circuit board waste have been successfully tested in a smaller bench scale reactor.

The demonstration plant does not yet have the necessary environmental permits to handle hazardous materials such as medical "red bag" waste, hydrocarbon contaminated sludges and soils, petroleum refining waste sludges, and waste chemical solutions and solvents; however, these materials can be gasified by the process and should also be considered as candidate feedstock.


## ADDITION OF VITRIFICATION TO THE SYNGAS PROCESS

In late 1993 Proler decided to proceed with the addition of vitrification to the Proler Syngas Process according to the following guidelines:

(1)  Solid residue (non-metallic) from the gasification process should be conveyed by a closed circuit system to the vitrifier.
(2)  Remaining fuel value in the residue should be an asset in the vitrifier.

48

(3) The hot process gas from the vitrifier must be beneficially integrated with syngas to avoid having a "smoke stack" with the vitrifier.

(4) The vitrifier and gasifier must be able to operate either in unison or separately.

(5) If possible, existing vitrification technology should be used.

Vitrification trials were successfully conducted at demonstration plant levels in mid-1994, and engineering studies directed toward integrating the vitrification process with the gasification process to achieve the above listed goals indicated excellent feasibility. Plans are currently under way to install a vitrifier with the Proler Syngas Demonstration Plant in Houston.

The addition of vitrification to the Proler Syngas Process is illustrated schematically in Figure 2.

## SCALE UP DESIGN

Proler contracted HYL Engineering to do the conceptual design engineering for the first full-scale plant. Design parameters involved such considerations as handling of low density solids, heat transfer and heat flux, gas flow velocities, retention times, gas cooling, cleaning, and handling, etc. HYL engineers have excellent scale-up experience in designing progressively larger HYL Process Plants for direct reduction of iron.

Based on size considerations for the rotary reactor, it was decided that the first full-scale plant should be designed to produce and process between 425,000 scf/hr and 550,000 scf/hr of syngas.

A typical Mass Balance for a full-scale ASR gasification plant is illustrated schematically in Figure 3.

## SYNGAS APPLICATIONS

Four basic uses have been identified for Proler Syngas as follows:

**Produce Electric Energy**
Turbine
Combustion Engines
Fuel Cells

**Produce Thermal Energy**
Boiler Fuel
Duct Firing
Furnace Firing

**Chemical Feedstock**
Hydrogen Gas
Carbon Monoxide Gas
Methanol
Ammonia
Paraffin/Olefin

**Direct Reduction of Iron**
HYL Process (DRI)
Iron Carbide
Non-Ferrous Metal Oxides

# Proler SynGas Process with Vitrification

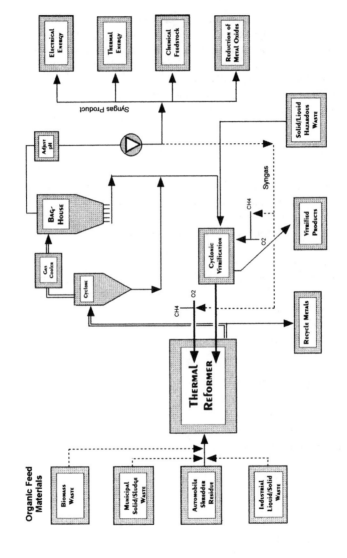

*Figure 2*

50

# TYPICAL MASS BALANCE
## ASR Material

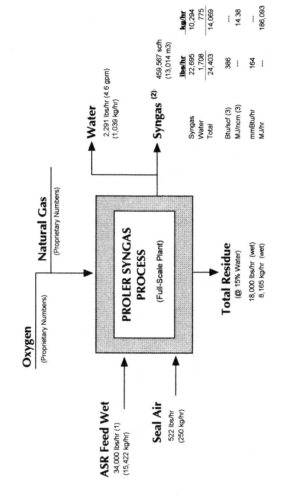

**Water**
2,291 lbs/hr (4.6 gpm)
(1,039 kg/hr)

**Oxygen**
(Proprietary Numbers)

**Natural Gas**
(Proprietary Numbers)

**PROLER SYNGAS PROCESS**
(Full-Scale Plant)

**ASR Feed Wet**
34,000 lbs/hr (1)
(15,422 kg/hr)

**Seal Air**
522 lbs/hr
(250 kg/hr)

**Total Residue**
(@ 15% Water)
18,000 lbs/hr (wet)
8,165 kg/hr (wet)

**Syngas** (2)
459,567 scfh
(13,014 m3)

| | lbs/hr | kg/hr |
|---|---|---|
| Syngas | 22,695 | 10,294 |
| Water | 1,708 | 775 |
| Total | 24,403 | 14,069 |
| Btu/scf (3) | 386 | --- |
| MJ/ncm (3) | --- | 14.38 |
| mmBtu/hr | 164 | --- |
| MJ/hr | --- | 186,093 |

*Figure 3*

(1) Normally contains 10% H2O
(2) 7% H2O, without Secondary Reactor
(3) H.H.V.

51

This list of possible uses is not intended to be all inclusive and numerous other uses may be economically attractive.

## PROCESS ECONOMICS

The capital costs for each plant will depend upon cost factors relative to the exact plant configuration, location and type of organic waste materials to be processed. Preliminary considerations include engineering, construction, land, site preparation, utility connections, management, and taxes.

The Proler Syngas Process is designed to minimize operating costs. The ongoing revenue stream received through the production of syngas, glass, and metals along with tipping fees (via direct revenue or offset costs from feedstock disposal fees), will pay operational costs and amortization of a typical plant over a six to ten years period.

Proler's objective is to maintain a 50% or more ownership in projects by offering its extensive management and operations experience in industrial waste recycling, as well as thermal and electrical energy, to reduce costs, improve efficiencies and minimize any potential risk to industrial users.

## CONCLUSIONS

The Proler Syngas Process has undergone six years of process and equipment development; and the demonstration plant has gasified millions of pounds of combined ASR, MSW, Recycled Cardboard Residue, and chipped tires. The operating thermal balance of the process is easy to control, and the syngas product quality is uniform with carbon monoxide and hydrogen gases comprising between 70% and 80% of the total volume of the gas. The quantity of syngas produced is relative to the richness of the raw feedstock.

The addition of vitrification to the syngas process does not add significantly to the capital or operating cost; and the combined processes are able to accept a wider range of organic as well as inorganic feedstocks, including materials classified as hazardous. The combined technologies are capable of processing a wide variety of both non-hazardous and hazardous waste organic materials, as well as fine inorganic materials and baghouse dust.

The combined technologies consume almost 100% of all fuel value contained in the waste organic feedstocks and convert it to syngas and glass.

### References

1. Investigation of Energy Value of ASR. DOE/ID/1255-1

# WASTE MANAGEMENT PRACTICES AT AN INTEGRATED STEEL MILL

JOHN D. LYNN, JOSEPH R. KOCH
Bethlehem Steel Corporation
Bethlehem, Pa. 18016

R. DOUGLAS LANE
Waylite, a subsidiary of International Mill Service, Inc.
R.D. #5 Easton Road
Bethlehem, PA 18015

As anyone who has ever been to an integrated steel plant knows, and as anyone who has even seen such a steel plant from the street knows, integrated steel manufacturing--from ore to steel--is a large-scale operation. Integrated steelmaking requires a tremendous amount of raw materials on the input end, and produces, besides steel, a tremendous amount of materials on the output end. These end products, historically seen as wastes, have been a problem for the steel industry for many years and have become even more so as environmental regulations make economical disposal a thing of the past. Because of these regulations and the economics of disposal, the industry has looked for new ways to minimize the generation of wastes and for ways to recycle or develop useful coproducts from the remaining materials.

This paper offers an overview of some steps that the Bethlehem Structural Products Corporation, a subsidiary of Bethlehem Steel Corporation, has taken to convert materials once considered as troublesome and costly waste disposal problems, into economically recyclable and saleable products and coproducts.

## RECYCLING REVERTS AS CEMENT-BONDED AGGLOMERATES

Bethlehem Steel contracted with the Pennsylvania Recovery Corporation (PRC), a subsidiary of International Mill Service, Inc., to agglomerate or cold bond various very fine plant reverts, including BOF precipitator dust, Blast Furnace scrubber sludge and other plant materials, so they could be recycled to the Blast Furnace to replace iron ore. PRC developed an extrusion process to blend various sludges and dusts with cement and then extrude them. The extrusions are 1 inch in diameter by about 4 inches long. Once formed, the extrusions are cured for about 1 to 2 weeks, the way concrete is cured, after which they are strong enough to be added to the Blast Furnace. The following is a typical composition of the extrusions:

74.8% BOF precipitator fume
8.8%  Blast Furnace filter cake
2.2 % PreLimer clarifier sludge
2.2%  Lime grit
12.2% Portland cement
100.0% Total

In its process, PRC initially premixes and stockpiles the plant reverts, and then mixes and extrudes the premix with cement in a two-step operation. After curing, the agglomerates exhibit a minimum 70 percent +1/4-inch ASTM Tumble Index comparable to the sinter that is commonly used as a Blast Furnace feedstock. These agglomerates are added to the Blast Furnace at a rate of about 5,000 NNT/ month or about 170 NNT/day. Approximately 50,000 NNT of stockpiled BOF dust has been recycled to the Blast Furnace using this process. Since there is a small amount of zinc in the BOF dust, it is necessary to take measures to remove the zinc from Blast Furnace waste and sludges prior to reuse or discharge.

## FIGURE I--SIMPLIFIED SCHEMATIC DIAGRAM OF RECIRCULATED BLAST FURNACE GAS SCRUBBER WATER CIRCUIT

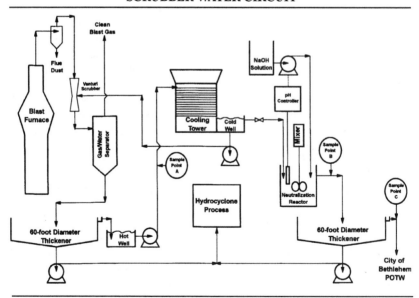

# BLAST FURNACE WASTE WATER PROCESSING

When raw materials that contain zinc are added to the Blast Furnace, most of the zinc is volatilized and removed from the furnace in Blast Furnace gases that leave the top of the furnace. These gases are cleaned by a high-pressure-drop venturi scrubber, which captures very fine particulate matter, including zinc, in waters injected into the scrubber. Then these venturi scrubber waters must be processed to remove particulate matter and zinc prior to discharge. Figure I shows a diagram of the Blast Furnace gas cleaning system and the Blast Furnace waste water processing facility at the Bethlehem Plant's Blast Furnaces.

## VENTURI SCRUBBER WATER PROCESSING

The processing of the venturi scrubber discharge consists of first separating most of the suspended solids from the Blast Furnace waters, followed by cooling and recirculating most of the flow to the venturi. A small portion of the total flow is further processed to remove additional suspended solids and zinc, and the water is then discharged to the Bethlehem City Public Owned Treatment Works (POTW).

From the venturi scrubber, the water flows by gravity to a 60-foot-diameter thickener for removal of suspended solids. A polyelectrolyte is added to the water entering the thickener to aid in flocculation of the solids, which improves the effi-

## TABLE I--CHEMICAL ANALYSIS OF BLAST FURNACE RECIRCULATED WATER SYSTEM BEFORE, DURING AND AFTER TREATMENT TO REMOVE ZINC

| Date | Scrubber Water To Cooling Tower (Sampling Point A) | | | | | Effluent from Zinc Precipitation (Sampling Point B) | | | | | Final Effluent (Sampling Point C) | | | | |
|------|------|-----|-----------|-------|------|-------|-----------|-------|------|-------|-----------|-------|
| | pH | TSS | Zinc | | pH | TSS | Zinc | | pH | TSS | Zinc | |
| | | | Soluble | Total | | | Soluble | Total | | | Soluble | Total |
| | | ppm | ppm | ppm | | ppm | ppm | ppm | | ppm | ppm | ppm |
| 11-7 | 7.57 | 25 | 4.9 | 5.8 | 9.50 | 1,155 | <.10 | 5.1 | 9.19 | 2.8 | <.10 | 0.10 |
| 11-9 | 7.57 | 32 | 12.0 | 16.0 | 9.29 | 996 | <.10 | 16.0 | 9.17 | 2.8 | <.10 | 0.14 |
| 11-11 | 7.54 | 32 | 14.0 | 18.0 | 9.53 | 1,118 | <.10 | 17.0 | 9.24 | 6.4 | <.10 | 0.15 |
| 11-14 | 7.69 | 14 | 4.1 | 4.6 | 9.55 | 1,397 | <.10 | 5.0 | 9.16 | 3.2 | <.10 | 0.13 |
| 11-16 | 7.75 | 46 | 3.4 | 4.7 | 9.44 | 1,212 | <.10 | 2.9 | 9.15 | 3.6 | <.10 | <0.10 |
| 11-18 | 7.61 | 25 | 6.7 | 7.7 | 9.39 | 1,213 | <.10 | 7.6 | 9.20 | 3.2 | <.10 | 0.10 |
| 11-21 | 7.58 | 22 | 4.5 | 5.7 | 9.12 | 1,285 | <.10 | 5.2 | 9.16 | 0.8 | <.10 | <0.10 |
| 11-23 | 7.61 | 28 | 4.3 | 5.4 | 9.27 | 969 | <.10 | 4.1 | 9.23 | 4.0 | <.10 | <0.10 |
| 11-28 | 7.43 | 23 | 7.3 | 7.9 | 9.30 | 1,035 | <.10 | 7.2 | 9.10 | 4.4 | <.10 | 0.10 |
| 11-30 | 7.67 | 19 | 9.6 | 14.0 | 9.47 | 1,249 | <.10 | 14.0 | 9.22 | 4.0 | <.10 | 0.13 |
| 12-2 | 7.58 | 26 | 6.4 | 7.7 | 9.24 | 902 | 0.12 | 7.6 | 9.18 | 8.4 | <.10 | 0.17 |

ciency of the thickener to remove solids. The thickener underflow slurry is sent to a hydrocyclone system to separate the zinc from the particulate matter that contains mostly iron and carbon, which are recovered and sent to Bethlehem's Sparrows Point Division for recycling in its sinter plant. The thickener overflow is pumped to a cooling tower prior to being recirculated to the venturi scrubber. A small amount, typically 50 to 75 gallons per minute, is taken from the cooling tower cold-well as a blowdown to control buildup of dissolved solids in the recirculated venturi scrubber water system. This must be done to prevent mineral scale buildup on the pipes and equipment.

## ZINC REMOVAL SYSTEM

Blowdown from the recirculated water system is processed to remove zinc consists of raising the pH and then separating the solids and liquid by gravity. The process facility comprises a stirred-tank reactor, pH control circuit, sodium hydroxide addition system, and thickener. The first step of the process consists of increasing the pH of blowdown water to 9.5 to precipitate any soluble zinc as zinc hydroxide. Increasing the pH also causes precipitation of much of the calcium carbonate in the water. This carbonate precipitation aids in removing the zinc by coprecipitation. This pH-adjusted blowdown flows to a thickener by gravity for clarification and removal of the solids and the clarified thickener overflow is discharged to the Bethlehem POTW. Table I lists typical analyses of the blowdown waters at various stages in the zinc removal process. The final effluent is well below the applicable effluent limitations.

TABLE II--TYPICAL DISTRIBUTION OF IRON, CARBON AND ZINC VALUES FOLLOWING HYDROCYCLONE PROCESSING OF BLAST FURNACE WASTE WATER SLUDGE,
(Weight Percentage Distribution)

|  | Hydrocyclone Underflow | Hydrocyclone Overflow |
|---|---|---|
| IRON | 85 % | 15 % |
| CARBON | 82 % | 18 % |
| ZINC | 10 % | 90 % |

56

## FIGURE II--BETHLEHEM PLANT BLAST FURNACE SCRUBBER SOLIDS COMPOSITION BY SIZE

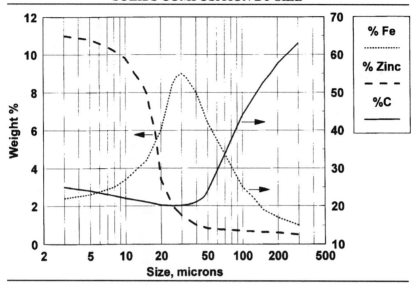

## RECYCLING BLAST FURNACE WASTE WATER SLUDGE

If iron-bearing Blast Furnace sludge is continually recycled without processing, unacceptably high levels of zinc build up in a Blast Furnace and affect its performance. This limits the amount of sludge that can be recycled. To overcome the zinc problem at its plants, Bethlehem Steel developed and patented a new technology in which sludge is beneficiated by a hydrocyclone process that recovers iron and carbon values, while rejecting constituents such as zinc and alkalis.

While characterizing the sludge solids, we found that these metals are concentrated in the finest size fractions (See Figure II), due, most probably, to a vaporization-condensation process within the Blast Furnace, which concentrates the metals on the highest-surface-area iron and carbon particles. This natural partitioning of the zinc provided a simple means of size-classification for removing these metals and recycling the beneficiated sludge solids. However, our studies showed that to effectively beneficiate this material, a size separation at about 10 microns was required, which precluded conventional screening operations for this application.

In studying this problem, we found that operating small-diameter hydrocyclones at inlet pressures typically two to three times that used in conventional hydrocyclone operations resulted in zinc and lead rejections of 80 to 90 percent to the hydrocyclone overflow, and 80 to 90 percent recovery of iron and carbon in the hydrocyclone underflow. Table II shows typical iron, carbon and zinc

## FIGURE III--BETHLEHEM PLANT BLAST FURNACE SLURRY HYDROCYCLONE FACILITY

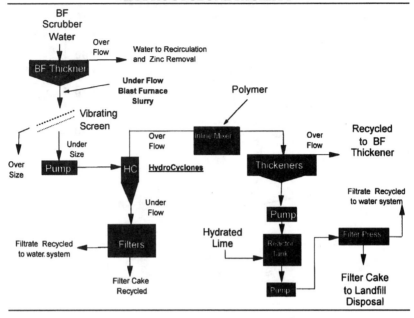

recovery rates.

Since Bethlehem began looking into this, several hydrocyclone configurations have been evaluated for Blast Furnace operations. These include both single- and double-stage hydrocyclones. As a result, we are installing a two-stage hydrocyclone at Bethlehem's Burns Harbor plant in Indiana. The second-stage hydrocyclone re-treats the primary hydrocyclone overflow to improve recovery of iron and carbon units.

Figure III presents the flow sheet of the hydrocyclone and related slurry-processing equipment installed at the Bethlehem Structural Products Corporation. In this application, we installed only one-stage hydrocyclones. This facility treats about 35 DNT/day of scrubber sludge, recovering about 80 percent of the solids-- primarily iron and carbon units--as hydrocyclone underflow. This slurry is dewatered and eventually recycled to the Blast Furnace. Hydrocyclone overflow containing the bulk of the zinc is thickened and processed by a lime-enhanced pressure filtration process.

Taken together, hydrocyclone and extrusion processes have allowed Bethlehem to replace an equivalent amount of purchased iron ore. Moreover, thousands of dollars have been saved by reducing the cost of disposing of plant reverts in a landfill.

58

## TABLE III
## BETHLEHEM STRUCTURAL PRODUCTS CORPORATION
## PRODUCTS AND CO-PRODUCTS RECYCLED OR SOLD

| SOURCE | AMOUNT Tons/Day | USES |
|---|---|---|
| **BLAST FURNACE** | | |
| Dry flue dust | 35 | Recycled for iron recovery |
| Scrubber sludge-- HydroClone product | 28 | Recycled for iron recovery |
| Slag | 350 | Pavement subbase<br>River bank and roadside stabilizer<br>Railroad ballast<br>Anti-skid aggregate<br>Roofing aggregate<br>Asphalt-concrete aggregate<br>Cement<br>Portland cement concrete aggregate<br>Mineral fiber |
| **BASIC OXYGEN FURNACE** | | |
| Dust | 100 | Agglomerated/recycled for iron recovery<br>Sold for iron content |
| Slag | 350 | Recycled to BF for flux content<br>Asphalt-concrete aggregate<br>Subbase<br>Anti-skid aggregate<br>Railroad ballast<br>Bituminous sand |
| **OTHER MATERIALS** | | |
| Spent silica refractory bricks | 25 | Processed for use as a filler in vitrious china, toilets, water basins, artificial slate, roofing tile and other products |
| Spent clay refractory bricks | -- | Processed for use in baseball infields, running tracks and decorative aggregate |
| Spent foundry sands | 8 | Mixed with soil and compost to make quality topsoil |
| Demolition concrete | -- | Subbase |

## PROCESSING SLAGS, SANDS AND REFRACTORIES FOR SALE

The steelmaking industry historically has generated a tremendous amount of slag, spent refractories and sands, as well as waste concrete and asphalt (from reconstruction sites), that have been placed in landfills despite their potential for

59

reuse. At the Bethlehem Structural Products Corporation, however, these materials have been diverted from landfills through a highly successful partnership with an industry that prepares and sells them as high-quality materials for a variety of uses in the construction industry. (See Table III)

## SLAG

Bethlehem Structural Products Corporation produces seven railroad cars (700 tons) of slag daily from its Blast Furnace and Basic Oxygen Furnace (BOF), slag that once was placed in landfills. Seven hundred tons a day amounts to 255,500 tons a year and 2,555 railroad cars, enough slag to make a 28-mile-long train. Today, however, instead of being placed in a landfill, this slag is transported to a nearby independent industry, Waylite, for processing. Waylite is a subsidiary of International Mill Service, Inc.

Waylite crushes and grades all this slag for a variety of uses including use as a pavement subbase, river bank and roadside stabilizer, railroad ballast, anti-skid ice-control aggregate, roofing aggregate, component in Portland cement concrete mixes and concrete masonry, and raw material for conversion to mineral fiber. BOF slags contain extra lime and are recycled to the Blast Furnace for reuse as a flux and partial replacement for limestone or are used as construction aggregate.

For Waylite, however, the 700 tons of daily slag represents only 15 percent of its daily work load. Waylite also *mines* another 4,000 tons of slag daily from old Bethlehem Steel slag banks--another 40 railroad cars a day.

## REFRACTORIES

Besides slag, Bethlehem generates 9,000 tons of refractories (clay brick, Coke's silica brick, and high magnesium brick) yearly from its Blast Furnace, BOF, coke ovens and other furnaces. This too is sent to Waylite, which crushes and sizes it. For Waylite, processing the refractories was the easy part; developing markets was the hard part. Once potential markets were identified by Waylite, Bethlehem Steel had to change its style of managing refractories by separating them before shipping them to Waylite to make it easier for Waylite to process them into different products. Silica brick, for example, is ground to a flour-like consistency for use as a filler in the china used in making toilets, artificial slate roofing tiles, and certain bricks.

Crushed clay brick mixed with crushed slag is successfully used for baseball infields because its high absorption rate makes infields resist rain saturation. This infield mix product is specifically formulated to Major League specifications. Crushed clay bricks are also used as a decorative landscape aggregate.

## SANDS

Bethlehem's Blast Furnace and foundries (Ingot Mould, Iron and Brass) produce 3,000 tons of sands a year. Waylite takes these too and has found a variety of uses for these sands, including use as an additive for blending with yard and leaf compost to make a high quality organic topsoil sold to landscapers and building contractors. Another use is as an additive in asphalt.

## CONCRETE AND AGGREGATE

Because Bethlehem is constantly razing or rebuilding parts of its facility, it generates a lot of concrete and asphalt waste material. Historically these materials would have been sent to a landfill, but again the partnership with Waylite found other uses for these "resources." Waylite crushes the concrete, which is blended with BOF slag for use as subbase.

## CONCLUSION

Bethlehem's experiences with finding ways to use materials that were once discarded shows how one industry has met its environmental responsibilities, generated income, worked in partnership with smaller industries to create new products and jobs, and acted as a responsible neighbor in the community.

The environmental impact of Bethlehem's waste management program is significant in several ways that include (1) reducing the need for landfill space, (2) eliminating disposal problems related to troublesome wastes, (3) conserving natural resources, and (4) reducing energy consumption and the cost of controlling pollution.

Given Bethlehem's experiences, there is little to suggest that similar ventures could not be undertaken by other companies.

# WASTE MINIMIZATION IN ELECTROPOLISHING: PROCESS CONTROL

ALLEN P. DAVIS and CAIROLE BERNSTEIN
Environmental Engineering Program
Department of Civil Engineering
University of Maryland
College Park, Maryland 20742

PAUL M. GIETKA
Technology Extension Service
University of Maryland
College Park, Maryland 20742

## ABSTRACT

Electropolishing of metals produces high dissolved metal concentrations in the polishing baths and an eventual buildup of metallic sludge. Consequently, hazardous waste management is a significant operating cost for the electropolisher. A study of the removal of metal during the electropolishing process has found that this removal varied greatly and is a function of the age of the bath. In all cases, work quality was acceptable. Thus contaminant input to the electropolishing bath may be reduced by controlling electropolishing conditions. Regulating electropolishing time based on the age, or metal content of the electropolishing bath (a solution low in metal content polishes a piece more quickly) as opposed to maintaining a constant polishing time regardless of bath age has promise to reduce waste generation and to increase useful bath life.

## INTRODUCTION

Electropolishing is the process of passing a current through a metal work piece, in this case, stainless steel, removing a thin layer of the work. Metals are electropolished to produce a bright, corrosion-resistant finish and for efficient sterilization. The stainless steel workpiece is the anode and a thin layer of the work is oxidized, dissolving the piece and thus releasing iron, nickel, and chromium, the constituents of the stainless steel alloy:

$$\text{Stainless Steel} \rightarrow Fe^{3+} + Ni^{2+} + Cr^{6+} + n\ e^- \qquad (1)$$

At the cathode (usually copper), the iron and chromium are reduced to the divalent and trivalent species, respectively [1]:

$$Fe^{3+} + Cr^{6+} + n\ e^- \rightarrow Fe^{2+} + Cr^{3+} \qquad (2)$$

The current density (current per unit area of anode) and corresponding dissolution are largest in rough areas and sharp edges, thus smoothing or polishing the work. The most effective polishing is achieved in an acid bath, of which several mixtures are used.

Pier-Sol, Inc. is a small electropolishing company in Baltimore, MD. Pier-Sol currently operates a primary electropolishing line consisting of an electropolish bath mixture of sulfuric and other proprietary acids, followed successively by two dead water rinse tanks and one spray rinse tank (Figure 1). In each tank, daily liquid losses are replenished from the succeeding tank, with losses from the second rinse tank made up with fresh water. This recycle process creates a closed-loop system that conserves water and minimizes waste load to the wastewater treatment system, but causes the accumulation of dissolved metals in the first two tanks. In fact, on occasion, the primary rinse tank has been measured to have a *higher* concentration of iron and copper than the electropolishing bath. This may result from a high concentration of metals in the slime attached to parts when dragged out from the electropolish bath to the primary rinse tank.

The removed metals accumulate in the electropolishing bath until either the buildup of metal sludge interferes with polishing, or the concentration of metals in solution reduces operational efficiencies to unacceptable standards. The dense sludge is removed from the polishing tanks and disposed of as a hazardous waste due to the heavy metals present. However, much of this waste is iron.

The common technologies used for waste reduction for electroplaters (i.e., ion exchange and evaporation) are not applicable to electropolishers since they have no use for concentrated metal solutions and the majority of their waste is non-valuable iron. Electrolytic treatment of electropolishing wastewater was unsuccessful due to the difficulty in plating iron and the low pH of the solution [2].

Consequently, an investigation of waste production by Pier-Sol and possible waste minimization by varying plating line operational procedures with associated process control was initiated.

The chemical makeup of the electropolishing baths change considerably with use over time. Metal analyses for the electropolishing tank and the subsequent two static rinse baths are presented in Table I. Metal concentrations in the baths are a function of the age of the electropolishing baths and material throughput. Usually iron is present in the largest quantity (in one case at 155 mg/L in the primary rinse tank). Copper results from some corrosion of the anodes in the strong acid media.

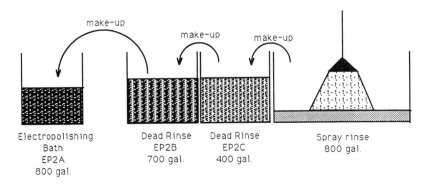

**Figure 1.**   Schematic of Electropolishing Line at Pier-Sol, Inc., Baltimore, MD.

## PROCESS ANALYSIS AND MODELING

The primary parts electropolished by Pier-Sol are mouse cage lids for medical research supply companies. A detailed investigation was conducted on quantifying the amount of metal removed during the electropolishing process by weighing the stainless steel part before and after electropolishing. Mouse cage lids were collected and monitored over a 4-month period encompassing portions of 4 electropolishing bath cycles. The cages were weighed and the metal removal calculated based on the average difference in mass between the polished and unpolished cages. Table II shows the results from this investigation. Two types of cages were processed by Pier-Sol over this time period.

It is found that the amount of stainless steel removed by a typical electropolishing treatment strongly depends on the age of the polishing bath. Mouse cage lids polished using an old bath ready for desludging (sludge removal with subsequent addition of fresh acids) removed as little as 0.3% of the total cage mass during electropolishing, with little variation among lids. In contrast, electropolishing the same lids using a fresh bath removed over 4% of the stainless steel from the cages with a variation from 2 to 6%. All of the cages were polished to acceptable quality, although those from the newest and oldest baths are considered as extremes. In all cases, the electropolishing time is approximately 10 minutes, regardless of the bath age.

A plot of metal removal as a function of bath life is presented in Figure 2.

**Table I.** Analyses of Electropolishing Bath and Rinsewaters (as measured by private laboratory).

| | Concentration (mg/L, except for pH) | | | | |
|---|---|---|---|---|---|
| | pH | Chromium | Copper | Iron | Nickel |
| *EP Bath* | | | | | |
| March 1991 | <0 | 42.8 | 0.39 | 37.3 | 13.69 |
| November 1991 | <0 | 21.0 | 0.71 | 32.4 | 10.9 |
| *Primary Rinse* | | | | | |
| March 1991 | <0 | 10.11 | 1.43 | 154.6 | 3.4 |
| November 1991 | <0 | 18.3 | 3.46 | 37.3 | 9.2 |
| January 1992 | <0 | 11.2 | 5.6 | 26.4 | 5.84 |
| *Secondary Rinse* | | | | | |
| March 1991 | <0 | 4.15 | 0.82 | 6.9 | 1.37 |
| November 1991 | 0.3 | 4.77 | 2.50 | 13.2 | 2.54 |

The large cage masses are presented on the right axis of the plot in the same proportion as the small cages. On average, 17 grams of stainless steel from the cages are removed in new baths, while this value approaches 1 gram in nearly-spent baths. This trend is evident even though the data involve two types of cages and four different bath cycles.

A large deviation in cage mass is noted in the cages with the most metal removal (Figure 2). This may be due to short variations in electropolishing times which, in the new baths, would result in large differences in metal removal. The variability may also be caused by the location and placement of the lids on the anode rack, where again, small differences would be magnified in the fresh electropolishing baths. The number of cages loaded into the bath on each cycle was always the same.

In order to estimate the effect that altering the electropolishing operational characteristics will have on bath life, a simple model describing the electropolishing process is developed. The mass of metal removed from an electropolished piece (M) is given as the rate of removal (r) times the electropolishing time, $\Theta$:

$$M = r\Theta \tag{3}$$

**Table II.** Masses of Electropolished Mouse Cage Lids from Polishing Baths of Various Ages.

| Bath Cycle | Weeks after Desludge | Cage Mass (g) | | Metal Removed | |
|---|---|---|---|---|---|
| | | Average | Ave. ± 1 Std. | (g) | (%) |
| *Small Cages* | | | | | |
| Unpolished Cages | - | 409.1 | 408.7 - 409.5 | - | - |
| A | approx. 9 | 408.0 | 407.2 - 408.8 | 1.1 | 0.27 |
| B | approx. 1/2 | 391.8 | 384.2 - 399.4 | 17.3 | 4.2 |
| C | 1.9 | 393.6 | 386.6 - 400.6 | 15.3 | 3.7 |
| C | 4.6 | 399.7 | 396.6 - 402.8 | 9.4 | 2.3 |
| C | 5.7 | 397.4 | 395.1 - 399.7 | 11.7 | 2.9 |
| D | 1.6 | 395.0 | 392.7 - 397.4 | 14.1 | 3.4 |
| D | 9.1 | 404.6 | 403.8 - 405.4 | 4.5 | 1.1 |
| *Large Cages* | | | | | |
| Unpolished Cages | - | 619.5 | 614.9 - 624.1 | - | - |
| D | 3.3 | 604.0 | 601.8 - 606.2 | 15.5 | 2.5 |

The rate of polishing is a function of the bath life, as determined above. It is assumed that this rate can be described as an initial electropolishing rate ($r_o$), with a first-order dependence on the concentration of soluble and insoluble metal impurities (C) in the bath that decrease the rate:

$$r = r_o - kC \qquad (4)$$

where k is a rate constant. C varies and increases over the bath life due to the accumulation of metals in the bath as a result of metal removal during the electropolishing process:

$$C = \int_0^t \frac{n\,M}{V}\,dt' \qquad (5)$$

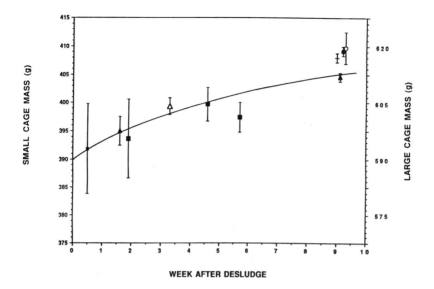

**Figure 2.** Mass of Cage Lids as a Function of Electropolishing Bath Age.

where n is the number of pieces processed, V is the volume of the electropolishing bath and t is the age of the bath.

These three expressions are interrelated can be evaluated based on the mode of operation used. Substituting Equations 3 and 5 into 4 results in:

$$M = r_o\Theta - \frac{kn\Theta}{V} \int_0^t M \, dt' \tag{6}$$

For example, under the current operating conditions, electropolishing time, $\Theta$, is held constant at 10 minutes; however the mass of metal removed varies as a function of bath life. The solution to this integral equation for mass removal is:

$$M = r_o\Theta \exp\left(\frac{-kn\Theta t}{V}\right) \tag{7}$$

thus predicting an exponential decrease in the amount of metal removed as the bath ages.

Equation 7 is fit to the data on metal removal (Figure 2) using a non-linear, least-squares program. This best-fit is produced by $r_o$ = 1.93 g/min-cage and kn/V = 0.0155 $min^{-1}$-$weeks^{-1}$. For typical Pier-Sol values of n = 3500 cages/week and V = 800 gallons, k = $3.54 \times 10^{-3}$ gal/cage-min. The equation fit is given by the solid line in Figure 2, showing that the model adequately describes the polishing process.

The total amount of metal input to the bath at any time is given by:

$$n\Theta \int_0^t M \, dt' = \frac{r_o V}{k} [1 - \exp(\frac{-kn\Theta t}{V})] \qquad (8)$$

For a typical 9.5 week bath life, this corresponds to 336 kg of steel per bath.


## PROCESS MODIFICATION

A modification is suggested to decrease the amount of metal added to the bath, and thus increase bath life and minimize waste disposal. Currently, electropolishing time is fixed at about 10 minutes. Thus if polishing time is decreased during the initial weeks of the bath, the metal removal will be less and the bath life extended, with no reduction in final product quality. It is proposed that the minimum acceptable electropolishing time be used in fresh baths. This time can be extended up to the full 10 minutes (or longer) as the bath ages.

The rate of metal removal (as estimated from Figure 2) is approximately 1.9 g/min for the small cages in a fresh bath . A removal target of, for example, 4 grams would reduce the initial electropolishing time to about 2.1 minutes. Consequently, during this initial week, the metal input to the bath would decrease from about 17 to 4 g/cage. At 3500 cages per week, this corresponds to a reduction from 60 to 14 kg of stainless steel accumulated in the polishing bath, with comparable reductions in sludge production and correspondingly, an increase in bath life. Less significant reductions would result in subsequent weeks of the bath life.

It is proposed to operate the electropolishing line using a fixed metal removal mass and vary the electropolishing time. Thus with a fixed M, Equation 6 gives:

$$M = r_o\Theta - \frac{k M n \Theta t}{V} \qquad (9)$$

and thus the electropolishing time required to produce this selected mass of stainless steel removal can be calculated as a function of bath age:

$$\Theta = \frac{M}{r_o - k M n \, t/V} \qquad (10)$$

Recommended polishing times for the electropolishing line at Pier-Sol can be found by using the numerical constants determined for the constant time operation. Thus:

$$\Theta = \frac{M}{1.93 - 0.0155 \, M \, t} \qquad (11)$$

where $\Theta$ will be given in minutes, M in grams and t in weeks.

Table III presents the electropolishing time calculated for the removal of both 4 and 6 grams of stainless steel from all cages as a function of bath age. The polishing times become very large after the removal of 336 kg of steel by the bath and at this point desludging is required. Such a process control system based on metal removal is more efficient than one based on electropolishing time, however the control is more difficult. A precision balance may be used to monitor removed stainless steel from the lids.

## ANALYSIS OF WASTE MINIMIZATION

Two desludgings at Pier-Sol have each produced four and five drums of sludge, respectively. Thus at 5.5 desludges per year (52/9.5), 4½ drums per desludge, the hazardous waste production is approximately 25 drums per year from this electropolishing line. From above, it is estimated that 336 kg of steel is removed for each bath life, resulting in 1850 kg of steel removed per year.

If, via a process control procedure, all lids were equally polished to a 4 g/lid removal, at 336 kg/bath, the bath would polish 84,000 lids and, consequently, last 24 weeks at the same 3500 cage/week throughput--approximately 2.5 times the current bath life. The desludging would be required only 2.2 times a year (which in itself saves manpower, down-time, and laborious work), resulting in a yearly waste generation of 9.8 drums per year--less than half of the current amount. This estimate assumes that electropolishing throughput remains constant. Presently, throughput is limited by manpower and not by electropolishing time.

## TESTING

Subsequent testing of modifications of the polishing time has been inconsistent with predicted results. The metal removal did not correspond to trends determined earlier. Apparently during the previous desludge, the acid bath mixture was altered. Thus the control of the bath composition is also

**Table III.** Calculated Electropolishing Time for Constant Metal Removal as a Function of Electropolishing Bath Age (Eqn 11).

| Bath Age (Weeks after Desludge) | Electropolishing Time (min) | |
|---|---|---|
| | 4 g Metal Removal | 6 g Metal Removal |
| 0 | 2.1 | 3.1 |
| 2 | 2.2 | 3.4 |
| 4 | 2.4 | 3.9 |
| 6 | 2.6 | 4.4 |
| 8 | 2.8 | 5.1 |
| 10 | 3.1 | 6.0 |
| 12 | 3.4 | 7.4 |
| 14 | 3.8 | 9.6 |
| 16 | 4.3 | 13.6 |
| 18 | 4.9 | - |
| 20 | 5.8 | - |
| 22 | 7.1 | - |
| 24 | 9.0 | - |
| 26 | 12.6 | - |

important in order to optimize electropolishing efficiency. Further study is needed to evaluate this parameter and to determine if either an optimal bath composition exits, or bath composition can be incorporated as a refinement to the process control model.

## SUMMARY

Controlling electropolishing time based on the age, or current metal content, of the electropolishing tank (a solution low in metal content polishes a piece more quickly) as opposed to maintaining a constant polishing time regardless of bath age has promise to reduce waste generation and to increase useful bath life. Further evaluation of the operating conditions at Pier-Sol will determine the feasibility of process control by monitoring metal removed or by adjusting electropolishing time based on calculated and predicted metal removals.

## ACKNOWLEDGEMENT

This work was supported by the Maryland Industrial Partnerships which is administered by the University of Maryland Engineering Research Center, and the U.S. Environmental Protection Agency's Pollution By and For Small Business Program which is administered by the Center for Hazardous Materials Research at the University of Pittsburgh.

## REFERENCES

1. Faust, C.L. 1982. "Electropolishing: Stainless Steel," *Metal Finishing*, 80(9):89-93.

2. Bernstein, C. and Davis, A.P. May 1992. "Waste Minimization in Electro-polishing: Electrodeposition and Process Control," Final Report to Center for Hazardous Materials Research, through Pier-Sol, Inc., Baltimore, MD.

# APPLICATION OF PYROLIZED CARBON BLACK FROM SCRAP TIRES IN ASPHALT PAVEMENT DESIGN AND CONSTRUCTION

TAESOON PARK
BRIAN J. COREE
Indiana Department of Transportation
Division of Research
1205 Montgomery Street
P. O. Box 2279
West Lafayette, IN 47906
C. WILLIAM LOVELL
School of Civil Engineering
Purdue University
West Lafayette, IN 47907

## INTRODUCTION

According to EPA reports (1991) of the over 242 million waste tires generated each year in the United State, 5 % are exported, 6 % recycled, 11 % incinerated, and 78 % are landfilled, stockpiled, or illegally dumped. A variety of uses for these tires are being studied. Among these is pyrolysis which produces 55% of oil, 25% of carbon black, 9% of steel, 5% of fiber and 6% of gas. Pyrolized carbon black contains 9 % of ash, 4 % of sulfur, 12 % of butadine copolymer and 75 % of carbon black.

The objective of this research is to investigate the viability of using PCB as an additive in hot mix asphalt. The use of PCB in asphalt pavement is expected not only to improve the performance of conventional asphalt, but also to provide a means for the mass disposal of waste tires.

## MATERIALS USED

### Aggregate

Following Indiana Department of Transportation (INDOT) specifications, #9 binder aggregate was used for the target gradation. This target gradation is shown in Table I. The bulk specific gravity and apparent specific gravity value of the coarse aggregate are 2.47 and 2.51, respectively, and for the fine aggregate, 2.742 and 2.797, respectively.

Table I. The Gradation of Aggregate Used

| Sieve Size | % Passing (Controlled) | Spec. Range % Passing |
|---|---|---|
| 3/4 " (19 mm) | 100 | 100 |
| 1/2 " (12.5 mm) | 81 | 70 - 92 |
| 3/8 " (9.5 mm) | 63 | 50 - 76 |
| # 4 (4.75 mm) | 40 | 40 ± 5 |
| # 8 (2.36 mm) | 25 | 18 - 45 |
| # 16 (1.18 mm) | 16 | 10 - 36 |
| # 30 (0.6 mm) | 10 | 6 - 26 |
| # 50 (0.3 mm) | 6 | 2 - 18 |
| # 100 (0.15 mm) | 4 | 0 - 11 |
| # 200 (0.075 mm) | 2 | 0 - 4 |

## Asphalt

Two grades of asphalt, AC-10 and AC-20 were used. The physical properties of AC-10 and AC-20 comply with the INDOT specifications.

## Pyrolized Carbon Black

The pyrolized Carbon Black (PCB) was provided by Wolf Industries, Brazil, Indiana. Yields from tire pyrolysis vary with the factory and the process used. Tire pyrolysis typically yields 55% of oil, 25% of carbon black, 9% of steel, 5% of fiber and 6% of gas. Pyrolysis is also called destructive distillation, thermal depolymerization, thermal cracking, carbonization, or cooking. Pyrolized carbon black used in this study is obtained from the most common process which is reductive (retort) pyrolysis.

As previously stated, pyrolized carbon black contains a maximum of 9% ash content, and 4% sulfur content, 12% of minimum butadine copolymer content(nitrile rubber), and 75% of carbon black. This type of carbon black could partially replace commercial blacks for the preparation of low-grade rubber products (Roy et al., 1990). The particle size and surface area of mill ground pyrolized carbon black are shown in Table II. General properties of a similar carbon sample produced during vacuum pyrolysis of waste is also summarized in Table III. From the environmental point of view, the pyrolized carbon black may form toxic materials, carbon dioxide and carbon monoxide, however as PCB is insoluble and stability is high, this may not easily occur.

Table II. The Particle Size and Surface Area of Mill Ground PCB

| Name of the Products | Rg (Å), Particle Size | Specific Surface Area, $gm/cm^3$ | Large Scale df | Large Scale Rg ($\mu$m) |
|---|---|---|---|---|
| BC 100 | 430 | 157 | 1.9 | 0.87 |
| BC 200 | 343 | 188 | 2.3 | 0.52 |
| BC 500 | 439 | 159 | 2.4 | 0.70 |
| WC 500 | 230 | 338 | 1.7 | N/A |
| NC 339 | 304 | 187 | N/A | (> 0.49) |

NOTE : NC 339 is a pure carbon black and listed for comparing to pyrolized carbon black.

Table III. General Properties of Carbon Black Produced during Vacuum Pyrolysis of Used Tires. (After Roy et al, 1990)

| | |
|---|---|
| Iodine Index (mg/g) | 144.2 - 151.4 |
| DPB Adsorption (ml/100g) | 84.6 - 93.0 |
| Heat Loss at 105 °C (%) | 0.4 - 1.0 |
| Tint Strength | 57.1 - 60.6 |
| Ash (%) | 15.5 - 17.0 |
| Volatile Matter | 4.9 - 3.3 |
| S ( %) | 2.5 - 3.0 |

NOTE : Ultimate temperature was 525°C and total pressure varied between 1.5 and 4.5 kPa. (Feedstock included both regular and steel belt used tire samples.)

The pyrolized carbon black is blended with asphalt as received. Specific gravity of pyrolized carbon black is 1.486. The particles of pyrolized carbon black are much coarser than high structure HAF type carbon black, however, most of the coarse particles are easily broken down by normal pressure. The color is lighter than HAF type carbon black.

Carbon Black
Carbon black (CB) was purchased from CABOT Industry, Boston, Massachusetts. The carbon black used in this study is high structure high abrasion furnace (HAF) type carbon black. Specific gravity of carbon black is 1.967. The reason for choosing the high structure HAF type is that several researchers have reported that the carbon black modified asphalt cement using the high structure HAF type resulted in improvement of temperature susceptibly, rutting and cracking resistance (Khosla, 1991, Yao and Monismith, 1986, Vallerga and Gridley, 1980, Rostler et al, 1977)

**EXPERIMENTAL WORK**

Preparation of Binder
Heated pyrolized carbon black and carbon black were blended separately with heated asphalt cement. The content of pyrolized carbon black and carbon black based on the weight of asphalt, was chosen as 5%, 10%, 15%, and 20%, because previous study (Yao and Monismith, 1986) showed that carbon black contents between 10% and 15% produced enhanced rutting resistance and less cracking.

Mix preparation and Compaction for Marshall Method
Mix preparation of Marshall specimens used in this study is outlined in ASTM D1559. A total of 18 sets of mixtures were prepared. Three samples were prepared per each mixture. The aggregate used for this study evidenced very little absorption, therefore, compaction was carried out immediately after mixing. The level of compaction followed was in accordance with INDOT Specification, 75 blows per side, and other procedures were followed according to ASTM D1559 and MS-2.

## Gyratory Testing Machine (GTM)

The gyratory testing machine has been accepted as an effective and practical tool in the evaluation of characteristics and performance of bituminous mixtures. The gyratory testing machine produces test specimens by a kneading compaction process. Specimens prepared by the gyratory testing machine seem to provide the most representative stress-strain properties.

## Mix Preparation and Compaction for GTM

Mix preparation for the gyratory compaction specimens was the same as for the Marshall test method. Other procedures were followed in accordance with ASTM D 3387 and a GTM manual provided by McRae (1993). A four inch (101.6 mm) diameter mold was used. A 1.25° angle of gyration and 120 psi (827.4 Kpa) normal pressure were selected to produce the specimens. The number of revolutions was selected at 250 for the ultimate compaction effort. The variation of the roller pressure and the height of sample were recorded at every 50 revolutions, to indicate the effects of traffic loads for different ratios of pyrolized carbon black and carbon black. The height of sample, gyratory angle, and applied pressure was recorded by the gyrograph. After the compaction was completed, the sample was extruded from the mold. The compacted samples were cooled in the laboratory temperature (67°F to 72°F) for 12 hours for the bulk specific gravity test.

## TEST RESULTS AND DISCUSSIONS

### Marshall Test Method

The Marshall test results define the characteristics of bituminous mixtures in relation to their binder contents. The optimum binder contents were determined from the test data. The INDOT Marshall criteria were used to determine the optimum binder content. The void relationships and the mechanical properties of each mixture were measured and compared.

## Air Voids (VTM)

The air voids are estimated by comparing the average bulk specific gravity ($G_{mb}$) to the theoretical maximum specific gravity ($G_{mm}$) for that asphalt content. The air voids can be calculated by the following relationship.

$$VTM = \left(1 - \frac{G_{mb}}{G_{mm}}\right)100 \qquad (1)$$

Figure 1 shows the air voids versus the PCB content for AC-10 mixtures. The air voids increase almost linearly with increasing PCB contents. This trend is the same as compared to the conventional mixtures and the CB mixtures.

## Voids in Mineral Aggregate (VMA)

The VMA can be neither low nor too high. A low VMA indicates that a satisfactory asphalt film thickness can not be provided, while a high VMA indicates a low stability for the mixture. The AC-10 PCB mixtures show increases in the VMA with increase of the PCB content, except for the 20% mixture. The AC-20 PCB mixture shows the same trend.

75

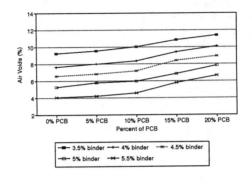

Figure 1. Air Voids vs. PCB Content (AC-10 Mixtures)

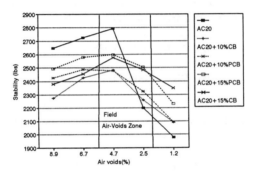

Figure 2. Stability vs. Air Voids (AC-20 Mixtures)

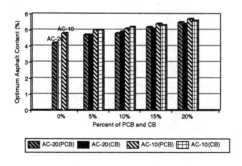

Figure 3. Optimum Binder Contents

## Marshall Stability

The mechanical properties of the asphalt mixture are indicated by the Marshall stability. While the Marshall stability is decreased by the inclusion of the PCB in both grades of asphalt mixture, the stability increases with increasing PCB contents. The stability values of the PCB mixtures and CB mixtures are almost identical. The stability of the AC-10 mixture increases up to 15 percent of PCB, and then decreases. However, the stability of AC-20 mixtures increases with increasing PCB content. Stability is roughly proportional to PCB content for AC-20 mixtures.

The relationship between air voids and stability was examined in order to compare the sensitivity of stability to the variation of the air voids. Figure 2 shows the stability versus the air voids for AC-20 mixtures. As can be seen in Figure 2, decrease of stability in the conventional mixture is much more severe than for CB and PCB mixtures. The PCB mixture shows less sensitivity and has higher strength than the CB mixture with decreasing the air voids for the AC-20. The AC-10 mixtures show the same trend.

## Marshall Flow

The flow is essentially independent of the PCB content with AC-10. However, the AC-20 mixtures show a significant decrease in flow with increasing PCB content. The latter indicates a potentially enhanced resistance to plastic flow.

## Optimum Binder Content

The optimum binder content is the most important factor in bituminous pavement design. Figure 3 compares the optimum binder contents and shows that the optimum binder content varies somewhat depending on the grade of the asphalt. The optimum asphalt content decreases with increasing the PCB and CB content in AC-10 mixtures. However, the optimum binder content increases with increasing the PCB and CB contents in AC-20.

## Gyratory Testing Machine (GTM)

The GTM provides stress-strain and plastic deformation relationships for asphalt mixtures. The Gyratory Compactibilty Index (GCI), the Gyratory Stability Index (GSI), Gyratory Shear Index (Sg), the Gyratory Shear Factor (GSF) and the unit weight versus the GTM revolution can be obtained through the GTM test data

## Gyratory Compactibilty Index (GCI)

The closer the index approaches unity, the easier the mixture is to compact. The GCI value for both grades of asphalt with PCB is close to one. Thus the compactibilty of the mixtures is not effected by the inclusion of PCB.

## Gyratory Stability Index (GSI)

The stability of a mixture can be estimated by the GSI. The GSI is related to the plastic deformation of pavements. As the GSI value is closer to unity, the mixture becomes more stable and less plastic deformation is likely to occur. The GSI can be obtained from the following relationship, the ratio of the maximum gyratory angle ($\theta_{max}$) to the minimum gyratory angle ($\theta_{min}$).

$$GSI = \frac{\theta_{max}}{\theta_{min}} \qquad (2)$$

The official criterion for a GSI value has not yet been finalized. McRae (1993) suggested that a GSI close to unity is typical of a stable mix. Robert et al. (1991) specified that a value above 1.1 usually indicates unstable mixtures. According to Zhang et al. (1994), research by the Maine DOT suggested that the GSI should be less than 1.15 after 300 revolutions to prevent rutting, Illinois studies suggested that the GSI should be less than 1.25 after 300 revolutions. A GSI value of 1.2 is selected as the criterion in this study.

The relationship between GSI and number of revolutions for the AC-10 mixtures is provided in Figure 4. The GSI increases with an increasing number of revolutions. It is observed that the PCB mixtures perform much better than the CB mixtures, and both conventional mixtures also show stable conditions in terms of GSI.

Gyratory Shear

Gyratory Shear (Sg) is related to the shear resistance of mixtures. A reduction of this value during the compaction process indicates loss of stability. Currently, an official criterion for Sg is not available. Research from the Maine DOT (1992) recommends the minimum Sg value as 35 psi (241.32 KPa) after 300 revolutions. A gyratory shear value of 40 psi is used in this study.

In Figure 5 the effects of the PCB in AC-10 mixtures are shown. The conventional AC-10 mixture maintains a constant Sg after 100 revolutions in spite of increase of the GTM revolutions. While the PCB mixtures initially exhibit higher Sg values, the Sg is decreased with increasing the GTM revolutions. The same general trend is maintained in the AC-20 mixtures. Based on the parameter Sg, the shear resistance of the mixture can be increased by the inclusion of PCB as an additive.

Gyratory Shear Factor

The GSF is a factor of safety type index. When the GSF value is less than unity, inadequate shear strength for the anticipated maximum shear in the pavement is expected. McRae (1993) states in the GTM manual that GSI and GSF values should be considered at the same time, because the GSF value is not valid if the GSI is greater than unity. Table V shows the GSI and GSF value at 250 revolutions. The 5 % and the 10% PCB mixtures show desirable performance for both grades of asphalt mixture. From the GSI and GSF analysis, it can be concluded that the inclusion of the PCB in the asphalt mixture reinforces the shear resistance of the pavement and the plastic deformation can be controlled by using an appropriate amount of PCB.

Unit Weight

The initial increase rate of unit weight, during 50 to 150 revolutions, is very significant. Unit weight of CB mixtures steadily increases with the CB content for increases of 50 to 250 revolutions. However, PCB mixtures show a different behavior in unit weight, and the variation of unit weight is not so significant as for CB mixtures. The unit weight in terms of PCB contents is almost the same from 5 percent to 15 percent PCB, and decrease slightly after 15 percent PCB.

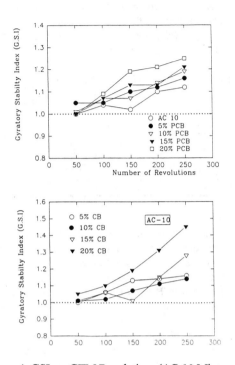

Figure 4. GSI vs. GTM Revolutions (AC-10 Mixtures)

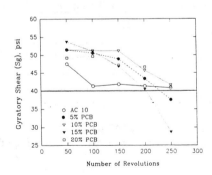

Figure 5. The Effects of PCB on Sg (AC-10 Mixtures)

The difference in unit weight for PCB mixtures could be attributed to the difference in specific gravity. The specific gravity of CB is approximately 1.948 and that of PCB is 1.486. Another reason could be differences in the particle sizes. The PCB involves much coarser particles than carbon black., therefore, when the asphalt is replaced by the PCB, more volume is occupied by the PCB, so that the unit weight of PCB mixtures is less than that of CB mixtures.

Table V Analysis of GSI and GSF at GTM 250 Revolutions

| Content | AC20+PCB | | AC20+CB | | AC10+PCB | | AC10+CB | |
|---------|------|------|------|------|------|------|------|------|
| | GSI | GSF | GSI | GSF | GSI | GSF | GSI | GSF |
| 0 % | 1.12 | 1.10 | 1.12 | 1.10 | 1.12 | 1.07 | 1.12 | 1.07 |
| 5 % | 1.17 | 1.07 | 1.12 | 1.19 | 1.16 | 0.98 | 1.16 | 0.81 |
| 10 % | 1.20 | 1.17 | 1.28 | 1.08 | 1.19 | 1.09 | 1.14 | 0.70 |
| 15 % | 1.29 | 0.97 | 1.41 | 0.94 | 1.21 | 1.09 | 1.28 | 0.45 |
| 20 % | 1.18 | 0.84 | 1.33 | 0.38 | 1.25 | 1.08 | 1.45 | 0.18 |

Summary of Discussion

The most significant improvement of the PCB mixtures is that the rate of reduction of stability with decreasing air voids is not as severe as that observed in the conventional mixtures. The same trend is observed with the CB mixtures, however, the rate of reduction of the stability is slightly higher than the PCB mixtures.

The relative effects of PCB and CB, when the GTM is use, are summarized in Table VI. As shown in Table VI, the PCB mixtures generally exhibit better performance.

The raw material cost of PCB is 16 cents/lb (Wolf Industries, 1995) and CB costs 71 cents/lb (CABOT, 1995). Therefore, cost-effectiveness can be achieved in addition to providing the improved performance of the pavement.

TABLE VI. The Effects of PCB and CB in AC-10 and AC-20 mixtures

| | PCB Mixture | | CB Mixture | | Conventional Mixture | |
|-----|---------|---------|---------|---------|---------|---------|
| | AC - 10 | AC - 20 | AC - 10 | AC - 20 | AC - 10 | AC - 20 |
| GCI | o | o | o | o | o | o |
| GSI | * | * | x | x | o | o |
| GSF | o | o | x | x | * | * |
| Sg | o | o | * | * | * | * |

Note : o : Good, * : Reasonable, x : Poor

## CONCLUSION AND RECOMMENDATIONS
The following conclusions and recommendations are drawn from the test results and analysis.

1) The characteristics and performance of PCB modified asphalt mixtures is somewhat dependent on the type of the asphalt.
2) The addition of PCB shows equal or improved performance for all the tests.

3) The inclusion of PCB in both grades of asphalt mixtures improves the shear resistance of the pavement. The plastic deformation can be controlled by the appropriate amount of PCB, which seems to be 5% to 10% of weight of asphalt.

4) The PCB additive should be studied in the laboratory for other aggregate types, such as slag.

5) Finally, test road verification will be required for PCB additives in the amounts determined by laboratory studies.

## ACKNOWLEDGMENTS

This study was sponsored by the Indiana Department of Transportation (INDOT) and the Federal Highway Administration (FHWA). The authors also appreciate the help of Wolf Industries, Brazil, Indiana, and Koch Materials, Terre Haute, Indiana.

## REFERENCES CITED

1)CABOT Industries (1995), Telephone Conversation with Christ Cornille on April 14, 1995.

2) EPA (1991) "Markets for Scrap Tires," Report No. EPA/530-SW-90-074A, United States Environmental Protection Agency, Office of Solid Waste, Washington , D.C..

3) Khosla, P.N. (1991), "Effects of the Use of Modifiers on Performance of Asphaltic Pavements." TRB Record 1317.

4) McRae, J. L. (1993), "Gyratory Testing Machine Technical Manual," Engineering Developments Co. Inc., Mississippi, 103 pp.

5) Maine Department of Transportation (MEDOT), (1992), "Evaluation of Gyratory Testing Machine Mixture Design Results for Work Performed in 1991", Research Report No. Problem Solving 92-21.

6) Robert, F.L., Kandhal, P.S., Brown, E.R., Lee, D.Y. and Kennedy, T.W. (1991), "Hot Mix Asphalt Materials, Mixture Design and Construction" 1st Edition, The National Asphalt Pavement Association (NAPA) Education Foundation, Lanham, Maryland.

7) Rostler F.S. (1977), "Carbon Black as a Reinforcing Agent for Asphalt." AAPT, Vol.46.

8) Roy, C., Labrecque , B. and Caumia de, B. (1990), "Recycling of Scrap Tires to Oil and Carbon Black by Vacuum Pyrolysis," Resources, Conservation and Recycling, No.4, pp.203-213.

9) Vallerga B.A. and Gridley P.F. (1980), "Carbon Black Reinforcement of Asphalt in Paving Mixtures" ASTM, STP 724.

10) Wolf Industries (1995), Telephone Conversation with Mr. Donald Foster on April 14, 1995.

10) Yao, Z and Monismith C.L.(1986), "Behavior of Asphalt Mixtures with Carbon Black Reinforcement." AAPT, Vol.55.

11) Zhang, X., Gress, D. and Eighmy, T., (1994) "Bottom Ash Utilization as an Aggregate Substitute in Hot Mix Asphalt" Proceedings, The 2nd Annual Great Lake, Geotechnical/Geoenvironmental Conference, May 20, 1994, Purdue University, West Lafayette, Indiana.

CATEGORY II:
# Technology, Regulations and Ethics

# TECHNOLOGY TRANSFER
## HOW IT CAN BEST BE ACCOMPLISHED

JAMES H. SALING
National Institute for Environmental Renewal
1300 Old Plank Road
Mayfield, PA 18433

## INTRODUCTION

The National Institute for Environmental Renewal (NIER) was formed through a Cooperative Research and Development Agreement (CRADA) between the Lackawanna Heritage Valley Authority and Martin Marietta Energy Systems representing the U.S. Department of Energy.

The goals of the NIER are; (1) to identify and prioritize existing environmental problems, (2) to identify appropriate technologies and systematic approaches to improve the condition of environmentally disturbed areas, (3) to facilitate the development of government, academic and private industry partnerships, (4) to export these technologies, methods and information to a variety of users through teleconferencing and long distance learning, and (5) to transfer these technologies to the private sector by creating alliances involving government agencies, private industries, and academic institutions to conduct field demonstrations. To achieve these goals, four separate programs have been established as shown in figure 1.

1.  An environmental research and development program to examine systematic approaches to improve the environment, and the specific technologies necessary to implement these approaches.
2.  A public policy and action program to develop alliances between government, academic, and private sectors to test, implement, and manage the systematic approaches and technologies.
3.  An education and training program to research, develop, and demonstrate technologies, methods, and information to distribute to a variety of users using a diverse set of training and education-delivery mechanisms.
4.  An incubator and technology transfer program among federal agencies, academia, and the private sector to exchange innovative, practical, and proactive technologies.

Figure 1 - Strategic Business Areas

## NIER APPROACH TO TECHNOLOGY TRANSFER

The NIER was established to restore the environmental and economic vitality of communities and industrial sites in the United States and internationally. Although we are presently focused on technology demonstrations and reclamation of properties in the Lackawanna Valley, our long range goals are for the NIER to be the environmental technology clearinghouse for the entire world. This would be accomplished by a test-bed for innovation and research for emerging technologies designed to remediate exploited lands such as coal mines and industrial sites throughout the Lackawanna Valley and the nation. The overall mission of the NIER entails not only the development and demonstration of existing and emerging technologies to solve environmental problems but also the subsequent education of a broad audience on the results of the institute's projects. The NIER's mission also mandates the adoption of public policy programs to facilitate interjurisdictional partnerships whose goal is to implement systematic approaches to environmental reclamation.

The NIER expects to accomplish its goals by creating partnerships among academic institutions, the private sector, and federal, state, and local governments on a project-specific basis. Consequently, the knowledge and experience gained through these projects will arise directly to the public and private sector partners as well as to international organizations.

### CURRENT PROGRAMS

Some of the programs that are currently in progress at the NIER are shown in the next three figures. Figure 2 shows the current programs in the areas of technology transfer and in research, development, and demonstration. The Environmental Monitoring and Management Systems (EMMS) program will collect existing environmental data for the Lackawanna Rivershed, model the data, and display the results in a Geographic Information System. This will then identify the need for and the type of additional data needed to properly characterize the environmental problems in the area. This program will also identify, procure, and emplace additional sensor throughout the watershed that will collect additional data and monitor all aspects of the environment in the watershed. The National Oceanic and Atmospheric Administration (NOAA) program will design and demonstrate a sensor platform that can be placed in streams or bodies of water to monitor all aspects of water quality. These sensors will relay near realtime environmental data continuously by radio or cellular phone back to the NIER operations center to be organized, displayed, and evaluated. The DOD/SERDP program is split into two separate programs. One is to demonstrate concepts to monitor and manage vast quantities of environmental data, to build engineering models of environmental conditions, and to support decisions related to remediation plans and progress. This effort will involve sensor data analysis and handling, image understanding, signal processing, data fusion, command and control, and sensor and computer system simulation and modeling. The other program will demonstrate four of the most promising technologies for removal of heavy metals from acid mine drainage pools.

- Environmental Monitoring and Management (EMMS)
- National Oceanic and Atmospheric Administration (NOAA)
- Department of Defense/Strategic Environmental R&D Program (DOD/SERDP)
  - Environmental Quality Monitoring and Assessment
  - Heavy Metals Removal

Figure 2 - Program Areas
Technology Transfer and R&D

Figure 3 illustrates some of the programs in education and training that the NIER is involved in. The "Environmental SMARTS" document is a proposed curriculum document for the introduction of environmental studies into the curriculum of 19 high schools in Northeastern Pennsylvania. This approach would weave environmental information into each and every subject that is offered instead of developing an environmental course which many students would not take. The NIER is also involved in international environmental education which involves a high school in Belarus and a high school in the Lehigh Valley. These two schools are collecting and sharing data on nuclear fallout in each of their local areas. Also, the NIER and ECOLOGIA hosted 16 visitors from the newly independent states of the former Soviet Union to discuss environmental issues. The NIER conducted a short training session with the visitors on the use of Technology Logic Diagrams. The NIER also has a trained and certified instructor for lead-based paint abatement. When funding becomes available, we will be able to train others in this important area.

- Environmental "SMARTS" Document
- International Environmental Education
- Technology Logic Diagram
- Lead-Based Paint Initiative

Figure 3 - Program Areas
Education and Training

In the area of public policy, the NIER has been and will continue to be heavily involved. Figure 4 shows some of the areas where we have been or are currently involved. In May of 1994, the NIER hosted a conference on Geographic Information Systems (GIS). Representatives from Government, academia and industry were in attendance at this conference. The NIER also worked with the city of Duryea to make available to them information on technologies that could be used to treat or dispose of used automobile and truck tires. The NIER is also sharing information with PADER, the U.S. Army Corps. of Engineers, the U.S. Office of Surface Mining, the Lackawanna County Commissioners, the Scranton City Storm Water Management Task Force, the Lackawanna River Corridor Association, and others on environmental technologies and issues.

- Geographic Information Systems Conference
- City of Duryea Tire Disposal
- PA Department of Environmental Resources
- U.S. Army Corps. of Engineers
- U.S. Office of Surface Mining
- Lackawanna County Commissioners
- Scranton City Storm Water Management Task Force
- Lackawanna River Corridor Association

Figure 4 - Program Areas
Public Policy

As discussed earlier, the NIER expects to accomplish its goals by creating partnerships and alliances with government, academic, and industrial partners. Figures 5 through 8 provides a list of academic institutions, private companies, non-profit organizations, and government agencies that we have met with, shared information with, and have a working relationship with.

Figure 5 is a list of academic institutions that the NIER has met with and has continuing contacts and exchanges of information with. Many of these institutions are actively involved in current NIER programs and others are expected to be involved in future programs.

Bloomsburg University
East Stroudsburg University
Keystone Junior College
King's College
Lackawanna Junior College
Lehigh University

Marywood College
NEIU - #19
Penn State University
University of Pittsburgh
University of Scranton

Figure 5 - Strategic Partnerships and Alliances
Academic Institutions

The private companies that are involved in our current programs are illustrated in figure 6. We are, of course, interested in involving private industry in our efforts to the maximum extent possible. We will, therefore, continue in our efforts to identify other industrial participants and to help new companies get started in a variety of environmental areas.

- AAI/SMI
- Booz-Allen & Hamilton, Inc.
- Centech Corp.
- Compost America, Inc.
- Hazleton Environmental Products, Inc.
- Martin Marietta Energy Systems, Inc.
- PAR Government Systems
- Phoenix Systems and Technologies
- Synergist

Figure 6 - Strategic Partnerships and Alliances
Private Companies

Some of the non-profit organizations that we are working with and sharing information with are shown in figure 7. Some of these organizations are actual team members on programs. For example, the National Environmental Technology Applications Center will provide assistance to the NIER in determining the feasability and cost of commercializing some of the leading edge technologies that are being used in our demonstration programs. ECOLOGIA is also providing support and key contacts in the Newly Independent States of the former Soviet Union in our efforts to work with them to demonstrate certain environmental technologies.

- Concurrent Technologies Center
- Conversion for the Environment
- Earth Conservancy
- ECOLOGIA
- Lackawanna River Corridor Association
- National Environmental Technology Applications Center

Figure 7 - Strategic Partnerships and Alliances
Non-Profit Organizations

Figure 8 lists some of the government agencies that we are either working for to demonstrate innovative technologies for cleaning up the environment or are sharing information on the results of those demonstrations.

- EPA
- Lackawanna Heritage Valley Authority
- National Park Service
- PA Department of Community Affairs
- PA Department of Environmental Resources

- PA Department of Education
- TVA
- U.S. DOA
- U.S. DOE
- U.S. DOD
- U.S. DOT

Figure 8 - Strategic Partnerships and Alliances
Government Agencies

We obtain technical support from the National Laboratory System (principally ORNL) and we have access to all governmental labs in our search for environmental technologies that can be used to address environmental problems of interest to the Institute. Figure 9 is a list of the laboratories that are most likely to provide information on the most promising types of technologies.

- Agonne National Labs (ANL)
- Hanford Engineering Development Labs (HEDL)
- Idaho Nuclear Engineering Labs (INEL)
- Oak Ridge National Labs (ORNL)
- Sandia National Labs (SNL)
- Savannah River Labs (SRL)
- Waterways Experiment Station (WES)

Figure 9 - Federal Laboratories

**CONCLUSION**

In conclusion, the NIER is a new institution with our first program being a little more than one year old. We have gotten off to a flying start. However, this has only been possible because of cooperative efforts of government, (federal, state, and local) private industries and academic institutions. We firmly believe that continued success can only be achieved by continuing to cooperate and share information in this important effort. We, therefore, invite all parties who are interested in improving the environment to work with us to this end.

# ENVIRONMENTAL CONSIDERATIONS OF CONSTRUCTION PROJECTS

H. SCOTT LAIRD, P.G., PROJECT MANAGER
Woodward-Clyde Consultants
5120 Butler Pike
Plymouth Meeting, PA 19462

PETER R. JACOBSON, P.G., VICE PRESIDENT
Woodward-Clyde Consultants
5120 Butler Pike
Plymouth Meeting, PA 19462

## INTRODUCTION AND BACKGROUND

Construction projects, whether commercial, industrial, or residential, often encounter unexpected conditions. Increasingly in recent years, these unexpected conditions are of environmental concern -- waste materials, contaminated soils, ground water or surface water, or related debris or structures. These features are encountered at construction sites in urban and suburban settings, and are often surprises that could not be accounted for in the planning of the construction work, despite extensive due diligence efforts (e.g., Phase I Environmental Site Assessments [1]). At a minimum, these surprises result in delays; most often, there are other costs accrued that were not in the original construction budget.

Increased awareness of the possible presence of contamination and the regulations regarding health and safety and waste management is responsible for a steadily  increasing number of these types of interruptions. When they do occur, before a decision on how to proceed is taken, there is often confusion as to what information is needed to assess the risks, whether regulatory agencies require the situation to be reported, and what regulations apply to actions that might be taken. Appropriate reactions and responses by the contractor, owner, and the engineer are needed to avoid significant and unnecessary delays to the project schedule while the best course of action is decided. The courses of action available can result in a wide range of additional costs to the owner and/or contractor depending on numerous factors. A possible worst case scenario is that expensive remediation can lead to implementation of overly conservative protective measures to human health and environment at high cost to the owner and sometimes to the contractor.

The purpose of this paper is to identify, from an environmental engineering perspective, some of the more significant and appropriate factors to be considered when potentially contaminated fill, soil, surface water, and ground water are unexpectedly encountered during the course of construction. It is important to keep in mind that each project has its own unique characteristics, requirements, and constraints. There is no single decision process applicable to any project, and the authors make no such implication. Also, since there are possible legal concerns in any decisions that are made with regard to contaminated materials, this paper is intended simply to heighten the awareness of potential owners, engineers, and contractors of the factors. Decision priorities are not proposed herein because they are most appropriate on a site-specific basis.

To illustrate the above points, the authors include several cases as examples. The particular cases were selected to represent typical situations, although not exactly similar. What is similar about them, however, are the types of deliberations needed to effect resolution so the project could proceed.

## FACTORS AFFECTING PROJECT COST

### Problem Definition

When an anomalous occurrence of material is initially encountered on the site, it is often difficult to know what data are necessary to characterize the problem, or, in other words, one must decide how much information about the contaminated material (i.e., nature and extent) is necessary to define the problem impacting the construction project. The decisions required need to be clearly articulated and data collection activities should be focused on only the information needed to make those decisions. Obtaining too much detailed information during the initial characterization stage can lengthen the project delay time as well as prove to be a disadvantage later. The reason is that subsequent decisions may have to take account of all data acquired separate and aside from the actual cost of obtaining and evaluating irrelevant data. This process is known as the Data Quality Objectives Process [2, 3], and it would be streamlined for most construction projects. For example, if contamination was noted during soil excavation for shallow footings, the process would focus on the areas to be excavated. Conversely, if deeper excavation were planned, and de-watering is necessary, it would be important to know in advance whether treatment of the extracted water would be necessary.

## Characterization/Classification of Material

If the nature of the problem constituents is unknown, it is generally good practice to consider whatever information is available about the site history as quickly as possible when deciding what analytical tests or characterization must be done, and to focus the testing on the most appropriate range of constituents. Once the composition of the constituents of interest is determined, a number of regulations may come into play which will influence the decision process depending on (a) what is the concentration level of the constituent of concern in the material, (b) whether the material is a solid or liquid waste, or soil or water containing waste thereby requiring special management requirements, and (c) whether the material is controlled or uncontrolled and a potential release to the environment is imminent. For example, the foundation of a building addition at a chemical plant was designed so that excavation would not be required because it was known that excavation spoils would have to be disposed offsite as hazardous waste. In this case consideration was given to identifying the process which generated the waste at a specific location. When construction is unexpectedly interrupted, this approach may be helpful in limiting the stop-work condition to smaller parts of the project site. It is also relevant to waste management because knowing the nature of the material may govern how it is temporarily stored onsite. The same information is also required for disposal profiles.

## Evaluation of Worker/Public Protection

Measures to protect workers who initially discovered the problem are often implemented immediately when the project is halted by bringing in industrial hygiene professionals to assess the nature of the chemical hazards associated with potential exposure to the material. When the material has been satisfactorily classified, one needs to know whether ongoing special health and safety measures for site workers are required before the project may resume, as well as whether there are potential public exposure concerns associated with resumption of work. In the authors' experience, resolution of these questions is not always straightforward. Sometimes the normal working industrial environment that surrounds the project site may seem to be harsher than conditions encountered during the construction project, and implementation of a conservative health and safety plan causes unnecessary concerns among other site workers. This can trigger additional delays and costs to the project until an appropriate balance of health and safety protection can be achieved. For example, at one construction site, a delay was resolved when a pile of soil emitting vapors from an organic solvent in the soil was stockpiled in an isolated area where workers did not need to go onsite so that safety-trained personnel were not required to proceed with the foundation construction.

### Regulatory Notifications

Federal, state, and local regulations may exist that require notification to the agency if a release to the environment of a regulated substance has occurred. State and local regulations vary, so it is generally good practice to check these regulations as part of the decision process. If notification is required, the alternatives for addressing the contaminated material are often constrained by regulatory requirements. This is illustrated below in Case History No. 2. This factor, therefore, tends to increase the difficulty of controlling additional cost associated with the discovery. However, agencies are becoming increasingly anxious to avoid adding to their list of problem sites, and it is often possible to satisfy their requirements with minimal additional effort.

### Standards and Guidelines

When fill, soil, or surface or ground water at the site appear impacted by residual contaminants, it may be helpful, or even necessary, to consider whether soil or water quality standards are in force for the impacted media. There are a wide range of regulatory initiatives whose primary purpose is to define acceptable levels of contaminants in the environment. Promulgated federal and state water quality standards exist for both surface water and ground water, whereas soil and sediment quality guideline values, based on a conservative level of additional incremental risk, are used to evaluate soil/sediment impacts. When the anomalous finding occurs at the construction site, it is good practice to determine whether the residual constituents of interest have standards or guidelines and factor this information into the decision process for data collection.

### Materials Storage and Disposal

When contaminated materials are involved, regulations exist regarding how those materials may be stockpiled, stored, handled, transported, or disposed. Knowledge of these regulations is essential to avoid unnecessarily creating a waste which can only be managed at high cost. Where generation of waste is unavoidable, commonly, cost effective construction objectives in terms of materials handling are incompatible with cost effective waste management. It is also possible to contravene certain regulations inadvertently by proceeding with normal construction practices. Examples of these issues are presented below in Case History Nos. 1 and 2.

### Assessment of Potential Future Liabilities

The decision process needs to evaluate what potential future contact may occur with the residual constituents encountered at the site. This includes not only land, water, and property uses, but also potentially any impacted material that is removed from the site. The original construction project objectives and approach may be incompatible with the nature and extent of the residual constituents with respect to long term future use. For example, if non-toxic odors are encountered, will they seep into the buildings over time and create an aesthetically unacceptable condition? These situations can sometimes be corrected at minimal cost compared to post-construction mitigation. Community relations may further influence the decision process. In the authors' experience, future liabilities are better managed where good relations exist with the neighbors.

## Understanding the Construction: Options and Constraints

The engineer needs to understand how the project is normally constructed and incorporate that into the decision process when anticipating how to manage the affected media at the site. Sometimes contingency plans can be developed. For example, a contractor may plan to stockpile soil from the site for later re-use as compacted backfill. If the contractor knows that some of the soil to be excavated may contain constituents which cannot be used as backfill, but could have an alternate onsite use, such as landscaping berms, then this may be incorporated into the work plan, even with agency approval. By knowing how the foundation is constructed and other features of the project, and the material quantity requirements, the engineer can build in ways to avoid extra costs like offsite disposal in this example.

The factors discussed above are illustrated by the collection of case histories that follow.


## CASE HISTORY NO. 1

A residential housing project is being built on a site that was formerly used as farmland. Prior site usage also included the application of municipal wastewater treatment sludges throughout the property under a state-issued permit. The permit had been officially discontinued by the state five years prior to the start of construction, with the recognition of two important facts: (1) elevated levels of certain heavy metals had accumulated in the topsoils as a result of the sewage sludge application; and (2) the intended land usage after closure of the sludge application permit was residential. All of the above facts were documented in a Phase I Environmental Site Assessment that was used to obtain financing for the project before construction started.

One of the initial site activities was to strip the topsoil from the site, stockpiling it in a "convenient" location and allowing site grading. Once stockpiled, the contractor inquired the owner if some of the soils could be used off-site, as there would likely be a surplus. Remembering the Phase I data demonstrating the high metals levels, the owner did not allow any off site movement. The contractor, through his environmental consultant, then claimed that the soil could not be re-used on site because of the elevated metals (above residential criteria) and the absence of a state-approved soil re-use plan.

This standoff was resolved with some alterations of the final grading plans and a confirmation from the state agency that the five-year-old permit closure was still valid and applied to soils being re-used on site. Before this was achieved, however, there were several weeks of contractor standby (the location of the soil stockpile was not so convenient, after all), consultants, lawyers, laboratory fees, and many sleepless nights by an owner who envisioned negative publicity about his development being declared unfit for residential use.

Some claim that these issues may have been unavoidable, and that when they arose, they were addressed appropriately and efficiently; however, there appears to have been a disconnect between the contamination issues (elevated metals), the regulatory issues (permit closure), and the construction issues (need to stockpile and re-spread topsoils). Better and earlier coordination between the environmental personnel performing the Phase 1 assessment and the construction engineers planning the earthworks might have identified this issue and obtained the state's approval before construction started. On the other hand, part of the uncertainties about the validity of the original closure arose because the state's regulatory framework and criteria had changed significantly during the five years after the closure.

## CASE HISTORY NO. 2

A public water purveyor was conducting routine maintenance on a 60-inch potable water main. The work required excavation of a 10 ft x 10 ft x 6 ft trench to completely expose the pipe so a section could be cut to allow access for cleaning and leak repairs. The contractor intended to continuously de-water the trench during the four to six weeks the line was expected to be serviced. The water was to be discharged to a nearby railroad ditch that drained to a brook. Although the water company had been previously apprised that the area was currently under a remedial investigation (RI) as part of a Consent Order, they did not realize that the work site was included within the area of the RI, but near the limits of the affected area where there was only very limited information regarding the extent of contamination. Both soil and ground water had been impacted by heavy metals, and the contractor had already excavated the trench when the consultant for the responsible party informed the contractor of the situation. The excavated soils were tested and found to be below state cleanup guidelines except for what had been the surface soils. The state was notified and they agreed to a replacement of the unaffected soils in the trench upon completion of the work. The elevated soils were disposed offsite. The water company brought in an OSHA-trained contractor to perform the work. A more cost effective plan for groundwater control was developed. Extracted water was field screened and discharged under a temporary permit if it tested below the surface water quality standards. If the water tests indicated it could not be discharged, it would have to be transported offsite for treatment and disposal. To limit the potential water quantity for treatment, methods were planned to minimize trench volume to be de-watered. Offsite treatment turned out not to be necessary because all the water could be discharged to surface water.

Knowledge of other environmental conditions in the area would have caused the water company to plan the work differently. Cooperation from the state agency allowed the work to be completed within the water company's critical time period. Waste minimization methods controlled incremental cost increases associated with the contaminated conditions, although the requirement to notify the agency affected the overall schedule and incremental cost increase.


## CASE HISTORY GROUP

Three small case histories, taken from portions of recent projects, provide examples of a range of responses relative to regulatory interaction by owners upon discovery of contaminated conditions.

1.  During the excavation of footings for a warehouse expansion, contaminated fill materials, tentatively identified as a slag product, were encountered. The contractor and the owner jointly called in the state agency, who responded quickly with a site inspection and sample collection. Analysis of the materials confirmed elevated arsenic concentrations, with the suspected source being a former smelter at an adjacent property (now a Superfund site).

    Several meetings were held with the owner, contractor, consultants, state, and architects, and procedures were developed for sampling in areas of excavation, analysis, including speciation, and decision-making regarding the ultimate disposal needs of the materials. In the end, all of the excavated, contaminated slag materials were buried under the building floor slab. No materials were removed from the site for disposal. This highly cost-effective solution was achieved in large part due to the cooperative teamwork established by all parties concerned.

2.  A thirty-foot extension to a retail store in an urban setting was being constructed, when a concrete vault was encountered by the excavator for the last footing being installed. The vault turned out

to contain lead and acid wastes, presumably a remnant of the site's former use as a battery manufacturing facility. (This former use had not been identified until the vault was discovered.) Laboratory testing and regulatory agency discussions were initiated promptly. The state agency representative confirmed that the vault would need to be cleaned and closed properly, but not subject to UST rules. Subsequent testing confirmed that the vault contents were hazardous wastes, which were then disposed of at a RCRA incinerator.

3.  During construction of caisson foundations for an addition to an existing office building, construction was stopped when strong odors were encountered in one of the caissons. The odor was identified as belonging to an organic solvent believed to be used in radio manufacturing prior to construction of the existing office building. The caisson had to be vented to permit inspection to be safely completed by an individual having the requisite OSHA Hazwoper (40 hour) training. However, the constituent in the caisson soil spoils (about 50 cubic yards) also emitted vapors above recommended ceiling limits of exposure (there was no time-weighted average threshold limit value for this compound). Both water and state soil criteria exist for the constituent, the soil concentration was below the state soil criterion, so water impacts were not an issue. The extent of the affected soils was known based on the limited number of caissons where the odors were detected. Available alternatives included (a) spread the soil spoils as backfill beneath the slab of the addition; (b) envelop the soils in a geomembrane for re-use as fill beneath the building; (c) level the pile and cap in place; and (d) dispose offsite. The owner elected alternative (c). Had this project occurred a few years ago, the owner would likely have involved the agency in resolving the impasse. Conceivably, many more samples could have been required, and mitigation may have been necessary.

## CONCLUSIONS

When contaminated fill, soil, ground water, surface water, or structures are encountered unexpectedly during a construction project, the decision process which is invoked to resolve the interruption must consider a number of objective and subjective factors to arrive at the most cost-effective outcome. Traditional approaches can result in unnecessarily expensive remedial investigations and resultant corrective action, whereas less expensive approaches not requiring remediation are acceptable. Agencies are showing increasing tendencies to work cooperatively with owners in developing reasonable solutions to these types of problems.

**REFERENCES**

1.  American Society for Testing and Materials. 1993. "Standard Practice for Environmental Site Assessments: Phase I Environmental Site Assessment Process", ASTM E 1527 -93, 24 p.

2.  U.S. EPA. 1987. "Data Quality Objectives for Remedial Response Activities - Development Process". EPA/540/G-87/003.

3.  U.S. EPA. 1993. "Data Quality Objectives Process for Superfund - Interim Final Guidance". EPA/540/R/93/071.

# REGULATORY AND LEGAL ISSUES IN INDUSTRIAL AND HAZARDOUS WASTE MANAGEMENT

ALFRED A. SIESS, JR.
C E RESEARCH GROUP
P.O. Box 39
Coopersburg, PA  18036

## ABSTRACT

This paper discusses the failure of the existing regulatory process, at all levels of government, to adequately address problems of managing industrial and hazardous wastes.

In order to formulate rational public policy which leads to optimum solutions to environmental and other societal problems it is necessary to employ an objective, or "scientific," approach to problem-solving.  This approach requires that one must first identify the problem, and the cause(s) of the problem, to be solved.  Next, the desired outcome, or goal, of the intended policy must be objectively established.  Only then is it possible to promulgate policies which will lead to appropriate solutions, which may or may not be primarily technological solutions. To the extent that they are, it is necessary to use appropriate technology.

Using case studies in waste management, the author will explain why past and present waste management policies have failed to adequately address the social, economic and environmental problems. A formula for correcting past failures will be presented.

## THE DECISION-MAKING PROCESS -- OVERVIEW

The decision-making process for establishing public policy may be diagramed as shown below. [Figure 1]

If the process is rational, then the results will be beneficial to society.  A win-win situation.  If the process is subjective, i.e., distorted to serve a vested private interest at the expense of the public, then the results will exacerbate rather than solve the real problem.  A zero-sum or win-lose situation.

A rational outcome requires honest and objective input at each step of the

100

process. For example, GOALS must be chosen based upon unbiased research and an honest effort to pursue only those goals which meet the legitimate needs of the public-at-large.

The translation of goals into LAWS or REGULATIONS must not be subverted by private vested interests. (i.e., laws must be promulgated in the open with all interested parties afforded equal access to present their views. The input from all participants must be honestly considered. If the regulation requires a technological solution, appropriate technology must be utilized. The law must provide sufficient authority for enforcement, must not be ambiguous or contain "loopholes" which invite evasion, etc.).

The conduct of the regulated community in RESPONSE must be lawful and comply with both the intent and the "letter" of the law. Compliance should be de facto as well as de jure.

A flawed process will result in laws or regulations that exacerbate the problem rather than mitigate it.


STEP 1

ESTABLISH **GOAL** BASED ON PERCEIVED NEED


|  | STEP 2 | LEGISLATURE OR |
| --- | --- | --- |
|  |  | REGULATORY AUTHORITY |
| GOAL BECOMES | **REGULATION** | RECEIVES INPUT |
|  | **OR LAW** | FROM VARIOUS |
|  |  | OFTEN CONFLICTING |
|  |  | INTERESTS |


| INTENT OF LAW IS | STEP 3 |
| --- | --- |
| OBSERVED OR |  |
| CIRCUMVENTED | **RESPONSE BY** |
|  | **REGULATED COMMUNITY** |


| RESULT SOLVES | STEP 4 |
| --- | --- |
| PROBLEM OR |  |
| EXACERBATES | **RESULT** |
| PROBLEM |  |


**FIGURE 1 - DIAGRAM OF DECISION-MAKING PROCESS**

Failures in the regulatory process occur at one or more of the steps in the process. Most often the failure occurs at the very first step as a result of an entity, with the power to influence policy, lobbying for the wrong goals. Occasionally this happens because of genuine ignorance of the cause(s) of the problem to be ameliorated. Most often, however, and particularly where industry is involved, the process is knowingly subverted for personal or corporate profit. The latter action is particularly egregious because it demonstrates a willingness to deliberately harm others in order to serve a private vested interest.

Real or feigned ignorance of the basics of our free-market economic system and of the serious adverse environmental and health effects of pollution is at the root of many of our most harmful policy failures.

The author presented a paper in 1983 at a "Conference on Disposal of Solid, Liquid and Hazardous Waste" held at Lehigh University which provided the economic and social rationale for the need for laws regulating the handling and disposal of wastes. That paper was written, in large part, to rebut the "misguided deregulation rhetoric" of the Reagan Administration and the administration's supporters in private industry.

Industry's "deregulation" arguments were examined objectively from the economic perspective of classical free-market theory and from the political perspective of this country's ideal historic model of representative government to demonstrate that there is no dichotomy between a healthy environment and a healthy national economy.

The free-market model utilized in the 1983 paper showed that costs of not regulating were incalculably higher than the costs of government regulation. Furthermore, industry's deregulation arguments were shown, by their economist's admission, to be inapplicable to pollution control laws [1].

Economic arguments aside, it is clear that the existing regulatory process has failed to satisfactorily address the exceedingly serious problems caused by pollution. It is equally clear that many of the worst failures have occurred in the resource and waste management industries and that decision-makers in those industries are primarily to blame.

The following section describes in detail cases where public policy intended to clean up pollution has been subverted.


## CASE STUDIES OF REGULATORY FAILURES

CASE I - The response of the electric utility industry to the Clean Air Act and other regulations designed to reduce air pollution.

Title IV of the 1990 Clean Air Act Amendments addresses the acid rain problem by requiring electric utilities to reduce their $SO_2$ and $NO_x$ emissions. Compliance with the $SO_2$ requirements can be achieved under the act in any of three ways: switching to low sulfur coal, installing scrubbers or purchasing allowances from other sources.

The utilities embrace the allowance provision because it provides the opportunity to meet the letter of the law without seriously trying to reduce emissions. For example, if the utility has plants of varying age where the cost of additional emission reduction will also vary significantly, the opportunity exists for the company to be very selective in deciding what, if any, pollution will be eliminated. For example, at Plant A it may cost $200 /Ton and at Plant B it may cost $100/Ton to reduce emissions with improved scrubbers. If sufficient reduction can be achieved at Plant B to bring the company into compliance the company is able to do absolutely nothing about cleaning up the emissions from Plant A. If $SO_2$ pollution credits can be purchased for less than $100/Ton then the company would be able to avoid reducing emissions altogether. The sad fact is that many, perhaps all, utility companies are willing to do just that. It was the utility industry that led the lobbying effort to include this loophole in the act. The industry lobbied for the trade-off provisions on the grounds that the higher cost of doing the job properly, i.e., by eliminating emissions, would hurt their competitive position. They pretended not to understand the fact that pollution control costs are passed on to their residential customers, just as other rate increases are ultimately paid by their residential customers. [Industrial and commercial customers add the rate increase to the price of their products. School districts and municipalities obtain the needed revenue from property taxes.] As regulated monopolies , the utilities cannot use competition as an argument for not spending money on pollution control--except in the very special case where they have, through past lobbying effort, created a thorny problem for themselves. (See PURPA below).

Another case illustrating the utility industry's short-sightedness with respect to reducing sulphur dioxide ($SO_2$) emissions was recently aired by National Public Radio. The report told how a class of sixth graders from New York State outbid a Cleveland-based utility company in the $SO_2$ allowances market. ($SO_2$ credits are traded on the open market at the Chicago Board of Trade.) Concerned with the effects of acid rain, which the students had found to be "horribly acidic" (as low as ph 3.0), the class had raised over $3000 and used it to purchase the rights to 21 Tons of $SO_2$ pollution. The spokeswoman for the Cleveland electric utility company which was outbid by the middle school students had this to say about the incident. "If, for whatever reason, sufficient emission allowances were not available for us to continue to use our coal plants as they're currently configured, then we would have to invest in a more expensive technology, such as a scrubber or such as burning natural gas or something like that. And, if the allowance market went away that would merely drive up the price of what it would cost us to generate the energy for our customers. So, in the long run the customers are the ones who pay." (emphasis added) [2].

One must question, if the industry recognizes that their costs of doing business are paid by their customers, and they are truly concerned for their customers' welfare--why they supported the perversion of the Public Utilities Regulatory Policy Act (PURPA) by the incinerator industry, why they support the attempts to have hazardous incinerator ash regulated as "special waste," why they are lobbying to gut the Resource Conservation and Recovery Act (RCRA) by amending it to allow hazardous waste oils to be burned in utility boilers, why they have formed non-regulated companies to build municipal solid waste (MSW) incinerators in other states, why they became and remain champions of nuclear power, and perhaps the most important question of all, why do electric utility companies still promote CONSUMPTION of energy instead of conservation?

Let's consider the consequences of each of the above policies (discussed in reverse order for clarity).

## CONSUMPTION VS. CONSERVATION & THE BURDEN OF NUCLEAR POWER

The following observations, contained in a 1991 magazine article about Amory Lovins explain why, at one time several decades ago, electric utilities viewed growth in consumption and the need for massive additional generating capacity as never-ending. It even explains why private investor-owned utilities succumbed, at that time, to the idea that nuclear plants could supply so much power, so inexpensively, that it would be "too cheap to meter."

What the reporter wrote in 1991 and the quote which follows, made by Amory Lovins himself in 1994, do NOT explain is why the utilities continued to support nuclear power long after the nuclear power plant dream became an economic and environmental nightmare. It is also difficult to understand why many utility companies continue to promote increased consumption some two and one-half DECADES after it became obvious that sustainable energy and energy savings were the order of the day.

"In the 1950s and 1960s it had been impossible to overestimate future demand, because electricity obeyed the Field of Dreams phenomenon: If you built a power plant, people would come to use it--irrigators, aluminum companies, airplane manufacturers. And thanks to the economics of scale, every time a new plant came on line the cost of electricity went down. Then about 1970 the energy order was jolted by a series of economic, political, and technical shocks . . . "Increasing supply now meant increasing the price the customer paid"; when the cost of a commodity went up, common sense made people look for ways to use less. Suddenly there were terrible penalties for over-estimating demand: canceled plants, irate stockholders, even bankruptcy. " . . . Already some prudent utilities were canceling nuclear projects as too expensive. (By 1984 they would have lost more than $20 billion on abandoned plants alone.)" [3].

"Since 1979 this country has gotten four and a half times as much new energy from savings as from all net increases in energy supply put together. . . . You can now save twice as much electricity as you could five years ago at only a third the cost" [4].

## UNREGULATED UTILITY ENTRY INTO WASTE DISPOSAL BUSINESS

Delmarva Power and Light operates a non-regulated business segment in Pennsylvania which has been trying for nine years to build a municipal solid waste (MSW) incinerator in Northampton County. The same company owns the Pinegrove Landfill in Schuylkill County. The subsidiary of another, Maryland based, utility tried but failed to build a MSW incinerator in Monroe County, Pennsylvania.

As regulated monopolies in their home states, electric utilities enjoy tremendous competitive advantage in the fact that their profits are guaranteed. Allowing these companies to move their wealth to other states and operate without regulation has serious regulatory implications. This is particularly so when their investments are in incinerators which generate electricity which must be purchased at premium rates by the home state (regulated) utilities.

## WASTE OILS AND RCRA

Energy intensive industries such as steel, aluminum and cement producers have always sought less costly alternatives to coal and oil for their energy needs.

One unfortunate result is the loophole in federal law which allows the burning of hazardous wastes in boilers and industrial furnaces (BIFs).

This has led to an entire new industry to supply hazardous solvents, waste oils, tires and other alternative "fuels" to any producer unscrupulous enough to burn toxic wastes in facilities, such as cement kilns, which are not properly designed or sited to safely control toxic releases.

In the Lehigh Valley (PA) the local utility has taken the leadership position in lobbying Congress to allow the burning of hazardous wastes in BIFs. Pennsylvania Power and Light (PP&L) President and CEO William F. Hecht headed a coalition of Lehigh Valley business leaders--the Rational RCRA Reauthorization Coalition (RRRC) which had the avowed purpose of developing "mutual goals and a strategy" for the reauthorization of RCRA . Among RRRC's "Priority Concerns" was to exempt toxic wastes, such as cement kiln dust and used (waste) oil, from regulation as hazardous wastes under Subtitle C of RCRA.

## UTILITY INDUSTRY'S SUPPORT FOR INCINERATORS

Incinerator ash, from municipal solid waste (MSW) as well as from commercial hazardous waste burners, is toxic--containing hazardous dioxins and heavy metals. It is properly regulated as hazardous waste under Subtitle C of RCRA. The law requires that incinerator ash which tests hazardous, as does about 99% of fly ash, must be sent to licensed hazardous waste facilities for disposal.

However, the proper designation of toxic ash as hazardous is cost-prohibitive and would deal a fatal blow to the waste incineration industry. Consequently, incinerator proponents have sought to declare this toxic waste legally non-hazardous by changing the definition to "special waste' or "residual waste"; (a ploy referred to by environmentalist Dr. Paul Connett as "linguistic detoxification").

Industry lobbied Congress to have ash designated a special waste in the 1990 revisions to the Clean Air Act, a move as hypocritical as it was unsound-- using the law designed to protect air quality to promote incineration, one of the worst generators of toxic air emissions.

The "special waste" provision was knocked out of the 1990 Clean Air Act but industry succeeded in replacing it with the following: "Incinerator ash shall not be handled as hazardous waste under Subtitle C of RCRA for two years." This gave the incinerator industry a two-year reprieve and postponed consideration of the "ash-as-special-waste" issue until the time when RCRA was to be reauthorized.

The utility industry supported the incinerator revisions to the Clean Air Act in 1990 and also supported the ash revisions in the ensuing RCRA debates.


PURPA LEGISLATION

The Public Utilities Regulatory Policy Act (PURPA) was designed to encourage the development of non-polluting alternative energy sources such as solar and wind. A major provision of the act was to require public utilities to purchase excess power from alternative energy producers at premium rates.

The worthy intent of the legislation was subverted, however, when major polluters, principally MSW incinerators, became the principal beneficiaries of the incentive subsidies.

A newspaper story dated April 8, 1995, headlined "Glendon Energy accepts Met-Ed's Contract Buyout," illustrates the serious impact on utility industry stockholders and ratepayers of this major public policy mistake, here compounded by company mismanagement: "Glendon Energy Co. has accepted Metropolitan Edison Co.'s $1.65 million offer to terminate its 25-year power purchase contract. . . . Joseph Reibman, president of Glendon Energy . . . said the Reading based utility needed to reduce its potential liability to remain competitive. Med Ed has asked the Public Utility Commission to pass the $1.65 million buyout cost on to its customers (emphasis added). . . . In the long run, the deal will save Met-Ed customers about $16 million, said spokesman Gary Plummer, adding that the contract with Glendon would have been too expensive" [5].

Perhaps it occurs to you, as it does to me, that Met-Ed should not have entered into a contract, with an unbuilt incinerator developer that would have spent unnecessarily "about" $18 million of ratepayers money, in the first place.

An additional case example which demonstrate the widespread subversion of policy-making in the areas of environmental and public-health and safety legislation follows:

CASE II -- LANDFILLS AND INCINERATORS

The viability of using modern "sanitary" landfills for disposal of municipal solid waste is dependent on several factors:  The facility must be properly 1. Designed, 2. Constructed, 3. Operated, 4. Sited, and, the most important consideration of all, 5. Toxic or Hazardous Wastes must NEVER be landfilled.

The last parameter is the most important of all because all landfills leak. Permeability of ground or synthetic liners is part of the design.  Unplanned leakage at much higher than design rates is inevitable.

Incinerators compound the problem in two major ways.  They convert ordinary household refuse, such as tin cans, newspapers and plastics, which may be relatively benign when landfilled, into extremely toxic air emissions.  (These are made much more readily available to the environment, and to ingestion, because of the fine particle size of combustion products).  Additionally, incinerators do not eliminate the need for landfills; the ash created by burning is highly toxic and the disposal method of choice is landfilling.

The use of incineration, ostensibly for treatment and volume reduction of garbage, was promoted as an easy solution to the "crisis" in available landfill capacity.  Industry defined the problem:  "Increasing amounts of garbage and the rapid loss of landfill capacity to bury the garbage."

Industry also defined the solution:  "Build more landfills and build incinerators to reduce the volume of trash and stretch out the life of landfills."

Industry knew that they were not promoting a viable solution because they knew that they had not properly defined the problem.  They had a hidden agenda--There was no real crisis involving disposal of municipal waste.  The problem was that industry had huge amounts of industrial waste, 80% of which was improperly stored or disposed on site, to get rid of.  Because much of the industrial waste was hazardous or toxic, DISPOSAL was absolutely the worst answer to their problem.

Even if there had been a genuine shortage of landfill capacity, the rational solution was to REDUCE the amount of waste, not INCREASE DISPOSAL CAPACITY.

Incinerators and double-lined landfills were a ploy designed to enrich waste companies' consultants, bond counselors and lawyers, while enabling the generators of toxic industrial waste to get rid of their problem.  (Incinerators serve the added benefit to the toxics generator of destroying the evidence.)  All of this has been done at huge cost to the environment and public health, not to mention the tax monies wasted in the process.  The result: The waste disposal problem gets larger and industry constantly invents new ways of insuring that it will never be solved.

# REFERENCES

1. Siess, Alfred A., Jr. 1983. "Economic and social Rationale for Regulation of Wastes--A Free-Market Perspective," *Proceedings-- Conference on Disposal of Solid , Liquid and Hazardous Waste.*

2. Feinsmith, Robin (Reporter). April 1995. Excerpts from "Living on Earth," April 23, 1995 news report, National Public Radio--WHYY, Philadelphia. (Transcribed from audiotape.)

3. Brown, Chip. 1991. "High Priest of the Low-Flow Shower Heads," *Outside* (magazine) p. 64.

4. Lovins, Amory. 1994. Excerpted from his Keynote address tot he Campus Earth Summit in "Blueprint for a Green Campus" *The Heinz Family Foundation*, January, 1995 p. 29.

5. Hay, Bryan (Reporter). April 8, 1995. Newspaper story, *The Morning Call*, Allentown, PA.

# EVOLUTION OF ENVIRONMENTAL RESPONSIBILITY IN CIVIL ENGINEERING

C. W. "BILL" LOVELL
Professor Emeritus and Research Engineer
School of Civil Engineering
Purdue University
West Lafayette, IN 47907

## INTRODUCTION

Change in environmental concepts over the past half century have been profound in my specialty of geotechnical engineering. As the technology available to sculpture the earth's surface developed, the engineer became smitten with its ability to overwhelm all natural obstacles. No river valley was too deep or wide to be dammed; no water body was too wide to be bridged above or tunneled beneath; no mineral resource too remote to be exploited. The ruggedness or expansiveness of the natural scene might pose enormous technological challenges, but they could be overcome. Nature was an adversary to be bested, and losses accruing to the loser were...well, natural.

Consciable technologists of course had canons of ethics which demanded that such activity be "for the benefit of Human-kind". However, short term and localized benefits were the focus. There was little attention given to the potential unfavorable effects on future generations or of natural effects which transcended the local scene.

The healing and restoring capacities of Nature were overestimated or ignored. Solid wastes were dumped, piled and buried; liquid wastes were flushed into waterways; combustion products, both particulate and gaseous, spewed into the air. Such activities could not be sustained; environmental quality deteriorated badly over large geographic regions...there were even global effects.

Of necessity, the image of Nature was slowly, even reluctantly, changed from Adversary (to be beaten) to Partner (deserving consideration and cooperation). The notion of building with Nature and of building to achieve total and longstanding social benefit gained acceptance [8]. Assessments of environmental effects (even impacts) of new building activities became routine and were often mandated. Lest the argument that the building activity should simply be avoided always prevail, the social cost of <u>not</u> building (the no-build alternative) was also estimated.

Developments continued at a rate and in a manner that accrued environmental damage, until the set of environmental issues reached the current (1995) critical level. The need for drastic rethinking of economic policy has led to the definition and beginning implementations of <u>Sustainable Development</u>. In this philosophy, development is adjusted to that level/kind which admits both economic growth and environmental preservation...even some restoration.

In terms of the human species, Sustainable Development is meeting the needs of the present without compromising the ability of future generations to meet their own needs [3]. In terms of the global ecosystem, Sustainable Development is the middle ground where humans thrive in the same landscape with healthy ecological processes, characteristic vegetation and biological diversity.

Sustainable Development recognizes that an economy which provides new jobs must continue, and rejects the irrational argument that development must be abruptly and drastically curtailed to "save the environment". Sustainable Development as well rejects the position that environmental losses are imaginary or overstated and that we can proceed with "business as usual". Sustainable Development challenges all of us to reevaluate and strengthen our fundamental values and to rethink our operational procedures, including routine actions that are based primarily on short-term individual convenience.


ENVIRONMENTAL ISSUES

We are accustomed to seeing environmental issues one at a time, and they are formidable enough on that basis. Recently, ecologist Norman Myers was asked to list the top 10 problems. He responded with the following list, [9]:

1. Global Warming
2. Biodepletion
3. Consumption
4. Population
5. Third World Poverty
6. Soil Erosion
7. Water Shortages
8. Ozone Depletion
9. Synergisms (compounded impacts)
10. Unknown Unknowns (impacts as yet undiscerned)

Space does not permit a detailed discussion of each of these issues, but a few may warrant particular attention.

Changes of only a few degrees in the planet's average temperature are able to produce profound changes in its living environment. For example a few degrees decrease in the average temperature was apparently able to produce an ice age. The accumulation of carbon dioxide and other greenhouse gases certainly has the capacity to raise the average temperature a few degrees. These accumulations, if continued, will undoubtedly cause global warming. It is not a question of whether, only of when.

Many of the consequences of warming are well appreciated. As a specialist in the cold regions of the planet, I need to emphasize the effects of warming on the accumulations of snow and ice, and the perennially frozen ground (permafrost). Ocean levels will rise dramatically and runoff will erode and drastically alter the terrain near streams. As the permafrost is thawed over enormous areas in the North, massive ground subsidences will occur, along with countless landslides. Roads, airfields, pipelines and other human infrastructure will be rendered inoperable as the supporting frozen ground beneath them thaws and settles.

Energy conservation and development of alternate fuels seem to be required to control the greenhouse effect.

Biodepletion, or the reduction of biodiversity, is proceeding at a rate which alarms biological scientists. Are we about to lose one third to two thirds of the Earth's species, before we have even identified their value? The natural habitat in which these species exist must be protected from human destruction, and from external factors such as the global warming and acid rain. To replace lost species through evolution requires millions of years.

"Recycle" is a much emphasized modern activity, but it is actually the final environmental step in the 3 "R's"...Reduce, Reuse, Recycle. Consumption of non renewable resources must, of course, be limited not only for conservation, but also to reduce the quantity of wastes which must be returned to the environment. To "reuse" can mean to use with care and with proper maintenance to extend life. It also means storing infrequently used items so that they may be periodically retrieved, used, and then stored again.

How easy is it to recycle, once further reuse is not practical? A few car manufacturers are already fabricating to facilitate recycling through disassembly and separation of materials. The same concept can be applied to structures, replacing the "wrecking ball" by tools which cut, pry, and pull components apart in the demolition process.

Much attention is directed toward recycling domestic wastes, but quantity wise, domestic waste rates well below mining wastes and industrial wastes. Areas of carelessly disposed mining wastes must now be remediated and reclaimed for a secondary use. An excellent example is the coal waste "gob pile" and the acid mine drainage which originates in it. These areas must be graded, drained and "capped" to prevent infiltration of rainwater which would continue to generate the acid drainage. Modern regulations require that mined material which is not marketed be placed back in the mined area, and that the mined area be appropriately graded for reuse.

Population control seems to be the most necessary element in controlling environmental problems. Even the most avid proponent of Sustainable

Development will readily concede that the concept will fail unless population growth rates are sharply reduced.

Some 90% of the population growth occurs in developing nations, where it tends to be most difficult to control. There, political power resides in numbers, and outnumbered groups (minorities) may resist efforts viewed as an effort to simply maintain their minority status. Cultural ethics and religious beliefs may effectively mitigate against controlling birth rates. Large numbers of women are currently unable to make responsible decisions about family size.

Certainly there are signs of hope. The Peoples Republic of China has an effective control plan, one child per marriage. However, minorities are exempt, and the plan is not well enforced in rural areas.

The most pervasive argument in favor of control may be the ability to sustain a well-fed and healthy population. Were the world on a diet deriving 25% of calories from animal protein, as in North America, a population of less than three billion could be sustained. For people on a vegetarian diet, and with present agro-technologies, the present population of 5.5 billion is sustainable [9]. While our ability to raise and distribute food will undoubtedly continue to increase, the trend will lag behind the projected population growth.

Many of us are in the physical/chemical sciences and their applications, and we relate better to the next three items on the critical symptoms list: soil erosion, water shortages, and ozone depletion.

Soil erosion..."is as bad in parts of Indiana, one of America's agricultural states, as in India" [9]. Loss of top soil results in loss of fertility; a downstream effect is the compromising of downstream aquatic environments and water supplies through siltation.

Economic incentives to protect topsoil work well in the U.S., and will probably work elsewhere. Hilly terrain is best kept in vegetation, including trees, where rainfall allows.

Water sources are either surface or underground, with flow connections of course. Where water is scarce it is expensive and conservation measures are strict. Overuse of surficial supplies is obvious and tends to be controlled, however, excessive groundwater use is hidden, and depletion of major aquifers is common. Agricultural irrigation is needed for a substantial part of global food production. And, as more food production is required, water supplies will be further stretched.

Careless and unregulated underground disposal of toxic materials has contaminated large quantities of groundwater. Industrial and domestic sewage has polluted surface waters. Environmental laws and regulations have sharply reduced such pollution in the U.S., but in the former Soviet bloc of nations, abatement has scarcely begun.

Contaminated waters produce enormous health problems. Today, more than a billion people suffer shortages of clean water. It is estimated [9] that this number will triple in the next two decades.

The ozone layer shields the earth from lethal solar rays. Thus ozone loss is a grave threat to most living creatures. Ozone losses occur due to the release in the atmosphere of chemical compounds containing chlorine...including coolants used in refrigerators and air conditioners. These chlorine molecules combine with the ozone to form acid compounds. Substitution for the harmful chemicals is taking place; unfortunately the replacement materials are more expensive. Will the reductions in ozone-depleting refrigerants in North America and Europe be countered by increases in Asia and Africa?

## THE NEW ETHIC

In the face of all of this, governments, institutions, agencies, families, and individuals require a philosophy, which is realistic, positive and adequate for the challenge. The Sustainable Development approach will work, if population growth rates can be sharply reduced.

Our philosophy, and our activities in behalf of it, are undergirded by a set of behavioral principles and guidelines. As a Civil Engineer with 50 years of experience, the author is a member of and subscribes to the Canon of Ethics of the American Society of Civil Engineers. The seven canons in this list will, after a healthy debate, be joined by an 8th, which will likely read: Engineers shall perform services in a manner as to sustain the world's resources, and protect the natural and cultural environment.

The implications of this pledge are far-reaching, and already have significant beginnings. More and more structures are being planned and designed with a predetermined life cycle of construction, maintenance, and deconstruction [1].

The "built environment" is made compatibility with the "natural environment". The modern structure also conserves natural building material resources, while maximizing recycled content and the recyclability of the materials in the structure. The structure is not only "healthy" for those using it, but also avoids being an environmental liability when destructed.

Certain institutional and commercial structures, e.g., churches, public buildings, financial institutions, are properly built for a very long life, with periodic remodelling/renovation. Other structures with a limited service life, e.g., motels, strip malls, fast food restaurants, should be built so that upon decommissioning, the materials can be simply separated for recycling.

## WASTE TYPES AND VOLUMES

The effort to reuse/recycle begins with an assessment of waste types and rates of generation. Our data are of course better for the U.S., and we will use them for purpose of comparison. The annual rate of production of waste types is in order from greatest to least is: agricultural, mining, industrial, and domestic (including scrap tires).

Agricultural waste tends to stay close to the points of generation, and is

113

not a high-profile waste in the U.S. However, in small but livestock oriented countries like The Netherlands, manure production must be legally controlled, and even plans to export the waste are under consideration [5]. In developing countries, dried manure is often used as fuel.

Second on the list is mining waste. This is comprised of the earthen material which must be removed to access the commercial mineral, the host rock, if any, and the waste from extraction/refinement of the mineral. Improper disposal of the past has seriously compromised the quality of the surface and groundwater in the vicinity. Current disposal is highly regulated. The third waste category is industrial. A primary example is coal ash or coal combustion byproducts (CCBPs). This grouping includes the cementitious fly ashes (Class C) and others (Class F) and the boiler slags and bottom ashes. Improvements in combustion processes and the desire to burn higher sulphur coals are leading to increasing amounts of fluidized bed residuals and flue gas desulfurization products. Interestingly enough, as combustion and clean air technologies are improved, more and more solid wastes are generated.

Class C fly ash results from the combustion of lower grades of coal, and the product is in demand as a cement replacement in concrete. The more abundant Class F fly ash (from the better grades of coal) has no established market. The same is true of the coarser residues, i.e., all bottom ash and some boiler slag.

A key to the use of these by-products, as soil or mineral aggregate replacements, is their method of disposal. The finer fly ash and the coarser bottom products are collected separately, but may be commingled in disposal. When this happens, a complex depositional mixture of sizes of material occurs, which complicates retrieval and use.

The noncementitious fly ashes and the bottom ashes may be used as a soil replacement when building fills for structures, including highways [6]. They may also be used as fillers in asphaltic mixes. While the use of these wastes to replace soil in fills may seem to be of little importance, remember that in flat areas, soil from adjacent areas must be excavated for the fill. This often reduces tillable land. Use of the CCBPs for the fill not only conserves the adjacent land for agriculture, but also reduces the disposal costs for the generators.

Another industrial byproduct waste of importance, particularly in the Midwest, is spent foundry sand [7]. Foundries cast a variety of metals, but iron is often involved. As the greensand process is used to cast ferrous metals, the spent sands are often environmentally suitable for a number of engineering applications.

One of the uses of the spent foundry sands is again as a soil substitute in fills. Another is as a constituent in a low-strength cementitious flowable material called "flowable fill" [4]. This material is superior to compacted soil in and around pipes and other "tight" places. It is purposely designed to have low strength for utility cuts in streets, so that it may be easily excavated and replaced. (We are all familiar with the scenario of periodic trenching operations in nearly all streets.)

The materials in buildings and pavements should be largely reused. This is perhaps easiest in an asphaltic pavement, where, with the addition of a rejuvenator, the materials may be reused for the new paving. The same is true of old concrete paving, except that it is more expensive to break up the concrete pieces and remove the reinforcing steel, prior to constituting a new concrete mix. Reuse of products from demolished buildings is complicated by two factors: the technology of construction; and the technology of destruction. We have all observed the "wrecking ball" process and the impossible jumble being loaded into trucks for disposal. A number of the materials can be reused if separated. Among these are concrete, steel, brick, wood, and rubber/asphalt roofing. Disposal should not be in a municipal solid waste landfill, but in a construction debris facility. Separation can occur before disposal of ultimate residuals in a processing yard adjacent to the landfill.

Certain domestic wastes are commonly recycled, viz., aluminum cans, bimetal cans, glass, paper, cardboard and some plastics. A waste which needs increased attention is the scrap tire. This discard is generated in the U.S. at the rate of about 0.8 tire/ person/year. Since these objects have been presumed by "speculators" to have ultimate value, they have been stored. This has often been accomplished in a most unsightly and unsanitary manner, producing disease and fire hazards.

Scrap tires must now be stored inside, which means that they are usually cut or shredded, to save space. Because whole tires cannot be compacted with the other wastes, and actually tend to "float" upward in the mass, tires must be cut into pieces for disposal in solid waste landfills.

Tire rubber is a strong and durable material that lends itself to a variety of reuse potentials [2]. For example, whole tires can be used to build an earth retaining structure, a floating breakwater, or an underwater reef. Sidewalls and beads can be cut out and linked in a mat to stabilize low cost roads or earthen masses. Tire shreds, a few inches wide on a side, form an excellent lightweight substitute for soil in a fill. Reducing the tire rubber to small crumbs allows it to be reused to mold new rubber products or to produce a new variety of asphalt.

All of the reuse/recycle examples listed in this paper are simplistic ones and represent very low technology. These procedures can be expected to advance and develop enormously in the near future.


SUMMARY

Environmental responsibility has evolved slowly and only after abundant evidence of damage to the earth. The global issues constitute a formidable list, all of which require immediate attention and remediation. A basic principle which can unify and cause scientists and engineers to cooperate and synergize is that of Sustainable Development. In this strategy, development takes place with appropriate environmental sensitivity.

Unless population growth rates are sharply decreased, we will undoubtedly exhaust food supplies, even given great technology developments.

Sustainable technology will involve many ideas and approaches, but an important one is reuse/recycle of current wastes such as scrap rubber tires, coal combustion ash, and spent foundry sands. Paving should be recycled, as well, and products of building demolition should also be separated and reused. The author has significant personal interest in this topic, and has given some details in the paper.

## LITERATURE CITED

1.  Abraham, D. M., Lovell, C. W. and Kim, Jongmin (1994) "Recycling of Destructed Building Products", Proceedings of 1st International Conference on Sustainable Construction", CIB T1616, International Council of Building Research Task Group 16, Nov. 6-9, Tampa, Florida, pp. 755-764.

2.  Ahmed, I. (1993) Use of Waste Materials in Highway Construction, Noyes Data Corporation, Park Ridge, New Jersey, 114 pp.

3.  American Association of Engineering Societies (1994), The Role of Engineering in Sustainable Development, AAES, Wash., D.C., 106 pp.

4.  Bhat, S. T., Lovell, C. W., Scholer, C. F. and Nantung, T. E. (1995) "Flowable Fill Using Waste Foundry Sand", Proceedings, 11th International Symposium on Use and Management of Coal Combustion By-Products, Vol. 2 (CCBs), EPRI TR-104657-V2, Project 3176, Orlando, FL, Jan., pp. 39-1 to 39-14.

5.  Dahlburg, J.-T. (1994) Los Angeles Times, Los Angeles, California, August 26.

6.  Huang, W.-H. (1990) "The Use of Bottom Ash in Highway Embankments, Subgrades and Subbases", Joint Highway Research Project, FHWA/IN/ JHRP 90/4, Purdue University, W. Lafayette, Indiana, 315 pp.

7.  Javed, S. and Lovell, C. W. (1993) "Use of Waste Foundry Sand in Highway Construction", 44th Highway Geology Symposium, Tampa, Florida, May, pp. 19-34.

8.  McHarg, Ian (1969), Design with Nature, Doubleday & Co., Inc., Garden City, NY, 197 pp.

9.  Myers, Norman (1994), "What Ails the Globe?" International

_Wildlife_, Vol. 24, No. 2, March-April, National Wildlife Federation, Vienna, Va, pp. 34-41.

# RISK ASSESSMENT AND COMMUNICATION - TOOLS FOR SITE CLOSURE

MARK MYRICK
Delta Environmental Consultants, Inc.
6701 Carmel Road
Charlotte, NC 28226
(704) 541-9890
(704) 543-4035 fax

## ABSTRACT

The costs of remediation at sites with contaminants in soil or ground water have increased by more than a factor of ten in the last ten years. Annually billions of dollars are spent by the government, industry, and individuals in an attempt to restore sites that have been impacted by accidental or intentional disposal of wastes. Despite the large expenditures, few if any of the sites have been restored to pristine conditions.

The Science Advisory Board for the EPA has recommended that environmental hazards be dealt with based on the severity of the risks to human health and the environment. Despite EPA's attempts to account for risks in its priority setting strategy, few state regulatory agencies allow the concept of risk to be factored into the remediation of contaminated, non superfund or hazardous waste sites. While many regulatory agencies are wary of the concept of acceptable risks, some states do accept risk assessments in order to establish remediation closure goals above the current federal and state maximum allowable concentrations of contaminants.

Site specific risk assessment has the ability to determine the risks associated with compounds of concern, establish remediation goals for these compounds that are practical and achievable. By establishing achievable remediation goals environmental professionals are able to significantly reduce costs to the client, while protecting the health of the population surrounding the site.

A risk assessment (exposure, toxicity, and ecological) was completed for a gasoline contaminated site in southwest Virginia. The original closure goals that were set by the Commonwealth of Virginia were renegotiated after the risk assessment was completed. The clean up levels for ground water were raised from "natural ground water quality" ($<1$ ug/L) for benzene, toluene, ethylbenzene, and xylenes to 1690, 1980, 4950, and 19,080 ug/L respectively. The determination of the risks associated with the potential exposure to and toxicity of the specific compounds detected in the ground water was used to establish a precise remediation goal that suited the hazards at this site. The increase in the remediation end point concentrations allowed the site to be closed after twenty months of ground water remediation. The risk assessment significantly shortened the time needed to remediate the site, reduced project costs, while not adversely affecting the public's health or the environment.


## INTRODUCTION

In the early 1970s, after decades of under regulated industrial activity, the United States Environmental Protection Agency (EPA) began assessing the contamination associated with industry. The EPA set as an early goal to address, curb the damages from, and clean up gross pollution of the air and water. Vast amounts of money were spent in these attempts. While there were many successful achievements, it soon became clear that sites where gross soil and ground water contamination had occurred would not be easily restored, even with significant funds available. The term "risk" was rarely used or thought of in the early 1970s (EPA Journal, 1991, p. 13). By the late 1970s, the EPA had narrowed its focus to toxic pollutants (those thought to affect the public's health), (Paustenbach, 1989 p. ix). An issue that the EPA faced with this new focus was determining which environmental chemicals were actually toxic and which were not. In 1980, the Comprehensive Environmental Response Compensation and Liability Act (CERCLA) was enacted in order to provide adequate resources, technology, and political pressure to clean up the worst hazardous waste sites. Risk assessment (RA) was developed in response to CERCLA in order to develop criteria for determining the severity of the contamination at these sites and establishing appropriate clean-up levels. By 1983 the EPA had begun using RA to aid in making intelligent decisions about the risks posed by the many pollutants of concern (Paustenbach, 1989, p. ix). In the Fall of 1990 the Science Advisory Board recommended to the EPA that they deal with all environmental hazards based on the severity of risks to human health and the environment (EPA Journal, 1991 p. 29).

Many of the environmental laws that we now have were developed prior to the introduction of the concept of the evaluation of risks posed by specific chemicals. Many of the laws were created through piecemeal environmental policies that better reflected public perception of environmental risks than they did actual risks (EPA Journal, 1991, p. 18). Little, if any, of the legislation addresses the overall effects of the policy, reduction of risks to the public, or overall costs.

It is currently estimated that it will cost approximately one trillion dollars (not counting legal fees) for remediation of the 32,000 known and 43,000 estimated hazardous waste sites across the United States (Abelson, 1992 p. 901). Between the years of 1982 and 1989 the costs of remediating a given amount of contaminated media increased by more than a factor of ten (Environmental Restoration and Waste Management, 1991). Even with this vast expenditure, it is unlikely that any of the sites will be restored to pristine conditions. Many of the sites will never even be restored to a condition that will allow for unrestricted public access.

With the costs of remediation increasing rapidly and technology not capable of achieving pristine conditions, it is likely that the role of risk assessment and risk management will become increasingly important at contaminated sites. This paper will address some of the aspects to consider when using RA and the importance of quality field data collection during the site characterization of a project. An example of a successful use of RA will then be presented.

## OVERVIEW OF RISK ASSESSMENTS

Risk assessments have been utilized by scientists in assessing the hazards posed by chemicals for approximately 28 years (Paustenbach, 1989 p. 27). The RA process in the environmental field became formalized after CERCLA was enacted in 1980. If used properly, RA has the ability to provide the environmental professional with sufficient data to define the nature and potential threat of any substance of concern. While there is a large difference in the site characterization detail that is required for a RA at a superfund site as compared to one conducted for a leaking underground storage tank site, all RAs generally consist of an estimation of the magnitude and characteristics of a studied risk. This is commonly achieved through the use of several or all of the following steps: site characterization, contamination identification, hazard identification, exposure assessment, toxicity assessment, risks characterization, ecological assessment, and environmental fate determination (Lehr, 1990 p. 4-5) and (Paustenbach, 1989 p. 30-32).

The detail required in RA varies, but the goal of all should be to provide the risk manager with the data needed to inform his client of actual risks associated with a contaminated site. The client can then weigh remediation options against the costs and benefits of eliminating or reducing the risks. With this approach, the best site specific option can be selected and implemented.

It should be pointed out that while conducting RA, a clear distinction between assessment and risk management should be maintained. The assessment should take all pertinent scientific data and informed judgements into account and present them in a written document communicating the risks at a site. The risk assessor should not consider costs, benefits, political or social ramifications of the outcome of the investigation. On the other hand, the risk manager should take the information presented in the risk assessment and use it in combination with political, social, and economic information to help the client determine the best site specific solution (Paustenbach, 1991 p. 29).

## DIFFICULTIES ASSOCIATED WITH RISK ASSESSMENTS

There are numerous difficulties encountered when a RA is applied at a site. The assessment itself attempts to quantify and communicate the risks posed in a complex and often poorly defined environment. Numerous assumptions and estimations are used in the assessment which make it an inexact process. The assumptions used, by design cause an overestimation of risk (Johnson, 1992 p. 37). The final product represents a range of potential impacts and risks rather than a precise measurement of the actual risks (EPA Journal, 1991 p. 39). The nature of the RA itself requires that the public, whose health may be impacted, accept a certain amount of uncertainty with regard to the assessment.

Communicating risk with the public may be the biggest challenge facing the risk manager. The scientists conducting the assessment view the process as a technical endeavor. However, the public view it as a personal decision process that directly affects them (Cardinal, 1991 p. 1984). The complex task of assessing personal risks by the layperson is most often reduced to simple judgmental operation based on prior experience and beliefs (Johnson, 1992 p. 37). Recent studies suggest that the growing environmental movement is not so much due to the publics' concern for natural resources, but more a factor of worries about personal health (EPA Journal, 1991 p. 42). The risk manager is likely to face public resentment, and possible outrage, if anything less than total cleanup is proposed regardless of the risks.

The language of the RA poses another difficulty. The "technical experts" who conduct the assessments and general public quite often use language that is vastly different. The use of words such as "toxic", "hazardous", or "contamination" can have a dramatic impact on the way the public evaluates what is presented to them. If the message about environmental risks is improperly or poorly conveyed to the pubic, a phenomenon known as social amplification can occur (Kasperson, 1988). With this phenomenon perceived risks become exaggerated and distorted from actual risks. The jargon that scientists so often use only helps confuse the public and may cause distrust of what is being presented.

State government agencies often do not have adequate technical staff to review and evaluate RA that are presented to them. To compound this problem, the concept of allowable environmental risk is not a commonly accepted public belief, even though everyday living is full of acceptable risks (driving, smoking, flying). The prospect of approval of RA when contaminants are left in the ground is not a politically attractive option for those making the final decision. The challenge for the regulatory agency and those conducting the assessment is to develop a balance between allowable environmental risk and public values and opinions.

While RA has become an essential tool of the EPA, many state governments have been reluctant in their application to leaking underground storage tank (LUST) projects. All states require RA at superfund or hazardous waste sites due to the guidance that has been provided by the EPA (USEPA 1989). The slow acceptance of state agencies in applying RA at LUST sites is in part a factor of the current legislation that is in place in many states.

Much of the legislation that states have adopted is not based on sound scientific assessment and prevention of risks, but reflects the public's perception of environmental risks. Public opinions have been shaped more by the amount of attention that the media gives an issue than by the true understanding of the magnitude to the risk (EPA Journal, 1991, p. 41). Studies have proven that there is a clear correlation between the amount of coverage a health issue is given by the media and the public's perception of the importance of that particular issue (Sofalvi and Airhihenbuwa, 1992 p. 298). After public concern of an issue is raised by the media, the public's beliefs (no matter how based in fact they are) have the tendency to influence public policy through a pattern of leaders following followers (Page and others, 1983 p. 180) and (Lipset, 1981 p. 367). To compound the problem the public's general "chemophobia" has also helped to create some of the laws that do not allow for risk to be considered at LUST sites.

However, as remediation costs continue to rise, and more and more sites become eligible for state trust fund monies, legislatures have begun to reevaluate to use of RA in determining which sites to clean up (and thus pay for) and which sites to not pursue active remediation on. States have begun to realize that the objectives of all remediation activities are to protect human health and the environment as well as to reach an end-point. Thus the use of RA in the future will become a more common occurrence on many sites with environmental issues.

## FIELD DATA REQUIREMENTS IN RISK ASSESSMENT

The proper use of RA requires a good understanding of the risk evaluation process and an abundance of site specific data. Most of the information used to prepare a RA is gathered during the site characterization phase of a project. The quality of data gathered during this phase of the project will have a direct effect on the magnitude of the remediation end points that are later established. A lack of detail in the characterization of subsurface conditions will result in a speculative RA. Speculation in the RA process often necessitates the use of conservative "worst case" assumptions. The worst case assumptions often result in more stringent remediation end-points when compared to end-points establish with quality field data. More stringent closure goals result in an increase costs for the client and a longer time needed to reach the goals.

It is common to find sites where soil contamination is present, but is covered with buildings, asphalt, or too deep to come in contact with receptors (dermal contact or ingestion). In these cases the contaminants in the soil do not pose significant exposure risks from direct contact. However, risks may be present if the contaminants have the potential to leach from the soil and reach the ground water table. Contaminant leachability depends on the way the compound partitions between the soil particles and the pore water present in the subsurface. The leachability of the contaminants can either be established analytically or empirically by determining the soil diffusion coefficient ($K_d$). $K_d$ is defined as

122

the ratio of contaminant on solid phase (mg/kg) to the contaminant in liquid phase (mg/L). The higher the $K_d$ the more likely a chemical is to bind to the soil than to remain in water (USEPA, 1989).

The empirical method of determining $K_d$ involves establishing the organic content (OC) of the soil. The OC should be determined for a soil sample or samples collected from the contaminated zone. Once the OC is determined, $K_d$ can be calculated by multiplying OC by the carbon partitioning coefficient ($K_{oc}$) of the contaminant. $K_{oc}$ data are available from reference books for a wide range of organic chemicals. The empirical method is relatively straight forward, but tends to yield overly conservative $K_d$ values, which in turn result in overly conservative remediation goals (Huggins and Money, 1994, p. 11).

A more realistic $K_d$ value can be established by directly testing soil samples that are collected in the field. This procedure involves collecting an undisturbed soil sample from the zone of contamination using a Shelby tube sampling device. Samples should also be collected for laboratory analysis from this zone. The Shelby tube sample should be tested with a laboratory permeameter in order to establish the order of magnitude vertical hydraulic conductivity. The soil samples collected in conjunction with the Shelby tube sample should then be analyzed for the target compounds. The leachate from the Shelby tube sample should also be analyzed for the target compounds in order to establish which compounds leach from the soil. The data from these analyses can then be used to establish a realistic $K_d$.

$$K_d = \text{soil contamination (mg/kg)} / \text{leachate contamination (mg/L)} \qquad (1)$$

Once the $K_d$ is established for the site, the soil remediation level for each contaminant can be determined by multiplying the ground water maximum contaminant level (MCL) with the $K_d$.

$$\text{soil remediation goal (mg/kg)} = K_d \times MCL \qquad (2)$$

This method assumes ground water will not be adversely affected by soil contamination above the established MCL. Solute transport modeling can also be used to determine the affects of different soil contamination levels on ground water quality if remediation goals for ground water are above the MCL (Huggins and Money, 1994).

There are numerous additional parameters that are required for risk exposure and analysis model simulation. These data for soil include defining: soil bulk density, intrinsic permeability, soil porosity, organic content, pH, grain size distribution, percent sand / clay / silt, stratification or structures present, migration pathways, thickness of the unsaturated zone, depth to rock, contaminants present, concentrations of contaminants, horizontal and vertical extent of contaminants, and background soil quality.

Data requirements with regard to ground water include defining: aquifer characteristics (perched, confined, water table), depth to water, height of the capillary fringe, seasonal variations, horizontal and vertical gradients, flow direction, flow rates, aquifer thickness, defining if confining units are present, discharge points, contaminants present, concentration of contaminants, horizontal and vertical extent of contaminants, background water quality, and potential receptors.

It is important for the environmental professional in the field to understand the importance of quality data collection while in the site characterization phase of a project. Documentation of how and why certain data were gathered is vital since the information collected during this phase of the investigation will be used in the RA which is often conducted years after the site characterization has been completed.

Soil bulk density, intrinsic permeability, soil porosity, organic content, pH, grain size, and percent clay are parameters that geotechnical and analytical laboratories can easily determine. One drawback of laboratory derived values is that a single sample is assumed to be representative of the entire site. This assumption may or may not be true. This is where the judgment and experience of the environmental professional are crucial. Field personnel in charge of collecting laboratory samples should be aware of horizontal and vertical variations in soil conditions across the site. If conditions are homogeneous, one set of samples may be sufficient to define the subsurface conditions. However, if the soil is found to be highly variable, either vertically or horizontally, several sets of samples will need to be submitted for laboratory analyses. A range of values would then be derived for the subsurface conditions and used in the RA.

The identification of stratification or structures present in the subsurface, migration pathways, thickness of the unsaturated zone, depth to rock, and in place hydraulic conductivity can only be determined in the field. The field investigator should be aware of the need to identify the type of structures present in the regolith while conducting the site investigation. Split spoon samples should be collected continuously from the surface to the termination of each boring in order to define the variation in soil type, presence of stratification, foliation, fractures, joints, or other structures such as quartz veins, which could aid in contaminant migration. The type of structure, depth, orientation, and frequency should be noted on the boring logs. Laboratory data can not determine any of these soil qualities. Only the field investigator, using observations and knowledge can define the nature of the subsurface conditions in this manner.

The estimated thickness of the unsaturated zone and depth to rock should also be noted on boring logs. The unsaturated zone thickness can be confirmed once monitoring wells are installed. Using the monitoring wells, in place hydraulic conductivity can then be determined. In place hydraulic conductivity values may be more representative of the subsurface conditions than laboratory derived values since the tests conducted in monitoring wells extend over a greater thickness of the aquifer than samples that are analyzed in the laboratory. Hydraulic conductivity values obtained the in field will be affected by structures present in the regolith while the laboratory sample collected may or may not have intersected any of the structures present in the subsurface. Depending on the site conditions, single well permeability test (slug tests) or aquifer tests may be used to derive the range of hydraulic conductivities across the site.

In sandy soils, an estimate of the hydraulic conductivity can be derived from the grain size analysis using the Hazen method. This method used the data from grain size distribution curves to establish order of magnitude K values. The K estimate (cm/sec) is derived using the following equation.

$$K = C (D_{10})^2 \qquad (3)$$

The equation parameters are defined as follows:

K = hydraulic conductivity (cm/sec)
$D_{10}$ = ten percent finer than size (cm)
C = a coefficient based on the following table

| | |
|---|---|
| very fine sand, poorly sorted | 40-80 |
| fine sand with appreciable fine | 40-80 |
| medium sand, well sorted | 80-120 |
| coarse sand, poorly sorted | 80-120 |
| coarse sand, well sorted, clean | 120-150 |
| (Fetter, 1988, p. 80-81) | |

Note that the Hazen method of K estimation is only appropriate when the $D_{10}$ grain size is between 0.1 and 3.0 millimeters.

Several phases of site characterization may be need to define all of the contaminants present, concentrations of contaminants, horizontal and vertical extent of contaminants, and background soil quality. It is important that the characteristics of the site are well define before attempting to conduct the RA.


RISK ASSESSMENT IN USE

In Virginia, the Department of Environmental Quality (DEQ) has legislation that clearly requires the use of risk assessment at each site where a release of a regulated substance has occurred. State leaking underground storage tank regulation VR 680-13-02 Section 6.4 requires that a risk assessment be included as apart of the initial site characterization. The risk assessment must include "evidence that wells of the area have been affected; use approximate locations of wells potentially affected by the release; identification of potential and impacted receptors; migration routes; surrounding populations; potential for additional environmental damage;" (VR 680-13-02 Section 6.4 2b, p. 32). The application of the risk assessment at each individual site allows for site specific data and information to be used in developing remediation criteria. The subsequent corrective action plan (CAP) includes a copy of the risk assessment. The CAP then establishes site specific remediation goals and end points based on the risks posed to all identified receptors around the site. The DEQ will

125

approve the CAP only after ensuring that implementation of the plan will adequately protect human health, safety, and the environment.

In October 1988, an estimated 1200 gallons of premium unleaded gasoline was lost from a 4000 gallon underground storage tank at a retail gasoline station. Site assessment activities were conducted during 1989. A risk assessment was conducted in 1990 by a team of hydrogeologists and toxicologists for the site which is in southwest Virginia. The objectives of the assessment were to provide an estimate of the potential risks to public health and the environment due to exposure to site-specific contaminants. The risk assessment was conducted in general accordance with guidance outlined in the "Risk Assessment Guidelines For Superfund, Human Health Evaluation Manual Part A", U.S. EPA, Washington, D.C. (USEPA, 1989) since the DEQ could give little guidance as to exactly what was required at the time it was conducted.

The risk assessment was conducted in five steps as follows:
1. Site Characterization; 2. Contaminant Identification; 3. Exposure Assessment; 4. Toxicity Assessment; 5. Risk Characterization.

Site characterization provided a physical description and an historical review of the site. The physical description included the site location, immediate surrounding area and population, construction details of buildings on site, geology, hydrogeology, and soil characteristics. The historical review included the ownership history, operational/management history and regulatory history.

Contaminant identification determined which compounds were of potential concern. The contaminants of potential concern were determined to be chemicals that could pose potential risk to human health or the environment.

An exposure assessment estimated the magnitude and frequency of exposure of human and environmental receptors to site-specific contaminants. The exposure assessment included on-site and off-site pathway analysis, receptor evaluation, and estimation of contaminant intake.

The toxicity assessment (TA) was conducted to weigh available evidence regarding the potential for contaminants of concern to cause adverse effects in exposed individuals. The TA also estimated the relationship between the extent of exposure to a contaminant and the increased likelihood and/or severity of adverse effects. Toxicity values (numerical expressions of a substance's dose-response relationship) also were obtained.

The final step of the assessment was the risk characterization. In this step the exposure and toxicity assessments were summarized and integrated into qualitative and quantitative expressions of risk (Kinsler and Bell, 1991 p. 1-2).

Site specific remediation end points were developed with the RA data. The original closure goals posed by the DEQ were "natural ground water quality" (<1 ug/L) for benzene, toluene, ethylbenzene, and xylenes (BTEX). The risk assessment proposed to raised the remediation end points to 1690, 1980, 4950 and 19,080 ug/L respectively for BTEX. The proposed end points were accepted by the DEQ.

Ground water remediation activities began in September 1991. The remediation system was operated for approximately twenty-one months. During this period the level of dissolved gasoline compounds detected in ground water decreased substantially. Quarterly ground water quality monitoring indicated that BTEX compounds had been remediated below the proposed closure goals for over six months in June of 1993 so the remediation system what shut down at this time. Post operational ground water monitoring was conducted for one year after the treatment system was shut down. The levels of contaminants left in the subsurface never exceeded the established risk based closure levels after the remediation system was turned off. DEQ granted site closure on July 26, 1994.

## SUMMARY AND CONCLUSIONS

Risk assessment can be a valuable and useful tool when used correctly. Although RA is routinely used at superfund and hazardous waste sites, many states are still reluctant to apply them to LUST sites. The reasons for the current reluctance is a factor of the current laws that are in place and the general public attitude that no level of environmental risk is acceptable. With time it is likely that laws and attitudes will change and more and more states will begin to use risk assessment at LUST sites. As more risk assessments are conducted, there will be various applications to protect human health and the environment.

In Virginia a RA was used to develop site specific remediation end points well above ones that were originally proposed by the DEQ. The application of risk assessment allowed the site to be closed after twenty-one months of ground water remediation. Site closure resulted in significant cost savings to the client, while not adversely affecting the health of the population surrounding the site.

## ACKNOWLEDGEMENTS

Thanks to Mike Carey and Marilyn Gauthier for their skillful review of this manuscript and the constructive comments along the way to its completion.

## REFERENCES

Abelson, P.H., (1992) "Remediation of Hazardous Waste Sites", Science, vol. 255, no. 5047.

Cardinal, A.E., (1991) "Risky Business Communicating Risk for the Government", Environmental Science and Technology, vol. 25, no. 12.

Environmental Restoration and Waste Management, (1991) Five Year Plan, U.S. Department of Energy, Washington D.C.

EPA Journal, (1991) Setting Environmental Priorities: The Debate About Risk, vol. 17, no. 2 21K-1007.

Fetter, C.W. (1988) Applied Hydrogeology, 2nd edition, Charles E. Merrill Publishing Co, Columbus, Ohio.

Huggins, B. and Money, B., (1994) "Reaching the End-Points: Risk and Remediation", Virginia's Environment, Volume 1, Number 9.

Johnson, S.D., (1992) "Public Response to Environmental Health Risks", The National Environmental Journal, July/August 1992.

Kasperson, R., (1988) "The Social Amplification of Risk: A Conceptual Framework", Risk Analysis, vol. 8, no. 2.

Kinsler, S.K., and Bell, J.U., (1991) "Risk Assessment Report Tyler's Auto Service Roanoke Virginia" unpublished report by Delta Environmental Consultants, Inc.

Lehr, J.H., (1990) "Toxicological Risk Assessment Distortions: Part I", Ground Water, vol. 28, no. 1.

Lipset, S.M., (1981) The Wavering Polls, in N.R. Luttbeg (Ed.) Public Opinion and Public Policy. Models of Political Linkage, Edward F. Peacock, Itasca, IL.

Page, B. I., Shapiro, R.Y., and Dempsey, G.R., (1983) "Effects of Public Opinion on Policy", The American Political Science Review, vol. 77.

Paustenbach, D.J., (1989) The Risk Assessment of Environmental Hazards, John Wiley & Sons, Inc. New York.

Sofalvi, A.J. and Airhihenbuwa, C.O., (1992) "An Analysis of the Relationship Between News Coverage of Health Topics and Public Opinion of the Most Important Health Problems in the United States", Journal of Health Education, July/August 1992, vol. 23, no. 5.

USEPA, (1989) Risk Assessment Guidance for Superfund, Human Health Evaluation Manual, Part A., Solid Waste and Emergency Response, U.S. Environmental Protection Agency, Washington, D.C.

VR 680-13-02, (1989) Virginia Underground Storage Tank Technical Regulations.

# REMEDIAL ACTION SUITABILITY FOR THE CORNHUSKER ARMY AMMUNITION PLANT SITE

SUJITHKUMAR NONAVINAKERE
Plexus Scientific Corporation
980, Awald Drive, Suite 202
Annapolis, MD 21403

PHILIP RAPPA III
Plexus Scientific Corporation
980, Awald Drive, Suite 202
Annapolis, MD 21403

## INTRODUCTION

Numerous Department of Defense (DOD) sites across the nation are contaminated with explosive wastes due to munitions production during World War II (1942-1945), Korean Conflict (1950 to 1957) and Vietnam Conflict (1965 to 1973).    Production activities included explosives manufacturing, loading, packing, assembling, machining, casting and curing.  Contaminants often present at these sites include TNT, RDX, HMX, Tetryl, 2,4-DNT, 2,6-DNT, 1,3-DNB, 1,3,5-TNB and nitrobenzene.  Manufacture of TNT in the United States was banned in the mid-1980s (Hercules, 1991).  The Cornhusker Army Ammunition Plant (CAAP) is one such DOD site that has determined to be contaminated with explosives (USAEC, 1993).  The CAAP is located approximately 2 miles west of the City of Grand Island in Hall County, Nebraska.   The plant was put into operation in 1942 and was operated intermittently over a period of 30 years. The most recent operations were terminated in 1973.    The plant produced artillery, bombs, boosters, supplementary charges and various other experimental explosives.  In addition, the plant produced ammonium nitrate that was used as a fertilizer.  The CAAP site was placed on the National Priorities List (NPL) by the USEPA Region VII in October 1984 with Hazard Ranking System (HRS) score of 51.13.

The purpose of this paper is to provide an overview of the site background, review of the remedial alternatives evaluation process and rationale behind the selection of present remedial action.

## SITE BACKGROUND

An Environmental Site Assessment (ESA) of CAAP performed was by the United States Army Toxic and Hazardous Materials Agency (USATHAMA) in 1989. Based on the assessment report, USATHAMA concluded that the surrounding soil sub-strata was contaminated and a potential for groundwater contamination and migration of contaminants existed. A Field Screening Program (FSP) was intiated involving rapid-turnaround soil sample analyses using off-site laboratories to delineate the areal extent of surface soil contamination. It was identified that approximately 3400 cubic yards (USACE,1994) of soil was to be remediated. The concentrations of contaminants found in the CAAP soil is presented in Table 1.

**Table 1. Concentration of Contaminants at the CAAP Site**

| Contaminant | Highest Levels Detected in mg/kg |
|---|---|
| *Organic Contaminants* | |
| Hexahydro-1,3,5-trinitro-1,3,5-triazine (RDX) | 890 |
| Octohydro-tetranitro-tetrazocine (HMX) | 83 |
| 2,4,6-trinitro-toluene (TNT) | 2,400 |
| 1,3,5-trinitro-benzene (TNB) | 4.18 |
| *Inorganic Contaminants* | |
| Mercury | 0.70 |
| Cadmium | 64 |
| Chromium | 7,798 |
| Lead | 5,217 |
| Aluminum | 13,700 |
| *Others* | |
| Asbestos (Friable & Non-Friable) | Unknown* |

*The presence of asbestos in the soil was confirmed, the quantities present was unavailable at the time this paper was written.

# REGULATIONS IMPACTING THE DECISION PROCESS

The contaminants found at the CAAP site (see Table 1) are considered to be hazardous substances as defined in Section 101(14) of CERCLA. Even though, the contaminants do not pose an immediate threat, they need to be addressed by implementing an appropriate response action to prevent the threat of release of hazardous substances into the environment. The remedial action selected should be able to prevent, minimize, or mitigate damage to the public health or welfare or to the environment that may otherwise result from a release or threat of release.

USEPA classifies the removal actions at hazardous waste sites into three different categories based on the immediate hazard. Depending on the type of situation, threat or potential threat of release and the time frame for initiating removal action, the three different classifications are as follows:

1. Emergency Removal Actions
2. Time-Critical Removal Actions
3. Non-Time-Critical Removal Actions

The guidelines for emergency and time-critical removal actions (USEPA, 1993) calls for response to releases that requires removal action within six months. Non-time critical removal actions respond to releases requiring actions that can start later than six months. Based on the quantities and types of contaminants and evaluation of exposure routes, the removal actions at CAAP site can be readily classified as non-time critical removal action.

Regardless of the classification of the removal action at a site or the type of removal action chosen as per the National Oil and Hazardous Substances Contingency Plan (40 CFR Part 300) and CERCLA Section 104 & Section 106 are required, to the extent practicable considering the exigencies of the situation, to attain applicable or relevant and appropriate requirements (ARARs) under Federal, State or facility siting laws. The following three types of ARARs (OSWER Dir.9834.11, 1987) are applicable to removal actions at hazardous waste sites including the CAAP site:

1. *Chemical-specific ARARs*: These requirements are based on the health or risk-based amounts or concentrations of the contaminants that are found in or discharged to the ambient environment at the CAAP site. The chemical-specific ARARs applicable to the CAAP site are as follows:
   - The National Drinking Water Standards (40 CFR Part 141)
   - EPA Procedures for Approving State Water Quality Criteria (40 CFR Part 131)
   - Fish and Wildlife Coordination Act (16 USC, 40 CFR Part 6.302(g))
   - Nebraska Surface Water Criteria (NDEQ, Title 117)

2. *Location-specific ARARs:* These requirements are based on the restrictions placed on the removal activities execution at the CAAP site due to its geographic location. The location-specific ARARs relevant to the CAAP site are as follows:
   - General Information, Regulations & Information (49 CFR Part 171)
   - Hazardous Materials Table & Communication (49 CFR Part 172)
   - General Requirements, Shipment/Packaging (49 CFR Part 173)
   - Carriage by Public Highways (49 CFR Part 177)
   - Packaging Requirements (49 CFR Part 178)

3. *Action-specific ARARs:* The requirements triggered by the particular technology chosen or the activities occurring on site related to management of hazardous substances, pollutants or contaminants. The action-specific ARARs applicable to CAAP site are as follows:
   - National Emission Standards for Hazardous Air Pollutants (40 CFR Part 61)
   - The National Pollutant Discharge Elimination System (40 CFR Part 122)
   - Standards Applicable to Transporters of Hazardous Waste (40 CFR Part 263)
   - Standards for Owners and Operators of Hazardous Waste Treatment, Storage, Disposal Facilities ( 40 CFR Part 264)
   - Land Disposal Restrictions (40 CFR Part 268)
   - Designation, Reportable Quantities (RQs) and Notification (40 CFR Part 302)
   - Emergency Planning and Notification (40 CFR Part 355)
   - General Pretreatment Regulations for Existing and New Sources of Pollution (40 CFR Part 403)
   - Emergency Response and Contingency Procedures (29 CFR Part 1926.65(1))
   - Occupational Safety and Health Administration (29 CFR Part 1910)

## IDENTIFICATION & SCREENING OF REMOVAL ACTIONS

The clean-up alternatives reviewed in this section are based on potential applicability to the site, nature of contaminants, extent of contaminant migration and geography of the site. The removal action alternatives are screened on the following criteria:
- Extent of Human Health protection and Environmental Impact
- Compliance with the ARARs and acceptance by concerned agencies
- Overall, Long-term and Short-term effectiveness
- Technical and Administrative Implementation Feasibility
- Implementation Cost

## TABLE 2

## SCREENING OF REMOVAL ACTION ALTERNATIVES AT THE CAAP

| General Response Action | Associated Technologies | Human Health & Environmnetal Protection | Time Limit Compliance | Technical Implementation Feasibility | Concerned Agencies Acceptance |
|---|---|---|---|---|---|
| On-site Treatment | Solidification & Stabilization | No | Yes | Uncertain[1] | Yes[2] |
| | *In-Situ* Bioremediation | No | No | Uncertain[1] | No |
| | *Ex-Situ* Soil Washing | No | No | Uncertain[1] | No |
| | *In-Situ* Soil Vetrification | No | Yes | Uncertain[1] | No |
| Removal | Soil Excavation | Yes | Yes | Yes | Yes |
| Offsite Treatment | Off-Site Incineration | Yes | Yes | Yes | Yes |
| | Stabilization | Yes | No | Uncertain | Yes |
| Offsite Landfilling | Permitted Landfill | Yes | Yes | Yes | Yes |

[1] These technologies have not been implemented on full-scale basis. They have been shown to be effective through Bench Scale Studies or Pilot Studies. Therefore, their technical implementation on a full-scale basis is considered to be uncertain.

[2] Solidification & Stabilization is acceptable by the U.S EPA if the final product passes the TCLP tests.

*General Notes:*

Technologies were considered to be capable of protecting the human health and the environment if the action detoxified the contamination. Work safety during the implementation of the technology was also a deciding factor.

- Community Acceptance

Based on the currently available and widely accepted technologies, the following technological alternatives (VISITT, 1994) were identified as having demonstrated effectiveness or potential effectiveness:

1. Solidification and Stabilization
2. In-Situ Bioremediation
3. Ex-situ soil washing and replacement of washed soil
4. In-situ soil vitrification
5. Off-site incineration or thermal destruction
6. Disposal of excavated soil in an off-site landfill

Due to space restrictions the selected technologies are not discussed in detail in this paper, rather a summary of all the technologies is presented in Table 2. A comparative analysis evaluating the relative performance of all the removal actions mentioned above indicated Disposal of excavated soil in an off-site landfill as being the most suitable removal action alternative for the removal or remediation of contaminated soil at the CAAP site.

## REFERENCES

Hercules (1991). *A Petition for the Reclassification of TNT Process Red Water to a Secondary Material,*. Radford Army Ammunition Plant, Radford VA, January 1991.

OSWER (1987). *Revised Procedures for Implementing Off-site Response Actions*, OSWER Directive 9834.11, November 13, 1987.

USAEC (1993). *Site Characterization Document*, Cornhusker Army Ammunition Plant Remedial Investigation and Feasibility Study, Volume IX: Appendix C-1, Section 1-7, Contract #DAAA15-90-D-0018, June 1993.

USEPA (1993). *Guidance on Conducting Non-Time-Critical Removal Actions Under CERCLA*, Office of Solid Waste and Emergency Response, Washington, D.C 20460, EPA540-R-93-057, August 1993.

USEPA (1984). *National Priorities List Site*, Hazardous Waste Site Listed under the CERCLA Act of 1980 ("Superfund"), Adjusted Final, NPL-U2-2-169, United States Environmental Protection Agency, October 1984.

VISITT (1994). *VISITT - Vendor Informatin System for Innovative Treatment Technologies, Version 3.0*, USEPA, Solid Waste and Emergency Response, EPA 542-R-94-003, July 1994.

# A MODEL FOR THE PERMITTING OF REACTIVE CHEMICAL EMERGENCY PROJECTS: CLOSING THE GAP BETWEEN REGULATION, SAFETY AND COMMON SENSE

April 1995

MICHAEL F. RICHTER, DIRECTOR
Safety and Environmental Health
Advanced Environmental Technology Corporation
3 Gold Mine Road
Flanders, New Jersey 07836
(201) 691-7332

RICHARD HARTNETT, DISTRICT CHIEF
Planning Department
Boston Fire Department
115 Southampton Street
Boston, Massachusetts 02118
(617) 343-3390

## INTRODUCTION

The City of Boston is host to many industrial facilities, private research institutions and universities. Many of these facilities are in close proximity to densely populated residential neighborhoods, large office buildings, sports arenas and hospitals. These facilities all utilize and store many different potentially reactive chemicals either as "raw materials", finished products or hazardous waste in industrial process or research activities. In the vast majority of cases, these chemicals are used safely and without incident. In Boston, however, a series of minor incidents involving reactive chemicals several years ago, prompted the Boston Fire Department (BFD) to develop a departmental procedure and city regulation for the safe management of unstable/reactive or explosive materials discovered in Boston. BFD's purpose in developing the regulation was to act decisively to protect the Health and Safety of the Boston Public from the hazards of Reactive Chemicals.

The procedure provides for a standardized BFD response, evacuation of the public at risk and requires the facility involved to immediately arrange for a licensed hazardous waste contractor to inspect, stabilize and remove the designated material. The paper briefly summarizes the evolution of the program and it's effective implementation within Boston, citing two recent case studies as examples.

135

## THE DEPARTMENTAL PERSPECTIVE

The authority for the City of Boston to regulate reactive chemicals is derived from the Massachusetts Fire Prevention Regulations (527 CMR 1.03). In 1943, in a progressive response to the Coconut Grove incident, the BFD established the Fire Department Chemist's Office. Similarly, in 1992 the Fire Prevention and Planning Departments, together with the Chemist's Office instituted Standard Operating Procedure 53A (Reactive Chemicals and Unstable Materials). The regulation itself calls for the following sequence of events once an unstable reactive chemical is discovered in the city:

1. Any facility that discovers an unstable reactive chemical is to immediately report the discovery to the Fire Department Alarm Office and evacuate the affected area or building. The affected facility must identify the exact street address, building name or number, floor, area or room number and describe fully, the quantity, condition and type of unstable material.

2. Upon receipt of the call, the fire department initiates an incident response, with the District Fire Chief, and appropriate fire/rescue equipment dispatched to the scene.

3. The incident is under control of the District Fire Chief, who makes an assessment of the potential hazard after consulting with the facility representative and other resources such as the Fire Department Chemist, Health and Environmental Officials, the Boston Police Bomb Squad or Private Experts. An immediate evacuation of the potentially impacted area is ordered and maintained until the hazard is abated. The safety and welfare of the public is the prime consideration. Fire officials are not permitted to enter the scene without the approval of the District Fire Chief.

4. Containment, desensitization or neutralization and removal of the unstable materials is accomplished by a contractor, licensed by the City of Boston to handle reactive and unstable materials. An emergency removal permit is issued to the contractor after submission of the following documentation:

   (a). Site Specific Reactive Chemical Safety Plan
   (b). Specific Reactive Chemical Handling Procedures
   (c). Request for Specific Fire Detail Assistance Outlining
        Contingency Procedures.
   (d). Credentials Summary

5. The project is permitted and scheduled by the District Fire Chief. The District Chief will frequently discuss the project at length with the contractor and the facility management prior to issuing the permit. During this iteration, project logistics and emergency/contingency measures are discussed. Typically the Fire Prevention Office and Fire Department Chemist is also consulted during the permit review.

6. At the time designated by permit, the contractor must conduct a pre-project safety and emergency contingency meeting. During this brief meeting, all participants roles are discussed. The project commences only after the District

Chief is satisfied with safety plan implementation and that all permit conditions have been met.

## THE CONTRACTOR'S PERSPECTIVE:

As a licensed contractor serving numerous clients in the City of Boston, AETC approached the BFD regarding review and pre-approval of AETC's Safety Program and Reactive Chemical Stabilization procedures. Our goal in requesting this review was two-fold: Maximize project Safety for AETC and BFD personnel; and second, minimize the potential impact of the project to the public and our client's facility. BFD agreed that such a review would enhance project safety and agreed to review our procedures almost immediately following our request.

The first step in the review process was to clarify the phrase "Reactive Chemicals". To insure a conservative approach to these incidents, BFD discussed openly, their interpretation.

BFD uses the term "reactive chemicals" in a very broad sense to help insure that all incidents are properly reported. Discussions with the BFD Planning Department, Chemist's office and Department of Fire Prevention, resulted in AETC's submittal of detailed procedural information for the handling and stabilization of the following classes of reactive chemicals:

| Category: | Example: |
|---|---|
| 1. Air Reactive | Triethyl Aluminum in Hexane |
| 2. Water Reactive | Chlorinated Silanes |
|  | Sodium Metal |
| 3. Temperature Sensitive | Organic Peroxides |
|  | Dibenzoyl Peroxide |
| 4. Shock, Friction, or Static Sensitive | Picric Acid |
|  | Polynitrated Alkyl compounds |
| 5. Contaminated Strong Oxidizers | Nitric Acid |
|  | Perchloric Acid |
| 6. Extremely Toxic Gases | Hydrogen Cyanide |
|  | Phosphine |
|  | Arsine |
| 7. Peroxidizing Liquids | Ethers |
|  | Furans |
| 8. Unstable Monomers | Uninhibited Styrene |
|  | Acrylic Acid |

Clearly many of these are common industrial chemicals, however, under certain conditions such as improper storage, handling, or they have the potential to release dangerous amounts of energy or plumes of toxic gases. It is under these circumstances that the BFD would initiate emergency response procedures for reactive chemicals. According to Standard Operating Procedure 53A (SOP 53A).

In addition to the procedural information, AETC provided BFD with a Corporate Health and Safety Program Plan, Training Documentation, Personnel Resumes,' training transcripts and examples of Site Specific Reactive Chemical Safety Plans for review. Several weeks following the submittal AETC met with the department to provide clarifications regarding the submittals and to listen to the department's perspective.

The results of this informal process certainly justified the investment of time and resources. During these meetings AETC personnel learned exactly what our response personnel could expect from the BFD during a reactive chemical incident and vice versa. Probably the most practical piece of information developed during this process was a standardized format for reactive chemical safety plans, the outline of which appears below:

1. General Information
2. Project Procedures
3. Personal Hygienic Measures
4. Hazard Assessment
5. Permits
6. Medical Monitoring
7. Personal Protective Equipment
8. Site Entry/Work Areas
9. Decontamination
10. Training Requirements
11. Emergency Response Procedures

AETC then took this information back to our clients to help them plan internally for reactive chemical emergencies. Subsequent projects utilizing this system were approved more quickly and conducted more swiftly. In all cases, because both BFD, client facilities and AETC personnel better understood each-other's capabilities and project roles, the projects were conducted with an even greater margin of safety.

**CASE STUDIES:**

Two case studies have been selected. To illustrate the function of BFD SOP 53A, the authors have selected two case studies. The first went very smoothly, the other initially, did not. The purpose in describing these case studies is to illustrate the importance of standardizing the approach to reactive chemical emergencies.

## CASE STUDY #1:   AGED HYDROGEN CYANIDE CYLINDER: PRIVATE RESEARCH LABORATORY

### SCENARIO:

A lecture bottle of Hydrogen Cyanide was discovered encased in ice in a very old laboratory freezer when the freezer was being removed from service. The age of the cylinder was approximately 20 years, and it appeared to be lab-fabricated.

### CHRONOLOGY

### "DAY 1: DISCOVERY"

Upon discovery of the cylinder, the facility, familiar with the SOP 53A immediately reported the incident concurrently to the BFD and AETC, the facilities' hazardous materials contractor.  BFD and the Research Laboratory, initiated their respective emergency/contingency plans, evacuating the building and mobilizing response equipment.  Concurrently, AETC was directed to inspect the cylinder and develop a removal plan for the material.  Having conducted an inspection of the cylinder, AETC reported that the material was indeed a lecture bottle labelled Hydrogen Cyanide.   The cylinder was approximately 20 years old, constructed of Stainless Steel, laboratory fabricated and, possibly bulged at one end.  The Research Laboratory reported that the cylinder did not appear on their existing inventory of chemicals and that they had no accurate information with regard to the amount of the contents or presence of stabilizers or inhibitors in the material.  Based on this information and the inherent instability and toxicity of uninhibited Hydrogen Cyanide, the facility, together with the District Fire Chief, made the decision to maintain the evacuation of the laboratory and have AETC place the cylinder immediately into an emergency cylinder containment vessel.

### "Day 2: CONTAINMENT"

AETC submitted a Reactive Chemical Safety Plan to the Boston Fire Department and obtained an emergency Permit to place the cylinder into a containment vessel.  Scheduling occurred immediately with AETC, and the Boston Health Department providing Risk Assessment Information to the BFD. The District Chief established an evacuation radius based on this information and scheduled the removal for 1 AM to minimize the population at risk in the downtown area.   The procedure was completed without incident, with the cylinder containment vessel placed into temporary storage.   The vessel was continuously monitored for pressure buildup and leakage through a combination of video monitors, and portable monitoring equipment which was hard-wired into the facilities' alarm system.  The storage room ventilation was isolated with exhaust directed directly through a laboratory hood system to insure there would be no contamination of the facility should a release occur.

## "DAY 3 - DAY 50: PROJECT PLANNING"

With the cylinder safely contained, planning and risk assessment began to determine the best way to handle the cylinder. Uninhibited HCN is a DOT Forbidden Material, and further, no RCRA facilities accept Uninhibited HCN. BFD, the Boston Health Department, AETC and the Facility all agreed that the cylinder could not be removed from the containment vessel onsite due to the extreme hazards present. AETC submitted a Reactive Chemical Safety Plan and Risk Assessment to perform a remotely operated controlled detonation and treatment of the cylinder contents. The procedure called for a 500 foot exclusion zone, with AETC personnel performing the treatment garbed in Level A protection together with fire protective coveralls and blast fragmentation suits.

The selection of an appropriate location to conduct the project and transportation of the containment vessel proved to be the most difficult aspect of the project. The following government organizations participated in the development and approval of the project: Boston Fire Department, Boston Health Department, Mass. Department of Environmental Protection, Mass. State Fire Marshall's office, Boston Police Department, Boston Bomb Squad, USEPA, and the US Coast Guard. Project location selected was a remote Island in Boston Harbor. The cylinder was moved to the island under the following conditions:

From Research Facility to Pier
*Police Department  Security Escort
*Fire Department Escort
*Ambulance with paramedics: HCN antidote equipped
From Pier to Island:
*Pushed Barge and Tugboat with AETC Emergency Response
 Team and Continuous Monitoring
*Coast Guard Escort

At the island, the cylinder was removed from the containment vessel and placed in a shape charge cradle assembly. This unit was then placed in a bunker, packed in dry slurry of Sodium Hydroxide and Sand, and punctured. Treatment occurred safely, and without incident. Cyanide salts and excess Sodium Hydroxide from the treatment process were manifested to RCRA licensed facilities.

## CASE STUDY 1: COMMENTARY:

Clearly, this is an extreme example from the standpoint of permitting, agency participation and hazard. This example however serves to illustrate the importance of a well defined procedure and clear chain of command. Further, the pre-approval of the contractor's procedures and familiarity of the regulators with the contractor's Reactive Chemical and Safety Procedures streamlined the project planning process. Had this relationship not been previously established, the approvals process could have stretched out for months, possibly placing the public at increased risk.

## CASE STUDY 2: ABANDONED INDUSTRIAL FACILITY:

### SCENARIO:
An abandoned Cold Storage Facility, owned by the Boston Public Facilities Department is being investigated by an Environmental Consultant. During a site-walk, numerous large unknown cylinders, waste containers and tanks are discovered. The Facility is bordered by a large day care center on one side, an elementary school on the opposite side and is located in the middle of a densely populated residential area.

### CHRONOLOGY:
Upon discovery, the City Agency owning the property notifies the Boston Fire Department. Acting in accordance with departmental policy, the District Fire Chief orders the evacuation of the day care and school. The Public Facilities Department then enlisted the services of a private Hazardous Materials Contractor to conduct the waste removal activity. The emphasis from both the Boston Fire Department and the Public Facilities was to remove all potentially dangerous or reactive materials from the facility, rendering the neighboring School and Day Care "safe" for re-occupation.

Several weeks later, the Boston Fire Department and the Public Facilities Department, dissatisfied with progress at the site requested AETC's services at the site based on our prior experience on similar projects in the city.

AETC conducted a site-walk with BFD representatives, and delivered a project work scope, together with a Reactive Chemical Project Safety Plan the following morning. The submittal included an estimated time-frame for the specified goal of re-occupation of the day-care and School of only 96 hours. AETC was authorized to proceed immediately with approval from the BFD, Public Facilities Department, Mass. Department of Environmental Protection and Boston Department of Health. The Department of Education was also consulted to insure their concerns were being met.

Work began immediately upon approval with the following tasks being completed. 1). A safety meeting with all involved parties was conducted upon AETC's arrival at the site. Tasks, emergency measures and each group's role was clearly defined at this time. Since the project work included two work shifts/day, around the clock coverage and security assignments were also finalized. 2). Approximately 30 unknown and decaying cylinders were removed from the building, sampled, and characterized. Nearly half of these were returned to the original manufacturers for recycling or re-processing. 3). In addition, nearly 100 additional containers of Hazardous wastes were removed, characterized and manifested off-site for proper disposal. 4). The final task called for the evaluation and stabilization of an Ammonia based refrigerant system with a total volume of approximately 3500 gallons. This final task was accomplished with the cooperation of the BFD utilizing water walls and fog lines during a series of confined space entries, ammonia scrubbing and line-

breaking procedures. All work was completed as planned, with the school certified safe for occupancy by Boston Public Safety Officials nearly a full day ahead of the projected 96 hour schedule.

## CASE STUDY 2: COMMENTARY:

The key to the ultimate success of this project originated in its eventual organization. During the first segment of work by the original contractor, progress occurred slowly because communications were inneffictive. The project chain-of-command, approved procedures and activities, project safety plans and goals were not clear to all parties involved.

Despite time constraints and the perceived hazard, AETC developed the plans and documentation required by all regulatory authorities involved. Permitting submittals were prepared in the pre-arranged format, dramatically reducing the review and approvals process. Concurrently with the review process, a clear chain-of-command was established in accordance with BFD SOP 53A. With permits in place, a safety meeting was conducted on site with all involved parties. Task leaders were identified with specific task goals clearly explained. All questions were addressed at this time.

The resultant project went very smoothly. Work interruptions were minimized because all parties; AETC, BFD, MADEP, Boston Health Department and the Education Department all understood their respective roles and were working together. When problems did arise, the established communications format facilitated their swift resolution.

## CONCLUSION:

The goal of the Boston Reactive Chemical Emergency Program (SOP53) is simple. Protect the safety and health of the public to the greatest extent possible. Since the BFD program forbids the handling of reactive chemicals by BFD responders, the most common response tactic is conservative evacuation ordered by the District Fire Chief, often for an indefinite period. This action provides for the immediate security and protection of the population at risk. Clearly, such evacuations, while safe, are extremely costly to private facilities involved, since their business is effectively closed during this process. Other intangible costs are associated with the disruption of private residences and city services. The wisdom of delaying the response until procedural safety documentation is complete has been questioned by affected industries, primarily due to the negative potential business impact. BFD's response to this concern is simply that their single most important concern is the public safety, with property and business concerns secondary. However, SOP53 also empowers the responsible District Fire Chief to make decisions on site and provides substantial resources to aid in the decision process. Those facilities who have planned for potential emergencies have experienced minimal impacts from the implementation of SOP53. Generally speaking, reactive chemical emergencies

in these facilities have been handled swiftly and effectively. In cases where facilities have not pre-planned, the resolution of emergency situations has been conducted safely, however much more slowly. It is imperative therefore, that facilities implement detailed contingency plans with qualified contractor support familiar with the local authorities and their procedures.

From the municipal perspective, the need for establishing internal procedures and communication with commercial enterprises is also clearly a necessity.

Boston's experience demonstrates that it is possible to implement a response program that protects the public safety, establishes clear lines of authority and communication, and minimizes negative impacts on commerce.

# COMPARATIVE ANALYSIS OF RISK-BASED CLEANUP LEVELS AND ASSOCIATED REMEDIATION COSTS USING LINEARIZED MULTISTAGE MODEL (CANCER SLOPE FACTOR) VS. THRESHOLD APPROACH (REFERENCE DOSE) FOR THREE CHLORINATED ALKENES

LINDA J. LAWTON
Geraghty & Miller, Inc.
One Corporate Drive
Andover, MA 01801

JOHN P. MIHALICH
Geraghty & Miller, Inc.
3000 Cabot Boulevard West
Suite 3004
Langhorne, PA 19047

## ABSTRACT

The chlorinated alkenes 1,1-dichloroethene (1,1-DCE), tetrachloroethene (PCE), and trichloroethene (TCE) are common environmental contaminants found in soil and groundwater at hazardous waste sites. Over the past few decades, concerns about chlorinated solvents being released into the environment have increased and regulatory agencies have tightened regulations on the use and release of these compounds. Site cleanups costing industry millions of dollars have been driven by health concerns associated with potential exposures to these compounds. These health concerns stem largely from positive outcomes in laboratory cancer bioassay studies with rodents. Recent assessment of these data and the data from epidemiology and mechanistic studies indicates that although exposure to 1,1-DCE, PCE, and TCE causes tumor formation in rodents, it is unlikely that these chemicals are carcinogenic to humans. Nevertheless, many state and federal agencies continue to regulate these compounds as carcinogens through the use of the linearized multistage model and resulting cancer slope factor (CSF).

The available data indicate that 1,1-DCE, PCE, and TCE should be assessed using a threshold (i.e., reference dose [RfD]) approach rather than a CSF. This paper summarizes the available metabolic, toxicologic, and epidemiologic data that question the use of the linear multistage model (and CSF) for extrapolation from rodents to humans. A comparative analysis of potential risk-based cleanup goals (RBGs) for these three compounds in soil is presented for a hazardous waste site. Goals were calculated using the USEPA CSFs and using a threshold (i.e., RfD) approach. Costs associated with remediation activities required to meet each set of these cleanup goals are presented and compared.

## INTRODUCTION

Current state and federal guidelines and regulations establishing allowable soil concentration for 1,1-DCE, PCE, and TCE are largely based upon studies which demonstrate tumor formation in rodents exposed daily to high concentrations of these compounds. USEPA Regions III [1] and IX [2] and a number of state regulatory agencies publish risk-based soil concentrations for these compounds that are based on the use of a CSF calculated using the linearized multistage model. The CSF is combined with a defined acceptable risk level (i.e., $1 \times 10^{-6}$) and a set of exposure assumptions to calculate a RBG. The basic equation for calculation of the cancer RBG is shown below:

$$RBG = \frac{Risk\ (1 \times 10^{-6})}{Exposure \times CSF} \qquad (1)$$

Use of the CSF to calculate soil RBGs is based on the assumption that human risk associated with environmental exposure to these constituents can be estimated by a straight line extrapolation from high-dose rodent data. A review of the metabolic, toxicologic, and epidemiologic data for 1,1-DCE, PCE, and TCE call this approach into question [3, 4, 5]. The weight of available evidence on the mechanism of tumor production by TCE and PCE in rodents indicates that tumor production is related to rodent-specific responses triggered as a result of tissue toxicity associated with extremely high exposure levels. The rapid metabolism of 1,1-DCE in mice results in greater quantities of reactive metabolites being formed at the same dosages compared to humans. Thus, the metabolic data indicate that humans are less sensitive to 1,1-DCE, TCE, and PCE than either rats or mice [4]. This is coupled with the fact that humans are exposed to much lower concentrations of these compounds in the environment compared to the levels to which rodents are exposed in laboratory studies.

Exposures of human populations to 1,1-DCE, PCE, or TCE have typically been associated with simultaneous exposures to other chemicals. As a result, the available epidemiologic studies are difficult to interpret. Nevertheless, the available epidemiology data, like the mechanistic data, do not provide compelling evidence to suggest that these constituents are human carcinogens [3, 4]. The available data thus indicate that the current state and federal guidelines for calculating soil RBGs for 1,1-DCE, PCE, and TCE are too stringent.

It have been suggested that exposure limits (i.e., RBGs) for 1,1-DCE, PCE, and TCE would more appropriately be calculated using procedures based on a threshold-type dose-response relationship (i.e., based on an evaluation of

145

noncancer effects) [3, 4, 5]. The use of a threshold approach results in soil RBGs that are two or three orders of magnitude higher than those calculated using the linearized multistage model. Risk-based goals are often used as soil cleanup goals; thus, these values dictate the most appropriate remedial alternative and the volume of soil requiring remediation. Both of these factors have a profound effect on the ultimate cost of soil remediation. We emphasize that the risk assessment approach used to calculate RBGs for 1,1-DCE, PCE, and TCE is an important factor that should be more closely considered by industry and regulating agencies.


## METABOLISM, TOXICOLOGY, AND EPIDEMIOLOGY

1,1-Dichloroethene

The available data suggest that the liver and kidney toxicity, the mutagenicity, and the apparent carcinogenicity of 1.1-DCE are due to reactive intermediates formed during its metabolism [4]. These adverse effects do not appear to be due to the parent compound. The metabolism of 1,1-DCE is dose dependent, and saturation of metabolic systems results in changes in toxic potency at different doses. Toxic effects manifested at high doses may be very different than those observed at much lower exposure levels, such as those to which humans may be exposed in the environment. Mice show a greater sensitivity to 1,1-DCE toxicity compared to humans, as a result of their greater metabolism of 1,1-DCE [6]. The results of a pharmacokinetic model suggest that humans would likely be exposed to lower tissue doses of the reactive metabolite of 1,1-DCE compared to rats, assuming that administered doses were equal, due to metabolic and physiological differences [7]. It is this reactive metabolite that is thought to be responsible for the apparent carcinogenicity of 1,1-DCE. This questions the validity of classifying 1,1-DCE as a potential human carcinogen based on results of rodent studies.

Several epidemiology studies have evaluated the relationship between exposure to 1,1-DCE and cancer. No significant increases in mortality due to cancers were observed in several studies of petroleum workers with potential exposure to 1,1-DCE [3,4]. Two cohort studies showed no excess risk of cancer among subjects exposed to 1,1-DCE plus a mixture of other chemicals [4]. There were a number of limitations to these studies; therefore, the results are considered inconclusive.

The USEPA has classified 1,1-DCE as a possible human carcinogen (Group C) under the USEPA weight-of-evidence classification. It should be noted that the International Agency for Research on Cancer (IARC) (1986) has concluded that the evidence is inadequate to classify 1,1-DCE as a human carcinogen [8]. Nevertheless, as discussed below, the CSF is still used by most state and federal regulatory agencies to calculate acceptable environmental exposure levels for 1,1-DCE.

146

## Tetrachloroethene

In a 1986 NTP bioassay, a marginal, nonsignificant increase in the incidence of kidney tumors was observed in treated male rats [4]. The kidney tumors were associated with high exposure levels that also led to increased mortality and overt kidney toxicity. The low incidence of kidney tumors may also have been related to the accumulation of a male rat-specific urinary protein [4]. Based on the available data on the proposed mechanism of action of PCE, the male rat kidney tumors observed in the NTP study are not likely relevant to humans since accumulation of the urinary protein does not occur in humans and since the cytotoxicity is likely to occur only after prolonged exposure to concentrations of PCE substantially greater than those to which humans are typically exposed [4].

Liver tumors were reported in mice exposed to high doses of PCE by gavage and inhalation [4]. The tumors were associated with concomitant liver toxicity and appear to be related to the induction of peroxisome proliferation. (Peroxisomes are ultramicroscopic particles that contain enzymes that generate hydrogen peroxide in cells, particularly in liver cells). Peroxisome proliferation does not appear to occur in humans; therefore, the liver tumors observed in mice may not be relevant to the assessment of risks in humans.

The epidemiology data relating PCE exposures to certain forms of cancer have been inconclusive. A number of studies have reported some positive associations between PCE exposure and an elevated risk for cancer; however, individuals in these studies were exposed to a variety of other chemicals in addition to PCE and confounding exposure to other carcinogenic agents were clearly identified. The epidemiology data do not provide any compelling evidence to suggest that PCE is a human carcinogen [4].

The USEPA has withdrawn the carcinogenicity assessment of PCE in response to the data that the mechanism of action by which PCE induces tumors in rodents may not be relevant to the assessment of cancer risks in humans. The lack of relevance of the tumors observed in rodents prompted ACGIH to classify PCE as "Group A3-Animal Carcinogen" or "Available evidence indicates that the agent is not likely to cause cancer in humans, except under uncommon or unlikely routes or extreme rates of exposure" [9].

## Trichloroethene

Studies demonstrate that TCE induces liver tumor formation in mice exposed to daily high doses of the compound (in the range of 1,000 mg/kg/day) [4]. These tumors are formed on a background of liver cell necrosis (i.e., cell death). TCE itself is minimally mutagenic and, like 1,1-DCE, the carcinogenicity in mice appears to require its metabolism. Humans, dogs, rats, and cats have not been found to develop liver tumors when chronically dosed with large quantities of TCE. Mice metabolize TCE to the toxic metabolite (trichloroacetic acid) much faster than rats or humans and data indicate that this metabolite induces liver tumors in mice through peroxisome proliferation [4]. As mentioned earlier, peroxisome proliferation appears not to occur in humans.

Epidemiology studies, covering a number of years of exposure to TCE, have not demonstrated an excess number of cancers [4]. These studies, however, have been limited by a number of factors that limit the conclusions that can be drawn from the results. A more recent study of over 14,000 workers at an aircraft maintenance facility that used various solvents, including TCE, showed no significant associations between several indicators of TCE exposure and any excess mortality due to cancer. Overall, the authors concluded that occupational exposure to TCE is not associated with an increased risk for cancer. Based on the results of the epidemiology studies and, in particular, the negative results reported in the large study of aircraft workers, the ACGIH has classified TCE as "Group A5" or "Not Suspected as Human Carcinogen".

The IARC concluded that TCE was unclassifiable as to its carcinogenicity in humans based upon limited evidence of carcinogenicity in one animal species [4]. The USEPA has withdrawn the carcinogenicity assessment of TCE (the USEPA had classified TCE as a probable human carcinogen[Group B2]) in response to the weight of evidence that the mechanism by which TCE induces liver tumors in mice may not be relevant to the assessment of carcinogenic risks in humans. Nevertheless, many regulatory agencies, including the USEPA, continue to use the CSFs derived from the mouse liver rumor data to calculate RBGs for hazardous waste sites.

## CALCULATION OF RISK-BASED GOALS (RBGs)

Risk-based goals for soil were calculated for 1,1-DCE, PCE, and TCE using USEPA CSFs derived from the application of the linearized multistage model (i.e., RBGs for cancer effects), and using a threshold (RfD) approach (i.e., RBGs for noncancer effects). Reference doses and CSFs used in the derivation of the RBGs are published by the USEPA and are shown in Table I. The RFDs for 1,1-DCE; PCE, and TCE were all derived based on studies which reported liver effects in exposed rodents [4].

### TABLE I - TOXICITY VALUES

| Constituent | Oral RfD (mg/kg/day) | Oral CSF $(mg/kg/day)^{-1}$ |
| --- | --- | --- |
| 1,1-DCE | 0.009 | 0.6 |
| PCE | 0.01 | 0.052 |
| TCE | 0.006 | 0.011 |

Risk-based goals were calculated based on a standard exposure scenario involving a 30-year resident being exposed to affected soils via incidental soil ingestion. The goals for cancer effects were calculated based on a target excess lifetime cancer risk of $1 \times 10^{-6}$. Noncancer goals were calculated based on a target hazard index of 1. These risk goals are specified in USEPA's Risk Assessment Guidance for Superfund: Volume 1 - Human Health Evaluation Manual (Part B, Development of Risk-Based Preliminary Remediation Goals) [10] and are typically used by regulatory agencies in calculating RBGs. The equations used to calculate RBGs for cancer (based on a CSF) and for noncancer effects (based on a RfD) are shown below [10]:

$$RBG \text{ (Cancer)} = \frac{1 \times 10^{-6} \times 70 \text{ years} \times 365 \text{ days/year}}{CSF \times 10^{-6} \text{kg/mg} \times 350 \text{ days/year} \times 114 \text{ mg-yr/kg-day}} \quad (2)$$

$$RBG \text{ (Noncancer)} = \frac{1 \times 30 \text{ years} \times 365 \text{ days/year}}{1/RfD \times 10^{-6} \text{ kg/mg} \times 350 \text{ days/year} \times 114 \text{ mg-yr/kg-day}} \quad (3)$$

The calculated cancer and noncancer (i.e., threshold) RBGs are shown in Table II.

## TABLE II-RISK-BASED GOALS

| Constituent | Noncancer RBG (mg/kg) | Cancer RBG (mg/kg) |
|---|---|---|
| 1,1-DCE | 2,500 | 1 |
| PCE | 2,700 | 12 |
| TCE | 1,600 | 58 |

The RBGs calculated using a threshold-type approach (i.e., using the RfDs for noncancer effects) are 28 to 2,500 times greater than the RBGs calculated using the CSFs derived from the linearized multistage model.

## REMEDIATION COSTS

Costs associated with remediating soils at an actual hazardous waste site to the levels shown in Table II for noncancer effects and cancer effects were calculated. Regulatory agencies often require that industrial sites be cleaned up to residential use RBGs based on the assumption that the site could one day be developed for residential use; therefore, we have calculated remedial costs associated with cleanup to residential use RBGs. The total cost to remediate soil is dependent on several factors, such as the volume of soil to be remedied, the characteristics

149

of the contaminant(s), the soil characteristics (e.g., clay vs. sand), the physical conditions of the site, and the timeframe during which remediation must be completed. However, the driving factor for soil remediation, and therefore the basis for selecting the most appropriate remedial alternative, is the required cleanup level. Ultimately, it is the required cleanup level that determines the cost of soil remediation.

The cost comparison is based on an actual site located in an industrial area. The site consists of a 20-foot by 30-foot concrete pad which was historically used to store hazardous waste (solvents) containing 1,1-DCE, PCE, and TCE. For the purpose of the cost comparison, it is assumed that a total of 1500 pounds of solvent have contaminated a 20-foot by 30-foot by 10-foot deep volume of soil (approximately 222 cubic yards of soil) underlying the concrete pad. The soil is characterized as unsaturated, rather homogeneous, silty sand and silty clay.

The selection of a remedial alternative at this site is complicated by the presence of underground sewer lines and the proximity of nearby buildings, both of which limit the use of soil excavation as a form of remediation. An insitu remedial technology, such as soil vapor extraction, may be appropriate for this site due to the volatile characteristic of the contaminants and the physical constraints of the site. However, the success of soil vapor extraction may be limited by the fine-grained nature of the underlying soils. Given a time-frame of two years, cleanup to below cancer levels using soil vapor extraction is unlikely. Cleanup to below non-cancer levels using soil vapor extraction, alternatively, is estimated to take six months. The cost to cleanup to below cancer levels using excavation must be compared to the cost to cleanup to below non-cancer levels using an insitu technology such soil vapor extraction before a remedial alternative is selected.

Costs associated with excavation include soil removal, transportation, and disposal. Additional costs associated with excavation include backfilling and repavement. There are also engineering constraints associated with excavation, such as excavation under a building. Costs associated with soil vapor extraction include system installation, aboveground soil vapor treatment (e.g., using activated carbon), operation and maintenance of the system, and permitting. Costs associated with these two alternatives for the site described above are summarized in Table III.

## TABLE III-COMPARISON BETWEEN EXCAVATION AND SOIL VAPOR EXTRACTION

| Excavation | | Soil Vapor Extraction (for six months) | |
|---|---|---|---|
| Tasks | Cost | Task | Cost |
| work plan preparation, permitting | $10,000 | design, installation, permitting | $20,000 |

150

| removal | $16,000 | | | electricity | $2,000 |
|---|---|---|---|---|---|
| backfill/repavement | $6,000 | | | activated carbon | $20,000 |
| contractor oversight | $10,000 | | | vapor monitoring | $5,000 |
| transportation, disposal | $132,000 | to | $416,000 | confirmation sampling | $6,000 |
| confirmation sampling | $2,000 | | | reporting | $5,000 |
| reporting | $5,000 | | | | |
| Total | $181,000 | to | $465,000 | Total | $58,000 |

A review of the costs associated with each of the alternatives for this site, in light of the required cleanup levels, demonstrates the economic impracticability of cleanup to below cancer levels in an industrial area. The cost to remediate to below cancer levels using excavation as a remedial alternative ranges from $181,000 to $465,000, depending on factors such as the method of disposal. The cost to remediate to below non-cancer levels, however, is estimated to be 58,000, assuming six months of soil vapor extraction. Thus, the remediation costs to reach the cancer RBGs for 1,1-DCE; PCE, and TCE are three to eight times higher than the remediation costs required to reach the noncancer RBGs for these compounds.

## CONCLUSION

Cleanup goals for soils and other media at hazardous waste sites contaminated with 1,1-DCE, PCE, and TCE are typically based on a CSF derived from the application of the linearized multistage model to animal tumor data. Given the metabolic and mechanistic data that indicate that the tumors observed in rodents are in response to extreme rates of exposures to these compounds that are not likely to be relevant to human health, and given the available epidemiology data which do not indicate that these compounds are human carcinogens, the RBGs are more appropriately determined based on mechanisms involving threshold-type dose-response relationships (e.g., an RfD). Thus, the available data indicate that the methods that many state and federal agencies are using to calculate RBGs for these compounds are too stringent.

This paper calculated RBGs for 1,1-DCE, PCE, and TCE in soil using the USEPA CSFs and using a threshold (i.e., RfD) approach. The RBGs calculated using a threshold-type approach (i.e., using the RfDs for noncancer effects) were 28 to 2,500 times greater than the RBGs calculated using the CSFs derived from the linearized multistage model. Costs associated with remediating soil at a hazardous waste site to the cancer and noncancer RBGs for 1,1-DCE, PCE, and TCE also were calculated. Estimated soil remediation costs required to meet the cancer RBGs were three to eight times higher than the remediation costs required to reach the noncancer RBGs. The results of this comparison indicate that the risk assessment approach used to calculate RBGs for 1,1-DCE, PCE, and TCE is an important factor that has a major impact on potential site remediation costs.

## REFERENCES

1. United States Environmental Protection Agency (USEPA), 1995. Region III Risk-Based Concentration Table, January-June 1995. March 7.

2. United States Environmental Protection Agency (USEPA), 1995. Region IX Preliminary Remediation Goals (PRGs) First Half of 1995. February 1.

3. Steinberg, A.D. and J.M. DeSesso, 1993. Have Animal Data Been Used Inappropriately to Estimate Risks to Humans from Environmental Trichloroethylene? Regulatory Toxicology and Pharmacology 18: 137-153.

4. Regulatory Toxicology and Pharmacology, 1994. Interpretive Review of the Potential Adverse Effects of Chlorinated Organic Chemicals on Human Health and the Environment. F. Coulston, and A.C. Kolbye, eds. 20(1).

5. Brown, L.P., D.G. Farrar, and C.G. de Rooij, 1990. Health Risk Assessment of Environmental Exposure to Tricholoroethylene, Regulatory Toxicology and Pharmacology 11: 24-41.

6. Jones, B.K., and D.E. Hathway, 1978. Differences in Metabolism of Vinylidene Chloride Between Mice and Rats, BR.J. Cancer 37: 411-417.

7. D'Souza, R.W. and M.E. Andersen, 1988. Physiologically Based Pharmacokinetic Model for Vinylidene Chloride, Toxicol. Appl. Pharmacol. 92(2): 230-240.

8. International Agency for Research on Cancer (IARC), 1986. IARC Monographs on the Evaluation of Vinylidene Chloride. IARC, Lyon, France.

9. American Conference of Governmental Industrial Hygienists (ACGIH), 1993. Results of Database Printout on Tetrachloroethylene (March 25, 1991).

10. United States Environmental Protection Agency (USEPA), 1991. Risk Assessment Guidance for Superfund: Volume I-Human Health Evaluation Manual (Part B, Development of Risk-Based Preliminary Remediation Goals). Office of Emergency and Remedial Response.

CATEGORY III:

# *Air Pollution: Modeling and Control*

# SIMULATION OF VOC EMISSIONS FROM TREATMENT OF AN INDUSTRIAL WASTEWATER IN A BIOLOGICAL TREATMENT SYSTEM

RAJIV K. AGARWAL
   Principal Research Engineer
   Air Products and Chemicals, Inc.
   7201 Hamilton Boulevard
   Allentown, PA 18195-1501

MARTIN A. LARSON
   Senior Research Technician
   Air Products and Chemicals, Inc.
   7201 Hamilton Boulevard
   Allentown, PA 18195-1501

## ABSTRACT

At an Air Products' chemical plant, wastewater is treated biologically prior to its discharge to a nearby river. The wastewater treatment system consists of a large aeration basin with gravity clarification for solids-liquid separation. The aeration basin utilizes floating surface aerators for providing oxygen and mixing energy and treats an average of 1.2 MGD of wastewater flow. The biological system operates at a 4 day HRT and F/M (TOC basis) of 0.06 and consistently produces a final effluent low in BOD and TSS.

A number of pertinent organic compounds such as acetaldehyde, methanol, methyl acetate and vinyl acetate are present in the waste feed to the bio-system. EPA modeling data for a completely mixed biological system suggests that a large fraction of the organic compounds in the feed may be air stripping causing high reportable SARA emission numbers. It is believed that EPA model overstates the VOC emissions to air from a well operated biological system. This project was initiated to simulate the biological wastewater treatment system and to determine relative emission of the four named compounds to air.

A closed completely mixed bench top bioreactor was setup and operated under field conditions to simulate the plant's aeration basin. Air flow was regulated to maintain a DO value between 2.0 and 4.0 mg/l. Synthetic feed containing aforementioned compounds as the sole carbon source was fed to the reactor. The addition of the feed was controlled to enable acclimatization of the biomass to the compounds. After acclimation, a portion of the off gas was trapped on sorbent tubes and analyzed. The results of off gas analysis indicated

155

that no methanol was evaporating and less than 1% of the incoming acetaldehyde, methyl acetate, and vinyl acetate were evaporating. These emissions results are much lower than predicted by the EPA models. The study results will be presented in this paper.

## INTRODUCTION

At one of Air Products and Chemicals Inc.'s facilities, a number of specialty chemicals are manufactured. Wastewaters from different plant areas are collected and treated in a wastewater treatment (WWT) system. The WWT consists of a mixing area, an equalization basin and an activated sludge system. The treated water is discharged to the local river. The activated sludge system at the plant consists of a large aeration basin with gravity clarification for solids-liquid separation. Aeration and mixing is provided by floating surface aerators. The plant operates at a mixed liquor biomass of approximately 4,000 mg/l. The hydraulic load of the biological system is about 1.2 MGD. The biological system operates at a low F/M of < 0.06 (TOC/VSS) and consistently produces a final effluent quality low in BOD and TSS.

A number of volatile organic compounds are present in the wastewater fed to the aeration basin. The compounds of interest for this project are acetaldehyde, methanol, methyl acetate and vinyl acetate. Modeling data for a completely mixed biological system indicates that some of the organic compounds in the feed are air stripping from the aeration basin during normal loading periods. For example, in 1993, the plant reported 96.6 TPY of volatiles emissions from the wastewater treatment system. 89% of the total emissions from the WWT were due to acetaldehyde, methanol and vinyl acetate. The SARA reported amounts are shown in Table I. In Table I, estimated emissions from the aeration basin are also reported. These emissions were estimated by Surface Impoundment Modeling System (SIMS).

Methanol is the biggest source of emissions from WWT and aeration basin. Methanol is highly biodegradable and not very strippable under ambient conditions. It is believed that modeling overstates the emissions of methanol and other compounds. This project was initiated to provide quantitative data on the off-gassing of the four targeted organics under controlled biological treatment conditions. It was hoped that this project will provide direction for short-term and long-term regulatory needs and for estimating organic emissions from a completely mixed bio-system handling acetaldehyde, methanol, methyl acetate and vinyl acetate.

## PROJECT OBJECTIVES

The project had the following objectives:
- Perform biotreatability under dynamic conditions in a completely mixed laboratory activated sludge system that simulates plant operating conditions, using synthetic feeds containing all four of the aforementioned volatile compounds.
- Sample the air in the headspace of the bioreactor to quantify the off-gassing of the targeted four organic compounds of interest.

## EXPERIMENTAL DETAILS

A schematic of the experimental system is shown in Figure I. A covered aeration reactor with a volume of 45 liters was used. Air stones placed at regular spacing at the bottom of the reactor provided air to the reactor. Air flow was controlled via a flow controller. Air flow rate was set by measuring the DO in the reactor which was regulated at 2 mg/l. Reactor hydraulic residence time was set at four days which approximates the current plant operation. A recycle sludge line continuously pumped the recycle sludge back to the reactor from a clarifier. The clarifier had a holding time of approximately 15 hours.

The reactor was operated at F/M values of 0.05-0.06. The feed to the reactor was synthetic and made from DI water and pure compounds. 10 ppm of nitrogen and 2 ppm of phosphorus were added as nutrients. During acclimation phase, feed TOCs were sometimes lower than those necessary to maintain the desired F/M. After acclimation, as measured by steady low effluent TOCs, the TOC of the feed was increased to obtain the desired F/M. Sludge population was maintained between 4,000-5,000 mg/l by wasting it regularly. Experiments for each compound took about three weeks. Acclimation time for a given compound varied from 7-14 days.

Air sampling in the headspace was done once a steady state operation as evidenced by low effluent TOCs was noted. An air sample was collected from about three inches above the liquid level using SKC, Inc. personal pumps. Analytical methods were developed to measure the amount of organics in the air in the headspace. Details of the analytical methods have been provided in Table II. Control experiments were done to ensure that the selected analytical methods were suitable for analysis as desired. In the control sampling, a sample with a known amount of the given chemical was sparged with air. The air sample was collected by the methods outlined in Table II and analyzed to determine the amount of compound lost due to sparging. From the TOC loss of the sparged sample, efficiency of the chosen method was determined.

Single and mixed substrate experiments were conducted for the four compounds of interest. In single substrate experiments, a given compound was used as sole carbon source. Typically, the feed TOC was maintained close to 1,000 mg/l to obtain the desired F/M values. In mixed substrate experiments, each of the four compounds was mixed in equal TOC amount to obtain the desired TOC loading.

## RESULTS

In Table III air sampling results are presented. Air samples were taken when the reactor had acclimated to the feed and was resulting in low effluent TOC values. The results show that in both single and mixed substrate cases, methanol was not detected in the headspace. Acetaldehyde was found at a fraction of a percent (<0.003%) in both the single and mixed substrate cases. For methyl acetate, in single substrate case where more than 1,200 mg/l as TOC was present in the feed, it was found that about 0.1% of the incoming feed concentration air stripped. In the mixed substrate case, methyl acetate was seen in the air sample but its concentration was too low to be measured. For vinyl acetate, less than 0.04% of the incoming feed was found to have stripped in the single substrate case while nothing was seen in the mixed substrate case.

## DISCUSSION OF RESULTS

Physical and chemical properties of acetaldehyde, benzene, methanol, methyl acetate and vinyl acetate are shown in Table IV. Benzene data are included for comparative purposes. Henry's Law constant is a measure of strippability of a compound. These values were calculated from Air Products' thermodynamics data base. According to Perry and Eckenfelder, compounds having Henry's constant (H) greater than $10^{-3}$ atm.m$^3$/mol are considered easy to strip, those with H between $10^{-5}$ and $10^{-4}$ atm.m$^3$/mol are difficult to strip and compounds with H less than $10^{-5}$ atm.m$^3$/mol are generally non-strippable [1]. Therefore benzene is highly strippable, vinyl acetate is moderately strippable, acetaldehyde and methyl acetate are difficult to strip and methanol is not-strippable at 25°C. Vinyl acetate hydrolyzes slowly to acetaldehyde and acetic acid and this chemistry is likely to reduce its strippability.

Biodegradability of a compound is determined by its BOD to TOD ratio [2]. Typically, values between 0.4-0.7 indicate high biodegradability. The respective BOD and TOD values are shown in Table IV. BOD value for methyl acetate could not be found but we know from experience that it is biodegradable. Based on the BOD/TOD ratio and from experience, all the compounds listed are therefore quite biodegradable.

In an activated sludge system, both biodegradation and stripping would be expected to occur simultaneously. As benzene has the highest Henry's Law constant, we would expect it to strip most from the aeration basin. In a similar work, Kincannon and coworkers determined that 16% of influent benzene stripped from an activated sludge system [3]. As Henry's Law constants for the four compounds of interest are at least an order of magnitude lower, it is expected that they will also strip in substantially reduced amounts than 16%

observed for benzene. This is indeed seen in the results of this study. Methanol, which is not strippable was not seen in the headspace of the activated sludge system while acetaldehyde, methyl acetate and vinyl acetate were all seen at substantially less than 1% of their influent feed concentrations. These experiments were done when the feed contained these compounds as the sole source of organics. In actuality, these compounds are present with a number of other biodegradable compounds. The biodegradation rates are typically higher in mixed substrate cases than in single substrate cases. Therefore, in such a situation, stripping should be further reduced.

## CONCLUSIONS

These results show that biodegradable volatile compounds that were studied in this work strip in negligible amounts from a well operated biological system operating at low F/M and high hydraulic residence time (HRT). Most of the TOC reduction of these chemicals is due to biodegradation. Modeling overestimates emissions of these compounds from an activated sludge system. At the plant, this can potentially impact 60% of the reported air emissions from wastewater treatment system. Based on this work, real time monitoring of the aeration basin is under review to quantify air emissions from WWT. It is expected that this will result in lower reported emissions from the plant.

## BIBLIOGRAPHY

1. P. W. Lankford and W. W. Eckenfelder, "Toxicity Reduction in Industrial Effluents," Van Nostrand Reinhold, 1990, Chapter 8.

2. P. Pitter and J. Chudoba, "Biodegradability of Organic Substances in the Aquatic Environment," CRC Press, 1989, pp. 267-289.

3  D. F. Kincannon, E. L. Stover, V. Nichols, D. Medley, Journal WPCF, 1983, Vol. 55 (2), pp. 157-163.

## TABLE I

## SARA 313 REPORTED EMISSIONS FROM WWT SYSTEM IN 1993

Total Emissions reported from WWT = 96.6 TPY

| Compound | WWT Emissions TPY | % Emissions from Aeration Basin |
|---|---|---|
| Acetaldehyde | 8.40 | 27.0 |
| Methanol | 59.04 | 1.4 |
| Vinyl Acetate | 17.13 | 46.8 |

# TABLE II

## ANALYTICAL METHODS FOR GAS SAMPLING AND ANALYSIS

### Acetaldehyde

Analytical Method Used:  OSHA #2538
- Pass air through solid sorbent tube 2-HMP on XAD-2
- Flow rate set at 50-60 ml/min for 300 or more minutes
- Extract oxazolidine derivative of acetaldehyde in toluene
- Analyze derivatized acetaldehyde using GC

### Methanol

OSHA Method #2000 was not found suitable; control experiments did not show any detectable methanol in silica gel tubes.

Passed air through cold water traps to solubilize methanol.  The water samples were analyzed by direct injection using GC.

### Methyl Acetate

Analytical Method Used:  OSHA #S42
- Pass air thru coconut shell charcoal tube
- Flow rate set at 60-200 ml/min for 300 or more minutes
- Extract methyl acetate using carbon disulfide
- Analyze methyl acetate using GC

### Vinyl Acetate

Followed a non-agency method developed by Union Carbide.  This method uses a sorbent tube with a large desiccant presection followed by a hydroquinone inhibited carbon adsorbent.  Carbon disulfide was used to desorb vinyl acetate from the carbon tubes.  Vinyl acetate in carbon disulfide was measured using GC methods.

161

# TABLE III

## RESULTS OF AIR SAMPLING

### Reactor Operating Parameters

HRT = 4 days
Reactor Volume = 45 liters
DO = 2.0-4.0 mg/l

| Compound | Feed TOC mg/l | Effluent TOC mg/l | F/M | # Data Points | Air Emission as % of Influent Concentration |
|---|---|---|---|---|---|
| **Single Substrate Testing** | | | | | |
| Acetaldehyde | 1,356 | 120 | 0.065 | 4 | 0.0003 |
| Methyl Acetate | 1,211 | 46 | 0.053 | 3 | 0.12 |
| Methanol | 1,130 | 162 | 0.057 | 4 | 0 |
| Vinyl Acetate | 1,091 | 65 | 0.052 | 3 | 0.025 |
| **Mixed Substrate Testing (Each Compound added as approx. 300 ppm TOC)** | | | | | |
| Acetaldehyde | 1,288 | 72 | 0.084 | 2 | 0.0022 |
| Methyl Acetate | 1,277 | 64 | 0.080 | 2 | 0 |
| Methanol | 1,277 | 64 | 0.08 | 2 | 0 |
| Vinyl Acetate | 1,288 | 72 | 0.084 | 2 | 0 |

## TABLE IV

## SPECIFIC ORGANIC COMPOUND CHARACTERISTICS

| Compound | H (@25 C) atm.m³/mol | BOD g/g | TOD g/g | Strippability | Biodegradable |
|---|---|---|---|---|---|
| Acetaldehyde | $9.2 \times 10^{-5}$ | 1.27 | 1.82 | Difficult | Yes |
| Benzene | $4.0 \times 10^{-3}$ | 1.65 | 3.08 | High | Yes |
| Methanol | $5.4 \times 10^{-6}$ | 0.85 | 1.50 | Non-strippable | Yes |
| Methyl Acetate | $8.6 \times 10^{-5}$ | | | Difficult | |
| Vinyl Acetate | $5.9 \times 10^{-4}$ | 1.00 | 1.67 | Moderate | Yes |

163

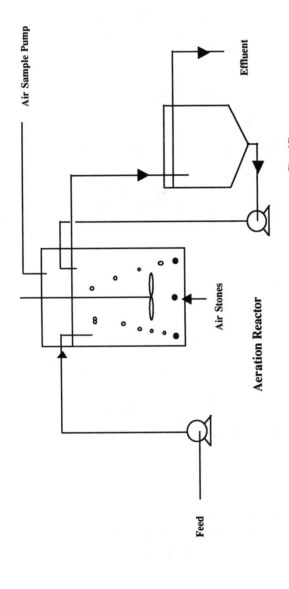

**OPERATING CONDITIONS**
Reactor Volume = 45 liters
HRT = 4 days
DO = 2 ppm

**FIGURE I**

**SCHEMATIC OF THE VOC/BIOTREATMENT SYSTEM**

# QUANTIFYING AND MODELING VAPOR PHASE DIFFUSION OF HAZARDOUS SUBSTANCES IN SURFACE SOILS - IMPACT ON AIR EMISSION

**GEORGE P. PARTRIDGE**
**ANINDYA DASGUPTA**
**JOHN P. KLINE**
School of Science, Engineering and Technology
Pennsylvania State University at Harrisburg
Harrisburg, PA 17057

## INTRODUCTION

The current emphasis on air quality as a result of the Clean Air Act Amendments of 1990 has increased the focus on site environmental risk assessment and remedial activities. There is a need to accurately quantify the potential for fugitive air emissions from the soil-air interface due to volatilization of organics in contaminated surface soils at hazardous waste sites.

Techniques for data collection at hazardous waste sites range from head space analysis, soil-vapor surveys, to modeling [1]. Considering the number of sites and the magnitude of chemical species that may be present, industry has frequently incorporated modeling techniques into the decision making process regarding site management and the need for remediation. Models like the CHEMDAT air emission model series have been used to predict air emissions from hazardous waste treatment, storage, and disposal facilities. A study using CHEMDAT 7 indicated that emissions could be over predicted by greater than 200 times the actual emission rate from a wastewater disposal unit [2]. There has been equal concern regarding the accuracy of the land treatment and landfill models incorporated into CHEMDAT. The predicted effective diffusivity used by models, significantly affects model predictions. Effective diffusion coefficients are typically calculated from the diffusivity of the chemical in air considering such factors as tortuosity of soil matrix, soil partitioning, soil moisture content, and soil void volume [3].

To accurately quantify the effective diffusion coefficient an experimental apparatus and methodology was developed at Penn State Harrisburg to quantify and model the effective soil gas diffusion coefficient for hazardous chemicals in soil matrices [4]. An additional factor that affects vapor phase transport in the soil is water movement (infiltration) which has not been investigated. The apparatus includes capabilities to evaluate the effect of water movement in soil on the effective diffusion of the volatile soil contaminant to the air-soil interface. Research work currently in progress involves the measurement of BTEX components through synthetic and soil matrices.

This paper presents an overview of modeling approaches for chemical vapor transport in soils and the approaches to predict the effective diffusion coefficient. The apparatus and methodology for measuring soil gas diffusion rates that was developed is also presented.


## VAPOR TRANSPORT IN SOILS

Fugitive air emissions from contaminated soils result from volatilization of chemicals in the soil matrix followed by subsequent loss of the chemical to the atmosphere. The process involves vaporization of the chemical to the soil gas phase followed by diffusion to the soil-air interface. This is then followed by movement to the atmosphere. The rate at which a chemical migrates away from the soil surface is diffusion controlled [5]. There is very little movement of air next to the surface. The rate of air flow over the surface affects the thickness of the boundary layer. Mass transfer studies have correlated the transport rates through boundary layers next to surfaces and accounted for such affects as surface texture or surface roughness. Due to the complexity of the soil matrix there has been limited modeling of the diffusion rate through the soil to the soil-interface. An understanding of vapor transport in the surface soil is critical to the quantification of the overall fugitive air emission rates.

Vapor phase transport will typically lead to an enrichment of the gas phase with the more volatile organic compounds. To assess the subsurface contamination, to conduct source identification, predict contaminant fate, and ultimately design an appropriate remedial program, it is important to be quantify the vapor transport rate in the soil gas. The behavior of VOCs in a heterogeneous soil matrix is affected by: chemical, biological and photo-degradation/transformation, soil sorption and leaching, capillary flow through the soil system, and vapor transport within and volatilization from the soil surface [6].

Upon introduction of a VOC into a soil, the initial mechanism of transport in the soil subsurface is liquid flow. Once the liquid flow ceases, a region of residual saturation of VOC liquid remains in the soil [7]. The VOC will begin to partition into the liquid and vapor phases and become dissolved in soil moisture and adsorbed onto the surfaces of soil minerals and organic matter. The degree of partitioning of the VOC among these four components will depend on the volatility and water solubility of the VOC, the soil moisture content, and the soil solids, i.e., the minerals and

166

organic matter [8]. Due to low water solubility, low sorptive capacity of many moist soils towards VOCs and high vapor pressure which is typical of many VOCs, VOCs will tend to partition into the vapor phase [9]. Therefore, since vapor-phase diffusion values are approximately $10^4$ greater than solution-phase diffusion values, vapor-phase diffusion will be the dominant transport mechanism for VOCs in the vadose zone of soil [10].

The measured in-soil diffusion coefficient referred to as an effective diffusion coefficient, $D_{eff}$, is generally not available. Traditionally, effective diffusion coefficients have been derived empirically from the in-air diffusion coefficient correcting for the porosity and tortuosity of the soil. They do not include secondary effects of adsorption or interactions with soil moisture [10]. The removal of VOCs from the soil gas by partitioning into the soil moisture and soil organic matter results in a reduction in the apparent diffusion rate, and consequently, the effective diffusion coefficient [8].

The addition of water to the soil can have at least four major influences on the vapor-phase diffusion in soil. Physically, the water may block air passages in the soil, resulting in a more tortuous path or even a barrier condition for the gas. Secondly, water may absorb the contaminant, depending on the VOC's solubility and Henry's law constant. Thirdly, the soil water content may interfere with or enhance the ability of the soil to absorb the VOC [10]. Finally, the water can displace the vapor inducing a localized pressure head within the soil matrix and subsequently creating a convective flow.

Thus, a more appropriate approach is to develop an algorithm for an effective diffusion coefficient considering soil-gas partitioning, soil moisture, tortuosity of the soil matrix, and the soil organic content [3]. The effective diffusion coefficient, therefore, would correct an in-air diffusion coefficient for the presence of a soil matrix and account for the decreased cross-sectional area for flow, increased length of diffusion path and competing mechanisms such as adsorption and absorption.

## MODELING SUBSURFACE VAPOR PHASE TRANSPORT

Models that have been developed for prediction of contaminant movement in the subsurface usually involve one-dimensional analytical solutions to concentration profiles in homogeneous soils [11]. Fick's first law [12,13,14]

$$J = -D \, dC/dZ \qquad (1)$$

and second law [8,11]

$$dC/dt = D \, d^2C/dZ^2 \qquad (2)$$

where:  $J$ = diffusive flux of gas, $(g/cm^2 \, s)$
$D$ = in-air diffusion coefficient of chemical, $(cm^2/s)$

$C$ = concentration of chemical in the vapor phase, $(g/cm^3)$
$Z$ = distance travelled, (cm)

have been incorporated into models to predict the vapor phase diffusion of a chemical. A modeling approach commonly used is to combine a mass transport relationship based on Fick's Law with a algorithm to predict the effective diffusion coefficient accounting for the presence of the soil matrix [15]. The diffusion coefficient for the chemical in soil vapor space must reflect the microenvironment through which the chemical must move to reach the air-soil interface [16]. There are several approaches that have been used and are summarized below.

To account for the effect a soil matrix has upon soil gas diffusion, a soil gas diffusion coefficient can be obtained using the Millington-Quirk formulation [8,11,14].

$$D_G = Da^{10/3} / n^2 \qquad (3)$$

where: $D_G$ = soil gas diffusion coefficient, $(cm^2/s)$
$D$ = in-air diffusion coefficient, $(cm^2/s)$
$a$ = volumetric air content of soil, $(cm^3/cm^3)$
$n$ = total soil porosity, $(cm^3/cm^3)$

Jury et al. [8,11] has also developed a relationship for an effective diffusion coefficient which incorporates the soil gas diffusion coefficient and chemical partitioning

$$D_E = D_G/[(bK_D/K_H) + w/K_H + a] \qquad (4)$$

where: $D_E$ = the effective diffusion coefficient in soil gas corrected for partitioning, $(cm^2/s)$
$D_G$ = soil gas diffusion coefficient, $(cm^2/s)$
$b$ = bulk dry density of the soil, $(g/cm^3)$
$K_D$ = soil partitioning coefficient, $(cm^3/g)$
$K_H$ = Henry's law constant of the chemical, (dimensionless)
$w$ = volumetric soil moisture content, $(cm^3/cm^3)$
$a$ = volumetric air content, $(cm^3/cm^3)$;

where $a + w$ = total porosity (n).

To date experimental data and model validation is limited due to prior work focusing on individual factors affecting transport and fate. As a result there is a need for a comprehensive model that considers all the competing mechanisms in a heterogeneous soil matrix.

To obtain the experimental data necessary to quantify the mechanisms that contribute to the effective diffusivity of a contaminant through a soil, an apparatus and an experimental methodology was designed and is presented below.

## EXPERIMENTAL APPARATUS AND METHODOLOGY

The apparatus for this project was designed for a long-term experimental program which will be conducted over the next several years. The entire experimental apparatus is presented in Figure 1. It incorporates a gas conditioning system, the diffusion cell, and various methods of sampling. The diffusion cell is presented in Figure 2. For diffusion measurements, it is important that parameters such as temperature, pressure, and relative humidity be monitored and maintained constant through the diffusion cell in order to negate any concerns of transport due to pressure gradients or convective movement. The diffusion cell was placed in a constant temperature environmental chamber to maintain isothermal conditions. A thermocouple, relative humidity probe, and manometer are placed both directly before and after the diffusion cell to measure and monitor the temperature, relative humidity, and pressure, respectively, of the carrier gas entering and exiting the diffusion cell.

The objective of the experimental methodology is to develop an algorithm for calculating the effective diffusivity based on a pure gas phase diffusivity incorporating the physical characteristics of a well defined soil matrix. This algorithm is then incorporated into a fugitive air emission model such as CHEMDAT8. The experimental methodology systematically investigates the different components that should be included in predictive algorithms to calculate effective diffusivities from in-air diffusivities. By comparing measured diffusion rates incorporating various types of media under different experimental conditions, the components that have the greatest significance can be incorporated into a refined algorithm for prediction of the effective diffusivity. The experimentally determined effective diffusivity is then compared to existing algorithms that predict effective diffusion coefficients for soils. Based on comparisons of algorithms published in the literature and the current algorithm used in CHEMDAT8, a predictive algorithm will be developed.

The experiments for each component are performed utilizing different cell heights (soil depths). This permits the evaluation of the effect that increased soil depth has on the transport rate calculations incorporating the effective diffusivity coefficient. The experimentation focuses on one component or one set of components at a time throughout the procedures of this methodology. All experimentation is performed at steady-state conditions. The following is an example of how this systematic methodology is applied for vapor phase diffusion measurement of the chemical(s):

1) measure in-air diffusion rate (calibration of diffusion cell)
2) measure diffusion rate in inert synthetic media

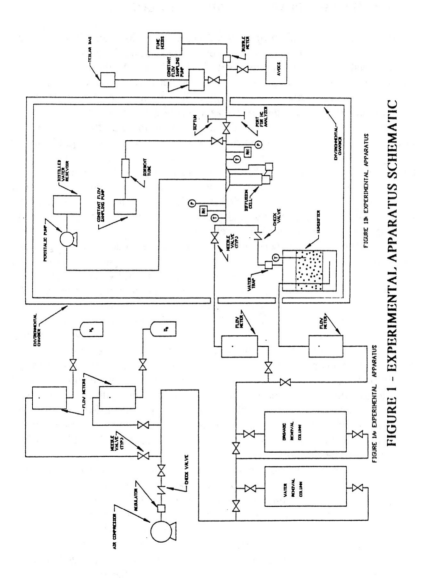

FIGURE 1b EXPERIMENTAL APPARATUS

FIGURE 1a EXPERIMENTAL APPARATUS

FIGURE 1 - EXPERIMENTAL APPARATUS SCHEMATIC

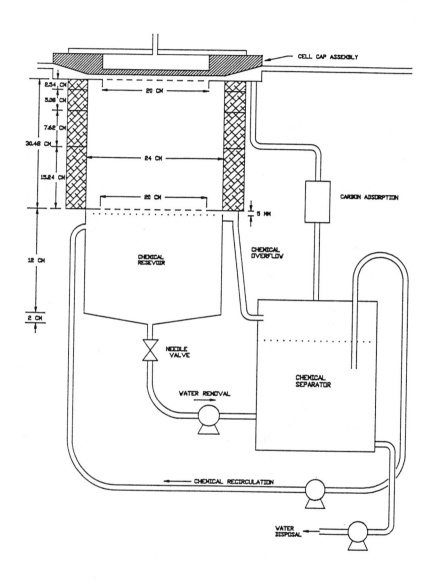

FIGURE 2 - DIFFUSION CELL

3) measure diffusion rate in inert synthetic media with water addition
4) measure diffusion rate in adsorbent synthetic media
5) measure diffusion rate in adsorbent synthetic media with water addition
6) measure diffusion rate in well characterized soil
7) measure diffusion rate in well characterized soil with water addition
8) repeat steps 1-5 for each cell height

**Effective Diffusivity Measurements**

**Inert Synthetic Media.** Experimentation is performed using a synthetic inert homogeneous media (spherical glass beads) in the cell. The diffusion rate of the chemical(s) through the inert media is measured. Knowing the size and shape of the inert synthetic media, the stacking geometry of the column is well defined. The interparticulate void fraction and the tortuosity is mathematically calculated from the orientation of the individual media particles based on geometry. The measured diffusion rate through the synthetic media is compared to the in-air diffusion rate to determine the effect that void space and tortuosity has on the vapor phase diffusivity.

**Adsorbent Synthetic Media.** Experimentation is performed using a synthetic adsorbent type homogeneous media in the cell. The diffusion rate of the chemical(s) through the adsorbent type media is measured. Since the geometry of the adsorbent type synthetic media is also well defined, the measured diffusion rate through the media is compared to the diffusion rate through the inert synthetic media to determine the effect adsorption has on the vapor phase diffusivity.

**Well Characterized Soil.** The diffusion rate of the chemical(s) through the well characterized soil media is measured. The experimental procedure includes soil analyses to evaluate the following soil characteristics:
- porosity
- void fraction
- moisture content as a function of temperature
- particle size distribution
- organic content
- chemical adsorptive capacity as a function of temperature
- indication of geometry of particles (microscopic examination).

**Water Percolation in Soil Matrix**

Experimentation is performed to evaluate the effect that infiltration through a soil matrix has on the effective diffusivity. The experimentation is performed with the synthetic media, adsorbent synthetic media, and well characterized soil. The effective diffusivity with water addition to the soil is compared to the effective diffusivity without water addition. This comparison gives insight into the effects that water has due to changes in the soil void fraction, soil gas movement, and absorption.

## CONCLUSIONS

A design of an experimental apparatus and a methodology to quantify the effective diffusivity of a volatile organic compound in a soil matrix is presented. The project is unique in its systematic approach for evaluating the components of an algorithm to predict effective diffusivity. The methodology identifies which components have the greatest significance and allows a step-wise approach toward constructing a refined model for predicting the effective diffusivity coefficient. A unique feature of the experimental apparatus is the allowance for water to percolate through the diffusion cell to simulate the effect of water infiltration in the soil. This feature will help clarify the effect subsequent downward movement of water through a soil has on soil gas vapor movement and chemical diffusion.

## REFERENCES

1. *Air Pathway Analysis - Engineering Bulletin.* EPA/540/S-92/013, November, 1992.

2. Comey III, Kenneth R., Christopher G. Rabideau, and Raymond T. Willingham. 1993. "Evaluation of a Flow-Through Flux Chamber Method for Measuring Volatile Organic Emissions from a Wastewater Surface Impoundment," Unpublished.

3. Silka, Lyle R. 1988. "Simulation of Vapor Transport Through the Unsaturated Zone - Interpretation of Soil-Gas Surveys," *Ground Water Monitoring Review*, 8(2).

4. Kline, John P., George P. Partridge, and Kenneth R. Comey III. 1995. "Experimental Apparatus and Methodology for Quantifying and Modeling the Effective Soil Gas Diffusion Coefficient in Heterogeneous Soil Matrices," Unpublished.

5. Spenser, William F., Walter J. Farmer, and William A. Jury. 1982. Behavior of Organic Chemicals at Soil, Air, Water Interfaces as Related to Predicting the Transport and Volatilization of Organic Pollutants." *Environmental Technology and Chemistry*, Vol. 1, pp. 17-26.

6. Gan, D. R., and R. R. Dupont. 1989. "Multiphase and Multicompound

Measurements of Batch Equilibrium and Distribution Coefficients for Six Volatile Organic Compounds," *Hazardous Waste and Hazardous Materials*, 6(4): 363-383.

7. Amali, S., and D. E. Rolston. 1993. "Theoretical Investigation of Muliticomponent Volatile Organic Vapor Diffusion: Steady-state Fluxes," *Journal of Environmental Quality*, 22: 825-831.

8. Silka, L. R. 1986. "Simulation of the Movement of Volatile Organic Vapor Through the Unsaturated Zone as it Pertains to Soil-Gas Surveys," *Proc. of the NWWA/API Conference on Petroleum Hydrocarbons and Organic Chemicals in Ground Water - Prevention, Detection, and Restoration*, National Water Well Association, pp. 204-224.

9. Shonnard, D. R., R. L. Bell, and A. P. Jackman. 1993. "Effects of Nonlinear Sorption on the Diffusion of Benzene and Dichloromethane from Two Air-dry Soils," *Environmental Science & Technology*, 27: 457-466.

10. Hutter, G. M., G. R. Brenniman, R. J. Anderson. 1992. "Measurement of the Apparent Diffusion Coefficient of Trichloroethylene in Soil," *Water Environment Research*, 64 69-77.

11. Jury, W. A., W. F. Spenser, and W. J. Farmer. 1983. "Behavior Assessment Model for Trace Organics in Soil: I. Model Description," *J. Environ. Qual.*, 12(4): 558-564.

12. Troeh, et. al. 1982. "Gaseous Diffusion Equations for Porous Materials," *Geoderma*, 27(3): 239.

13. Lappala, E. G., and G. M. Thompson. 1984. "Detection of Ground Water Contamination by Shallow Soil Gas Sampling in the Vadose Zone," *Proc. of the National Conference on Management of Uncontrolled Hazardous Waste Sites*, Hazardous Materials Control Research Institute, Silver Spring, Maryland, pp. 20-28.

14. Jury, W. L. and Valentine, R. L. 1986. "Transport Mechanisms and Loss Pathways for Chemicals in Soil," *Vadose Zone Modeling of Organic Pollutants*, S. C. Hern and S. M. Melancon (Eds.), Lewis Publishers, Inc.

15. Thibodeaux, L. J. 1981. "Estimating the Air Emissions of Chemicals from Hazardous Waste Landfills," *Journal of Hazardous Materials*, 4: 235-244.

16. Thibodeauz, L. J., and H. D. Scott. 1985. "Air/Soil Exchange Coefficients," *Environmental Exposure From Chemicals - Volume I*, Brock, W., and Gary E. Blau, eds. CRC Press, pp. 65 - 89.

# EVALUATION OF THE EFFECT OF MOBILE SOURCE EMISSIONS ON OZONE FORMATION IN MEMPHIS, TENNESSEE

NANDITA BANDYOPADHYAY
State of Tennessee
Department of Environment and Conservation
Division of Air Pollution Control
9th Floor L & C, Annex
401 Church Street
Nashville, TN - 37243.

WAYNE T. DAVIS
The University of Tennessee
Department of Civil and Environmental Engineering
73 Perkins Hall
Knoxville, TN - 37996

TERRY L. MILLER
The University of Tennessee
Department of Civil and Environmental Engineering
73 Perkins Hall
Knoxville, TN - 37996

## INTRODUCTION

Under the 1970 Clean Air Act Amendments (CAAA), a National Ambient Air Quality Standard (NAAQS) of 0.12 ppmv was established for ozone. The NAAQS for ozone is defined as a one hour average concentration of 0.12 ppmv, not to exceed more than three times in any consecutive three year period.

Volatile organic compounds (VOCs), and nitrogen oxides ($NO_x$, including all oxides of nitrogen expressed as $NO_2$) are the most important precursors for ozone formation. The main sources of VOCs are automobile exhaust and vegetation (mobile and biogenic sources), whereas the main sources of $NO_x$ are power plants and automobile exhaust (point and mobile sources). High ozone concentration may result either from the reactions between these locally emitted precursors or from transportation of ozone from the upwind direction. Uncontrolled biogenic emissions are known to be responsible for high

175

concentration of ozone (photochemical smog) in areas where anthropogenic $NO_x$ emissions are high [2,5].

In 1991, there were three ozone nonattainment areas in Tennessee, of which the Memphis area was designated as 'marginal' by the CAAA of 1990 and had a compliance date of 1993. The memphis area violated the ozone NAAQS ten times during 1986-1988. A comprehensive emission inventory of ozone precursors was developed followed by modeling the ozone formation in order to enhance the assessment of the future compliance with the ozone standard. Computer models are generally used in order to explain and predict the effects of different types of emissions and also of changes in these emissions on ozone formation.

A modeling study was done by the University of Tennessee, Knoxville, Department of Civil and Environmental Engineering using EPA's Urban Airshed Model (UAM). The UAM is an air quality simulation model, developed by Systems Application, Inc., used to analyze the ozone air quality. This modeling approach assumes a three dimensional grid-based modeling domain. The application of UAM requires a large database of ozone precursor emissions based on gridded emission inventory.

The UAM was run for the Memphis area using a 1988 emission inventory and representative meteorological data based on the ten highest ozone violation days during 1986-1988. The ozone concentrations predicted by the model were analyzed and compared to actual observed concentration data to validate the model. A sensitivity analysis was conducted to determine the contribution of specific source categories to ozone formation and to gain insight into the probable reasons for the violations of the ozone standards.

The main objective of this paper is to present different steps of UAM model application development and to understand and analyze the model results. This paper includes the analysis of the effect of emissions from mobile sources, which include large amounts of VOC and $NO_x$ emissions, on ozone formation in Memphis area.

Qin [6] used Empirical Kinetic Modeling Approach (EKMA) recommended by U. S. EPA. This is a single trajectory model using the Carbon Bond-IV mechanism of ozone formation. The author examined the sensitivity of ozone to biogenic and anthropogenic emissions in the Middle Tennessee area using the EKMA approach and concluded that ozone NAAQS in Middle Tennessee area could be attained by a 50% reduction in anthropogenic VOC emissions, a 50% reduction in $NO_x$ emissions, or a 30% reduction in both VOC and $NO_x$ emissions.

Zhou [8] simulated ozone formation in the Middle Tennessee area using a 1988 base year inventory. The author used both the EKMA and the UAM models to compare the predicted and observed ozone concentrations. The simulation results were in good comparison with the observed data. The predicted ozone profiles from both models were similar, but the EKMA model resulted in higher ozone values. The author explains further that the UAM considered the turbulent diffusion in addition to the consideration of transport of ozone and chemical reactions of chemical species by the EKMA model. The

176

diffusion effect on ozone formation in the UAM was proposed as the reason behind the lower ozone concentration predicted by the UAM than that by the EKMA.

## MODELING INPUT DATA COLLECTION

In this work, a specific modeling domain was chosen which was located within the boundaries of the emission inventory. It consisted of a gridded, 20 X 20 matrix with a 5 km X 5 km resolution. The gridded area for modeling was chosen in such a way, that the locations at which maximum ozone concentrations occurred in 1986-88 were within the modeling domain.

The meteorological inputs required in the modeling such as temperature, atmospheric pressure, water concentration, and solar intensity were obtained by analyzing the data for the ten highest ozone days in the period of 1986-88 to arrive at a composite or average "ozone day". Table I provides the data about the ten highest ozone days between 1986 and 1988.

Ozone precursors are emitted from different types of sources including point, area, mobile and biogenic sources. Each source has a different temporal and spatial distribution with specific chemical composition. An emission inventory for the Memphis area had been completed for the year of 1988 prior to the initiation of the work reported herein. This inventory included Shelby, and Tipton counties in Tennessee, Crittenden county in Arkansas, and Desoto county in Mississippi. Emissions from point, area, and mobile sources were

Table I : Ten Highest Ozone Days (1986 - 1988).

| DAY | HIGHEST OZONE CONCENTRATION (ppb) | TIME OF OCCURRENCE (hr) | MONITORING STATION |
|---|---|---|---|
| 05/13/86 | 0.130 | 14:00 | MUDDVILLE |
| 07/24/86 | 0.165 | 15:00 | FRAYSER |
| 08/05/86 | 0.159 | 14:00 | MUDDVILLE |
| 06/09/87 | 0.127 | 16:00 | FRAYSER |
| 06/24/87 | 0.146 | 16:00 | FRAYSER |
| 07/30/87 | 0.134 | 12:00 | MUDDVILLE |
| 06/15/88 | 0.144 | 13:00 | FRAYSER |
| 06/21/88 | 0.140 | 16:00 | FRAYSER |
| 06/22/88 | 0.125 | 15:00 | FRAYSER |
| 06/23/88 | 0.140 | 15:00 | FRAYSER |

Table II : Ozone Precursor Emissions in the Study Area - 1988
Base Case (tons/day).

| SOURCE | | POLLUTANT | |
|---|---|---|---|
| | | VOC | NO$_x$ |
| POINT | TVA (T) | 0.14 | 75.60 |
| | OTHER SOURCES (P) | 55.71 | 14.90 |
| MOBILE (M) | | 169.14 | 90.24 |
| AREA (A) | | 92.74 | 83.61 |
| BIOGENIC (B) | | 155.88 | 0.00 |
| TOTAL | | 473.61 | 264.35 |

obtained from the 1988 emission inventory prepared by the University of Tennessee, Knoxville. The daily emissions of VOC and NO$_x$ from different sources in the modeling domain are summarized in Table II.

The emission inventory included all the point sources which emit more than 100 tons per year of VOC or NO$_x$. Point source data for the Tennessee areas (Shelby County, Tipton County) were obtained from the State EIS (Emission Inventory System) computerized data format, while data for Arkansas (Crittenden County) and Mississippi (Desoto County) were obtained from U. S. EPA's standard NEDS (National Emissions Data System) computerized format.

Area source emissions included numerous small point, area, and non-highway mobile source emissions. Countywide area source emissions were calculated using the U. S. EPA emission estimating procedure [3]. These emissions data were allocated to each grid within a county using spatial surrogates. Surrogate indicators included information such as population, specific source type, and land use pattern.

The emissions from automobiles and trucks travelling on different types of roads including interstate, major arterial, minor arterial, collector, and local roads were considered as mobile source emissions.

The emissions of VOCs from trees and other vegetation have a significant effect on ozone formation, thus requiring an estimation of biogenic emissions. The U. S. EPA recommended personal computer program PC-BEIS was used to estimate the hourly biogenic emissions.

The VOC emission data were speciated according to SCC code and distributed spatially and temporally. For the point sources, the grid numbers for each source were obtained from the exact location (latitude and longitude) of the plant, and the temporal distribution was done according to the operating hours of each point source.

For the area sources, countywide VOC emissions data were obtained for population based categories and source specific categories. The SCC code was available from the U. S. EPA database for each source type. The area source population based emissions were spatially distributed by the gridded population data in each county. Temporal distribution was made considering an 8-hr emission (8 a.m. - 4 p.m.) for each source, except for railroads and forest fires, which were distributed over 24 hours.

To calculate the mobile source emissions, the vehicle miles travelled (VMT) data based on different road types were required for each county. These data were obtained for Shelby County from the Memphis and Shelby County Office of Planning and Development, for Tipton and Crittenden County by calculating the VMT data based on Average Daily Traffic (ADT) data provided by the Tennessee and Arkansas Departments of Transportation, and for Desoto County from the Mississippi Department of Transportation. Hourly emission factors for emission of different pollutants from different vehicles on different types of roads were calculated by using U. S. EPA's MOBILE 4.0 model. These emission factors were applied to the VMT data of different road types in order to calculate the mobile source emissions. The speciation profiles recommended by U. S. EPA were used to speciate the mobile source VOC emissions.

The hourly biogenic VOC emissions were speciated [1] using EPA'a recommended profile to various chemical species used in the Carbon Bond-IV Mechanism including Isoprene, Olefines, Paraffins, and higher aldehydes.

The $NO_x$ emission data were expressed in terms of Nitrogen Dioxide ($NO_2$). The $NO_x$ data were speciated into NO (90%) (Nitric Oxide) and $NO_2$ (10%). Temporal and spatial distributions were done in the same way as for VOCs.


## MODEL INPUT DATA PREPARATION

The UAM model run requires eleven input files for defining various meteorological and emissions data. A first input file was required to initialize the concentration of every species at the beginning of the model simulation. For a one day simulation, the prediction is highly sensitive to the assumed initial concentration. The concentration of each species representing the initial condition was obtained from EPA's recommended default value [4] except for the concentration of ozone. The monitoring station data were analyzed to obtain the initial concentration of ozone. A second input file defining the boundaries of the modeling domain and hourly pollutant concentrations in each of the boundary cell was created. The concentration of each species representing the boundary was obtained from U. S. EPA's recommended default value [4] except for the concentration of ozone. The monitoring station data were analyzed to obtain the boundary concentration of ozone. A third input file was created defining the mixing height for the modeling domain at each hour of the day. The hourly mixing height data were very important in order to predict the ozone concentration as the dilution of ozone precursors in the atmosphere depends on the mixing height. A fourth input file was created containing all the data for spatially and temporally distributed emissions from all the sources including area, mobile, and biogenic sources. A fifth input file was created defining time-varying meteorological data for the modeling domain. A sixth input file was

179

created to define the horizontal location of the point sources in the modeling domain and calculate and allocate the time-varying emission data from elevated point sources into specific vertical cells. A seventh input file was created to define the region top heights for the modeling domain. The regiontop height is defined as the top of the modeling domain above the ground. An eighth input file was created to define the hourly average temperature data for the modeling domain. A ninth input file was created to define the surface roughness and vegetation factors for the modeling domain to calculate the vertical diffusivity and surface deposition. A tenth input file was created to define the pollutant concentration at the top of the region during the period of simulation. The concentration of each species representing the concentration at the top of the region was obtained from the U. S. EPA'a recommended value [4] except for the concentration of ozone. The monitoring station data were analyzed to obtain the boundary concentration of ozone. An eleventh input file was created to define the horizontal wind vector values in the modeling domain and overall maximum and average values at the boundaries. For modeling, a constant wind speed of 3.5 km/hr was chosen with a calm period of 6 - 9 A.M.

## ANALYSIS OF RESULTS

The UAM was run for a base case condition after which sensitivity analyses were conducted in order to determine the effect of emissions from an entire source category or the emission of a particular pollutant from a source category on ozone formation.

## BASE CASE EVALUATION

In order to evaluate the validity of the model, the model was run for a base case condition and compared to the actual observations at a monitoring station in the Memphis area. The base case was defined as the scenario which included emissions from all sources (Scenario PTAMB).

The model was run with the input data as described above. A comparison of actual observed peak ozone at the Mudville monitoring station located northeast of the urban area and that predicted by the model is shown in figure 1. The highest observed ozone concentration profile in figure 1 was obtained by plotting the highest ozone concentration at each hour for the ten highest ozone days. Similarly, the second highest ozone concentration profile was obtained by plotting the second highest ozone concentration at each hour for the ten highest ozone days. The ozone concentration profile from the model output shows a peak of 152.2 ppb at 19:00, whereas the observed peak was 165.0 ppb at 14:00 as shown in figure 1. The model predicted the peak concentration within 7% of the observed peak, This is in agreement with previous UAM studies.

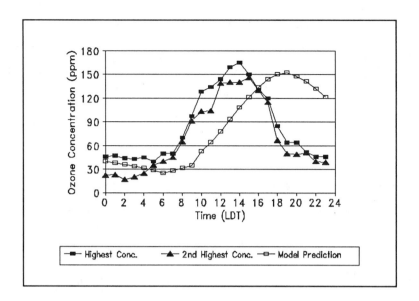

Figure 1.    Comparison of Observed and Predicted Ozone Peak : Base Case
Evaluation.

Seinfield [7] showed that the grid-based photochemical models generally predict the peak to within 15-20% of the measured value. He also stated that the measured ozone concentrations contain some uncertainty including measurement errors and the naturally random character of the atmosphere. Again, the model predicts the volume average concentrations over the grid, whereas observations represent point measurements at the monitoring station. The base case model run predicted a peak concentration which occurred at time 18:00 - 20:00 as compared to the observed peak ozone concentration which occurred at time 13:00 - 15:00. Preliminary evaluations of the discrepancies have not identified the reason for this shift. It should be noted that Zhou [8] found a similar shift in the modeling of the Middle Tennessee area.

## SENSITIVITY ANALYSIS

In order to determine the effect of different pollutant emissions from mobile sources on formation of ozone, the UAM was run with various scenarios containing different combinations of amount of pollutant emissions from mobile sources. The mobile source emissions accounted for about 36% of total VOC and 34% of the total $NO_x$ emissions in the modeling domain as shown in table II. The ozone concentrations at the location of the Mudville monitoring station at time 19:00 are summarized in table III for various scenarios. Columns

Table III :   Contribution of Mobile Source VOC Emissions to the Ozone Formation.

| Row No./ Column No. (1) | Base Case (2) | Scenario containing all other sources and the Mobile Source containing | | | | |
|---|---|---|---|---|---|---|
| | | 100% NO$_x$ | | | 100% VOC | |
| | | 50% VOC (3) | 200% VOC (4) | 300% VOC (5) | 50% NO$_x$ (6) | ZERO NO$_x$ (7) |
| Ozone Concentration (ppb) (2) | 152.2 | 147.4 | 151.1 | 148.8 | 141.9 | 128.8 |
| Differential Ozone Concentration (ppb) (3) | | 4.8 | 1.1 | 3.4 | 10.3 | 23.4 |

(3), (4), and (5) of table III show the contribution of various levels of mobile source VOC emissions. By keeping the NO$_x$ emissions constant at 100%, the change of VOC emissions in the input showed very little effect on ozone formation. In particular, the column (5) of table III showed that increasing mobile source VOC emissions by 300% resulted in a lower ozone concentration at the Mudville monitoring station location. On the other hand, column (6) and (7) of table III show a significant contribution of mobile source NO$_x$ emissions on ozone formation. The Row (3) of table III presents the differential ozone concentration for different combinations of pollutant emission as compared to the base case model run at the Mudville monitoring station at time 19:00.

To understand the contribution of the mobile source emissions to the ozone formation in the overall modeling domain, a modeling scenario was conducted in which mobile source emissions were completely removed (Scenario PTAB). The result for this scenario is presented in figure 2. In this scenario peak ozone concentration occurred at the Mudville monitoring station. At this location, ozone concentration decreased from 152.2 ppb with the mobile source emissions down to 128.0 ppb without the mobile source emissions at time 19:00. The contribution of the mobile source emissions was 15.9% (24.2 ppb) of the peak ozone concentration.

It was observed that at the Mudville monitoring station (25,26), the ozone concentration profile was lower (ozone scavenging) up to 10:00 in the scenario containing all sources (PTAMB) than in the scenario without mobile source emission (PTAB). After 11:00, an incremental increase in ozone concentration

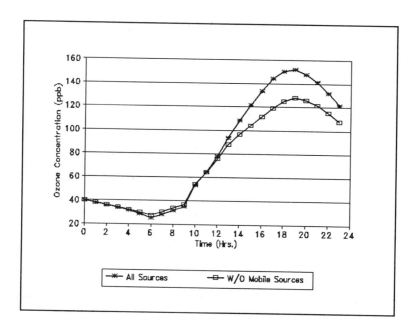

Figure 2.    Comparison of Ozone Concentration Profiles at the Location of Mudville Monitoring Station : with or without the Mobile Source Emissions.

was observed at the mudville monitoring station up to the end of the simulation period with the addition of the mobile source in the modeling domain.

## CONCLUSION

The effects of mobile sources on ozone formation were evaluated by using the Urban Airshed Model. The Model predicted the peak ozone concentration within 7% of the observed peak ozone concentration.

A series of modeling run were conducted by removing emission of a particular pollutant from the mobile source or removing the total mobile source emission at a time. It was observed that the emissions from mobile sources had the single largest influence on the ozone formation. The VOC and $NO_x$ emissions from mobile source were also greater than any other category. The contribution of the mobile source category to the ozone formation was found to be highest in the downwind direction in the late afternoon.

The overall study showed that the ozone formation was more dependent on the $NO_x$ emissions than VOC emissions. In other words, in the scenario where

NO$_x$ emissions were reduced, ozone reductions were higher than in the scenario where VOC emissions were reduced, with the conclusion that the ozone formation was NO$_x$ limited for those modeling scenarios and meteorological conditions studies.

## REFERENCES

1. Causley, M. C., J. L. Fieber, M. Jimnez, and L. Gardner. 1990. *User's Guide for the Urban Airshed Model : Volume IV: User's Manual for the Emissions Preprocessor System. EPA-450/4-90-007D*. United States Environmental Protection Agency. Research Triangle Park. NC. (NTIS No. : PB91-131250).

2. Chamedies W. L., Lindsay R. W., Richardson J. L. and Kiang C. S. 1988. "Role of Biogenic Hydrocarbon in Urban Photochemical Smog," *Science*, 241, pp. 1473-1475.

3. Davis W. T., T. L. Miller. 1990. *Comprehensive Emission Inventory for Precursors of Ozone in the Nashville and Memphis Areas*, Civil Engineering Department, The University of Tennessee, Knoxville, TN.

4. EPA. 1991. *Guideline for Regulatory Application of the Urban Airshed Model. EPA-450/4-91-013*. United States Environmental Protection Agency. Research Triangle Park. NC.

5. Lamb, B., A. Guenther, D. Gay and H. Westberg. 1987. "A National Inventory of Biogenic Hydrocarbon Emissions," *Atmospheric Environment*, 21, No. 8, pp. 1695-1705.

6. Qin, H. 1992. "The Role of Biogenic and Anhropogenic Air Pollution Emissions in the Production of Tropospheric Ozone in Middle Tennessee," *A Thesis presented for the MS Degree*. The University of Tennessee, Knoxville.

7. Seinfield, J. H. 1988. "Ozone Air Quality Models: A Critical Review," *JAPCA*, 38, pp. 616-645.

8. Zhou, N. 1992. "Simulation of Ozone Formation in the Middle Tennessee Area," *A Thesis presented for the MS Degree*. The University of Tennessee, Knoxville.

CATEGORY IV:
# *Biotreatment and Bioremediation*

# ANAEROBIC BIODEGRADATION OF NITROGLYCERIN BY DIGESTER SLUDGE

CHRISTOS CHRISTODOULATOS

NIRUPAM PAL

SYAMALENDU BHAUMIK
Center for Environmental Engineering
Stevens Institute of Technology
Hoboken, NJ 07030

## INTRODUCTION

Nitroglycerin (NG) is an energetic compound primarily present in gun and rocket propellants as a primary explosive. It is also abundantly found in spent wastes from several chemical or pharmaceutical industries and in the wastewater of munitions manufacturing facilities causing significant environmental pollution.

NG or glycerol trinitrate, $C_3H_5(ONO_2)_3$, is a pale yellow viscous liquid with a vapor pressure of 0.00026 mm of Hg at 20°C and a solubility of 1.25 gm/liter of water. It gives off nitrous yellow vapors at 135°C and explodes at 218°C with a heat of combustion of 1580 cal/gm. NG is readily adsorbed through the skin, lungs, and mucous and it can be harmful to humans and other living organisms. Prolonged exposure to NG may cause nausea, vomiting, cyanosis, palpitations of the heart, coma, cessation of breathing, and even death. Because of its adverse effects on the environment nitroglycerin must be properly treated and disposed of in an environmentally sound manner.

Open burning-open detonation of energetic materials and explosive mixtures, such as obsolete rocket and gun propellants, has been a common practice for many years. However, as more stringent environmental regulations are enacted at the state and federal level open burning-open detonation does not appear to be a viable option for the destruction of energetic materials. Incineration, other thermal processes, and chemical treatment such as acid or alkaline hydrolysis can effectively destroy these high energy compounds but they are associated with high treatment costs. Moreover, chemical processes may generate waste streams which require further treatment prior to their discharge in the environment. For

187

instance, alkaline hydrolysis of NG produces nitrites and nitrates which may need to be removed by denitrification techniques. There is therefore, a pressing need for the development of new technologies that can economically and effectively deal with the disposal of energetic compounds. Biological treatment of energetic compounds amenable to microbial degradation provides an alternative to costly thermal and chemical methods. It has recently received considerable attention and it may be the technology of choice for various types of munitions wastes and propellants.

Biodegradation is a process where naturally occurring microorganisms under aerobic or anaerobic conditions utilize the organics as carbon and energy sources to produce environmentally safe products such as carbon dioxide and water. NG can be aerobically biodegraded by several fungal and bacterial consortia in the presence of co-substrates. The decomposition proceeds through a number of intermediate products whose formation is catalyzed by extra-cellular enzymes.

The anaerobic biodegradation of NG by a mixed bacterial culture from digester sludge was investigated in this study. The study focused on the ability of anaerobic bacteria to degrade NG and utilize it as sole carbon source, the identification of possible intermediates, and the effect of co-substrates on the rates of transformation.

Wendt et al. [1] demonstrated the feasibility of microbially mediated decomposition of NG working with activated sludge and pure cultures. The two isomers of dinitroglycerin namely, 1,3-dinitroglycerin (1,3 DNG), and 1,2-dinitroglycerin (1,2 DNG), and possible mixtures of 1-mononitroglycerin, and 2-mononitroglycerin were detected during biodegradation. However, the mononitroglycerin isomers were not separated. Based on their study, a pathway for the biodegradation of nitroglycerin was proposed in which the parent molecule is decomposed, through successive denitrations, to glycerol being widely used by all researchers later. But the intermediates were not quantified during the study and the postulated was confirmed.

Biodegradation of this compound by activated sludge was studied by Kaplan et al. [2] where the intermediates were quantified during the experiments. However, the complete mineralization of NG was proven and only a single intermediate, 1 MNG, was detected during their study. They also studied the biodegradation of glycerol, a complete denitrified product of glycerol trinitrate. The results showed that glycerol can be degraded easily with a higher rate compared to nitroglycerin or its intermediates, both aerobically and anaerobically as a sole carbon source.

Ducrocq et al. [3] studied the biodegradation of NG by the fungus Geotrichum candidum. The dinitro isomers were individually identified in this study, but not the mononitro isomers. The data presented by these investigators did not show complete mineralization of NG and degradation appeared to cease after the formation of the mononitroglycerin isomers. Servent et al. [4] studied the metabolism of NG by Phanerochaete chrysosporium and discussed several enzymatic pathways for the degradation of NG and the intermediates which were identified but they were not completely mineralized. Servent et al. [5] documented the formation of nitric oxide and nitrates during the metabolism of nitroglycerin by *Phanerochaete chrysosporium*. Presari and Grasso [6] working with mixed wastes containing NG in sequencing batch reactors characterized NG as an inhibitory non-growth substrate. However, no intermediates were reported although high destruction efficiency of NG was observed in these systems. In recent study, White et al. [7] reported the metabolism of this compound by bacterial cultures as a sole nitrogen source where glycerol was used as a primary carbon source.

The results of the present study indicate that anaerobic bacterial consortia are able to degrade NG by successive denitration to dinitroglycerin and mononitroglycerin and their corresponding isomers and utilize it as a sole carbon source. The experiments were conducted in microcosm vials that were amended with the required nutrients. The various intermediates were identified and quantified. All intermediates are biodegradable but at lower rates than the parent compound, and are converted to glycerol which is completely mineralized to carbon dioxide and water. Based on these results a degradation pathway is proposed for the anaerobic degradation of NG.


## MATERIALS AND METHODS

The anaerobic experiments were carried out in 160 ml microcosm vials which were inoculated with microorganisms obtained from an anaerobic digester. The chemical analysis, for determination and identification of nitroglycerin and the produced intermediates, was performed by high pressure liquid chromatography (HPLC).

NG and its biotransformation intermediates were quantified by HPLC (Varian Instruments Co., Palo Alto, CA) using a Partisil 10 ODS-3, 4.6 x 250 mm (Whatman Inc., Clifton, NJ) column at room temperature. For the analysis of nitroglycerin, a mobile phase containing 70% acetonitrile and 30% water at a flow rate of 1 ml/min was used and the highest absorbance was observed at 210 nm. For the intermediates, a mobile phase of 5% acetonitrile and 95% water at a flow rate of 1 ml/min was used and the highest absorbance was at 205 nm. Four possible intermediates (1,3-dinitroglycerin (1,3-DNG), 1,2-dinitroglycerin (1,2-DNG), 2-mononitroglycerin (2-MNG), and 1-mononitroglycerin (1-MNG)) and nitroglycerin standards were purchased from Radian Corporation (Austin, TX). The nitroglycerin used in these experiments was obtained from ICI. Calibration curves were prepared for nitroglycerin as well as its intermediates using the methods as indicated above.

Since anaerobic microorganisms work at an optimum pH of around 7.0 (Grady and Lim, [8]), growth media was prepared using a strong phosphate buffer to maintain the pH around 7.0. Other nutrients were added as needed for typical bacterial cultures. The final pH of the growth media was 7.1. The composition of the growth media prepared for these experiments is given in Table 1. Anaerobic digester sludge with a mixed bacterial population used in these experiments was brought from a nearby municipal wastewater treatment plant.

TABLE I - Composition of growth Media.

| Items | Amount |
|---|---|
| Sodium phosphate | 11.2 gms |
| Potassium phosphate | 5.7 gms |
| Ammonium sulfate | 500 mg |
| Magnesium sulfate | 100 mg |
| Manganous sulfate | 10 mg |
| Ferric chloride | 5 mg |
| DI Water | 1 liter |

The anaerobic experiments were conducted in 160 ml serum bottles in batch mode. Samples were prepared in an anaerobic glove-box under nitrogen atmosphere. Two sets of experiments were carried out in this study. First experiments were carried out to evaluate the effect of the presence of co-substrates on the biodegradation process and then experiments were performed to study the degradation pathway. Glucose was used as a co-substrate and it was added in different concentrations. In our first experiment, five microcosm vials were prepared each with 160 ml growth medium to which 50 mg/l of nitroglycerin were added. In two of these flasks, 1 gm/l of glucose was added to assess the effect of co-substrates. The microcosms were inoculated with anaerobic digester sludge. The fifth bottle was used as a control to which no culture was added. Kaplan *et al.* [2] have shown that no significant photodegradation of nitroglycerin occurs. However, the control was necessary for this experiment in order to check the stability nitroglycerin in presence of growth media chemicals. The bottles were sealed with Teflon lined septa to prevent adsorption and incubated at 30°C in a gyratory incubator at 180 rpm. Samples were withdrawn once a week using a syringe, and analyzed in HPLC for nitroglycerin only. In the second set of experiments the microcosm vials were prepared in a similar manner but they were supplemented with 1 gm/l of glucose and the initial NG concentration was 160 mg/l. Samples were periodically withdrawn for nitroglycerin and its intermediates.

## RESULTS AND DISCUSSION

The concentration profiles of nitroglycerin in the absence and presence of co-substrate are given in Figure 1. From these results one can observe that nitroglycerin can be biodegraded anaerobically by digester sludge in presence of glucose as a co-substrate but it can also be utilized as a sole carbon source. It was observed that nitroglycerin was completely disappeared within week in the bottles where glucose was used as a co-substrate, but the degradation rate was slower in the unamended microcosms. No significant change in NG concentration was observed in the control vials indicating that the degradation was mediated by the anaerobic microorganisms.

The second experiment, as indicated in the experimental procedure, was carried out to determine the biodegradation pathway of nitroglycerin. The experiment was continued until mineralization of all di- and mono-isomers of nitroglycerin was complete. The intermediates were identified by comparing their retention times against the standards. The concentration profiles for nitroglycerin as well as its biotransformation intermediates are given in Figure 2.

From the profiles of the biotransformation intermediates, one observes that nitroglycerin first degrades to 1,3-DNG and 1,2-DNG and then it is converted to 2-MNG and 1-MNG. Since there is a possibility of formation of 1-MNG from both dinitroglycerin isomers, there should be more 1-MNG formed at any transient time during degradation. The aerobic degradation pathway proposed by Wendt et al.[1] appears to be followed also during the anaerobic degradation of NG. In the aerobic studies reported by Kaplan et al. [2], Ducrocq et al. [3], and Servent et al. [4, 5] the decomposition of NG appears to stop after the formation of either 1-MNG or 2-MNG and complete mineralization was not documented. The results of this study showed that both 1-MNG and 2-MNG degrade further, possibly to glycerol which readily converted to carbon dioxide and water.

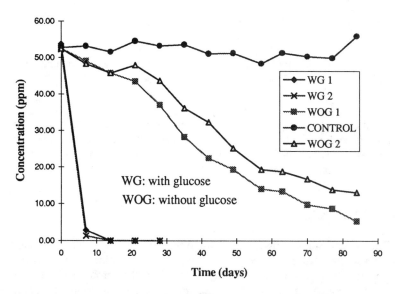

FIGURE 1. Anaerobic degradation of nitroglycerin in the absence and presence of co-substrates.

191

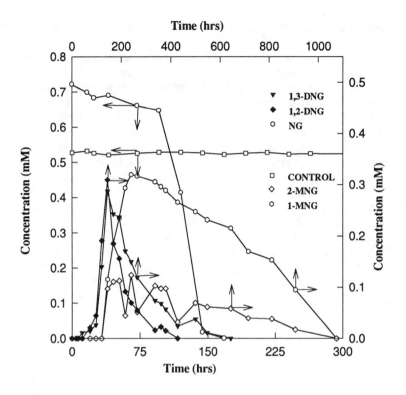

FIGURE 2. Concentration profiles of intermediates formed during anaerobic
degradation of nitroglycerin.

Head space gas samples from the unamended microcosm vials were injected
into a GC equipped with TCD detector and carbon dioxide was detected. Since
there was not carbon dioxide present initially in the system it must have been the
product of complete mineralization of NG.

## CONCLUSIONS

The present study showed that nitroglycerin can be degraded by anaerobic
bacteria in presence or absence of a co-substrate. The nitroglycerin anaerobic
biodegradation pathway appears to be identical to the aerobic pathway which
proceeds through successive dinitration of the parent molecule to produce

glycerol. Glycerol is readily degraded to produce carbon dioxide and water thus resulting in complete mineralization of nitroglycerin. The intermediates degrade much slower than nitroglycerin and require longer reaction times to completely disappear. The reaction rate is limited either by the conversion of dinitroglycerin to mononitroglycerin or by the conversion of mononitroglycerin to glycerol. Which step is the slowest and therefore rate controlling is not clear and further research is necessary to clarify the mechanism.

## ACKNOWLEDGMENTS

This work was performed at the James C. Nicoll Jr. Laboratory at Stevens Institute of Technology for Picatinny Arsenal, Dover, New Jersey, and it was sponsored by the U.S. Army AMCCOM under contract No. DAAA21-93-C-1018. The authors express their gratitude to Bruce Brodman from Picatinny Arsenal for his support and guidance throughout this project.

## REFERENCES

1. Wendt., T. M., J. H. Cornell, and A. M. Kaplan. 1978. "Microbial Degradation of Glycerol Nitrates," *Appl. and Env. Microbiology*, 36:693-699.
2. Kaplan, D. L., J. H. Cornell, and A. M. Kaplan. 1981. "Biodegradation of Glycidol and Glycidyl Nitrate," *Appl. and Env. Microbiology*, 43:144-150.
3. Ducrocq, C., S. Claudine, and M. Lenfant. 1989. "Bioconversion of Glycerol Trinitrate into Mononitrates by *Geotrichum candidum*," *FEMS Microbiology Letters*, 65:219-222.
4. Servent, D., Ducrocq, C., Henry, Y., Servy, C., and M. Lenfant. 1992 "Multiple Enzymatic Pathways Involved in the Metabolism of Glyceryl Trinitrate in *Phanerochaete chrysosporium*," *Biotech. and Biochem.*, 15:257-266.
5. Servent, D., C. Duerocq, Y. Henry, A. Guissani, and M. Lenfant. 1990. "Nitroglycerin Metabolism by *Phanerochaete chrysosporium* : Evidence for Nitric oxide and Nitrite Formation," Biochem et Biophysica Acta, 1074:320-325.
6. Presari, H., and D. Grasso. 1993. "Biodegradation of an Inhibitory Nongrowth Substrate (Nitroglycerin) in Batch Reactors," *Biotech. and Bioengg.*, 41:79-87.
7. White, G. F., J. R. Snape, and S. Niklin. 1993. "Bacterial Biodegradation of Glycerol Trinitrate," in *The 9th International Biodeterioration and Biodegradation Symposium*, The University of Leeds, UK.
8. Grady, Jr., C. P. L., and Henry C. Lim, 1980. *Biological Wastewater Treatment, Theory and Applications*, New York: Marcel Dekker, Inc.

# BIOREMEDIATION: A SYSTEMATIC, TIERED APPROACH

John V. Forsyth III
   Dames & Moore, Inc.
   2325 Maryland Road
   Willow Grove, PA 19090

Richard Bleam
   Bioscience, Inc.
   1550 Valley Center Parkway Suite 140
   Bethlehem, PA 18017

Neil Wrubel
   Dames & Moore
   2325 Maryland Road
   Willow Grove, PA 19090

A potential bioremediation project must be approached with the knowledge that each site will exhibit different and unique characteristics and will require a specific bioremediation plan. Therefore, a tiered response is appropriate, starting with a screening level determination (Tier I: feasibility of bioremediation and development of a rough estimate of remediation time and costs), continuing with pilot scale studies (Tier II) and ending with a complete work plan, budget and project management plan for a full-scale bioremediation effort which is conducted in the field (Tier III). Information gained from work performed in the lower tiers is utilized in higher tiers to determine the feasibility of bioremediation, method(s) of bioremediation selected, location of treatment (in place or in above ground treatment cells), to seek alternative remedial technologies or to develop more specific information before any decision can be made. By following a systematic and tiered approach, costly bioremediation project delays and failures can be averted.

The potential effectiveness of bioremediation is affected by technical, economic and regulatory factors. As progress is made from one tier to the next, more information becomes available and, therefore, the bioremediation project becomes more well-defined. Generally, larger sums of money are committed as the project moves to higher tiers. However, there is an accompanying increase in the level of confidence that the remediation will be completed accurately and cost-effectively.


## TIER 1

Tier I is comprised of preliminary screening analyses, a preliminary design plan and a rough estimate of costs involved in the project.  These three areas include:
- A review of all prior environmental investigations conducted at the site.
- Implementation of biofeasibility screening and interpretation of data (soil and/or groundwater physico-chemical and biological parameters).  Depending upon the site and processes which are investigated, biofeasibility screening may include -among other techniques- laboratory analyses of available nutrients, contaminants present and soil characteristics.  Characterizations of existing microbial populations, if any, are made in this tier.
- Selection of process technology (one or a combination of several options may be chosen as appropriate Tier II or Tier III processes).

The feasibility of bioremediation can be determined from this information.  Additional information needs are then defined to develop a conceptual design for the project, to further define anticipated costs and procedures and to make recommendations for subsequent project activities.


## TIER II

Tier II procedures confirm and quantify the potential of soils and/or groundwater at the site to support bioremediation.  The confirmation and quantification is accomplished through biotreatability studies (laboratory work) and/or process confirmation (pilot work).  Biotreatability studies are generally designed to determine the degradation rates of contaminants, the extent of degradation and optimal conditions in which biodegradation can occur.  Biotreatability studies can be used to represent both in-situ and ex-situ bioremediation projects.  The results of these tests will determine the applicability of employing in-situ bioremediation technologies at the site, the approximate length of treatment and general soil conditions such as pH, nutrient and moisture contents.

Treatability studies provide information to estimate the timeframe for full-scale bioremediation and to estimate the costs involved in the full-scale project.  To perform the treatability study, soils, sediments, slurries and/or groundwater can be treated in place or collected from the site and be treated aboveground or in a laboratory setting.  Treatability studies may be conducted with these media using the following techniques/technologies:


## ELECTROLYTIC RESPIROMETRY

Electrolytic respirometry equipment can be used to determine the biodegradation of organic compounds present in solid, slurry and aqueous phases.  The most common function of this equipment is to measure the amount of oxygen consumed during bacterial utilization of organic substrates as food sources. Electrolytic respirometers rely upon the electrolysis of water to provide the oxygen necessary for biodegradation of a sample which is seeded with microorganisms. Each respirometric cell consists of a closed reaction vessel and an electrolytic cell for oxygen supply. As oxygen is consumed by the microbes in the reactor vessels (and metabolically produced carbon dioxide is adsorbed by a potassium hydroxide scrubber solution), a slight vacuum lowers the electrolyte level in the electrolysis cell triggering oxygen production. Oxygen is produced at the positive electrode and added to the reactor vessel until the original internal pressure is restored. An electrical control unit monitors the amount of oxygen required to equalize the pressure. A continuous readout of cumulative oxygen demand is obtained and expressed as a concentration. Ultimately, respirometry studies can estimate degradation rates of contaminants and delimit the requisite environmental conditions to maintain optimal biodegradation rates in the field.

## MICROCOSMS

Microcosms are laboratory-scale systems which simulate environmental conditions found in the field. Tests conducted in microcosms can provide 1) an indication of the tested media to support biodegradation, 2) reasonable estimates of contaminant removal rates through mass balance determinations and 3) controlled conditions for experimentation to determine, for example, biogeochemical processes involved in the breakdown of constituents of concern and associated metabolic pathways involved in biodegradation. The following sections briefly describe common microcosm techniques in, generally, increasing order of sophistication and cost:

## SOIL SLURRY REACTORS

Soil slurry reactors can be utilized in bioremediation applications where bench-scale models are required for large-scale conceptual designs. These bench scale-models have been used to design slurry systems for waste lagoon closures and for saturated soils where a solid phase bioremediation approach is not feasible.

## SOIL BOXES

Soil box studies are conducted on soils obtained from the site and evaluated for solid phase bioremediation applications. Applications for soil box biotreatability studies are from ex-situ projects where soil has already been excavated, for shallow surface bioremediation or for soil to be remediated in place in lifts. These studies may simulate -among other processes- actual mechanical mixing of soil, nutrient addition and moisture control.

## FLASK STUDIES

Flask studies are used to analyze biodegradation of contaminants in water, soils or slurries. These studies are inexpensive and are commonly used to determine degradation rates and the biotoxicity of compounds. Flask studies are best used as indicators of a medium's ability to support biodegradation.

## SOIL COLUMNS

Soil columns can be bench-scale or small, mobile biological reactors for low flow treatment of contaminated soils, sediments and aquifer material. Vogel *et al* (1987) characterized abiotic and biotic loss mechanisms of water-borne halogenated aliphatic compounds which was passed through glass columns in which biofilms were attached to glass beads. The authors demonstrated the importance of the availability of electron acceptors and retention time of flowing water in such a column for degradation of halogenated aliphatic compounds. Decreases in retention times resulted in the maintenance of semi-batch conditions within the columns. The authors also characterized the requisite conditions within the columns to maintain optimal rates of biodegradation of the compounds of concern [1].

The results of soil column studies are expected to represent only the medium(a) which is tested. However, as the number of column studies which are performed continues to increase, general information should become available to accurately anticipate biodegradation of contaminants in a wide variety of subsurface conditions.

## MODELS

Models which characterize bioremediation are best employed as one of many tools consulted during the selection of remedial alternatives. The use of models is most powerful when little is known about subsurface conditions at the specific site and intrusive, investigative procedures are too costly. In general, models can provide oxygen transfer requirements, soil permeability, nutrient requirements and nutrient and oxygen transport requirements for applications where soil cannot be moved. In general, certainty about the results of model simulations increases in proportion to decreases in subsurface heterogeneities.

Semprini and McCarty (1991) describe favorable comparisons between short-term non steady-state model simulations and field results of an in-situ biorestoration of aquifer material containing chlorinated aliphatics. However, the authors encountered varying results between observed and long -term simulations of the transport of methane and dissolved oxygen. The dispersion was thought to result from vertically distributed heterogeneities in the aquifer material [2].

## PILOT-SCALE STUDIES

The application of pilot-scale studies conducted in the field are becoming more prevalent. Coover *et al* (1994) describe a pilot-scale demonstration of liquid solids treatment (LST), a biological treatment process similar to conventional activated sludge, which can be utilized to remediate petroleum refinery sludges in surface impoundments. The authors constructed an LST bioreactor by retrofittting existing wastewater treatment units. Weathered petroleum sludge had accumulated for several years in a surface impoundment and was to be treated. The primary constituents of the sludge were oil and grease which contained low molecular weight polyaromatic hydrocarbons (PAHs) and benzene, toluene, ethylbenzene and xylenes (BTEX). The concentration of PAHs ranged from less than 2 parts per million (ppm) to 24 ppm.

The sludge was excavated from the surface impoundment and placed in an abandoned concrete clarifier which was retrofitted with a float-mounted mixer. Water was added to the parent material to form a slurry which was subsequently aerated and seeded with petroleum degrading microorganisms. The final slurry volume was approximately 947,000 gallons.

The slurry was treated for 8 weeks and periodically amended with calcium hydroxide, ammonia nitrate and prilled superphosphate to buffer the pH and provide nutrients to the activated slurry. The slurry was periodically sampled for constituents of concern and solids content. In addition, the temperature, pH, dissolved oxygen and oxygen uptake of the slurry was periodically monitored. The results of the pilot-scale demonstration included an estimated 0.21% of nonmethane hydrocarbons were emitted from the approximately 425,000 kg of sludge. However, the reduction in residual semivolatile and nonvolatile hydrocarbons was attributed primarily to biological degradation [3].


## PILOT-SCALE BIOVENTING

Bioventing is a proven active method of bioremediation which enhances the abilities of existing soil microbial populations to metabolize and degrade organic compounds in the subsurface into simpler compounds and, ultimately, to carbon dioxide and water. This technology can be readily installed as a pilot-scale system in the field.

Bioventing uses the same process units as soil vapor extraction (SVE), but the cycle of operation is modified. Rather than being used to actively remove organic constituents, the SVE unit is used to periodically introduce fresh air to the microbes and to remove excess carbon dioxide generated during the degradation of organic compounds. Fresh air is drawn to the extraction well via passive air inlet wells installed either in the subsurface or piles of soil or sediment. The rate of movement of air through the treated medium(a) can be adjusted to balance emissions generated during microbial respiration so that microbial degradation is optimized. Nutrients and moisture may be introduced to the air stream to facilitate degradation of the contaminants.

Bioventing equipment is readily available and capital costs for this process option are low. Requisite equipment and installations for the bioventing system include a vacuum blower unit and associated piping, passive air inlet or outlet wells, air extraction or air injection well(s), off-gas treatment units (e.g., activated granular carbon units) and, if necessary, a subsurface amendment infiltration system to inject bacteria or nutrients.

Pilot-scale bioventing treatability studies are expected to yield the following information: 1) radius of influence of the injected air, 2) air injection system operating parameters (used to design a full-scale bioventing system) and 3) microbial growth response to supplemental air, moisture and nutrients (if applied).

198

## COMPARISON OF LABORATORY AND FIELD TREATABILITY STUDIES

Laboratory treatability studies are more common than those conducted in the field. The main difference between laboratory microcosms and field conditions relates to the exchange of matter across system boundaries. Laboratory experiments and field "reactors" are, generally, closed and open systems, respectfully. There are also several practical limitations encountered in the lab. Small laboratory microcosms may not represent the behavior of processes which require larger scales. Microcosms cannot capture heterogeneous conditions which are present in anaerobic sediments, soils and aquifers. Microcosms can, however, indicate the potential for the representative medium(a) to support biodegradation. In addition, extrapolations of results obtained from microcosms have improved with use of multi-media microcosms, such as sediment and soil columns.

The main advantage of laboratory microcosms is the maintenance of greater control over experimental conditions. The representative media can be subjected to a variety of experimental manipulations. The environmental conditions maintained in the treatability study(ies) are either 1) maintained at levels which are similar to those observed in representative media collected from the field or 2) maintained at levels which attain optimal conditions to conduct the treatability test with the intent of achieving efficient full-scale bioremediation. Treatment of the representative media is generally conducted in accordance with processes being considered for full-scale remediation at the site.

## CAUTION WITH TREATABILITY STUDIES

The rates, patterns and extent of biodegradation observed in microcosms may only generally represent actual rates which occur in the field. Workers often compare results in experimental batches or columns to sterile controls. However, they may not have characterized the affect biota have on physico-chemical processes occurring in the experimental media. For example, they ignore the affect(s) biota (e.g., extracellular polymers) can have on contaminant partitioning to organic and inorganic materials in the supporting media. Workers commonly focus on one or a few isolated contaminants in their experiments. Thus, little is known about the effects of co-contamination on biotransformation of many contaminants. In summary, strong attention should be paid to contaminant losses in microcosms that could arise from processes other than biodegradation.

## TIER III

Tier III is comprised of the implementation and performance of the full-scale site bioremediation. The process chosen for full-scale remediation is dependent upon the contaminant characteristics, site characteristics, regulatory requirements and economic factors as determined in Tiers I and II. These processes may include single methodologies or a combination of methodologies. The goals, degradation rates of contaminants, material requirements, estimated time frame for cleanup and design for full-scale bioremediation have been realized.

Biological treatment can be accomplished by bacteria, fungi, and higher plants. Methodologies which incorporate biota during treatment of contaminated soils, slurries, sediments and groundwater include passive bioremediation (monitoring only) and active (bioaugmentation and bioventing) bioremediation methods. Passive bioremediation of contaminants generally occurs naturally and at slower rates than active methods. This method can be an option where natural attenuation of contaminants can be documented and effectively contains the contamination and contaminated medium(a).

Bioremediation is commonly segregated into two categories, ex-situ and in-situ treatment biotreatment. Ex-situ biotreatment can be described as the process of using either indigenous or augmented microorganisms to remediate soil or slurry which has been removed from the original zone of contamination and treated elsewhere. In-situ bioremediation is the in-place treatment of soil, slurries and or water. This may include:

- Solid phase biotreatment of shallow, contaminated soils.
- Solid phase biotreatment of soil in layers.
- Slurry treatment such as in waste lagoons.
- Soil treatment through recirculation of water/air through the zone of contamination.
- Groundwater treatment through contacting contaminated groundwater with introduced oxygen, nutrients and/or exogenous microorganisms.

## A CASE HISTORY INCORPORATING THE TIER APPROACH

Leaking underground storage tanks at a formulating plant contaminated surrounding soil with gasoline and a C-13 - C-15, branched-chain, aliphatic hydrocarbon solvent. The underground storage tanks were removed and 1,200 cubic yards of surrounding soil were excavated. Analytical monitoring of soil samples which were collected from the excavated soil determined that the excavated soil contained an average of 230 mg/kg Total Petroleum Hydrocarbons (TPH). Levels of other compounds were below RCRA regulatory limits. "Hot spots" in the excavated soil showed TPH levels of greater than 600 mg/kg. An initial screening of the site characterized the amount of available nutrients, quantity of microorganisms present, moisture levels and TPH.

The excavated soil was placed on an impermeable liner and covered while tests were conducted to determine the feasibility of bioremediation by means of a solid-phase process. Solid phase bioremediation was chosen after an analysis of remedial options, their associated costs and the available space for treatment on-site. Although gasoline is a readily biodegradable substrate, little was known about the C-13 - C-15 branched chain, hydrocarbon solvent. There was no information on whether it could be biodegraded by either indigenous microorganisms or commercially available, natural hydrocarbon degrading species.

To determine the ability of the indigenous microorganism to degrade the constituents of concerns, Bioscience, Inc., of Bethlehem, Pennsylvania, performed two types of treatability studies. The first study was conducted using electrolytic respirometry to measure the oxygen consumption of microorganisms during biodegradation of the contaminants under a variety of conditions. The second experiment consisted of two bench-scale reaction vessels which contained composites of soil from the site, nutrients, water and bacteria.

Both of the reactors containing the C-13 - C-15 solvent showed lower oxygen uptake than the controls during the first 80 hours of incubation. The reactor containing a higher concentration showed a significant lag in oxygen uptake for the first 16 hours. Between 70 and 150 hours into the experiment, oxygen uptake for both reactors was similar to that of the seed blank (control containing). After 150 hours, both reactors began to show a consistent increase in oxygen uptake compared to the seed blank, but not enough to indicate complete biodegradation of the hydrocarbon solvent substrate. Analysis of the reactor contents after incubation indicated a greater than 90 percent reduction in TPH.

TPH reduction in the reactors with indigenous microbes were only 15 and 45 percent after 14 and 28 days, respectively. Where the commercial inoculum had been added, the reductions were 39 and 78 percent after 14 and 28 days, respectively.

Based on the test results, which indicated the inhibitory nature of the hydrocarbon solvent and a doubling of the biodegradation rate by bioaugmentation, the formulator selected a solid-phase process, using nutrients and a commercial microbial inoculum, as the most timely and cost-effective treatment alternative for the site. Based on the laboratory results, reduction of TPH levels to the target of less than 100 mg/kg was expected to take approximately three months, depending primarily upon varying environmental conditions such as soil temperature.

Solid-phase bioremediation of the site began in mid-October. The soil was divided into three treatment cells which were placed on a 20-mil plastic liner. Each cell was covered with a 10-mil cover to prevent volatilization of hydrocarbons and the infiltration of surface water. The soil was initially conditioned with a commercial, hydrated microbial product and the required levels of specialized commercial nutrients for microbial growth. The first cell was spread to a depth of one foot and inoculated with nutrients and microbes. Another foot of soil was added and the inoculum procedure was repeated therein. Since this method was time-consuming, the remainder of the soil was inoculated by spray through a tank sprayer as it was being spread on the liner by a track-hoe. The soil was mechanically distributed to provide sufficient mixing and aeration for microbial growth.

One month after inoculation, each cell was sampled and tested for TPH. Each cell was divided into four sections and a composite of three samples from each section was analyzed. The average concentration of TPH in all cells was below 100 mg/kg. However, some individual hot spots remained as a result of incomplete mixing. The site was left untreated over the winter and specific hot spots were re-inoculated and rototilled in the following spring.

Final laboratory analysis of soil samples collected in April showed that the TPH levels were well below the permit levels of 100 mg/kg in every cell section. TPH concentrations in soil samples collected from 7 of the 12 sections were below 10 mg/kg TPH. Supplemental analyses were conducted on the same soil samples for BTEX and for total organic halogens (TOX). Each of the samples contained BTEX and TOX levels which were below detectable limits.


## SUMMARY

Adherence to the tiered approach to a bioremediation project allows the engineer and the client to evaluate the project at each step to determine the proper courses of action. Each step contains options of evaluating experimental and remedial alternatives and evaluating the project. The nature of the project may dictate that biotreatability studies may not be economical or may not be warranted because the contaminant is known to be highly biodegradable. However, all projects should base this decision on evaluations from Tier I. While it may not be feasible to incorporate all of the steps above, the initial tiers should be evaluated for the site prior to specifying a bioremediation application. Adhering to the steps outlined in Tier I and Tier II will limit the time to start-up and completion of the full-scale phase. When problems are encountered in the full-scale phase, sufficient information is available to overcome the problem and improve the selected bioremediation process(es).

## REFERENCES

1. Vogel T.M., Criddle C.S., and McCarty P.L. 1987. "Transformations of Halogenated Aliphatic Compounds," *Environmental Science & Technology* 26(8):722-736.

2. Semprini L. and McCarty P.L. 1991. "Comparison Between Model Simulations and Field Results for In-Situ Biorestoration of Chlorinated Aliphatics: Part 1. Biostimulation of Methanotrophic Bacteria," *Ground Water* 29(3):365-374.

3. Coover M.P., Kabrick R.M., Stroo H.S., and Sherman D.F. 1994. "*In Situ* Liquid/Solids Treatment of Petroleum Impoundment Sludges: Engineering Aspects and Field Applications." in, *Bioremediation Field Experience*, Flathman, P.E., Jerger D.E., and Exner J.H, eds. Boca Raton, Florida:CRC Press, Inc., pp. 197-224.

# INVESTIGATION OF THE BIOTRANSFORMATION OF TNT BY A RHODOCOCCUS SP.

John P. Tharakan and Glaister Welsh
Department of Chemical Engineering
Howard University
2300 6th Street, NW, Washington D.C. 20059

James H. Johnson, Jr.
Department of Civil Engineering
Howard University
2300 6th Street, NW, Washington, D.C. 20059

## INTRODUCTION

Tri-nitro-toluene (TNT) has contaminated waterways and soils as a result of its widespread military and non-military uses. Toxic effects of TNT pollution including liver damage and anemia have been widely reported; TNT is also toxic to fish and wildlife. The U.S. Government is aware of at least one million tons of explosives-contaminated soils awaiting clean-up; a logical technology to implement for such a large-volume source is bioremediation. Unfortunately, TNT has proven to be quite recalcitrant to mineralization through direct biodegradation.

Several investigators, however, have shown that TNT can be biologically transformed [1,2]. Biological degradation of nitroaromatic compounds have been demonstrated to occur by one of three pathways: (1) the initial conversion of nitro groups to hydroxyl groups, (2) the reduction of a nitro group to an amino group, and (3) the release of nitro groups as nitrite with the concomitant formation of a phenol.

Reduction of the nitro groups usually proceeds stepwise with the *para* being the most susceptible, followed by the *ortho* groups; the third nitro group is reduced only under strictly anaerobic conditions[3]. Formation of a diphenol is a necessary prerequisite to ring cleavage but no phenolic intermediates have been found [2].

Several researches have isolated or constructed strains of microorganisms that are able to utilize TNT as sole carbon, nitrogen or energy sources[4]. Researchers have also reported high TNT concentrations ($>50$ mg/l) as inhibitory to bacteria, yeast and fungi[5].

Fernando et al.[6], reported biodegradation of TNT by *Phanerochaete chrysosporium*, which mineralized 35% of $^{14}C$ TNT after 12 days of incubation; another study with *P. Chrysosporium* showed that the highest $^{14}CO_2$ release (10%) was at 5 ppm of TNT, and no mineralization occurred with TNT levels greater than 15 ppm[7]. However, the use of a matrix, such as corn-cobs may immobilizes and protect the fungus [6].

In activated sludge systems, no $^{14}C$ or $^{14}CO_2$ could be detected in an aerated reactor following 3 to 5 days of incubation with labelled TNT[5]. In another series of studies, a total of 66 microorganisms were isolated from a munitions waste discharge point with 15 demonstrating 100% degradation of TNT[8,9]. *Pseudomonas sp. strains* were also isolated that oxidized TNT [10,11]. Enzymatic preparations from isolates have also reduced TNT [2].

This study examines the degradation of TNT and pyrene using a *Rhodococcus sp.* isolated based on its ability to survive on pyrene as the sole carbon source[12]. The experiments were designed to study both the direct and cometabolic transformation of TNT by the microbes in both batch and continuous modes, and to determine the optimum conditions under which TNT would be degradable.

**Materials & Methods**

Minimal Salt Media Preparation: Minimal salt media (MSM) was prepared as described earlier [12]. The pH was adjusted to 7.0 and then pyrene, TNT, or both were added to obtain minimal salt media with pyrene (MSP), TNT (MST) or both (MSTP).

PAH and TNT Solution Preparation: Stock solutions of non-labeled pyrene (Eastman Kodak Co., Rochester, NY) and TNT (Chem Service, West Chester, PA) were prepared in HPLC grade acetonitrile (Fisher Scientific, Pittsburgh, PA) at a concentration of 5000 mg/L.

Microorganism: The *Rhodococcus sp.* was isolated from a creosote contaminated wood treatment site and was characterized as aerobic, gram positive, and with optimum growth conditions at 38°C and pH 7.0 [12].

Batch System: Thirty ml amber tubes were used as the batch culture vessel. One hundred and twenty of these, incubated at 38°C served as the batch system. Inocula were obtained from stock cultures in MSP, which were repeatedly washed in MSM and then resuspended in MSP, MST, or MSTP, as required. Two abiotic control were maintained, one without microbes and the other with cells killed with 0.5% $NaN_3$. Six triplicate sets of culture tubes were prepared for each media formulation, and one set was sacrificed for cell and substrate analysis at days 0, 3, 7, 21, 35 and 50.

Continuous System: A benchtop fermentor (Bioflo I Model C30, New Brunswick Scientific, New Brunswick, NJ) with a 2-liter aerated (1.51 liter air/min) culture vessel was utilized for the chemostat studies. After sterilization, filling, and system integrity checks, the continuous fermenter was inoculated by adding a 100-ml seed culture raised under the appropriate conditions (MSP or MST). One run was conducted with MST as the feed and another with MSTP. Samples were collected every 24 hours and analyzed for cell number and then extracted and analyzed for TNT and pyrene on an HPLC. In the continuous cultures, the dilution rate was increased step-wise from steady state to steady state (defined as constant cell and substrate concentration over a 24 hour period) to determine kinetics of transformation.

Instrumentation and Sample Analysis Heterotrophic plate counts (EPA 9215C) and optical density readings (540 nm) were used to measure biomass. Samples were then liquid extracted with methylene chloride and analyzed for pyrene and TNT by HPLC on a 25 cm x 4.6 mm C18 reversed phase column (Rainin Instrument Co. Inc, Emeryville, Ca)

## RESULTS AND DISCUSSION

TNT concentrations in the *Rhodococcus* culture in MST initially spiked with 5, 10, 20 and 30 mg/L TNT, respectively, were monitored over a period of 50 days. Similar cultures spiked with the TNT concentrations above and with a total of 50 mg pyrene/L culture, were also monitored over the 50 day period. Data points were obtained using triplicate samples. The standard deviation at each data point for cultures in MST ranged from $\pm$ 0.7 mg/L at an initial concentration of 5 mg/L TNT to $\pm$ 2.0 mg/L at an initial concentration of 30.0 mg/L TNT. In the cultures supplemented with pyrene, the standard deviation at each point was approximately $\pm$ 2.0 mg/L.

In Figure 1, the log of *Rhodococcus sp.* cell concentration is shown for batch cultures in MST at 5, 10, 20 and 30 mg/L of TNT. Maximum cell growth ($2.5 \times 10^{18}$ cfu/ml) was achieved in the MST with 5 mg/L of TNT, compared to $7.5 \times 10^{14}$ cfu/ml at 30 mg/L TNT. With *P. chrysosporium*, good fungal growth at low TNT concentrations (0.36 mg/L) were obtained, but growth was limited at 20.36 mg TNT/L[13].

When pyrene was added to the media, the difference between growth at low and high TNT concentrations was significantly reduced. Growth of the organism on pyrene alone demonstrated an increase from $1.90 \times 10^8$ to $8 \times 10^{19}$ CFU/ml of cell culture over a 50 day period, similar to that obtained for the organism during the initial isolation[12].

Figure 2 and 3 show the percentage reduction of TNT in MST and MSTP culture, respectively, over 50 days. Percentage TNT reductions in the MST of 94.4%, 85%, 94.5% and 94% were obtained for initial TNT concentrations of 5, 10, 20 and 30 mg/L, respectively. In the MSTP media the percentage reductions of TNT were 92.4%, 88.5%, 94.4% and 92.0%, respectively. The percentage reduction in total pyrene concentrations for the two substrate cultures were 18.1%, 26.5%, 19.3% and 17% for cultures with initial TNT levels of 5, 10, 20 and 30 mg/L, respectively. The percent reduction in pyrene in MSP was approximately 24% over the 50 day incubation period, comparable to the ranges obtained earlier[12].

Figures 4 and 5 show the time course of TNT transformation in batch cultures for low (5 mg/ml) and high (30 mg/ml) concentrations of TNT, respectively. Data are shown for four separate conditions[descriptive key on figures shown in parentheses]: (1) abiotic control in MST(TNT Control), (2) killed cell control (Dead Cells + Media), (3) cultures in MST (TNT + cells), and (4) cultures in MSTP (TNT + pyrene + Cells). The concentration of TNT in the controls remained relatively unchanged, suggesting the absence of abiotic mechanisms of TNT disappearance. In the culture with 30mg/L TNT (Fig. 5), TNT reductions of 94% was obtained for cultures in MSTP, and 92% reduction for cultures in MST.

The data suggest that higher TNT concentrations had no significant effect on the rate or extent of biotransformation. Initial TNT concentrations of 5, 10, 20 and 30 mg/L were reduced to 0.24, 1.12, 0.94 and 1.4 mg/L, respectively over 50 days. Similar results were obtained for a *Desulfovibrio sp.* B strain [14] and a *Pseudomonas sp.* strain C151[4], both of which could grow on and transform 100 mg/L of TNT within 7 days of incubation at 37°C. Interestingly, cell growth appeared inhibited at higher concentrations of TNT. The yield coefficient decreased with increasing TNT concentration for cases where TNT was the sole carbon source. However, the yield was essentially constant in media with both TNT and pyrene present.

The changes in yield may be related to implementation of adaptive strategies. Cells that are exposed to high nutrient concentrations develop high cell quotas for common bioelements such as nitrogen, phosphorous, silicon and coenzymes, such as vitamin $B_{12}$. When these cells are exposed to large nutrient concentrations, the progressively depressed rate of nutrient uptake observed may be a result of utilization of a smaller quantity of cell material as a base for rate determination, which could cause declining yields.

Data for the continuous culture of *Rhodococcus* in a bench-top fermentor are shown in Figure 6, where TNT concentration and cell count are shown as a function of dilution rate. The fermentor inlet TNT concentration was 30 mg/L. The TNT concentration increased with increasing dilution rate while the cell concentration decreased, eventually reaching the point close to wash out[15]. Duque *et al.*,[4] found that *Pseudomonas sp.* clone A in a chemostat reduced TNT concentration from an inlet of 70 mg/L to a steady state concentration of 0.202 mg/L at a dilution rate of 0.01 hr$^{-1}$. The lowest effluent concentration achieved in these experiments was 3.5 mg/L at a dilution rate of 0.02 hr$^{-1}$.

The data obtained when the continuous culture was repeated with MSTP are shown in Figure 7. TNT inlet concentration was 30 mg/L and pyrene was supplied at a total level of 50 mg/L. TNT concentration increased with dilution rate while the cell population decreased. The initial concentrations of TNT in the culture vessel was 22.0 mg/L, while total initial pyrene was measured at 8.0 mg per liter of culture volume. The steady state concentration of TNT at the lowest dilution rate ( 0.02 hr$^{-1}$) was 4.0mg/L, while chemostat pyrene content was measured at 1.40 mg per liter of culture volume. This represents an 82% reduction in TNT and an 83% decrease in available pyrene.

Using conventional fermentation and Monod kinetic analysis, the batch and continuous culture data can be utilized to obtain kinetic coefficients assuming straightforward Monod kinetics[15]. Table 1 summarize the results of the calculations for $\mu_{max}$ and $K_S$ for batch culture in MST and Table 2 summarizes calculations of $\mu_{max}$ and $K_S$ for culture in MSTP, with calculations of both dual substrate and single substrate maximum specific growth rates.

TABLE 1: Monod Constant for Batch Growth of *Rhodococcus* in MST

| TNT Concentrations (mg/L) | $\mu_{max}$ (hr$^{-1}$) | $K_S$ (mg/l) |
|---|---|---|
| 5 | 0.029 | 0.203 |
| 10 | 0.040 | 4.28 |
| 20 | 0.027 | 1.35 |
| 30 | 0.025 | 2.18 |

TABLE 2:Monod Constant for Batch Growth of *Rhodococcus* in MSTP

| TNT Conc. mg/L | Pyrene Conc. mg/l | $\mu_{m1}$[1] hr$^{-1}$ | $\mu_{m2}$[2] hr$^{-1}$ | $\mu_{max}$[3] hr$^{-1}$ | $K_S$ mg/l |
|---|---|---|---|---|---|
| 5 | 50 | 0.029 | 0.069 | 0.043 | .430 |
| 10 | 50 | 0.041 | 0.046 | 0.041 | .804 |
| 20 | 50 | 0.027 | 0.043 | 0.041 | .783 |
| 30 | 50 | 0.025 | 0.044 | 0.033 | 1.37 |

[1]$\mu_{m1}$: Calculated for growth with respect to TNT
[2]$\mu_{m2}$:Calculated for growth with respect to pyrene
[3]$\mu_{max}$: Single substrate model calculation results.

The calculations demonstrate that the value of $\mu_{max}$ remains relatively unchanged regardless of the presence of both TNT and pyrene, and also for increasing concentrations of TNT. On the other hand, the $K_S$ increases for increasing concentrations of TNT, both for the cultures with and without pyrene. This behavior is typical of microbial systems that are inhibited by higher concentrations of a particular substrate. In the continuous culture, the data revealed a $\mu_{max}$ of 0.071 hr$^{-1}$ and a $K_S$ of 5.5 mg/l for MST growth and 0.167 hr$^{-1}$ and 8.11 mg/l for pyrene.

## CONCLUSIONS

In this study we have demonstrated the ability of a *Rhodococcus sp.*, isolated from a wood treatment site based on its ability to degrade pyrene as a sole carbon source, to transform TNT when it was supplied as the sole carbon source. When cultured in the presence of pyrene, the percent of TNT reduction remained similar. The extents of TNT reduction were far greater than that of pyrene in the batch culture experiments. Interestingly, the continuous culture experiments revealed similar extents of reduction for both compounds. TNT was also found to be inhibitory to cell growth at high concentrations. Future directions of research will focus on the identification of metabolites of TNT biotransformation and on the use of $^{14}$C-labeled TNT; this will facilitate identification of the metabolic pathway of TNT disappearance.

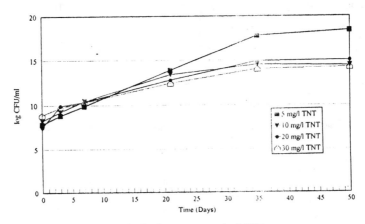

FIGURE 1: Growth of *Rhodococcus sp.* in MST

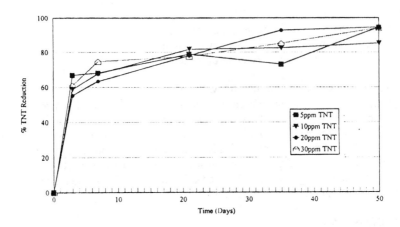

FIGURE 2: Percentage reduction in TNT concentration in MST.

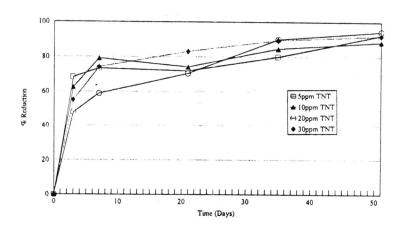

FIGURE 3: Percentage reduction in TNT in MSTP.

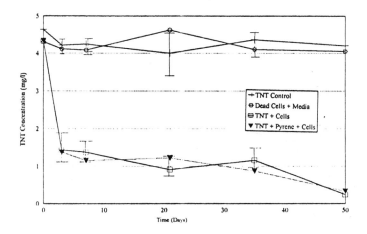

FIGURE 4: TNT Disappearance in MST (5 mg/l), MSTP (5 mg/l TNT, 50 mg/l pyrene) and Abiotic Controls.

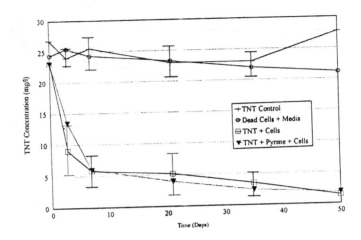

FIGURE 5: TNT Disappearance in MST (30 mg/l), MSTP (30 mg/l TNT, 50 mg/l pyrene) and Abiotic Controls.

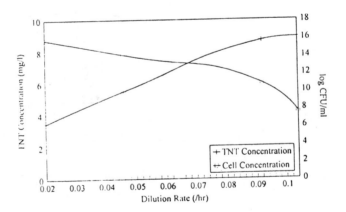

FIGURE 6: TNT and Cell Concentration in Continuous Chemostat Culture in MST (Feed TNT Concentration: 30 mg /L).

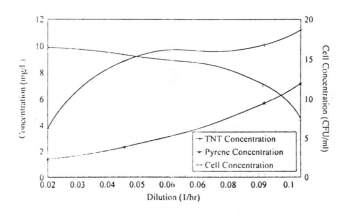

FIGURE 7:   TNT, Pyrene and Cell Concentration in Continuous
Chemostat Culture in MSTP (Feed TNT Concentration: 30
mg /L; Chemostat Pyrene level: 50 mg/L).

## REFERENCES

1  Kaplan, D.L. and Kaplan, A.M.,"2,4,6-Trinitrotoluene-Surfactant
   Complexes: Decomposition, Mutagenicity and Soil Leaching Studies,"
   Environmental Science and Technology, Vol.16, No. 9, pp.566-
   571,(1982).
2  McCormick, N.G., Feeherry, F.E. and Levinson, H.S.,"Microbial
   Transformation of 2,4,6-Trinitrotoluene and Other Nitroaromatic
   Compounds," Applied and Environmental Microbiology, Vol. 31, No. 6,
   pp. 949-958 (1976).
3  Funk, S.B., Roberts, D.J., Crawford, D.L. and Crawford, R.L.,"Initial-
   Phase Optimization for Bioremediation of Munitions Compound-
   Contaminated Soils," Applied and Environmental Microbiology, Vol. 59,
   No. 7, pp.2171-2177, (1993).
4  Duque, E., Haidour, A., Godoy, F. and Ramos, J.,"Construction of a
   Pseudomonas Hybrid Strain That Mineralizes 2,4,6-Trinitrotoluene,"

Journal of Bacteriology, Vol. 175, No. 8, pp.2278-2283, (1993).

5  Carpenter, D.F., McCormick, N.G., Cornell, J.H., and Kaplan, A.,"Microbial Transformation of $^{14}$C-Labeled 2,4,6-Trinitrotoluene in an Activated-Sludge System," Applied and Environmental Microbiology, Vol. 35, No. 5, pp.949-954,(1978).

6  Fernando, T., Bumpus, J.A. and Aust, S.A.,"Biodegradation of TNT (2,4,6-Trinitrotoluene) by Phanerochaete chrysosporium," Applied and Environmental Microbiology, Vol. 56, pp. 1666-1671, (1990).

7  Spiker, J.K., Crawford, D.L. and Crawford. R.L.,"Influence of 2,4,6-Trinitrotoluene(TNT) Concentration on the Degradation of TNT in Explosive-Contaminated Soils by the White Rot Fungus Phanerochate chrysosporium," Applied and Environmental Microbiology, Vol. 58, No. 9, (1992).

8  Kanekar, P. and Godbole, S.H.,"Microbial Degradation of Trinitrotoluene (TNT)," Indian Journal of Environmental Health, Vol. 26, No. 2, pp. 89-101, (1984).

9  Kanekar, P. and Godbole, S.H.,"Pseudomonas trinitrotoluenophila,sp. nov. Isolated from Soil Exposed to Explosive Waste," Biovigyanam, Vol. 9, pp. 5-8,(1983).

10 Won, W.D., Heckly, R.J., Glover, D.J. and Hoffsommer J.C.,"Metabolic Disposition of 2,4,6-Trinitrotoluene," Applied Microbiology, Vol. 27, No. 3, pp. 513-516 (1974).

11 Won, W.D., Disalvo, L.H. and James, N.G.,"Toxicity and Mutagenicity of 2,4,6-Trinitrotoluene and its Microbial Metabolites," Applied and Environmental Microbiology, Vol. 31, No. 4, (1976).

12 Cheeks, J , "Isolation and Characterization of a Microbe Capable of utilizing Pyrene as a Sole Carbon Source and Its Growth Kinetics in Batch and Continuous Culture," M.S. Thesis, Howard University, Washington D.C. (1994).

13 Michels, J., and Gottschalk, G., "Inhibition of the Lignin Peroxidase of P. Chrysosporium by Hydroxylamino Dinitrotoluene, an Early Intermediate in the Degradation of 2,4,6-Trinitrotoluene," Applied and Environmental Microbiology, 60 (1), 187-194 (1994).

14 Boopathy, R., C.F. Kulpa, and M. Wilson, "Metabolism of 2,4,6-trinitrotoluene by Desulfovibrio sp. B Strain," Appl. Env. Microbiol., 39, 270-275 (1993).

15 Bailey, James E. and Ollis, David F., eds., Biochemical Engineering Fundamentals, McGraw-Hill, New York, 1986. pp. 373-456.

# BIOLOGICAL REDUCTIVE DECHLORINATION OF CHLORINATED ETHYLENES: IMPLICATIONS FOR NATURAL ATTENUATION AND BIOSTIMULATION

## THOMAS D. DISTEFANO
Assistant Professor, Environmental Programs
The Pennsylvania State University — Harrisburg

## INTRODUCTION

Chlorinated organic compounds are the most frequently found contaminants at many hazardous waste sites and industrial facilities. Numerous industries use chlorinated organics such as tetrachloroethylene — also known as perchloroethylene (PCE) — and trichloroethylene (TCE), as degreasing agents, paint strippers, and in textile processing. These solvents are often detected as soil and ground water contaminants due to improper storage and disposal practices. Laboratory and full-scale investigations have proven that complete biological transformation of PCE and TCE is possible under anaerobic conditions. Biological treatment of chlorinated ethenes has received much interest due to the prevalence of these contaminants and the need to develop technologies that destroy contaminants rather than transfer them to other media. The purpose of this paper is to give an overview of the biological process by which anaerobic bacteria biodegrade chlorinated ethylenes. The benefits of this process are discussed along with key findings that may be employed to determine if dechlorination is occurring under natural conditions. Requirements of these bacteria are described and an assessment of future research needs is provided.

## POTENTIAL BENEFITS OF BIOTRANSFORMATION OF PCE AND TCE

Biological transformation of PCE and TCE offers the following potential benefits:

- In general, biological treatment offers the potential transformation of PCE and TCE to ethylene (ETH), an environmentally acceptable compound. Competing technologies, such as air strippers and activated carbon columns, merely transfer contamination from one environmental media to another. As long as the chemicals exist, so too will potential liability.

- Biological treatment of PCE and TCE may cost less than competing technologies.

- For sites with contaminated ground water, in situ biological treatment may achieve lower concentrations faster than conventional pump-and-treat methods. The rate that an aquifer can be remediated by a pump-and-treat system is often mass transfer limited— that is, getting the contaminants to the above-ground treatment system is often dependent on the rate of desorption from aquifer material. Although chlorinated ethenes are relatively soluble, sorption of these contaminants can significantly increase the duration of clean up for pump-and-treat systems.

- Some sites with PCE and TCE contaminated ground waters have experienced biological dechlorination of these contaminants by bacteria that are naturally present in an aquifer. This process of naturally occurring biodegradation is termed "natural bioattenuation". Amendments may be added to stimulate this naturally occurring dechlorination and shorten cleanup times — this is known as biostimulation.

## BACKGROUND

Due to its highly oxidized nature, biodegradation of PCE occurs only in anaerobic environments. Under anaerobic conditions, PCE and TCE are biotransformed by a reductive dechlorination mechanism [1]. Although TCE biodegradation under aerobic conditions is possible, naturally occurring biotransformation favors anaerobic mechanisms due to negligible or absence of molecular oxygen in biologically active aquifers.

During the 1980s, anaerobic biodegradation of PCE and TCE was reported by many researchers. Specifically, partial biological reductive dechlorination of PCE and TCE has been observed in laboratory cultures [2-6], soil microcosms [1,7] and in the natural environment [1,8,9]. Nearly all researchers reported incomplete dechlorination of PCE and TCE in that lesser chlorinated dichloroethylene isomers (DCE) or vinyl chloride (VC) were identified as end products. Because all chlorinated ethenes are known or suspected carcinogens, partial dechlorination is of little benefit; complete transformation to ethylene is required.

Figure 1 describes the biological reductive dechlorination of PCE to ETH under anaerobic conditions as determined by Freedman and Gossett [10]; some investigators have reported further reduction of ETH to ethane (ETHA), as observed by DeBruin et al. [11]. Because complete dechlorination of PCE to ETH was first observed in a mixed culture that was primarily methane producing, it was believed that methanogenic organisms were directly responsible for PCE dechlorination. In support of this, some researchers have reported partial dechlorination of PCE by a pure methanogenic culture [12]. However, more recent work has demonstrated that non-methanogenic, PCE-reducing bacteria are responsible for complete dechlorination to ETH [11,13]. The biotransformation pathway depicted in Figure 1 has been observed in batch and continuous-flow anaerobic reactors and in aquifers contaminated with PCE or TCE. The following points are made with respect to Figure 1:

- PCE is sequentially dechlorinated to TCE, DCE, VC, and then ETH; ETH may be further reduced to ETHA.
- An electron donor, designated as "H", must be available to drive the dechlorination. Many organic compounds have been employed as a source of electron donor. Addition of electron donor may represent a major operating cost for full-scale applications.
- Each dechlorination step results in the production of $H^+$ and $Cl^-$ ions.
- Transformation of PCE to VC is fairly rapid; the final dechlorination step of VC to ETH is slowest. In some systems, dechlorination may be incomplete and lesser chlorinated intermediates, such as DCE and VC may be detected.

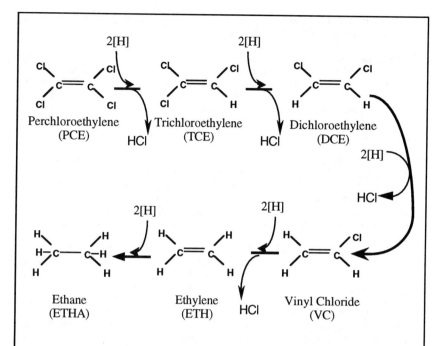

**Figure 1.** Biological transformation of PCE to ETH under anaerobic conditions (adapted from [10,11]).

## APPLICATION TO SITE REMEDIATION

Biological reductive dechlorination has been demonstrated in the laboratory and observed at field locations. Partial dechlorination has been reported at many field locations under natural conditions [1,8,9]. For example, at some facilities, TCE, DCEs, and VC have been identified as contaminants at sites where only PCE was disposed. This process of biotransformation under natural conditions has been termed natural or intrinsic bioattenuation.

**Indicators of natural bioattenuation.** Based on Figure 1, the following observations at a contaminated site would suggest that natural reductive dechlorination is occurring.

- Detection of TCE, DCE, VC, and ETH at sites where only PCE was disposed. Lesser chlorinated ethylenes are "daughter products" from the reductive dechlorination of PCE and TCE.

- Elevated chloride ion concentrations coinciding with ETH or less chlorinated ethylenes. Each dechlorination step produces 1 mole each of $H^+$ and $Cl^-$ ions.

215

A decrease in pH (due to hydrogen ion production) may also be expected if the buffering capacity of the ground water has been exceeded.

- A reduction in chemical oxygen demand (COD) coinciding with increasing occurrence of ETH or less chlorinated ethylenes. A source of reducing power (an electron donor) must be provided for reductive dechlorination to occur. If natural bioattenuation of PCE or TCE is occurring, organics in the aquifer are being consumed and serving as electron donors for reductive dechlorination. Biodegradation of organics in an aquifer would coincide with reductions in COD.

- Presence of elevated concentrations of methane coincident with lesser chlorinated ethylenes. Biological reductive dechlorination of PCE and TCE occurs under conditions suitable to methane producing bacteria. If organics in an aquifer are biodegraded as a source of electron donor, a consortium of anaerobic bacteria is actively degrading the organics. Under such conditions, methane is often an end product of the biodegradation of aquifer organics.

- Spatial distribution of ETH and chlorinated intermediates may also occur at sites where natural reductive dechlorination is significant. As PCE is sequentially dechlorinated, mobility of each resulting product increases. In other words, less chlorinated ethylenes sorb less readily to aquifer material; therefore retardation — the movement of a contaminant through an aquifer relative to ground water — decreases as dechlorination progresses. Field observations of compound mobility have been rationalized based on the log $K_{ow}$ (octanol water coefficient) of each compound. It follows that PCE with log $K_{ow} = 2.60$ would be least mobile compared to TCE (log $K_{ow} = 2.38$), DCE (log $K_{ow} = 1.9$) and VC (log $K_{ow} = 0.9$). Observations by Major et al. [8] are depicted in Figure 2. As shown in Figure 2, lesser chlorinated daughter products were found at the leading edge of a contaminant plume.

**Potential Benefits of Natural Bioattenuation.** Natural bioremediation of PCE, TCE, and other chlorinated solvents offers the following potential benefits:

- At sites with no known receptors of contamination, natural bioremediation may preclude the need for active treatment systems, such as pump-and-treat technology. Site management costs and potential liability would thus be reduced.

- Sites that are currently using active treatment may realize decreased times to project completion by exploiting natural bioremediation because in situ biodegradation increases the rate of contaminants removed. This is because in situ biodegradation is much less dependent on contaminant transport than pump-and-treat methods.

# REQUIREMENTS FOR BIOLOGICAL DECHLORINATION

Design of a successful biodegradation system must consider the growth requirements of the microbial community to be employed for reductive dechlorination. This section gives a general description of nutrition and growth requirements in the context of mixed anaerobic cultures because complete dechlorination of PCE has not been achieved by a pure culture of dechlorinating bacteria. It must be noted that we lack a complete understanding of the activities

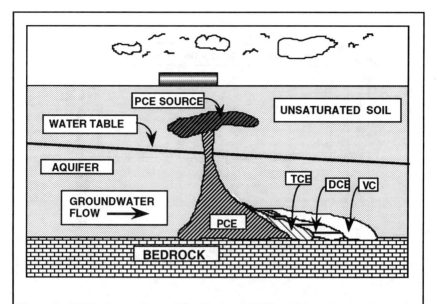

**Figure 2.** Field observations of distribution of PCE and daughter products of reductive dechlorination as observed by Major et al [8].

and relationships of individual species within a PCE-reducing culture; therefore on-going studies are critical to understand the nutritional requirements necessary to achieve successful field-scale dechlorination.

In general, all biological processes require a carbon source, electron acceptor, electron donor, and various macro- and micronutrients. PCE-reducing bacteria utilize inorganic carbon and small quantities of acetate to satisfy carbon requirements [14-16]. During reductive dechlorination, PCE and lesser chlorinated ethylenes serve as electron acceptors. Nitrogen and phosphorous may be added in various forms to satisfy macronutrient requirements. Electron donor and micronutrient requirements are described in the following paragraphs.

**Electron Donor.** Numerous electron donors have been used in laboratory and field applications to supply reducing equivalents. Such electron donors include glucose, acetate, formate, methanol, hydrogen [10], ethanol, lactate [11], benzoate [17], and toluene [18]. These electron donors also provide a carbon source, which is required for growth of bacteria (other than PCE-reducers) in the mixed culture.

Recent work with a mixed anaerobic culture capable of degrading high concentrations of PCE (91 mg/L nominal dose) indicates that hydrogen is the direct electron donor for reduction of chlorinated ethylenes [19]. This finding is supported in that pure PCE-dechlorinating cultures have been identified that use only hydrogen or formate as electron donors [14,15]. These observations likely explain why numerous organic electron donors have been successfully used to reduce PCE and TCE. The organic electron donors listed previously are fermented to various end products and contribute to a hydrogen pool that PCE-dechlorinating organisms can utilize. Figure 3 illustrates the concept of hydrogen serving as the direct electron donor for reductive dechlorination.

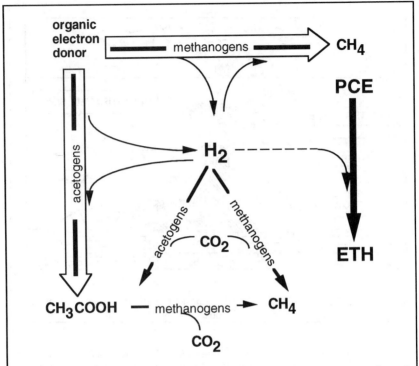

**Figure 3.** The Role of Hydrogen as the Direct Electron Donor for Reduction of PCE (adapted from [13]).

From Figure 3, note that an organic electron donor is fermented to methane or acetate (other fermentation products may be observed, depending on the bacterial culture and electron donor employed). During fermentation, molecular hydrogen ($H_2$) is produced and utilized by PCE-dechlorinating bacteria to reduce PCE to ETH. Therefore, the best electron donors may prove to be those organic compounds that provide a steady supply of hydrogen for the PCE-reducing bacteria. Note that $H_2$ may be used by other bacteria in the mixed culture as well.

   **Electron Donor Efficiency.** Many investigators have employed electron donor doses over 50 times that needed for complete dechlorination of PCE to ETH. As in all natural microbial systems, competition for substrates (i.e., the electron donor) is common; therefore doses of electron donor greater than required for complete PCE reduction should be expected. An efficient reductive dechlorination system would maximize the fraction of electron donor used for reductive dechlorination and therefore minimize electron donor usage for competing reactions. To compare donors and products, it is convenient to express all chemicals on an electron equivalent (eq) basis; details on conversion of chemical concentration to electron equivalents are given elsewhere [13]. In this culture, 984 µeq of methanol was added as the electron donor. Approximately 91 mg/L PCE was added. After incubation, 712 μeq acetate, 206 μeq VC, and 149 μeq ETH were measured; other products were insignificant. Therefore, roughly two-thirds of the added electron donor was used by acetate-forming organisms, whereas the remaining one-third was utilized by PCE-dechlorinating bacteria. Because addition of electron donor represents an operational expense, efficient use of the added electron donor by PCE-reducing bacteria is necessary.

**Micronutrients.** Currently, a complete understanding of the complex nutritional requirements of PCE-dechlorinating bacteria is lacking. In this context, micronutrients refers to essential organic and inorganic compounds that are necessary for microbial growth. Essential organic micronutrients may be various vitamins; inorganic requirements may include various metals. The term "micronutrients" indicates that low concentrations are required relative to electron donor, electron acceptor, carbon, nitrogen, and phosphorous sources.

Possibly, production of micronutrients by bacteria other than PCE dechlorinators may explain the fact that complete dechlorination has not occurred in pure culture. Indeed, the noted successful complete dechlorination of PCE by mixed cultures may be due, albeit indirectly, to the activity of non-dechlorinating bacteria. Numerous examples have been documented as to the syntrophic activities of mixed microbial cultures. It may be that methanogenic, acetogenic, and possibly other genera produce various micronutrients and growth factors that support the activity of PCE-dechlorinating bacteria. Note that Figure 3 illustrates the competition for the electron donor. As mentioned previously, supplying an electron donor could represent a major operational cost in a full-scale in situ reductive dechlorination system. To minimize this cost, a high proportion of the added electron donor should be utilized for PCE or TCE reduction. That is, efficient use of electron donor means that the mass of microbial end products such as methane or acetate should be minimized relative to the mass of ethylene produced from reductive dechlorination. However, efforts should focus on minimizing competition (i.e., production of other end products) — not excluding non-dechlorinating bacteria completely because they may be providing required micronutrients to PCE-dechlorinating bacteria.


## RATES OF DECHLORINATION

Very little information is currently available regarding the rates of biological transformation of PCE. Because evidence exists that some bacteria can grow from the energy yield of reductive dechlorination [13,15,16], rates of transformation are dependent on the quantity of PCE-dechlorinating biomass present. Therefore, reported rates of PCE dechlorination are most useful when reported per unit biomass. Work completed by Tandoi et al. on a high-rate PCE dechlorinating mixed culture indicated a PCE biotransformation rate of 0.76 mg PCE / mg VSS-day (35°C) [20]. This culture transformed PCE, TCE, and DCE with near zero-order kinetics. Significant VC transformation to ETH occurred with first-order kinetics after PCE was gone. The half-life of VC was estimated at 17.3 hours. Note that these results portray the rate of a mixed culture in which PCE reduction was a significant activity. If expressed in terms of the PCE-dechlorinating biomass, the rate would be higher. For comparison, Neumann et al. reported a rate of 20 to 150 nmol Cl$^-$ / min-mg protein (25°C) by a pure culture that dechlorinated PCE to cis 1,2-DCE [16]. This converts to roughly 1 to 9 mg PCE / mg VSS-day if one assumes 0.5 mg protein / mg VSS and 2 mol chloride ion released per mol PCE transformed to DCE. Such high transformation rates are indicative of a metabolic transformation as opposed to a cometabolic process.

Cometabolic transformations are those that involve biotransformation of a non-growth substrate — rates of cometabolic transformation are normally much less than those reported herein.

## REQUIRED FUTURE WORK

Many studies are currently underway ranging from fundamental laboratory investigations to full-scale biostimulation of contaminated aquifers. Whereas biological reductive dechlorination of PCE and TCE is a promising technology, further studies must be completed to consistently achieve successful application of this technology. The following items are suggested:

- For sites in which natural bioattenuation is occurring, required additives must be identified to stimulate this natural process so that complete dechlorination to ethylene may be achieved in reasonable a time. Pending identification of required amendments, uniform delivery of additives may be a major challenge, depending on aquifer conditions.

- The effect of environmental factors, such as temperature and pH, must be determined to define the practical limits of this technology. Successful lab studies have demonstrated reductive dechlorination at temperatures ranging from 10°-35°C and neutral pH. Field observations have confirmed that dechlorination to ETH is possible; however, chlorinated intermediates often persist.

- The effect of mixtures of other chemicals must be assessed. Very little work has been done to identify the effect of various chemicals on the biological reductive dechlorination process. This is significant because many sites are contaminated with a variety of organic chemicals and inorganics, such as heavy metals.

- The efficiency of the electron donor must be improved. Some studies have indicated that efficient use of an added electron donor is possible. However, numerous studies have employed electron donors at doses that would be economically infeasible under full-scale conditions. Inexpensive and efficiently used electron donors must be identified because electron donor addition may likely represent a major operating cost of such a system.

- Environmentally acceptable electron donors must be identified. Organic electron donors such as glucose, acetate, formate, lactate, methanol, ethanol, butyrate, benzoate, and toluene have been employed. Obviously, addition of some of these electron donors to aquifers may be unacceptable to regulatory authorities.

# REFERENCES

1. Parsons, F., P.R. Wood, and J. DeMarco. 1984. "Transformations of tetrachloroethene and trichloroethene in microcosms and groundwater,". *Journal American Water Works Association*, 76 (2): p. 56-59.

2. Bouwer, E.J. and P.L. McCarty. 1983. "Transformations of 1- and 2-carbon halogenated aliphatic organic compounds under methanogenic conditions,". *Appl. Environ. Microbiol.*, 45 (4): p. 1286-1294.

3. Bouwer, E.J. and P.L. McCarty. 1985. "Utilization rates of trace halogenated organic compounds in acetate-grown biofilms,". *Biotechnol. Bioeng.*, XXVII 1564-1571.

4. Bouwer, E.J. and J.P. Wright. 1988. "Transformations of trace halogenated aliphatics in anoxic biofilm columns,". *J. Contam. Hydrol.*, 2 155-169.

5. Gossett, J.M.. 1985. *Anaerobic degradation of $C_1$ and $C_2$ chlorinated hydrocarbons*, Engineering & Services Laboratory, U. S. Air Force Engineering and Services Center: Tyndall AFB, Fla.:

6. Vogel, T.M. and P.L. McCarty. 1985. "Biotransformation of tetrachloroethylene to trichloroethylene, dichloroethylene, vinyl chloride, and carbon dioxide under methanogenic conditions,". *Appl. Environ. Microbiol.*, 49 (5): p. 1080-1083.

7. Parsons, F., G.B. Lage, and R. Rice. 1985. "Biotransformation of chlorinated organic solvents in static microcosms,". *Environ. Toxicol. Chem.*, 4 739-742.

8. Major, D.W., W.W. Hodgins, and B.J. Butler. 1991. "Field and laboratory evidence of in situ biotransformation of tetrachloroethylene to ethene and ethane at a chemical transfer facility in North Toronto," in *On-site bioreclamation.*. San Diego, CA: Butterworth-Heinemann Stoneham.

9. McCarty, P.L. and J.T. Wilson. 1992. "Natural anaerobic treatment of a TCE plume, St. Joseph, Michigan, NPL site,". *In Bioremediation of hazardous wastes*, U.S. EPA (EPA/600/R-92/126): p.

10. Freedman, D.L. and J.M. Gossett. 1989. "Biological reductive dechlorination of tetrachloroethylene and trichloroethylene to ethylene under methanogenic conditions,". *Appl Environ Microbiol*, 55 (9): p. 2144-2151.

11. De Bruin, W.P., *et al.* 1992. "Complete biological reductive transformation of tetrachloroethene to ethane,". *Appl. Environ. Microbiol.*, 58 (6): p. 1996-2000.

12. Fathepure, B.Z., J.P. Nengu, and S.A. Boyd. 1987. "Anaerobic bacteria that dechlorinate perchloroethene,". *Appl. Environ. Microbiol.*, 53 (11): p. 2671-2674.

13. DiStefano, T.D., J.M. Gossett, and S.H. Zinder. 1991. "Reductive dechlorination of high concentrations of tetrachloroethene to ethene by an

anaerobic enrichment culture in the absence of methanogenesis,". *Appl. Environ. Microbiol.*, 57 (8): p. 2287-2292.

14. Zinder, S.H., *et al.* Characterization of an anaerobic enrichment culture which rapidly converts tetrachloroethylene to ethene. in *American Society for Microbiology Conference on Anaerobic Dehalogenation and Its Implications*. 1992. Athens, GA., Aug 30-Sept. 4, 1992.:

15. Holliger, C., *et al.* 1993. "A highly purified enrichment culture couples the reductive dechlorination of tetrachloroethene to growth,". *Appl. Environ. Microbiol.*, 59 (9): p. 2991-2997.

16. Neumann, A., H. Scholz-Muramatzu, and G. Diekert. 1994. "Tetrachloroethene metabolism of *Dehalospirillum multivorans*,". *Arch. Micobiol.*, 162 295-301.

17. Scholtz-Muramatsu, H., *et al.* 1990. "Benzoate can serve as a source of reducing equivalents for PCE reduction,". *FEMS Microbiol Lett*, (66): p. 81.

18. Sewell, G.W. and S.A. Gibson. 1991. "Stimulation of the reductive dechlorination of tetrachloroethene in anaerobic aquifer microcosms by the addition of toluene,". *Environ Sci Technol*, 25 (5): p. 982-984.

19. DiStefano, T.D., J.M. Gossett, and S.H. Zinder. 1992. "Hydrogen as an electron donor for dechlorination of tetrachloroethene by an anerobic mixed culture,". *Appl. Environ. Microbiol.*, 58 (11): p. 3622-3629.

20. Tandoi, V., *et al.* 1994. "Reductive dehalogenation of chlorinated ethenes and halogenated ethanes by a high-rate anaerobic enrichment culture,". *Environ. Sci. Technol.*, 28 (5): p. 973-979.

# BIOTREATMENT OF CHROMIUM (VI ) EFFLUENTS

Teresa Tavares
  Universidade do Minho
  Engenharia Biológica
  P-4709 Braga Codex
  Portugal

Cecília Martins
  Universidade Lusíada
  Edifício da Lapa
  Largo Tinoco de Sousa
  P- 4760 Vila Nova de Famalicão
  Portugal

Paulo Neto
  Universidade do Minho
  Engenharia Biológica
  P-4709 Braga Codex
  Portugal

## ABSTRACT

The presence of heavy metals in industrial wastewaters is still a serious problem for some of our local small and medium size industries. Particularly electroplating and tanneries produce highly concentrated chromium effluents, which are treated by traditional physico-chemical processes. Those are able to reduce the total chromium concentration from some hundreds of mg.l$^{-1}$ to very low concentrations, but the allowable final value of 0.1 mg.l$^{-1}$ is hardly obtained as the referred processes become too costly for those small and medium size industries.

The aim of these studies is the definition of an efficient system, economically attractive and friendly to the environment, based on the ability of some microorganisms to concentrate heavy metals. This system would be used as a final treatment step to remove low concentrations of hexavalent chromium.

Three different bacteria were used in batch systems to evaluate their resistance to Cr (VI) and their ability to reduce it to the trivalent form. The results were compared with those obtained with microorganisms isolated from sludge of treatment plants receiving wastewater loaded with chromium.

One of those bacteria was supported on granular activated carbon and the biofilm was optimized in terms of adhesion and removal efficiency. The chromium adsorption capacity of the support was also studied as albeit it is known that adsorption is not used for heavy metals removal, granular activated carbon is an excellent immobilization support for the biofilm and certainly has some responsibility on the chromium fixation process.

## INTRODUCTION

The effect of heavy metals in human health and in the environment is still a motive of concern for researchers. Chromium, in particular, appears in the hexavalent and in the trivalent form in industrial and municipal wastewaters and due to its toxicity the threshold limit is quite demanding but not always fulfilled. The potential environmental hazards are usually avoided by treatment of the industrial effluents by the traditional physico-chemical processes: reduction-precipitation, ion exchange, solvent extraction, reverse osmosis [1]. Usually the treated effluent has a final metal concentration less than 10 $mg.l^{-1}$ [2], but although very efficient such techniques require large capital reagent and/or energy costs.

The potential of live or dead microorganisms to accumulate certain ions is already recognized and biosorption, defined as the capture of metal ions by solid materials of natural origin [3] is an alternative technology to the ones referred above. There are now several biosorption systems in operation treating low metal contamination [4]. The microorganisms to be used may be metabolically inactive and still able to retain metallic ions [5].

Considering the toxic effect that heavy metals have on microorganisms, several studies were made in order to define the conditions in which bacteria, algae and fungi can stand those elements [6,7,8]. It is known that the ratio of the weight of metal in the biosystem to that in the surrounding aqueous phase for mercury, cadmium and zinc at equilibrium ranged from 4000 to 10000 [9].

Besides chromium and other metallic ions, there are organic and inorganic compounds in the waste streams of the electroplating unites and tanneries, so it would be desirable to co-extract several pollutants at the same time. The utilization of granular activated carbon as the support of the biofilm seems to be advantageous, as it can retain the other substances while the biofilm remove the heavy metal [10,11].

Among the microorganisms able to retain the referred metallic ions, bacteria are specially interesting as some of them excrete enormous quantities of polysaccharides allowing a good adhesion to the support, implementing the retention capacity of cells and protecting them from the xenobiotic effect of the

224

heavy metal ions. The bacteria used in these studies were pure cultures of *Pseudomonas fluorescens*, *Pseudomonas putida* and *Escherichia coli*. One of the purposes of this work is the isolation of microorganisms present in the sludge of treatment plants receiving liquid effluents loaded with chromium, that somehow developed resistance to unfriendly environment [12]. Those microorganisms were used in biosorption studies as a comparative basis.

## MATERIALS AND METHODS

*Batch cultures*

The bacteria used in these studies were *Pseudomonas putida*, *Pseudomonas fluorescens* and *Escherichia coli* from the Spanish Type Culture Collection. They were seeded in 500 ml Erlenmeyer flasks with two different culture media. The first one included a minimum salts solution (125 ml) - $NH_4Cl$ (10g), $NH_4NO_3$ (2g), $Na_2SO_4$ (4g), $K_2HPO_4$ (6g), $KH_2PO_4$ (2g), $MgSO_4.7H_2O$ (0.2g), $dH_2O$ (500 ml); glucose 20% (5 ml); $dH_2O$ (370 ml). The same bacteria were grown in rich medium with yeast extract (0.5% w/v), glucose (2% w/v) and peptone (1% w/v). $K_2Cr_2O_7$, was added to the media before the seeding and the resulting Cr (VI) concentrations ranged from 4 to 25 mg.l$^{-1}$. The pH was kept 7.2, with NaOH or $H_2SO_4$. The cultures were incubated for periods between 6 hours and 24 hours in a rotary shaker, 200 rpm, at 27$^0$C for the *Pseudomonas* and at 37$^0$C for the *Escherichia*.

A microorganisms consortium, resistant to Cr (VI), was isolated from sludge of treatment plants receiving tannery wastewaters. The sludge was diluted in three 500 ml Erlenmeyer flasks with 100, 200 and 300 mg.l$^{-1}$ of Cr (VI). The culture medium was composed of yeast extract, glucose and peptone in the same composition was before and after several days of incubation the viability of the resistant microorganisms was confirmed by seeding in Petri plates with nutrient agar.

The Cr (VI) concentration in the supernatant of the centrifuged samples was determined by a colorimetric method using diphenylcarbazide and a spectrophotometer (JASCO 7850) at 540 nm. The total Cr concentration was determined by Atomic Absorption Spectroscopy (VARIAN SPECTRA AA-250 PLUS), after exposure of the samples to ultra-violet radiation and centrifugation.

*Continuous systems*

The granular activated carbon used as support for the biofilm was charcoal with an average particle diameter of 1.5 mm. It has a density of 2.34 g.cm$^{-3}$, a Langmuir specific area of 1270 m$^2$.g$^{-1}$ and an average pore diameter of 20 angstrom.

The columns have a height of 30 cm and a diameter of 0.9 cm. They are partially filled with the activated carbon. *Pseudomonas fluorescens* cells, in the exponential growth phase and in the stationary growth phase, as well as nutrient broth were circulated through the carbon slowly enough to allow biofilm attachment but expanding the bed enough to avoid the gluing of the granules. The adhesion of the biofilm was characterized by stopping the cells suspension feed, washing the system gently with distilled water and quantifying the amount of cells being detached.

This optimized biofilm, supported on the GAC, was then used in biosorption studies allowing the contact of this system with aqueous solutions of Cr (VI) of concentrations between 4 and 20 mg.l$^{-1}$, at 27$^0$C. Determinations of the concentration of the hexavalent and of the total chromium were systematically made, by the methods referred above. Some of the runs were repeated with the same support after this has been heated for 12 hours at 200$^0$C, in order to improve biofilm and metal fixation. A new and metabolically active biofilm was, then, developed over the old one.

Samples of GAC with biofilm and metal ions were observed by Scanning Electron Microscopy (LEICA S 320) and elementary analysis was made by Energy Dispersive Spectrometry. Samples to be observed were prepared by fixation of biomass with glutaraldehyde (5%) and chilled cacodylate buffer (0.1M), dehydration in graded ethanol series and drying in airtight container for 12 hours.

Similar studies were made with granular activated carbon without the biofilm so the participation on the process of this adsorbent may be quantified. An adsorption isotherm was determined in batch studies where 5 g of GAC were put in contact with 200 ml of Cr (VI) solutions with concentrations ranging from 40 to 1000 mg.l$^{-1}$. These solutions were placed in Erlenmeyer flasks , in a rotary shaker at 27$^0$C, and the concentration of the metallic ion in the liquid phase was followed for several days.

## RESULTS AND DISCUSSION

*Batch studies*

The determination of the limit threshold of hexavalent chromium for viability of bacteria aimed the selection of the best biosystem, between those who are known to be good exopolysaccharides producers, for the removal purpose. As may be seen in Table I most of the suspensions were able to remove or reduce Cr (VI), exception for the microbial consortium, that although could stand high concentrations of the metallic ion (it has been exposed to concentrations of 300 mg.l$^{-1}$, and survived) was not able to accumulate or reduce any of the hexavalent chromium present in the liquid phase. Among the bacteria that were able to live with the xenobiotic metal, *Pseudomonas fluorescens* showed the best removal efficiency.

TABLE I - Behavior of different microorganisms in presence of Cr (VI).

| Microorganism | Culture Medium | Initial Cr $^{6+}$ Conc. (mg.l$^{-1}$) | Growth | Removal % |
|---|---|---|---|---|
| Pseudomonas fluorescens | Minimum | 10 | + | 77 |
| | Rich | 10 | + | 62 |
| Pseudomonas putida | Minimum | 4 | - | 0 |
| | Rich | 4 | + | 22 |
| Escherichia coli | Minimum | 25 | + | 36 |
| | Rich | 10 | + | 12 |
| Sludge Microorganisms Consortium | Minimum | 25 | - | 0 |
| | Rich | 25 | + | 0 |

Although the minimum culture medium would be advisable, the continuous biosorption studies were carried out with the peptone/ yeast extract/glucose medium as it is easier to grow the bacteria in this last one.

*Adsorption on GAC*

As a powerful adsorbent the activated carbon participates in the metal ions fixation even if the main retention process would be promoted by the biofilm that will cover it, in the biosorption systems. As may be concluded from Table II the removal efficiency is not really high and although increasing the retention time may seem advisable in order to implement the contact between the solid and the liquid phase, such small rate flow favors the development of preferential paths and does not take full advantage of the whole GAC surface. The pre-washing with NaOH aimed the maintenance of ionic strength of the system increasing the H$^+$ concentration on the surface and raising the efficiency of the entrapment of the anionic metallic species. In fact that did not happened again due to the fact that the low flow rates would not allow the complete utilization of the whole adsorbent surface. The isotherm for this specific activated carbon, determined at 27$^0$C, indicates a maximum Cr (VI) uptake of 400 mg per gram of GAC, corresponding to a concentration in the aqueous phase of 1000 mg.l$^{-1}$.

TABLE II - Adsorption of Cr (VI) on GAC in packed bed columns

| Retention time (min) | Pre-Washing NaOH, 1M | Initial Cr (VI) Conc. (mg.l$^{-1}$) | Efficiency (%) | Metal Uptake (mg Cr/g GAC) |
|---|---|---|---|---|
| 0.460 | + | 50 | 24.5 | 15.6 |
| 0.460 | - | 50 | 32.0 | 56.0 |
| 0.954 | - | 50 | 28.0 | 27.4 |

*Biosorption experiments*

The adhesion of the *Pseudomonas fluorescens* biofilm was studied in terms of the percentage of initial number of cells that are retained by the GAC bed as a function of time of residence of the bacterial suspension in the column. This determination indicated an optimal residence time of 20.6 sec, which corresponds to a flow rate of 31 cm$^3$.min$^{-1}$. It was also observed that a better adhesion is obtained when the suspension is fed to the GAC column during the exponential growth phase of the microorganism. That was concluded by washing out two biofilms, one produced during the exponential growth phase and another produced in the stationary growth phase of the *Pseudomonas* and verifying that the first one resisted around 6 hours to detachment, while the second one took just 4 hours to suffer from the same effect.

Chromium solutions of concentrations ranging from 4 to 20 mg.l-1 were fed to the system biofilm / GAC and the concentration of the total and of the hexavalent form was measured in the exit stream. The residence time seems to have a double effect on the total Cr removal efficiency. On one hand a longer contact between the metallic solution and the biosorption system will implement the removal process, Table III, on the other hand the toxicity of chromium is more evident when low flow rates are used by the slow detachment of the biofilm.

Figure 1 presents the results of the experiment with highest total Cr removal, i.e. the initial Cr concentration of 4 mg.l$^{-1}$ at a flow rate of 7 cm$^3$.min$^{-1}$. For this run it was observed that 50 min would be enough to reach a constant total Cr concentration of 0.5 mg.l$^{-1}$ in the exit stream. From that time onwards some Cr$^{6+}$ appeared, reaching a constant concentration of 0.13 mg.l$^{-1}$ Similar patterns were obtained during other experiments, with different flows and initial metallic concentrations, with an initial period smaller with increasing flow rates. This indicates that, although the major part of the initial metal ions is retained by the biosystem, there is always a partial reduction of Cr$^{6+}$ to Cr$^{3+}$,

TABLE III - Total chromium removal efficiency of the biosorption system for different flow rates and concentrations of the metallic solution.

| Flow rate ($cm^3.min^{-1}$) | 7 | 15 | 25 | 31 |
|---|---|---|---|---|
| Concentration (4 mg.$l^{-1}$) | 87% | 74% | 61% | 58% |
| Concentration (10 mg.$l^{-1}$) | - | 68% | - | - |
| Concentration (20 mg.$l^{-1}$) | - | 60% | - | - |

which diminishes as the fixation process approaches a stationary level. The biosorption systems is still far from the saturation, as may be observed in Table III for the other concentration influents and respective removal efficiency, but a steady state is reached that will last as long as the biofilm survives. The reduction of the hexavalent to the trivalent form should be avoided as this last one may precipitate and may raise the pressure drop in the system. The heat treatment of the GAC covered with biofilm and loaded with chromium for

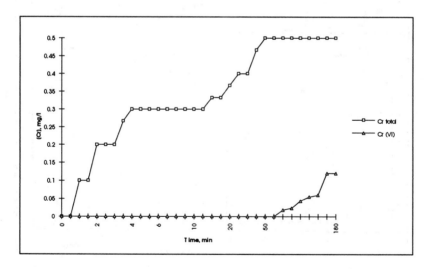

Figure 1 - Total Cr and hexavalent Cr in the biosorption system effluent .

12 hours and 200⁰C, for metal fixation allowed the growth of a new biofilm which was used for another Cr solution treatment. The efficiency of the process was as high as in the first run , i.e. 87%, but the steady state was reached much faster, in 20 min in contrast with the 50 min of the first run.

Some samples of GAC with biofilm and chromium were observed in SEM. The polysaccharide matrix covers completely the whole surface not allowing the viewing of the bacteria which are trapped between the carbon surface the referred matrix. Sample collected in the earlier phase of the biofilm formation showed the typical configuration of the *Pseudomonas fluorescens* attached preferentially to the kinks and edges of the surface. The elementary analysis by EDS of GAC loaded with Cr and GAC/Biofilm loaded with Cr allowed the conclusion that the ratio between the signals obtained at 30 kV and at 12 kV increases with the presence of the biofilm, Figure 2. The depth of the analysis is, for this material 8.5 μm at 30 kV and 1.7 μm at 12 kV.

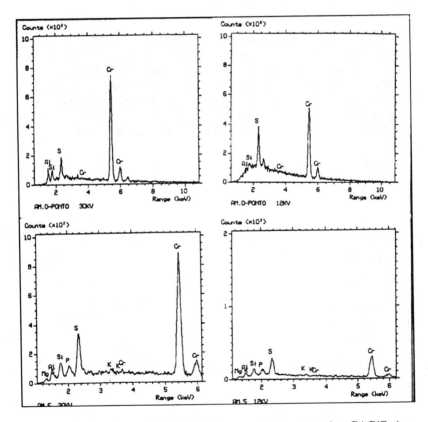

Figure 2 - Elementary analysis by EDS for two different samples: GAC/Cr (top two) and GAC/Biofilm/Cr (bottom two).

Although no quantitative conclusion may be inferred from this analysis, it may be observed that the fixation of the Cr ions occurs at a deeper level than the first micrometers (two top graphs of Fig.2), and that profile of concentrations is accentuated in the presence of a biofilm (two bottom graphs of Fig.2).

## CONCLUSIONS

This work reports just the preliminary results of a research project that aims the definition of a biosorption system able to reduce to acceptable values the concentration of hexavalent chromium in industrial and municipal wastewaters. These results indicate that: i) there are microorganisms able to develop some resistance to the xenobiotic effect of the metallic ion and this resistance may be induced by the right choice of the culture medium; they are also able to reduce $Cr^{6+}$ to $Cr^{3+}$ ; ii) as supported biological systems have some advantages over suspension systems, a proper choice of the immobilization matrix will enhance the efficiency of the treatment process and the participation of GAC, the support in question, in the metallic ions adsorption was determined; iii) the biofilm enhances the removal process and the optimization of the biofilm build-up as well as of the biosorption parameters are still under consideration.

## ACKNOWLEDGMENTS

This work was financially supported by Junta Nacional de Investigação Científica, Project: PEAM/TAI/261/93.

## REFERENCES

1. Mahajan, S.P., 1989, *Pollution Control in Process Industries*, Tata McGraw-Hill Publishing Company Limited, pp. 82-89.

2. Eckenfelder, W.W., 1989, *Industrial Water Pollution Control*, NY McGraw-Hill Book Company, pp. 98-110.

3. Volesky, B., 1986, "Biosorbent Materials", *Biotechnology and Bioengineering Symp.*, 16, 121-126.

4. Mattuschka, B., Straube, G., 1993, "Biosorption of Metals by Waste Biomass", *J. Chem. Tech. Biotechnol.*, 58, 57-63.

5. Tsezos, M., 1980, *Biosorption of Uranium and Thorium*, Ph.D. Thesis, McGill University, Montreal.

6. Gadd, G.M., 1990, *Microbiology of Extreme Environments*, C. Edwards, ed., Milton Keynes: Open University Press, pp. 178-210.

7. Gadd, G.M., 1992, *Microbial Control of Pollution*, C.J. Fry, G.M. Gadd, R.A. Herbert, C.W. Jones, I.A. Watson-Craik, eds., Cambridge University Press, pp. 59-88.

8. Gadd, G.M., 1993, "Interactions of Fungi with Toxic Metals", *New Phytol.*, 124, 25-60.

9. Neufeld, R.D., Hermann, F.G., 1975, "Heavy Metal Removal by Acclimated Activated Sludge", *JWPCF*, 47, 310.

10. Scott, J.A., Karanjkar, A.M., 1994, "Decontamination Biosorption to Concentrate Metals on Granular Activated Carbon - Influence of pH and Temperature and use of this Material to Adsorb Chloroform", *Progress in Biotechnology*, L. Alberghina, L. Frontali, P. Sensi, eds., Elsevier Science, 1231- 1234.

11. Scott, J.A., Karanjkar, A.M., Rowe, D.L., 1994, "Exploitation of Biofilm Covered Granular Activated Carbon - Influence of pH and Temperature", *Resources Conservation and Environmental Technologies in Metallurgical Industries*, CIM, Montreal.

12. Baldi, F., Vaughan, A.M., Olson, G., 1990, "Chromium VI-Resistant Yeast Isolated from Sewage Treatment Plant Receiving Tannery Wastes", *Applied and Environmental Microbiology*, 4, 913-918.

# DESORPTION AND BIODEGRADATION BEHAVIOR OF NAPHTHALENE SORBED TO SOIL COLLOIDS

## ARUP BANDYOPADHYAY
Environmental Engineer,
State of Tennessee
Department of Environment & Conservation
Division of Water Pollution Control
401 Church Street
L & C Annex, 6th Floor
Nashville, TN - 37243-1534

## KEVIN G. ROBINSON
Assistant Professor,
Department of Civil & Environmental Engineering
219A Perkins Hall
The University of Tennessee
Knoxville, TN - 37996

## INTRODUCTION

Groundwater and soil have been widely contaminated by a variety of man-made chemicals. In recent years, special attention has been given not only to minimize or prevent environmental pollution, but to restore contaminated environments.

Organic pollutants, especially Polycyclic Aromatic Hydrocarbons (PAH) pose special problems for soil-groundwater remediation. PAH's are of growing concern from a public health perspective. They are produced both by natural and man made processes. Contributing sources include petroleum, creosote and coal-tar products, atmospheric deposition of combustion products, natural oil seepage, stormwater runoff, and petroleum spills [7].

Many PAH's are hydrophobic in nature; once sorbed to the soil and sediment, they are extremely difficult to remove and act as a source of low level, long term contamination. Their fate and transport are tremendously influenced by their sorptive behavior with soils and sediments.

Soil colloids can influence the mobility of organic contaminants, including PAH's, via sorptive interactions thereby facilitating PAH transport in groundwater systems [1, 4]. The binding of hydrophobic contaminants is controlled by the hydrophobicity of the pollutant and the organic matter content of the colloids. In addition, colloid-PAH interactions can impact bioavailability. Contaminants present in the aqueous phase (unbounded and uncomplexed) may be bioavailable [3], whereas contaminants bound to the complex soil matrix are not readily available for degradation [5]. Desorption of bound contaminants into the aqueous phase can enhance bioavailability.

As compared to sorption behavior, desorption of PAH's and other similar organic compounds has been less thoroughly investigated. Desorption data are often extrapolated from adsorption data, assuming the sorption process as reversible. Both sorption and desorption rates are reported to follow a two phase pattern: an initial fast rate, followed by a slow rate.

This paper details research performed to determine the influence of soil colloids on the desorption and bioavailability of naphthalene (target PAH). The goal of this research was to determine if sorption or mineralization controls naphthalene biodegradation.

## MATERIALS AND METHODS

Batch adsorption experiments were conducted with naphthalene and soil colloids. Sorption was measured as a function of time and apparent equilibrium was determined (data not presented). Desorption experiments were performed after apparent equilibrium was reached and was measured as a function of time. Finally, mineralization experiments were conducted using a naphthalene degrading strain of bacteria (pseudomonas fluorescens 5R). Mineralization was measured by $^{14}CO_2$ collection as a function of time in a system initially at apparent equilibrium with soil colloids.

Potting soil was selected as a source of colloids due to its high soil organic matter content. A stock colloid solution was generated from potting soil mixed in distilled, deionized water (DD Water). Larger solids were allowed to settle and colloids were collected from the supernatant. The final concentration of stock colloids was 350 mg/L ($\pm$ 5 mg/L). The particle size distribution of the stock colloids was measured using a Coulter Multisizer II Counter. The mean size distribution was measured as 4.265 $\mu$m (standard deviation 0.093).

A stock solution of naphthalene (Mallinckrodt Chemical Company) was prepared (average concentration 32 mg/L). Radiolabeled naphthalene crystals (Sigma Chemical Company; specific activity = 10.1 mCi/mmole) were dissolved in 0.5 ml of 0.1 M methanol and the final working radiolabeled naphthalene solution had a concentration of 0.115 mg/L. The methanol concentration in the naphthalene solution was 0.8 mg/L. A Beckman LS 5000 TD Liquid Scintillation Counter was used to measure radioactivity.

A sodium azide solution (Mallinckrodt, Practical grade) was used in desorption experiments to inhibit biodegradation. The final azide concentration in each microcosm was 10 mg/L.

Hexane and isopropanol (Mallinckrodt, as AR grade), (4:1 v/v), was used to extract naphthalene from the colloids. Phosphate buffer ($10^{-4}$ M) was used to maintain a constant pH in all samples. All chemicals were used as supplied, without purification.


## DESORPTION PROCEDURE

In the batch experiments, teflon centrifuge tubes with teflon-silicon-lined screw caps were used. The stock colloid solution was well mixed by hand before addition to each centrifuge tube (9.2 ml). Stock sodium azide (0.2 ml) was then added to each tube. In addition, 0.1 ml of stock, autoclaved buffer was added to the system. Tubes were then mixed for approximately 15-20 minutes to ensure complete inhibition of biological activity by the azide solution. Finally, 0.5 ml stock, labeled naphthalene solution was transferred to each tube. This brought the total volume in each tube to 10 ml. Each tube was then capped, wrapped with parafilm, inverted to minimize volatilization losses through the cap, and placed on a Psychrotherm Incubator Shaker at a constant temperature (25 °C) and speed (100 strokes/min).

Samples were continuously shaken until analyzed. All tubes containing samples and blanks were removed and centrifuged. Aqueous phase radioactivity was directly measured while the sorbed phase was extracted with an organic solvent mixture before analysis.

A multiple extraction procedure was used to measure naphthalene remaining on the colloids. Radioactivity extracted from the colloids was determined in duplicate, and reported as a percent of initial radioactivity present on the colloids. Each tube served as a single data point for desorption evaluation. Similarly, the mass of naphthalene desorbed at a specific time point was also calculated. From replicate samples, the total amount of radioactivity present on the colloids at each sample point was evaluated. Desorption experiments were continued until the concentration of naphthalene in solution fell below the detection limit (40 dpm). From these data, desorption rates were determined.

In later mineralization experiments, a headspace gas was maintained inside the centrifuge tubes such that glass center wells containing NaOH could be attached to the cap of each tube to trap $^{14}CO_2$. Tubes containing the attached glass traps were rotated horizontally. This desorption procedure was identical to that used in mineralization.


## MINERALIZATION PROCEDURE

A strain of pseudomonas fluorescens 5R (pKA3) bacteria was used to degrade naphthalene. This organism was recovered from a PAH-degrading inoculum, developed from a manufactured gas plant soil slurry reactor [6]. A pure culture of this strain was grown on liquid and/or agar yeast extract, peptone, and glucose media (YEPG media). Microorganisms were grown on the agar plates, refrigerated at 4 °C and restreaked every seven days. A full-strength YEPG media was directly inoculated from plate cultures, and placed on a shaker table. Flasks were shaken overnight at 150 rpm at 28 °C in an Environ-Shaker incubator (Lab-Line Instruments, Inc.).

Once in the log phase of growth, 25 ml of the microorganism cell suspension was transferred to a 50 ml polypropylene centrifuge tube. The cells were centrifuged at 10,000 rpm for 15 minutes at 25 °C. The supernatant was decanted, the microorganisms collected as a pellet, then resuspended in 25 ml of buffered media. This wash process was repeated twice before resuspending the culture in 5 ml of buffered media. This served as the stock microorganism culture used to inoculate each tube.

In this research, biodegradation experiments were based on mineralization only. Radioactive $^{14}CO_2$ evolved was trapped in a NaOH trap and measured using a liquid scintillation counter. Mineralization experiments were conducted to determine the rate of mineralization of colloid-associated naphthalene.

Colloids were brought to apparent equilibrium using radiolabeled naphthalene, as described earlier. Aqueous phase radioactivity was measured after centrifugation. In order to measure the exact amount of naphthalene present on the colloids, one set of tubes with colloids at apparent equilibrium were multiple extracted with organic solvent. In a set of replicate tubes, the stock microorganism suspension was added along with buffered media (total volume = 5.0 ml). At the top of each tube a small glass vial containing 0.2 ml of 0.4 N NaOH solution was attached to the cap to collect $CO_2$ evolved during mineralization. Each trap with its constituents was transferred directly into a scintillation vial over time containing 10 ml of cocktail. This method resulted in minimum sample quenching. Blanks containing naphthalene, but no microorganisms, were analyzed with each sample to determine non-mineralized naphthalene in the trap.

## RESULTS AND DISCUSSIONS

### DESORPTION STUDIES AT APPARENT EQUILIBRIUM

Desorption data presented represents only the naphthalene mass recovered in the extract. That is, each extract (amount desorbed) was directly measured and desorption was calculated based on reduction of soil-bound naphthalene concentration. Further, apart from the mass desorbed, mass remaining bound to the colloids was also directly measured.

Results of the desorption experiment are presented in figure 1. It was observed that the entire desorption process could be divided into two phases :

1)    Initial desorption rate was extremely high and desorption occurred very fast, starting from time zero and lasting approximately up to 24 hours.

2)    A second, slow desorption rate, starting from around 24 hours and lasting up to 150 hours was observed.

It was also noted that approximately 45% of the total naphthalene desorption occurred within the initial phase. During the slow phase, an additional 41% could be removed via desorption.

The initial fast desorption rate is believed due to the concentration gradient which

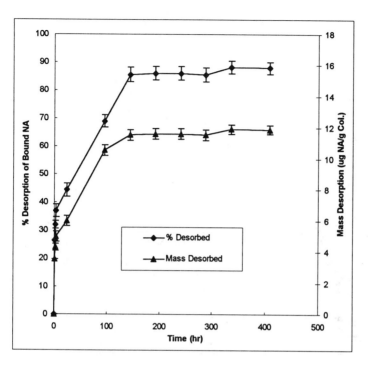

Figure 1. Desorption of Naphthalene using Distilled Deionized Water.

exists between the aqueous and soil colloid phases. The second, slower desorption rate is believed due to desorption occurring from internal micropores of the colloids.

In the later stages of the desorption experiment, it was observed that desorption equilibrium was attained; that is, there was little desorption after a period of 150 hours.

## BIOAVAILABILITY OF SORBED NAPHTHALENE

In mineralization experiments, radiolabeled naphthalene and colloids were first brought to apparent equilibrium (120 hours). The system was then centrifuged and the aqueous phase decanted and analyzed. Colloidal-bound naphthalene was measured after extraction in a series of replicate tubes.

A fresh solution of microorganisms in buffered media was added to each tube resulting in a final microorganism concentration of approximately 300 mg/L ($\pm$ 5%). The top of each tube contained a small glass vial, with 0.2 ml of 0.4 N NaOH solution, to trap the $^{14}CO_2$ evolved during mineralization. Samples were analyzed over time in duplicate. Net mineralization of bound naphthalene was plotted and presented in figure 2, after netting out all other contributing factors.

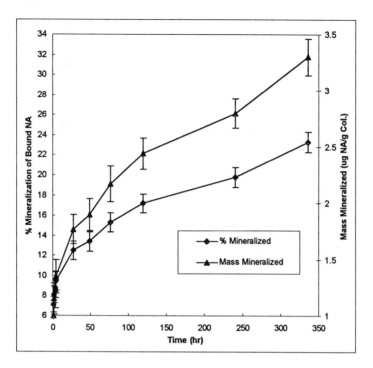

Figure 2. Mineralization of Naphthalene Sorbed to Colloids.

Mineralization was measured directly as the mass of radioactive carbon ($^{14}C$) collected as $CO_2$ in the center wells. Naphthalene mass on colloids at apparent equilibrium was determined by sacrificing replicate samples over time. Mineralized mass and percentage of naphthalene was calculated based on initial naphthalene loading on the colloids at apparent equilibrium.

It was observed that colloid-bound naphthalene mineralization was only approximately 23% (a corresponding mass of 3.3 mg naphthalene/g colloids) after a period of 336 hours. In the batch system used in this research, substrate concentration was very low in the system corresponding to the colloid-bound naphthalene mineralization. In this research, only mineralized mass of total $^{14}C$ was measured. From figure 2, it may be observed that mineralization had two distinct rates, one initial fast rate followed by a slow rate. The fast rate was observed for about 27 hours and approximately 12.5% mineralization was recorded. The slower rate was observed for the next 309 hours and it resulted in an additional 10.7% mineralization.

The reason for slow mineralization of bound naphthalene may be that, though naphthalene became constantly available for degradation after desorption, microbial metabolism was limited. Naphthalene degradation intermediates or biodegradation by-products might have formed [2], but they were not separately measured.

Desorption and mineralization rates were calculated using linear regression analysis. It may be observed that the initial desorption rate (2.5 $\mu$g NA/g colloids/hour) was approximately five fold higher than the corresponding initial mineralization rate (0.5 $\mu$g NA/g colloids/hour). The slower phase desorption rate (0.061 $\mu$g NA/g colloids/hour) was much higher than the corresponding mineralization rate (0.0046 $\mu$g of NA/g colloids/hour). However, for the period when no further desorption was observed, mineralization continued, though at a much slower rate.

## SUMMARY AND CONCLUSIONS

Desorption was measured as the change in colloid-bound naphthalene concentration over a period of 120 hours. Desorption rates followed a two fold pattern, a very fast initial desorption phase, lasting only a few hours, followed by a slower desorption phase, lasting a few days. The desorption rates were measured as 2.5 and 0.061 $\mu$g NA/g colloids/hr using DD water. These desorption patterns were in close agreement with reported results.

Mineralization of sorbed naphthalene followed an initial fast phase and a later slower phase. These two rates were measured as 0.5 and 0.0046 $\mu$g NA/g colloids/hr respectively.

Mineralization rates of colloid-bound naphthalene were then compared to the corresponding desorption rates. The initial mineralization rate was approximately 5 times lower than the initial desorption rate, under the identical operating conditions. However, it was observed that mineralization continued at a slower rate for a longer duration than that measured for desorption alone. Microbial metabolism might have been limiting under these experimental conditions. For this reason, microorganisms failed to degrade all of the aqueous labeled 14C, which increased in concentration over time.

## LIST OF REFERENCES

[1]     Backhus, D. A., and P. M. Gschewend. 1990. "Fluorescent Polycyclic Aromatic Hydrocarbons as Probes for Studying the Impact of Colloids on Pollutant Transport in Groundwater," *Envir. Sc. & Tech.*, Vol. 24, No. 8, pp 1214-1223.

[2]     Cerniglia, C. E., and M. A. Heitkamp. 1989. "Microbial Degradation of Polycyclic Aromatic Hydrocarbons (PAH) in the Aquatic Environment," *Metabolism of Polycyclic Aromatic Hydrocarbons in the Aquatic Environment*, Editor Usha Varanasi, CRC Press, Inc., Boca Raton, FL, pp 41-68.

[3]     Heitkamp, M. A., and C. E. Cerniglia. 1989. "Polycyclic Aromatic Hydrocarbon Degradation by a Mycobacterium Sp. in Microcosms Containing Sediments and Water from a Pristine Ecosystem," *Appl. & Envir. Microbl.*, Vol. 55, No. 8, pp 1968-1973.

[4]     McCarthy, J. F., B. D. Jimenez, and T. Barbee. 1985. "Effect of Dissolved Humic Material on Accumulation of Polycyclic Aromatic Hydrocarbons: Structure-Activity Relationships," *Aquatic Tox.*, Vol. 7, pp 15-24.

[5]     Pignatello, J. J. 1989. "Sorption Dynamics of Organic Compounds in Soils and Sediments," B. L. Sawhney and K. Brown (eds.), Reactions and Movements of Organic Chemicals in Soils, SSSA Special Publication No. 22, *Soil Sc. Soc. of America, Madison*, WI, pp 45-80.

[6]     Sanseverino, J., C. Werner, J. Fleming, B. M. Applegate, J. M. H. King, G. S. Sayler, and J. Blackburn. 1992. "Molecular Analysis of Manufactured Gas Plant Soils for Naphthalene Mineralization," Unpublished Data.

[7]     Sims, R. C., J. Keck, M. Coover, K. Park, and B. Symons. 1989. "Evidence for Cooxidation of Polynuclear Aromatic Hydrocarbons in Soil," *Water Res.*, Vol. 23, No. 12, pp 1467-1476.

# A CERAMIC MEMBRANE BIOREACTOR FOR AEROBIC TREATMENT OF VERY HIGH COD AND BOD DAIRY EFFLUENTS

**Karen L. Smith**
Wessex Water plc
Kingston Seymour
Avon BS21 6UY, UK

**John Ashley Scott**
School of Chemical Engineering
University of Bath, Bath
Avon BA2 7AY, UK

**John A. Howell**
School of Chemical Engineering
University of Bath, Bath
Avon BA2 7AY, UK

## INTRODUCTION

The progressive increase in the stringency of legislated discharge constraints is reflected in steady rise in external industrial effluent treatment charges by municipal and other water treatment companies. As a consequence, a growing number of industries now consider provision of in-house waste treatment as a more economic and effective solution to disposal problems. For the dairy and related industries, membrane bioreactors, a coupled biological treatment and separation process to tackle high strength wastes at source, offer great potential.

The membrane bioreactor is a very straightforward concept and involves a continuous biological treatment stage combined with a cross-flow membrane filter for separation. The membrane selected should be robust and act as an effective barrier to prevent biomass loss from the reactor. This enables high microflora concentrations to be retained without need to rely on settlement for separation, as in conventional biological treatment systems (*e.g.* activated sludge).

Processes, such as activated sludge, rely on flocculation of a mixed microflora to enable biomass separation from the treated liquor by sedimentation.

Flocculation is primarily brought about by exopolysaccharide produced by certain bacterial species [1]. Excreted exopolysaccharide is thought to provide cell protection (*e.g.* against heavy metals) and act also as a food source in nutrient deficient conditions.

To an extent, the level of exopolysaccharide production reflects organism stress, and hence medium "quality" and dissolved oxygen supply. If environmental factors are of high quality, then net production levels are often reduced as the cell metabolism "focuses" on internal activity, rather than extra-cellular products. A very dense, metabolically active microbial population would provide, therefore, enhanced rates of effluent treatment, but a subsequent decline in exopolysaccharide, leading to reduced flocculation and poor separation characteristics. Consequently, activated sludge plants are not necessarily operated at maximum potential in order that adequate separation through sedimentation of aggregated flocs can be achieved.

With a membrane bioreactor, the need for flocculation is removed and also, development (and retention) of waste specific microorganisms can be achieved. The outcome is that much higher biomass levels can be obtained, leading to more efficient removal of COD/BOD per unit reactor volume. An additional benefit is that most dairy and other food processing wastes contain large quantities of low grade waste heat. By locating a bioreactor on-site, better exploitation of this heat through enhanced microbial activity can be achieved (along with the potential provision of on-site recyclable water).

Applications of membranes in waste-water treatment has been mainly restricted to polymeric units [2]. Whereas, in the bioreactors under study, we have been looking at ceramic membrane filters, which have been employed for many years in industrial separations [3] and provide good thermal, pressure and chemical resistance. Previous investigations into use of ceramics in bioreactors, however, have been restricted to unusual design, small-scale systems [4]. For municipal effluent treatment, under normal aerobic (activated sludge) bioreactor operating conditions, the relatively high unit costs of ceramics are currently considered prohibitive.

COD and BOD are generally the most significant factors in cost calculations for treating trade effluent by municipal water companies. Hence, for high level, large volume producers, significant savings can be made by investing in effective on-site waste treatment. Such a solution could be also of benefit to municipal water treatment companies, as receiving sewage works would have a reduced loading, thereby possibly delaying need for expansion.

To illustrate the principal, presented is work on a very high COD/BOD dairy waste (COD typically 4000-9000 mg/l, BOD 2000-5000 mg/l), treated using a ceramic membrane based bioreactor.

## MATERIALS AND METHODS

The membrane used was a 0.2 μm ceramic (Adams Hydraulics Ltd., York, UK). A module was 0.6 m long and 0.02 m diameter, with a surface area of 0.06 m² (7 internal star-shaped channels of approximately 2 mm diameter).

Temperature controlled fermenters (7-20 l working volumes) were coupled via a re-cycle loop to a ceramic membrane module operated at a constant cross-flow rate of 160 l/h. System temperature was maintained at 25±1°C, which was similar to the effluent discharge temperature. Figure 1 shows the experimental rig layout.

The rig was operated in a cyclic backwash/aeration mode. Air was normally fed directly into the permeate side of the membrane. This provided a high degree of aeration and also back-flushing through the membrane's porous structure. At regular intervals, the aeration was stopped, permeate (treated effluent) drawn off and an equivalent volume of raw effluent feed pumped in (by means of a level controller). At the same time as permeate was drawn off, samples were also taken of the bioreactor liquor.

Changing the frequency of permeate removal/effluent feed addition, was used to vary the hydraulic residence time (HRT). The average HRT within the bioreactor could be maintained between 12-48 h, in order to compare levels of treatment.

Figure 1: Ceramic Membrane Bioreactor Arrangement

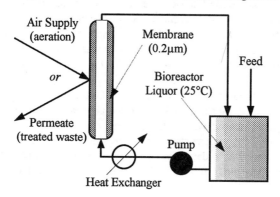

## RESULTS AND DISCUSSION

Fluxes obtained using the ceramic membrane in cross-flow filtration of activated sludge were around 170 l/m$^2$/h [5]. This was 2-3 times higher than those achieved with polymeric membranes of equivalent pore size in both flat sheet and hollow fibre modules. The star-shaped profile of the ceramic membrane channels appeared to enhance turbulence and surface scour to limit fouling. However, on a surface area basis, the costs of the ceramic units are typically 4-6 times greater than polymeric systems. As a consequence, their use in municipal treatment is generally considered, within the industry, as economically unattractive.

A limited number of studies have been carried out using a cross-flow membrane as an alternative aerator [6], but this has been at the expense of them acting as a filter. But, with the ceramic units operated in a dual role mode (*i.e.* aeration and filtration) and exposed to very high-strength wastes, then they can offer real advantages.

Amongst wastes under investigation, we are looking at treating effluent from an ice-cream and dairy products factory. At the factory site, waste-streams from different process units are combined and passed through a fat-trap before discharge into the local sewerage net-work. The waste was typically 23-30°C at the sample collection point (which was up-stream of the fat-trap). This heat would be normally dissipated in the sewerage system, whereas by operating a bioreactor at source, positive use in terms of enhanced metabolic activity, could be achieved.

A clear indication of the importance of temperature and aeration in achieving BOD/COD reduction is illustrated in Figure 2. In this example, fresh effluent was placed in incubated 2 l glass-jar fermenters (with and without aeration), and decline in COD and BOD followed over six days. Not surprisingly, introducing aeration and increasing the temperature both significantly enhanced degradation rate of the organic load in the raw feed. For example, at 15°C (typical sewage treatment plant temperature), reduction in COD was 54% and BOD, 67%. At 25°C (on-site temperature), reductions were COD, 82% and BOD, 91%.

A number of bioreactor runs lasting from 10-40 days have been performed. Figure 3 shows a typical set of feed and treated permeate COD and BOD values for a 27 day run. The feed, with a COD that varied between 3880-9990 mg/l, was kept refrigerated (5°C) to prevent degradation prior to use (each sample was fed to the bioreactor over 7 days, before replacement with a fresh waste-water). Hydraulic residence time (HRT) in the bioreactor was approximately 24 hours.

The first 8-10 days of operation represent a stabilisation period in the bioreactor, over which the feed's indigenous microflora acclimatised and increased. Thereafter, despite a variable quality in the raw effluent feed, bioreactor permeate COD was maintained at around 150-200 mg/l (a 98% reduction at the

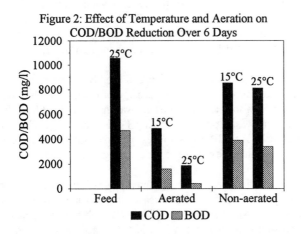

Figure 2: Effect of Temperature and Aeration on COD/BOD Reduction Over 6 Days

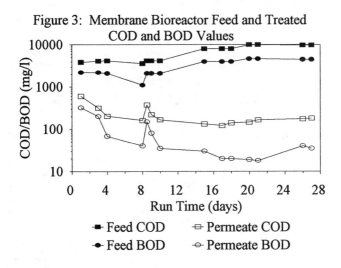

Figure 3: Membrane Bioreactor Feed and Treated COD and BOD Values

peak feed load of 9990 mg/l). of around 10 kg/m³/day. This compares very favourably with typical activated sludge process and also COD removal efficiencies of 80-98% reported for a membrane bioreactor operating with municipal sewage (feed COD of 60-200 mg/l ) [3].

For BOD₅, input levels over the 27 days varied between 2200-4700 mg/l. Figure 3 also shows the decline of both BOD in the permeate after the microbial population in the bioreactor had established. Once the bioreactor was stable, the BOD was maintained below 50 mg/l and often down to 20 mg/l.

An additional problem to very high COD and BOD levels, were periodic high pH values (often above 10) due to discharge of cleaning in place (CIP) storage tanks. A benefit of the bioreactor system, was the level of pH buffering achieved. Food processing wastes often have high pH due to discharge of cleaning in place (CIP) chemicals. As CIP tends to be periodic, rather than continuous, any treatment system must be capable of withstanding "shock loads".

Untreated effluent frequently registered pH >9.5, but in a 16 day run, despite a consistently very high feed pH, bioreactor liquor and permeate pHs were held below 8.1 (Figure 4). The membrane bioreactor enabled an indigenous population of microorganisms to build up, which included lactic acid bacteria. Growth of these bacteria is inhibited at high pH, but at around 5-7 they are capable of generating high levels of lactic acid. This helped "buffer" the effluent to an acceptable pH discharge level.

Figure 4: Membrane Bioreactor Feed, Reactor Liquor and Permeate pH Values

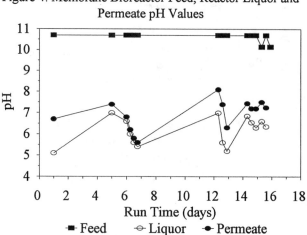

## CONCLUSION

A membrane bioreactor is a flexible means of providing waste-specific effluent treatment. With the dairy waste illustration, a ceramic membrane based unit provided a system that enabled the natural indigenous microflora to build-up, and provide effective and stable treatment. The membrane also fulfilled a dual role, combining the functions of aerator and filter. Furthermore, with the bioreactor located on-site, full advantage can be taken of waste-heat through significant enhancement of metabolic activity.

## ACKNOWLEDGEMENT

The authors are very grateful to Stephen Churchouse, Lakshan Saldin and Kate Parker for their assistance on the project. They also would like to thank the Wessex Water plc/SERC/DTI Teaching Company Scheme for financial assistance.

## REFERENCES

1.  Novak, J. S., Tonenbaum, S. W. and Nakes, J. P.. 1992. Heteropolysaccharide formation by A.viscosus grown on xylose and xylose oligosaccharide. *Appl. Env. Microbiol.*, 58:3501-3507.

2.  Chiemchaisri, C., Yamaoto, K. and Vigneswaran, S.. 1993. Household membrane bioreactor in domestic wastewater treatment. Wat. Sci. Tech., 27(1):171-178.

3.  Lahiere, R. J. and Goodboy, K. P.. 1993. Ceramic membrane treatment of petrochemical wastewater. *Env. Prog.*, 12 (2):86-96.

4.  Yamagiwa, K., Ohmae, Y., Dahlan, M. H. and Ohkawa, A.. 1991. Activated sludge treatment of small-scale wastewater by a plunging liquid jet bioreactor with cross-flow filtration. *Biores. Technol.*, 27:215-222.

5.  Scott, J. A., Smith, K. L. and Howell, J. A.. 1995. A membrane bioreactor for aerobic treatment of high COD effluents. In *Towards hybrid membrane and biotechnology solutions for environmental problems*. Eds. Noworyta, A. and Howell, J. A., 195-201, Wroclaw University Press, Wroclaw

6.  Ahmed, T. and Semmens, M. J.. 1992. Use of sealed end hollow fibres for bubbleless membrane aeration: experimental studies. *J. Memb. Sci.*, 69:1-10.

# ACTIVATED CARBONS AS BIOFILM SUPPORTS FOR DECONTAMINATION OF STREAMS POLLUTED WITH HEAVY METALS AND ORGANIC RESIDUES

**John Ashley Scott**
School of Chemical Engineering
University of Bath, Bath
Avon BA2 7AY, UK

**Atul M. Karanjkar**
School of Chemical Engineering
University of Bath, Bath
Avon BA2 7AY, UK

## INTRODUCTION

Environmental hazards from toxic heavy metal contaminants in mine tailings, industrial effluents and dump site leachates, are often best tackled at source. A range of physical/chemical methods are available, such as sulphide and hydroxide precipitation, solvent extraction, ion-exchange, adsorption and reverse osmosis. Whilst usually effective, they can require large capital, reagent and/or energy costs, and are normally most efficient at metal concentrations above 50-100 mg/l. There is an identifiable niche for alternative technologies, particularly at the sub 10 mg/l range, of which biosorption is now a recognised option [1].

Invariably waste streams, however, are likely to be a cocktail of organic and inorganic chemicals, in which metals are only one, albeit important, constituent. The option to CO-extract more than one pollutant can, therefore, provide a significant process advantage. On this theme, we are investigating the possibilities of granular activated carbon (GAC) covered with defined bacterial biofilms, as a multi-component waste-water treatment system [2-4].

GAC is in general not noted as a particularly efficient adsorber of metals, but its high loading capacity for organic compounds, finds extensive application in decontamination of liquid and gaseous streams. With a suitable biofilm immobilized over GAC particles, the metal uptake level can be increased several fold [5]. A combined biofilm/GAC system has been devised, therefore, to exploit the abilities of the biofilm to tackle metals and the GAC other contaminants.

An effective group of organisms are bacteria which generate exopolysaccharides to form extensive and cohesive biofilms. It is the excreted polymer and not metabolic activity that accounts for almost all subsequent biosorption [6]. The ideal biofilm, will be one that retains bacteria through entrapment, but is open enough not to smother the GAC surface (i.e. so that it remains accessible for adsorption of organic residues etc.). Use of a support for immobilization of micro-organisms can also provide greater resistance against potentially toxic agents and increase metabolic activity [7].

## MATERIALS AND METHODS

Microbial selection, culture conditions and analytical techniques are described in detail elsewhere [4]. Biofilms were developed using an exopolysaccharide producer obtained from the National Collection of Industrial and Marine Bacteria, Aberdeen, Scotland (NCIMB), *Pseudomonas sp.* (NCIMB 11592).

To develop an immobilized biofilm, 2 l of seeded culture medium was circulated (up-flow, 25°C, pH 7.4±0.1) for two days through 15 mm id glass columns containing 10 g of GAC (Filtrasorb 400, Chemviron Carbon). After two days, the culture medium was replaced with a solution containing various metals and/or organic residues for uptake studies by either plain GAC, or biofilm-GAC. A - +schematic of the process is shown in Figure 1.

Figure1: Biofilm-GAC Decontamination Column
(Up-flow Mode)

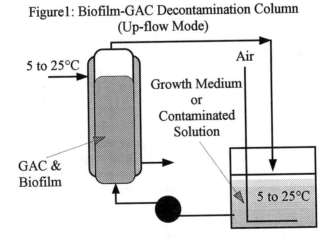

5 to 25°C

Air

Growth Medium
or
Contaminated
Solution

GAC &
Biofilm

5 to 25°C

## RESULTS AND DISCUSSION

### The Biofilm

A fully developed and immobilized biofilm (estimated 10-100 μm thick) consists of rod-shaped bacteria bound in an exopolysaccharide coating [ ]. It is the nature of this coating that provides sites for attraction of positive metal ions, as it acts essentially as an ion-exchange material. It is important to preserve film integrity and, therefore, necessary to keep the bacterial population viable in order that it is able to "repair and maintain". If the organisms cease to metabolise, the biofilm will very rapidly break-down (over 6-12 hours), and slough off the GAC surface.

### Influence of Temperature and pH on Biosorption of Metals

Figure 2 illustrates both the effectiveness of an immobilized biofilm in taking up cadmium (25 mg/l), along with the influence of solution temperature on equilibrium metal loading levels. That is, the presence of the biofilm, estimated at around only 20-100 mg (dry weight) per g of GAC [8], results in a several-fold increase in metal uptake when compared to plain (non-biofilm) GAC. Furthermore, over a temperature rise of 5-25°C, the slight increase in metal uptake indicates physical adsorption, rather than metabolic activity as the prime factor in metal accumulation by the biofilm-GAC system.

Biofilms were developed at pH 7.4, but many metal containing wastes are acidic in nature. Figure 3 shows both the affinity of the biofilm-GAC to three metals and also the influence of solution pH on equilibrium uptake levels. As a general rule, for the range of metals we have investigated (*e.g.* Cu, Zn, Cd, Ni, Cr, Ag), the biofilm-GAC system demonstrated a marked loss in performance below pH 3.5-4. Under these acidic conditions, other than detrimental stress placed on the micro-organisms, the net surface charge over the exopolysaccharides is likely to be unfavourably (*i.e.* a progressive shift to positive charge).

### Kinetics of Metal Uptake

Biofilms immobilized over GAC clearly enhance the uptake of metals, but previous work [4], has shown that the biofilm does not to have the same affinity for all species. In terms of process design, the adsorption rate constants, $k_{ad}$, provides an indication of the preferential uptake order and also importantly, a means of comparing biofilm-GAC and with other systems. Rate constants have been reported [4], for a variety of metals using a Lagrergren type expression for metal adsorption onto virgin GAC [10]. These values for six different metals, adsorbed from 25 mg/l solutions onto biofilm-GAC or plain GAC, are presented in Figure 4.

This data confirms that with biofilm-GAC, consistently higher rates of adsorption were achieved for all metals. The importance of this lies in the design of a process, as residence times in contact columns can be reduced by increasing

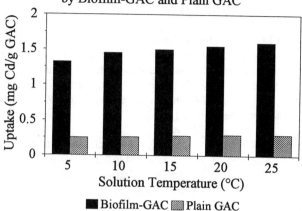

Figure 2: Effect of Temperature on Cadmium Uptake by Biofilm-GAC and Plain GAC

Figure 3: Effect of Solution pH on Biosorption

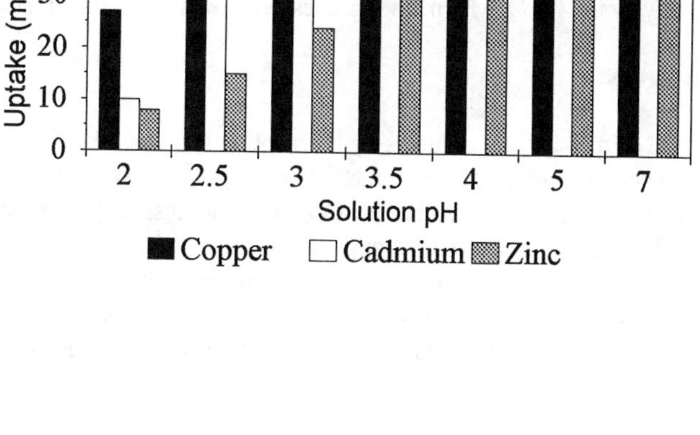

uptake rate. Another observation from Figure 4, is that the order in which the metal adsorption rates are ranked, is modified by the biofilm. With regards plain GAC, metal uptake is generally low, but a particular affinity to silver is found. Whereas, with an immobilized biofilm in place, silver drops to the bottom of the ranking (although still better than achieved by plain GAC). Conversely, cadmium, which had a relatively low rate of uptake by the carbon studied, was found to be very efficiently removed by the biofilm.

### Co-Uptake of Metal and Organic Residue

GAC has a well established role in potable and waste-water treatment for removing organic contaminants. As discussed in the introduction, the philosophy behind using GAC as a biofilm support, is of course to exploit this excellent adsorptive capacity. Otherwise, other, more economic immobilizing media (e.g. sand, anthracite) could be employed.

With regards the ubiquitous pesticide residue, atrazine (a common herbicide), we have shown that the capacity of biofilm-GAC exposed to 25 mg/l cadmium solutions, was unaffected by the presence of various low levels of inorganic salts and 8 mg/l atrazine, [2]. Also, if plain GAC was used to take-up atrazine before biofilm development, atrazine concentrations over 30 mg/l were required to bring about a reduction in subsequent metal uptake performance (e.g. even after pre-exposure to 100 mg/l atrazine, cadmium biosorption was only reduced by 25%).

Trials have been also carried out with solutions containing high levels of nitrate (up to 100 mg/l), with the GAC again either pre-loaded, or exposed to the nitrate after biofilm development. There was with either approach, no discernible effect on either biofilm development, or subsequent rate of metal uptake.

Finally, in Figure 5, uptake from a solution containing both 25 mg/l of cadmium and 100 mg/l benzene is shown for both plain GAC, and biofilm-GAC. As expected, the biofilm dramatically improved metal uptake, but encouragingly, also did not appear to inhibit too much, despite the need to diffuse through the film, simultaneous benzene removal. The data also indicates, as before, metal was predominately taken up by the biofilm, whereas benzene adsorption was predominantly by the GAC, with neither process interfering with the other.

### CONCLUSION

Removal of metals from effluents by biosorption is an established alternative method for decontamination of heavy metal bearing waste streams. However, for operating a practical process, two factors have to be addressed. (i) the biosorbent should be physically immobilized/entrapped, such that it is large enough to be retained within the treatment zone (i.e it can be prevented from flowing out of the system loaded with metals). The most common approach is to use encapsulated or fixed dead biomass. Whilst this may be effective at removing metals, it does not

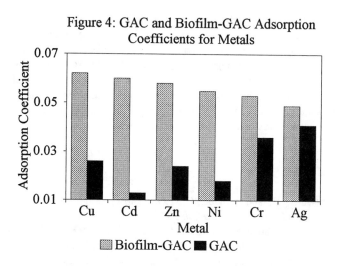

Figure 4: GAC and Biofilm-GAC Adsorption Coefficients for Metals

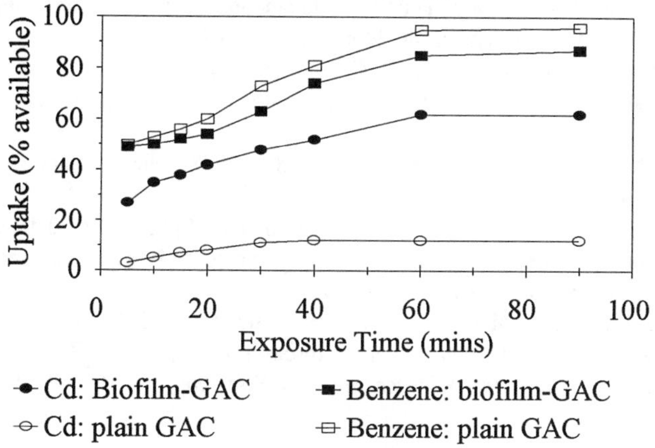

Figure 5: Co-Uptake of Cadmium and Benzene by Biofilm-GAC and GAC

offer any opportunity to exploit microbial metabolism (*e.g.* to neutralize/degrade contaminating organic residues). (ii) the system should be ideally multi-functional, with an ability to tackle more than one type of pollutant.

We are looking at immobilizing viable biofilms over granular activated carbon (GAC). The advantages of good metal removal can be then combined with an excellent adsorber for inorganic salts and organic residues (which may in turn be metabolised by micro-organisms colonising the surface). GAC by itself, is not in general an effective adsorbent for heavy metals, but with a biofilm attached which consists of bacteria held in an expansive excreted polysaccharide coating, rate and quantity of metal extraction can be significantly increased. At the same time, an ability is maintained to co-adsorb inorganics and organics.

## REFERENCES

1.  Mattvschka, B. and Straube, G. 1993. Biosorption of metals by waste biomass. *J.Chem.Tech.Biotechnol.* 58:57.
2.  Scott, J.A., Karanjkar, A.M. and Rowe, D.L. 1994. Exploitation of biofilmcovered granular activated carbon for enhanced removal of metals, In *Resources Conservation and Environmental Technologies in Metallurgical Industries*, Eds. Mahant,P., Pickles, C. and Lu, W.-K., 345, CIM, Montreal.
3.  Scott, J.A. and Karanjkar, A.M. 1994. Decontamination biosorption to concentrate metals on granular activated carbon - influence of pH and temperature, and use of this material to adsorb chloroform. In *Progress Biotechnology 9*, Eds. Alberghina, L., Frontali, L. and Sensi, P., 1231, Elesvier Science, Amsterdam.
4.  Scott, J. A., Karanjkar, A. M. and Rowe, D. L. 1995. Biofilm covered granular activated carbon for decontamination of streams containing heavy metals and organic chemicals. *Min. Eng.* 8(1/2):221
5.  Scott, J.A. and Karanjkar, A.M. 1992. Repeated cadmium biosorption by regenerated *E.aerogenes* biofilm attached to activated carbon. *Biotechnol.Letts.* 14:737.
6.  Scott, J.A., Sage, G.K. and Palmer, S.J. 1988. Metal immobilisation by microbial capsular coatings. *Biorecovery* 1:51.
7.  Morsen, A. and Rehm, H.J. 1990. Degradation of phenol by a defined mixed culture immobilized by adsorption on activated carbon and sintered glass. *Appl. Microbiol. Biotechnol.* 33:206.
8.  Scott, J.A., O'Reilly, A.M. and Karanjkar, A.M. 1992. Heavy metal ion accumulation over activated carbon through biosorption. In *Waste Processes Recycling: Metal Waste Processing and Recycling*, Ed. Rao, R., 157, CIM, Montreal.

# BIOLOGICALLY CATALYZED REDUCTION OF NITROUS OXIDE TO NITROGEN GAS USING WASTE WATER TREATMENT SYSTEMS

Drew Endy
  Biotechnology & Biochemical Engineering
  Thayer School of Engineering
  Dartmouth College
  Hanover, NH 03755-8000
  endy@dartmouth.edu

## ABSTRACT

Nitrous oxide is one of several gases in the atmosphere which contribute to the greenhouse effect and ozone depletion. Atmospheric concentrations of nitrous oxide have risen at 0.2-0.3% per year over the last thirty years [1]. A documented anthropogenic source of nitrous oxide is the nitrification/denitrification process of waste water treatment plants [2]. One method of eliminating these emissions is to facilitate the biologically catalyzed reduction of nitrous oxide to nitrogen gas using the activated sludge from a municipal waste water treatment plant. This study details the feasibility of such a reaction. The effect of using a supplementary electron donor to encourage the reduction as well as the possible inhibition and/or competition of $NO_x$ is examined. Experimental results allow for the estimation of rate constants describing the reaction $-d[N_2O]/dt$ and indicate that an electron donor such as methanol or dextrose increases the rate of reduction significantly. These findings could be used to develop biological treatment processes for the elimination of nitrous oxide emissions from industrial sources and waste water treatment plants or for the remediation of ground waters.

## INTRODUCTION

Reported atmospheric concentrations of nitrous oxide ($N_2O$) have increased at a rate of 0.2-0.3% per year over the last thirty years to a current concentration of approximately 330 ppb by volume [1,3,4]. In the troposphere $N_2O$ is extremely stable  but once in the stratosphere high intensity radiation results in its oxidation. The stability of the $N_2O$ molecule results in an average life in the atmosphere of 100 to 150 years. Since the other oxides of nitrogen are scrubbed out in the troposphere, $N_2O$ is the principle vector for nitrogen transmission to the stratosphere. Once in the stratosphere, about 10% of the $N_2O$ is oxidized to nitric oxide (NO) in a series of reactions involving solar radiation at $\lambda < 310$ nm. In turn the free radical NO catalyzes the conversion of ozone to $NO_2$ and $O_2$ [5]. Although the concentration of $N_2O$ is low when compared to other greenhouse gases ($CO_2$ and $CH_4$) a molecule of $N_2O$ is 290 times as potent as a $CO_2$ molecule and should account for 4% of global warming over the next century [6]. The combination of a long life span with global warming and ozone destroying properties makes it important to understand the reasons behind increasing $N_2O$ concentrations and explore possible control mechanisms.

Removal of nitrogen in waste water treatment plants is accomplished by bacteria which catalyze several reactions. First, organic nitrogen is broken down via ammonification to ammonia ($NH_3$). The $NH_3$ is oxidized to nitrite ($NO_2$) which in turn is oxidized to nitrate ($NO_3$) through the nitrification reactions. The $NO_3$ is reduced to $NO_2$, NO, $N_2O$, and finally nitrogen gas ($N_2$) by the denitrification reactions. Although $N_2O$ is produced as a free intermediate in the denitrification process, nitrogen was believed to be removed from the waste stream as $N_2$ (low solubility in water) left the liquid phase. Recent studies have shown however that $N_2O$ is an end product of both the nitrification [7] and denitrification [2] stages in waste water treatment. Other work has shown that during denitrification the ratio of $N_2O:N_2$ produced increases with the concentration of $NO_3$ and $O_2$ [8]. Additionally the effect of organic carbon, temperature, pH, and solids retention time (SRT) on $N_2O$ production has been studied [2]. Despite this work and the significant effort devoted to understanding the global $N_2O$ balance, the emission of $N_2O$ from waste water treatment plants is not well understood quantitatively.

The elimination of $N_2O$ emissions from waste water treatment plants can be accomplished by either preventing $N_2O$ formation or reducing $N_2O$ once it has been generated. By providing an electron donor such as dextrose or methanol, and an anaerobic environment, the reduction of $N_2O$ to $N_2$ should be possible (equations 1 & 2). Here, the viability of catalyzing this reaction with the biomass from a municipal waste water treatment plant is studied. These experiments were conducted as part of a larger study examining the effect of

256

$$2N_2O + CH_2O \rightarrow 2N_2 + H_2O + CO_2 \qquad (1)$$

$$3N_2O + CH_3OH \rightarrow 3N_2 + 2H_2O + CO_2 \qquad (2)$$

an anaerobic storage period on the concurrent biological nitrification/denitrification (CBND) process. During previous CBND investigations a significant quantity of $N_2O$ was generated [10].

## MATERIALS AND METHODS

**Biomass**  The biomass used in this study was obtained in October, 1993 from the recycle of Line 2 in the Largo Waste Water Treatment Plant (5100 150th Avenue North, Clearwater, FL 34260). After shipment (1 day) in anaerobic containers to the Environmental Studies Center, Lehigh University, Bethlehem, PA 18015, the biomass was placed in four separate sequencing batch reactors and maintained as described below.

**Inorganic and Organic Feeds**  The inorganic feed comprising of $MgSO_4$ (11.4 ppm-Mg), $NaH_2PO_4$-$H_2O$ (7.64 ppm-P), $KHCO_3$ (3.40 ppm-K), $CaCl_2$ (15.5 ppm-Ca), $FeCl_3$-$H_2O$ (0.25 ppm-Fe), $NaHCO_3$ (12.93 ppm) and $NH_4Cl$ (variable concentration) was made weekly in 25 L carboys. Organic feeds comprising of dextrose (2521 mg/L) and yeast extract (10 g/L) were prepared in 3.8 L and 0.5 L carboys as required. All reagents were supplied by Fisher Scientific with the exception of the yeast extract supplied by Difco Laboratories. In between fillings all carboys were contacted with household bleach and rinsed vigorously to prevent bacterial growth in the feed solutions.

**Sequencing Batch Reactors**  Operation of the sequencing batch reactors (SBR's) began in October, 1993 and continued throughout these experiments (May, 1994). The program controlling the SBR's was dictated by the results of the parental CBND studies and may have provided a biomass for these experiments that was not at 'steady state' [9]. A detailed investigation of the CBND process is provided elsewhere by Spector and Kugelman [10]. Figure 1a shows the schematic of a SBR. A single operating cycle started with approximately 6 g of biomass in the 0.6 L reactor heel. 1.3 L of inorganic feed was fed into the reactor and the stirrer activated. 0.1 L of organic feed was added to provide 252 mg dextrose and increase the total volume to 2.0 L. During the initial anaerobic stage 3.2 ml of the yeast extract solution was added. Oxygen was provided at 2.0 ppm during the subsequent aeration stage. After aeration the solids were allowed to settle and begin an anaerobic storage period while 1.4 L of

257

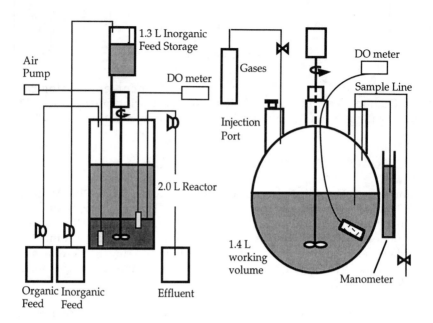

**Figure 1a.** Sequencing Batch Reactor (SBR)   **b.** Closed Reactor (CR)

supernatant was pumped to an effluent carboy. The effluent carboys were treated with 3 ml of concentrated $H_2SO_4$ to prevent further biodegradation of $NH_4$ or $NO_x$. Solids were kept at 3000 mg/L by periodic wasting of excess biomass.

**Closed Reactor**     Once the SBR's reached equilibrium as defined by stable effluent $NO_x$ and $PO_4$ concentrations, the biomass from a SBR was transferred into the closed reactor (CR) for a batch experiment (figure 1b). During a typical closed reactor experiment, the biomass was taken from a SBR immediately after feeding. Once settled, approximately 1.4 kg of sludge and supernatant were weighed and transferred into the CR which was then sealed and pressure tested with 100% argon. The reactor head space was flushed with argon for ten minutes to remove any nitrogen and oxygen from the CR. Throughout the run a slight positive pressure was maintained to prevent $N_2$ from leaking into the CR. At the start of the experiment $N_2O$ was injected through the sampling port with 10cc syringes. During the experiment, head space pressure was monitored and gas samples withdrawn for analysis. Pressure was maintained by the addition of inorganic feed *sans* $NH_4$. At the conclusion of the run, the biomass was returned to the SBR and it's program reinstated.

**Analytical Methods**    $NO_2$, $NO_3$, $PO_4$, and $SO_4$ concentrations were determined with a Dionex Ion Chromatograph 16 run at 800 psi using 0.025 N $H_2SO_4$ as the regenerant and 0.25 g/L $Na_2CO_3$ + 0.25 g/L $NaHCO_3$ as the anion eluent. $NH_4$ concentrations were found in accordance with the procedure listed in section 417b of Standard Methods [11]. Nessler reagent and Rochelle salt from Fischer and a Bausch & Lomb Spectronic 70 set at $\lambda = 412$ nm were used. All mass measurements were found using a Mettler H20T balance. TKN analysis was performed on 25 ml samples in accordance with section 420a of Standard Methods. Gas samples from the CR were taken with Hamilton Gas Tight syringes. A Fisher Gas Partitioner Model 1200 with a helium carrier (Linde Specialty Gases) was used for the analysis of gas samples. Gas standards were provided by calibrated mixtures (2.53% $N_2O$ in $O_2$ and 5.23% $CO_2$ + 4.96% $N_2$ in $O_2$) obtained from Air Products, Trexlertown, PA. Dissolved oxygen was measured with Yellow Springs Instrument Co. Model 54ARC D.O. meters.

## RESULTS

**Continuous Operation**    Starting on October 27, 1993 the four SBR's were operated continuously. During this period the initiation of CBND and the characterization of the biomass was monitored by a series of material balances. The average values describing the biomass during the period of CR experiments is given in Table I and the range of operating programs in figure 2. Over this study, an average of 26 mg $NH_4$-N entered with the feed and an average of 3.4 mg N/L was removed via CBND. Previous studies [10] found nitrogen removal levels of 10.0 to 15.0 mg N/L via CBND. Throughout this study nitrogen removal due to CBND was not realized at the levels obtained previously.

### TABLE I    AVERAGE BIOMASS PARAMETERS

| | |
|---|---|
| Yield | 0.43 gram/gram |
| Food:Biomass | 0.28 /day |
| MCRT | 5-7 days |
| %N | 7.3-9.5 %wt |
| [$O_2$] | 2.0 ppm |

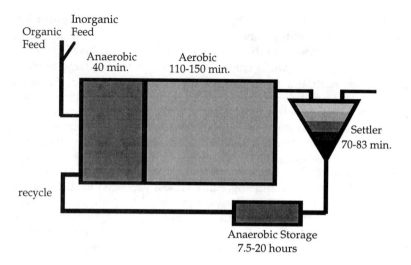

**Figure 2.** Operating Program for SBR's during CBND studies

**Batch Experiments**    A series of eight experiments (RAS 201-208) attempted to detail the reduction of $N_2O$ using a biomass exhibiting minimal CBND under various conditions.  RAS 201 served as the 'proof of concept' experiment.  In this case the biomass was placed in the CR after feeding with 252 mg dextrose.  $N_2O$ was added to maintain the CR pressure while an increase in the $N_2$ concentration was recorded, thereby indicating a reduction in $N_2O$.  After RAS 201, $N_2O$ was injected  only at the start of the run.  RAS 202 was conducted five days later (about one MCRT) using the biomass from RAS 201 prior to feeding with dextrose.   In this run, the $N_2O$ level remained constant until the addition of 67 mg methanol (4 x stoichiometry, eq. 2) after which it dropped rapidly.  RAS 203 repeated RAS 201 but with more precise experimental control.  Here the  biomass  was fed initially with 252  mg dextrose and  50 cc $N_2O$ (1 atm, 70F). Figure 3a details the loss of $N_2O$ and corresponding increase of $N_2$.  A least squares regression performed on the natural log of the $N_2O$ levels versus time (figure 3b) allowed for the determination of the rate constant, k, describing the pseudo first order reduction.  For this experiment k = 0.9+-0.08/hr.  RAS 204 (repeat of RAS 202) used an unfed biomass and resulted in a slow but steady decrease in $N_2O$ levels (1.3 mg $N_2O$-N/g MLSS/hr) with a corresponding increase in $N_2$ concentration.  RAS 205 used a biomass which had been fed and aerated    for    two    hours    in    order    to    simulate

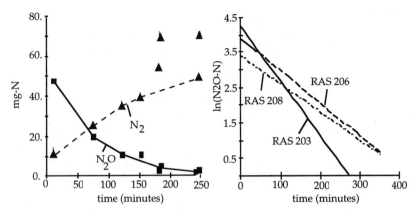

**Figure 3a.** RAS 203 $N_2O$ and $N_2$ data     **b.** 1rst order regression for k

the reduction of $N_2O$ in the later stages of a waste water treatment plant. Again, a slow but steady decrease in $N_2O$ was observed (1.2 mg $N_2O$-N/g MLSS/hr).

RAS 206 repeated RAS 203, using the biomass immediately after feeding (252 mg dextrose and 50 cc $N_2O$ (1 atm, 70F)). The decrease in $N_2O$ followed the same form as RAS 203 but at a slightly slower rate (figure 3b). RAS 207 was effected with biomass taken directly from Line 2 of the Largo plant. This run examined the ability of a non-CBND biomass to reduce $N_2O$. Initially the biomass was fed with 163 mg dextrose and 50 cc $N_2O$ (1 atm, 70F) resulting in a moderate decrease in $N_2O$ levels (1.7 mg $N_2O$-N/g MLSS/hr). After three hours 223 mg of methanol was added which increased the rate of $N_2O$ reduction (2.8 mg $N_2O$-N/g MLSS/hr). RAS 208 probed the effects of adding 5 ppm $NO_3$ at the start of a run identical to RAS 203. Analysis showed that the $NO_3$ was completely removed within the first ten minutes. Subsequent $N_2O$ destruction paralleled that seen in RAS 203 and RAS 206 (figure 3b). Table II summarizes these experimental results. A complete listing of all experimental results, including mass balances for the SBR's is available elsewhere [12].

### TABLE II     SUMMARY OF RESULTS

| RAS 203 | 9.5->1.7 mg-N/g/hr | k = 0.9/hour (0.96) |
|---------|--------------------|---------------------|
| RAS 204 | 1.3    mg-N/g/hr   | (R squared) |
| RAS 205 | 1.2    mg-N/g/hr   | |
| RAS 206 | 5.7->1.1 mg-N/g/hr | k = 0.5/hour (0.99) |
| RAS 207 | 1.7->2.8 mg-N/g/hr | |
| RAS 208 | 3.2->0.6 mg-N/g/hr | k = 0.5/hour (0.99) |

## DISCUSSION

In a first order model the rate of substrate destruction is controlled by its concentration (equation 3). This model has been

$$[N_2O]_t = [N_2O]_0 \exp(-k_{N2O}t) \qquad (3)$$

used widely to predict the rates of biologically catalyzed reactions where the exact species and concentrations are not precisely known (eg. BOD models). Because of the complex population catalyzing the reaction, application of this model to $N_2O$ reduction merits consideration. The underlying assumption is that the substrate itself, or the metabolic response to its concentration, controls the reaction rate. In the case of RAS 203, 206, and 208 the electron donor was provided at a dosage of about three times stoichiometry (based on equations 1 & 2). Since little or no oxygen was available for the aerobic oxidation of the carbon source, most of the carbon should have been available as an electron donor for $N_2O$ reduction. This could imply that the $N_2O$ concentration is indeed the limiting factor for its own reduction. Proceeding with this analysis showed that the experimental results fit a first order rate model remarkably well (figure 3b and table II).

The rate developed for RAS 203 is approximately 180% that found for RAS 206 & 208. The most probable explanation for this difference is the reduction in the length of the anaerobic storage stage (20 to 8 hours) which took place immediately after RAS 203. This change might have produced either a shift in the fraction of denitrifying organisms, or conditioned the existing bacteria so that they became less adept at reducing $N_2O$.

RAS 204 attempted to reproduce the lack of N2O reduction seen in RAS 202 prior to the addition of methanol. While the rate was slow, the steady reduction may have been due to an oxidative couple provided by a stored carbon source (eg. endogenous metabolism). RAS 207 detailed the feasibility of $N_20$ reduction with an unconditioned biomass and demonstrated that the addition of a supplementary carbon source increase the reaction. Overall, the rates generated in this study are four to five times faster then those reported by Hanaki et al [2]. Again, this may be due to the conditioning of the bacteria during anaerobic storage.

These experiments clearly demonstrate the ability of a mixed culture from a municipal waste water treatment plant to reduce $N_2O$. The rate of reduction appears to follow a pseudo first order model in the presence of excess electron donors. Additional work should be performed to detail the effect of $NO_x$ and various anaerobic storage times on the reaction rate. Regardless, the elimination of $N_2O$ emissions from various sources via biological reduction seems feasible.

## ACKNOWLEDGMENTS

The author would like to acknowledge the support and guidance provided by Dr. Irwin J. Kugelman and Marshall Spector. This work was supported by the Department of Civil & Environmental Engineering, Lehigh University and the USEPA.

## REFERENCES

1. Rasmussen, R.A. and M.A.K. Khalil, 1986, "Atmospheric trace gases: trends and distributions over the last decade", *Science* 232:1623-1624.
2. Hanaki, K., Hong, Z., Matsuo, T., 1992, "Production of nitrous oxide gas during denitrification of wastewater", *Wat. Sci. Tech.* 26(5-6):1027-1036.
3. Weiss, R.F., 1981, "The temporal and spatial distribution of tropospheric nitrous oxide", *Journal of Geophysical Resources* 86:7185-7195.
4. Crutzen, P.J., "Atmospheric chemical processes of the oxides of nitrogen, including nitrous oxide", National Center for Atmospheric Research, Atmospheric Quality Division, Boulder, Colorado.
5. Cicerone, R.J., 1987, "Changes in stratospheric ozone", *Science* 237:34-42.
6. *Greenhouse gas emissions, the energy dimension*, 1991, International Energy Agency, 2 Rue Andre-Pascal, 75775 Paris CEDEX 16. France.
7. Ueda, S., Ogura, N., Yoshinari, T., 1993, "Accumulation of nitrous oxide in aerobic groundwaters", *Water Res.* 27(12):1787-1792.
8. Firestone, M.K., Davidson, E.A., 1989, "Microbial basis for NO and $N_2O$ production and consumption", in *Exchange of trace gases between terrestrial ecosystems and the atmosphere*, M.O. Andreae and D.S. Schimel, eds., Dahlem Konferenzen, pp 7-21, Wiley, Chichester.
9. Cassell, E.A., Sulzer, F.T., Lamb, J.C., 1966, "Population Dynamics and selection in continuous mixed cultures", *Journal of the Water Pollution Control Federation*, 38(9).
10. Spector, M., Kugelman, I.J., March 31, 1992, "Final Report", USEPA Project-R-814838-01-0.
11. *Standard methods for the examination of water and waste water*, 16th edition, 1985, American Public Health Association, 1015 15th St. NW, Washington D.C. 20005.
12. Endy, Drew, 1994, "Biologically catalyzed reduction of nitrous oxide to nitrogen gas", *M.S. thesis*, Civil & Environmental Engineering, Lehigh University, Bethlehem, PA 18015.

# REMOVAL OF VOLATILE ORGANIC COMPOUNDS (VOCs) USING BIOFILTERS

## PATRICK E. CARRIERE
Department of Civil and Environmental Engineering
West Virginia University,
Morgantown, WV-26506.

## SHAHAB D. MOHAGHEGH
Department of Petroleum and Natural Gas Engineering
West Virginia University,
Morgantown, WV-26506.

## BABU SRINIVAS MADABHUSHI
Department of Civil and Environmental Engineering
West Virginia University,
Morgantown, WV-26506

## INTRODUCTION

One of the most significant air pollution control challenges being faced by the Federal and State agencies and the chemical process industries is the control of emissions of volatile organic compounds(VOCs).  VOCs are discharged from process industries as major components of  mixed organic wastes which contaminate the environment.  Among these wastes, benzene, toluene, ethyl benzene and xylene are classified as major pollutants with high frequencies of occurrence on the EPA list of priority pollutants.  Biofiltration, a recent air pollution control technology, is the removal and decomposition of contaminants present in emissions of non hazardous substances using a biologically activated medium.  Biofiltration involves contacting the contaminated emission gas stream with microorganisms in a filter media.  Biofiltration utilizes microorganisms immobilized in the form of a biofilm layer on an adsorptive filter media.  Compared to other technologies, biofiltration is inexpensive, reliable and requires no post treatment.  The main objective of this study was to compare the performance of both Granular Activated Carbon (GAC) and Biologically Activated Carbon (BAC) for the removal of benzene and toluene.

## BACKGROUND

Biofiltration is the removal and oxidation of organic gases, Volatile Organic Compounds (VOCs) from contaminated air by beds of compost or soil, wherein microorganisms in a moist, oxygen rich environment oxidize organic

264

compounds to $CO_2$ and $H_2O$. While a successful technology in Europe, biofiltration in the United States is held at arms length by EPA and industry, mostly because engineering data is limited. There are a few installations in the U.S. and Canada and their primary application is odor control.

Biofiltration has been applied to remediate gaseous contaminants like VOCs and other gases since the early sixties. Pertinent literature on the development of soil bed biofiltration and the various researchers include, Smith et al [1], who studied the removal of sulfur bearing compounds, $CO_2$, acetylene and ethylene. Bohn [2], Bohn [3] employed biofiltration for odor control. Kampbell et al [4] studied the removal of VOCs such as aliphatics which consist of butane and isobutane.

A variety of biofilters using peat and compost systems have also been studied. Ottengraf and Van Den Oever [5], Ottengraf et al [6], studied biofiltration by using peat/compost as the support media for the microorganisms on a variety of easily biodegradable VOCs. Don and Feenstra [7] studied biofilters with peat/compost as the support media for treating toluene, ammonia and aldehydes. Leson et al [8] employed a compost biofilter and observed a significant removal of ethanol. Engras et al [9] studied biofiltration for the removal of chlorinated organics, hydrogen sulfide and aromatics in a packed biofilter with a mixture of pearlite and crushed oyster shells as the support media. Jang Young Lee et al [10] have studied biodegradation of toluene and xylene and observed that the optimization of the process requires the proper selection and efficient control of operating conditions such as moisture content in the bed and the pH. As the contaminants are discharged in a mixed state, they also suggested that a single microorganism with a catabolic activity on all of the components in the mixture is to be selected for the complete degradation of the mixture of contaminants.

In the case of GAC the removal mechanism is adsorption of the contaminants on to granular activated carbon. But in the case of BAC the removal mechanism is a combination of both adsorption and biodegradation of the contaminants.

## EXPERIMENTAL METHODOLOGY

To accomplish the objective, column studies were performed by investigating different parameters such as (i) the effect of low initial contaminant concentration on removal efficiency and (ii) the effect of flow rate. The experimental setup is presented in Figure 1. The methodology followed is as given below:

### Packing the reaction column:

The reaction column was packed with activated carbon (Vapure 6 x 16) with placing glass wool and glass beads for uniform distribution. The support media was packed with 100% saturation. The microbial seed which was already acclimated to the contaminant environment was pumped upward through the column at 50 ml/min. Nutrient solution was also pumped upward through the column. Air was supplied continuously to the column.

### Contaminant stock preparation:

Prior to injecting the contaminant, the gas reservoir was filled with air at atmospheric pressure and room temperature and sealed gas-tight. The contaminant, in liquid phase was injected into the gas reservoir through a gas tight septum port. As the contaminant is known for its volatility and the concentration being very low, the contaminant changes into gas phase in a short span of time. The mixture of contaminant and air was conditioned to obtain consistent concentration, by recirculating the mixture prior to passing through the reaction column. The gas samples were drawn at regular intervals from the reservoir by a gas tight syringe and were analyzed by using a GC. This process was carried out till obtaining a consistent concentration. This reservoir serves as the stock for the contaminant After attaining the consistent concentration, the mixture was passed through the reaction column using a gas pump. The contaminant concentrations were expressed in mg/l of mixture.

### Acclimation of Microorganisms:

This process enables to avoid the normal lag phase that the microorganisms have to undergo to get acclimated to the environment of activated carbon and contaminant. This process involved making the microorganisms acclimated to use contaminants as the substrate. This was carried out in an aqueous reactor. The microbial stock was prepared by adding the microbial culture stored as slants to the nutrient media. 5 ml of the microbial stock solution was added to the aqueous reactor which has 300 ml of deionized water. To this mixture 5 ml of nutrients solution was also added. The mixture was kept stirring with the help of a magnetic stirrer to facilitate air supply to the microorganisms. The contaminants were gradually introduced to the microorganisms in small increments so as to make the microorganisms use contaminants as their substrate. These microorganisms were supplied with gradual increments of contaminant, since higher doses of contaminant could be toxic to the microorganisms and might inhibit their growth. This mixture was the stock of microorganisms. Once after the microorganisms were acclimated to the contaminant environment, these were grown on the support media.

### Column Study:

The contaminants were pumped to the column through a humidifying column. The humidifying column was used to increase the moisture content of the contaminants which otherwise might have dropped the water content of the column bed. In the humidifying column, the water was sprinkled from the top to maintain the water content of the mixture. The contaminant mixture was recycled to get a consistent concentration before passing through the column. Once after achieving the consistent concentration, the air contaminant mixture was then passed through the column and the mixture is recycled to till the desired concentrations were attained. The necessary nutrients for the microorganisms were supplied from the top of the column. The drainage from the column was collected at the bottom of the column. The initial concentrations of benzene injected for the constant flow rates were, 9.75 and 19.5 mg/l. The flow rates studied for the constant initial concentration of 19.5 mg/l were 330 ml/min and 155 ml/min. The initial concentrations of toluene injected as the contaminant in both GAC and BAC was 21.75 mg/l. The flow rates were 255 ml/min and 330 ml/min. The removal efficiencies were calculated with respect to the contaminant concentrations in the reservoir. The bed volumes required to reach the desired levels of contaminant concentrations were calculated. The precise chemical analysis of the contaminant mixture was carried out by employing the Gas Chromatography.

## RESULTS AND DISCUSSION

### *Effect of flow rate on the removal efficiency of benzene using GAC:*
The results of the study on the effect of flow rate on benzene removal are presented in Figure 2. When the flow rate was increased from 155 ml/min to 255 ml/min and then to 330 ml/min, keeping the benzene concentration constant at 19.5 mg/l, the number of bed volumes has increased from 32 to 37 to achieve 90% removal. This can be attributed to insufficient contact time with the adsorption sites.

### *Effect of initial concentration on the removal efficiency of benzene using GAC:*
The results of the study on the effect of initial concentration of benzene on its removal are presented in Figure 3. When the concentration was increased from 9.75 mg/l to 19.5 mg/l, keeping a constant flow rate of 255 ml/min, the pertinent data suggested that the number of bed volumes required to attain about 90% reduction was increased from 20 to 34.

### *The effect of flow rate on removal efficiency of benzene in BAC*
The flow rates studied for the constant initial concentration of 19.5 mg/l were 415 ml/min, 330 ml/min, 255 ml/min and 108 ml/min. The removal efficiencies were calculated with respect to number of bed volumes required to achieve the desired concentration of 0.5 mg/l.

The relation between the number of Bed Volumes and contaminant concentration for different flow rates is shown in Figure 4. This data indicated that the maximum removal efficiency was obtained when the flow rate was 255 ml/min, which has given the optimum retention time required for the microorganisms to achieve the biodegradation.

### *Comparison of removal efficiencies using GAC and BAC*
The removal efficiencies were compared with respect to the bed volumes required to reach the desired concentrations in GAC and BAC and the results are presented in Figure 5. The data indicated a lesser number of bed volumes

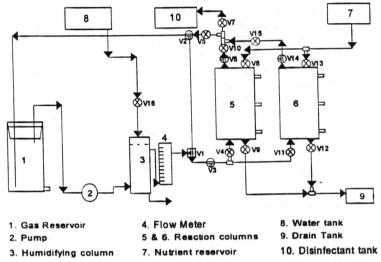

1. Gas Reservoir    4. Flow Meter    8. Water tank
2. Pump    5 & 6. Reaction columns    9. Drain Tank
3. Humidifying column    7. Nutrient reservoir    10. Disinfectant tank

**Figure1.Biofiltration Process Experimental Setup**

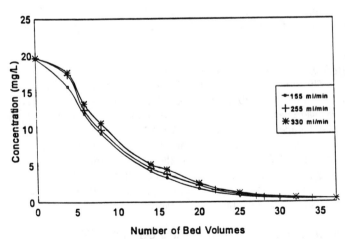

**Figure 2.** **Effect of flow rate on removal efficiency of benzene (GAC)**

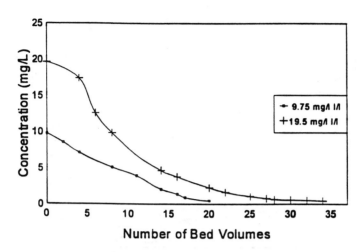

**Figure 3.** Effect of initial concentration on removal efficiency of benzene (GAC)

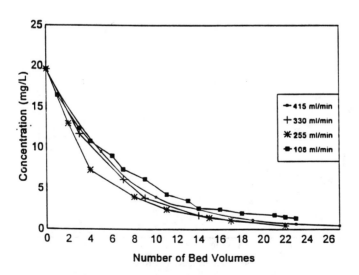

**Figure 4.** Effect of flow rate on removal efficiency of benzene (BAC)

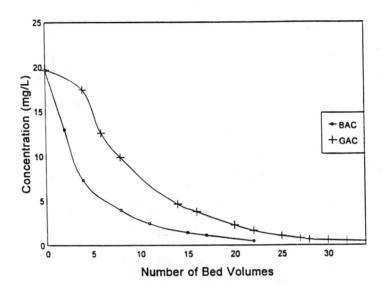

**Figure 5.  Comparison of removal efficiencies using GAC and BAC (Benzene)**

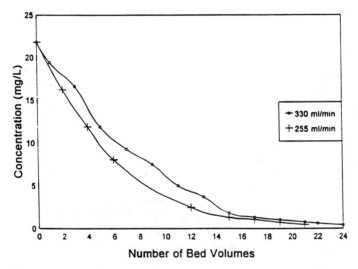

**Figure 6.  Effect of flow rate on removal efficiency of toluene (BAC)**

(22) required in case of BAC than in the case of GAC (35) which can be attributed to biodegradation of the contaminants in addition to adsorption and bioregeneration of adsorption sites.

## *The effect of flow rate on removal efficiency of toluene in BAC*

The results of the study on the effect of flow rate on toluene removal are presented in Figure 6. The flow rate was increased from 255 ml/min to 330 ml/min keeping the initial concentration constant at 22 mg/l. The number of bed volumes required, to achieve the desired concentrations of 0.5 mg/l, was increased from 21 to 24 with the flow rate. This is due to insufficient retention time to achieve the biodegradation of the contaminants in the column. In this case only adsorption and partial biodegradation were achieved.

## *The effect of flow rate on removal efficiency of toluene in GAC*

The results of the study on the effect of flow rate on toluene removal are presented in Figure 7. The flow rate was increased from 255 ml/min to 330 ml/min keeping the initial concentration constant at 22 mg/l. The data indicated that the difference in bed volumes required was marginal 32 to 33. This shows that the increased flow rate from 255 to 330 ml/min has no effect on the removal efficiency in GAC.

## *Comparison of removal efficiencies using GAC and BAC*

The removal efficiencies by using GAC and BAC were compared and the results are presented in Figure 8. The number of bed volumes required in BAC and GAC, operated for a given concentration and flow rate were compared. A considerable difference of number of bed volumes required 21 to 30, was observed when the initial concentration and flow rate were 22 mg/l and 255 ml/min respectively. This indicated the better performance of BAC over GAC for the current concentration and flow rate.

## CONCLUSIONS

It can be concluded that the removal of VOCs using biologically activated carbon (BAC) column was more significant than the granular activated carbon (GAC) column. The biological degradation rate of VOCs using BAC decreased with increase in initial VOC concentration and flow rate. The number of bed volumes required to reach the desired levels of VOC concentration increased with the increase of the initial concentration. Under the same conditions, the degradation of VOCs using BAC column required lesser number of bed volumes to reach the desired concentration than the GAC column due to adsorption and biodegradation processes.

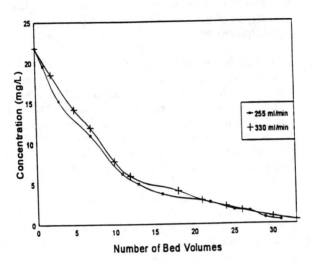

**Figure 7.** Effect of flow rate on removal efficiency of toluene (GAC)

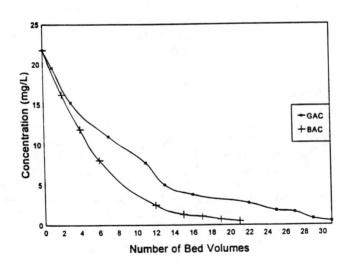

**Figure 8.** Comparison of removal efficiencies using GAC and BAC (Toluene)

# REFERENCES

1. Smith, K.A., J.A. Bremmer and M.A. Tatabai. 1973. "Sorption of gaseous atmospheric pollutants by soil" *Soil Science*, 116:313
2. Bohn H.L. 1975. " Soil and Compost filters of malodorant gases" *JAPCA*, 25, 953.
3. Bohn H.L and R.K. Bohn. 1986. " Soil bed scrubbing of fugitive gas releases" *Journal of Environmental Science and health*, A 21: 561
4. Kampbell D.H., J.T. Wilson., H.W. Read., and T.T. Stockdale. 1987. " Removal of volatile aliphatic hydrocarbons in a soil bioreactor" *JAPCA*, 37: 1236.
5. Ottengraf, S.P.P and H.C Van den Oever, 1983. "Kinetics of organic compound removal from waste gases with a biofilter, *Biotechnology and Bioengineering*, 25:3089
6. Ottengraf S.P.P., H.C Van den Oever., F.J.C.M Kempenaars, 1984. " Waste gas purification in a biological filter bed" in *Innovations in Biotechnology*, E.H. Houwink and R.R. Van der Meer, eds Elsevier, Amsterdam.
7. Don J.A., and L. Feenstra, 1984. "Odor abatement through biofiltration" Symposium on characterization and control of odoriferous pollutants in process industries". Belgium.
8. Leson G., A.M.Winer and D.S Hodge, 1991. "Application of biofiltration to the control of air toxics and other VOC emissions" *84 th annual meeting of AWMA*, Vancouver.
9. Ergas, S.J., E.D Schroeder., Y.Chang and R.Morton, 1992. "Control of volatile organic compound emissions from a POTW using a compost biofilter" 85th Annual meeting of Air and Waste management Association, Kansas City, Missouri.
10. Jang-Young Lee., Yong-Bok Choi and Hak-Sung Kim, 1993. "Simultaneous biodegradation of Toluene and Xylene in a novel bioreactor: Experimental and Mathematical analysis" *Biotechnology Progress*, 9: 46.

# DESIGN OF ACTIVATED SLUDGE SETTLING TANKS FOR OPERATING WITH A BULKING SLUDGE

YOUNGCHUL KIM AND WESLEY O. PIPES
Environmental Studies Institute
Drexel University
Philadelphia, PA 19104

## INTRODUCTION

Activated sludge bulking is a common operating problem. Sludge bulking is measured by the sludge volume index (SVI) which is the volume occupied by a gram of solids after 30 minutes of settling and compaction. Bulking activated sludge occupies too much volume (has too much bulk). This means that the suspended solids concentration in the underflow ($S_u$) from the settling tanks will be relatively low.

The settling test used to determine the SVI is usually performed in a one liter graduated cylinder. A sample of the mixed liquor from the aeration tanks is allowed to settle, the volume occupied by the sludge after settling for 30 minutes is measured ($SV_{30}$) and the mixed liquor suspended solids ($S_m$) are measured. The SVI is calculated from

$$SVI = 1000(SV_{30}) / S_m \tag{1}$$

Since $SV_{30}$ is measured in mL/L and $S_m$ in mg/L, the units of SVI are mL/g.

The average concentration of suspended solids in the sludge after 30 minutes settling is called the sludge density index (SDI):

$$SDI = (1000)(S_m)/SV_{30} = 10^6/SVI \tag{2}$$

SDI is an estimate of the maximum suspended solids concentration which can be attained by settling. With a SVI of 100 mL/g, the SDI will be 10,000 mg/L; with the SVI = 200 mL/g, the SDI = 5,000 mg/L and so forth. Of course, one liter graduate cylinders are not designed as settling tanks and the suspended solids concentration reached in the activated sludge process settling tanks should be somewhat higher that the value calculated from equation (2). The point is that the laboratory settling test provides the information that the sludge will thicken to only slightly more than to the SDI in the settling tanks.

Under normal flow conditions, an operator can often compensate for the low underflow suspended solids concentration by using a higher return sludge pumping rate and thereby avoid having the settling tanks fill up with sludge which then overflows into the effluent. However, there is an upper limit to the return

274

sludge pumping and, during times of increased hydraulic loading during rainstorms, it may be impossible to increase the return sludge pumping rate enough so that the sludge is removed from the bottom of the settling tanks as rapidly as it is carried in from the aeration tanks. When activated sludge bulking occurs, the most likely time for high effluent suspended solids concentrations are during rainstorms.

A wastewater treatment plant in Pennsylvania which has experienced bulking sludge for a number of years and has sometimes had high effluent suspended solids concentrations during rainstorms is the Western Regional Treatment Plant (WRTP) of the Delaware County Regional Water Quality Administration (DELCORA). Studies of this plant showed that the high effluent suspended solids concentrations resulted from a failure of the settling tanks to provide adequate volume for storage of sludge solids during these periods of hydraulic overloading [1].

The purpose of this paper is to provide an illustration of how this requirement for storing activated sludge solids during hydraulic overloading can be formulated as a design criterion.

## CURRENT DESIGN PRACTICES

Parameters which are now used for the design of activated sludge settling tanks are the overflow rate which is related to the clarification function and the solids loading which is related to the thickening function. The most important feature of the settling tank for both of these parameters is the surface area for settling, $A_s$. The depth of the tank also influences both the clarification and thickening functions but becomes a much more important design feature when the need for storing of suspended solids during hydraulic overloads is considered.

**Total Area of Settling Tanks** - Circular settling tanks are very common in activated sludge processes and the emphasis of this discussion is on circular upflow settling tanks. Figure 1 is a settling tank diagram with the solids inflow and outflows indicated. The overflow rate (O.R.) is defined by

$$O.R = (Q-W)/A_s \qquad (3)$$

The solids loading (S.L.) is given by

$$S.L = (Q+R)(S_m)(f)/A_s \qquad (4)$$

where Q = plant inflow (MGD), R = return sludge flow rate (MGD), W = waste sludge flow (MGD), $S_m$ is mixed liquor suspended solids concentration (mg/L), and f is a unit conversion factor.

Settling tanks are usually sized on the basis of overflow rates and solids loading rates. According to the EPA manual [2], the average overflow rate for design of an air activated sludge settling tanks should be in the range of 0.68-1.36 m/hr (400-800 gpd/ft2) and the average solids loading should be 98-146 kg/m2/d (13.9-20.73 lb/ft2/d). The overflow rate and the solids loading established for peak flow are 1.70-2.04 m/hr (1,000-1,200 gpd/ft2) and 244 kg/m2/d (34.6 lb/ft2/d), respectively. These criteria are interrelated and each affects both functions of settling tanks. Similar values for the parameters are recommended in the WEF and ASCE design manuals [3]. Daigger et al [4] and Keinath [5] have suggested settling tank operation and design nomograph based on the solids flux

275

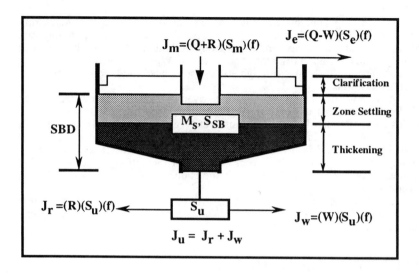

Figure 1  A definition sketch for an upflow, center-feed settling tank

analysis of various activated sludges as a function of sludge volume index.

**Side Water Depth -** In addition to the area required for clarification and thickening, a certain depth (or retention time) must also be provided. The volume requirement commonly has been believed to be associated with the need for time to help settling of flocculent sludge particles in the upper part of settling tanks and to accomplish compression of the sludge solids in the bottom. Analysis of the volume requirement for thickening has not been developed on a rational basis, but several empirical methods have been proposed [6,7].

Metcalf and Eddy [6] suggested the method of sizing the side water depth of settling tanks including respective depth for clarification and thickening function. In Germany, the side water depth is determined by the four-parts of depth: clear water depth, separating depth, storage depth, and thickening depth [7]. Few people considered an entire settling tanks as a solids storage space.

Primarily for reasons of excavation cost, the settling tank depth is limited. Criteria for the side water depth of settling tank by the US Environmental Protection Agency (EPA) is 3.66 - 4.57 m [2]. The trend is toward increasing depth for the improvement of performance [3].

## SOLIDS STORAGE FUNCTION

One function of the settling tanks is to provide space to store sludge solids during hydraulic overloads which occur during rainstorms. The total mass of sludge solids in an activated sludge process ($M_T$) includes the sludge in the settling tanks ($M_s$) as well as the sludge in the aeration tanks ($M_a$).

$$M_T = M_a + M_s \qquad (5)$$

A materials balance of the suspended solids into and out of the settling tanks can be written as

$$dM_s/dt = J_m - J_u - J_e = \{(Q+R)(S_m) - (R+W)(S_u) - (Q-W)(S_e)\}(f) \qquad (6)$$

where f is a unit conversion factor and other terms are defined in Figure 1. If it is assumed that $S_e$ is very small compared with $S_m$ and $S_u$, $J_e$ can be neglected and the condition for $M_s$ to be constant is

$$(Q+R)(S_m) = (R+W)(S_u) \qquad (7)$$

From equation (7) it is possible to define a thickening ratio, k, as

$$k = S_u/S_m = (Q+R)/(R+W) \qquad (8)$$

A thickening ratio is largely determined by sludge settling characteristics and hydraulic loading conditions in the settling tanks. Unless, prevented by its settling characteristics, the sludge will thicken in the settling tanks until the ratio of $S_u$ to $S_m$ is equal to the ratio of inflow to the underflow. Also, the operator can control the thickening factor by selection of the underflow rate (R+W) during a normal flow conditions.

In municipal wastewater treatment plants, Q may increase to two or three times greater than normal dry weather flow during a rainstorm. When Q is very high, R+W, will not carry sludge out of settling tanks as fast as they are carried in with Q+R and $dM_s/dt > 0$. In this case, the settling tanks has to store the excess sludge solids until the rainstorm is over and the sludge can be pumped back to the aeration tanks. The total solids accumulation in the settling tanks at $t = t$ from $t = 0$ during hydraulic overloads can be evaluated from

$$M_{st=t} = M_{st=o} + \Sigma (J_m - J_r - J_w - J_e)\Delta t = M_{st=o} + \Sigma \Delta M_s \qquad (9)$$

If the generation of new activated sludge solids is balanced by sludge wasting, $M_T$ can be considered to be constant and the increase in the amount of solids stored in the settling tanks will balance the decrease in the amount of solids in the aeration tanks.

$$dM_a + dM_s = 0 \qquad (10)$$

This storage function of activated sludge settling tanks during hydraulic overloads is recognized in the literature by a few people [8,9].

## PROPOSED DESIGN PARAMETER

The important questions to answer are 1] how to define the solids storage capacity of settling tanks, 2] how much storage capacity should be provided to prevent storage failures during hydraulic overloads and 3] what is the relationship between the solids storage parameter and the presently used design parameters.

**Definition** - In Figure 1, the total amount of sludge in the settling tanks is represented by $M_s = (V_s)(S_{SB})(f)$ where $V_s$ is the volume of the sludge layer, $S_{SB}$ is an average suspended solids concentration in the layer and f is a conversion factor (f=0.001 kg/g). Using an average $S_{SB}$ concentration greatly simplifies the

277

estimation of solids storage in the settling tanks. The solids storage capacity of the settling tanks can be defined as the maximum amount of sludge solids which can be stored with a maximum return sludge flow (R) without greatly increasing the effluent suspended concentration. Wash-out of sludge solids into the effluent occurs during hydraulic overloads if the solids storage capacity is exceeded. These observations have been made by several investigators [1,8,9,10,11,12]. The solids storage capacity of the settling tanks ($M_{sc}$) depends on the surface area ($A_s$), the critical SBD, and the sludge blanket concentration.

$$M_{sc} = (A_s)(SBD_c)(S_{SB})(f) \qquad (11)$$

**Design Parameter** - The critical sludge blanket depth ($SBD_c$) may be determined by the side water depth of settling tank ($h_w$). The trend to improved clarifier performance with increased depth was found for various operating conditions. The common parameter involved in designing for the three different functions of activated sludge settling tank is the surface area ($A_s$). A dimensionless ratio of maximum solids storage in the settling tanks before wash-out starts to the amount of solids in the process during normal operation is proposed for the design of the settling tank area.

$$M_{sd} / (M_a + M_s) = \Phi \qquad (12)$$

where $M_{sd}$ is a design solids storage capacity (kg) of the settling tanks, $M_a$ is a desired amount of activated sludge solids in the aeration tanks under normal flow conditions, $M_s$ is an initial storage of sludge solids in the settling tanks which depends on the process operation and solids inventory management, and $\Phi$ is a dimensionless storage ratio which is controlled by the severity of storm flow and the operating conditions.

The designer could establish the value of $\Phi$ needed to protect against storage failure during rainstorm flows. The design methodology for aeration tanks is well developed. The mass of sludge solids in the aeration tanks is given by

$$M_a = (V_a)(S_m)(f) \qquad (13)$$

From equation (12), the design solids storage of the settling tanks (this would be solids storage capacity during storm flow conditions) is given by

$$M_{sd} = \Phi (M_a + M_s) \qquad (14)$$

Designing for a higher $M_a$ of the aeration tanks means that the volume provided for solids storage must be increased for a given value of $\Phi$, just as designing for a higher $M_a$ requires that the thickening capacity of the settling tanks be increased. The interaction between aeration tanks and settling tanks in terms of thickening and clarification function was previously studied [13]. In that study, they allow consideration of the effect of mixed liquor concentration ($S_m$) upon the sizing of both types of tanks.

According to equation (14), if the initial sludge blanket depth (at the start of the hydraulic overload) is higher, more storage capacity would be required. The initial $M_s$ will be higher if the process is normally operated with a high SBD. It was found that the higher $M_s$ before storm flow starts greatly increases the chance of storage failure of settling tanks during hydraulic overloads [1].

The larger design $\Phi$ value would prevent storage failure of settling tank for larger storm flow but require more construction cost. The $\Phi$ is considered a greatly simplified factor which describes the complicated sludge transport phenomena between aeration tanks and settling tanks by various types of hydraulic overloads (equation 9). Equation (14) can be combined with equation (11) to give

$$M_{sd} = (A_s)(SBD_c)(S_{SB})(f) = \Phi (M_a + M_s) \qquad (15)$$

The critical sludge blanket depth ($SBD_c$) can be expressed in terms of side water depth of the settling tank ($h_w$) as

$$SBD_c / h_w = z \qquad (16)$$

where z is the dimensionless ratio of the critical SBD to the side water depth of the settling tanks. The ratio for the circular upflow and center-feed settling tanks studied by Authors [1] was found to be about 0.7 (3m/4.57m). The different ratio (z) values can be obtained from other studies [9,10,11]. Substituting equation (16) into equation (15) and rearranging gives equation (17).

$$M_{sd} /\{(A_s)(z)(h_w)\} = S_{SB} \qquad (17)$$

**Estimation of Solids Storage** - Some investigators have studied distribution of suspended solids concentrations in the settling tanks [8,12,14]. Vitasovic's model of thickening [8] predicts solids concentration profiles by dividing the depth of settling tanks into layers of constant thickness and applying solids balance to each layer. The currently available technique for the estimation of the average $S_{SB}$ concentration is averaging the mixed liquor suspended solids ($S_m$) and the underflow suspended solids ($S_u$) concentration [6].

$$S_{SB} = (S_u + S_m)(0.5) \qquad (18)$$

The use of this relationship assumes that the solids concentration gradients from top to bottom of the sludge blanket is linear.

The $S_{SB}$ for the upflow, center feed circular settling tanks at the WRTP, was measured on 56 separated days from September 1992 to March 1994 and those data were compared with the calculated values from equation (18) [1,15]. The settling tank has a side wall depth of 4.57 m, and a 4 m diameter center feed-well projecting down 1.47 m. The results showed that averaging the $S_m$ and $S_u$ frequently overestimated the $S_{SB}$ by a large amount. A different relationship between $S_{SB}$ and parameters which are normally measured for operational purposes was developed as an improved method for estimation of the $S_{SB}$.

$$S_{SB} = S_u \exp [-K_f (Q/A_s)(SV_f^2)/S_m] \qquad (19)$$

where $S_u$ = underflow suspended solids concentration (mg/L), $SV_f$ = sludge volume fraction (dimensionless) after 30 minute settling in an one liter graduated cylinder and $S_m$ = operating mixed liquor suspended solids concentration in the aeration tanks (dimensionless). The relationship shows that an increasing overflow rate at times of hydraulic overloading will result in an exponential decrease in the ratio of $S_{SB}$ to $S_u$. Also, the $SV_f$ has a marked effect on the ratio of $S_{SB}$ to $S_u$ and thus on the amount of suspended solids which can be stored in the settling tanks. The increase of $S_m$ tends to increase $S_{SB}$ concentration in the settling tanks.

The area needed for solids storage can be obtained from equation (19) and (17).

$$M_{sd}/\{(A_s)(z)(h_w)\} = S_u \exp[-K_f(Q/A_s)(SV_f^2)/S_m] \qquad (20)$$

The area of settling tanks ($A_s$) required for the design solids storage ($M_{sd}$) can be calculated from equation (20) by using a trial and error method.

**Illustration of Settling Tank Design Based on Solids Storage** - In order to illustrate the design procedures, the following hypothetical example is presented. The treatment conditions are the same as those of WRTP plant. The average wastewater flow rate (Q) is 124,905 m³/d (33 MGD). The volume of aeration tanks ($V_a$) which is mechanically aerated is 37,850 m³ (10 MG). The representative operating $S_m$ value at the WRTP was 1,800 mg/L (0.0018) and the typical SVI value was about 125 mL/g. The linear regression of equation (19) for 56 sample data from WRTP plant resulted in $K_f = 0.00076$ d/m.

1)  The underflow suspended solids concentration ($S_u$) during hydraulic overloading period is not significantly different from the value for normal flow condition. During a normal flow condition, the return sludge flow rate (R) is set 50 % of influent flow rate (Q) and then $dM_s/dt = 0$. The $S_u$ can be calculated by

$$S_u = (Q+R)(S_m)/R = [(33 + (0.5\times33))(1,800)] / (0.5 \times 33) = 5,400 \text{ mg/L}$$

2)  The designer can establish the value of design storage ratio ($\Phi$) needed to protect against storage failure. The amount of sludge solids in the aeration tanks is

$$M_a = (37,850 \text{ m}^3)(1,800 \text{ mg/L})(0.001) = 68,130 \text{ kg}$$

The initial amount of solids in the settling tanks can be calculated by using equation (11). The initial SBD which was usually observed at the WRTP during normal flow condition is about 1 m and the area of the settling tanks is 4,930 m². For SVI = 125 mL/g, $S_u$ = 5,400 mg/L, and O.R. = 25.3 m/d,

$$S_{SB} = (5,400 \text{ mg/L}) \exp[(-0.00076)(124,905/4930)(0.225^2/0.0018)]$$
$$= 4,349 \text{ mg/L}$$

and $M_s = (4,930 \text{ m}^2)(1\text{m})(4,349 \text{ mg/L})(0.001) = 21,441 \text{ kg}$.

For the value of $\Phi = 0.4$, the design solids storage of the settling tanks ($M_{sd}$) is obtained by

$$M_{sd} = \Phi(M_a+M_s) = (0.3)(68,130 \text{ kg} + 21,441 \text{ kg}) = 35,829 \text{ kg}$$

3)  The ratio (z) of a critical sludge blanket depth to the side water depth of settling tank ($h_w$) is about 0.7 and the side water depth of settling tanks is 4.57 m. The design peak-flow rate is 249,810 m³ (66 MGD) which is two times higher than the average flow rate (33 MGD).

4) The settling tank area required for design $\Phi=0.4$ is obtained by solving equation (20).

$$\{(35,829)(1,000)]/[(A_s)(0.7)(4.57)\}$$
$$= \exp\{(-0.00076)(249,810/A_s)(0.225^2/0.0018)\}$$

The calculated $A_s$ for $\Phi = 0.4$, SVI = 125 mL/g and peak-factor = 2 is 5,488 m$^2$, which suggests that 558 m$^2$ of more storage space of settling (5,488 - 4930) should be provided so that storage in the settling tanks would be adequate to prevent failure during design storm flow.

## DISCUSSION

Various values of the storage ratio ($\Phi$), SVI, $S_m$, and initial sludge blanket depth (SBD$_i$) may be assumed and the corresponding area of settling tank ($A_s$) can be calculated while holding the design storm flow constant. For a given value of SVI and storm flow, equation (20) can be used to construct a family of curves for the area of the settling tanks ($A_s$) as a function of dimensionless storage ratio ($\Phi$). Also, the corresponding overflow rate ($Q/A_s$) may be plotted as a function of $\Phi$ with a different SVI value. Figure 2 would correspond to SVI values from 100 to 150 mL/g and to $\Phi$ values from 0.2 to 0.6.

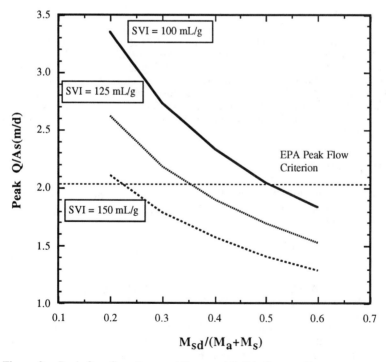

Figure 2    Peak Overflow Rate and Proposed Solids Storage Parameter

The horizontal line at $Q/A_s$ = 2.04 m/hr [1,200 gallon/ft²/d] on Figure 2 represents a criterion required for the clarification during peak-flow rate established by the EPA.

There will be different settling tank areas required for each storage factor ($\Phi$) and SVI value. Figure 2 shows that the higher design $\Phi$ value requires more storage area of settling tanks and also more area should be provided for the increased SVI value (this is shown by the decrease of overflow rate for the given peak flow). The overflow rate corresponding to $\Phi$= 0.4 and SVI = 125 mL/g is about 1.8 m/hr [1,060 gallon/ft²/day] which is below the horizontal criteria line for clarification. This means that the area required for solids storage is larger than that required for clarification.

As the design $\Phi$ value increases, the area requirement for solids storage becomes greater compared with that required for clarification. The intersection point between the area ($A_s$) and the overflow rate lines represents the solids storage ratio which can be achieved when the settling tank is designed by the peak overflow criteria.

The current area of the settling tank ($A_s$) at the WRTP is 4,930 m² and overflow rate = 2.11 m/hr which corresponds to 0.33 of storage ratio when the operating SVI is assumed to be 125 mL/g. The WRTP plant plans to build another settling tank which would increase the total area of settling tanks from 4,930 m² to 6,163 m². This project would increase the solids storage capacity of settling tanks from $\Phi$ = 0.33 to $\Phi$ = 0.47 which results in much safer operation during storm flow conditions.

## SUMMARY

A dimensionless ratio ($\Phi$) of maximum solids storage in the settling tanks before wash-out starts to the amount of solids in the process during normal operation is proposed for the design of the settling tanks. The designer should establish the value of $\Phi$ needed to protect against storage failure. In order to illustrate design procedures, the hypothetical design problem for existing activated sludge process is presented. As the design $\Phi$ and SVI value increases, the area requirement for solids storage becomes greater compared with that for clarification.

## REFERENCES

1.  Kim, Youngchul. 1995. "Solids Storage Function of Activated Sludge Settling Tanks During Hydraulic Overloads," *Ph.D. dissertation*, Drexel University.

2.  EPA, U.S. 1975. *Process Design Manual For Suspended Solids Removal*, EPA 625/1-75-003a.

3.  WEF and ASCE. 1991. *Design of Municipal Wastewater Treatment*, WEF   Manual of Practice. No.8, Vol.1.

4.  Daigger,G.T., and Ropper, R.E. 1985. " The Relationship between SVI and Activated Sludge Settling Characteristics," *Journal Water Pollution Control Federation*, 57(8):859-866.

5.  Keinath, T.M. 1990. "Diagram for Designing and Operating Secondary Clarifiers According to the Thickening Criterion," *Journal Water Pollution Control Federation*, 62(3):254-258.

6.  Metcalf and Eddy 1991. *Wastewater Engineering: Treatment and Disposal*, MaGraw Hill.

7.  Gunthert, F.W. 1984. "Thickening Zone and Sludge Removal in Circular Final Settling Tanks", *Water Science and Technology*, 16:303-316.

8.  Vitasovic, Z.Z. 1989. " An Integrated Control Strategy for the Activated Sludge Process," *Ph.D. dissertation*, Rice University.

9.  Thompson, D. 1988. "Activated Sludge: Step Feed Control to Minimize Solids Loss during Storm Flow," *M. Eng. Thesis*, McMaster University.

10. Parker, D.S. 1983. "Assessment of Secondary Clarification Design Concepts," *Journal Water Pollution Control Federation*, 55(4): 349-359.

11. Ghobrial, F.H. 1977. "Importance of the Clarification Phase in Biological Process Control," *Water Research*, 12:1009-1016.

12. Pflanz, P. 1969. "Performance of Secondary Sedimentation Basin," In *Advances in Water Pollution Research* (edited by Jenkins, S.H.). pp. 569-581. Pergamon Press, London.

13. Mynhier, M.D. and Grady, C.P.L. 1976. "Design Graphs for Activated Sludge Process," *Journal of Environmental Engineering*, ASCE, 101 (EES): 829-846.

14. Patry, G.G. and Takacs, I. 1992. "Settling of Flocculent Suspensions in Secondary Clarifiers,"*Water Research*, 26:473.

15. Kim, Youngchul, Pipes, W.O. and Chung, P.G. 1995. "Activated Sludge Process: Estimation of Sludge Blanket Suspended Solids Concentration in the Settling Tanks," In *Proc. of I.A.W.Q-Asian Regional Conference on Water Quality and Pollution Control*, pp. 221-229, Manila, Philippine, February 5-9

# ENHANCED DEGRADATION OF DINITROTOLUENE BY *PHANEROCHAETE CHRYSOSPORIUM* IN PACKED-BED BIOREACTOR

NIRUPAM PAL
CHRISTOS CHRITODOULATOS
SAI KODALI
Center for Environmental Engineering
Stevens Institute of Technology
Hoboken, NJ 07030

## INTRODUCTION

Due to recent demilitarization activities in United States, Eastern Europe, and former soviet union, a large quantity of propellants, explosives, and pyrotechnics are to be disposed in an environmental safe manner [7]. The nitroaromatic compound 2,4-Dinitrotoluene (DNT) is a precursor for the manufacture of 2,4,6-trinitrotoluene and polyurethane foam and extensively used in manufacturing propellants and explosives [5]. Wastes from these manufacturing processes contain DNT, a priority pollutant, and found to be toxic to various forms of life including mammals [3]. The manufacture of explosives, cleaning and repackaging requires large quantities of water, which become contaminated with explosives during explosive processing. For years, explosive-contaminated wastewater (pink water) was discharged on ground or lakes. Approximately 9% of the organic content of the pink water come from DNT [3]. Due to its recalcitrant nature, this toxic chemical contaminated groundwater and soil. A number of treatment technologies have been proposed to convert DNT to innocuous products or to separate it before it reaches the waste stream. The current method of remediation for explosive contaminated wastes is Open burning and Open detonation (OB/OD) [7]. This method is costly, energy intensive, and not suitable for explosive containing wastewater. In addition, due to production of NOx, OB/OD may be discontinued in future. Other alternative technologies such as adsorption of DNT on activated carbon as a means of separating DNT from wastewater and anaerobic degradation of DNT in presence of a co-substrate has been proposed. Spanggord et al. [7] identified a *Pseudomonas* sp. obtained from DNT contaminated waste stream that uses DNT as the sole carbon source.

Recently a basidiomycete, *Phanerochaete chrysosporium*, commonly known as a 'white rot fungus' have been shown to be a potential microorganism for mineralizing DNT [5]. Use of this fungus has been tested for decontaminating explosive contaminated soil [5]. Valli et al. [9] reported degradation mechanism of DNT by *P. chrysosporium* and showed that it can completely mineralize DNT. The extraordinary ability of *P.chrysosporium* to mineralize a large number of xenobiotic compounds over a wide range of pH and temperature is well documented [1,2,6]. Until now, the major drawbacks for utilizing this fungal culture for commercial application was (i) its slower degradation rate compared to bacterial process; (ii) high dependability of degradation rate on nutrient conditions and the optimum nutrient conditions were not known; (iii) sensitivity to shear stress; and (iv) incomplete information on mechanism of degradation. Recently Pal et al. [6] have shown that in case of trichlorophenols, the degradation rate can be substantially increased using a continuous reactor system where the shear-stress effects can be eliminated and continued supply of energy source can be maintained. These researchers have shown that the optimal carbon to nitrogen ratio (C:N) varies between 60 to 80, and the volumetric flux around $1.0 \ mL/(cm^2 \ x \ min)$. The complete mechanism for mineralization is yet to be revealed. Recent studies showed that the first step in the mineralization process is initiated by extracellular enzymes or free radicals forming a complex [2]. The complex is subsequently mineralized by biomass to innocuous products [1]. Although a number of studies have been performed on biodegradation of DNT by both bacterial and fungal cultures, but objectives of all these studies were the microbial or biochemical aspect of degradation processes rather than engineering aspect of implementing the technology.

The objective of this study is to enhance the degradation rate of DNT by selecting proper reactor configuration and subsequently to model the degradation kinetics for scale-up purpose to treat wastewater containing DNT.

## EXPERIMENTAL PROCEDURE

### Growth of Fungal Mycelium in Batch Fermenter

*Phanerochaete chrysosporium* (ATCC 24725) was obtained from the American Type Culture Collection (ATCC) and maintained on yeast malt agar. The fungus was grown in growth medium (Table I), in Erlenmeyer flask on a gyratory shaker at 30°C.

TABLE I: Composition of Growth Medium and Induction Medium

| Components | Growth Medium | Induction Medium |
|---|---|---|
| Glucose | 6.0 g | 0.5-0.9 g |
| $KH_2PO_4$ | 2.0 g | 2.0 g |
| $NaNO_3$ | 0.2 g | 15-40 mg |
| $MgSO_4$ | 0.5 g | 0.5 g |
| $CaSO_4$ | 0.1 g | 0.1 g |
| Mineral Salt Solution* | 5 mL | 5 mL |
| Thiamin Hydrochloride | 5 mg | 5 mg |
| Deionized Water | 1.0 L | 1.0 L |

*The composition of mineral salt solution is given elsewhere [1].

After growing for five days, the glucose and nitrogen concentration dropped to about 30 mg/L and less than 1 mg/L respectively. The pH dropped to about 3.0. This fungal culture was used as inoculum as described later.

## Batch Experiments in Shaker Flasks

Batch experiments were conducted in shaker flasks. A stock solution of $200 \pm 10$ mg/L DNT (98% purity, Sigma Chemical Co.) was prepared. Aliquots of the stock solution were added to the shaker flasks containing growth medium and as required. Control flasks containing growth medium, inoculum and acid/base as pH adjusters were autoclaved, and supplemented with DNT. All flasks (including controls) were incubated at 30°C on a gyratory shaker rotating at 150 rpm. Samples of different volumes (2 or 5 mL) were taken from the shaker flasks to measure DNT, nitrogen, and glucose concentrations.

## Experiments in Continuous Packed-Bed Reactors

A schematic of the set up is shown in Figure 1 (length 45.7 cm and internal diameter 10 cm). The reactor was equipped with a feed pump and air sparger at the bottom, and provided with an external recirculation loop (including a circulation pump) was used in each experiment. Clear polyethylene terephthalate (PET) flakes (irregular in shape and size, cross-sectional area = 2-15 mm$^2$; thickness $\cong 0.5$ mm) were obtained from the Polymer Recycling Plant, Rutgers University, New Brunswick, NJ.

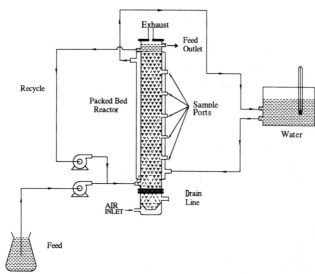

FIGURE 1: Schematics of the Fixed-film Bioreactor Setup

After cleaning and autoclaving these flakes were used as random packing material (void volume: 50-60%). The temperature in the reactor was maintained at 30° C by circulating water from a water bath through the reactor jacket. The reactor was continuously fed from a feed tank at feed rates between 0.5 and 3.0 mL/min. The external recirculation rate was between 6 to 20 mL/min, and was always much larger than the feed rate to ensure that the liquid content of the reactor was always well mixed. To confirm this, a separate study was conducted in which a step change in the concentration of a tracer in the feed stream was imposed while the tracer concentrations at three sampling ports along the reactor were measured. The port concentrations were compared with the theoretical response of a corresponding continuously stirred tank reactor (CSTR) and found to behave like a CSTR (results not shown).

To immobilize the fungus in the packed-bed reactor, a 5-day old fungal culture broth from shaker flask was transferred to the reactor with the simultaneous addition of PET flakes. The fungus was allowed to grow for an additional period (2 to 10 days depending on the experiment) while continuously feeding growth medium and maintaining external recirculation. Then, the feed was switched from growth medium to induction medium. These media differed only in their glucose and $NaNO_3$ compositions (Table I). The induction medium was supposed to stimulate the secretion of extracellular enzymes by limiting the availability of nutrients to the fungus [1,2,6]. The induction medium fed to the reactor was supplemented with DNT. Samples from different ports along the length of the reactor were analyzed for DNT, nitrogen, glucose, pH, nitrate, and total protein concentrations.

The attainment of steady state condition was confirmed when no significant change in outlet concentrations was observed throughout a period of at least 3 to 4 days.

## ANALYTICAL METHODS

### Determination of Nitrogen, Chloride, Glucose and Protein Concentrations

The concentrations of nitrogen and chloride ion were measured using specific electrodes (Orion Inc., Model Nos. 93-07, 96-17B, respectively) placed directly in the samples collected from the fermenter, reactor, or shaker flasks. Fixed nitrogen was measured as nitrate ion concentration but reported as N. Glucose was assayed using the ortho-toluidine reaction method [1]. The protein content was measured as described elsewhere [6].

### Determination of DNT and Transformation Intermediate Concentrations

Aqueous phase concentration of DNT and other intermediates was determined using an HPLC (Varian, Model 9050) provided with a tunable detector (Varian, Model 9065) and autosampler (Varian, Model 9095) with a C8 Ultrasphere column (Alltech Associates Inc.). A mixture of methanol and deionized water in gradient method (Water is to methanol ratio varied from 70:30 to 30:70) was used as the mobile phase and the absorbence was measured at 254 nm. During analysis, aqueous samples from the reactors were spun for 10 minutes at 12,000 rpm in an ultracentrifuge (IEC Centra-M, International Equipment Co.) to separate the biomass. Then 60 µl of each sample was injected for analysis via auto-sampler. The reproducibility of the analysis was within ±0.37 mg/L.

## PROPOSED REACTION SCHEME AND MATHEMATICAL MODEL

Armenante et al.[1] have recently shown that the degradation of 2,4,6-TCP by *P. chrysosporium* is a process that requires the simultaneous presence of both the mycelium and the extracellular enzyme system released by the mycelium to initiate degradation. These investigators have also shown that the rate limiting step in the process is the concentration of extracellular protein, and that the concentration of mycelial biomass does not affect the rate of degradation, although the presence of some biomass is necessary for degradation to occur. Based on this reaction scheme, Pal et al. [6] proposed the following reaction scheme:

$$C + E_{ex} \Leftrightarrow C\text{-}E_{Ex} \text{ (Complex)} \tag{1}$$

$$C - E_{ex} \Rightarrow E_{Ex} + \text{Intermediate (s)} \tag{2}$$

$$\text{Intermediate } + E_{Cell\,bound} \Leftrightarrow \text{Intermidiate-}E_{Cell\,bound} \text{ (Complex)} \tag{3}$$

$$\text{Intermidiate-}E_{Cell\,bound} \Leftrightarrow \text{Products } + E_{Cell\,Bound} \tag{4}$$

A justification for this scheme is the following. It has been reported by many investigators that the fungus is able to break down large molecules, such as lignin, having molecular weights in excess of 600 kilodaltons [2]. Such large molecules are unlikely to be transported across the cell membrane. This leads to the conclusion that the first step in the degradation process is typically an extracellular enzymatic or free radical reaction. Since DNT can not support fungal growth as the carbon source (it can support its nitrogen requirement), one can assume that the second step in the reaction scheme occurs on surface bound enzyme or by biomass itself. However, the mechanism of nitrogen transfer is not clear yet, which may involve intracellular transfer of the intermediate(s). The schematic mechanism is given in Figure 2.

Therefore, a Michaelis-Menten model was used here to describe the rate of degradation of the target compound according to the above reaction scheme.

$$r_c = \frac{V_p\, P\, C}{K_m + C} \tag{5}$$

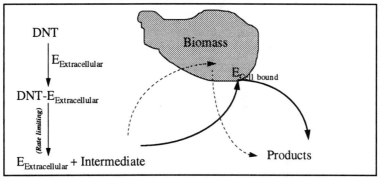

FIGURE 2: Schematics of Degradation Mechanism of DNT by *P.chrysosporium*

Where $r_C$ is the degradation rate, $C$ is the concentration of the target compound, $P$ is the concentration of extracellular protein(s) (proportional to the extracellular enzyme concentration, $E_{Ex}$), and $V_P$ and $K_m$ are the Michaelis-Menten parameters.

Therefore for a well mixed-reactor, the steady-state mass balance results:

$$\frac{P\tau}{C_0 - C} = \frac{K_m}{V_P C} + \frac{1}{V_P}$$

(6)

The above equation is valid when the enzyme deactivation is neglected or the retention time ($\tau$) is much less than the life time of the enzyme [6]. Equation (6) can be used to find the Michaelis-Menten parameters, $V_P$ and $K_m$, from a regression of the experimental data.

## RESULTS AND DISCUSSIONS

### Results of Batch Studies in Shaker Flask

The results of a shaker flask study (conducted in triplicate) are depicted in Figure III. The results show that DNT can be degraded by *P.chrysosporium* with glucose as the carbon source however it can not utilize DNT as the sole carbon source. During the batch study, a number of intermediates were observed and two of these intermediates were stable, identified as 4-methyl 3-nitro aniline and 2-amino 4-nitrotoluene. Similar intermediates were reported by Valli et al. [9] and they have elucidated degradation pathway.

From the results of the batch studies one can observe that the rate of degradation of DNT is significantly slower compared to the degradation rate by bacterial species observed by other researchers [7]. Slower rate of degradation has been seen by other researchers and attributed to depletion of energy source and accumulation of inhibitors [1,2].

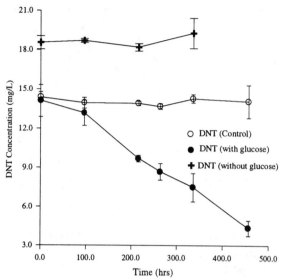

FIGURE 3: Concentration Profile of DNT in Shaker Flask Study

The initial step in this degradation process is initiated by a extracellular enzyme(s) released by this fungus. The generation of these enzymes is an energy consuming process, hence in a substrate depleted system the fungal culture can not secrete any enzymes [6]. This was further justified by the fact that in case of trichlorophenol degradation by *P.chrysosporium*, when glucose was added to a substrate depleted system, the degradation restarted [6]. This confirms that for continued degradation, continuous supply of energy source is essential.

On the other hand, when the substrates are in abundant, the fungi shift its metabolic activity from secondary to growth phase and no lignolytic enzymes are produced [2]. These two contradicting parameters must be optimized to attain continued high degradation rate which can not be achieved in conventional batch reactors. In addition, conventional stirred tank reactors are unsuitable since it exerts significant shear-stress on this filamentous microorganism during mixing operation [6].

**Results of Packed-Bed Reactor Studies**

A typical result for a packed-bed reactor is shown in Figure 4. At steady state 96.5% of the 2,4 DNT initially fed to the reactor was mineralized in 19.1 hours. The biomass loss from the bed was negligible (0.36 mg/h to 0.78 mg/h). Small concentrations of glucose (10-60 mg/L) and nitrogen (0.2-1 mg/L) were detected in the outlet. Because the system was not buffered, the pH of the feed and the outlet differed significantly (4.6 and 5.1 respectively).

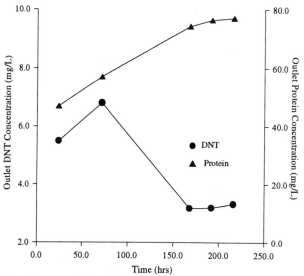

FIGURE 4: Development of Steady State Outlet Concentrations in Continuous Packed-bed Reactor (Inlet DNT concentration = 106 ± 0.9 mg/L, glucose = 600 mg/L, nitrogen 3.0 mg/L, retention time 19.1 hours)

A comparison of this results with that with batch study results shows that the specific degradation rates were very different in the two systems examined here (0.0208 mg DNT/(L × h) in shake flasks vs. 5.34 mg 2,4,6-TCP/(L × h) in the packed-bed reactor).

This increase in degradation rate can be attributed to the preference of the fungus to be attached to a solid support [6], and the constant availability of low levels of carbon and nitrogen in the packed-bed reactor (which likely enable the fungus to synthesize fresh extracellular enzyme to initiate the attack on the target molecule).

In addition, any effects related to shear-stress was eliminated in this reactor configuration. Therefore one can observe two orders of magnitude increase in degradation rate in packed-bed reactor compared to suspended growth system in shaker flask.

The degradation rate in packed-bed bioreactor is a strong function of residence time, pH and nutrient conditions [6]. Further studies have shown that improper parameters not only reduce the degradation rate but a large number of undegraded intermediates appear in the outlet stream during chromatographic analysis (results not shown).

**Results of Modeling Studies**

Series of experiments were conducted in packed-bed reactor at a fixed pH of 5.6. Figure 5 shows the regressions of obtained at a pH of 5.6 using equation 6.

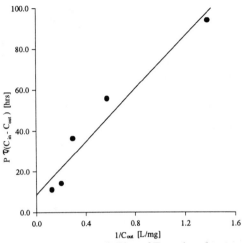

FIGURE 5: Plot of Equation 6

Regressions of the experimental degradation data from the set of results were used to determine the Michaelis-Menten parameters $V_P$ and $K_m$ in equation 6. Since variables such as shear stress, nutrient concentration, and pH can affect the value of these parameters only selected data sets were used in the regressions.

Specifically, the data regressed came from experiments in which shear stress effects were negligible (i.e., the volumetric flux was below 1.0 mL/(cm$^2$ × min)), the glucose concentration was less than 50 mg/L, the nitrogen concentration was low enough to prevent inhibitory effects (less than 2 mg/L), and the inlet C:N ratio was between 60 and 80. As for pH, different data sets were considered in which the pH was kept constant throughout the experiment.

Each data point in this figure represents a steady-state condition. The correlation coefficient for the line was 0.93. This figure shows that the model adequately fits the experimental data. The values of the Michaelis-Menten parameters were calculated from the slope and intercepts of such plots. Similar plot was obtained at pH 4.6 and the resultant data shows to match the model. The results show that $V_P$ is a strong function of pH however $K_m$ remains unchanged for all practical purposes. These results indicate that a single enzyme is involved in the degradation process [6].

**CONCLUSION**

The following conclusions can be drawn from this study:
- The use of a continuous packed-bed reactor containing the immobilized fungus can enhance the degradation rate of 2,4-dinitrotoluene by two orders of magnitude as compared to a batch reactor containing suspended biomass.

- A Michaelis-Menten kinetic model, based on a reaction scheme involving a rate controlling step of formation of the reaction intermediate, appears to correlate the experimental data well. The value of the Michaelis-Menten parameter $V_P$ is strongly dependent on the system pH, however pH has no effect on $K_m$.

REFERENCES

1. Armenante, P.M., N. Pal, and G. Lewandowski. 1994. "Role of Mycelium Protein and Extracellular Protein in Biodegradation of 2,4,6-Trichlorophenol by *Phanerochaete chrysosporium,*" *Appl. & Env. Microbiol,* 60:1711-1788.

2. Barr, D.P., and S. Aust. 1994. "Mechanisms White Rot Fungi Use to Degrade Pollutant," *Environ. Sci. Technol,* 28:78A-87A.

3. Keith, L.H., and W.A. Telliard. 1979. "Priority Pollutant-A Perspective View," *Env. Sc. & Tech,* 13:416-423.

4. McCormick, N.G., J.H. Cornell, and A.M. Kaplan. 1986. "Identification of Biotransformation Products of 2,4-Dinitrotoluene," *Appl. & Env. Microbiol,* 35:945-948.

5. McFarland, M., S. Kalaskar, and E. Baiden. 1992. "Composting of Explosives Contaminated Soil Using the White Rot Fungus *Phanerochaete chrysosporium,*" Final Report by Utah Water Research Laboratory, Utah State University, Logan, UT. Contact No. DAAL03-91-C-0034.

6. Pal, N., G. Lewandowski, and P.M. Armenante. 1995. "Process Optimization and Degradation Modeling for Biodegradation of Trichlorophenols by *Phanerochaete chrysosporium,*" *Biotec. & Bioeng,* 46:599-609.

7. Spanggord, R.J., J.C. Spain, S.F. Nishino, and K.E. Mortelmans. 1991. "Biodegradation of 2,4-Dinitrotoluene by a *Pseudomonas Sp.,*" *Appl. & Env. Microbiol,* 57:3200-3205.

8. Spontareli, T., G.A. Buntain, J.A. Sanchez, and T.M. Benziger. 1992. "Destruction of Waste Energetic Materials Using Base Hydrolysis," *Waste Energetic Materials,* 787-791.

9. Valli, K., B.J.Brook,.D.K. Joshi, and M.H. Gold. 1992. "Degradation of 2,4--Dinitrotoluene by the Lignin-Degrading Fungus *Phanerochaete chrysosporium,*" *Appl. & Env. Microbiol,* 58:221-228.

# ENHANCED BIODEGRADATION OF CREOSOTE CONTAMINATED SOIL USING A NONIONIC SURFACTANT

PATRICK P.E. CARRIERE
FEHMIDAKHATUN A. MESANIA

Department of Civil and Environmental Engineering
West Virginia University
Morgantown, West Virginia 26506-6103

## INTRODUCTION

There is a growing concern in the U.S. about the increasing number of industrial sites containing concentration of polynuclear aromatic hydrocarbons (PAHs) in their soil and waste sludge above background levels. PAHs , neutral and non-polar organic compounds, consist of two or more fused benzene rings which are generated from industrial activities such as creosote wood treating, gas manufacturing, coke making, coal tar refining, petroleum refining, and aluminum smelting. Low molecular weight PAHs are generally considered as extremely toxic compounds, whereas the higher molecular weight PAHs are carcinogenic in nature. The aqueous solubility and volatility of PAH's decrease with increasing molecular weight.

Bioremediation, a viable option for treatment of PAH contaminated soils, can reduce PAH concentration to acceptable levels. It is primarily a water-based process influenced by sorption (absorption/desorption), diffusion and dissolution mechanisms which serve to control the accessibility of the organics to the bacteria that are present in water. In most cases, sorption is the rate limiting step controlling both the rate and extent of biodegradation. The process of bioremediation can be enhanced by application of surfactant by increasing the availability of the organic compounds to the microorganisms.

In previous bioremediation studies, the use of several kinds of surfactant was found to enhance the solubility of hydrophobic compounds. The main objective of this study was to evaluate the effect of a nonionic surfactant on biodegradation of creosote contaminated soil.

## BACKGROUND

### Bioremediation

Bioremediation has been used since the 1970's to treat soils contaminated with petroleum hydrocarbons [1]. Many researchers have contributed to the development of this area. Extensive work has been performed by the Gas Research Institute [2] to study the mechanisms of sorption and biological oxidation during the process of bioremediation. Wu and Gschewend [3] have described the sorption kinetics in terms of a radial diffusive penetration model modified by a retardation factor that reflects micro scale partitioning of the sorbate between the pore fluids and the solids which make up the aggregate. This mathematical model has been tested against experimental data for lower ring PAHs and moderately hydrophobic organic chemicals, but not for very large soil aggregates or very hydrophobic substances due to the extended time frame required for complete equilibration. In addition, the model suggests that the physical/chemical properties of the PAH and the size of the soil aggregate are the major variables which influence the rate of PAH adsorption/desorption. PAHs may take considerably longer to become available to bacteria if the soil aggregates are larger and/or the number of aromatic rings in the PAHs are greater than three. This time will be extended further by aging of soil-waste matrices providing more time for diffusion of the PAHs into the aggregates. It has also been noted that sorption of hydrophobic organic compounds into soils is highly correlated with the soil organic fraction and the rate of soil mineralization [4].

Various methods of enhancing biodegradation of heavier PAH compounds have been investigated by Sherman *et al.,* [5] in Manufactured Gas Plant residues. These include the addition of microbial enhancements such as nutrients, and organic supplements. The enhancements were tested in pan studies and the preliminary results of the analyses indicated that the more recalcitrant PAHs may not be biologically available with some of the simpler forms of enhancement due to high sorption of the contaminant to the soil matrix.

## Surfactants

Several investigations in the last few years have assessed the potential for surfactants to assist in the treatment of contaminated soils. Most of this work has consisted of laboratory screening studies using soil-water suspensions or small column apparatus. Generally, surfactant solutions in the range of 1 to 4 % have been suggested as promising for enhancing removal of scrubbing fuel components, PCBs and other chlorinated hydrocarbons from soil [6].

The use of surfactants in bioremediation of petroleum contaminated soil has been extensively studied by the Texas Research Institute [7]. TRI investigated a combination of nonionic and anionic surfactants application to a sandy soil. The combination of a lower tension surfactant (Richonate-YLA) and a nonionic surfactant (Hyonic PE-90) proved to be the most effective with an efficiency of 80% removal of the gasoline residual when used for flushing gasoline from sandy soil. Other surfactants used were : Alfonic 1412-S, Poly-Tergent B-500, and Alrosol O. These surfactants resulted in formation of viscous emulsions and consequently, in a low flow rate and low removal efficiency.

Volk Field researchers in 1985 conducted a field test of in situ washing of a fire training pit. The subsurface soil was determined to be contaminated primarily with a medium weight oil (2,000 - 25,000 mg oil/ Kg soil) and (600 - 3,500 mg/Kg soil) of chlorinated compounds. Two types of surfactants were used as wash solutions:

1) Synthetic surfactants that are commercially available, and 2) natural surfactants that had their origins at the fire training pit and were the by products of biological activity. These solutions were added to a series of test cells and it was found that the use of surfactants on in situ soil washing did not prove successful due to by-passing of contaminated zones by the wash solutions.

Luthy and coworkers [8], have demonstrated that the addition of surfactants to soil-water systems solubilizes PAH compounds by incorporation of the contaminant in the surfactant micelles. The solubilization of anthracene, phenanthrene and pyrene was evaluated in soil-water suspensions with several nonionic and anionic surfactants in batch equilibrium extraction tests. According to experimental results obtained by Luthy et al., [8] the most effective surfactant was nonionic octylphenylethoxylate with 9 to 12 ethoxylate units (Triton X-100).

## MATERIALS AND METHODOLOGY

To evaluate the effect of the nonionic surfactant on degradation of creosote contaminated soil, abiotic and respirometric experiments were performed on soil samples with and without surfactant. Triton X-100 was the surfactant selected for the study. It is a commercial name for polyethylene glycol p-tert-octylphenyl ether with chemical formula : $C_8H_{17}(C_6H_4)O(CH_2CH_2O)_xH$ and a non biodegradable product with a specific gravity of 1.07 g/mL. Abiotic experiments were conducted to determine the kinetics of release of the contaminants in the aqueous phase, while respirometric experiments were performed to monitor the oxygen uptake by the microorganisms. The oxygen uptake is an indication of the degradation of the contaminant. A sandy soil was collected and contaminated at contamination levels of 112.5, 275.6, 551.5, and 1,102.5 mg of creosote per kilograms of soil. The contaminated soil was placed on a shaker for two days. For the abiotic experiments, 3-, 4-, 5-, and 6-ring PAHs which are acenaphthylene(ACL), benzo(a)anthracene(BAA), benzo(b)fluoranthene(BBF), and indeno(1,2,3-cd)pyrene(ICDP), respectively were selected and monitored.

### Abiotic Experiments

Samples without surfactant, consisted of 15 grams of contaminated soil, 45 mg of mercuric chloride (HgCl₂), and 45 mL of buffered water solution ( 200 mg/L of NaHCO₃), were placed on a shaker for contact time of 24, 48, 72, 96, and 120 hours. At each contact time, duplicate samples were removed from the shaker and centrifuged at 15500 rpm for 25 minutes using Sorvall RC-5C Plus. The solid part was extracted by using a soxhlet extraction (EPA Method 3540), and the extract analyzed for chemicals of interest (ACL, BAA, BBF, and ICDP). The soxhlet extraction method consisted of mixing the soil sample with anhydrous sodium sulfate, placing it in an extraction thimble, and extracting with using methylene chloride. A sample set with blank samples with uncontaminated soil were performed under abiotic conditions by following the same procedures.

296

For samples with surfactant, the same procedure was followed, but an amount of nonionic surfactant (Triton X-100) at a concentration of 0.35mg/Kg and 0.71mg/Kg was added to each sample bottle. A sample set with blank samples with uncontaminated soil and corresponding amount of surfactant was performed under abiotic conditions by following the same steps.

## Respirometric Experiments

An N-Con respirometer monitoring the oxygen uptake by bacteria in a closed system was used to investigate the effect of surfactant on degradation of creosote. The respirometer consists of an oxygen cylinder, an isothermal bath, 6 reactors equipped with solenoid valves and vacuum switches, and a personal computer equipped with an I/O interface. The reactors consist of 500 mL Schoth Duran glass bottles equipped with magnetic stirrers and KOH traps. The magnetic stirrers were used to provide an uniform oxygen distribution in the solution. During the biological reaction, oxygen is consumed and carbon dioxide ($CO_2$) is produced. The oxygen uptake is an indication of the degradation rates of creosote contaminated soil and the oxygen supply rates required. $CO_2$ removal from the headspace is facilitated by KOH. The oxygen consumption causes a negative pressure inside the reactor which is detected by a vacuum switch by comparing with the reference pressure. Then the vacuum switch opens the solenoid valve to deliver a measured amount of oxygen to the reactor. The amount of oxygen delivered is monitored by the computer and is reported as mgBOD/L versus time.

For each level of contamination, respirometric experiments were performed in duplicate for samples with and without surfactant. The batch reactors were prepared in Schoth Duran glass bottles equipped with magnetic stirrers and KOH traps. The biological reactors were prepared by adding 10 mL of microbial seed (acclimated activated sludge with creosote), 2.5 mL each of nutrient solutions phosphate buffer, magnesium sulfate, calcium chloride, and ferric chloride (Standard Methods 507, 1985), and 15 grams of contaminated soil. Deionized water was added to make up a volume of 300 mL. The samples were placed in the respirometer for approximately 480 hours. After preparing the reactors, KOH traps were filled with KOH pellets to remove $CO_2$ from the headspace in the closed system. Failure to do so may invalidate the test and kill the bacteria.

## RESULTS AND DISCUSSIONS

### Abiotic Experiments

Desorption tests were performed in duplicate to determine the kinetics of release of selected PAH compounds from the soil matrix to the aqueous phase. The release percentage of the PAHs at different contact time without and with 0.35 mg/kg of surfactant are presented in Figures 1 and 2. As observed, comparison of samples with and without surfactant for the selected PAHs showed higher desorption when surfactant was present. Faster desorption was noticed after 24 hours of contact time when surfactant was present.

The 3- and 4-ring PAHs showed higher desorption after 24-hr contact time than the 5-, and 6-ring PAHs for samples without and with surfactant. As noticed, approximately 90% of the total percentage removal, for 3- and 4-ring PAH for samples with surfactant concentration of 0.35mg/Kg, were desorbed after a contact time period of 24 hours while approximately 85% for the 5- and 6-ring PAH were desorbed. For samples without surfactant the 3- and 4-ring PAH showed that about 24% of the total percentage removal were desorbed after a contact time period of 24 hours. The 5- and 6-ring PAH presented approximately 29% of the total percentage removal.

**Respirometric Experiments**

The amount of oxygen monitored by the respirometer for samples with different level of creosote contamination is presented in Figures 3 and 4. As observed in Figure 3 for samples without surfactant, increase in contamination level from 112.5 mg/kg to 1,102.5 mg/kg showed increase in oxygen uptake. For samples with surfactant concentration of 0.35 mg/kg, Figure 4(a) indicated that increase in contamination level from 112.5 mg/kg to 771.8 mg/kg showed increase in oxygen uptake. But for contamination level of 1,102.5 mg/kg, the oxygen uptake was similar to the contamination level of 771.8 mg/kg. This may be due to inhibition of biological activity at high level of contamination.

As noticed in Figure 4(b), increase in surfactant concentration from 0.35 to 0.71 mg/Kg showed an increase for contamination level of 112.5 mg/kg, but a 80% decrease in oxygen consumption for contamination level of 275.6 to 1,102.5 mg/kg. The inhibition of biological activity was more significant. As mentioned previously, creosote is composed at nearly 300 compounds, any of which is potentially toxic to microorganisms at certain concentrations. It is possible that at a certain surfactant concentration level above 0.35 mg/Kg soil, creosote availability in the aqueous phase becomes toxic to the microorganisms, thus causing inhibition and reducing metabolic oxygen demand.

**CONCLUSIONS**

Surfactant addition can be considered as a possible method to enhance bioremediation of contaminated soils with PAHs compounds, showing increase in contaminant availability to the microorganisms when compared with samples without surfactant. The abiotic kinetics results demonstrated that PAH compounds can desorb at different rates. The 3- and 4-ring PAH compounds achieved a greater removal and desorption rate than the 5- and 6-ring PAHs. For the 3- and 4-ring PAHs, 90% of the total contaminant removed occurred after 24 hours of contact time, while 30% for samples without surfactant. For the 5- and 6-ring PAHs, 80% of the total contaminant removed occurred after 24 hours of contact time, while 15% for samples without surfactant.

# REFERENCES

1. Roberts, Eve Riser. 1992. *Bioremediation of Petroleum Contaminated* Sites, CRC Press, Inc., NY.

2. Gas Research Institute. 1992. "The GRI Acceleration Biotreatability Protocol for Accessing Conventional Biological Treatment of Soils: Development and Evaluation Using Soils from Manufactured Gas Plant Sites". Report No. GRI-92/0499, Chicago, IL.

3. Wu, S.C., and Gschewend, P.M. 1986. "Sorption Kinetics of Hydrophobic Organic Compounds to Natural Sediments and Soils", *Environmental Science Technology*, 20:717-725.

4. Laha, Shonali, and Luthy, Richard. 1991. "Inhibition of Phenanthrene Mineralization by Nonionic Surfactant in Soil-Water Systems". Department of Civil Engineering, Carnegie Mellon University, Pittsburgh, PA, 15213.

5. Sherman, Donald F., Stroo, Hans F., and Bratina, James. 1990. "Degradation of PAHs in Soils Utilizing Enhanced Bioremediation", Remediation Technologies, Austin, TX, 78705.

6. Liu, Zhongbao, Laha, Shonali, and Luthy, Richard. 1990. "Surfactant Solubilization of Polycyclic Aromatic Hydrocarbons Compounds in Soil-Water Suspensions", *15th Biennial International Conference*, Kyoto, Japan, July.

7. Green, Guy. 1989. "The Use of Surfactant in the Bioremediation of Petroleum Contaminated Soils", *Environmental Protection Agency*, Washington, DC, EPA/PB90-256546.

8. Luthy, Richard G., Dzombak, David A., Roy, Sujoy B., Ramaswani, Anuradha, Nakles, David V., and Nott, Babu R. 1994."Remediating Tar-Contaminated Soils at Manufactured Gas Plant Sites", *Environmental Science Technology*, 28(6):266-276.

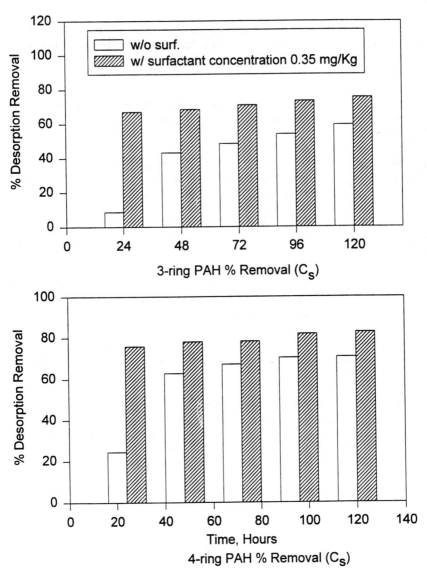

Fig. 1: Percentage of Desorption for 3- and 4- ring PAH Compounds

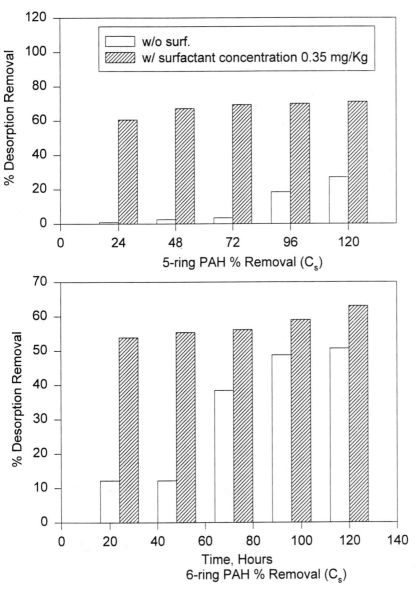

Fig. 2: Percentage Desoprtion for 5- and 6-ring PAH Compounds

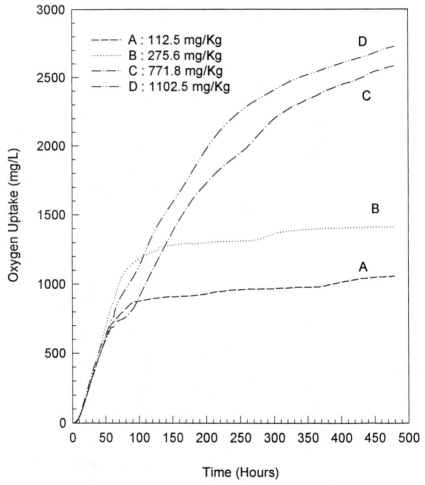

Fig. 3: Oxygen Uptake for Samples w/o Surfactant

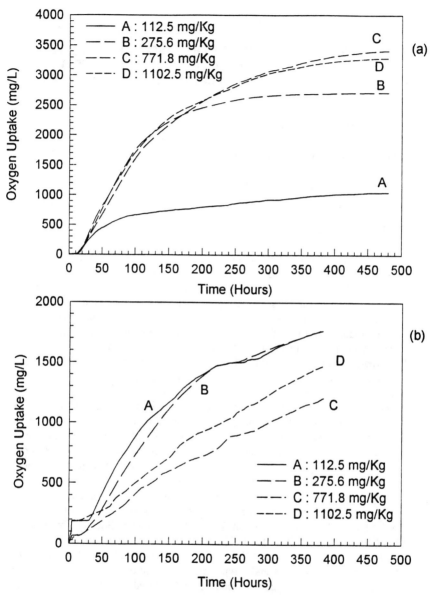

Fig. 4 : Oxygen Uptake for Samples With Surfactant
(a) 0.35 mg/Kg Soil
(b) 0.71 mg/Kg Soil

# A SEQUENTIAL CHECKLIST FOR THE ASSESSMENT OF NATURAL ATTENUATION OF DISSOLVED PETROLEUM CONTAMINANT PLUMES FROM LEAKING UNDERGROUND STORAGE TANKS

NICHOLAS DE ROSE, P.G.
Langan Engineering and Environmental Services, Inc.
Georgetown Crossing, Suite 225
3655 Route 202
Doylestown, Pennsylvania 18901-1699

## INTRODUCTION

Estimates of the number of leaking underground storage tanks (UST) are measured in the hundreds of thousands in the United States alone [1]. Commercial products and chemicals stored in USTs contain some of the most common and notorious environmental contaminants. This includes benzene in gasoline and trichloroethylene in commercial solvents. The promulgation of UST regulations in the 1980's was the result of the perceived large number of these potential contaminant sources and the lack of appropriate requirements for UST management. As a result of experiences gained by complying with the new Federal and State regulations, we are now beginning to understand the dynamics of UST releases.

The most common types of products stored in USTs include petroleum products. Therefore, the release of petroleum products to the environment, including diesel fuel, heating oil, gasoline and other petroleum distillates have been subject to the greatest amount of study by industry and government agencies. Through the gained experience and conducted research, new investigation and remediation approaches have been developed and accepted by the regulatory, professional and business communities. This is evidenced by the publication of an "Emergency Standard Guide for Risk-Based Corrective Action Applied at Petroleum release Sites" by the American Society for Testing Materials (ASTM) in July, 1994 [2].

The concept of "Risk-Based Corrective Action" is based upon the assessment of contaminant properties and environmental conditions to identify potential exposure routes to the general population. Based on the results of the assessment, appropriate corrective action requirements are determined. Researchers have found that most petroleum products will naturally attenuate in the environment [3]. Therefore, in many situations, dissolved petroleum contaminant plumes will naturally attenuate in the environment prior to creating a potential exposure hazard. This is generally attributed to the volatility of

304

many of the lighter distillates including petroleum based solvents and motor fuels, and the biodegradability of most of the "simpler" petroleum compounds which make up motor fuels and "lighter" heating oils. Coincidentally, the shallow groundwater environment often provides the required ingredients for biodegradation, including indigenous petroleum degrading bacteria, and oxygen as a metabolizer and election receptor [3, 4].

The discussion in this article largely pertains to the lighter motor fuels which contain aromatic petroleum hydrocarbons. These include benzene, toluene, ethylbenzene and xylenes (BTEX). The occurrence of dissolved BTEX groundwater contaminant plumes is most commonly associated with leaking gasoline USTs. However, their association with other petroleum products is not uncommon. This article and checklist provide guidance for completing UST assessments, which can support the decision-making process presented in the ASTM Emergency Standard Guide. Following the checklist will ensure that the initial site assessment results in an accurate and functional characterization of the details and subtleties of the UST Source Impact Zone. This includes evaluating the UST area to identify release pathways for residual contamination in soil and groundwater. Based on the results obtained from the Source Impact Zone Evaluation, natural attenuation can be assessed for it's applicability and performance.

Recent experience with petroleum hydrocarbon releases from UST sites has demonstrated that by completing an accurate and detailed Source Impact Zone Evaluation, the assessment of natural bio-attenuation can be readily completed. This is a result of the integral relationship between the Source Impact Zone and the resulting dissolved contaminant plume. Often by completing an effective Source Impact Zone Evaluation and plume delineation only limited supplemental geochemical analysis will be required to demonstrate natural attenuation. This article, therefore, does not present detailed geochemical procedures or techniques for characterizing natural attenuation processes.

**OBJECTIVE**

The assessment of natural attenuation requires combining fundamental hydrogeology with the analysis of the plume's biological and geochemical dynamics. Because the lateral and vertical extent of these plumes is generally limited, spatial requirements for data collection points can be relatively detailed. This requires a comprehensive characterization of the unsaturated zone and underlying shallow aquifer to determine the relationship between the plume and contaminant source.

The subtle and varying hydrogeologic conditions associated with the underground storage tank environment often must be defined to a scale that is measured in inches, not feet or tens of feet. For example, understanding the hydraulic interaction of a backfilled tank excavation with surrounding soils and underlying bedrock may be critical to ensuring the success of the project.

The checklist developed for this article underscores the necessity to ensure that basic geologic and hydrogeologic information is collected and analyzed. This information must be collected, interpreted and analyzed prior to assessing the potential for natural attenuation.

## FIRST GENERATION UST SITE ASSESSMENT APPROACHES

In the early days of UST management, site assessments were performed using limited post-closure soil sampling often followed by the immediate installation of a groundwater monitoring system. The monitoring system was typically located immediately surrounding the former UST location. The detailed interaction of the former UST excavation (Source Impact Zone) with the overall site and regional subsurface conditions was not considered. This often resulted in incorrect determination of site groundwater flow conditions and conclusions regarding the associated contamination. Furthermore, if results obtained from these investigations indicated exceedances of standards, active groundwater remediation may have been initiated without a complete or accurate hydrogeologic assessment. This would result in ineffective and sometimes unnecessary remedial programs.

To some extent, this site assessment approach was justified as a cost savings effort for owners of large UST populations. However, anticipated long-term cost savings of this "boiler plate" approach were often not realized as a result of excessive remedial costs. The current second generation of risk based site assessments, which support natural attenuation as the preferred remedial option, have been shown to provide a more cost-effective UST management approach [5].

## SOURCE IMPACT ZONE EVALUATION

In order to assess the potential for natural attenuation as a remediation option, an accurate evaluation of the Source Impact Zone must be performed. This requires characterizing features and processes which are relatively small components of the geologic and hydrogeologic framework. This characterization is identified herein as a Source Impact Zone Evaluation and is considered as significant as effective source control in order to achieve effective remediation.

The following broad data gaps can result from not completing an accurate Source Impact Zone Evaluation:

- Incomplete/incorrect hydraulic characterization of the impacts of former UST excavation on the site groundwater flow regime.

- Incomplete/incorrect assessment of unsaturated zone soils, geology and flow conditions.

306

- Incorrect interpretation of shallow groundwater flow direction and gradient by improper well design (e.g. installing well screens through perched zones).

- Incorrect interpretations of shallow groundwater flow direction and gradient by not considering temporal fluctuations in groundwater elevations.

- Incomplete/incorrect assessment of impacts of shallow fill deposits, site construction modifications, subsurface barriers or preferential flow paths on contaminant or product migration pathways from the source.

From each of these data gaps, it would not be possible to accurately assess natural attenuation as a remedial option. Limitations to the assessment process include (1) not determining reliable data trends (temporal or spatial) from the monitoring well system, (2) not identifying all occurrences of separate phase petroleum product, (3) not having completed the plume delineation and, (4) having missed a release pathway from the source area.

## THE SEQUENTIAL CHECKLIST FOR ASSESSING NATURAL ATTENUATION

A sequential checklist has been developed in order to ensure that the initial site assessment and hydrogeologic characterization does not overlook the technical features and processes that control the impacts from a leaking UST. This is accomplished by providing a relatively simple sequential framework for source characterization, evaluation of the Source Impact Zone, analysis of migration pathways, natural attenuation assessment and groundwater monitoring.

Completion of each sequential level of assessment ensures that subsequent activities will be planned accurately and related to the dynamics which characterize the Source Impact Zone. More than one level of assessment can be combined and efficiently completed within a limited investigation program.

The following levels of site assessment for natural attenuation are suggested:

Level 1:    Source Control/Removal Verification
Level 2:    Source Impact Zone Evaluation
Level 3:    Contaminant Migration Analysis
Level 4:    Natural Attenuation Assessment
Level 5:    Groundwater Monitoring

The checklist is presented as Table I and includes a summary of the technical issues to be evaluated within each level of the assessment process. Completion of each level will satisfy the minimum requirements to confirm source removal (Level 1), evaluate the release pathways to the environment from the Source

Impact Zone (Level 2), assess potential impacts of dissolved groundwater contamination (Level 3), assess the appropriateness of natural attenuation as a remedial approach (Level 4) and monitor the natural attenuation process results (Level 5).

## CHECKLIST FOR THE ASSESSMENT OF NATURAL ATTENUATION OF DISSOLVED PETROLEUM CONTAMINANT PLUMES
### TABLE I

### Level 1:  Effective Source Control/Removal

☐ Removal of all recoverable separate phase product.

☐ Elimination of all releases of separate phase product.

☐ Removal of all soils visibly saturated with petroleum product.

### Level 2:  Source Impact Zone Evaluation

☐ Delineation of lateral and vertical extent of separate phase product.

☐ Delineation of lateral and vertical extent of soils visibly saturated with product.

☐ Determination of complete lateral and vertical extent of former UST excavation.

☐ Identification and location of preferential migration pathways (e.g., utility trenches, foundations, pavement sub-base layers) which may intersect former UST excavation.

☐ Identification of surface/near surface drainage features that may impact former UST excavation.

☐ Determination of shallow groundwater hydraulics of former UST excavation in relationship to natural subsurface conditions.

☐ Determination of natural subsurface soils and geologic formations in immediate vicinity of former UST excavation.

☐ Determination of depth and type of bedrock underlying former UST excavation.

□ Evaluation of potential short and long term fluctuations in surface water and groundwater hydrologic conditions.

□ Complete vertical characterization of unsaturated zone stratigraphy beneath former UST excavation.

□ Field screening or sampling of entire thickness of unsaturated zone vertical section beneath former UST excavation.

**Level 3: Containment Migration Analysis**

□ Lateral and vertical delineation of unsaturated zone contamination.

□ Lateral and vertical delineation of dissolved groundwater contaminant plume quality.

□ Determination of background soil and groundwater quality.

□ Determination of site groundwater flow conditions and relationship to regional conditions.

□ Identification of groundwater recharge and discharge features.

□ Identification of nearby (one mile) water supply wells and/or surface water bodies.

□ Determination of groundwater flow velocities.

**Level 4: Natural Attenuation Assessment**

□ Complete plume delineation achieved.

□ Dissolved oxygen levels profiled for plume.

□ Contaminant levels profiled for plume.

□ Lateral trends established for D.O. and contaminant levels.

□ Maximum plume travel distance and duration estimated.

□ Travel time calculated to potential receptors.

□ Additional geochemical analysis performed (e.g. bacteria, electron receptors, pH).

### Level 5: Groundwater Monitoring Requirements

☐ Evaluate seasonal groundwater quality trends (minimum two years).

☐ Establish decreasing contaminant levels over time (minimum two years).

☐ Confirm estimated maximum plume travel distance and plume duration.

☐ Monitor potential receptors, as appropriate.

☐ Monitor Geotechnical Trends (natural conditions and contaminant plume parameters)

## LEVEL ONE:  SOURCE CONTROL/REMOVAL VERIFICATION

The removal of leaking underground storage tanks as a source removal effort is typically the first action to be taken.  The removal of obvious separate phase petroleum product and of grossly contaminated soils is also typically accomplished during, or soon after tank removal.  In all UST removals, site assessment activities should include confirmation of the absence of all petroleum product and petroleum saturated soils.

In many cases, the reliance solely upon post-closure soil samples and limited groundwater monitoring may not be adequate to confirm the effectiveness of the source removal.  For example, this includes sites where former USTs were situated within bedrock formations.  In these settings, the investigation of soil underlying the former UST may not reveal petroleum contamination or separate phase petroleum product.  This could lead to the conclusion that source removal has been effective.  However, it is possible that the petroleum product may have migrated through the soils into underlying fractured bedrock.

## LEVEL TWO:  SOURCE IMPACT ZONE EVALUATION

The objective of the Source Impact Zone Evaluation (SIZE) is to characterize the immediate environment of the former UST location and to evaluate release pathways to the environment.  Through this process, an evaluation of the effectiveness of the source removal/control activities will also be accomplished.  Upon completion of the Level 2 assessment, a determination should be made as to whether additional source removal is required.  In addition, the findings should be utilized to determine an effective approach for characterizing potential impacts from residual contaminants in Level 3.

In the case of the UST situated within bedrock, the SIZE might determine that the former excavation has created a relatively impermeable sump within the top of bedrock.  In this case, effective removal of all contaminated material from within the former excavation may be all that is required to complete site remediation. In other cases, open fractured bedrock may not create an impermeable sump, but instead may act as a permeable dry well. In these cases, the SIZE should identify the likely release pathways from the former source location.

## LEVEL THREE: CONTAMINANT MIGRATION ANALYSIS

The goal of the Level 3 Contaminant Migration Analysis is to establish the lateral and vertical extent of the contaminant plume and to identify potential receptors. Subsequent levels of the attenuation assessment will determine the appropriateness of natural attenuation as a remedial approach and evaluate the natural attenuation process at the site.

The assessment of natural attenuation requires complete plume delineation. This can be accomplished through the Level 3 Contaminant Migration Analysis (CMA) which is designed based upon the results of the SIZE. Tue efficiency of the plume delineation process will be directly affected by the accuracy of the SIZE. The use of screening techniques including non-destructive site-wide soil gas surveys may be effective in early stages of the CMA. In the immediate vicinity of the source zone, the effects of hydraulic mounding, the occurrence of preferential release and migration pathways, and fluctuating shallow saturated zone conditions may require relatively detailed analysis.

Useful techniques to incorporate in the CMA include soil gas surveys and continuous water level measurements. In addition, the installation of piezometers to aid in distinguishing between regional (site vicinity) groundwater flow patterns and localized hydraulic effects from the former tank excavation should be considered.

## LEVEL FOUR: NATURAL ATTENUATION ASSESSMENT

Upon complete plume delineation, the assessment of natural attenuation may be performed. The appropriateness of the natural attenuation approach is determined by characterizing plume conditions and assessing the risk to potential receptors. If source removal has been accomplished, accurate plume delineation achieved, and impacts to potential receptors eliminated, then natural attenuation should be an appropriate remedial approach.

In general, based upon the calculation of travel times using groundwater flow velocities and contaminant properties, the downgradient extent of plume migration can be estimated and the duration of the contaminant in the environment can be predicted. This information is used to assess potential impacts to identified receptors by determining the extent and duration of the plume.

Favorable conditions for natural attenuation include dissolved oxygen levels in excess of 2 ppm and moderate hydraulic conductivities and flow gradients. Spatial trends which indicate an inverse relationship between dissolved oxygen

311

levels and contaminant concentrations are typically indicative of active natural bio-attenuation. Temporal trends of increasing dissolved oxygen levels and decreasing contaminant levels are also indicative of natural bio-attenuation [4].

Plume dimensions generally are limited by aerobic biodegradation which occurs at plume fringes. Over time this will be evident by a limit to the lateral downgradient and transverse migration of the plume. In addition, varying rates of migration and degradation for the individual BTEX contaminants will occur resulting in varying plume chemistry over time [4, 6]. Geochemical indicators of natural bio-attenuation can also be evaluated by conducting sophisticated monitoring programs and analysis of several other electron acceptors including iron, sulfate and nitrate [6].

## LEVEL FIVE: NATURAL ATTENUATION MONITORING

The establishment of spatial and temporal trends in groundwater quality and geochemistry provides the confirmation of effective source removal/control and of ongoing attenuation. Effective monitoring systems must account for short-term and long-term fluctuations in groundwater elevations at a site, variations in natural groundwater quality, the distinction of multiple contaminant plumes and background impacts.

Trends cannot be reliably established and interpreted without sufficient data points. This typically requires a period of regular semi-annual monitoring of several well locations. Incorporation of statistical techniques is also occasionally warranted. Based upon favorable monitoring results which support determined estimates of maximum plume extent and the expected duration of the plume, the effectiveness of natural attenuation can be demonstrated. These results would also demonstrate safe protection of any identified potential receptors. For final site closure, monitoring should be proposed as necessary to confirm complete plume attenuation.

## SUMMARY AND CONCLUSIONS

Natural attenuation has been identified as the preferred risk based corrective action for petroleum hydrocarbon releases. This is based on the identification of a combination of favorable conditions which frequently occur at leaking UST sites. The objective of the assessment of natural attenuation is to document these conditions as they relate to the release at the site. This includes accurate and complete delineation of the plume by building upon the findings of the Source Impact Zone Evaluation.

The completion of a comprehensive Source Impact Zone Evaluation is a critical requirement for the assessment of natural attenuation. The process should focus on both confirming the effectiveness of source removal and identifying the release pathways to the environment at the source location.

Source Removal and Source Impact Zone Evaluation both must be completed in order to assess the feasibility of natural attenuation as a remedial approach.

Demonstration of natural attenuation relies upon establishing reliable trends of decreasing contaminant concentrations within the delineated plume. This can be supported by geochemical analysis to document specific attenuation mechanisms occurring at the site.

The sequential checklist for the assessment of natural attenuation illustrates the levels and objectives that define the process to demonstrate natural attenuation. By emphasizing the objectives to be accomplished and relating these to the Source Impact Zone, a thorough, yet efficient assessment process can be accomplished.

## REFERENCES

1. U.S. Environmental Protection Agency. Jan 1987. *Underground Storage Tank Corrective Action Technologies*, Cincinnati, OH: Hazardous Waste Engineering Research Laboratory. EPA/625/6-87-015.

2. American Society of Testing Materials. July 1994. "Emergency Standard Guide for Risk-Based Corrective Action Applied at Petroleum Release Sites," Philadelphia, PA: ES38-94.

3. Salanitro, Joseph P. 1993. "The Pole of Bioattenuation in the Management of Aromatic Hydrocarbon Plumes in Aquifers," Groundwater Monitoring Review, pp. 150-161.

4. McAllister, P.M. and C.Y., Chiang. 1994. "A Practical Approach to Evaluating Natural Attenuation of Contaminants in Groundwater," Groundwater Monitoring Review, pp. 161-173.

5. Marqis, S.A. Jr. and D., Dineen. 1994. "Comparison Between Pump and Test, Biorestoration and Biorestoration/Pump and Treat Combined: Lessons from Computer Modeling," Groundwater Monitoring Review, pp. 105-119.

6. Borden, R.C., C.A. Gomez and M.T. Becker. 1995. "Geochemical Indicators of Intrinsic Bioremediation," Groundwater, 33(2): 180-189.

# PERFORMANCE OPTIMIZATION OF BIOLOGICAL WASTE TREATMENT BY FLOTATION CLARIFICATION AT A CHEMICAL MANUFACTURING FACILITY

BELA J. KERECZ
Lead Group Chemist
Air Products and Chemicals, Inc.
7201 Hamilton Boulevard
Allentown, PA 18195-1501

DONALD R. MILLER
Application Specialist
Komline-Sanderson
Holland Avenue
Peapack, NJ 07977

## INTRODUCTION

Air Products and Chemicals, Inc., utilizes a deep-tank activated sludge wastewater treatment system with a dissolved air flotation clarifier (DAF) to effectively treat amine wastes. The bio-system, a deep tank aeration system, produces a high quality final effluent low in biochemical oxygen demand (BOD), ammonia and organic nitrogen, turbidity and total suspended solids. Prior to installing the DAF, treatment performance was at risk with a gravity clarifier. Waste treatment performance was jeopardized by poor settling bio-flocs and uncontrollable solids-liquid separation problems within the gravity clarifier. The solids settleability problems resulted primarily from mixed liquor nitrogen supersaturation degassing in the clarifier. As a result of the degassing, biomass floated on the gravity clarifier or overflowed the effluent weir. As a result of biomass loss periodically organic carbon and total Kjeldahl nitrogen loadings had to be reduced in order to maintain optimal food-to-mass ratios. As biomass levels dropped within the aeration basin, waste treatment performance was at risk and waste loads had to be decreased causing waste inventories to increase in storage tanks.

## BACKGROUND AND DISCUSSION

Air Products and Chemicals, Inc., owns and operates the world's largest higher amine production facility at St. Gabriel, Louisiana. The facility, located on 667,590 m$^2$ about 16 km east of Baton Rouge and 70 km west of New Orleans, produces ethyl amines and isopropyl amines. Product end-uses include agricultural chemicals, herbicides, pharmaceuticals, surfactants, lube oil additives, fabric softeners, epoxy curing agents, asphalt additives and corrosion inhibitors. The plant operates on a batch campaign basis making one amine product during each 20 to 40 day campaign. The facility started-up as a green field plant in September of 1976.

During production, the amine reaction co-produces process wastewaters containing residual organics, ammonia-nitrogen and organic nitrogen. The plant typically produces a total daily wastewater volume of 450 to 700 m$^3$ containing high BOD and TKN levels. These wastewaters are readily assimilated biologically on-site in the activated sludge wastewater treatment system. The existing treatment system consists of wastewater equalization, pH adjustment, aeration, clarification, filtration and sludge digestion in aerated lagoons. See Figure I for process flow diagram of existing wastewater treatment system. The effluent from the biological system, typically containing 20 mg/L or less TSS, <10 mg/L BOD, and <1 mg/L ammonia-nitrogen, is polished through a mixed-media filter before final discharge to the Mississippi River.

Because of the deep tank aeration basin (7.3 m water depth) and type of aeration equipment, nitrogen gas supersaturation of the mixed liquor occurs routinely [1]. Degassing would then occur in the gravity clarifier (4.3 m water depth), due to reduced partial pressure. The resulting nitrogen gas bubbles formed within the bio-mass flocs, caused sludge buoyancy problems and chronic high effluent turbidity and TSS prior to the installation of the DAF.

Nitrogen gas supersaturation occurs in the 1625 m$^3$ (429,000 gallon) deep tank aeration basin even when dissolved oxygen levels are low (<3 mg/L). The wastewater treatment system for aeration utilized a 125-horsepower motor-driven submerged flat blade turbine mixer with an air diffuser sparger ring. The turbine is located near the bottom of the 7.3 m deep aeration basin about 0.5 m above the sparger ring (1.3 m diameter). Two 40-horsepower blowers, each capable of supplying 15 m$^3$ (550 SCFM) of air, provide oxygen for waste treatment. Nitrogen gas supersaturation can occur within this type of system at liquid levels greater than 4.3 m negatively impacting gravity clarifier performance [2]. Two 60-horsepower submerged induced air aerators were installed in late 1992 and two more in 1994 to supplement the blower turbine aeration system and to provide increased oxygen capacity for additional BOD (Biochemical Oxygen Demand) and NOD (Nitrogen Oxygen Demand) throughput.

Periodically during severe degassing events and prior to installing a DAF clarifier, effluent TSS levels ran in the 300 to 500 mg/L range and a floating mat of bio-solids which would not settle appeared on the surface of the gravity clarifier. In order to prevent biomass loss, waste loadings and wastewater flows were regularly cut back during these high solids loss periods. Nitrification performance was at risk as sludge ages were significantly reduced. Wastewater storage tank inventories swelled. At times production capacity was jeopardized because waste storage tank inventories were at or near capacity levels. As a result of the performance problems and capacity limitations, various upgrade options for improving the system were evaluated.

DAF technology was selected after extensive laboratory and pilot studies demonstrated preliminary technical feasibility and cost-effectiveness of the approach. The installation of a 23 square meter DAF in November of 1991 with a hydraulic design capacity of 1400 $m^3$/day accomplished the targeted goals.

Performance to date shows:
1. Two-fold increased hydraulic capacity.
2. Doubling of BOD and NOD capacity.
3. Two-fold increase in concentration of biomass (mixed liquor).
4. Superior effluent quality low in TSS <20 mg/L, <10 mg/L BOD, <1 mg/L ammonia.
5. Less requirement for operator attention, resulting from a more controllable and reliable operation.
6. Reduced biomass (sludge) wasting requirements.
7. Better tolerance to changes in influent waste strength and waste types.

## DISCUSSION AND RESULTS

DAF Process

Dissolved air flotation (DAF) is a liquid-solid separation process that takes advantage of the fact that if air is dissolved in water under pressure and the pressure is later released, the dissolved air will be released from solution in the form of vast numbers of micro-bubbles. "Flotation" occurs when these bubbles become attached to or entrained in suspended solids in the DAF, thus increasing their buoyancy and causing them to rise to the surface of the liquid. Floated material can then be skimmed from the surface and collected. Clarified liquid, or product water minus the solids, underflows the DAF through baffles and exits the unit.

Effective clarification is the result of one or more of the following phenomena: (1) physical attractions between the air bubbles and the particles; (2) entrainment, in which the rising air bubbles are entrapped by the particles; (3) impingement, in which the particles are "pushed" upward by the rising air bubbles. These associations of bubbles and particles result in a density of the combined particle (oil/solid/air agglomerate) substantially below that of water, causing the particle to float to the surface. Chemical additives (organic and inorganic coagulants, polymers) can be used or may be necessary to increase the effectiveness of the air flotation. Cationic polymer addition is required at St. Gabriel to flocculate the activated sludge for effective clarification by dissolved air flotation.

Figure II shows a simplified internal flow diagram of the installed DAF unit. The influent containing the biomass enters the inlet mixing chamber. After being pressurized (500 kPa) and aerated, a portion of clarified effluent is recycled to the inlet chamber where it is mixed with the influent. As the

316

pressure of the recycled water is reduced, air comes out of solution in the mixing chamber in the form of vast numbers of microbubbles. A cationic polymer is added to the influent to promote coagulation and flocculation of the bio-solids.

The microbubbles become attached to or entrained in the flocculated bio-solids. This attachment of bubbles "reduces" the density of the bio-solids resulting in increased buoyancy thus effecting flotation of the bio-solids. The float-solids are collected by a countercurrent float skimmer device which skims the solids into a sludge hopper. The float solids at 2 to 3 % total solids are then pumped back into the aeration basin or blowndown to the aerated sludge lagoons. Clarified product water, at less than 20 mg/L TSS, is removed from the unit via an underflow baffle.

## LABORATORY AND PILOT SCALE TEST DATA

Laboratory and pilot testing of the dissolved air flotation process was performed to qualify the technology. This qualification consisted of establishing values for performance parameters and costs for equipment and operations. Table I shows the values determined from the test work, together with engineering design basis for the installed equipment.

The installed DAF, while not preventing nitrogen gas supersaturation, is able to effectively handle the poorly setting biomass producing product water going to the mixed-media filter low in TSS and turbidity and a concentrated sludge float. The sludge float containing 2 to 3% solids is easily recycled back to the aeration basin to maintain a food-to-mass ratio of 0.2 or less (kg BOD/kg MLVSS), or wasted to the aerated sludge lagoons.

Waste treatment performance improved dramatically and the need for operator attention was reduced when the 23 m$^2$ DAF was installed. As a direct result of the DAF, the system now operates with two to three times the biomass in the same aeration volume and produces a consistent quality effluent low in BOD, ammonia-nitrogen, suspended solids and turbidity. In addition to requiring less operator attention, the upgraded system can process twice the BOD and TKN waste loads, resulting in the elimination of additional storage capacity and expensive waste treatment alternatives.

## TABLE I

### DAF TREATABILITY AND DESIGN DATA

| Parameter | Laboratory Data | Pilot Plant Data | Design Data |
|---|---|---|---|
| DAF Surface Area, m2 | 0.1 | 1.25 | 23 |
| Flow Rate, m3/hr | 0.45 | 6.1 | 9-64 |
| Solids Rate, kg/hr | 2.3 | 22.5 | 27-678 |
| Pressurized Recycle Rate, m3/hr | 0.23 | 3.1 | 57-114 |
| Effluent TSS, mg/L | <50 | ≤10 | <50 |
| Effluent BOD, mg/L | <10 | <10 | <10 |
| Operating Cost, $/m3 | -- | -- | .13 |
| Flocculant Demand, mg/L | 30 | 15-25 | 25 |

## CONCLUSIONS

Dissolved air flotation is a cost-effective means of effluent clarification for Air Products and Chemicals St. Gabriel deep tank activated sludge system. Four years of operation with the DAF has provided a final effluent low in BOD, TSS, ammonia, and the wastewater treatment systems capacity has been increased more than two-fold. In addition, plant production is no longer jeopardized by episodic liquid/solids separation problems which previously occurred with gravity settling.

## REFERENCES

1. Japanese Conference on Sewage Treatment Technology (1977). Development of the Deep Aeration Tank, Nagaharu Okuns.

2. Deep Tank Aeration Experience at New York City's North River Plant, Lin Lu, J. A. Turcotte, J. Donnellon, T. E. Wilson, presented at the 63rd Annual Conference of the Water Pollution Control Federation, Washington D.C. 11 October 1990.

## ACKNOWLEDGMENTS

The authors would like to express their thanks to the following people who helped make this project a success.

Mr. R. Pophal, Mr. Rhett Levins and Mr. F. Grunewald (formerly of Air Products, now with Great Lakes Chemical) and the many unnamed plant operators at the St. Gabriel plant.

In addition, special thanks to Mr. Paul G. Daly, Deep Shaft Technology, Inc., who assisted the project team in providing technical expertise and in helping to solve the solids-liquid biomass separation problems associated with deep tank biological systems.

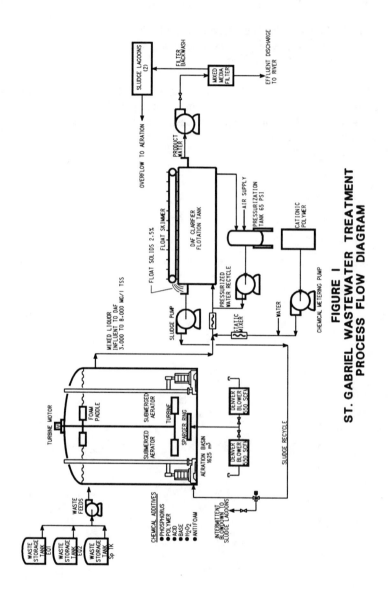

**FIGURE I**

**ST. GABRIEL WASTEWATER TREATMENT
PROCESS FLOW DIAGRAM**

320

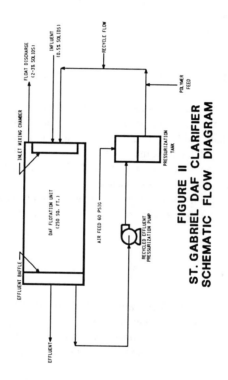

FIGURE II
ST. GABRIEL DAF CLARIFIER
SCHEMATIC FLOW DIAGRAM

# TREATMENT OF A DILUTE WASTE OIL EMULSION FROM ALUMINUM ROLLING MILL OPERATIONS BY A BIOLOGICAL AERATED FILTER

PATRICK E. CARRIERE
BRIAN E. REED
JEFF BROOKS

Department of Civil and Environmental Engineering
West Virginia University
Morgantown, WV 26506-6103

## INTRODUCTION

Wastewater from aluminum industry contains oil used primarily during the aluminum rolling process. In this study, oily wastewater from the manufacturer's processes was transferred to two holding ponds having a total capacity of about 5 million gallons. A detention time of 10 days was maintained to allow the free oil to rise to the surface and the solids to settle. Wastewater from the ponds was withdrawn at a depth of 9 feet and used as influent to the pilot-scale biological aerated filter. The main objective of this study was to evaluate the performance of the pilot-scale biological aerated filter on oil and grease(O&G) and total suspended solids (TSS) removal.

## BACKGROUND

Biological aerated filters (BAF) are attached-growth processes. Biological aerated filters remove contaminants from wastewater through both physical and biological processes. The reactor functions as a filter to retain suspended solids, while the biomass consists of a highly concentrated community of bacteria and microorganisms that stabilize contaminants through biological processes. Most of the soluble matter is transformed through biochemical oxidation, while smaller portions are absorbed in the biomass. The biological community must have a continuous source of carbon, nutrients, and oxygen to sustain an optimal population.

The pilot-scale BAF system used in this study consists of four reactors operating in series. The system is designed to operate from a flowrate of 1 to 3 gpm and an oil grease concentration from 1,000 to 8,000 mg/L. The media in the reactors is known as Durapore. This media has a high surface/volume ratio. The microorganisms used are Pseudomonas fluorescens. Each reactor is equipped with a recirculation pump to increase the detention time, promote mixing, and sloughing. Oxygen and nutrients are applied to enhance the ability of the microorganisms to remediate the wastewater. Each reactor is equipped with a porous stone to provide aeration. Nutrients are supplied only to the first reactor, and consist of ammonium sulfate, nonbasic sodium phosphate, and dibasic potassium phosphate. The wastewater is introduced to the first reactor in a downflow mode.

## EXPERIMENTAL DESIGN

The performance of the pilot-scale BAF system was evaluated under different operating conditions. Different ranges of initial O&G and TSS concentration, hydraulic loading rates of 294, 440 and 587 gpd/ft2, and different particle sizes of the media were investigated. During the initial stage of this study, the particle size of the media size in all four reactors ranged from 3/8" to 1/4". The particle size of the media in reactor 1 was increased to the 1/2" to 1" range. The increase in particle size in reactor 1 provided larger void space, but smaller specific surface area for microbial growth. Each reactor was equipped with a recirculation pump and air source. The total air flowrate to all four reactors was approximately 12 cfm. A nutrient solution was injected into the first reactor. Each experiment was divided into 4 experimental runs. An experimental run consisted of five consecutive days of wastewater treatment followed by a backwash event.

Samples were collected from each of the four reactors and analyzed for oil and grease, total suspended solids, dissolved oxygen, pH, temperature, and bacterial counts. Also, samples were taken for chemical oxygen demand, potassium, phosphorus, and the various forms of nitrogen.

### Backwash Procedure

The backwash consists of utilizing both air and water to remove accumulated solids. The addition of air in the backwash procedure greatly reduced the flowrate requirement of the backwash water. A backwash sequence was performed in two steps. The first step of the backwash sequence was an air scouring step, which fluidized the media. The air was applied at 18 cfm and 30 psi for a period of ten minutes for each sequence. In the second step, air and water were simultaneously applied, removing accumulated solids from the media bed. The water was applied at 40 gpm and 45 psi, while the air was applied at 10 cfm and 20 psi. The second step was performed for a duration of 4, 3, 2.5, and 2.5 minutes for sequence 1, 2, 3, and 4, respectively. After completion, the sheared solids floating at the top of the reactor were flushed out with a water rinse. The water was applied at 10 gpm for a period of five minutes.

A composite sample of each sequence was collected for total solids and total suspended solids analysis. The analytical results were used to determine the mass of solids removed. The effectiveness of each backwash event was evaluated based on the amount of solids removed, hydraulic capacity of the system, and the condition of the media bed. The hydraulic capacity of the system was evaluated by determining the maximum recirculation flowrate through each reactor. The condition of the media bed was evaluated by sampling each reactor for oil and grease after completion of the backwash.


## RESULTS

The performance of the BAF technology was evaluated by investigating the effect of initial O&G concentration, hydraulic loading, and media size on O&G and TSS removal.

### Effect of Feed O&G Concentration

The effect of initial O&G concentration on the performance of the BAF technology is illustrated in Figure 1. The fractional distance through the BAF technology represents the effluent from each of the four reactors. For example, the effluent from reactor 2 is located at 0.5 on the x-axis. As shown in Figure 1, O&G concentration exponentially decreased as the wastewater moved through the system. The decrease in removal rate with depth was the result of the low O&G concentration within the system, since the available O&G controls the rate of biological reactions. It was also found that the percent O&G removal declined as the influent concentration was increased. As observed in Figure 1, a much smaller fraction of O&G was removed when the influent O&G concentration exceeded 3,000 mg/L.

### Effect of Hydraulic Loading Rate on O&G and TSS Removal

The effect of hydraulic loading rate on O&G and TSS removal is illustrated in Figures 2 and 3, respectively. As observed in Figures 2 and 3, the removal of O&G and TSS decreased as the hydraulic loading was increased. The removal percentage of O&G and TSS was similar for both hydraulic loading rates of 294 and 440 gpd/ft$^2$. However, the removal of O&G and TSS significantly decreased when the hydraulic loading was increased to 587 gpd/ft$^2$.

### Effect of Particle Size of the Media on O&G and TSS Removal

The effect of the particle size of the media in reactor 1 is illustrated in Figures 4 and 5 for O&G removal and Figures 6 and 7 for TSS removal, respectively. As noticed in Figure 4, the removal percentage of O&G for the largest particle size was approximately 80%, while 60% for the smallest size under the same operating conditions. As shown in Figure 5, it was found that the removal efficiency of O&G was similar for both particle size under the same operating conditions, but the feed O&G concentration was much higher for the largest particle size.

As shown in Figure 6, for an initial TSS concentration ranging from 1,000 to 1,500 mg/L and hydraulic rate of 294 gpd/ft2, the removal percentage of TSS for the largest size was approximately 90%, while 80% for the smallest. However, the difference in removal of total suspended solids was negligible when the initial TSS concentration ranged from 500 to 1000 mg/L as shown in Figure 7.

**SUMMARY**

Three experiments were carried out to examine the effects of feed O&G and TSS concentration, hydraulic loading rates, and particle size of the media on the performance of the BAF. When the feed O&G concentration exceeded 3,000 mg/L, the results indicated that the percent of O&G removal declined. It was observed that the removal of O&G and TSS significantly decreased when the hydraulic loading was increased to 587 gpd/ft$^2$. It was found that the removal efficiency of O&G was similar for both particle size under the same operating conditions, but the feed O&G concentration was much higher for the largest size. When the initial TSS concentration varied from 1,000 to 1,500 mg/L and the hydraulic rate was 294 gpd/ft$^2$, the removal percentage of TSS for the largest particle size was approximately 90%, while 80% for the smallest. However, the removal percentage was similar for both size at low initial TSS concentration.

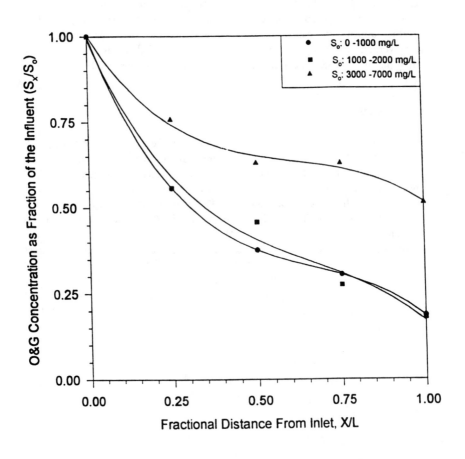

Figure 1 : Effect of Feed on O&G Concentration Removal

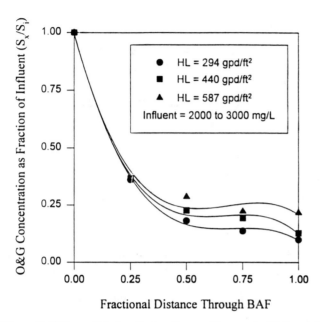

Figure 2: Effect of Hydraulic Loading on Removal of Oil and Grease

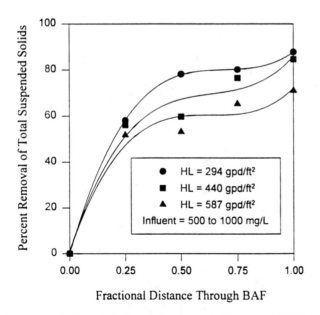

Figure 3 : Effect of Hydraulic Loading on Suspended Solids Removal

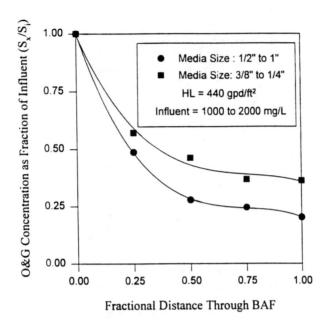

Figure 4: Effect of Media Size on O&G Removal

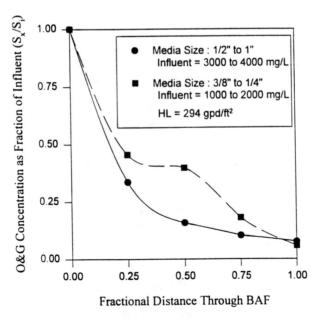

Figure 5 : Effect of Media Size on O&G Removal

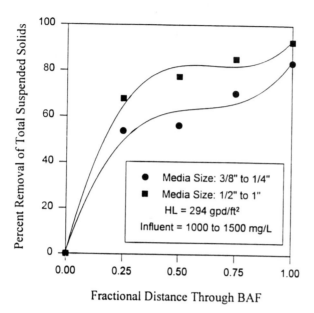

Figure 6 : Effect of Media Size on TSS Removal

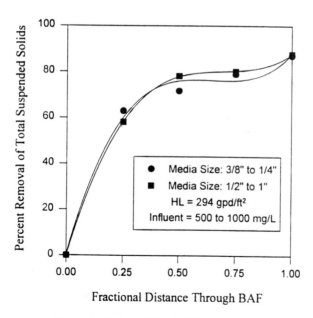

Figure 7 : Effect of Media Size on TSS Removal

CATEGORY V:
# *Advanced Oxidation Process*

# UPDATE ON WET OXIDATION TECHNOLOGIES FOR WASTEWATER AND SLUDGE MANAGEMENT

David L. Bowers
Frank J. Romano
John E. Sawicki
  Air Products and Chemicals, Inc., 7201 Hamiton Boulevard, Allentown, PA 18195-1501

## INTRODUCTION

Over the last several years, many companies have been evaluating wet oxidation processes for the treatment of industrial wastewater and management of sludge to meet new regulatory limits. Subcritical, near-critical and supercritical wet oxidation processes are currently being operated in pilot-plant scale as well as in full-scale plants to treat a wide variety of waste streams. Increasingly stringent environmental regulations and concerns about future regulations have lead academicians and industries to expand the data base and improve the fundamental understanding of wet oxidation.

## BACKGROUND

Wet oxidation occurs at elevated temperatures and pressures in the presence of an oxidizing agent to convert organic compounds to water and carbon dioxide. The exothermic reaction occurs in an aqueous ("wet") phase. This technology generally results in high organic removal efficiencies, clean vapor phases and, in the case of organic solids, a substantial volume reduction. Inert inorganic solids present in the waste streams pass through the system. Waste streams containing more than 3% organic material can sustain the required temperature without adding energy. Excess energy produced in the reaction can be recovered in some wet oxidation processes (Figure 1).

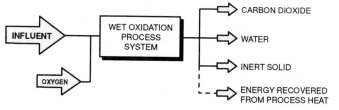

Figure 1.  General Wet Oxidation Flow Schematic

Control and management of accumulated inorganic material and organic scaling in the reaction and hot zones of a wet oxidation process continues to be one of the key challenges facing the development of this technology. The scale characteristics vary with the operating conditions of the process (sub or super-critical) and with changes in the influent material being treated. Several different approaches are being taken in the new wet oxidation techniques to effectively manage this problem.

## OVERVIEW OF PROCESS CONDITIONS

Wet oxidation processes can be divided into three (3) basic groups. The categories are based primarily on operating temperatures. In subcritical and near-critical processes, pressures are maintained sufficiently high to ensure a liquid phase in the reaction zones. In the supercritical process, the operating conditions exceed the critical point and the reaction fluid exhibits both liquid and gas-like properties that are well documented. Table I illustrates typical operating conditions for subcritical, near-critical and supercritical processes.

**Table I. Typical Wet Oxidation Process Conditions**

|  | Temp (°C) | Pressure (Bar) |
|---|---|---|
| **Subcritical** | 270 | 100 |
| **Near-critical** | 350 | 180 |
| **Supercritical** | >365 | >200 |

© Air Products and Chemicals, Inc. 1995

The choice of operating conditions generally depends on the required level of oxidation (removal efficiency) and the application. For example, processes designed to reduce the volume of organic solids will most likely operate at subcritical conditions while wet oxidation processes intended for maximizing removal of organic contaminants from water would operate at supercritical conditions. Table II shows approximate reaction times and their corresponding organic removal efficiencies for each of the three types of processes.

**Table II. Relative Reaction Times And Organics Removal**

|  | Reaction Time | Organics Removal (%) |
|---|---|---|
| **Subcritical** | 45 minutes | 90 |
| **Near-critical** | 5 minutes | 98 |
| **Supercritical** | 1/2 minute | 99.9 + |

© Air Products and Chemicals, Inc. 1995

Time-temperature effects on organic removal efficiencies is shown in Figure 2. These results were generated in a batch reactor using wastewater from a toluene nitration process to which 3% biomass was added. The removal of organic compounds from wastewater, measured in terms of chemical oxygen demand (COD) reduction, is obviously a very strong function of temperature. The dependency on residence time, although somewhat weaker, is also evident. In the subcritical region, residence time is substantially more important for high removal efficiency than in the supercritical region where the removal efficiency lines begin to converge. For the constituents studied, near complete COD removal occurs at approximately 500°C.

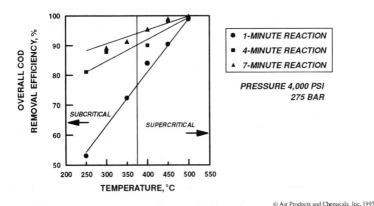

Figure 2. Nitration Wastewater Batch Reactor Work Results

## SUB-CRITICAL WET OXIDATION PROCESSING

Subcritical wet oxidation processes have been in operation for over thirty years for the treatment of industrial wastewater and for thermal conditioning of wastewater biosolids to improve dewatering. New facilities have come on-stream in recent years which have addressed the operating problems of earlier systems.

One facility located in Apeldoorn, the Netherlands is using the VerTech aqueous-phase oxidation (APO) process (Figure 3). The plant is currently processing nearly 31,000 dry metric tons (DMT) of municipal sludge per year, enabling local sewerage authorities to meet biosolids management regulations [1]. These new regulations require significant solid volume reductions, in addition to stringent air emissions and long term heavy metal stabilization.

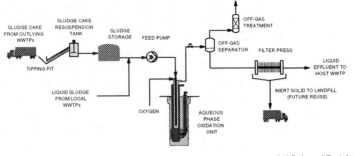

**Figure 3.  Apeldoorn VerTech APO Facility Flow Schematic**


## Full-Scale Commercial Operating Experience

Since the subcritical VerTech APO process began routine operations in August 1994, daily throughput of municipal sewage sludge has increased from 85 DMT to 110 DMT per day of dry solids at flowrates of 80 to 90 m³ per hour.  Weekly processing totals have increased from 300 to more than 600 DMT.  The facility has processed over 11,000 DMT of sludge as of April 1995.

Periodic cleaning of the facility's main processing unit (oxidation unit) to remove accumulated inorganic materials is necessary to maintain proper operating parameters.  The wash is performed weekly, generally after the facility has processed 600 to 650 DMT of sludge.  The wash frequency is dependent on the sulfur and calcium content of the incoming sludge stream.  The scaling material is nearly 100% calcium sulfate, and is deposited in the hotter regions of the reactor as a result of the inverse solubility phenomenon of calcium sulfate anhydrite.  The configuration of the oxidation unit in the VerTech APO process (two nested concentric pipes) allows a straight-forward washing procedure that can be completed in less than one day.  A wash solution of 15% nitric acid by weight is circulated through the unit which dissolves the calcium sulfate.  The scale is rapidly and completely removed from the reactor walls with minimal impact on onstream performance.

## Off-Gas Stream Characteristics

One of the benefits of wet oxidation processing is a small off-gas stream containing minimal pollutants.  This gas stream requires little additional treatment before it is vented to the atmosphere.  The process conditions do not produce NOx, SOx, particulate or the volated heavy metals that are normally produced in other thermal processes.  The off-gas stream from wet oxidation consists primarily of carbon dioxide saturated with water vapor.  The off-gas produced in Apeldoorn (Table III) requires treatment in a simple converter system to oxidize the carbon monoxide and minor amount of total hydrocarbons to carbon dioxide prior to atmospheric release.

**Table III.  Off-Gas Treatment System Performance, Apeldoorn**

| | Average Composition | |
|---|---|---|
| **Constituent** | **Before Treatment** | **After Treatment** |
| Carbon Dioxide | 90% | |
| Carbon Monoxide | 4% | < 25 ppm |
| Oxygen | 4% | |
| Nitrogen | 2% | |
| Total Hydrocarbons | ~ 100 ppm | < 2 ppm |
| Ammonia | | < 1 ppm |
| Nitrogen Oxides | | < 1 ppm |
| Sulfur Oxides | | < 1 ppm |

## Liquid Stream Characteristics

One of the historical drawbacks to subcritical wet oxidation has been the liquid effluent treatment required after waste processing.  The subcritical process conditions do not completely convert all of the waste COD to carbon dioxide. Typically, 10% to 20% of the incoming COD remains after processing, all of it now in the liquid phase. The biodegradability of the liquid stream generated in Apeldoorn can be seen in Table IV.  A major component of the residual COD is acetic acid, which is biodegradable.

**Table IV.  Liquid Effluent Biological Treatment System Performance**

| Parameter | Influent (mg/l) | Effluent (mg/l) | Reduction (%) |
|---|---|---|---|
| COD | 8,800 | 600 | 93 |
| BOD | 4,400 | 6 | 99.9 |
| TKN | 2,200 | 50 | 97.7 |
| $NH_4$-N | 1,750 | 1.5 | 99.9 |
| Total P | 21 | 3.6 | 83 |

A post-treatment system is usually not necessary with a VerTech APO process, and the liquid effluent can be returned to the host wastewater treatment plant for polishing.  The Apeldoorn facility requires post-treatment because it is a regional sludge processing plant, with over 80% of the sludge brought in from other wastewater treatment plants.  This results in a return stream with higher COD loads relative to the host facility's sludge contribution than would occur in a facility that treated only the host facility sludge stream.

## Solid Product Characteristics

The remaining solid product from the Apeldoorn process consists of inert soil materials. The solid phase also contains the heavy metals as well as the phosphorus from the wastewater sludge. The solid product meets all strict Dutch standards for environmental compatibility and can be landfilled or reused [2, 3]. The mineral composition of the end product is presented in Table V.

### Table V. Apeldoorn Solid Product Mineral Composition (by weight)

| | |
|---|---|
| Aluminum Phosphates | 38% |
| Kaolin | 16% |
| Quartz | 12% |
| Calcite | 9% |
| Portlandite | 5% |
| Feldspars | 4% |
| Gypsum | 2% |
| Balance fraction (amorphous) | 14% |
| | 100% |

## SUPERCRITICAL WET OXIDATION PROCESSING

New process concepts using supercritical wet oxidation (SCWO) have been developed in recent years, significantly advancing the knowledge of wet oxidation systems and design. Some applications have moved from the laboratory to large-scale pilot facilities in industrial locations. These facilities are intended primarily for removing organic pollutants from industrial wastewaters. A typical SCWO process schematic is presented in Figure 4.

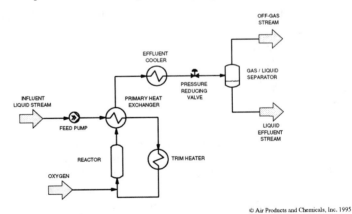

Figure 4. General Flow Schematic, SCWO Process

In 1994, Eco Waste Technologies (EWT), Austin, Texas, started up a commercial SCWO unit for Huntsman Corporation in Austin [4]. The Huntsman waste stream is generated from the production of long-chain alcohols, glycols, and amines. This liquid waste stream consists of 0.6 million gallons per year of wash water containing 1% to 4% total organic carbon (TOC) materials, up to 2 million pounds per year of concentrated organics containing 30% water, and 0.33 million pounds per year of amines.

The process is reported to use a plug-flow reactor. The reported aqueous effluent contains 2 mg/l ammonia, 20 mg/l COD, and 130 mg/l total dissolved solids. The effluent pH is between 7 and 7.5. The gaseous effluent stream consists of 90 - 95% carbon dioxide, 5 - 10% oxygen, less than 50 ppm carbon monoxide and less than 1 ppm $NO_x$. Information regarding inorganic scale formation and management procedures has not been disclosed at this time.

The effectiveness of the process in destroying organic compounds and the absence of pollutants in the gas vent are a testimony to the benefits of SCWO technology. There are technical challenges that need to be addressed to enable full commercialization of this developing technology. Key challenges include inorganic scale management and corrosion control.

Engineers at Modar, Inc., Natick, MA, have developed an SCWO reactor design that addresses the inorganic scale formed during processing. When water is brought to supercritical conditions, inorganic salts that are normally soluble in water become insoluble. Modar, Inc. developed a reactor that allows salts to form an aqueous brine that is continuously purged from the reactor. Removing this brine and not allowing the salts to accumulate allows continuous liquid waste processing.

## NEAR-CRITICAL WET OXIDATION PROCESSING

Engineers at Air Products and Chemicals, Inc., Allentown, PA, took a different approach to managing the scale formation inherent in SCWO. Faced with treating a liquid waste stream containing by-products of a nitration plant, they considered both subcritical and supercritical processes. The liquid waste stream constituents and the process treatment goals are presented in Table VI.

Laboratory work was conducted at several locations including the University of Texas at Austin, Modar, Inc., Natick, MA and at Air Products. Literature sources, including Dietrich [5] and Foussard [6], where reviewed. Based on these sources, the researchers settled on SCWO as the process most suited to meet the treatment goals. However, because of high concentration of inorganic compounds, notably sodium salts, precipitation from the supercritical fluid caused both scaling and plugging problems in the reactor. The scaling occurred rapidly, in several instances the pressure drop across the plugging section exceeding 80 bar before a complete flow blockage occurred. The salts were present in the original waste stream, and sodium hydroxide added for pH adjustment to mitigate corrosion resulted in increased sodium levels.

**Table VI. Waste Characterization and Treatment Goal, Nitration Wastewater [7]**

| Compound | Typical Concentration (ppb) | Treatment Goal (ppb) | Required Removal (%) |
|---|---|---|---|
| 2, 4-Dinitrotoluene | 630,000 | 113 | 99.98 |
| 2, 6-Dinitrotoluene | 250,000 | 155 | 99.90 |
| Nitrobenzene | 0 | 27 | --- |
| 2-Nitrophenol | 250 | 41 | 83.60 |
| 4-Nitrophenol | 350 | 72 | 79.43 |
| 2, 4-Dinitrophenol | 2,000 | 71 | 96.45 |
| 4, 6-Dinitrocresol | 9,000 | 78 | 99.13 |
| Phenol | 20,000 | 15 | 99.93 |
| Total cyanide | 55,000 | 420 | 99.24 |

Sufficient data were produced for the researchers to realize that the treatment goals could still be attained at conditions just below water's supercritical temperature. The relatively high temperature of a near-critical wet oxidation (NCWO) process would still benefit from rapid kinetics, and the pollutant removal levels would meet the treatment goals. The process temperature was lowered to 350°C where salt precipitation no longer occurred, and the effluent would still meet the treatment goals (Table VII).

**Table VII. NCWO Effluent Parameters (350°C, 4 minutes) [7]**

| Compound | Concentration in Effluent (ppb) | Treatment Goal (ppb) |
|---|---|---|
| 2, 4-Dinitrotoluene | < 200 | 113 |
| 2, 6-Dinitrotoluene | < 200 | 155 |
| Nitrobenzene | 5,220 | 27 |
| 2-Nitrophenol | < 5 | 41 |
| 4-Nitrophenol | < 5 | 72 |
| 2, 4-Dinitrophenol | < 5 | 71 |
| 4, 6-Dinitrocresol | < 5 | 78 |
| Phenol | < 5 | 15 |
| Total organic carbon | 331,000 | --- |

Notes:  Detection limits for DNT = 200 ppb
Detection limits for 2-NP, 2,4-DNP, 4,6-DNOC, Phenol = 5 ppb

The NCWO process was able to meet all treatment goals except for nitrobenzene. A small polishing system consisting of activated carbon could be used to reduce the nitrobenzene below the treatment goal. The total organic carbon load in the effluent is primarily acetic acid, which is biodegradable in a conventional wastewater treatment system.

## SUMMARY

Advancements in fundamental knowledge and improvements in process designs have enhanced the performances of wet oxidation systems. Processes in operation today can provide cost competitive solutions to wastewater treatment and biosolids management challenges confronting industrial and municipal managers. Continued developments and applications of near-critical and supercritical processes will result in improved fundamental understanding and advanced process technologies which will undoubtedly result in more environmentally sound solutions for complying with regulatory limits.

## REFERENCES

[1]  de Bekker, P., and K. Heerema. 1994. "Dutch Deep Shaft Takes the Pressure Off Wet Oxidation," *Water Quality International*, 3:28-29.

[2]  Brosnan, D. A., and W. A. Hochleitner. 1992. "Additional of Oxidized Sewage Sludge in Brick Manufacture as a Means of Revenue Generation," *Canadian Ceramics Quarterly*, 61(2):128-134.

[3]  Hochleitner, W. A., and F. J. Romano. 1995. "Solids Residue Used in Brick Manufacture," Water Environment & Technology, 7(1):16-19.

[4]  Caruana, C. M. 1995. "Supercritical Water Oxidation Aims for Wastewater Cleanup," *Chemical Engineering Progress*, April: 10-18.

[5]  Dietrich, M. J., T. L. Randall, and P. J. Canney. 1985. "Wet Air Oxidation of Hazardous Organics in Waste Water," *Environmental Progress*, 4(3):171-177.

[6]  Foussard, J. N., H. Debellefontaine, and J. Besombes-Vailhe'. 1989. "Efficient Elimination of Organic Liquid Wastes: Wet Air Oxidation," *Journal of Environmental Engineering*, 115(2):367-385.

[7]  Sawicki, J. E., and B. Casas. 1993. "Wet Oxidation System - Process Concept to Design," *Environmental Progress*, 12(4):275-283.

# STUDY ON THE PHOTOCATALYTIC DEGRADATION OF MONOCHLOROPHENOL POLLUTANTS BY TITANIUM DIOXIDE IN AQUEOUS SOLUTION

KUO-HUA WANG, KUO-SHU HUANG & YUNG-HSU HSIEH
Department of Environmental Engineering
National Chung Hsing University
250 Kuo-Kuang Road
Taichung, Taiwan

## INTRODUCTION

In recent years, the use of photocatalyic reaction to degrade non-biodegradable organic substances, especially the removal of toxic substances, is gradually being taken seriously. Using semi-conductors as the light-emitting sources for heterogeneous photocatalysis of organic substances in water not only completely degrades the organic substances into low molecular weight chemical compounds such as carbon dioxide and water, but also saves treatment costs because it is not necessary to add special chemicals. This is a controlled treatment method worth studying further[1].

Natural water contains only trace of phenolic compounds, mostly from industrial effluents discharged by such industries as petroleum refining, coal tar, steel, dyestuff, synthetic resins, coal gasification and liquefaction, surface runoff from coal mines and byproducts of agricultural chemicals[2,3]. The reaction of phenolic compounds with chlorine or other halogens produces unpleasant odor and taste, and becomes the main reason in controlling the concentration of phenolic compounds in water. Davis et al[3] used natural light/CdS, Matthews [5] and D'Ollveira[4] et al applied heterogeneous photocatalysis with $UV/TiO_2$ to study the possibility for the removal of chlorophenols, achieving favorable results.

In this investigation, batch experiments for photocatalytic decomposition of monochlorophenol either alone or co-existing were studied in a specially made glass reactor. The catalyst used was granulated titanium dioxide of the anatase type excited by UV lamp with monochromatic wavelength of 365 nm.. For single reactant, kinetic studies involved the variation of component concentrations, pH values and light intensities. For binary or ternary mixtures, the solid weight of titanium dioxide was changed to study the competitive reactions of such substances. In addition, for potential practical purpose, sunlight as natural light source was used to exploit the feasibility through single component experimentation.

## EXPERIMENTAL

The layout of the photocatalytic reaction system, shown in Fig. 1, consisted of a 5000 ml-glass reactor (27 cm × 17 cm i.d.) with an outer acrylic jacket for water circulation supplied from a constant temperature water bath (TU-16A, Techne) to keep the reactor isothermal. The reactor was fitted with a stirrer (N-04407-00 Cole-Parmer), pH controller (pH/ORP Controller 3675, Jenco). The UV light source was from a lamp fitted with a 365nm filter (TL-33, UVP Inc.). The $TiO_2$ used is anatase type (Janssen Chimica) with average particle size of 0.79 $\mu$m and specific surface area of 15.2 $m^2$/g.

The analysis of chlorophenols was performed with HPLC (Hitachi), column (25 cm × 4 mm i.d.) packed with LiChrospher 100RP-18, using a sample size of 20 $\mu\ell$. An excellent separation could be obtained when the composition of the mobile phase was $CH_3CN$: $H_2O$ : $CH_3COOH$=40 : 60 :0.1M and flowed by a rate of 0.7 ml/min. For a column pressure of 79-81 $kgf/cm^2$, and the retention time of around 10-12 min. the fastest effluent was 2-chlorophenol, followed by 4-chlorophenol and then 3-chlorophenol at last.

Table 1 shows the operational conditions including concentration of reactant, pH value and light intensity for the experiments of single reactant. The competitive reaction was conducted by changing the amount of $TiO_2$ added in the experiment of binary or ternary mixtures. At last, the possibility of using natural sunlight as light source in photocatalytic reactions was evaluated.

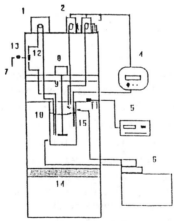

| 1.sampling pump | 6.water bath circulating basin | 11.photo-sensor |
| 2.acid pump | 7.sampling port | 12.Three-way valve |
| 3.base-addition motor | 8.stirring motor | 13.swith for sampling |
| 4.pH controller | 9.stirrer | 14.UV lamp |
| 5.luminous fluxmeter | 10.glass reactor | 15.pH electrode |

Figure 1-Layout of Reaction System

## TABLE I – EFFECT OF FACTORS ON OPERATING CONDITIONS

|  | Concentration ($\times 10^{-5}$M) | Light Intensity (mW/cm$^2$) | pH ($\pm 0.05$) |
|---|---|---|---|
| Controlled Factor | 38 | 2.95 | 4 |
| Changing | 38 | 4.55 | 10 |
|  | 15 | 2.95 | 8.5 |
|  | 7 | 1.39 | 7 |
|  | 3.8 | 0.87 | 5.5 |
| Factor | 1.9 | 0.36 | 4 |
|  | -- | -- | 2 |

Note:1. When one variable being changed the other two variables were under control.

2. pH values maintained to within $\pm 0.05$

3. TiO$_2$ : 10g/L
Temperature : 40°C
Stirrer speed : 200rpm
Aeration : Continuous aeration above 12 hrs.

## RESULTS AND DISCUSSION

1. For single Component

Figure 2 shows the curves for the change of five different initial concentrations of 3-chlorophenol with time. After 2 hrs of reaction time for the highest case of $38 \times 10^{-5}$ M, the concentration decreased to about $18 \times 10^{-5}$ M; Similarly, for initial concentrations of $15 \times 10^{-5}$ M and $7 \times 10^{-5}$ M, the respective residual concentrations were $3 \times 10^{-5}$ M and $0.1 \times 10^{-5}$ M. For the cases of 3.8 $\times 10^{-5}$ M and $1.9 \times 10^{-5}$ M, complete disappearance were observed at 80 and 40 minutes respectively. From these curves, the trend of slopes of concentration changes tends to be gradually flat, this might indicate the occurrence of ratarding effect. There could be two possibilities for such phenomenon one possibility could be the decrease of reactant concentration during its decomposition leading to the lowering of the concentration gradient between the solution and the surface of the semiconductor. Another cause might be due to the formation of chloride ion and other intermediates from monochlorophenol competing with the reactant for the adsorption sites on the surface of catalyst resulting the slow down of reaction rate.

344

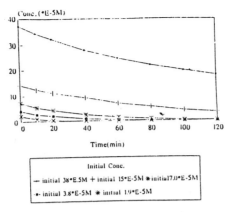

Figure 2–Time-Concentration change curves of 3-chlorophenol at different initial concentrations, pH=4, temp=40°C, lihgtintensity=2.95mW/cm², stirrer speed = 200rpm, continuous aeration above 12 hrs., $TiO_2$= 10g/L.


Figure 3 shows the time-concentration change of 3-chlorophenol under different light intensities. Faster reaction rate was favored by stronger illumination. With light intensities of 4.55, 2.95 and 1.39 mW/cm², the decomposition of the reactant could be completed in 60,100 and 120 minutes respectively. For light intensity of 0.39 mW/cm², however, there remained 0.5 × $10^{-5}$ M residual concentration. Photocatalysis in the presence of titanium dioxide can be regarded as photon energy transfer to affect oxidation on the solid surface. Stronger illumination means higher photon flux, which can provide more excitation by impact upon the surface of titanium dioxide to increase the probability for electron-hole pair production. Thus, the oxidation capacity of organic substances in water can be raised.

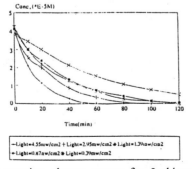

Figure 3–Time-Concentration change curves for 3-chlorophenol at different light intensities, pH=4, initial concentration=3.8 × $10^{-5}$M, temp.=40°C, stirrer speed=200rpm. TiO2=10g/L, continuous aeration above 12 hrs.

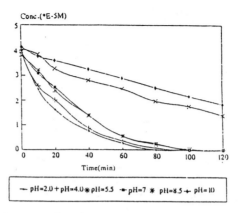

Conc.(*E-5M)

Time(min)

— pH=2.0 + pH=4.0 ⁎ pH=5.5 — pH=7 ⁎ pH=8.5 + pH=10

Figure 4–Time-Concentration change curves for 3-chlorophend at different pH
values, initial concentration=3.8 × $10^{-5}$M, light intensity=2.95
mW/cm$^2$, temp.=40 °C , stirrer speed=200rpm, continuous aeration
above 12 hrs, TiO$_2$ = 10g/L.

Figure 4 shows the time-concentration change curves of 3-chlorophenol at
different pH values. Obviously, the reaction rate was faster in acidic solution
than in alkaline solution. Within 2 hrs of reaction time, complete decomposition
of chlorophenol was obtained in acidic or neutral solution, however, only 50%
decomposition of 3-chlorophenol was observed at pH 10. According to the
reaction mechanism proposed by Okamoto[6], when pH value was above the
pka value of · HO$_2$ (4.88), opposite effect took place to lower the
concentration of · HO$_2$ and slow down the progress of reaction, as it could only
then depend on the supply of · OH radicals, generated from the hydroxyl group
and water molecule on the surface of TiO$_2$ . The amount of · OH radicals
generated in such reaction is much smaller than that by · HO$_2$ radical. This can
be verified by the obviously increasing reaction rate at low pH values although
the concentration of hydroxyl ions is diminished.

2. Modeling of Reaction Kinetics
    The photocatalytic reaction in the presence of titanium dioxide can be
described by the Langmuir-Hinshelwood model as following:

$$r_i = -\frac{dC_i}{dt} = k_r\,\theta_i = k_r\,\frac{k_iC_i}{(1+k_iC_i)} \tag{1}$$

where $k_r$ = reaction rate constant
       $k_i$ = adsorption coefficient
       $\theta_i$ = fractional surface coverage

346

For $t = 0$ and $C_i = C_0$, the integration of above equation leads to:

$$\ln \frac{C_o}{C_i} + k_i (C_o - C_i) = k_r k_i t \qquad (2)$$

For small $C_o$, Eq.(2) is reduced to:

$$\ln \frac{C_o}{C_i} = k_r k_i t = k't \qquad (3)$$

where $k'$ = pseudo-reaction rate constant

Table 2 gives the evaluation of $k'$ under different operating conditions. The linear regression coefficients for $k'$ were above 0.97, which showed the applicability of the Langmuir-Hinshelwood model for these experiments.

TABLE II — ESTIMATION OF $k'$ VALUES IN LANGMUIR-HINSHELWOOD MODEL UNDER DIFFERENT OPERATING CONDITIONS

| Factor | Changing Conditdion | 2-chlorophenol | 3-chlorophenol | 4-chlorophenol |
|---|---|---|---|---|
| Concantration | 38 | 0.00751 | 0.00346 | 0.00296 |
| | 15 | 0.01415 | 0.01216 | 0.01124 |
| | 7 | 0.03771 | 0.04503 | 0.06135 |
| $(\times 10^{-5}M)$ | 3.8 | 0.06236 | 0.06896 | 0.10690 |
| | 1.9 | 0.07857 | 0.09484 | 0.11880 |
| Light Intensity | 4.55 | 0.07980 | 0.07020 | 0.09959 |
| | 2.95 | 0.06236 | 0.06896 | 0.10690 |
| | 1.39 | 0.05273 | 0.04396 | 0.04498 |
| $(mW/cm^2)$ | 0.87 | 0.02554 | 0.04396 | 0.05972 |
| | 0.39 | 0.01718 | 0.02279 | 0.06173 |
| | 2 | 0.06198 | 0.04918 | 0.07417 |
| | 4 | 0.06236 | 0.06896 | 0.10690 |
| pH | 5.5 | 0.04621 | 0.07606 | 0.12300 |
| | 7 | 0.03363 | 0.05837 | 0.09613 |
| | 8.5 | 0.01884 | 0.01055 | 0.02596 |
| | 10 | 0.00565 | 0.09798 | 0.02092 |

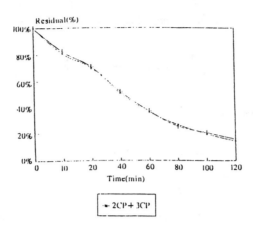

Figure 5–Time-Residual percentage curves for binary mixture of 2-chlorophenol and 3-chlorophenol, pH=4 , initial concentration=7 × $10^{-5}$M, temp.=40 °C , stirrer speed=200 rpm, light intensity=2.95mW/cm², continuous aeration above 12hrs, $TiO_2$=10g/L.

3. For binary and Ternary Mixtures of Chlorophenols

Figure 5 shows the curves of time-residual concentration relationship of 2-chlorophenol and 3-chlorophenol. After 2 hrs of reaction, both showed residual concentrations around 15%. In Fig. 5, the two curves appear to overlap as if obtained from only one specie. The pseudo-reaction rate constant k' for 2-chlorophenol and 3-chlorophenol were respectively 0.017095 and 0.015143 at 14 × $10^{-5}$ M of concentration and the average k' from experimentation with the binary mixture was found to be 0.01561, which differed from the average value calculated by interpolation by only 0.00041. Such a small deviation could be considered negligible, taking into consideration of experimental uncertainties in solution concentration and analytical errors. Thus, it could be concluded that the behavior of binary mixture of 2-chlorophenol and 3-chlorophenol may be treated as a single new specie, the reaction type and rate of which would not be affected by either parent specie and would not involve competing reactions but only parallel reactions among the two species. However, this conclusion has been deduced on the basis of sufficient amount of titanium dioxide in the reaction mixture, otherwise, with less amount different result would happen because of the competition of available adsorption sites on the solid surface by various species in reaction medium.

Figure 6 shows the curves of time-residual concentration profiles for a mixture of three isomeric chlorophenols. The residual percentages of all three reactants fell to around 20% after 2 hrs of reaction. As seen in Fig.6, the three curves are quite similar in shape and slope. In order to identify the competition among three individual chlorophenols, the amount of titanium dioxide was

348

reduced to 0.05g/L, as shown in Fig.7. The curves of 3-chlorophenol and 4-chlorophenol appear to overlap each other. After 2 hrs of reaction, the residual percentages of both fell to around 80%, while in the case of 2-chlorophenol, the residual percentage became lower than 60%. From this result, it indicated that 2-chlorophenol in a mixture of its isomers enjoyed much preference with faster reaction rate in the competing reactions. This result also differed completely with these from individual components

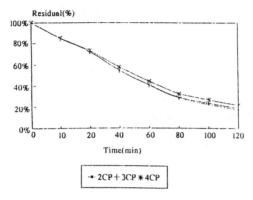

Figure 6–Time-Residual percentage curves of ternary mixture of chlorophenol isomers, pH=4, initial concentration = $7 \times 10^{-5}$ M, temp.=40°C, stirrer speed=200rpm, continuous aeration above 12hrs, $TiO_2$ = 10g/L.

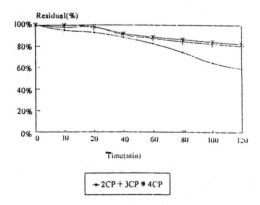

Figure 7–Time-Residual percentage curves of ternary mixture of chlorophenol isomers, pH=4, initial concentration=$7 \times 10^{-5}$M, temp.=40°C, stirrer speed=200rpm, light intensity=2.95mW/cm², continuous aeration above 12hrs., $TiO_2$= 0.05g/L.

349

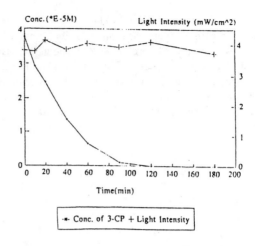

Figure 8–Time-Concentration change curve of 3-chlorophenol under irradiation by natural light, initial concentration=3.8 × 10⁻⁵M, continuous aeration above 12hrs., fixed speed of magnetic stirrer, uncontrolled temp. and pH.

4. Irradiation with Natural Light

Without controlling reaction parameters, the irradiation of $3.8 \times 10^{-5}$ M 3-chlorophenol resulted complete decomposition after three hours exposure under sunlight (Fig.8). The experiment was carried out in a day with continuous sunshine, the light intensity was measured with a luminous fluxmeter placed in the bottom of the reactor with readings in the range of 3.72 -4.12 mW/cm². Due to the initial low concentration, most decomposition took place within 90 minutes of irradiation. Using the average flux and empirical formula for k', its value was found to be 0.038598 for the initial concentration of $3.8 \times 10^{-5}$ M and only differed by approximately 0.001 from linear regression, i.e. k'=0.039758. From this calculation, it indicated the validity of the empirical formula for good prediction.

**CONCLUSIONS**

1. Using UV light excited titanium dioxide in photocatalysis, the decomposition of monochlorophenol in aqueous solution could achieve good result; with concentration less than $7 \times 10^{-5}$ M, a complete removal within 2 hours of reaction was observed.

2. From experimental results, it was found that the photocatalytic reaction proceeded according to the kinetic model of Langmuir-Hinshelwood.

3. In a reaction mixture containing two isomeric monochlorophenols, the reaction could be considered as that of a new single specie, its pseado-reaction rate constant could be estimated by doubling the concentration of a single component.

4. When the amount of titanium dioxide was below a certain level, the photocatalytic reaction of monochlorophenols involved competing reaction; the order diminished preference could be 4-chlorophenol, 3-chlorophenol and 2-chlorophenol.

5. Using natural light source, sunlight could be possibly considered as excitation source for photocatalytic reactions. This approach has significant potential in both practical and economic sense in treatment.

### REFERENCE

1. Gratzel, M.. 1983. *Energy Resources through Photochemistry and Catalysis*, Academic Press.

2. Milner, C. R., and R. Goulder. 1986. " The Abundance, Heterotropic Activity and Taxonomy of Bacteria in a Stream Subject to Pollution by Chlorophenols, Nitrophenols and Phenoxy Alkanoic Acids " , *Wat. Res.*, 20(1):85-90.

3. Davis, A. P., and C. P. Huang. 1990. "The Removal of Substituted Phenols by Photocatalytic Oxidation Process with Cadmium Sulfide " , *Wat. Res.*, 24(5):543-550.

4. D'Ollveira, J., G. Al-Sayyed and P. Pichat. 1990. "Photodegradation of 2- and 3-chlorophenol in $TiO_2$ Aqueous Suspensions" , *Environ. Sci. Technol.*, 24:990-996.

5. Mtthews, R. W.. 1987. "Photooxidation of Organic Impurities in Water Using Thin Film of Titanium Dioxide" , *J. Phys. Chem.*, 91:3328-3333.

6. Okamoto, K., Y. Yasunori, T. Hirok, T. Masashi and I. Akira. 1985. " Heterogeneous Photocatalytic Decomposition of Phenol over $TiO_2$ Powder" , *Bull. Chem. Soc. Jap.*, 58:2015-2022.

# INVESTIGATION OF H$_2$O$_2$ CATALYSIS WITH IRON OXIDE FOR REMOVAL OF SYNTHETIC ORGANIC COMPOUNDS

SHU-SUNG LIN
Graduate Student
Environmental Studies Institutes, Drexel University
Philadelphia, PA. 19104

DR. MIRAT D. GUROL
Professor
Environmental Studies Institutes, Drexel University
Philadelphia, PA. 19104

## INTRODUCTION

Hydrogen peroxide is not a strong oxidant for the majority of the organic water and soil contaminants, except for a few easily oxidizable compounds, e.g., phenols and cyanides. However, mixtures of hydrogen peroxide with iron salts have been effective in oxidizing a wide variety of organic substances in waters and wastewaters in acidic pH ranges [1-4]. It was first reported in 1894 by J. H. Fenton that the mixture of hydrogen peroxide with ferrous ion, which is referred to as "Fenton's Reagent" in the literature, strongly promotes the oxidation of tartaric acid. However, the catalytic effect of these mixtures was found to be limited to the low pH ranges of about 2 ~ 4. In addition, the colloidal precipitates of ferric hydroxides that usually form in the reaction mixtures make separation of the precipitates from solution difficult and expensive. Some recent studies [5-8] showed that mixtures of hydrogen peroxide with solid iron oxide surfaces can be used for removal of synthetic organic pollutants from soils and waters. The major advantage of using iron oxide particles rather than dissolved iron as a catalyst in water or wastewater treatment is that separation of mm size iron oxide particles is simpler than separation of colloidal precipitates.

The objective of this study is to investigate the feasibility of using hydrogen peroxide and iron oxide particles as a chemical oxidant for removal of organic pollutants from waters. In order to establish this objective, an organic compound was oxidized in a laboratory size reactor by hydrogen peroxide in the presence of iron oxide particles. The effect of several parameters on the oxidation efficiency of the compound was investigated. These parameters included the dosages of hydrogen peroxide and iron oxide, stirring speed, and the pH and bicarbonate ion concentrations in the solution. The experiments were conducted in a completely mixed slurry reactor operated in batch mode. Goethite ($\alpha$-FeOOH) was selected as the catalysts in this study since it is a common mineral in nature and also has a low solubility in water ($Ksp=10^{-39}$ [11]). The organic compound selected in this research was n-butylchloride (BuCl). BuCl serves as an OH•-probe due to its non-reactivity with hydrogen peroxide but high reactivity with OH•.

## EXPERIMENTAL METHODS

Solutions of hydrogen peroxide and BuCl were mixed and placed in the reactor which already contained iron oxide particles. Buffers were added for some experiments. The reactor was operated in batch mode and stirred vigorously to maintain complete mixing. The samples taken from the solution with certain time intervals were filtered through a #1 Whitman filter paper upon withdrawal to separate iron oxides from solutions. The filtrate was then analyzed for pH, and the concentrations of hydrogen peroxide, BuCl and total ferric and ferrous iron.

Goethite used in this study was purchased from Aldrich Chemical Company. Goethite particles have a diameter between 0.3~0.6 mm and density of 3.85 g/cm$^3$. Hydrogen peroxide concentration was measured by the permanganate titration method. The interference by organic matters for this method is not significant due to low concentrations of BuCl used in this study. Total ferric and ferrous iron concentrations were determined sepectrophotometrically by using the Phenanthroline Methods [9] which has a detection limit of 0.5 mg/l ($10^{-5}$ mole/l) as iron. Butylchloride was analyzed by a head-space gas chromatograph by using the internal standard, pentylchloride. The gas chromatography was a Shimadzu GC-mini 2 equipped with a flame ionization detector and a packed column of 10% SP-2100 on Supelco 80-100.

## RESULTS AND DISCUSSION

The experimental observations showed that varying the stirring speed in the reactor did not have any significant effect on the reaction rates of hydrogen peroxide and butylchloride at any level of the catalyst addition. Therefore it was concluded that the apparent reaction rates for this process were dominated by the intrinsic reaction rates on iron oxide surfaces rather than the mass-transfer rates of solution reagents to the surfaces. Preliminary experiments also showed that the initial pH of the solutions during the reaction reduced by less than 0.4 units from the original pH of 7.5. This indicates that no buffer was needed to control the pH of the solutions during these experiments.

### Effect of iron oxide dosage

The effect of variation of iron oxide dosage on oxidation rate of BuCl and decomposition rate of hydrogen peroxide is presented in Figures 1 and 2, respectively. In these figures, BuCl and hydrogen peroxide profiles were plotted as a normalized concentration ratio with respect to reaction time. It was clearly demonstrated that an increase in the iron oxide dosage has increased the removal of BuCl substantially. This is in accordance with our expectations since the increase of iron oxide dosage can enhance not only the binding (adsorption) but also the chemical reactions of BuCl on the iron oxide surfaces.

The results presented in Figures 1 and 2 also showed very similar behavioral patterns for hydrogen peroxide and BuCl as a function of iron oxide dosage. This implies a strong correlation between the rates of BuCl oxidation and decomposition of hydrogen peroxide, which is in accordance with our hypothesis that the organic compound would be oxidized by the hydroxyl radical upon decomposition of hydrogen peroxide over the surface of iron oxide.

### Effect of $H_2O_2$ dosage

In Figure 3, BuCl concentration as a function of reaction time was plotted for different hydrogen peroxide dosages, including the blank which was referred to as 0 ppm of peroxide. The results of the blank experiments showed about 35% removal of BuCl by adsorption on iron oxide. The presence of hydrogen peroxide significantly increased the removal of BuCl, indicating the dominance of an effective chemical oxidation process in the presence of hydrogen peroxide.

In addition, the application of higher hydrogen peroxide dosages caused a considerably higher removal of BuCl. This implies that the removal rate of BuCl is directly dependent upon the dosage of hydrogen peroxide as well as the rate of decomposition of hydrogen peroxide. Hence the initial reaction rates of hydrogen peroxide and BuCl were calculated based on the data and plotted as a function of hydrogen peroxide dosage in Figure 4. The results clearly show the dependence of the rate of BuCl oxidation on the decomposition rate of peroxide indicating that the hydroxyl radical which forms upon decomposition of hydrogen peroxide is responsible of the oxidation of BuCl

On the other hand, beyond about 170 ppm (5 mM) of $H_2O_2$ concentration, no appreciable effect of hydrogen peroxide dosage on BuCl oxidation is observable (Figure 3 and 4). These results suggest that under the operational conditions, the optimal removal rate of BuCl by this process was achieved at a hydrogen peroxide dosage of 170 mg/l for iron oxide concentration of 0.5 g/l, or at a molar ratio of 1 mol $H_2O_2$ / 1 mol FeOOH.

## Effect of pH

The effect of variation of pH on BuCl concentration is shown in Figure 5, in which the solid and dash lines represent the data obtained in the presence and absence of hydrogen peroxide, respectively. In the absence of hydrogen peroxide, the amount of BuCl adsorbed on iron oxide was independent of pH indicating that the pH of the solution does not affect the binding capacity of FeOOH for BuCl. In the presence of hydrogen peroxide, the oxidation rate of BuCl has increased slightly with increasing pH, but the effect was not substantially. These results indicate that neither the binding nor the oxidation rate of BuCl was dependent upon the pH. Hence, pH is not an important factor for this process.

## Effect of bicarbonate ion concentrations

In Figure 6, BuCl concentration at different reaction times is shown at various concentrations of bicarbonate ion. In the absence of hydrogen peroxide, the amount of BuCl adsorbed on iron oxide was independent of bicarbonate ion concentration indicating that the bicarbonate ion does not affect the binding capacity of FeOOH for BuCl. In the presence of hydrogen peroxide, significantly higher removal of BuCl was observed at all levels of the bicarbonate ion. Furthermore, the variation of BuCl measurements for different bicarbonate ion concentrations was within the experimental measurement error. In other words, no significant effect of bicarbonate ion was evident on oxidation rate of BuCl.

## CONCLUSIONS

The interaction of hydrogen peroxide with iron oxide was investigated as a potential treatment process for removal of synthetic organic pollutants from waters. The experimental results showed that the apparent rates of reactions for this process are determined by the intrinsic reaction rates on iron oxide surfaces rather than the mass-transfer rates of solution reagents to the surface. In addition, the observations obtained at various dosages of iron oxide and hydrogen peroxide showed that the oxidation rate of BuCl is directly dependent on the rate of hydrogen peroxide decomposition. An increase in iron oxide and hydrogen peroxide dosages can significantly enhance BuCl removal , however, an optimum molar ratio of $H_2O_2$/FeOOH was observed at 1/1.

The proposed process was tested for solutions at various pH and bicarbonate ion concentrations. It was observed that the rate of BuCl oxidation in this system was not significantly affected by the pH of the solution. This is a major advantage for this process as compared to the systems using Fenton's Reagents which are effective only at acidic pH. It was also observed that bicarbonate ions do not affect the oxidation and binding capacities of BuCl on iron oxide surfaces. Hence the alkalinity does not seem to be an important factor in this process as opposed to most other chemical oxidation processes, e.g. $UV/H_2O_2$ [10], $O_3/UV$ [12] and Fenton's Reagent, which are adversely affected by alkalinity.

In conclusion, this feasibility study clearly demonstrates that the mixture of hydrogen peroxide and iron oxide (goethite) particles can be used as an effective chemical oxidant for removal of synthetic organic compounds from contaminated waters.

## REFERENCES

1. Flaherty , K. A., an d Huang, C. P., (1992) "Continuous Flow Applications of Fenton's Regent for the Treatment of Refractory Wastewaters " in " 2nd International Symposium:Chemical Oxidation, Technology for the Nineties", Nashville, Tennessee

2. Munz, C., Galli, R. , Schlotz, R. and Egli, S.,(1992) " Oxidative Treatment of Process Water in a Soil Decontamination Plant: I. Laboratory Studies " Proceedings to the 2nd International Symposium: Chemical Oxidation, Technology for the Nineties , Nashville, TN.

3. Bowers, A. R., Gaddipati, P., Eckenfelder, W. W., Jr, and Monsen, R. M., ( 1989 )" Treatment of toxic or refractory wastewaters with hydrogen peroxide" Wat. Sci. Tech., 21, p.477

4. Walling, C., (1975) " Fenton's Reagent Revisited " Accounts of Chemical Research, 8, pp.12

5. Gurol, M. D. , and Ravikumar J. X,(1994) ""Chemical oxidation of chlorinated organics by hydrogen peroxide in the presence of sand."" Environ. Sci. Technol., 28, p.394

6. Wang, A.H., and Valentine, R. L.,(1993)" Hydrogen peroxide decomposition kinetics in the presence of Iron oxides " 3rd International Symposium: Chemical Oxidation, Technology for the Nineties , Nashville, Tennessee

7. Watts, R. J., Udell, M. D., and Monsen, R. M.,(1993) " Use of iron mineral in optimizing the peroxide treatment of contaminated soils " Water Environ. Res., 65,7,pp.839

8.Tyre, B. W., et al.,(1991) " Treatment of four biorefractory contaminants soils using catalyzed hydrogen peroxide ", J. Environ. Qual., 20, pp.832

9. Standard Method of Water and Wastewater Examination ,1985

10. Liao, C.-H., (1993) PhD thesis, Drexel University, Philadelphia, PA19104

11.Stumm, W. and Morgan,J. J., (1981) Aquatic Chemistry, 2nd edition, Wiley-Interscience (1981)

12. Akata, A., (1994) PhD thesis, Drexel University, Philadelphia, PA19104

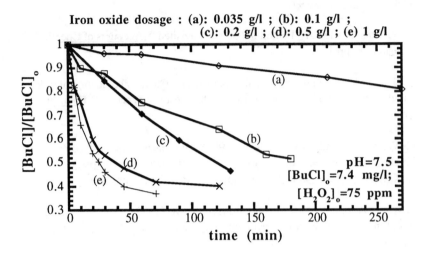

Figure 1. Effect of variation of iron oxide dosage on BuCl concentration

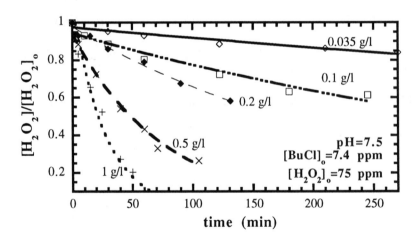

Figure 2. Effect of variation of iron oxide dosage on hydrogen peroxide concentration

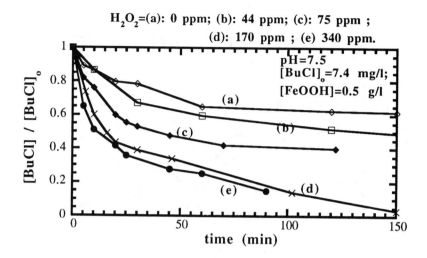

Figure 3. Effect of $H_2O_2$ dosage on the BuCl profile

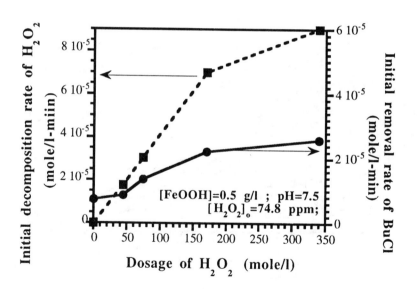

Figure 4. Effect of $H_2O_2$ dosage on the initial removal rate of BuCl and initial decomposition rate of $H_2O_2$

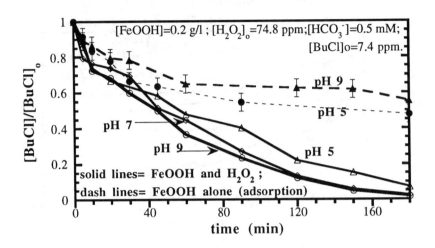

Figure 5. Effect of variation of pH on the BuCl concentration

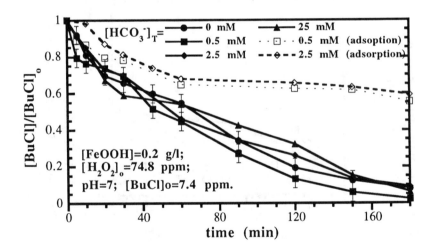

Figure 6. Effects of bicarbonate ion concentratio on BuCl concentration

# TiO$_2$-MEDIATED PHOTODEGRADATION OF 3-CHLOROPHENOL IN THE PRESENCE OF HYDROGEN PEROXIDE

CHENGDI DONG
   Department of Marine Environmental Engineering
   National Kaohsiung Institute of Marine Technology
   Kaohsiung, Taiwan 80005, R.O.C.

CHIU-WEN CHEN
   Department of Marine Environmental Engineering
   National Kaohsiung Institute of Marine Technology
   Kaohsiung, Taiwan 80005, R.O.C.

SHIOW-SHYAN WU
   Department of Marine Environmental Engineering
   National Kaohsiung Institute of Marine Technology
   Kaohsiung, Taiwan 80005, R.O.C.

## ABSTRACT

A method involving acetylation followed by gas chromatography/mass spectrometric (GC/MS) analysis has been applied to the study of the reaction products and pathways of the TiO$_2$-mediated photocatalytic oxidation of 3-chlorophenol (3-CP) in the presence of hydrogen peroxide. Results indicate that the photocatalytic degradation of 3-CP produces 3-chlorocatechol (3-CCA) and 4-chlorocatechol (4-CCA), the ortho hydroxylated products, as the main products. This disagrees with what was reported by other researchers who have proposed the formation of para hydroxylated products, chlorohydroquinone (CHQ) as the major intermediate. In contrast, direct photolysis of 3-CP produces resorcinol (RES), the dechlorination product, as the main reaction products. A reaction pathway has been proposed to describe the degradation of 3-CP by TiO$_2$/H$_2$O$_2$ oxidation process.

## INTRODUCTION

Recently, there has a great attention on the use of photocatalytic oxidation process with TiO$_2$ or other semiconductors in the field of hazardous wastes treatment [1-10]. The widely accepted mechanism of TiO$_2$-mediated photocatalytic oxidation proceess is the generation of hydroxyl radicals from water decomposition which attack the aromatic ring:

$$TiO_2 + h\gamma \longrightarrow h^+ + e^-  \qquad (1)$$

$$H_2O + h^+ \longrightarrow \cdot OH + H^+ \qquad (2)$$

$$(3)$$

Upon irradiation, the $TiO_2$ particle generates electron/hole pairs with positive holes ($h^+$) in the valance band (vb) and free electron ($e^-$) produced in the conduction band (cb) (equation 1). The holes migrate to the $TiO_2$ particle surface and react with the adsorbed water molecular to generate hydroxyl radicals (equation 2), which can further react with organic compounds (equation 3).

Many investigators have reported that this process can completely mineralize a variety of organic pollutants to carbon dioxide, water, or simple inorganic salts, e.g., chloride or phosphate [11-13]. However, only a few of these studies have dealt with the distribution or the nature of reaction products. For example, D'Oliveira et al. [1] have proposed an important reaction pathway for the oxidation of 3-chlorophenol based on HPLC analysis:

$$(4)$$

Only two intermediates, chlorohydroquinone and hydroxyhydroquinone, were detected. According to this pathway, the attack of hydroxyl radicals is preferential at the C4 (para) position and leads to the formation of chlorohydroquinone (CHQ). No explanation why hydroxyl radicals only prefer the para position is given.

From the viewpoint of water quality control, information on the degradation pathway and oxidation products is important. The objective of this study was to determine the intermediate products during the photocatalytic oxidation of 3-chlorophenol in $TiO_2/H_2O_2$ aqueous suspensions. Based on the resulted obtained, we proposed a reaction pathway for the oxidation of 3-chlorophenol by $TiO_2/H_2O_2$ process.

## MATERIALS AND METHODS

### Materials

Chemicals, 3-chlorophenol (98%), chlorohydroquinone (technical grade), 4-chlororesorcinol (98%), and hydroxyhydroquinone (99%), 3-chlorocatechol (99%) and 4-chlorocatechol (99%) were purchased from TCI Chemical Co., Tokyo, Japan. All chemicals were used without further purification. The photocatalyst, Degussa P-25 titanium dioxide (mainly anatase form; surface area 55 $m^2/g$; $pH_{zpc}$ 6.3).

### Experiments

All experiments were conducted in a fixed-bed batch pyrex reactor containing 100 ml $10^{-3}$ M 3-chlorophenol solutions and 0.05 M $NaNO_3$ as ionic strength. The reactor was coated with 0.014 g of $TiO_2$ catalyst at the inside surface of the reactor with a surface area of 113 $cm^2$. The initial solution pH was adjusted to 4.0 with sulfic acid (1 M) and sodium hydroxide (1 M) unless otherwise mentioned. The solution was magnetically stirred. Irradiation was done by a 3 watt lamp with a wavelength of 365 nm. At pre-selected reaction times, samples were taken and the reaction was quenched immediately in the absence of UV irradiation.

### Chemical Analysis

3-Chlorophenol and its intermediates were determined by GC/MS. The samples were acetylated in carbonate solution. 0.025 ml of 1 M potassium carbonate and 0.25 ml of acetic anhydride were added to 1 ml of sample and the acetylated derivatives extracted with 1 ml of hexane, and gently evaporated to dryness in a nitrogen stream. A GC (HP Model 5890) equipped with mass engine (HP Model 5989 II) and HP-5 capillary column (25 mm x 0.25 mm i.d. x 0.33 μm film thickness) was used. To separate different intermediate products, various oven temperature programs were performed. The GC/MS interface line was maintained at 300 °C. The mass selective detector was scanned at a rate of 1.2 seconds per decade and mass spectra were produced using the standard electron ionization (70 eV) at the electron impact mode.

## RESULTS AND DISCUSSION

### Effect of UV light, $H_2O_2$, and $TiO_2$

Figure 1 shows the effect of UV light, $H_2O_2$, and $TiO_2$ on the degradation of 3-chlorophenol (3-CP). UV light alone caused ~3% 3-CP degradation after 2 hrs. The presence of both $TiO_2$ and UV light resulted in ~30% 3-CP degradation and both $H_2O_2$ and UV light resulted in ~40% 3-CP degradation after 2 hrs. UV/TiO$_2$/H$_2$O$_2$ system shows the best efficiency on 3-CP degradation, over 65% of the initial 3-CP was removed after 2 hrs. $H_2O_2$ can trap electrons on TiO$_2$ surface to minilize the extent of elcetron/hole recombination, and then increase the chance of elcetron hole to react with adsorbed water molecular to generate hydroxyl radicals [14-15]. With UV radiation, $H_2O_2$ can also decompose to hydroxyl radicals. Both reactions will enhance the extent of 3-CP degradation.

## Formation of Intermediate Products

The derivatization technique and GC/MS analysis have been successfully used to detect organic pollutants in the aquatic environment. This same technique was chosen to identify the partial oxidation products of 3-chlorophenol in TiO$_2$ aqueous suspensions in the presence of $H_2O_2$. Figure 2 shows the ion chromatogram of the derivatized reaction products of 3-chlorophenol. Five derivatized peaks, **A**, **B**, **C**, **D**, and **E**, with retention times (RTs) at 11.37, 15.29, 15.62, 16.08, and 16.26 minutes, respectively, have been observed. Figures 3 show the GC mass spectra of these five peaks. Peak **A** is the acetylated parent compound, 3-chlorophenol. Peaks **B**, **C**, **D**, and **E** are acetylated intermediate products. The identification of peaks **B**, **C**, **D**, and **E** was based on the following observations:

(1) The difference in m/e value between M$^+$ peak and [M-Ac$_2$]$^+$ peak in peaks **B**, **C**, **D**, and **E** is due to the loss of two acetyl radicals (CH$_2$CO). This indicates the presence of two hydroxyl ions in these four intermediates because the acetyl groups are derivatized from the hydroxyl group of corresponding organic substrates after acetylation.

(2) The mass spectra of peaks **B**, **C**, **D**, and **E** have a near 3 : 1 ratio based on the isotope ion peaks at m/e 144 and 146. This indicates the possible presence of a chlorine atom, which also has a near 3 : 1 ratio based on the isotope ion peaks at m/e 35 and 37.

(3) The m/e value of the base peak of peak **A** (m/e=128) differs from the peaks **B**, **C**, **D**, and **E** by a value of 16. This indicates that compounds **B**, **C**, **D**, and **E** have one more OH molecular than compound **A**.

(4) A slight difference in other fragment ion peaks such as m/e 51, 61, 63, 79, 80, 115, and 128 between peaks **B**, **C**, **D**, and **E** is observed. This implies that these four intermediates have a similar chemical structure.

(5) According to the chemical structure of 3-chlorophenol, there are 5 possible ways of attack by ·OH radicals; all lead to the production of dihydroxybenzenes (equations 5 to 9):

$$\tag{5}$$

OH + ·OH → OH, OH (6)

OH + ·OH → OH, OH, Cl (7)

OH + ·OH → OH, OH, Cl (8)

OH + ·OH → OH, OH, Cl (9)

An attack at the C-2 position will generate 3-chlorocatechol with a replacement of a H atom by an OH molecular (eq. 5); an attack at the C-3 position will result in a dechlorination reaction to generate ressorcinol (eq. 6); an attack at the C-4 position will produce chlorohydroquinone (eq. 7); an attack at the C-5 position will produce 5-chlororesorcinol (eq. 8); an attack of C-6 site will form 4-chlorocatechol (eq. 9). These five dihydroxybenzenes can be identified using authentic standards.

(6) The GC/MS retention times and mass spectra of peaks **B**, **C**, **D**, and **E** are closely matched with the retention times and mass spectra of 3-chlorocatechol, 4-chlorocatechol chlorohydroquinone, and chlororesorcinol standards, respectively. The mass spectra of 3-chlorocatechol, 4-chlorocatechol chlorohydroquinone, and chlororesorcinol show the same $M^+$ peak at m/e 228 and $[M-Ac_2]^+$ at m/e 144 as peaks **B**, **C**, **D**, and **E**. That is, peaks **B**, **C**, **D**, and **E** correspond to acetylated 3-chlorocatechol, 4-chlorocatechol chlorohydroquinone, and chlororesorcinol, respectively.

Figure 4 shows the concentration change of 3-chlorophenol (reacted), 3-chlorocatechol (formed), 4-chlorocatechol, chlorohydroquinone (formed), and chlororesorcinol (formed) as a function of reaction time. Results show that both 3-chlorocatechol and 4-chlorocatechol products are present in significant amounts, whereas amounts of chlorohydroquinone and chlororesorcinol are at least ten times less than that of 3-chlorocatechol or 4-chlorocatechol. The maximun concentration of 3-chlorocatechol during the reaction is ca. 15 % of initial 2-chlorophenol concentration at about 1.5 hour of reaction time and the maximum concentration of 4-chlorocatechol is ca. 10 % of initial 2-chlorophenol concentration at about 1.5 hours of reaction time. After reaching the maximum concentration, these dihydroxybenzene intermediates gradually decrease with with further degradation reactions.

## Reaction Mechanisms

On the basis of the above analysis of intermediates, it is possible to propose reaction pathways describing the photocatalytic oxidation of 3-CP in $TiO_2/H_2O_2$ aqueous suspension. Figure 5 shows the proposed initial steps of photocatalytic oxidation of 3-CP. The oxidation of 3-CP is proposed to occur through an oxidative pathway which involves the initial addition of a hydroxyl radical to 3-CP to generate four possible chlorodihydroxycyclohexadienyl (ClDHCD) radicals (**I, II, III**, and **IV**) [16-17]. These ClDHCD radicals may further react with other hydroxyl radicals through an elimination of water to form 3-chlorocatechol (3-CCA), 4-chlorocatechol (4-CCA), chlorohydroquinone (CHQ), and 5-chlororesorcinol (5-CRE) (path **B**, Figure 5).

In the presence of air or oxygen, the oxidation reaction may also occur via an addition of oxygen to the ClDHCD radicals to form peroxyl radicals which then quickly form the same isomeric products, with an elimination of an $HO_2\cdot$ radical (not observed) (path **A**, Figure 5). The available data in the study does not allow us to discern whether path **A** or path **B** dominates. However, both pathways can happen simultaneously in the presence of an air or oxygen atmosphere.

Another potential minor pathway, which is not shown in Scheme I, is the bimolecular disproportionation reaction between two ClDHCD radicals, which, with an elimination of water, produces chlorophenols and chlorodihydroxybenzenes:

$$(10)$$

The disproportionation reaction usually occurs when organic substrates are at high concentration or when oxidants such as oxygen and hydroxyl radicals are at low concentration. Moreover, this kind of reaction commonly occurs simultaneously with the dimerization reaction which leads to the formation of hydroxylated biphenyl dimers . Since these conditions indicated above are not satisfied in this study, the disproportionation reaction is excluded as a major reaction pathway in the present system.

## Comparison with Direct Photolysis Process

For comparison, the direct photolysis of 3-CP was also examined under the same experimental conditions as the the photocatalytic oxidation process, except no $TiO_2$ catalyst and $H_2O_2$ was added. The degradation of 3-CP is initiated by the cleavage of the carbon-chlorine bond in 3-CP benzene ring [18]. The stronger polarity of the carbon-chlorine bond compared to the carbon-hydrogen bond causes UV photons to attack the C-Cl bond rather than C-H bond. As a result, the carbon-chlorine bond will be polarized then substitutied and hydroxylated:

$$(11)$$

## CONCLUSIONS

Results of this study indicate that the $TiO_2$-mediated photocatalytic oxidation process can readily degrade 3-CP in the presence of $H_2O_2$. The distribution of intermediates shows that the mechanism of $TiO_2$-mediated photocatalytic oxidation reaction can be described by the free radical mechanism involving $\cdot OH$ as the major reaction species. The reaction pathways follow the ortho and para hydroxylation pathways without dechlorination, so that the main intermediates found are 3-CCA and 4-CCA. In contrast, the direct photolysis of 3-CP follows the dechlorination pathway, which leads to the formation of non-chlorinated hydroxylated products, resorcinol. It has been suggested that the hydroxylated compounds can be more toxic to mammals than their starting mono-hydroxylated compounds such as monochlorophenols. Therefore, great care must be exercised in the application of photolysis process for the treatment of phenolic compounds.

## ACKNOWLEDGMENT

This work is supported by a grant (NSC-84-2211-E-022-002) from the National Science Council, Republic of China.

## REFERENCES

1. D'Oliveira, J. Al-Sayyed, G., and Pichat, P., "Photodegradation of 2- and 3-chlorophenols in $TiO_2$ Aqueous Suspensions," Environ. Sci. Technol., 24, 7, 990-996. (1990)
2. Barbeni, M., Morello, M., Pramauro, E., Pelizelli, E., Borgarello, E., Graetzel, M., and Serpone, N., "Photodegradation of 4-Chlorophenol Catalyzed by Titanium Dioxide Particles," Nouv. J. Chim., 8, 547-550. (1984)
3. Bard, A. J., "Photoelectrochemistry and Heterogeneous Photocatalysis at Semiconductors," J. Photochemistry, 10, 59-75. (1979)
4. Dong, C. and Huang, C. P., "Photocatalytic Degradation of 4-Chlorophenol in $TiO_2$ Aqueous Suspensions,", in Aqueous Chemistry, ACS Advanced Chemical Series 422. (1995)

5. Ho, P. C., "Photooxidation of 2,4-dinitrotoluene in Aqueous Solution in the Presence of Hydrogen Peroxide," Environ. Sci. Technol., **20**, 260-267. (1986)

6. Hussain, Al-Ekabi and Serpone, N. "Kinetic Studies in Heterogeneous Photocatalysis. 1. Photocatalytic Degradation of Chlorinated Phenols in Aerated Aqueous Solutions over $TiO_2$ Supported on a Glass Matrix," J. Phys. Chem. **92**, 5726-5731. (1988)

7. Ollis, D. F., Pelizzetti, E., and Serpone, N., "Destruction of Water Contaminants," Environ. Sci. Technol., **25**, 9, 1523-1529. (1991)

8. Ollis, D. F., Pelizzetti, E., and Serpone, N., "Heterogeneous Photocatalysis in the Environment: Application to Water Purification," in Photocatalysis-Fundamentals and Applications, Edited by Serpone, N. and Pelizzetti, E., John Wiley & Sons, New York, 6603-6637. (1989)

9. Schiavello, M., "Basic Concepts in Photocatalysis," in Photocatalysis and Environment- Trends and Applications, Edited by Schiavello, Kluwer Academic Publishers, The Netherlands, 351-356. (1987)

10. Turchi, C. S. and Ollis, D. F., "Photocatalytic Degradation of Organic Water Contaminants: Mechanisms Involving Hydroxyl Radical Attack," J. of Catalysis, **122**, 178-192. (1990)

11. Matthews, R. W., "Carbon Dioxide Formation from Organic Solutes in Aqueous Suspensions of Ultraviolet-Irradiated $TiO_2$. Effect of Solute Concentration," Aust. J. Chem., **40**, 667-675. (1987)

12. Matthews, R. W., "Photocatalytic Oxidation of Organic Impurities in Water Using Thin Films of Titanium Dioxide," J. Phys. Chem., **91**, 3328-3333. (1987)

13. Minero, C., Aliberti, C., Pelizzetti, E., Terzian, R., and Serpone, N., "Kinetic Studies in Heterogeneous Photocatalysis. 6. AM1 Simulated Sunlight Photodegradation over Titania in Aqueous Media: A First Case of Fluorinated Aromatics and Identification of Intermediates," Langmuir, **7**, 928-936. (1991)

14. Sclafani, A. Palmisano, L., and Davi E., "Photocatalytic Degradation of Phenol by $TiO_2$ Aqueous Dispersions: Rutile and Anatase Activity," New J. Chem., **14**, 265-268. (1990)

15. Sclafani, A., Palmisano, L., and Schiavello, M., "Influence of the Preparation Methods of $TiO_2$ on the Photocatalytic Degradation of Phenol in Aqueous Dispersion," J. Phys. Chem., **94**, 829-832. (1990)

16. Okamoto, K., Yamamoto, Y., Tanaka, H., Tanaka, M., and Haya, A., "Kinetics of Heterogeneous Photocatalytic Decomposition of Phenol over Anatase $TiO_2$ Powder," Bull. Chem. Soc. Japan., **58**, 2023-2028. (1985)

17. Okamoto, K., Yamamoto, Y., Tanaka, H., Tanaka, M., and Haya, A., "Heterogenous Photocatalytic Decomposition of Phenol over $TiO_2$ Powder," Bull. Chem. Soc. Jpn., **58**, 2015-2022. (1985)

18. Lipczynska-Kochany, E. and Bolton, J. R., "Flash Photolysis/HPLC Applications. 2. Direct Photolysis vs. Hydrogen Peroxide Mediated Photodegradation of 4-Chlorophenol As Studied by a Flash Photolysis/HPLC Technique," Environ. Sci. Technol., **26**, 2, 259-262. (1992)

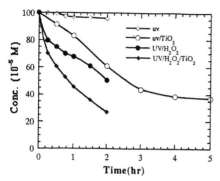

Figure 1 Effect of UV , H₂O₂, and TiO₂ on the degradation of 3-chlorophenol.

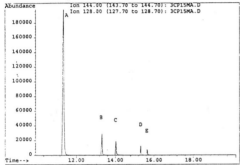

Figure 2 GC/MS ion chromatogram of the derivatized reaction products of 3-CP. Peaks are identified as the following: A: 3-CP, B: 3-CCA, C: 4-CCA, D: CHQ, E: 5-CRE.

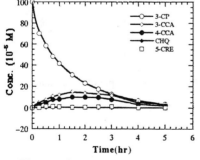

Figure 4 Concentration change of 3-CP, 3-CCA, 4-CCA, CHQ, and 5-CRE as a function of time.

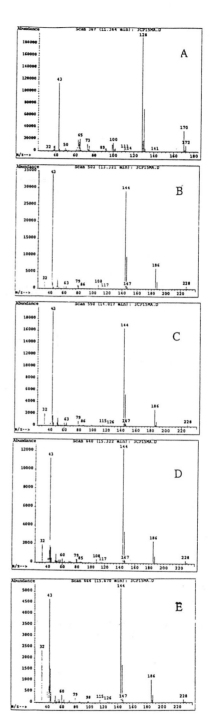

Figure 3 Electron impact mass spectra of peaks (A)-(E) in Figure 2.

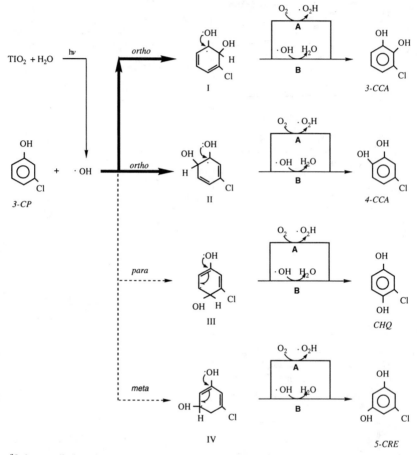

*Only nonradical species were detected.

Figure 5. Proposed routes for the initial steps of 3-chlorophenol degradation by TIO$_2$/H$_2$O$_2$/UV oxidation process.

# OXIDATION OF CHLOROBENZENE BY OZONE AND HETEROGENEOUS CATALYTIC OZONATION

NILESH N. BHAT
   Graduate Student
   Chemical Engineering Department, Drexel University
   Philadelphia, PA 19104

DR. MIRAT D. GUROL
   Professor
   Chemical Engineering Department & Environmental Studies
   Institute, Drexel University
   Philadelphia, PA 19104

## INTRODUCTION

Ozone is a chemical oxidant that is widely used for treatment of contaminated waters. Oxidation of organic compounds in water takes place by molecular ozone and/or by hydroxyl radicals, generated upon the decomposition of ozone in water. Molecular ozone may react with the electron rich sites of the organic compounds and this pathway is referred to as direct reaction of ozone. The selective nature of these reactions limit this pathway to only certain types of compounds which contain unsaturated bonds. Depending on the structure of the compound direct oxidation rate constants may range from less than 1 to about $10^8 M^{-1}s^{-1}$ [1-3].

The dissolved ozone also decomposes while it reacts with organic compounds in water. It is generally accepted that decomposition of ozone produces hydroxyl radical (OH·) which is a highly reactive and a non-selective oxidant. The rate constant of the reaction of OH· with the organic compounds is in the order of $10^9$-$10^{10}$ $M^{-1}s^{-1}$ [4]. The amount of OH· generated by self-decomposition of ozone is limited, therefore various catalysts including $H_2O_2$ and UV rays have been used for more efficient conversion of ozone into OH. Iron salts might also promote oxidation reactions by acting as promoters of the decomposition of ozone in water [8].

A few rare experiments on the heterogeneous catalysis of ozone using solid catalyst have been undertaken. For example ozone was used in the presence of $TiO_2$ to oxidize natural organic matter [5]. The efficiency of this system, as determined by the Total Organic Carbon (TOC) removal, was found to be better than both ozonation alone and ozone combined with hydrogen peroxide. Another application included the use of $O_3$ in the presence of alumina supported iron(III) [6] In this system both phenol and TOC were removed better than ozonation alone. It was suggested that the organic compounds were oxidized by generation of hydroxyl radicals by decomposition of ozone on the catalyst sites. A number of metallic oxides such as CuO, $Fe_2O_3$, $Cr_2O_3$, NiO were used in the presence of ozone showing an important catalytic effect [7].

Hence the intent of this study was to investigate the feasibility of heterogeneous catalytic oxidation using ozone and iron oxide particles. The study is based on the hypothesis that ozonation in the presence of solid catalyst particles such as iron-oxide involves the formation of hydroxyl radicals by the decomposition of ozone on the surface. In order to achieve the objective the process was examined in a semi-batch reactor by ozonating water samples with and without iron oxide particles. The effect of the dosage of iron oxide, size of iron-oxide particles, flow rate of ozone gas, concentration of ozone gas on the process rate was investigated. Monochlorobenzene was used as a model organic compound due to its fairly high water solubility, ease of quantification and high resistance to oxidation by molecular ozone (second order rate constant of $0.75M^{-1}s^{-1}$ [1] ). Geothite ($\alpha-FeOOH$) was selected as a catalyst in this study since it is a common mineral in nature and also has a very low solubility in water ($K_{sp}= 10^{-39}$) [9].

## EXPERIMENTAL METHODS

The experimental set-up used in this study is shown in Figure 1. The apparatus consists of an ozone generator, monitoring devices and a reaction vessel. Ozone gas was generated from oxygen using a Welsbach T-408 ozone generator. The amount of ozone produced was controlled by changing the power input to the generator. Ozone gas flow rate was controlled by a gas flow meter installed in the generator. The experiments were conducted in a reaction vessel which had openings for ozone gas inlet and outlet and for liquid sampling. A glass diffuser was used to sparge ozone gas into the solution. The influent and effluent ozone gas concentration was measured at 254nm by a Spectronic-1001UV spectrophotometer, which was calibrated by the Potassium Iodide method [10]. The solution in the reactor was mixed by a teflon coated stirrer at a rate to provide complete mixing. Teflon tubings were used for all connections from ozone generator to the reaction vessel. The experiments were conducted on solutions containing chlorobenzene at an initial concentration of 10 ppm and having alkalinity 0.5-1.0mM as bicarbonate. In ozonation experiments in presence of iron-oxide, a dosage of 0.2 g/L of FeOOH was used unless otherwise noted. The ozone gas was continuously sparged into the reactor at a rate of 0.2 L/min, and two different inlet concentrations of 3 mg/L and 22.4 mg/L were used.

The dissolved ozone concentration was determined by the Indigo Method [11]. There was no significant interference by iron oxide particles during the absorbance measurement. The absorbance measured both before and after filtration did not show significant difference. The concentration of monochlorobenzene was determined using a purge and trap gas chromatograph (Varian 3300). The column used was Carbopack B 60/80 mesh coated with 1% SP-1000. The total organic carbon was measured by a Dohrman DC-80 Carbon Analyzer according to the Standard Methods [10].

Chlorobenzene in the liquid form was obtained from Fischer Scientific Company and was of 99.98% purity. Aqueous chlorobenzene solutions used for the oxidation reactions were prepared from Millipore Milli-Q water and stock solution was prepared by stirring chlorobenzene for at least 1 week prior to use. Iron oxide ($\alpha$-FeOOH) particles were of catalyst grade(30/50 mesh) and purchased from Aldrich Chemical Company. The solutions from the catalytic system were filtered using qualitative Whatman filter paper of 11$\mu$m pore size and the filtrate was used for analysis of chlorobenzene.

## RESULTS AND DISCUSSION

Ozone concentration in solution in the semi-batch reactor was measured as a function of time in the absence of chlorobenzene in pure water and in the presence of iron-oxide particles. The results shown in Fig. 2 indicate significant ozone decomposition over the iron-oxide particles as expected. This decomposition is believed to generate hydroxyl radicals which can improve the oxidation of organic substances, including chlorobenzene. In the following sections the effectiveness of the process in the presence and absence of iron-oxide is discussed, in terms of removal of chlorobenzene as a function of time in the semi-batch reactor.

### I) Removal of Chlorobenzene by ozone alone.

The concentration of chlorobenzene in solution as a function of time for different treatments is depicted in Fig. 3. In order to quantify the effect of stripping in ozonation system, a control experiment was conducted by using oxygen gas to study the stripping of chlorobenzene. The results show that under the operating conditions the stripping contributed to the removal of chlorobenzene by about 50%. Although the direct ozone reactivity towards chlorobenzene was reported to be low ($k=0.75 M^{-1}s^{-1}$ [1] ), appreciable removal was obtained in the ozonation system. This can be attributed to hydroxyl radicals, that are generated by the decomposition of ozone.

### II) Removal of Chlorobenzene in the presence of iron-oxide.

Several experiments were performed in order to quantify the removal of chlorobenzene due to adsorption on iron-oxide particles alone. The results indicated that adsorption proceded rapidly in the initial period but gradually slowed down as the surface sites were saturated by chlorobenzene. This shows the limitation of relying on adsorption alone for removal of chlorobenzene. However in the ozone-iron-oxide system, chlorobenzene removal was observed to be significantly higher. This is probably because the sites are continuously renewed by oxidation of the bound chlorobenzene.

a) Effect of iron-oxide dosage.

The effect of variation of iron oxide dosage was studied by using three different dosages of iron oxide, 0.05 g/L, 0.2g/L, 1.0 g/L. The results presented in Fig. 4. indicate that the increase in iron oxide dosage resulted in an increase in the removal efficiency of monochlorobenzene. For example when iron oxide dosage was increased from 0.05 g/L to 1.0 g/L the chlorobenzene removal has increased from 68% to 83% at the end 30 minutes of reaction time. This implies that surface reactions play an important role in the removal mechanism of chlorobenzene, because increasing the mass of iron-oxide increases the surface area as well as the amount of iron available for the reactions.

b) Effect of size of iron-oxide particles.

The effect of the size of particles was studied by using two different sizes of particles. The original particles of size 0.6-0.3 mm were crushed to a size of 0.212-0.15 mm and were used in the reaction system with a dosage of 0.05 g/L. The experimental results are compared in Fig. 5. It can be seen that reducing the particle size had no effect on the removal of chlorobenzene. This suggests that the

374

apparent reaction rate is controlled by the reaction occurring on the iron oxide surface, rather than the mass transfer of chlorobenzene. The results also indicate that although the specific surface area was increased from 0.173 m-1 to 0.43 m-1 (by 2.5 times), there was no effect on the removal of chlorobenzene and hence the removal is controlled by the internal surface area of the pores rather than the external surface area of the iron-oxide particles.

c) Effect of Ozone Flow rate and Ozone concentration.

The effect of ozone flow rate was investigated by running the experiments at 0.2L/min and 0.4 L/min. It can be seen from the results presented in Fig. 6 that the increase in the flow rate leads to an increase in the removal of chlorobenzene from the system, all the other conditions being the same. The increased removal could be due to two factors: 1) an increased volatilization of chlorobenzene at higher gas flow rate, and 2) increased ozone dosage into the system (from 0.6 mg/min to 1.2 mg/min) which leads to more reaction with chlorobenzene. This issue was investigated further by changing the ozone concentration in the gas phase at the inlet from 3 mg/L to 4 mg/L by keeping the gas flow rate at 0.2L/min. The results presented in Fig. 7 indicate that increasing the ozone dosage even without increasing the gas flow would still improve the oxidation rate of chlorobenzene.

Oxidation of organic compounds always leads to formation of intermediates and by-products which have not been identified in this study. However, the effectiveness of the proposed process on the removal of intermediate organic compounds was checked by measuring the total organic carbon(TOC) removal. The information about the TOC removal is helpful in understanding whether the process can completely mineralize the organic matter. The TOC removals was measured for the cases of ozonation alone and ozonation in the presence of iron-oxide particles. It can be seen from Fig. 8 that the TOC removal obtained in the catalytic system was significantly greater than that obtained in the absence of the catalyst, under the same operating conditions. This increase in TOC removal can be explained by improved oxidation of the intermediate

compounds by OH˙ which forms on the catalyst surface. Although some degree of removal of the intermediates by simple adsorption on the surface is possible, but since the intermediate compounds of oxidation are usually more polar and more soluble compounds, such as, simple carboxylic acids and alcohols, not much contribution of adsorption to overall removal of the intermediates would be expected.

375

## CONCLUSIONS

The catalytic ozonation using iron oxide($\alpha$–FeOOH) particles was more effective than ozonation alone, for removal of chlorobenzene as well as TOC in aqueous solutions. The oxidation rate of chlorobenzene appeared to be controlled by the reaction on iron oxide surfaces, rather than mass transfer of ozone or chlorobenzene to the surface. The experimental results also showed that increasing the dosage of iron-oxide leads to an increase in the removal of chlorobenzene. An increased removal of chlorobenzene was also obtained by increasing the ozone gas flow rate and its concentration. The results on residual ozone concentration provided further evidence that the presence of iron-oxide particles leads to significant decomposition of ozone which produces radical entities more reactive than ozone itself resulting in increased removal of chlorobenzene and TOC.

## REFERENCES

1. Hoigné, J., and Bader, H., "Rate Constants of Reactions of Ozone with Organic and Inorganic Compounds in Water,I-Non-Dissociating Organic Compounds," Wat. Res., Vol 17, p 173, 1983.

2. Gurol, M.D. And Bremen, W.M., "Kinetics and mechanisms of Ozonation of Free Cyanide Species in Water." Environ. Sci. Tech. Vol 19,p 804, 1985

3. Gurol, M.D. And Nekouinaini, S. "Kinetic Behavior of Ozone in Aqueous Solutions of Substituted Phenols." Ind. Eng. Chem. Fund. Vol 23, p 54, 1984

4. Farhataziz, and Ross, A.B., "Selected Specific Rates of Reactions of Transients from Water in Aqueous Solution", National Bureau of Standards, NSRDS-NBS59, Washington D.C., 1977.

5. Allemane, H., Delouane, B., Paillard, H., and Legube, B., "Comparative Efficiency Of Three Systems ($O_3$, $O_3/H_2O_2$ and $O_3/TiO_2$) For the Oxidation Of Natural Organic Matter in Water". Ozone Sci. &Engg. Vol 15, p 419-432, 1993.

6. Al-Hayek, N., Legube, B., Doré, M. "Ozonation Catalytique (FeIII/Al2O3) du phénol et de ses produits d'ozonation", Environ. Technol. Letters, Vol 10, p 415-426, 1989.

7. Munter, R.R. ,Kamenev, S.B. Et al "Catalytic Treatment of wastewater with Ozone", <u>Khimiya i Technologiya Vody</u>, 7(6), p 17-19, 1985.

8. Hoigné, J. And Staehelin, J. "Decomposition of Ozone in Water in the presence of Organic Solutes Acting as Promoters and Inhibitors of radical Chain reactions", <u>Environ. Sci. Tehnol.</u> Vol 19, p 1206-1213, 1985.

9. Stumm, W. And Morgan, J.J., <u>Aquatic Chemistry</u>, 2nd edition, Wiley-Interscience 1981.

10. <u>Standard Method For The Examination of Water And Wastewater</u>, 16th. Edition, 1985.

11. Bader, H. And Hoigne, J. "Determination of Ozone in Water by the Indigo Method", <u>Wat. Res.</u>, Vol. 15, p 449, 1981.

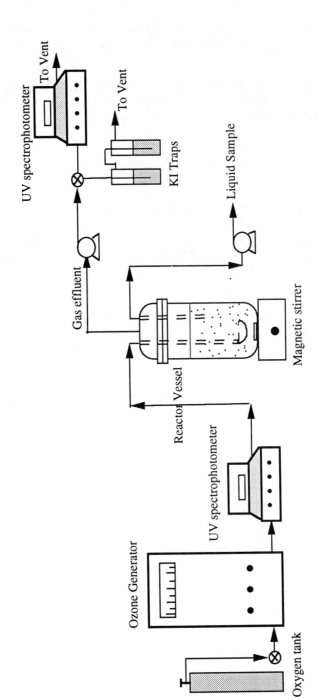

Figure 1. Experimental Set-up

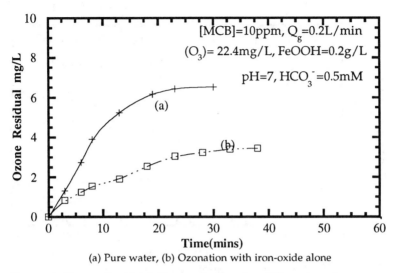

(a) Pure water, (b) Ozonation with iron-oxide alone

Figure 2. Ozone Residual Concentration for different systems.

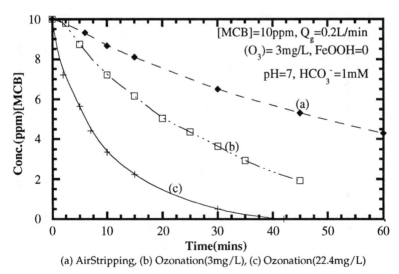

(a) AirStripping, (b) Ozonation(3mg/L), (c) Ozonation(22.4mg/L)

Figure 3. Removal of Chlorobenzene in the absence of iron-oxide.

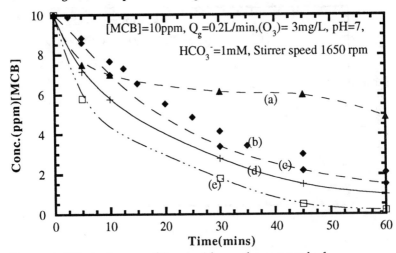

Figure 4. Effect of mass of Iron-oxide on the removal of Chlorobenzene

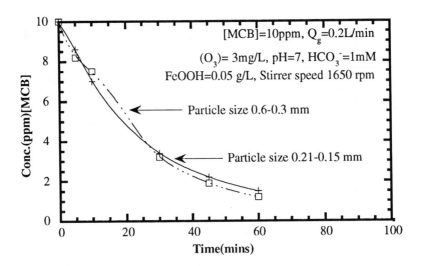

Figure 5. Effect of size of Iron-oxide on the removal of Chlorobenzene

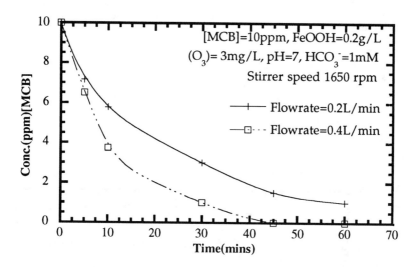

Figure 6. Effect of flow rate of Ozone gas on the removal of
Chlorobenzene

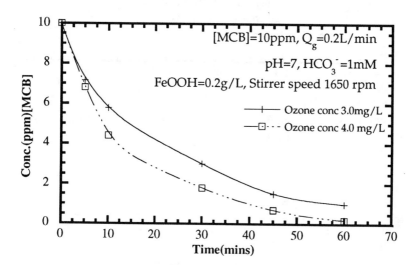

Figure 7. Effect of concentration of Ozone gas on the removal of
Chlorobenzene

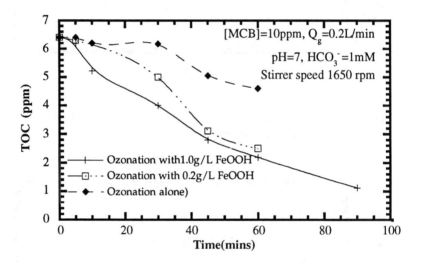

Figure 8. TOC removal for ozonation and catalytic ozonation
processes

# INVESTIGATION AND STUDY ON THE PHOTOCATALYTIC DEGRADATION OF POLYCHLOROPHENOLS IN AQUEOUS SOLUTION

YUNG-HSU HSIEH AND TING-SHAN CHIA
Department of Environmental Engineering
National Chung Hsing University
250 Kuo-Kuang Road
Taichung, Taiwan

## INTRODUCTION

Chlorination of phenolic compounds in aqueous solution leads to the formation of a mixture of chlorinated phenols over a broad range of pH values and with concentration of phenols as little as 0.05mg/L [1]. Most phenolic compounds are present in wastewaters from petrochemical, coal tar, plastic and pesticidal chemical industries. In addition, chlorinated phenols are used as intermediates in the manufacture of fungicides, herbicides, insecticides and preservatives. The toxicity of chlorinated phenols varies with the substitution position and number of chlorine atoms in the benzene ring. Beltrame reported that the more chlorine atoms attached in the phenol molecules, the more the power in suppressing the growth of micro-organisms[2]. Biological treatment of wastewaters containing chlorinated phenols requires longer retention time and in the case of higher concentration, the activated sludge formation can be substantially intoxicated and suppressed [3,4,5]. Since water treatment techniques have been developed adequately to remove toxic substances, chlorophenols can be eliminated completely by such methods as UV-initiated photo-oxidation with chemical reagents. Moza[6], Miller[7], and Sundstrom[8] reported the use of $UV/H_2O_2$ system and Khan et al with $UV/O_3$ treatment of wastewaters contaminated with chlorophenols[9]. Based on the photosensitization of semiconductor materials such as titanium dioxide, Matthews[10], D'Olivera [11] and Tseng [12] reported the heterogeneous photocatalytic degradation of chlorophenols for the possibility of their decontamination. The present investigation was to study the degradation of polychlorophenols in aqueous solution by the use of $UV/TiO_2$ system, including the batchwise treatment of 2,4-dichlorophenol, 2,4,6-trichlorophenol, and 2,3,4,6-tetrachlorophenol at various controlled pH-regimes.

## EXPERIMENTAL

The equipment for this investigation, shown in Figure 1, consisted of an acrylic jacketed glass reactor placed inside a black cloth covered cabinet, 167 cm(H) × 55 cm(L) × 28 cm(W), to avoid the interference of external light. Water from a constant temperature bath was circulated by a pump through the reactor jacket to keep the reaction temperature around 25°C. On top of the glass reactor was a movable cover with drilled ports for insertion of stirrer, pH electrodes, thermometer, sampling tube and acid/alkali feedlines into the reactor. UV lamp (UVP Inc. Model TL-33) was placed under the reactor bottom and projected upward. The intensity of UV light was controlled at 2.84 mW/cm² on the surface. Sample solution was prepared by aeration of ultrapure water for more than 12 hours and then mixed with the calculated amount of polychlorophenol (from Alderich) adjusted to a concentration of $7 \times 10^{-5}$M, followed by adding 10g/L of titanium dioxide (Janssen Chimica) with stirring at 200rpm. The sampling was performed with peristaltic pump connected to Teflon tubing(1/8 in. i.d.) immersed to the reaction mixture, taking a sample volume of 15ml for each time. The residual phenol content was analyzed by HPLC (Hitachi Model L-6000), the extent of degradation by TOC (O.I.Analytical Model 700), and chloride ion determination by ion chromatograph(Dionex Series 4500i).

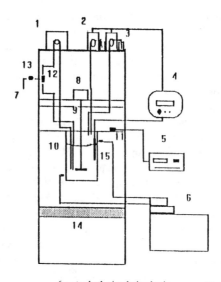

| | | |
|---|---|---|
| 1.sampling motor | 6.water bath circulating basin | 11.photo-sensor |
| 2.acid-addition motor | 7.sampling port | 12.Y-tube |
| 3.base-addition motor | 8.mixing motor | 13.swith for sampling |
| 4.pH controller | 9.mixing rod | 14.UV light source |
| 5.photometer | 10.glass reactor | 15.pH electrode |

Figure 1  Layout of experimental equipment

## RESULTS AND DISCUSSION

### 1. Effect of pH Value on Degradation

Figure 2 shows the time change of residual concentration of 2,4-dichlorophenol ($C/C_o$) during photocatalytic degradation at various pH values, indicating the order of reaction rate as pH 3>pH 5>pH 7>pH 9>pH 11. At pH 3, 2,4-dichlorophenol could be decomposed within 120 min. As shown in the curve, the residual concentration change becomes progressively flat, which could be the result of competing adsorption of reaction intermediates and chloride ion with the reactant molecules on the $TiO_2$ surface causing the decrease of reactant concentration on the active sites, or of its concentration gradient migrating toward the catalyst surface and slowing down the reaction. At pH 11, on the other hand, the residual concentration change in the curve remains descending possibly due to a higher concentration gradient around the reaction sites, and less affected by the interference of reaction intermediates.

The increasing reaction rate due to lower pH value could be attributed to three possible reasons. First, adsorption of 2,4-dichlorophenol on $TiO_2$ surface, by background experiment, was favored in acidic medium, thus, could promote better photocatalytic degradation. When the pH value is below the pK value of 2,4-dichlorophenol, it exists mainly in the molecular state. The zero point of charge of $TiO_2$ used in the investigation lies between 1 to 3, i.e. the functional groups on the $TiO_2$ surface would be in a deprotonated state and provide bonding formation in the transition complex for faster rate of photocatalytic conversion. Besides, coagulation of $TiO_2$ particles was found at controlled pH of 3 because of its closeness to the zero point of charge for $TiO_2$. This increased the transparency of the reaction medium for better transmission of high photon flux inducing faster reaction rate. Another reason for lower pH in increasing the reaction rate is related to the reaction mechanism, when the pH value being higher than the pK value (4.88) for the formation of $\cdot HO_2$ radicals, the formation of $\cdot OH$ radicals would be retarded and unfavorable to the photocatalytic degradation [13].

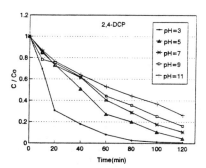

Figure 2  Residual concentration of 2,4-dichlorophenol vs. time of photodegradation at various pH values

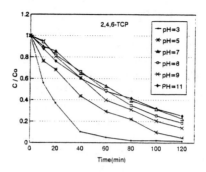

Figure 3  Residual concentration of 2,4,6-trichlorophenol vs. time of
photodegradation at various pH values

Figures 3 and 4 represent respectively the effect of pH on the photodegradation of 2,4,6-trichlorophenol and 2,3,4,6-tetrachlorophenol. Similar to the trend of 2,4-dichlorophenol, lower pH values increased the reaction rates, however, irrespective of either reactant, the reaction rate was faster at pH9 than that at pH 7. This trend could indicate that the pK values of these two compounds could be in the neighborhood of 9.

Taking pH value of 3 for comparing the residual concentrations of these compounds at the end of 120 min, 2,4-dichlorophenol had the least residual concentration, followed by 2,4,6-trichlorophenol and then 2,3,4,6-tetrachloro-phenol. The more number of chlorine atoms in the compounds, the higher the amount of released chloride ions, and in turn, the more suppression on the degradation.

## 2. Kinetic Model

The heterogeneous photocatalytic reaction involving $TiO_2$ can be described by the integral form of the Langmuir-Hinshelwood model [10,11,13]

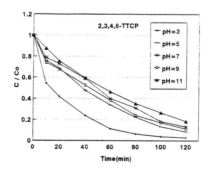

Figure 4  Residual concentration of 2,3,4,6-tetrachlorophenol vs. time of
photodegradation at various pH values

386

TABLE I  VALUES OF $K_a K_r$ (MIN$^{-1}$) AND K' (MIN$^{-1}$) FOR REACTANTS
AT VARIOUS pH VALUES

| Organic | pH | $K_a K_r$ | K' |
|---|---|---|---|
| | 3 | 0.03937 | 0.04145 |
| | 5 | 0.03932 | 0.02486 |
| 2,4-DCP | 7 | 0.03044 | 0.01814 |
| | 9 | 0.01852 | 0.01413 |
| | 11 | 0.01112 | 0.01036 |
| | 3 | 0.04521 | 0.03599 |
| | 5 | 0.03227 | 0.02420 |
| 2,4,6-TCP | 7 | 0.01923 | 0.01231 |
| | 8 | 0.02496 | 0.01433 |
| | 9 | 0.02440 | 0.01673 |
| | 11 | 0.00981 | 0.01120 |
| | 3 | 0.02493 | 0.02949 |
| | 5 | 0.02325 | 0.02001 |
| 2,3,4,6-TTCP | 7 | 0.02152 | 0.01648 |
| | 9 | 0.01929 | 0.01756 |
| | 11 | 0.01649 | 0.01370 |

$$\ln C_o/C + K_a(C_o-C) = K_a K_r t \qquad (1)$$

where $K_a$=adsorption constant
$K_r$=reaction rate constant
for small initial concentration, $C_o$, Eq.(1) can be reduced to

$$\ln C_o/C = K_a K_r t = K't \qquad (2)$$

where K' is the pseudo-reaction rate constant.

By fitting the experimental data to the Mathematica software for regression analysis, the calculated reaction rate constants, shown in TABLE I, followed closely the Langmuir-Hinshelwood model. Upon plotting $\ln C_o/C$ vs. time through linear regression, the slope of the straight line represents K'. Figure 5

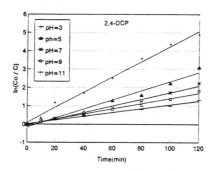

Figure 5  Plot of $\ln C_o/C$ vs. time for 2,4-dichlorophenol at various pH values

shows the plot of ln $C_o/C$ of 2,4-dichlorophenol against time at different pH levels, indicating the lower the pH value, the larger the slope with tendency to decrease systematically with increasing pH value.

Irrespective of pH levels, the reaction followed first-order. Prediction of residual concentration of 2,4-dichlorophenol at a given pH value can be accomplished with K' from Eq.(2) and $K_aK_r$ from Eq.(1). Figure 6 shows the

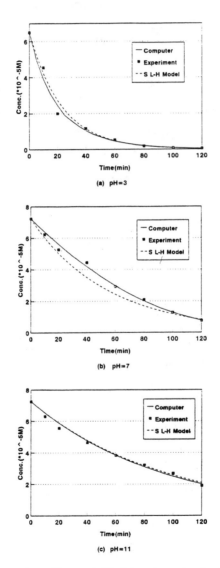

Figure 6  Comparison of Langmuir-Hinshelwood equation, its simplified version and experimental data for 2,4-dichlorophenol at various pH values

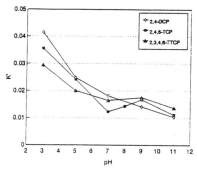

Figure 7  Relationship between K' and pH values of various polychlorophenols

simulation results from prediction, in which the solid line was derived from $K_aK_r$ and the dotted line from K'. These two lines, in general, matched the experimental data points closely.

Figure 7 shows the effect of pH on the pseudo-reaction rate constants for 2,4-dichloro-, 2,4,6-trichloro- and 2,3,4,6-tetrachloro-phenol. Observing these curves, the relationship could not be simply defined in quantitative terms. The complexity of photocatalytic reaction might be due to the functional groups on $TiO_2$ surface influenced by pH level, the chemical potentials of reactants, the respective concentration change of $H^+$ and $OH^-$ ions in solution, and the formation of · OH radicals. All these factors could alter the reaction paths in the formation of different reaction intermediates.

3. Exhaustive Decomposition

Figure 8 shows the result of residual concentration vs. time upon exhaustive decomposition of 2,4,6-trichlorophenol. It shows the formation of chloride ion at the beginning of the reaction through dechlorination and ring cleavage, being identical to the findings of D'Olivera[2] using $UV/TiO_2$ system in

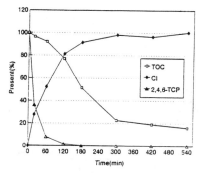

Figure 8  Exhaustive degradation of 2,4,6-trichlorophenol—TOC, Cl⁻
and residual concentration of reactant vs. time curves

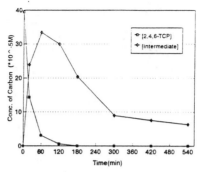

Figure 9  Exhaustive degradation of 2,4,6-trichlorophenol—carbon content
in reaction intermediates vs. time curve

the decomposition of 2-chloro- and 3-chloro-phenol. Following the initial
release of chloride ion, continuous dechlorination proceeded to decompose the
reactant. The removal of 2,4,6-trichlorophenol in 120 min reached 98.5% with
81% release of chloride ion, while TOC was only counted for 23%
decomposition. As mentioned previously [2], the toxicity of chlorinated phenols
depends on the substituted chlorine atoms in the benzene ring, this
decomposition could assist the detoxication of such compounds. After a
reaction period of 300 min, no more reactant could be detected, with 97%
chloride ion release but slow gains in TOC results. This could be explained by
the resistance of organic remains to photodegradation and/or deactivation of
$TiO_2$ surface from chloride ion and organic remains.

4. Concentration Change of Intermediates

Based on carbon balance, the concentration changes of carbon content in
the intermediates is shown in Figure 9, similar result of chlorine balance is given
in Figure 10. Both chlorine and carbon show maxima in the respective curve
followed by further reactions along with original reactant, which disappeared

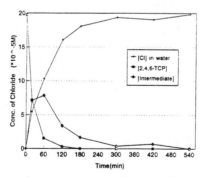

Figure 10  Exhaustive degradation of 2,4,6-trichlorophenol—chlorine content
in reaction intermediates vs. time curve

after 180 min leaving only the intermediates. At 60 min corresponding to the maximum concentration of intermediates, the carbon concentration was $33 \times 10^{-5}$ M and the chlorine content only $7.9 \times 10^{-5}$M indicating the production ratio of chloride ion was much larger than that of carbon dioxide. Following the course of reaction, the ratio of Cl/C became smaller in the intermediates, with both concentrations gradually diminished. Finally, the chloride ion disappeared after 540 min, but carbon-containing intermediates still remained, which might require more reaction time for complete removal.

## CONCLUSIONS

1. With titanium dioxide content of 10 g/L irradiated by UV light of wavelength 365 nm and intensity of $2.84$ mW/cm$^2$, the photocatalytic reaction can decompose polychlorophenols with good efficiency, especially at pH 3, achieving 96% removal in 120 min of irradiation.

2. The rate of photocatalytic degradation of 2,4-dichlorophenol became faster, the lower the pH level. However, 2,4,6-trichloro-and 2,3,4,6-tetrachloro-phenol were decomposed faster around pH9.

The pseudo-reaction rate constant, K', for the first-order kinetic model became larger, the lower the pH value. When the controlled pH value was close to the pK value of the reactant, the K' value was larger than these in nearby pH values, indicating the effect of pK on adsorptivity and also the reaction rate.

3. Using K' values to compare the reaction rates of polychlorophenols at different pH levels, the order was found to be 2,4-dichloro- >2,4,6-trichloro-> 2,3,4,6-tetrachloro-phenol in acidic medium, and reversed order in alkaline medium, but the influence on individual reactant was not significant.

4. Decomposition of 2,4,6-trichlorophenol took place by dechlorination followed by ring cleavage, and in exhaustive degradation, organic remains containing only carbon persisted after 9 hours of reaction time.

## REFERENCES

1. Environment Protection Ageny. 1980. "Ambient Water Quality Criteria for Chlorinated Phenols," EPA Report, 44015-80-032.
2. Beltrame, P. L., P. Carniti, and D. Guardione. 1988. "Inhibiting Action of Chlorophenols on Biodegradation of Phenol and Its correlation with Structural Properties of Inhibitors," Biotechnol. Bioeng., 31:821-828.
3. Schwien, U. 1988. "Degradation of Chloro-substituted Aromatic Compounds by Pseudomonas SP. Strain B13; Fate of 3,5-Dichlorocatechol," Arch. Microbiology, 150:78-84.
4. Haggblom, M. 1988. "Degradation and Tranformation of Chlorinated Phenolic Compounds by Strain of Rhodococcus and Mycobacterium," Water Research, 22(2): 171-177.

5. Dorn, E. 1988. "Isolation and Characterization of a 3-Chlorobenzoate Degrading Pseudomonas," Arch. Microbiology, 99: 61-67.

6. Moza, D. N., K. Fytianos, V. Samanidou, and F. Korte. 1988. "Photodecomposition of Chlorophenols in Aqueous Medium in Presence of Hydrogen Peroxide," Bull. Environ. Contam. Toxical., 41:678-682.

7. Miller, R. M., G. M. Singer, J. D. Rosen, and R. Bratha. 1988. "Sequential Degradation of Chlorophenols by Photolytic and Microbial Treatment," Environ. Sci. Technol., 22(10): 1215-1223.

8. Sundstrom, D. W., B. A. Weir, and H. E. Klei. 1989. "Destruction of Aromatic Pollutants by U. V. Light Catalyzed Oxidation with Hydrogen Peroxide," Environ. Progress, 8(1):6-14.

9. Khan, S. R., C. P. Huang, and J. W. Bozzelli. 1985. "Oxidation of 2-Chlorophenol Using Ozone and Ultraviolet Radiation," Environ. Progress, 14(4): 229-236.

10. Matthews, R. W. 1988. "Kinetics of Photocatalytic Oxidation of Organic Solutions over Titanium Dioxide," J. Catalysis, 111:264-272.

11. D'Olivera, J. C., G. Al-sayyed, and P. Pichat. 1990. "Photodegradation of 2- and 3-Chlorophenol in $TiO_2$ Aqueous Suspension," Environ. Sci. Technol., 24(7): 990-996.

12. Tseng, J. M., and C. P. Huang. 1991. "Removal of Chlorophenols from Water by Photocatalytic Oxidation," Wat. Sci. Technol., 23: 377-387.

13. Okamoto, K. 1985. "Heterogeneous Photocatalytic Decomposition of Phenol over $TiO_2$ Powder," Bull. Chem. Soc., 58: 2015-2022.

# Physical-Chemical Treatment

# SLUDGE THICKENING, DEWATERING AND DRYING: THE REMOVAL OF WATER BY MECHANICAL AND THERMAL PROCESSES*

P. AARNE VESILIND
Professor
Department of Civil and Environmental Engineering
Duke University
Box 90287
Durham, North Carolina 27708-0287

While the dewatering of wastewater sludges has received considerable research attention, we generally assume that if we could only be a little more clever in how we design mechanical sludge dewatering equipment, we should be able to dewater sludge to high solids concentrations. This notion rests on the assumption that the water continuum surrounding the sludge particles is mechanically removable and has the physical properties of bulk water. I suggest that this assumption may not be valid, and that there are at least four different identifiable types of water in sludge, two of which can not be removed by mechanical means and thermal energy is necessary to drive the water out.

## CHARACTERISTICS OF WATER IN SLUDGE

Sludge is made up of diverse solid particles suspended in an impure water continuum. Attempts to define a sludge 'particle' have been hampered by the problem that as soon as we try to measure or even look at these particles, they will change. Within the treatment system, sludge particles are dynamic -- dispersing and reforming depending on biological, chemical and physical conditions.

* Parts of this paper have been accepted for publication by *Waste Management and Research*, Journal of the International Solid Waste Association, Copenhagen, Denmark, co-authored with Thomas Ramsey. Other parts of this paper have been previously published in the *Water Environment Research* of the Water Environmental Federation, vol. 66, no. 1, p. 1, 1994.

Perhaps the only means of defining a sludge particle is to continually dilute the sludge with water under low shear conditions until the solids define their own size and structure. Sludge solids tend to flocculate, and these flocs include bacteria and larger animals, covered in slime and detritus, held together with filamentous organisms. Bacterial colonies, resembling small villages, are collected within this structure.

Water, with it various dissolved chemicals, is the continuum within and surrounding the sludge floc. The various water fractions associated with the sludge floc can be defined on the basis of the sludge floc structure. Although the names given to the various water fractions have evolved over the years, the following descriptions of different physical states of water in sludge (illustrated in Figure 1) appear useful:

**Free (or Bulk) water** — water not associated with and not influenced by the suspended solids particles.

**Interstitial water** — water trapped in the crevices and interstitial spaces of the flocs and organisms. Some of this water is held within the floc structure, and can become free water if the floc is destroyed. It is conceivable that some interstitial water is held within the structure of microbial cells, and can become free water only if the cells are destroyed.

**Vicinal water** — multiple layers of water molecules held tightly to the particle surface by hydrogen bonding. This force is believed to be short ranged but very intense close to the particle surface resulting in the appearance of highly structured water molecules. This water can be within cells as well, as long as it is associated with a solid surface. The major distinction between vicinal water and interstitial water is that the latter is free to move when the physical confinement is eliminated. Vicinal water is not free to move but adheres to solid surfaces.

The two properties of vicinal water that make it interesting in sludge dewatering are is lower density and higher viscosity than bulk water (Drost-Hansen, 1981). Small particles that are surrounded by vicinal water would have densities slightly less than the density of the particle itself since the surrounding water would act as part of the particle. If that particle moves through the fluid, the increased viscosity of the vicinal water would adversely affect its flow characteristics.

**Water of Hydration** — the water chemically bound to the particles and removable only by the expenditure of thermal energy. For example, water associated with the aluminum hydroxide floc can be removed only by thermal drying. Also, the thermal conversion of slaked lime [$Ca(OH)_2$] to quicklime [$CaO$] releases water of hydration.

The easiest water to remove from a wet sludge is bulk water, which can be eliminated by drainage, thickening or mechanical dewatering. Interstitial water, however, is trapped within the flocs and its release can be attained only by either the destruction or compression of floc structures using sufficient mechanical energy to squeeze the water out. Most dewatering processes are designed to remove both bulk water and interstitial water from sludge.

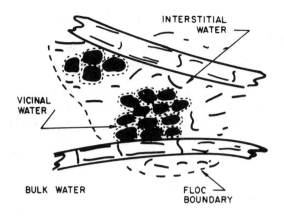

INTERSTITIAL
WATER

VICINAL
WATER

BULK WATER

FLOC
BOUNDARY

Figure 1. Representation of a sludge floc with the hypothetical types of water

Possibly the most interesting water fraction in dewatering is vicinal water, since it is quite difficult to remove mechanically unless some prior conditioning is affected. Similarly, mechanical dewatering can not remove the water of hydration which is chemically bound to the solids. It is of interest to know exactly how much vicinal water and water of hydration exists in a given sludge, since this represents the limit of mechanical dewatering.

## BOUND WATER IN SLUDGE

The application of the concept of different types of water to sludge technology is complicated by the historic definition of "bound water." Most past researchers have divided water in sludge into two categories—bulk water and bound water. By devising a measure of one water, the other is calculated by subtraction from the total water as measured by evaporation.

A widely accepted definition of bound water in sludges is water that does not freeze at some given temperature, usually -20 degrees C. The first researchers to define bound water in sludges in this manner were Heukelekian and Weisberg in 1956. Using dilatometers for their studies, and an activated sludge sample, they found that the quantity of bound water (unfrozen water in the dilatometer) is about 0.3 g/g of dry solids.

It seems reasonable to assume that "bound water", as the term is used in the environmental engineering and sciences literature, is actually a gross estimate of several forms of water, including some fraction of interstitial water, vicinal water and water of hydration. If this is true, what importance does this have in getting the water out of sludge?

Mechanical sludge dewatering probably removes mostly bulk and interstitial water from the sludge matrix. As pointed out by Richard Dick(1992), it may also be possible to remove some, but never all of the vicinal water as the particle surfaces are squeezed together. The water of hydration would of course not be affected by these processes. It would therefore by of interest to know, before a dewatering operation is planned, what the readily available water for dewatering (bulk plus interstitial water) is, recognizing that only a fraction of the vicinal water and none of the water of hydration can be removed by mechanical means. If it is possible to measure accurately the distribution of water in sludge, it might then be possible to estimate the highest possible cake solids concentration that can be achieved for a given sludge.

Using the dilatometric data of Katsiris and Kouzeli-Katsiri (1987), it is possible to show by calculation that the best cake solids that they could ever have achieved in their test tube centrifuge tests was 8% solids! It would have been unreasonable therefore to expect a mechanical dewatering device to achieve higher solids concentrations for this sludge. If a high solids concentration is required, the sludge would have had to be modified by chemical, biochemical, physical or thermal conditioning. The following section discusses what happens to water when sludge is subjected to high temperature drying. The experimental work for this study was conducted by Thomas Ramsey (1993).


## THERMAL DEWATERING (DRYING)

When sludge dries, there is a loss of mass. This loss is attributed to the evaporation of water. But is this actually true, or does the sludge lose mass by loss of volatile materials? This can be determined by measuring the fuel value of sludge when it has been dried under various temperatures.

Precise and accurate determination of the fuel value of wastewater sludges is possible using the oxygen bomb calorimeter. A measured sample of fuel is burned within a closed metal vessel (bomb) which retains all the products of combustion. Since water vapor is formed as a combustion product of hydrogen, its condensation within the bomb adds heat to the calorimetry results. Thus, these results are considered the gross, or higher heating value (HHV) of the fuel as defined by the American Society of Testing and Materials (ASTM). All data presented subsequently in this paper are HHVs.

Rudolfs and Baumgartner (1932) found volatile losses in heated sludge are minimal so long as water is present in the sludge. Their research indicates that regardless of the temperature of heating, little volatile matter is lost until 80 to 90 percent of the original moisture content of the sludge has evaporated. For a sludge with an initial total solids concentration of 25%, this corresponds to 63 to 77% total solids. Rudolfs and Baumgartner assumed that volatiles are not driven from a sludge sample until 63 to 77% total solids are reached given that residual moisture is evenly distributed throughout the sludge cake. In realty this condition is difficult to achieve, since exposed sludge surface areas tend to dry faster than those buried within the sludge sample.

Once a specific volume of sludge reaches a total solids concentration of approximately 70%, volatiles begin to distill from the sludge. If some of these volatiles are lost in the drying hearths of a multiple hearth furnace instead of in the combustion hearths, the loss represents a direct reduction in the fuel value of the sludge to the furnace, since the countercurrent flow of the MH furnace carries these volatiles away from the combustion zone.

Based upon laboratory studies by Tiery, *et al* (1990) and field studies of sludge incinerators by Vancil *et al*, (1991), the three most common volatile emissions identified from sludge incineration are benzene, toluene, and acrylonitrile. It is doubtful, however, that these compounds are present in large quantities as individual components in the sludge, given that hazardous solvents are specifically forbidden from effluents discharging into wastewater treatment plants. Previous study of concentrations of hazardous organic contaminants in sludge are typically less than 100 $\mu g/g$ (ppm) on dry weight basis (Dellinger and Mazer, 1990).

Since the most common volatiles emitted during sludge combustion are not initially present within the sludge feed, another mechanism must be responsible for their creation. Dellinger and Mazer (1990) theorize that the relatively simple volatile species, such as benzene and toluene, are formed from the thermal degradation of more complex organics in the sludge biomass during heating. Once formed, these species volatilize immediately and diffuse out of the sludge mass. Since benzene and toluene are stable at temperatures less than approximately 300°C, the lower temperatures of the drying hearths of the MHF (300°C to 450°C) cannot completely destroy these species before they escape the furnace.

Ideally, in order to obtain a true measurement of the HHV of a sludge sample, all of the water entrained within the sludge sample must be exclusively removed prior to its combustion in the bomb calorimeter. When sludge dries, water evaporation proceeds first from free water, followed by interstitial water, and finally vicinal water removal. Since free, interstitial, and surface waters are all chemically distinct from the sludge particle, their complete removal prior to combustion in the bomb calorimeter would theoretically provide the ultimate HHV for a sludge sample. In reality, however, once sludge dries beyond 70% total solids, volatiles cannot be prevented from distilling off the sludge as more water is lost. Therefore, the identification of weight loss due to water evaporation and weight loss due to volatile distillation are not easily distinguished in a sludge sample which has stabilized with regard to drying temperature.

In order to investigate this, different drying temperatures were used, ranging from ambient of 400°C. Unfortunately, no standard method exists for drying sludge under ambient temperature. Therefore, to determine if drying temperatures between ambient temperature and 105°C affect the HHV of a sample, a new procedure had to be developed.

Municipal wastewater sludge was collected from the Northside Treatment

Plant in Durham, North Carolina. The sludge is anaerobically digested and has not been conditioned with chemicals. Sludge was collected directly from outlet pipes leading from the anaerobic digesters to sludge drying beds used at the plant. Since the initial sludge was less than 5% total solids, some mechanical dewatering was desired prior to drying. Following sieving using a U.S. Standard #16 sieve (1.19 mm openings) for the removal of large particles, the sludge was centrifuged. Dewatering was performed using 500 ml cups centrifuged at approximately 2000 RPM for 5 minutes, resulting in a sludge cake of approximately 10% total solids. Centrate was discarded and the dewatered sludge cake was collected and mixed thoroughly to ensure that the initial characteristics of the sludge cake were as homogeneous as possible. The times used for heating the sludge were experimentally determined to allow the samples to attain essentially constant weight.

A closed apparatus was used which provided for extremely low atmospheric water vapor concentrations to ensure complete drying, and also allowed for rapid dewatering at ambient temperature. A schematic of the laboratory apparatus is shown in Figure 2.

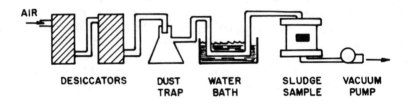

Figure 2. Apparatus used for sludge drying at low temperatures

Laboratory calorimetry was performed using a Parr Instruments Company model 1341 plain jacket, isothermal oxygen bomb calorimeter. Operating procedures used with the calorimeter were in accordance with ASTM procedures D3286-77 (1979) for the calorific determination of solid fuel and manufacturer recommendations. Details of the procedure used in this study are available elsewhere (Ramsey 1993).

Results indicate that the sludge investigated has a calorific value of approximately 3,000 calories per gram for solids dried at and below 105°C. This compares favorably with calorific results from similar sludges tested in previous studies. The results are shown graphically in Figure 3.

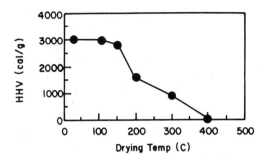

Figure 3. Higher heating values at various drying temperatures for the sludge sample used. Each data point represents the average of three experiments.

For sludges dried at 300°C and 400°C, the combustible portion remaining in the sludge sample was low enough that self-sustaining combustion did not occur in the bomb. As a result, benzoic acid was added to the sludge sample in order to ensure its complete combustion. Since the calorific value of the benzoic acid is known, thermochemical adjustments can be made to the subsequent calorimeter results in order to subtract out the heat addition due to benzoic acid.

High temperatures of combustion in the bomb caused the formation of "clinkers," or fused blobs from molten ash. The formation of clinkers is not unusual for sludge combusted under high temperatures. Because clinkers can cause operational difficulties in an incinerator, their formation is suppressed through limiting the highest operating temperature in a MHF to less than 980°C (U.S. EPA, 1985). Since combustion within the bomb calorimeter occurs vigorously and at high temperature, clinker formation is expected from the calorimeter ash. The ash from higher energy sludges tested (those above 1,500 cal/g) almost completely formed into clinkers. Sludges exposed to 300°C and 400°C temperatures had progressively fewer clinkers in their ash following combustion. This result can again be linked to the low fuel value of sludge samples exposed to temperatures above 300°C. Even with the addition of benzoic acid to these samples, comparatively little energy was released from these samples during their combustion. As a result, during combustion in the calorimeter, samples prepared at 300°C and 400°C could not maintain high temperatures for sufficient periods of time to completely fuse their ash into clinkers.

The data shown in Figure 3 indicate that HHVs for the sludge samples tested were strongly influenced by the temperature under which they were dried. Volatile losses affecting calorific value of the sludge appear in earnest at drying temperatures around 150°C, and at exposure to temperatures of 200°C the sludge has lost nearly half its HHV as compared to ambient conditions. The sample dried at 400°C exhibit sludge characteristics that are close to ash. At 400°C the sludge probably pyrolyzed within the drying furnace, since smoke was observed exiting the furnace during sludge drying at 400°, although no flame was visible. Previous research indicates that combustion for some sludges is problematic when drying is attempted at temperatures near 400°C (Rudolfs and Baumgartner, 1932).

Another result indicated by the plot is that drying temperatures between 25°C and 105°C apparently did not distill volatiles from the sludge samples in a quantity sufficient to be detectable within the limits of precision for this research. Since volatile losses will directly affect the final weight of the dried sludge sample, this result implies that if two identical samples are dried at 25°C and 105°C, respectively, the final weight loss between the two samples would be the same. Therefore, sludge drying under procedures outlined in *Standard Methods for the Examination of Water and Wastewater* for total solids determination at 105°C may be used without concern for the effect on the volatile content of the sludge.

Figure 4 shows a plot of percent HHV loss from the sludge samples versus drying temperature. This figure also shows the curve showing volatile solids loss as reported by Rudolfs and Baumgartner (1932). As can be seen from the data, the rate of HHV loss is proportional to temperature. This agrees with the loss of volatile solids as found by Rudolfs and Baumgartner.

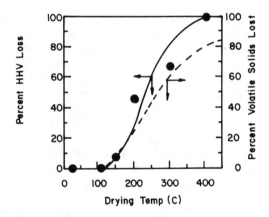

Figure 4. The data points and solid line indicate the higher heating value loss for sludge samples dried at different temperatures. The dashed line indicates the loss of volatile solids as reported by Rudolfs and Baumgartner(1932).

Although a host of volatile compounds are emitted during the thermal degradation of sludge, the primary species identified in incinerator emissions are acrylonitrile, benzene, and toluene (Dellinger and Mazer, 1990; Vancil, Parrish, and Palazzola, 1991). These compounds have boiling points of 78°C, 80°C, and 111°C, respectively. Therefore, if these compounds are initially present in a sludge sample prior to drying, we would expect to see HHV results fall between drying temperatures of 25°C and 105°C. Since the HHV data indicates that little, if any, chemical energy is lost at drying temperatures between 25°C and 105°C, it appears that acrylonitrile, benzene, and toluene are not initially present within the sludge. As a result, this observation also strengthens the argument that these volatile compounds are formed and released from the thermal degradation of organic matter within the sludge, thus implying that much of the organic matter in sludge is thermally unstable in air at temperatures well below normal combustion temperatures.

## CONCLUSIONS

One of the reasons that sludge dewatering is often unsuccessful is that some of the water tends to be attached to particle surfaces as vicinal water, and most of this can only be removed by thermal means or by changing the amount of surface that water can adhere to. Polymers will only influence the interstitial water, and will make it possible for the bulk and much of the interstitial water to be separated by mechanical means. These chemicals probably do not, however, affect most of the vicinal water, and thus mechanical devices cannot be expected to dewater polymer conditioned sludge past the point of removing the bulk and interstitial water. Only thermal means can drive out the remaining water.

Based on these experiments, we conclude that drying sludge at temperatures between ambient and 105°C does not measurably affect the subsequent HHV of the dried sludge, given the precision of the techniques used in this research. Thus the loss in mass at this temperature can be thought of as the removal of water.

Sludge held at temperatures above 105°C results in a measurable lowering of HHVs as compared to sludges dried at or below 105°C. The loss of heating value can be directly tied to the loss of volatile matter from the sludge.

The loss of volatile matter from the sludge is primarily the result of thermal degradation of biomass in the sludge. HHV results indicate that organic matter within sludges is thermally unstable at temperatures as low as 150°C, well below the point of active combustion for the sludge.

## REFERENCES

ASTM D3286-77. 1979. "Gross Calorific Value of Solid Fuel by the Isothermal Jacket Bomb Calorimeter," American Society of Testing and

Materials, Philadelphia, Pennsylvania.

Dellinger, B. and S. Mazer. 1990. "Laboratory Evaluations of the Thermal Degradation Properties of Toxic Organics in Sewage Sludge," U.S. EPA, Cincinnati, Ohio.

Dick, R. I., Cornell University, Personal Communication

Drost-Hansen, W. 1981. "The Occurrence and Extent of Vicinal Water," *Biophysics of Water*, F. Franks (Ed) John Wiley & Sons, New York

Huekelekian, H. and E. Weinberg (1956) "Bound Water and Activated Sludge Bulking," *Sewage and Industrial Wastes* 28(4):558-574.

Katsiris, N. and A. Kouzeli-Katsiri. 1987. "Bound Water Content of Biological Sludges in Relation to Filtration and Dewatering," *Water Research*, 24(11):1319-1327.

Ramsey, T. B. 1993. "Effect of Drying Temperature on the Fuel Value of Wastewater Sludge," MS Thesis, Duke University, Durham, NC.

Rudolfs, W. and W. H. Baumgartner. 1932. "Loss of Volatile Matter by Drying Sewage Sludge Before Incineration," *Water Works and Sewerage*, June:199-201.

*Standard Methods for the Examination of Water and Wastewater*, 17th ed., 1989. American Public Health Association, Washington, D.C., pp. 2-71 to 2-73.

Tirey, D., R. Striebich, S. Mazer, B. Dellinger, and H. Bostian. 1990. "Predicting Organic Emissions from Sewage Sludge Incineration," presented at the 16th Annual US EPA Research Symposium, Cincinnati, Ohio, April.

U.S. EPA. 1985. *Municipal Wastewater Sludge Combustion Technology*, Document # EPA/625/4-85/015, Cincinnati, Ohio.

Vancil, M., C. Parrish and M. Palazzolo, (1991) *Project Summary: Emissions of Metal and Organics from Municipal Wastewater Sludge Incinerators*, US EPA Document # EPA/600/S2-91/007, Cincinnati, Ohio, July.

# RESEARCH LABORATORY WASTEWATER TREATMENT PROCESS

Wilma Ann Jancuk. PE
CIH CSP CHCM CHMM
   AT&T Corp.
   PO Box 900
   Princeton, NJ 08542-0900

John. R. Fisher
   AT&T Corp.
   PO Box 900
   Princeton, NJ 08542-0900

## INTRODUCTION

Aqueous waste streams, described as laboratory wash waters, from the AT&T Engineering Research Center in Princeton, NJ are piped to a central facility for treatment prior to discharge to the receiving waters of the State of New Jersey per a NJPDES Permit within the limits as noted in Table I. The wastewater may contain acids, bases, metals such as copper and chromium from plating baths rinses, cyanides, ammonical and phosphorus compounds. The standard treatment technology used to treat these types of wastewaters include neutralization, cyanide oxidation, chromium reduction and metals precipitation as hydroxides followed by filtration to remove the solids.

Although hydroxide precipitation is generally sound it has its pitfalls [1]. The metals in the wastewater stream can be complex or chelated and hydroxide precipitation techniques may not remove the metals to the limits established on a NJPDES Permit [1][2]. The treatment pH for the formation of metal hydroxides is different for each metal [3]. Also, metals that are removed as metal hydroxides could redissolve during filtration if the pH is not carefully controlled[3].

The AT&T Engineering Research Center redesigned its Research Laboratory wastewater treatment process and treatment scheme using environmentally conscious techniques. Sulfide precipitation chemistry was selected to remove the dissolved metal contaminates to the lowest levels possible while using the least amount of treatment chemicals and generating the least amount of sludge. The treatment process includes multiple batch treatment tanks enabling a batch to be returned for treatment if it does not meet specifications.

405

## TABLE I. NJPDES PERMIT LIMITS

| Constituent | Concentration (mg/L) | Constituent | Concentration (mg/L) |
|---|---|---|---|
| Phosphorus | 1.0 | Lead | 0.15 |
| Cadmium | 0.1 | Nickel | 0.75 |
| Chromium | 0.5 | Silver | 0.1 |
| Copper | 0.4 | Zinc | 0.5 |
| Total Cyanide | 0.1 | Ammonia (as N) | 2.0 |

Several treatment schemes to remove the contaminates from the wastewater were studied and evaluated. There was an interest in using the sulfide precipitation chemistry for heavy metal removal. Sodium sulfide and sodium hydrosulfide have high solubilities and resulting high dissolved sulfide concentration causes rapid precipitation of the metals dissolved in the wastewater (Table II). The literature suggested that heavy metals can be removed to extremely low concentrations due to the low solubility of the metal sulfides and at a single pH value[1][4][5][6][7]. Also, sulfides allow precipitation of contaminants even in the presence of complexing agents[5][6][7]. The removal of a particular heavy metal is more effective when it is in a solution containing other heavy metals than when it is the only metal in solution[1].

Bench scale studies were conducted on the actual wastewater typically treated at the AT&T chemical wastewater treatment plant. The results of this study is listed in Table III. It was decided to design the plant to use the sulfide precipitation chemistry.

## TABLE II. SOLUBILITIES OF SULFIDES AND HYDROXIDES

| Metal | $K_{SP}$ Sulfide | $K_{SP}$ Hydroxide |
|---|---|---|
| Manganese | $4.7 \times 10^{-14}$ | $2.0 \times 10^{-13}$ |
| Iron (II) | $1.6 \times 10^{-19}$ | $4.8 \times 10^{-17}$ |
| Zinc | $2.9 \times 10^{-25}$ | $7.7 \times 10^{-17}$ |
| Nickel | $1.1 \times 10^{-21}$ | $5.5 \times 10^{-16}$ |
| Lead | $9.1 \times 10^{-29}$ | $1.4 \times 10^{-20}$ |
| Cadmium | $1.4 \times 10^{-29}$ | $5.3 \times 10^{-15}$ |
| Silver | $6.7 \times 10^{-50}$ | $1.5 \times 10^{8}$ |
| Copper (I) | $2.3 \times 10^{-48}$ | $2.0 \times 10^{-19}$ |
| Copper (II) | $1.3 \times 10^{-36}$ | - |
| Mercury (II) | $6.4 \times 10^{-53}$ | $3.1 \times 10^{-26}$ |

Handbook of Chemistry and Physics, 72nd Ed. 1991-1992, p. 8-43

## TABLE III. TREATABILITY STUDIES

| Constituent | Wastewater Conc. (mg/L) | Sodium Metabisulfite/ Lime | Sodium Metabisulfite/ NaOH | Ferrous Sulfate/ Sodium Sulfide |
|---|---|---|---|---|
| $Cr^{+6}$ | 1.0 | 0.01 | 0.01 | 0.01 |
| Cu (Chelated) | 2.0 | 0.92 | 0.93 | 0.01 |
| Pb | 3.2 | - | - | 0.07 |
| Ni | 1.3 | - | - | 0.69 |
| Ag | 0.17 | - | - | 0.01 |
| Zn | 0.90 | - | - | 0.11 |

## TREATMENT SCHEME

The new treatment scheme treats each batch as required depending on the contaminate and the concentration of the contaminate in the batch. Each batch is analyzed for selected substances, namely, total phosphorus (P), total copper (Cu), total chromium (Cr), cyanide (CN), and ammonia ($NH_3$).

The treatment scheme begins with cyanide destruction using alkaline chlorination[8]:

$$2CN^- + 5Cl_2 + 8OH^- \rightarrow 10Cl^- + 2CO_2 + N_2\uparrow + 4H_2O \qquad (1)$$

Ammonia is destroyed via breakpoint chlorination[9]:

$$NaOCl + H_2O \rightarrow HOCl + NaOH \qquad (2)$$

$$NH_3 + HOCl \rightarrow NH_2Cl + H_2O \qquad (3)$$

$$2NH_2Cl + HOCl \rightarrow N_2 + H_2O + 3HCL \qquad (4)$$

Heavy metals are removed using the insoluble sulfide process which precipitates dissolved metals by mixing the wastewater with ferrous sulfide (FeS)[2][5][6]. The sulfide (S=) is released from the ferrous sulfide (FeS) only when other heavy metals with lower equilibrium constants for their sulfide form are present in solution. The following reactions occur when FeS is introduced into a solution of dissolved metals (M) and metal hydroxides ($M(OH)_2$)[5][6].

$$FeS = Fe^{+2} + S^{-2} \tag{5}$$

$$M^{+2} + S^{-2} = MS \tag{6}$$

$$M(OH)_2 = M^{+2} + 2(OH)^- \tag{7}$$

$$Fe^{+2} + 2(OH)^- = Fe(OH)_2 \tag{8}$$

The unreacted ferrous sulfide is settled out with the metal sulfide precipitate. The FeS which is generated on-site in the reaction vessel as follows[6]:

$$FeSO_4 + Na_2S \rightarrow FeS + Na_2SO_4 \tag{9}$$

The $Cr^{+6}$ in the wastewater will not form a sulfide, however, the ferrous sulfate ($FeSO_4$) can be used to reduce the $Cr^{+6}$ to $Cr^{+3}$ and precipitate ferric and chromium hydroxides as follows[10]:

$$Na_2Cr_2O7 + 2FeS + 7H_2O = 2Fe(OH)_3\downarrow + 2Cr(OH)_3\downarrow + 2S^0 + 2NaOH \tag{10}$$

The process has been in operation for 15 months. Table IV. presents the influent and effluent concentration of dissolved chelated copper using this sulfide precipitation process. All batches treated removed the dissolved chelated copper to levels well below the required limit as established in the Permit.

### Table IV. Sulfide Precipitation of Copper in Wastewater
### (Actual Results)

| Influent Cu (mg/L) | Effluent Cu (mg/L) | Influent Cu (mg/L) | Effluent Cu (mg/L) |
|---|---|---|---|
| 0.49 | 0.01 | 0.37 | 0.13 |
| 0.21 | 0.09 | 0.7 | 0.03 |
| 1.02 | 0.16 | 0.33 | 0.03 |
| 0.24 | 0.06 | 0.31 | 0.04 |
| 0.37 | 0.04 | 0.72 | 0.05 |
| 0.37 | 0.06 | 0.20 | 0.02 |
| 0.32 | 0.11 | 0.25 | 0.01 |
| 3.57 | 0.04 | 0.22 | 0.01 |
| 0.24 | 0.15 | 0.21 | 0.02 |
| 0.30 | 0.01 | 0.20 | 0.05 |
| 0.28 | 0.07 | 0.48 | 0.10 |
| 0.32 | 0.17 | 0.21 | 0.04 |
| 0.88 | 0.03 | 0.30 | 0.06 |
| 0.36 | 0.08 | 0.27 | 0.02 |
| 0.44 | 0.20 | 0.44 | 0.02 |
| 0.30 | 0.01 | 0.24 | 0.07 |

**Average Influent  0.47 mg/L**          **Average Effluent  0.06 mg/L**

Phosphorus is removed via chemical precipitation with multivalent metal ions as follows[9]:

$$Al^{+3} + H_nPO_4^{3-n} \leftrightarrow AlPO_4\downarrow + mH^+ \tag{11}$$

## DESIGN

The new chemical wastewater treatment plant design included outfitting the existing three 23,000 gallon concrete underground equalization tanks with three 16,000 gallon fiber glass treatment tanks, tank within a tank, thereby providing secondary containment for the tanks. These three tanks are batch sequenced for a maximum treatment capacity of 48,000 gallons per day. The batch sequence involves one tank accepting the raw wastewater, another treating the wastewater, and the third discharging the treated wastewater to the filtration equipment and carbon adsorption system prior to discharge. The whole process , as illustrated in Figure 1, is controlled by a Distributed Control System (DCS).

The operator collects and analyzes each batch of raw wastewater and the concentration of the contaminate is entered into the DCS and a sequence of treatment steps is completed depending on the concentration of contaminate. The required amount of chemical for each treatment step, as described, is added based on the formulas programmed into the computer for each process step. The flow chart, see Figure 2, details the treatment methodology.

After treatment the batch is post qualified by filtering and analyzing a sample of the batch to verify that the treatment scheme has adequately removed the contaminates as required. After verification that the treatment phase has been successful the DCS switches the batch to the next phase called the filter/drain phase. In this phase the treated wastewater is filtered using a rotary drum filter followed by pH adjustment, passed through a carbon adsorption unit for organic removal and then discharged to the receiving waters of New Jersey. If the batch required further treatment during post qualifing, the DCS returns the process to the treatment phase. The design of the treatment plant and process considered many environmental and safety concerns. Each batch is analyzed (1) to ensure enough treatment chemical is added to remove the contaminates, (2) to add only the necessary amount of treatment chemical needed to the process to minimize the amount of sludge produced, and (3) to insure that the treatment phase adequately removed the contaminates.

The layout of the treatment plant, see Figure 3, included a unique design to add and control quantities of chemicals to the treatment tanks. A chemical feed room was designed to house all chemicals needed for the process. The DCS monitors and controls the pumping of chemicals which are delivered to a central chemical feed sink and then to the treatment tank. This procedure was developed to handle ten different chemicals for the process to treat three different tanks safely and with accuracy.

409

# Figure 1. Treatment Flowchart

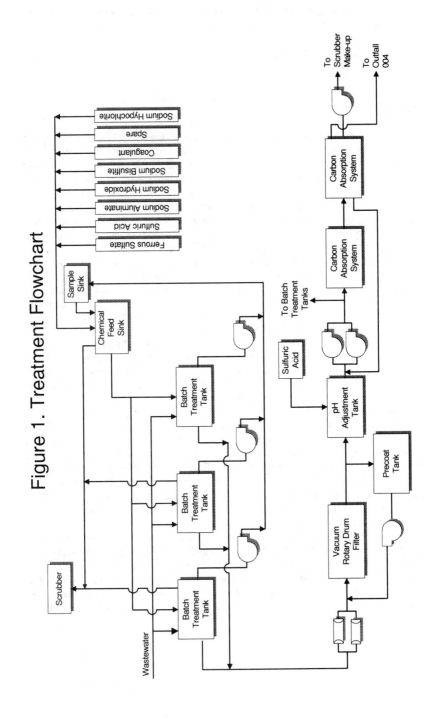

# FIGURE 2
## FLOW CHART OF THE
## WASTEWATER TREATMENT FACILITY

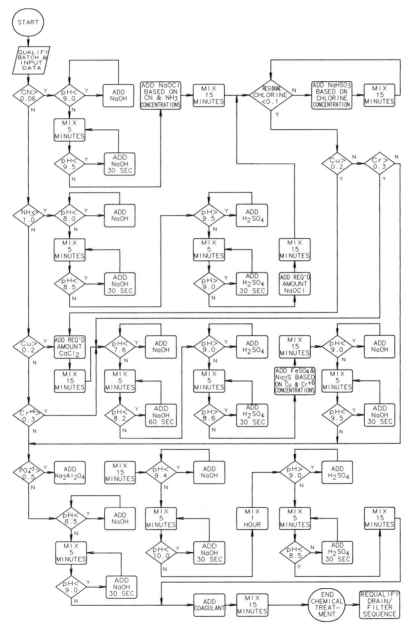

411

FIGURE 3

WASTEWATER TREATMENT PLANT
FLOOR PLAN

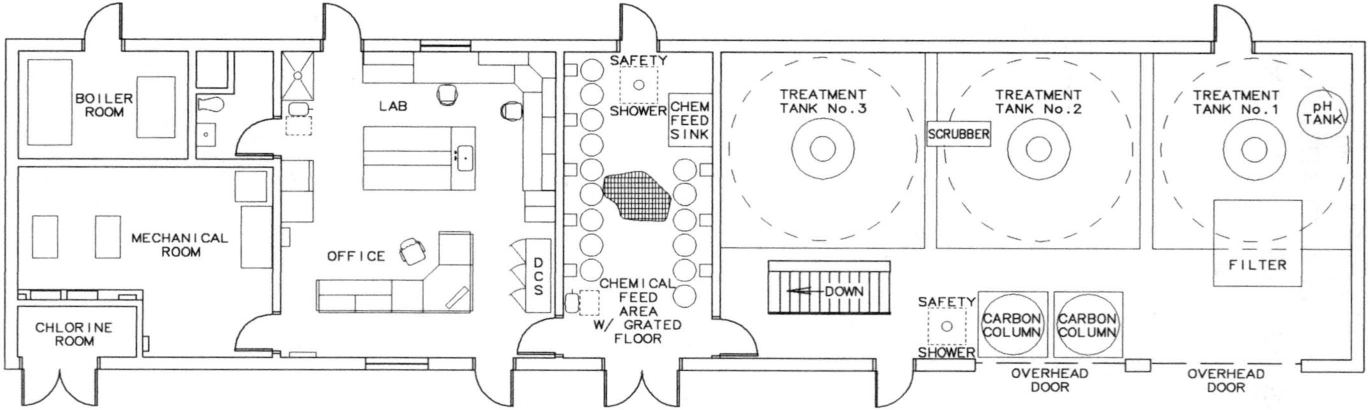

## CONCLUSION

As a result of this design and process control system installed at the AT&T Engineering Research Center the chemical wastewater treatment plant, the metals are removed from the wastewater to levels well below the permitted levels. The insoluble sulfide process, namely the addition of ferrous sulfate to form the FeS compound and control of the pH at about 8.5 during these reactions minimizes the emission of the hydrogen sulfide ($H_2S$) gas. The sulfide precipitation chemistry is a viable technology to remove complexed metals from wastewater. The use of a DCS to control and monitor the treatment plant has enabled the plant to operate efficiently and safely.

## REFERENCES

1. Clifford, D., T. J. Sorg. 1986, "Removing Dissolved Inorganic Contaminates from Water," *Environmental Science and Technology*," Vol.20 No. 11, pp 1072-1080.

2. Patterson, J.W. 1985, *Industrial Wastewater Treatment Technology*, Boston, MA:Butterworth-Heinmann, pp. 91-115.

3. Eckenfelder, W.W. 1989, *Industrial Water Pollution Control*, new York, NY: McGraw-Hill, p. 99.

4. Feigenbaum, H. N., 1977, "Removing Heavy Metals in Textile," *Industrial Waste*, March/April, pp32-34.

5. Scott, M. C. 1979, "Sulfide Process Removes Metals, Produces Disposable Sludge," *Industrial Wastes,* July/August, pp. 34-38.

6. Carpenter, C., D. Suciu, P. Wikoff. 1989, "Sodium Sulfide/Ferrous Sulfate Metals Treatment For Hazardous Waste Minimization," *Proceedings of the 44th Industrial Waste Conference, Purdue University,* pp. 617-624.

7. Ku, Y., R. W. Peters. 1988, "The Effect of Complexing Agents on the Precipitation and Removal of Copper and Nickel from Solution," *Particulate Science and Technology*, Vol. 6 No. 4, pp 441 -466.

8. Sawyer, C. N., P. L. McCarty, 1967, *Chemistry for Sanitary Engineers*, New York, NY"McGraw Hill, p.38.

9. Metcalf & Eddy, Inc., 1979, *Wastewater Engineering Treatment and Disposal/Reuse, New York, NY:McGraw-Hill, pp. 262,295.*

10. Eary, L. E., D. Rai. 1988, "Chromate Removal from Aqueous Wastes by Reduction with Ferrous Ion," *Environmental Science and Technology*, Vol.22 No.8, pp972-977.

# TREATMENT OF A DILUTE WASTE OIL EMULSION BY CHEMICAL ADDITION (CA)-DISSOLVED AIR FLOTATION (DAF)

BRIAN E. REED
PATRICK CARRIERE
XIAOFAN ZHU
TIM LORKOWSKI
Department of Civil and Environmental Engineering
West Virginia University
Morgantown, WV 26506-6103

## INTRODUCTION

Treatment of wastewater from aluminum rolling mill operations is an inherent problem in the aluminum fabrication industry. In this study, wastewater from the manufacturer's processes was transferred to two holding ponds having a total capacity of about 5 million gallons and a detention time of about 10 days. In the holding ponds, free oil was allowed to rise to the surface where it was periodically removed. Wastewater from the holding ponds was withdrawn from about a depth of 9' and used as influent to a variety of technologies. In this paper, results from the chemical addition (CA)-dissolved air flotation (DAF) portion of the treatability study are presented.

## BACKGROUND

Chemical addition (CA)- dissolved air flotation (DAF) is a two step process. In the first step, chemical, are added to the system to break the oil emulsion and alter the surface chemistry of the particles so that smaller particles can agglomerate into larger particles. In the flotation chamber, particles are physically separated from the liquid using buoyant forces. Very fine air bubbles are introduced into the wastewater and are attached to the particle surface. The buoyant force on the particle is increased so that it is greater than the summation of the gravitational and frictional forces. At this point, the particle-air bubble system will rise to the surface where it can be efficiently removed using a skimming apparatus.

## EXPERIMENTAL DESIGN

In this study, chemicals from two companies were investigated. Calgon Corp. supplied a cationic polymer (W-2923) to break the emulsion and an anionic polymer (POL-E-Z 2706) to enhance coagulation. KLAR-AID 2400, a cationic polymer, was supplied by Grace Dearborn. Jar tests were used to determine the required chemical(s) dosage prior to CA-DAF unit operation. A schematic diagram of the CA/DAF system is presented in Figure 1. The batch mixing tank (BMT) served as the chemical mix tank as well as a clarification unit for several experiments. Effluent from the BMT was sampled after settling to determine if the chemical addition-clarification process was a viable treatment method (*i.e.,* could the DAF unit be eliminated). The effluent from the BMT tank is then transferred to the DAF unit where it was mixed with recycled DAF effluent that has been supersaturated with air (air is dissolved into recycled effluent under high pressure). When the two streams are mixed at atmospheric pressure, the dissolved air comes out of solution, forming very fine bubbles. The cationic polymers were added directly to the BMT. The anionic polymer (Calgon only) was added at the transfer pump inlet and the turbulence from the pump provided the mixing energy. The CA-DAF unit was operated in a semi-batch and continuous mode. In the semi-batch mode, the BMT was filled with wastewater and the appropriate amount of cationic polymer and mixed for about 20 minutes. The contents of the BMT were then transferred to the DAF unit (Calgon anionic polymer added continuously). During the continuous operation the BMT was constantly being emptied/filled and all chemical additions were continuous. The turbidity of the BMT effluent (after 20 minutes settling) and DAF effluent were measured hourly. If the DAF effluent turbidity was greater than 30 NTU, a jar test was conducted to determine if a change in chemical dosage was warranted. Oil and grease concentration of the influent to the CA-DAF operation and of the BMT and DAF effluents were measured periodically. In Table 1, a summary of process conditions are presented.

**Table 1. Summary of Process Parameters**

| Parameter | Value | Parameter | Value |
|-----------|-------|-----------|-------|
| Flow rate, gpm | 8.9 | Air Pressure, psi | 42 |
| DAF recycle ratio | 0.70 | Detention Time, min | |
| | | BMT | 30 |
| | | DAF | 67 |
| Calgon Dosage[1], mg/L | | GD Dosage, mg/L | |
| Initial | 600/5 | Initial | 500 |
| Range[2] | 500-700 | Range | 500 |

[1]Cationic/anionic polymer concentration. [2]Only cationic polymer concentration was varied.

## RESULTS

### Calgon Chemicals

Effluent turbidity for the BMT and DAF unit versus operation time are presented in Figure 2. Turbidity's were highly variable and exceeded the maximum measurable value (200 NTU) on numerous occasions. Turbidity was used as a real-time indicator of process performance (*i.e.*, turbidity measure takes about 5 minutes and can be done on-site). When the DAF effluent turbidity exceeded 30 NTU a "quick" jar test was performed to determine if the chemical dosage required adjustment. A total of 16 of these tests were conducted during the 2/13-2/19 run. In 6 of the jar tests, a chemical adjustment was required. The new chemical dosage is indicated in Figure 2 by the vertically oriented numbers (*e.g.*, 600/50 ≡ cationic polymer dose/anionic polymer dose). The vertical arrows with no number designation refer to the "quick" jar tests in which a change in chemical dosage was not warranted. It would typically take 2 to 3 hours before the change in chemical dosage would effect.

In Figure 3, O/G concentration of the influent, BMT effluent and DAF effluent versus operation time are presented. Influent O/G ranged from 1490 to 3830 mg/L and averaged value 3130. Influent pH averaged $6.3 \pm 0.3$. BMT effluent O/G ranged from 60.5 mg/L to 688 mg/L (avg. $\pm$ std: $270 \pm 200$) in the semi-batch mode and from 327 to 1302 mg/L ($580 \pm 270$) in the continuous mode. DAF effluent O/G ranged from 11 mg/L to 278 mg/L ($60 \pm 80$) in the semi-batch mode and from 8 to 327 mg/L ($205 \pm 135$) in the continuous mode. The O/G concentration in the DAF effluent was significantly less than in the BMT effluent thus, including the DAF unit improved process performance over what was observed in the chemical addition-clarification process. During the semi-batch operation, between about 15 and 50 hours the DAF O/G ranged from 12 to 29 mg/L and averaged about 19 mg/L. Despite numerous chemical dosage adjustments, this was the only period of time where consistent and relatively low values of O/G were obtained. If CA/DAF was selected and Calgon chemicals were employed, it appears that the system would be operated in the semi-batch mode with frequent jar testing and chemical dosage adjustments.

For 72 hour semi-batch run, 25,870 gal of wastewater were treated and 190 gal of sludge were produced. The chemical cost the this portion of the run was for $117 which is equivalent to $4.54/1000 gal wastewater. The percent sludge produced was 0.73 %. For 72 hour continuous run, 38,448 gal of wastewater were treated and 262 gal of sludge were produced. The chemical cost the this portion of the run was for $156 which is equivalent to $4.05/1000 gal wastewater. The percent sludge produced was 0.68 %.

416

Effluent turbidity for the BMT and DAF unit versus operation time are presented in Figure 6.4. Turbidity's were lower and less variable compared with results from Calgon. For the DAF effluent, turbidity's were generally less than 20 except for a few spikes. The most noticeable turbidity spike occurred at about 70 hours. At this time there was a power outage and the chemical addition pump failed. The units were idle for about 12 hours and the turbidity's were high following the re-startup of the operation but decreased to < 30 NTU for the remainder of the run. Turbidity's were slightly higher during continuous operation compared with those observed during semi-batch operation. As in the Calgon chemical experiments, the turbidity was used as a real-time indicator of process performance (turbidity > 30 NTU a "quick" jar test was conducted). Only 2 of these tests were conducted during the Grace Dearborn run, and for both tests a change in chemical dosage was not warranted.

In Figure 5, O/G concentration of the influent, BMT effluent and DAF effluent versus operation time are presented. Influent O/G ranged from 2360 to 3240 mg/L with averaged 2950 ± 260 mg/L and influent pH averaged 7.5 ± 0.5. BMT effluent O/G ranged from 34 mg/L to 82 mg/L (55 ± 17) in the semi-batch mode and from 70 to 120 mg/L (94 ± 17) in the continuous mode. DAF effluent O/G ranged from 15 mg/L to 37 mg/L (28 ± 8) in the semi-batch mode and from 9 to 41 mg/L (30 ± 9) in the continuous mode. The O/G concentration in the DAF effluent was less than in the BMT effluent - the difference in average O/G values between the two units were 27 and 66 mg/L for the semi-batch and continuous operation modes, respectively. It remains to be determined if the improvement in O/G concentration by the addition of the DAF unit is warranted. For example, if the effluent was discharged to a wetland or sprayfield the slight increase in O/G concentration would most likely not affect the performance of the land systems.

For 70 hour semi-batch run, 24,404 gal of wastewater were treated and 325 gal of sludge were produced. The chemical cost the this portion of the run was for $264 which is equivalent to $10.8/1000 gal wastewater. The percent sludge produced was 1.33 %. For 60 hour continuous run, 32,040 gal of wastewater were treated and 742 gal of sludge were produced. The chemical cost the this portion of the run was for $347 which is equivalent to $10.8/1000 gal wastewater. The percent sludge produced was 2.3 %.

## SUMMARY

In Table 2, a summary of CA/DAF results for Calgon and Grace Dearborn chemicals is presented. O/G concentrations were significantly better for the Grace Dearborn Chemical and frequent jar testing was not required for the

Grace Dearborn chemical. However, depending on the mode of operation, there was between two and four time more sludge produced for the Grace Dearborn chemical compared with the Calgon chemicals. Assuming that both sludges cannot be burned at the facility and disposal of the sludge cost $0.26/gallon (a typical number) then the disposal cost per 1000 gallons of wastewater treated range $1.77 to $1.90 for the Calgon chemicals and $3.46 to $5.98 for the Grace Dearborn chemical. Additional experiments and an economic analysis is planned. Ultimately, the selection process will not only be based on the technical and economic results but also input from the state regulatory agency.

**Table 2 Summary of DAF Unit Operation Results
for Calgon and Grace Dearborn Chemicals**

| Parameter | Calgon | Grace Dearborn |
|---|---|---|
| Avg. ± std. O/G, mg/L | | |
| Semi-batch BMT | 270±200 | 55±17 |
| Continuous BMT | 580±270 | 94±17 |
| Semi-batch DAF | 60±80 | 28±8 |
| Continuous DAF | 205±135 | 30±9 |
| Sludge Produced, gal/1000 gal[1] | 7.3 | 13.3 |
| Semi-batch DAF | 6.8 | 23.0 |
| Continuous DAF | | |
| Costs, $/1000 gal | | |
| Chemical | | |
| Semi-batch DAF | 4.54 | 10.80 |
| Continuous DAF | 4.05 | 10.80 |
| Sludge Disposal[2] | | |
| Semi-batch DAF | 1.90 | 3.46 |
| Continuous DAF | 1.77 | 5.98 |

[1]gal sludge/1000 gal wastewater treated. [2]$0.26/gallon.

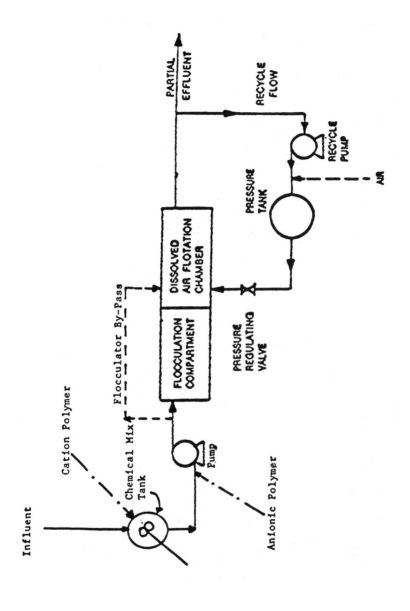

Figure 1. Schematic of the CA/DAF Unit

419

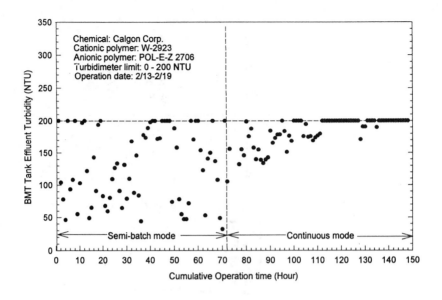

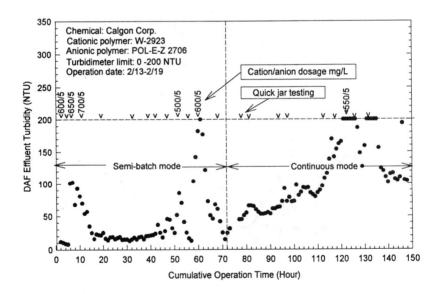

Figure 2.  Effluent Turbidity's for BMT and DAF vs. Time for Calgon
Chemicals.

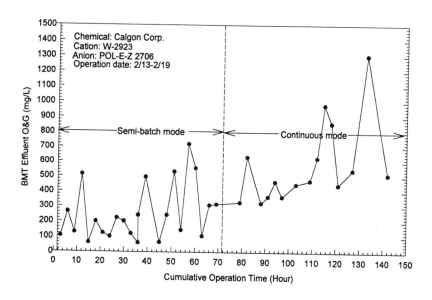

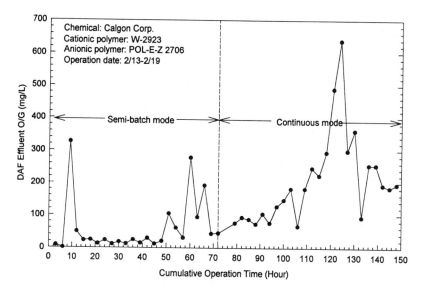

Figure 3. O/G Conc. for Influent, BMT & DAF Effluents for Calgon
Chemicals

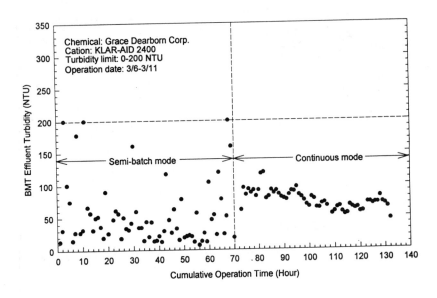

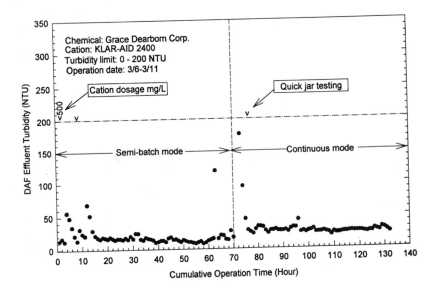

Figure 4. Effluent Turbidity's for BMT and DAF vs. Time for GD Chemical.

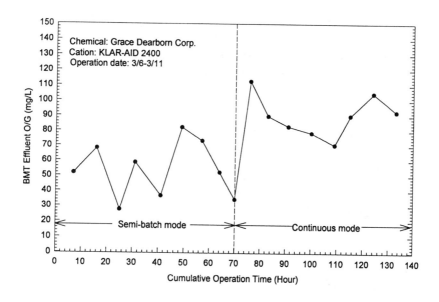

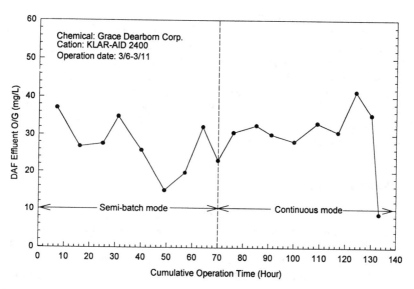

Figure 5. O/G Conc. for Influent, BMT & DAF Effluents for GD Chemical

# TREATMENT OF ALUMINUM MANUFACTURER COOLANT WASTE USING PILOT-SCALE ULTRAFILTRATION

BRIAN E. REED
PATRICK CARRIERE
CHRIS DUNN
   Department of Civil and Environmental Engineering
   West Virginia University
   Morgantown, WV 26506-6103

## INTRODUCTION

Treatment of oily wastewater from aluminum rolling mill operations is an inherent problem in the aluminum fabrication industry. In this study, wastewater from the manufacturer's processes was transferred to two holding ponds having a total capacity of about 5 million gallons and a detention time of about 10 days. In the holding ponds, free oil was allowed to rise to the surface where it was periodically removed. Wastewater from the holding ponds was withdrawn from about a depth of 9' and used as influent to a variety of technologies. In this paper, results from the ultrafiltration portion of the treatability study are presented.

## BACKGROUND

Ultrafiltration (UF) is a fluid/particle separation technology that utilizes porous membranes for the purification of liquid solutions. UF removes particulates from fluids by forcing the particle/fluid mixture through or alongside the porous membrane. The membrane allows the fluid to pass and rejects the particles. The result is a purified fluid stream and a concentrated particle stream, known as the permeate and the concentrate, respectively. The membrane system used for this study was a tubular system, a schematic of which is presented in Figure 1. Tubular systems employ membranes that are cast onto the inner walls of PVC tubes. The wastewater is pumped through the inside of the tubes. The concentrate exits the opposite end of the tube while the permeate passes through the walls of the membrane into effluent tubes.

Particles rejected at the membrane walls will form a "gel" layer on the membrane surface. This is due to the phenomenon known as "concentration polarization" which is the increase in concentration of solute on the membrane surface. This phenomenon causes the membranes to become fouled with waste material thus, the membrane must be cleaned periodically. Membrane cleaning is accomplished by pumping a proprietary detergent through the membranes to chemically remove the foulant. Spongeballs are also pumped through the tubes to mechanically remove foulant material.

Membranes are categorized by their molecular weight cut-off (MWCO). The MWCO designates the size of the compound that will be just rejected by the membrane. For example a membrane with a MWCO of 100,000 will remove compounds having a molecular weight greater than 100,000. Generally, as the MWCO increases the effluent quality decreases but the amount of wastewater treated increases. Membranes can also be constructed with a surface charge. For example, a negative surface charge will repel negative oil droplets resulting in a higher flux and less fouling and possibly a better quality effluent.

UF units are typically operated first in the semi-batch mode then the concentrate is or thickened in the batch mode ("batching-down"). In the semi-batch mode wastewater is added to the process tank periodically to replace the amount of permeate that was produced. In the batch mode, influent addition to the process is stopped but the UF unit is still operated (*i.e.*, permeate is being produced). The concentrate in the process tank is thickened. Concentration factors (CF) as high as 100 can be achieved.

The permeate flow rate or flux are the primary measures of how well the UF system is performing (*i.e.*, how much clean water is produced). The permeate flow rate is determined by measuring the volume of permeate produced per a given time. Flux is use more often than permeate flow rate because it can be used to scale-up pilot-scale results to a full sized system. Permeate flux is determined by dividing the permeate flow rate by the membrane area. The permeate flow rate and flux decrease with time due to the buildup of the gel layer on the membrane surface. The gel layer is removed by cleaning the membrane. Ideally, the flux after cleaning is equal to the flux observed at the beginning of the previous run (*i.e.*, membrane is not permanently fouled).

## EXPERIMENTAL DESIGN

The UF unit was leased from the manufacturer for 1 month in the summer (8 runs) and 1 month in the winter (4 runs). A schematic of the UF unit is presented in Figure 2. Two membranes, designated as "M" and "P", were evaluated in this study: Membrane "M" has a molecular weight cut-off (MWCO) of 100,000 and has no surface charge while membrane "P" has a

MWCO of 125,000 and a negative surface charge. For each membrane, 8 tubular membranes in series were used.

Pond effluent was fed directly to the UF unit during summer operation while in the winter the influent was first passed through a coalescing unit. The pressure drop across the membranes was maintained at 32 psi. O/G samples were taken at the start of the modified batch experiment and throughout the batch down process at several concentration factors. Temperature, pressure and permeate flow rate were monitored approximately every 2 hours for the first twelve hours of operation after which monitoring occurred every three to four hours. Permeate flow rate was monitored using a graduated cylinder and stopwatch. Permeate flux was calculated from permeate flow rate data. Following the completion of a run, the membranes were rinsed with a solution provided by the UF manufacturer (a mixture of surfactants and chelating agents). Following rinsing, spongeballs were sent through the system to physically remove particulates that accumulated on the membrane surface. Following rinsing and "spongeballing" the clean water flux (CWF) was determined. If the CWF was within the range specified by the manufacturer, the next run was started. If not, the cleaning procedure was repeated. Acid cracking of the UF concentrate using sulfuric acid was attempted three times during winter operation.

**RESULTS**

**Summer Operation**

In Table 1, the average process fluxes over the course of a given run for the M and P membranes are presented. Permeate flux for the M membrane ranged from 16 to 32 gal/ft$^2$-d (gfd) and averaged 26 gfd. Permeate flux for the P membrane ranged from 20 to 69 gal/ft$^2$-d (gfd) and averaged 39 gfd.

**Table 1 Process Flux During Summer Operation**

| Run Number | Process Flux, gfd | |
|---|---|---|
| | M Membrane | P Membrane |
| 1 | 26 | 43 |
| 2 | 30 | 63 |
| 3 | 58 | 58 |
| 5 | 16 | 25 |
| 6 | 17 | 20 |
| 7 | 26 | 31 |
| 8 | 32 | 69 |
| Average of 8 Runs | 26 | 39 |

Permeate and concentrate oil and grease (O/G) concentrations for the eight summer UF runs are presented in Table 2. Effluent O/G concentration for the M membrane ranged from of the < 5 to 47 mg/L and averaged 18 mg/L. Effluent O/G concentration for the P membrane ranged from of the < 10 to 46 mg/L and averaged 20 mg/L. O/G content of the concentrate varied from 1,200 to 65,000 mg/L depending on the CF.

**Table 2. Permeate and Concentrate Oil/Grease Content For Summer Operation**

| Run No. | Sample | O/G (mg/L) | Run No. | Sample | O/G (mg/L) |
|---------|--------|------------|---------|--------|------------|
| S1 | Concentrate 1X | 11000 | S4 con't. | Permeate M 14 | <5 |
|  | Permeate M 1X[1] | 47 |  | Permeate P 14X | 46 |
|  | Permeate P 1X | 14 | S5 | Concentrate 1X | 33000 |
| S2 | Concentrate 20X | 65000 |  | Permeate M 1X | 8 |
|  | Permeate M 20X | 15 |  | Permeate P 1X | 18 |
|  | Permeate P 20X | 14 | S6 | Concentrate 13X | 23885 |
| S3 | Concentrate 2X | 8700 |  | Permeate M 13X | 15 |
|  | Permeate M 2X | 9 |  | Permeate P 13X | 15 |
|  | Permeate P 2X | 10 |  | Concentrate 26X | 24803 |
| S4 | Concentrate 3X | 1200 |  | Permeate M 26X | 20 |
|  | Permeate M 3X | NT |  | Permeate P 26X | 21 |
|  | Permeate P 3X | NT |  | Concentrate 52X | 25126 |
|  | Concentrate 4X | 14000 |  | Permeate M 52X | 31 |
|  | Permeate M 4X | 7 |  | Permeate P 52X | 27 |
|  | Permeate P 4X | 14 | S7 | Permeate M 3X | 15 |
|  | Concentrate 5X | 25000 |  | Permeate P 3X | 19 |
|  | Permeate M 5X | 12 | S8 | Permeate M 9X | 20 |
|  | Permeate P 5X | 21 |  | Permeate P 9X | 18 |
|  | Concentrate 14X | 35000 |  | Concentrate 14X | 35000 |

NT: Not Tested. [1]Permeate at a concentration factor of __X.

The P membrane exhibited a higher average process flux compared to the M membrane while the oil and grease concentration for both membranes was approximately equal. Based on the summer data, the P membrane would be selected for use in full-scale operation.

**Winter Results**

In Table 3, a summary of O/G and permeate fluxes for the four winter runs are presented. In Table 4, O/G concentrations for the permeate and concentrate at various concentration factors (CF) are presented. The average influent O/G concentration ranged from 900 to 3,810 mg/L and averaged 2,300 mg/L over the four runs. For the M membrane, O/G concentration ranged from 15 to 196 mg/L and averaged 52 mg/L (99.4 % removal). O/G for the P membrane ranged from 16 to 299 mg/L and averaged 38 mg/L (99.4 % removal).

Average permeate flux over the four runs for the M membrane ranged from 17.2 gfd to 25.6 gfd and averaged 22.5 gfd. Permeate flux for the P membrane ranged from 14.6 gfd to 47.1 gfd and averaged 26.5 gfd. Thus, effluent quality, as measured by O/G, and permeate flux were slightly better for the P membrane compared to the M membrane. The O/G content of the concentrate ranged from about 1,350 to 25,790 mg/L and averaged 6,580 mg/L. The higher flux for the P membrane is due to the larger MWCO and the net negative charge on the membrane surface. Despite having the larger MWCO, the P membrane (MWCO = 125,000) produced a effluent with a lower O/G content compared with the M membrane (MWCO = 100,000). This was most likely due to the negative surface of the P membrane - the membrane net negative charge repels the negatively charged oil droplets. Based on the winter data, the P membrane would be selected for use in full-scale operation.

**Table 3. Summary of O/G and Permeate Fluxes From Winter UF Operation**

| Run | Range and Average[1] O/G Over The Duration of the Run, mg/L | | | | Average Flux gfd | |
|---|---|---|---|---|---|---|
| | Influent | Effluent | | Concentrate | | |
| | | M | P | | M | P |
| W1 | 1280-3081 (2350) | 15-43 (26) | 18-61 (34) | 2,670-5,590 (4,100) | 24.7 | 28.5 |
| W2 | 1010-2080 (1840) | 34-45 (39) | 30-36 (33) | 1,640-6,940 (4,290) | 22.4 | 47.1 |
| W3 | 1310-3810 (2910) | 28-139 (91) | 26-299 (20) | 1,350-18,330 (7,500) | 25.6 | 14.6 |
| W4 | 900-3230 (2110) | 21-196 (51) | 16-142 (65) | 2,290-25,790 (10,430) | 17.2 | 15.7 |
| Over All | 900-3810 (2300) | 15-196 (52) | 16-299 (38) | 1,350-25,790 (6580) | 17.2-25.6 (22.5) | 14.6-47.1 (26.5) |

[1] Average value in ( ).

**Table 4. Permeate and Concentrate O/G at Different Concentration Factors for Winter UF Operation**

| Run # | Sample | O/G mg/L | Run # | Sample | O/G mg/L |
|---|---|---|---|---|---|
| W1 | Concentrate 1X[1] | 2,670 | W2 | Concentrate 1X | 1,640 |
| | Permeate M 1X | 15 | | Permeate M 1X | 45 |
| | Permeate P 1X | 18 | | Permeate P 1X | 29 |
| | Concentrate 2X | 4,020 | | Concentrate 6.6X | 6,940 |
| | Permeate M 2X | 21 | | Permeate M 6.6X | 34 |
| | Permeate P 2X | 29 | | Permeate P 6.6X | 36 |
| | Concentrate 3.6X | 5,590 | | | |
| | Permeate M 3.6X | 43 | | | |
| | Permeate P 3.6X | 61 | | | |

[1] Permeate at a concentration factor of __X

**Table 4. Permeate and Concentrate O/G at Different Concentration Factors for Winter UF Operation, Con't.**

| Run # | Sample | O/G mg/L | Run # | Sample | O/G mg/L |
|---|---|---|---|---|---|
| W3 | Concentrate 1X[1] | 2,210 | W4 | Concentrate 1X | 2,290 |
| | Permeate M 1X | 28 | | Permeate M 1X | 27 |
| | Permeate P 1X | 26 | | Permeate P 1X | 16 |
| | Concentrate 3.8X | 1,350 | | Concentrate 4.5X | 6,600 |
| | Permeate M 3.8X | 100 | | Permeate M 4.5X | 78 |
| | Permeate P 3.8X | 300 | | Permeate P 4.5X | 138 |
| | Concentrate 4.3X | 4,720 | | Concentrate 7X | 8,620 |
| | Permeate M 4.3X | 126 | | Permeate M 7X | 93 |
| | Permeate P 4.3X | nd | | Permeate P 7X | 142 |
| | Concentrate 5.3X | 4,680 | | Concentrate 13X | 7,460 |
| | Permeate M 5.3X | 93 | | Permeate M 13X | 42 |
| | Permeate P 5.3X | 21 | | Permeate P 13X | 31 |
| | Concentrate 11X | 8,610 | | Concentrate 26X | 10,320 |
| | Permeate M 11X | 71 | | Permeate M 26X | 41 |
| | Permeate P 11X | 27 | | Permeate P 26X | 21 |
| | Concentrate 21X | 12,630 | | Concentrate 52X | 11,920 |
| | Permeate M 21X | 83 | | Permeate M 52X | 41 |
| | Permeate P 21X | nd | | Permeate P 52X | 31 |
| | Concentrate 42X | 18,330 | | Concentrate 104X | 25,790 |
| | Permeate M 42X | 139 | | Permeate M 104X | 196 |
| | Permeate P 42X | 38 | | Permeate P 104X | 75 |

[1]Permeate at a concentration factor of __X

In Table 5, the preliminary sizing of the UF system for summer and winter operation is presented. A 10 allowance for expansion has been incorporated into the capacity of the systems. Because of the lower flux experienced during winter operations, the winter design controls.

**Table 5 Preliminary Design of Full-Scale UF System**

| Design Parameter | Value | |
|---|---|---|
| | Summer | Winter |
| No. of Tubes | 1040 | 1,523 |
| Membrane Area/Tube, ft$^2$ | 2.2 | 2.2 ft$^2$ |
| Total Membrane Area, ft$^2$ | 2,287 | 3,352 |
| Average Flux, gfd | 39 | 26.5 |
| Capacity, gal/d | 89,248 | 88,800 gal |
| Est. Tube Life, yr. | 5 | 5 |
| Membrane Tube Cost, $ | 250 | 250 |

In Table 6, results from acid cracking experiments performed on concentrate generated during winter UF operations are presented. Two acid cracking experiments were performed on concentrate from Run W3 (CF = 42X). For the first attempt, the oil floc volume 60 mL and 140 mL of concentrated sulfuric acid was added. The turbidity of the supernate was > 200 NTU and pH was less than 1.0. No obvious cracking was evident based on visual observations and the high turbidity. In the second Run W3 cracking experiment a total of 180 mL of concentrated sulfuric acid was added. The waste temperature was 65°C and the floc volume was 75 mL. Floc size was relatively small. Supernatant turbidity was 124 NTU. The sample appeared to have been broken due to the relatively clear supernatant and floc formation in the sample. A third cracking experiment was conducted on concentrate (CF = 104X) generated from Run W4. Floc volume was 50 mL, the supernatant pH was < 1.0, and the supernatant turbidity was > 200 NTU. A total of 280 mL of concentrated sulfuric acid was used. No obvious cracking was evident based on visual observations and the high turbidity.
Acid cracking

**Table 6 Results of Acid Cracking Experiments**

| Run # | CF | Acid Added, mL | Floc Vol., mL/L | Supernatant Quality | | Comments |
|---|---|---|---|---|---|---|
| | | | | pH | Turbidity, NTU | |
| 3 | 42 | 140 | 60 | <1 | >200 | no break |
| 3 | 42 | 180 | 175 | <1 | 124 | break |
| 4 | 104 | 280 | 50 | <1 | >200 | no break |

**SUMMARY**

Treatment of an oily wastewater from an aluminum manufacturer by a pilot-scale ultrafiltration unit was studied. Two membranes, designated M and P were investigated. The M membrane had a molecular weight cut-off (MWCO) of 100,000 and had a neutral surface charge. The P membrane had a MWCO of 120,000 and had a negative surface charge. The UF unit was operated for 1 month in the summer and 1 month in the winter. For both seasonal operations, the P membrane was considered superior to the M membrane. For the summer operation, effluent O/G concentration for the P membrane ranged from of the < 10 to 46 mg/L and averaged 20 mg/L and the permeate flux the P membrane ranged from 20 to 69 gal/ft$^2$ membrane per day (gfd) and averaged 39 gfd. During winter operations, effluent O/G for the P membrane ranged from 16 to 299 mg/L and averaged 38 mg/L and the permeate flux for the P membrane ranged from 14.6 gfd to 47.1 gfd and averaged 26.5 gfd. The higher flux for the P membrane was due to the larger MWCO and the net negative charge on the membrane surface.  Acid cracking of the concentrate was attempted three times, but was only effective once.

## FEG TUBULAR MEMBRANE MODULE SPECIFICATIONS

Tubular Membrane Module Length          10 feet

Tubular Membrane Area                    2.2 square feet

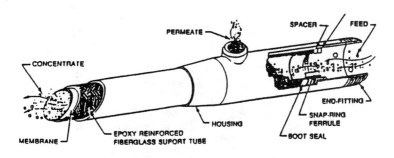

Figure 1.  Schematic Diagram of Tubular Membrane.

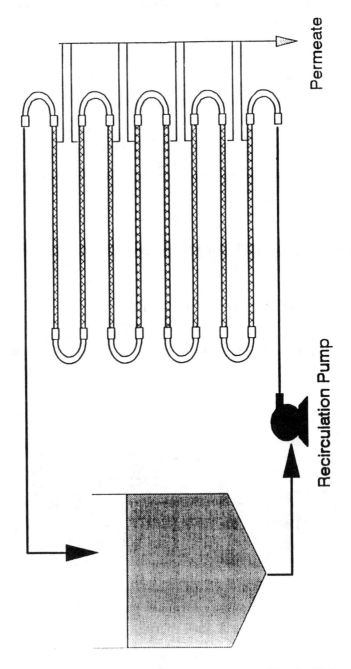

Figure 2. Schematic Diagram of Pilot-Scale Ultrafiltration Unit

# MECHANISMS THAT GOVERN THE SUCCESSFUL APPLICATION OF SPARGING TECHNOLOGIES

Stewart H. Abrams, P.E.
Michael Marley
Edward X. Droste, P.E.
   Envirogen, Inc.
   Princeton Research Center
   4100 Quakerbridge Road
   Lawrenceville, NJ 08648

## INTRODUCTION

Accidental releases of volatile organic compounds (VOCs) into the subsurface environment in the form of petroleum products or industrial solvents can result in costly remediation. Although virtually any form of remediation is expensive, developing a well planned, cost effective strategy at the onset of a spill or release can minimize expenses which accumulate throughout the duration of a cleanup project. Removal of the VOC source is the primary consideration to ensure effective remediation. Soil contamination which lies under and in the vicinity of a leaking underground storage tank (LUST) or a surface spill is a potential long term contributor/source to the migration of hazardous vapors in vadose zone soil and to dissolved VOCs in ground water. Frequently, contaminated soils exist below the ground water table (GWT) when free phase product mounds on the GWT and is transported vertically in response to seasonal GWT fluctuations or drawdown from pumping in nearby ground water/product recovery wells. Dense, non-aqueous phase liquids (DNAPLs) are frequently found on soils below the GWT as globules and/or residuals, due to their density driven vertical transport.

A few commercially applicable in-situ remediation technologies exist which can be applied as remedial alternatives at VOC spill sites although, generally no one technique can accomplish all the objectives of a complete site clean up. Utilizing pump and treat methods to remediate VOCs sorbed and/or trapped in saturated zone soils is considered to have significant limitations due mainly to standard pump and treat system designs, site specific soil heterogeneities and contaminant distribution and the kinetic limitations to the mass removal process [1]. Techniques such as soil flushing and augmented biodegradation provide potential in-situ enhancements to pump and treat based remediation methods.

Soil vapor extraction (SVE) has been demonstrated to be a successful and cost effective remediation technology for removing VOCs from vadose zone soils. This technique involves the controlled application of an air pressure gradient to induce an air flow through soils contaminated with VOCs. As soil gas is drawn towards the vacuum source (vapor extraction well), the equilibrium between the VOC phases as product, in soil vapor and soil moisture and on the soil is upset causing enhanced partitioning in the vapor phase. VOCs in the vapor phase are subsequently removed from the subsurface and treated utilizing a standard vapor extraction off-gas treatment system(s). One of the limitations of utilizing SVE is that it is not an optimal remediation technology to address contaminated soils within the capillary fringe and below the GWT.

A number of techniques have been developed and employed to expand the SVE process to include effective remediation of VOCs in saturated zone soils. Artifical water table drawdown is one approach that may be utilized to expose contaminated soils in the saturated zone to the advective air phase thereby increasing the efficiency of the SVE process. However, in some cases, this is not a practical, nor cost effective approach. An innovative, alternative approach is the application of sparging technology, which entails injecting a hydrocarbon free gaseous medium (standardly air) into the saturated zone below/within the areas of contamination. With air sparging, the VOC contaminants dissolved in the ground water and sorbed/trapped on the soil, partition into the advective gaseous phase effectively simulating an in-situ, saturated zone air stripping system. The stripped contaminants are subsequently transported in the air phase to the vadose zone, within the radius of influence of an operating soil vapor extraction system. The contaminant vapors are drawn through the vadose zone to the vapor extraction well(s) and are treated utilizing standard vapor extraction off-gas system(s).

Aeration of the ground water can occur or can be the primary objective of a sparging system. Under this condition, and assuming that additional relevant environmental parameters are or can be manipulated to be within an acceptable range, contaminant remediation via enhanced in-situ biodegradation can result (bio-sparging). A schematic depicting a typical sparging system configuration is presented in Figure 1.

Limited references exist in the literature as to the design and/or success of the laboratory or field application of the sparging process. Apparently, the process was first utilized as a remediation technology in Germany in the mid-1980's, predominantly to enhance the clean-up of chlorinated solvent contaminated ground water [2]. More recently the technology has been utilized in the enhanced remediation of gasoline contaminated saturated zone soils and ground water [3,4,5,6]. Apparently, in each of these cases the design of the air sparging systems have been empirically based.

The authors have performed sparging field pilot tests and implemented full scale sparging/SVE systems on numerous sites across the United States. Experience developed on these projects has demonstrated that there are numerous important criteria which must be considered when designing, installing, and operating sparging systems. Evaluation of these criteria are not only necessary to ensure effective remediation of saturated zone soils/ground water but also to preclude displacing and mobilizing potentially hazardous soil gas vapors or free phase product or dissolved phase contaminants in the saturated zone.

In developing a design process, it is necessary to better understand the mechanisms occuring during sparging. Air is injected into sparging wells at pressures in excess of the soil matrix air entry pressures. For coarse grained soils, minimum pressures are generally 1 to 2 pounds per square inch (psi) in excess of the hydrostatic head at the top of the injection well screen. For finer grained soils (with significant air entry pressures) the required minimum injection pressures can be significantly higher (factor of 2 or more) than the hydrostatic head at the top of the well screen. Sparging in highly stratified soils can require pressures that approach or exceed soil fracturing pressures. Fracturing of the soil matrix may or may not enhance the contaminant removal process. In many cases, the efficiency of mass removal may be significantly reduced due to limitations in the number and distribution of the fractures. Due to this potential, injection at pressures that may cause fracturing should be seriously reviewed.

Soils possess a relatively random distribution of pore sizes, pore scale heterogeneities and hydraulic conductivities. As air enters the saturated soil matrix, it will displace water from the highest conductivity pores, forming an air channel to the vadose zone. The displacement process is sensitive to the pore scale conductivity differences and results in an unpredictable distribution of air channels. This is contrary to the idealized perception of a uniform distribution of discrete air bubbles. With a limited

435

number of the air channels being formed, the sparging process would constitute contaminants being stripped or degraded in the immediate vicinity of the channels.

If this were the only mechanism of contaminant removal, the air sparging process would be significantly less efficient than it has been demonstrated to be in the field. It is believed that three major mechanisms may be contributing to the field observed enhanced removal efficiencies. First, the buoyant movement of the air through the saturated zone may tend to create convective movement of water in the vicinity of the channel, this will serve to induce movement of additional contaminants to the air channel for removal. Secondly, since the water displacement process is so sensitive to the pore scale hydraulic conductivities, a change in the hydraulic conductivity of a pore could effect the spacial location of the channel. To this degree, it is preferred to pulse air sparging systems during operation. The pulsing has a number of potential benefits. Pulsing causes a mixing of the water in the treatment zone effectively increasing the channel/contaminant contact. Pulsing also may cause a change in the air channel spacial distribution since the pore hydraulic conductivity can be effectively reduced due to the presence of residual air in the formerly preferred channels following a pulse event. Pulsing can also reduce capital and energy costs in the operation of this kinetically limited mass removal process.

Thirdly, the oxygen transfer into ground water and subsequent mixing from both the convective ground water movement and as a result of the pulsed mode of system operation, can produce contaminant removal through enhanced biodegradation (biosparging).

It is obvious that the sparging process still requires significant research to verify or to better describe the mechanisms occuring during system operation. As with most remediation technologies, it is not appropriate for all sites and is generally one of an integrated train of technologies required for a site clean up.


## SPARGING SYSTEM DESIGN

As previously stated, most sparging systems require the operation of a standard SVE system to capture and treat VOC laden air liberated by sparging activities. Major design considerations of sparging technology implementation (excluding a properly designed SVE system) are: site geology, contaminant type and distribution, gas injection pressures and flow rates, air injection interval (both areal and vertical), and site specific parameters that dictate the feasibility of enhanced biodegradation. Each of these design considerations are addressed briefly as follows.

## Site Geology

A thorough understanding of site geology is the most important design criteria of a sparging system. Sparging is generally more effective in uniform, coarse-grained soils where air-entry pressures (i.e. air injection pressures required to overcome the resisting hydrostatic head pressure) are relatively low. Generally, a more desirable air channel distribution is achieved in uniform coarse grained soils. Fine grained soils generally require higher air injection/entry pressures. Excess pressures may crate fractures in the soil formation which could reduce the sparging effectiveness.

Soil stratification is a critical geological feature to identify in terms of the expected performance of a sparging system. Air flow in the subsurface will follow the path of least resistance. The presence of less permeable soil stratas above the air injection point can create significant lateral air movement/channelling. Determination of the potential for preferred lateral air movement is critical to predict lateral displacement of ground water and capture of VOCs in the vadose zone due to sparging activities.

## Contaminant Type and Distribution

Sparging systems may be designed to remove volatile and semi-volatile contaminants through either or both volatilization and biodegradation processes.

The distribution of the contaminant in the subsurface (both areally and vertically) and the phases in which it is present (dissolved, non-aqueous phase liquid, or sorbed onto soil organic matter) is critical to predict the effectiveness and feasibility of sparging at a given site. One of the most difficult problems to address with sparging or any of the present commercially applicable technologies is the remediation of pools of DNAPL.

## Air Injection Pressures and Flow Rates

Air injection pressures are governed by the static water head above the sparge point, the required air entry pressure of the saturated soils, and the air injection flow rate. In the design process, the lowest effective air injection pressure will correspond to the pressure required to maintain a minimum continuous air flow through the saturated zone. Higher pressures will produce higher air injection flow rates and, due to the random distribution of air entry pressures in the soil, will likely produce additional air channels. The higher air injection pressures required in fine-grained soils can cause the formation of significant subsurface gas pockets. A gas pocket is essentially an unsaturated volume that expands from the air sparging well during the injection process until the pressure within the pocket is sufficient to overcome the vertical air entry pressure of the overlying soils. The pocket expansion will continue until a steady-state condition of air inlet flow to air escape flow is achieved. Too high an air injection pressure (0.8 psi per foot of normally consolidated overburden and 3 to 5 psi per foot of overconsolidated overburden) may create fractures in the sparging well annular seal or along weak joints in the soil. The creations of fractures may result in a loss of system efficiency or in some cases may actully improve channel distribution. However, when fracturing occurs, the effects are likely irreversible.

Air flow rates that are typically used in the field are in the range of 3 to 20 cubic feet per minute (cfm) per sparge point. As stated previously, pulsing of the air flow into the sparge points is considered to provide a more efficient approach to mass removal. Additionally, pulsed air flow is considered to provide a better distribution of air flow channels and ground water mixing over the project duration.

## Air Injection Interval

The areal locations of sparging wells will depend upon the areal delineation of the remediation area and the soil specific radii of influence. As previously mentioned, lateral displacement of the ground water is a potential concern at some sites. To alleviate this concern, traditional pump and treat methodologies may be required as a portion of the integrated remediation effort. Another option would be to install an array of defensive sparging wells or an intercepting sparging trench downgradient of the remediation area.

The air injected in saturated uniform soils is generally considered to travel upwards to the vadose zone in channels that describe a parabolic pathway. However, slight heterogeneities in the soil matrix can alter the upward flow path of injected air dramatically. To that extent, the site geology must be considered when determining the depth at which air will be injected. Based on the authors experience, the sparge point should be located no less that five feet

438

below the vertically delineated remediation zone. At shallower depths, it is expected that the radius of influence of the sparge point will be very limited, requiring excessive numbers of sparge points to enhance remediation of a unit volume of contaminated soil. On the other hand, placing sparge points at greater depths increases the risk of the site specific geology affecting the distribution of the air channels within the remediation soil volume.

Short screened intervals, on the order of one to three feet in length, are generally used in air sparging wells because most of the air exits through the top of the well screen where the pressure head is at a minimum. Use of longer screened intervals generally do not significantly add to the effectiveness of the air injection process, except in those cases where lateral migration of the air flow through the saturated zone is a design requirement.

## Enhanced Biodegradation

To determine the potential for enhanced biodegradation due to the ground water oxygenation, the sparging efficiency to oxygenate the saturated soil zone should be monitored. Increases in dissolved oxygen in nearby ground water monitoring wells or piezometers would indicate that the potential to increase biodegradation of applicable contaminants is present. However, care should be exercised to ensure that the dissolved oxygen level in the monitoring well is respresentative of dissolved oxygen in the formation and not as a result of short circuiting of the air flow to the monitoring well. Increases in measurement of $CO_2$ or $CO_2$ isotope ratios in the soil gas or dissolved phase after the sparging system is shutdown could indicate that indigenous soil microbes are consuming target VOCs. Analysis of soil samples to determine the population of indigenous soil microbes and the nutrient levels are recommended to gauge the potential effectiveness of biosparging.

## PILOT TESTING

Prior to full scale implementation of a sparging system, a pilot test is normally conducted. The purpose of the pilot test is to determine system design parameters: air entry pressures, flow rates and an effective radius of influence (ROI). Required air entry pressures and flow rates will dictate equipment requirements for a full scale system. Presently, sparging ROI determinations are empirically based by interpreting indirect measurements during the pilot test. A mathematical model has been recently developed by Marley and Li (1993) which predicts pressure distributions in the saturated soil zone induced by sparging. While initial model predictions have matched well with recently performed pilot tests, further research needs to be performed on the development and applicability of sparging models.

Sparging pilot testing generally includes the implementation of a SVE pilot test system and running of at least two sparging air flow rates. Measurements taken during the pilot test include: SVE off gas VOC $O_2$ and $CO_2$ concentrations, airflow rates and operating vacuum; soil vapor probe soil gas VOC, $O_2$ and $CO_2$ concentrations and pressures; sparging injection flow rates and pressures; water table mounding/piezometric pressures, dissolved oxygen concentrations, and the use of gaseous tracers to map the air channel distribution and SVE system capture effectiveness.

As previously stated, sparging ROI is still based somewhat on empirical interpretation of pilot test data. The analysis of collected data should be reviewed to predict the air flow channel distribution created during the sparging pilot test. In coarse soils, ROIs of 5 to 30 feet have been observed by the authors. However, in stratified soils, a ROI of 60 feet or greater has been observed. It is likely that in these cases, a uniform distribution of air channels was not present within a 60 foot radius from the injection source. Rather, preferential lateral air flow was likely occurring. ROI determination is also dependent on the objective of the sparging system (e.g., ground water oxygenation, VOC volatilization, or the development of a sparging cut off curtain or trench).

## SUMMARY

Sparging has the potential to significantly reduce the cost and duration for remediation of saturated, contaminated soils. The multi-fluid transport processes involved in sparging are complex. The technology is not appropriate for application at all sites. System control is best achieved in coarser grained soils without significant stratification. Pulsed operation of sparging system appears to provide a number of advantages over continuous system operation with respect to the efficiency of mass removal.

Volatile and semi-volatile contaminants may also be biodegraded in-situ through ground water oxygenation as a result of sparging processes. Oxygenation of the ground water can, under optimal environmental conditions, simulate the bioactivity of the indigenous bacterial population.

Major design considerations/concerns are the potential for uncontrolled contaminated vapor and water migration. System design generally requires ground water and vapor control through the proper positioning of sparging wells or trenches and/or ground water recovery wells, and, if appropriate vapor extraction wells.

Through the proper design and application of the technology, rapid and effective contaminant mass removal can be achieved. Under optimal conditions, ground water can be cleaned to MCL levels.

## REFERENCES

1. MacKay, D.M., and Cherry, J.A.: 1989. Ground Water Contamination: Limitations of Pump and Treat Remediation. *Environmental Science and Technology, Vol. 23, No. 6 pp 630.*

2. Gudemann, H. and Hiller, D.; 1988; In Situ Remediation of VOC Contaminated Soil and Ground water by Vapor Extraction and Ground water Aeration. *Proceeding of the Third Annual Haztech International Conference, Cleveland, Ohio.*

3. Ardito, C.P. and Billings, J.F.; 1990; Alternative Remediation Strategies: The Subsurface Volatilization and Ventilization System. *Proceeding of The Conference on Petroleum Hydrocarbons and Organic Chemicals in Ground Water: Prevention, Detection and Restoration,* NWWA, pp 281-296.

4. Marley, M.C., 1991. Air Sparging in Conjunction with Vapor Extraction for Source Removal at VOC Spill Sites. *Proceedings of the Fifth National Outdoor Action Conference on Aquifer Restoration, Ground water Monitoring and Geophysical Methods,* P89-103.

5. Marley, M.C., Hazebrouck, D.J., Walsh, M.T., 1992. The Application of In-Situ Air Sparging as an Innovative Soils and Ground water Remediation Technology. *Ground water Monitoring Review, Vol. 12, No.2,* P137-125.

6. Brown, R.; Herman, C. and Henry, E., 1991. The Use of Aeration in Environmental Cleanups. *Proceedings of Haztech International Pittsburgh Wast e Conference.*

# Air Sparging in Stratified Sands

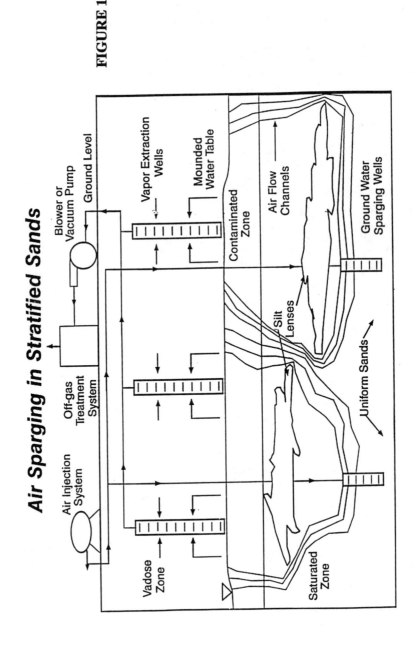

FIGURE 1

# CONSTRUCTION OF LOW PERMEABILITY SOIL-BENTONITE BARRIER CAPS AND LINERS FOR LANDFILLS

TOM WEBBER
Pyramid Environmental
P.O. Box 5532
Highpoint, NC 27262

MICK WILLIAMS, P.E.
IT Corporation
5754 Pacific Center Boulevard
Suite 203
San Diego, CA 92121

## ABSTRACT

A low permeability soil barrier layer is the usual regulatory requirement for both caps and liner systems on modern municipal, industrial, and hazardous waste landfills. This soil layer is either used as the sole barrier or as the soil component of a composite liner system. This paper presents construction experience for blending on site soils with sodium bentonite to produce a thick, low permeability soil barrier layer. The paper begins with a description of the components and construction of the barrier layer and discusses how soil-bentonite barrier layers meet or exceed the regulatory performance criteria for both State and Federal agencies.

## INTRODUCTION

Despite advances in on-site soil and groundwater remediation technology in recent years, the best and least expensive method of dealing with contamination is to prevent spills and leaks from ever occurring. The old adage "prevention is better than cure" could not be more true than for soil clean-ups. One prevention approach is to use a barrier layer, such as a low-permeability soil for the soil component of a composite liner system in a landfill. Unfortunately, native soils may be void of suitable low permeability soils in economical and available quantities. If so, one alternative is to

443

manufacture a soil liner by blending the local soils with imported high-grade sodium bentonite.

## REGULATORY REQUIREMENTS

Federal regulations (40 CFR Parts 264 and 265) for hazardous waste landfills, surface impoundments, land treatment units, and waste piles are applicable in the trust territories, territories, and commonwealths of the United States. Federal regulation (40 CFR Part 258) is applicable for municipal solid waste landfills.

For hazardous waste landfill bottom liner systems, RCRA Subtitle C (40 CFR 264.301(c)(1)(i) and 265.301(c)(1)(i)) requires a synthetic liner (geomembrane) on top of a leachate collection and removal system over a composite liner system. The composite liner system consists of another synthetic liner (geomembrane) overlying 900 mm (3 feet) of compacted soil material with a hydraulic conductivity of less than $1 \times 10^{-9}$ m/s. For municipal solid waste landfill bottom liner systems, RCRA Subtitle D (40 CFR 258.40(b)) requires a 1.5 mm (60-mil) thick HDPE geomembrane overlying 600 mm (2 feet) of compacted soil with a hydraulic conductivity of less than $1 \times 10^{-9}$ m/s. Generally, caps require a permeability less than the bottom liner system to avoid the "bathtub" effect.

## COMPONENTS OF SOIL-BENTONITE LINERS

There are three components of soil-bentonite liners:

- Soil
- Bentonite
- Water.

Soil

Soil is the building block of the soil-bentonite liner. For convenience and economic reasons, the engineer will first determine the suitability of the on-site soils. While a conventional soils analysis for Atterberg limits, gradation and soil classification gives most of the feasibility information, chemical compatibility of the soil with the bentonite is also important. For example, soils containing a high percentage of calcium carbonate can displace the sodium ions of the sodiumbentonite with the calcium ions and limit or severely reduce the bentonite's ability to swell, thereby increasing the permeability of the soil. However, this does not mean that the soil is

444

necessarily unsuitable for a soil-bentonite liner. The addition of soda ash or additional bentonite can counter this affect.

## Bentonite

Bentonite is a clay comprised primarily of the mineral montmorillonite, whose unique crystalline structure is responsible for the clay's properties. The clay structure, when dry, resembles a negatively-charged stack of plates. On contact with polar molecules such as water, the water molecules force the plates apart, exhibiting bentonite's characteristic swelling. Although there is more than one type of bentonite, the high-grade sodium bentonite clay (like those found in Wyoming) typically swell over 500 percent. With the addition of even more water the wetted bentonite acts like a highly viscous fluid and resists fluid flow, i.e., the bentonite lowers the permeability and seals the soil.

## Water

Often taken for granted, water is the component that makes bentonite swell. As a general guide, use only potable water but do a chemical analysis first. High concentrations of calcium, magnesium or chloride ions in the mix water may inhibit swelling and increases the resultant permeability of the completed liner. Brackish water from swamps or saline groundwater typical in the interstitial zone below the freshwater lens is probably not suitable for the mixing water.

## CONSTRUCTION OF SOIL-BENTONITE LINERS

Although the concept of mixing two soils together sounds simple, producing a low permeability soil-bentonite liner requires more quality control than conventional earthwork construction. Some of these elements are:

- Mixing equipment
- Expertise
- Placement
- Protection
- Construction Quality Assurance

445

## Mixing Equipment

There are two ways of mixing the bentonite into the soil: either spread the bentonite out over the native soil and till in place or blend the soil and bentonite in a pugmill. Design engineers prefer the latter method (USEPA,1988) because manufacturing the soil-bentonite blend is similar to a carefully controlled concrete and asphalt production process.

Specialty contractors now use the most modern mixing equipment available i.e., the on-site continuous mixing plant. This type of plant uses a high-speed, twin-shaft pugmill to accurately and consistently blend the soil, bentonite and water. The plant operates on a continuous basis, rather than the conventional batch process, which can produce a continuous, and repeatable blended soil-bentonite product of up to 500 tonnes per hour, depending on the soil type.

Having the best mixing equipment will not produce the best liner without the accompanying expertise. Without the right experience and equipment, the project is designed to go poorly, with project overruns, delays, and embarrassment for all involved. When writing the construction specifications, require the general contractor to use a specialty subcontractor with the appropriate expertise and equipment to mix the soil and bentonite for the liner.

## Placement and Compaction

Conventional earthmoving equipment such as scrapers or end-dump trucks haul the blended soil product from the mixing plant to the liner construction area. Since most liner projects typically cover an area of at least two hectares (five acres), there is a distinct advantage in using trailer mounted, self-erecting mixing equipment. Short haul distances and costs are dramatically reduced when the contractor locates the plant in the most convenient area with the shortest possible haul distance.

The most critical step governing a liner's performance is the construction. After spreading and leveling with dozers, smooth or padfoot rollers compact the soil-bentonite to meet the specified permeability requirements, typically $1x10^{-9}$ m/s. For designs requiring a geomembrane liner, use a smooth drum roller for the finish surface on the last lift. Typically, a 150 mm (6-inch) lift thickness of fine grained soil requires four to six passes with a padfoot roller. The operator needs to regulate the foot contact pressure to avoid shearing the soil on the third or fourth pass (USEPA, 1988).

446

One characteristic of soil liner construction is the creation of lift planes due to "springing" of the compaction equipment (USEPA, 1988). Lift planes can allow contaminants to communicate along the horizontal planes, until they find a vertical migration path or flaw in other lifts. The end result is a contaminant breakthrough time measured in months instead of years. Particularly vulnerable are steep sidewalls which usually require a horizontal lift construction method. Therefore, the compactive effort needs to tie the successive lifts together.

The problem can be minimized with good design and construction methods. With a little skill and a lot of experience, contractors can use a padfoot roller to knead the lifts together to produce as near a homogeneous layer as possible. Also, the specification should require the loose-lift thickness to not exceed the depth of the compactor pads.

## Protection

Desiccation or drying cracks are the main problems of clay liners, and to a lesser degree, soil-bentonite liners. However, current RCRA regulations require a geomembrane for the upper portion of the composite liner system. Like any synthetic low permeability cover, the geomembrane traps the moisture and reduces desiccation. However, even with the geomembrane in place, the diurnal temperature variations cause moisture to migrate from the soil liner and condense on the underside of the geomembrane. The effect can be so pronounced on sideslopes that the top of the slope can dry out resulting in deep cracks in the clay. Simultaneously, some engineers have noticed large volumes of condensed moisture at the toe of the slope (Basnett and Bruner, 1993).

One solution is to design the sideslopes no steeper than 3 horizontal to 1 vertical, then place the geomembrane, drainage layer and protective cover immediately after acceptance of the soil barrier construction. This scheduling procedure should effectively insulate the slope from desiccation effects.

However, if a soil-bentonite liner does desiccate and is rehydrated, then the montmorillonite crystalline structure causes the bentonite to act as a highly viscous fluid which moves in and fills the cracks.

## Construction Quality Assurance

The U.S. EPA recommends a detailed construction quality assurance (CQA) plan for each layer of the cover system (EPA, 1989). One tool of CQA is to use a prototype or test pad to demonstrate and test the construction practices and materials proposed for the actual liner of the soil barrier layer.

447

Here variables such as equipment and soil are compacted to an acceptable zone of moisture, density and permeability. The engineer and contractor then use this test pad experience to construct the actual soil liner.

## SETTLEMENT

Foundation movement or settlement is one of the characteristics of landfills. For liners, voids in the subgrade such as sinkholes in limestone, may collapse and tear the liner system. For unconsolidated landfills, waste decomposition will induce settlement in the cap system.

Two types of engineering controls to mitigate some of the settlement are to compact the waste and the other is to use materials that can tolerate small amounts of settlement. An engineer should select the appropriate and preferably best technology available; the correct construction material is a step in that direction. Studies of settlement using a water-filled bladder found that a soil-bentonite liner was the most effective of the clay liners in the laboratory test. The permeability of this liner did not increase after settlement was simulated by deflating the bladder. The authors of this study concluded that fine and well-distributed bentonite in the soil produced a more stable liner (USEPA, 1988).

## ALTERNATIVE DESIGNS

There are no official, published, alternative designs for bottom liner and cap systems. The EPA will review alternative designs on a case-by-case basis, however these designs must provide the same long-term performance as the recommended design. Furthermore, the Regional Administrator of the EPA must approve the proposed alternative design.

Soil-bentonite is a *soil*, therefore by definition, meets the requirements of both federal hazardous and municipal solid waste regulations.

## SUMMARY

If a soil clean-up project or remedial action options includes land disposal, federal regulations will require a low-permeability soil barrier layer as part of the cap or liner system. With a mixing plant and imported bentonite, a specialty contractor can blend the native soils and manufacture a low-permeability soil-bentonite barrier layer to regulatory standards.

# REFERENCES

1. Basnett, C.R., and Bruner, R., 1993. Clay Desiccation of a Single-Composite Liner System. Geosynthetics '93 Conference Proceedings, pp. 1329-1340

2. *Environmental Information LTD.*, 1993. Analysis Supports Access to Off-Site Treatment Centers. World Wastes, October, pp. 6-10.

3. Roy, K. A., 1993. EPA Report Outlines Market and Technology Trends in Site Cleanup. Industrial Wastewater, October/November.

4. U.S. Environmental Protection Agency, 1988. Design, Construction, and Evaluation of Clay Liners for Waste Management Facilities. November, EPA/530-SW-86-007-F, pp 5-71, 6-5 and 6-6.

5. U.S. Environmental Protection Agency, 1989. Final Covers on Hazardous Waste Landfills and Surface Impoundments. Technical Guidance Document, Office of Solid Waste and Emergency Response, Washington, D.C., EPA/530-SW-89-047, July.

# THE USE OF SOIL WASHING PROCESSES FOR THE RECLAMATION AND REUSE OF FOUNDRY WASTE SANDS

WALTER M. KOCHER
  Fenn College of Engineering
  Cleveland State University
  Cleveland, OH 44115

## INTRODUCTION

Soil washing processes are being investigated for possible use to reclaim foundry waste sands for beneficial reuse. The production of metal castings in foundry operations involves molds, coremaking, melting, pouring, cleaning and inspection. The molten metal is poured into molds cores made from sand during the casting process. After the castings harden, the metal product is separated from the molding sand and core materials in the shakeout process. These castings are then cleaned, inspected and prepared for shipment.

The coremaking, molding and shakeout operations usually generate more than 75% of the foundry's solid waste as spent cores and spent foundry sand. Some organic contaminants - phenolics in particular - might remain on these waste materials resulting from the use of molding binders and curing agents. However, it is most frequently the presence of leachable metals on the surfaces of these sands that cause these materials to be classified as a waste - and sometimes a hazardous waste. As a general rule, the iron and steel foundries do not generate hazardous spent sands, but leaded bronze and brass foundries frequently do find their spent sands classified as hazardous wastes [1].

The sweepings from the coreroom, the sand handling system dust, the dusts collected by cupola air pollution control devices, and the dust generated in the shakeout process in larger foundries are collected and accumulated in a slurry form. This slurry is usually stored in a settling lagoon and is called millpond sludge. The metals are frequently concentrated in this sludge and leachable concentrations may be particle size dependent. Such wastes still consist mostly of sandy material, but might also become classified as hazardous wastes.

Foundries overall have a high rate of spent sand reuse on-site, with about 93% of these sands being recycled. However, the remaining 7% of these waste sands which require disposal accounts for 7 to 8 million tons of waste generated annually in the United States alone [2]. The pollution prevention priorities of

450

source reduction and on-site recycle are not sufficient to fully address the issues of foundry waste sands generation.

Off-site beneficial reuse of these sands - both with and without reclamation treatment - are necessary to reduce the large quantities of these wastes being sent to disposal sites. Such reuses are generally limited by the environmental safety of the material and its application. A cost-effective reclamation treatment system could significantly reduce the amounts of both hazardous and non-toxic waste sands currently restricted to expensive disposal alternatives. Beneficial reuse will also be linked to normal market factors such as product quality, dependability and availability [3].

## REGULATORY ISSUES

The disposal requirements of these wastes and associated costs are very highly dependent upon the legal classification of the waste. Both federal and state laws play major roles in the appropriate management of these wastes.

Although foundry waste sands are not a RCRA listed waste, they may fall under RCRA regulations due to classification as a characteristic hazardous waste. As described in 40 CFR 261, foundry wastes might be considered hazardous due to reactivity (sulfides or cyanides content) or toxicity (TCLP test) [4]. The leachable heavy metals as measured by the TCLP criteria are most often the cause for a foundry waste to be classified hazardous.

If the characteristic hazardous waste criteria is not exceeded, then the waste becomes a non-hazardous exempt waste according to RCRA. Since the majority of these sands fall into that category, the waste is primarily regulated by state authorities as a non- toxic residual waste.

State laws, regulations and policies plus local ordinances generally dictate the feasibility of beneficial reuse of these materials [3, 5]. There is considerable variation in approval requirements for reclamation and beneficial reuse of foundry wastes between states and locales. For this reason, it is very difficult to effectively address this issue on a national basis. However, beneficial reuse of these materials can be expected to expand in scope and applications as the regulatory constraints, technical alternatives and cost factors become better defined.

## BENEFICIAL REUSE ALTERNATIVES

The regulatory constraints regarding the environmental safety of the reclaimed sand must be satisfied. However, the market constraints of material performance, cost (delivered) and availability will most often limit actual reuse applications. The market demand for the sands in the more severe waste categories is well below the supply - foundries today still need to pay to have these sands reused. Expanding the market demand - extending into other beneficial reuse categories - may be necessary to achieve a significant reduction

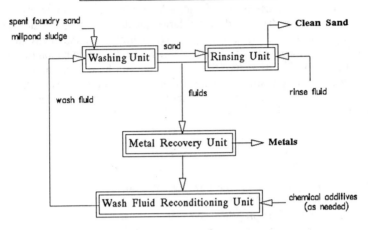

Figure 1. A Waste Reclamation System Model

in the amount of sands sent to disposal sites. The great majority of the spent sands and millpond sludges are currently shipped to disposal sites. Decreasing this disposal load to a significant degree will need an effective marketing strategy, and may require a reclamation treatment process to satisfy the environmental quality criteria of these additional beneficial reuses [2, 3].

## A RECLAMATION SYSTEM MODEL

The proposed reclamation system model (Figure 1) was developed with the intent of being a production unit rather than a treatment unit. The components would be separated into products which would have value, without generating a new waste stream.

The process would involve a type of soil washing/metal extraction process. Such processes are being investigated to determine the optimal metal/sand separation process. A weak acid solution, a complexing agent (Ethylenediaminetetraacetate - EDTA) and a solvent (toluene mixed with di-(2-ethyl-hexyl) phosphoric acid - HDEHP) extraction process have shown the potential to accomplish the separation in a cost- effective system.

These processes would be followed by a rinse unit, which would need to be compatable with the wash fluid selected.

Reclamation of the metals from the wash/ extraction fluids, would allow for the beneficial reuse of the metals and also the reconditioning and reuse of the wash and rinse fluids.

## Spent Sand Washing Using EDTA

Samples of spent foundry sands from a grey iron foundry were studied to determine the feasibility of an EDTA soil washing process to separate lead and chromium from the sand [1]. The equilibrium capacities of several solutions was compared to the standard TCLP test and the acid digestion test (EPA Method 3050) results in batch experiments. The kinetics of the extraction were also studied using sand columns.

Batch experiments were conducted comparing the extraction efficiency of distilled water, a solution of 5% nitric acid, and a range of EDTA concentrations (0.05, 0.10, 0.15, and 0.20 molar) in water (Figure 2). After 24 hours of agitation, the 0.10 M and higher EDTA solutions were effective compared to the acid digestion test for metals. It was noted that there was significant difference in the behavior of the lead and the chromium extractions by the various fluids, but the EDTA was potentially useful for capturing both metals.

The 0.10 M EDTA solution was selected as the column study wash fluid due to the batch test separation efficiency. This solution was compared to the nitric acid and the distilled water fluids using a 3-day column test (Figure 3). Each column contained 2700 grams of spent sand and was washed with 4 liters of

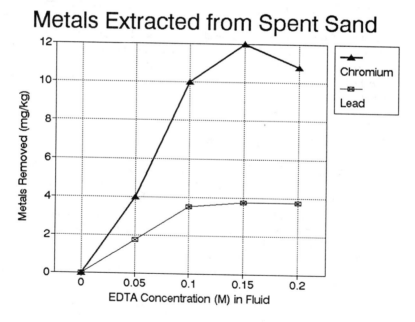

Figure 2. Batch Experiments of Metal Extraction Fluids

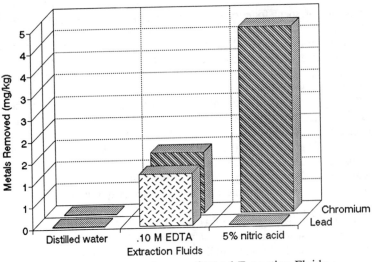

Figure 3. Column Experiments of Metal Extraction Fluids

fluid. As expected, the distilled water did not leach a measurable quantity of either metal. The nitric acid was most effective at removing the chromium, but was completely ineffective at removing lead. The EDTA providing a reasonable extraction of both chromium and lead from the sand. The very large difference between the chromium and lead removals by the nitric acid in one column suggests that the process was not necessarily limited by hydraulic factors. A fixed bed or column of sand might be effectively treated by an appropriate extraction fluid.

**Millpond Sludge Washing Using EDTA**

A concentration of 0.05M EDTA was selected as a wash fluid to separate the metals from dry millpond sludge. These experiments were run as a batch process with 6 hours of agitation [6]. The wash fluid was then collected and analyzed for lead, cadmium and chromium. The metal extraction results were then compared to the standard total metal digestion and TCLP tests performed on split samples of the sludge (Figure 4). The results indicate that the EDTA wash fluid extracted more metals from the sludge than the TCLP, but generally less than the total metal digestion results. The fluid was most effective at separating lead from the sludge.

The washed sludge was also rinsed with water and subjected to the TCLP test. The results were compared to both RCRA hazardous waste criteria and to

# Metals Extracted from Millpond Sludge

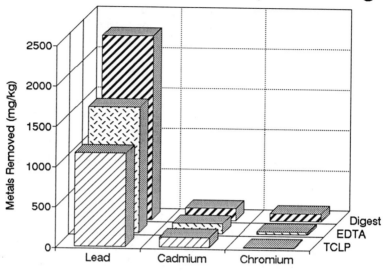

Figure 4. Comparison of Batch Process Extractions of Metals
from Millpond Sludge

# EDTA Treatment Effects on TCLP Results

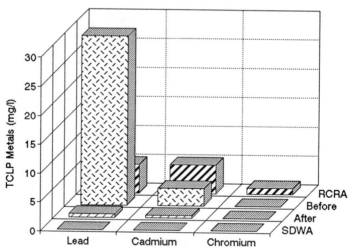

Figure 5. Effectiveness of EDTA Batch Extraction Process on Millpond Sludge

Safe Drinking Water Standards (Figure 5). The single cycle washing process was sufficient to convert the millpond sludge to non-hazardous waste. However, the results suggest that multiple wash cycles would be required to reduce leachate levels to Safe Drinking Water Standards, which would be required to allow for maximum beneficial reuse potential [5].

## Millpond Sludge Washing by Solvent Extraction

Toluene mixed with di-(2-ethyl-hexyl) phosphoric acid (HDEHP) was evaluated as a potential washing fluid to remove lead from millpond sludge [7]. The removal performance was compared to both TCLP and acid digestion analysis of the untreated sludge. Factors tested included the effects of HDEHP concentration, extraction (batch) contact time, and the fluid-to-sample ratio (FSR).

Results indicated that contact time was not a factor. The optimal treatment time is likely to be less than 30 minutes (the minimum time tested). The concentration of HDEHP and the FSR both had a very significant influence on the results (Figure 6). The optimal HDEHP concentration for any application is likely to be a function of estimated metal content of the untreated sludge. The optimal range for the sludge analyzed was 0.05 to 0.1 moles per liter of toluene. An FSR of 20 ml fluid per gram dry sludge was also found to be most effective.

An extraction using 0.05 mol/l of HDEHP and an FSR of 20 ml/g removed three fifths of the measured total lead in the sample and 125% of the measured TCLP lead. This suggests that a toluene-HDEHP extraction process has the potential to effectively reclaim metals from the millpond sludge.

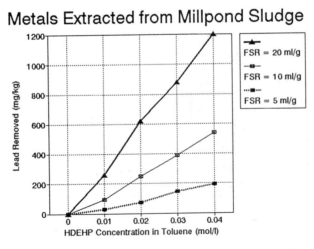

Figure 6. Batch Solvent Extraction of Lead from Millpond Sludge

## Soluble Metal Reclamation Using Onion Skins

Two types of onion skins (red and white) have been studied to determine the feasibility of soluble metal reclamation from aqueous wash fluids [8]. Acidic wash water containing a soluble concentration range of lead (0.25 to 50 mg/l), chromium (0.5 to 400 mg/l) and cadmium (0.001 to 40 mg/l) have been evaluated in a batch process. Results indicate that the removal rate correlates with the initial concentration of the metals in the fluid suggesting an equilibrium capacity relationship between the onion skins and soluble metal concentration (Figure 7).

The effects of pH were also studied (Figure 8). Although the optimal removal rates were obtained at pH8, very significatn removal rates (64 to 81%) were obtained for all three metals at pH2.

Recovery of the metals might require thermal destruction of the used onion skins. Such a process would also provide a waste reuse option for the onion processing industry.

## Ongoing and Future Work

A considerable amount of work is still needed to develop a feasible reclamation system. The washing fluids and processes must be studied for application to a wider variety of materials (bronze and brass spent sands plus

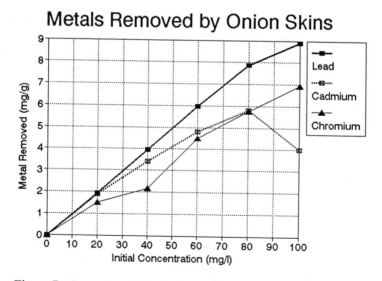

Figure 7. Recovery of Metals from a Weak Acid Extraction Fluid
by Batch Process

# pH Effects on Onion Skin Removal Rates

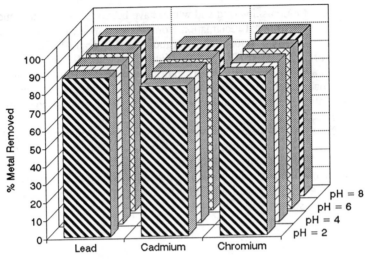

Figure 8. The Effects of pH on Metal Recovery

additional millpond sludges). The key to developing a reliable cost-effective reclamation system will likely involve the recovery of the soluble metals and reuse of the extraction fluid. Some of these studies are currently underway.

The EDTA wash fluid is being studied to evaluate the effects of multiple wash/rinse cycles and to find an effective metal recovery process. A modified catalytic electro-deposition process will be tested for metal recovery. The treated EDTA wash fluid will also be evaluated to determine necessary reconditioning and reuse criteria.

The toluene-HDEHP extraction fluid process is also under investigation, including the rinsing requirements of the sand material and the possible use of a modified ion exchange material for metal recovery from the fluid. Reuse of the treated fluid will also be studied.

The inter-dependency of the various units of the model reclamation system dictates that future experiments be designed based upon results not yet obtained.

**SUMMARY**

Significantly reducing the large quantities of foundry spent sands and millpond sludges which currently require disposal will require a combination of reasonable beneficial reuse regulations, cost- effective reclamation of the more highly contaminated materials, and an effective marketing strategy. Progress is being made in the development of both regulatory policies and marketing strategies.

A soil washing process is likely to be an integral part of a cost-effective reclamation system. Experimental results to date suggest that such a system is feasible, but further investigation is still needed.

## REFERENCES

1. Khare, J.S. and W.M. Kocher, 1994. "Extraction of Lead and Chromium from Spent foundry Sands Using EDTA," *Proceedings of the Air & Waste Management Association's Annual Meeting and Exhibition*, Cincinnati, Ohio, June 19 - 24.

2. Leidel, D.S., 1994. "Reduce, Reuse, Reclaim - What's the Scope," *Proceedings of the International Foundry Sand Reduce, Reuse, Reclaim Conference*, American Foundrymen's Society, Inc., Novi, Michigan, October 14 - 15.

3. Murray, R., 1994. "Paving the Way for Beneficial Reuse of Non-Toxic Foundry Sand," *Proceedings of the International Foundry Sand Reduce, Reuse, Reclaim Conference*, American Foundrymen's Society, Inc., Novi, Michigan, October 14 -15.

4. United States Environmental Protection Agency "Toxicity Characteristic Leaching Procedure," Code of Federal Regulations, Title 40, Part 261.

5. Ohio Policy No. 0400.007, November 7, 1994. "Beneficial Use of Nontoxic Bottom Ash, Fly Ash and Spent Foundry Sand and Other Exempt Wastes."

6. Armstrong, K. "Use of EDTA for the Extraction of Metals from Contaminated Millpond Sludge." Special Project, Cleveland State University.

7. Shah, D.B., A. Phadke, and W.M. Kocher, 1995. "Lead Removal from Foundry Wastes by Solvent Extraction," *Journal of the Air & Waste Management Association*, Volume 45, March.

8. Santapur, V., 1994. "Removal of Heavy Metals from Aqueous Solutions by Onion Skins," M.S.C.E. Thesis, Cleveland State University.

# LEACHABILITY OF ORGANIC COMPOUNDS FROM SODIUM SILICATE GROUTS CONTAINING ORGANIC REAGENTS

**J. MICHAEL MALONE**
**MORTON A. BARLAZ**
**ROY H. BORDEN**
North Carolina State University
Department of Civil Engineering
Box 7908
Raleigh, NC   27695-7908

## INTRODUCTION

The placement of chemical grouts is a proven technique for providing increased strength and reduced permeability of primarily granular soil deposits. Recently, concern has been expressed over the potential environmental impact of the chemicals involved in the chemical grouting process. Of the many types of chemical grouts available, sodium silicate based grouts are implemented in over 90% of the applications.

The ingredients of sodium silicate chemical grouts include aqueous sodium silicate, water and a neutralization reagent. Several types of reagents exist for sodium silicate grouts, including a variety of organic compounds. Organic reagents are widely used as they allow for greater control over the placement of the grout. More than 6 million liters of sodium silicate based grouts utilizing over 0.3 million liters of organic reagent were injected into the ground in 1993. Based on their present and former widespread field application, three organic reagents have been chosen for evaluation of organic compound leachability. The leachability of the organic reaction products has been characterized using methods similar to batch sorption/desorption tests. Total organic carbon analysis has been used to identify organic compunds in the aqueous phase leaching solution.

## SODIUM SILICATE GROUT CHEMISTRY

The chemical reaction of sodium silicate with three organic reagents, a dibasic ester combination known as DBE, ethyl acetate and formamide, is described. As detailed by equation 1, the equilibrium state of the sodium silicate solution is quite basic. The sodium silicate aqueous solution is in equilibrium with sodium hydroxide and silica at approximately pH 11.

$$SiO_2 * Na_2O + H_2O \Leftrightarrow 2NaOH + SiO_2 \qquad (1)$$
$$\text{sodium silicate} + \text{water} \Leftrightarrow \text{sodium hydroxide} + \text{silica}$$

Addition of the organic reagent causes a rate controlled reduction in the pH of sodium silicate. As the pH lowers, a polymer gel of silica acid forms. The gel binds soil grains together imparting a cohesive property to the overall soil-grout matrix. The reagent functions as a mechanism for pH reduction and does not appear to be a significant chemical constituent of the final gel [1]. Therefore, the final reaction products of the reagent are resident within the pore structure of the soil-grout matrix. For each reagent type, the detailed reaction chemistry along with description of reaction products is presented.

## DBE

The dibasic ester reagent (DBE) is composed of three organic dimethyl esters, dimethyl adipate, dimethyl glutarate, and dimethyl succinate. The esters of DBE serve as the principal reagent for several commercial based reagent types. Continued widespread placement of grouts using this ester combination is apparent. The percent of each dimethyl ester in the DBE reagent along with other relevant chemical data are presented in Table I.

In an irreversible base-promoted ester hydrolysis (saponification) the NaOH base is consumed and the ester is hydrolyzed to its dicarboxylate anion, in this case a sodium dicarboxylate (equation 2). Dimethyl succinate and sodium succinate, dimethyl glutarate and sodium glutarate, and dimethyl adipate and sodium adipate are represented by $n=2$, 3, and 4 respectively in equation 2.

$$H_3COOC(CH_2)_nCOOCH_3 + 2NaOH \Rightarrow NaOOC(CH_2)_nCOONa + 2CH_3OH \qquad (2)$$
$$\text{dibasic ester} + \text{sodium hydroxide} \qquad \text{sodium dicarboxylate} + 2\text{ methanol}$$
$$(C_{n+4}H_{2n+6}O_4) \qquad\qquad (Na_2C_{n+2}H_{2n}O_4) \qquad (CH_3OH)$$

TABLE I-- Components of DBE Reagent

| Compound | % of DBE (% wt.) | C.A.S. #[1] | Chemical Formula | M.W.[2] (g/mol) | Density[3] (g/ml) |
|---|---|---|---|---|---|
| dimethyl glutarate | 66 | 1119-40-0 | $C_7H_{12}O_4$ | 160 | 1.087 |
| dimethyl adipate | 17 | 627-93-0 | $C_8H_{14}O_4$ | 174.2 | 1.063 |
| dimethyl succinate | 16.5 | 106-65-0 | $C_6H_{10}O_4$ | 146 | 1.117 |
| Total | 99.5 | | | 159.3 | 1.09 |

[1]C.A.S. # - Chemical Abstracts Service Number
[2]M.W. - Molecular Weight
3 [2].

# ETHYL ACETATE

The result of a French patent in the late 1950's, ethyl acetate has been used extensively in the past for sodium silicate grouting [3]. Ethyl acetate, a monoester, reacts similarly to the dibasic esters with respect to the mechanism by which the NaOH is consumed. The reaction products of ethyl acetate saponification include sodium acetate and ethanol (equation 3). Note that the reaction of the dibasic esters neutralizes two moles of NaOH per mole of reagent (equation 2) whereas ethyl acetate neutralizes only one mole of base.

$$H_3CCOOCH_2CH_3 + NaOH \Rightarrow CH_3COONa + CH_3CH_2OH \qquad (3)$$

ethyl acetate + sodium hydroxide $\qquad$ sodium acetate + ethanol

$(C_4H_8O_2)$ $\qquad\qquad\qquad\qquad$ $(NaC_2H_3O_2)$ $\quad$ $(CH_3CH_2OH)$

# FORMAMIDE

The use of formamide, often in conjunction with ethyl acetate, as a reagent for sodium silicate grouts was widespread in the 1960's and 1970's. Labeled as a possible carcinogen, formamide use has dwindled in recent years with the advent of organic ester reagents [3]. However, the large quantities of formamide used previously warrant an investigation into their behavior and chemistry. Similar to DBE and ethyl acetate, formamide functions as a neutralizer of the NaOH base material. The products of the formamide reaction include sodium formate and ammonia (equation 4). Table II lists the properties of both the ethyl acetate and formamide reagent.

$$HCONH_2 + NaOH \Rightarrow HCOONa + NH_3 \qquad (4)$$

formamide + sodium hydroxide $\qquad$ sodium formate + ammonia

$(CNH_3O)$ $\qquad\qquad\qquad\qquad$ $(NaCHO_2)$

Consumption of the base is the mechanism which results in silica gelation. Sodium silicate is an aqueous solution having a pH of approximately 11. As the NaOH is consumed by ester hydrolysis, the pH is reduced to below approximately 10.5. At approximately pH 10.5, the silica is no longer soluble

TABLE II--Properties of Ethyl Acetate and Formamide Reagent

| Compound | C.A.S. # | Chemical Formula | M.W. (g/mol) | Density[1] (g/ml) |
|---|---|---|---|---|
| ethyl acetate | 141-78-6 | $C_4H_8O_2$ | 88.11 | 0.894 |
| formamide | 75-12-7 | $CNH_3O$ | 45.04 | 1.128 |

[1][2].

and forms polymeric groups of silicic acid. The gelation mechanism is identical for the three reagents and is detailed by equation 5a-c.

As the pH of the system lowers, the silica in solution ($SiO_2$) forms silicic acid ($SiOH)_4$ (Eq. 5a). The silicic acid condenses with itself to form the dimer of silicic acid (Eq. 5b). The dimer condenses again with silicic acid and forms a trimer of silicic acid (Eq. 5c). This pattern continues resulting in larger silicic acid polymers. These polymers continue to grow, eventually forming polymer aggregates and then polymer spheres. The polymer spheres grow until the hydroxyl groups attached to the surface of the silicon atoms of adjacent spheres condense with one another with the elimination of water. This condensation causes the polymer spheres to adhere to one another and at this point the solution begins to gel [4].

$$\textbf{a) } H_2O + SiO_2 \Rightarrow Si(OH)_4 \qquad (5)$$
$$\textbf{b) } Si(OH)_4 + Si(OH)_4 \Rightarrow Si(OH)_3OSi(OH)_3 + H_2O$$
$$\textbf{c) } Si(OH)_3OSi(OH)_3 + Si(OH)_4 \Rightarrow Si(OH)_3OSi(OH)_3OSi(OH)_3 + H_2O$$

The resulting gel binds the soil particles together thereby stabilizing the soil mass. The properties of the final gel are such that a cohesive property is introduced to the soil-grout matrix. In this manner, strength is improved, however, there is little change in the frictional response of the soil deposit.

The role of the reaction products in the overall gel is not completely clear. It appears that the reaction products are not chemical constituents of the final gel. This suggests that the organic compounds could be released to the environment along with the water from the polymerization of the silicic acid. Table III lists the organic reaction products for each reagent type.

TABLE III-- Properties of Organic Reagent Reaction Products

| Reagent Compound | | Reaction Product | Solubility[1] (g/l) | pKa1[1] | pKa2[1] |
|---|---|---|---|---|---|
| dimethyl glutarate | → | glutarate$^{2-}$ | 640 | 3.77 | 6.08 |
| dimethyl adipate | → | adipate$^{2-}$ | 15 | 4.41 | 5.41 |
| dimethyl succinate | → | succinate$^{2-}$ | 68 | 4.21 | 5.64 |
| | + | methanol | total | n/a | n/a |
| ethyl acetate | → | acetate$^{1-}$ | total | 4.74 | n/a |
| | + | ethanol | total | n/a | n/a |
| formamide | → | formate$^{1-}$ | total | 3.75 | n/a |

[1] solubility for protonated acid form [5,6]

## EXPERIMENTAL DESIGN

The use of organic reagents in sodium silicate grouts has been described above. The three organic reagents, DBE, ethyl acetate and formamide were selected for evaluation of TOC release. For each of the reagents, organic carbon is evident in the structure of the reaction products with the exception of the ammonia product from formamide. Therefore, total organic carbon (TOC) can be used to identify these products in aqueous phase leachate. Using the TOC analysis, the total mass of carbon leached can be calculated and compared with the total mass of organic carbon input as part of the mix. This computation allows for a quantitative evaluation of the fraction of organic material leached. Table IV lists the mass of organic compounds and the mass of organic carbon present in each of the mix designs of the experimental program. Each mix contained 50% N-grade sodium silicate. The column labeled "N %" refers to the base (NaOH) neutralization capacity of the mix design.

Specimens containing DBE and ethyl acetate reagent achieved gellation between 20 minutes to one hour after mixing of components. The formamide reagent specimens required the addition of calcium chloride to accelerate gellation. In the absence of calcium chloride, the specimens gelled only after a period of several days. The calcium chloride shortened the gel time to less than 24 hours. As calcium chloride ($CaCl_2$) is an inorganic chemical, no contribution to the total organic carbon results from its addition.


## EXPERIMENTAL METHODS

### SPECIMEN PREPARATION

All specimens were grouted using the "puddling method". The puddling technique has been shown to be repeatable and will ensure a more uniform

TABLE IV --Organic Content of Grouted Specimen Formulations

| Reagent Type | Reagent (% by volume) | N (%) | Organic Matter Mass (grams) | Organic Carbon Mass (grams) |
|---|---|---|---|---|
| DBE | 4 | 27 | 10.8 | 5.7 |
| DBE | 8 | 54 | 21.6 | 11.4 |
| Ethyl Acetate | 6 | 31 | 13.4 | 7.3 |
| Ethyl Acetate | 10 | 51 | 22.4 | 12.2 |
| Formamide | 2 | 25 | 5.6 | 1.5 |
| Formamide | 4 | 51 | 11.2 | 3.0 |

distribution of grout ingredients relative to grout injection methods [7]. The technique consists of pouring sand into a 3"x6" cylindrical mold containing the premixed grout components including the sodium silicate, reagent and water. The sand settles into the grout mix and is effectively saturated with the grout. The grout eventually reacts, binding the sand particles together into a stable mass which can be removed from the mold and tested. To prepare a 50% sodium silicate and 8% reagent mix, 1100 grams of dry Ottawa 20-30 standard sand are first poured into a graduated cylinder. Next, 125 ml of N-grade sodium silicate is mixed with 105 ml of tap water in a blender for two minutes. Using a graduated burette, 20 ml of reagent is then added to the sodium silicate-water solution and mixed for five minutes by closing the open end of a graduated cylinder with a stopper plug followed by shaking. Approximately 42 ml of the grout mix is then poured into an empty mold and one-sixth of the sand (183 grams) is also added. This procedure is repeated 5 more times creating the fully grouted specimen. The grout is uniformly distributed in each layer by tapping the mold on the table ten times. After hardening overnight, the specimen is removed from the mold. Sansalone has shown that the densities of puddled specimens are comparable to those of injected specimens [7].

## LEACHING TEST

Leachate from the grouted specimens was extracted using batch sorption/desorption type tests. Following the overnight hardening period, the grouted specimens were placed in 4 liter plastic containers. Each specimen was then pulverized in the open atmosphere using a standard Proctor type hammer until no visible clumps of grout remained. The container was then filled with approximately 3700 ml of deionized water. Due to the high pH of the grout leachate, typically near pH 10, biodegradation of the organics within the leaching cells was considered not to influence organic concentrations.

Other methods designed for the leaching of solid wastes, including the Toxicity Characteristic Leaching Procedure (TCLP) and the Extraction Procedure Toxicity Test (EP Tox) [8] were considered for their applicability to grouted soil specimens. These tests both use an acid buffered leaching media designed to simulate leaching present in landfills. The leachate is then evaluated and compared to an EPA list for the maximum allowable concentration of contaminants for the particular leaching procedure. The lists for TCLP and EP Tox tests do not include any chemicals or by-products of chemicals frequently used in sodium silicate grout. In addition, the acidic conditions of the TCLP and EP Tox tests would influence the gellation of the silica. Therefore, the TCLP and EP Tox tests were not chosen as methods for generating a leachate from a sodium silicate grouted soil mass.

## SAMPLING AND TESTING

The crushed specimen samples were collected by lowering a graduated pipette attached to a dispensing bulb into the leaching solution. In many cases, sampling was performed in duplicate. Samples were taken at time intervals of 1, 10 and 100 hours.

The presence of organics in a liquid grout leachate sample was detected using a Dohrmann DC-190 organic carbon analyzer. Through catalytic oxidation, the DC-190 completely oxidizes organic matter to carbon dioxide and water. After separating the water from the off gas stream, the $CO_2$ is measured using a non-dispersive infrared detector. This TOC procedure can detect organic carbon at levels as low as 1 mg/l (1 ppm). Analysis of the total organic carbon in the aqueous phase leaching solution consisted of sampling approximately 10 to 20 ml of grout leachate and injecting 30ul into the total organic carbon analyzer. In many cases, the samples required dilution to meet the detection range of the TOC analyzer. Prior to testing in the TOC analyzer, all samples were filtered with a 0.45um Gelman filter. Replicate injections were made until variation between successive readings was less than 5%.


## LEACHING TEST RESULTS

The results of the leaching tests indicate significant release of organic compounds from specimens regardless of reagent type. Figure 1 compares the measured TOC versus leaching time for a limited number of tests on each of the mix designs.

As evident from Figure 1-a, the release of organics from the DBE reagent specimens was more dependent on initial reagent content than on time. The 8% reagent specimens leached approximately twice the TOC of the 4% specimens. Also apparent from Figure 1-a, the majority of the organic carbon release occured by the first sample interval at 1 hour. Similar results for the ethyl acetate and formamide reagents are displayed in Figures 1-c and 1-e, respectively. Like the DBE specimens, initial reagent content influenced the release of organics more significantly than time.

Figures 1-b, 1-d and 1-f show the normalized TOC results for specimens containing DBE, ethyl acetate and formamide, respectively. Measured TOC values are normalized by the computed initial TOC contained in each specimen as shown in Table IV. The release of organics from the DBE specimens ranged from 70 to 95 percent of the initial TOC. This range was consistent for each of the tested specimens and no particular formulation released a greater or less percentage of organics. The ethyl acetate specimens released between 60 and 80 percent of the organics input as reagent. Specimens containing formamide reagent resulted in the release of between 80 and 100% of the available carbon.

466

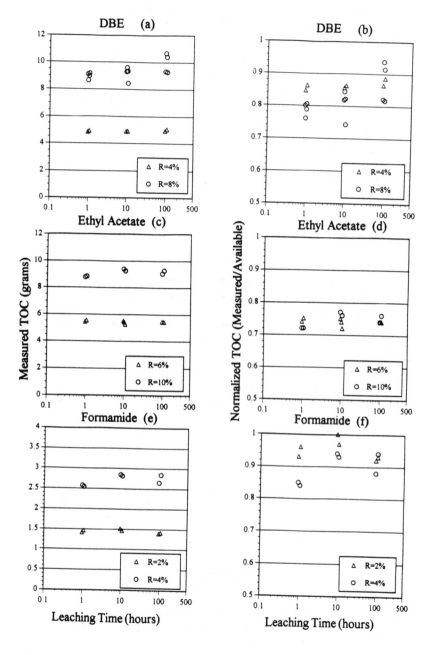

Figure 1 -- Measured TOC and Normalized TOC versus Leaching Time for
DBE (a,b), Ethyl Acetate (c,d) and Formamide (e,f)

Comparing the three reagent types, it can be said that the formamide specimens released approximately 10% more of the available organics than did the DBE, which in turn released approximately 10% more organics than did the ethyl acetate specimens. It is interesting to note that all carbon from the formamide reaction is represented by one organic by-product compound, sodium formate. Therefore, 100% of the carbon is contained in this compound. Whereas, the DBE and ethyl acetate reactions result in two organic compounds, one of which is an alcohol. The DBE specimens have 71% of the available organics in the sodium dicarboxylate form and 29% in the form of methanol. The ethyl acetate specimens have 50% of the carbon in the form of sodium acetate and 50% in the form of ethanol. Based on the test results, it appears that formulations resulting in greater alcohol production will release less organics than those producing no organic alcohol product.

## CONCLUSIONS

1. Reaction products from sodium silicate grouts using organic reagents do not appear to be chemically bound within the gelled grout.
2. In excess of 70% of the organic matter contained in the reagent was leached from all specimens.
3. Batch desoprtion type leaching tests are an effective method for quantifying the release of organics from grouted specimens within relatively short periods of time (1 hour).
4. Release of organics appears to be influenced by the chemistry of the reaction products. Among the three reagents tested in this preliminary study, the reagent not producing an alcohol reaction product leached more of the overall organic material.

## REFERENCES

[1] Malone, J. M., Barlaz, M. A., and R. H. Borden. 1995. "Methods to Evaluate the Environmental Impact of Sodium Silicate Chemical Grouts", Proc. of ASCE Specialty Conf. Geoenvironment 2000, ASCE, New Orleans, Feb. 24-26, 1995.

[2] Aldrich. 1991. Catalog Handbook of Fine Chemicals, Aldrich Chemical Company, Inc., Milwaukee, Wisconsin.

[3] Karol, R.H. 1982. "Chemical Grouts and their Properties", Proc. of Conf. on Grouting in Geotechnical Engineering, ASCE, New Orleans, Feb. 10-12, pp. 359-377.

[4] Scott, Raymond P.W. 1993. <u>Silica Gel and Bonded Phases; Their Production, Properties and Use in LC</u>, John Wiley and Sons, New York.

[5] Montgomery, John H. 1991. <u>Groundwater Chemicals Desk Reference</u>, Volume 2, Lewis Publishers Inc., Chelsea, Michigan.

[6] Verscheuren, Karel. 1993. <u>Handbook of Environmental Data on Organic Chemicals</u>, Second Edition, Van Nostrand Reinhold Company, New York.

[7] Sansalone, John J. 1992. "Unconfined Compressive Strength of Chemically Grouted Laboratory Sand Specimens as a Function of Particle Size and Injection Velocity", Thesis Submitted to North Carolina State University, pp. 34.

[8] EPA. 1989. "Stabilization/Solidification of CERCLA and RCRA Wastes, Physical Tests, Chemical Testing Procedures, Technology Screening and Field Activities," EPA/625/6-89/022, pp. 5.1-5.12.

# REMEDIATION AND REUSE OF CHROMIUM CONTAMINATED SOILS THROUGH COLD TOP EX-SITU VITRIFICATION

JAY MEEGODA, B. LIBRIZZI, G. F. MCKENNA, W. KAMOLPORNWIJIT
AND D. COHEN
New Jersey Institute of Technology
Newark, NJ 07102

DAVID A. VACCARI, S. EZELDIN AND L. WALDEN
Stevens Institute of Technology
Hoboken, NJ 07030,

BARRY A. NOVAL
Energy Products Enrichment, Inc.
Suite 903, One Montgomery Plaza
Norristown, PA 19401,

ROBERT. T. MUELLER AND S. SANTORA
New Jersey Department of Environmental protection
Trenton, NJ 08625

## INTRODUCTION

The origin of the chromite ore process residues (chromate chemical production waste) in the Jersey City area, the magnitude and extent of contamination, and the adverse health effects of chromium are discussed in depth in a recent publication [1]. Chromium is a metallic element that occurs in nature primarily in the trivalent form as the mineral chromite. However, most of the hexavalent chromium in the environment arises from processing of chromium. Health effects of the two common oxidation states of chromium trivalent and hexavalent are fundamentally different so they should be considered separately. The chromium compounds that are the most toxic and generally raise the greatest health concern are chromates (hexavalent chromium). The USEPA classifies hexavalent chromium as a Group A known human carcinogen. However, both hexavalent and trivalent forms of chromium are associated with allergic contact dermatitis.

470

The chromate chemical production waste has originated from the use of chromate waste obtained from local chromite ore processing industries. Over two million tons of chromate chemical production waste has been distributed throughout Hudson county. The chromate chemical production waste discharged at the residential, commercial, and industrial sites is a by-product of approximately seventy years, from 1905 to 1971, of chromate chemical production at three facilities located in Hudson county. The chromate chemical waste from the above facilities was distributed as fill material for use in construction and development projects at residential, commercial, industrial and recreational areas throughout Hudson county. The chromate production waste was also used for the backfilling of demolition sites, preparation for building foundations, construction of tank berms, roadway construction, the filling of wetlands and other construction and development related purposes. Chromate contamination has been found among other places, on the walls and floors of buildings, both interior and exterior, on the surfaces of driveways and parking lots and in the surface and subsurface of unpaved areas throughout Hudson county. These sites include residences, active work sites, public lands, commercial establishments and other populated areas of Hudson county. The New Jersey Department of Environmental Protection has identified over 150 areas in Hudson county where chromate chemical waste has been discharged and is present.

Chromium contamination in the range of a few PPM to 5 percent by weight has been measured at those sites. Some of the salt precipitates contain nearly 100% hexavalent chromium. In addition, the water solubility of hexavalent chromium compounds poses a potential threat to groundwater, and possible ingestion of particulate from contaminated soil poses a potential health threat in surrounding communities.

A solution to the problem is sought that is consistent with the requirements of SARA and the NJDEP technical rule for cost-effective, volume and toxicity reduction and permanent remediation methods. Several technologies such as vitrification, soil washing, bioextraction, stabilization/solidification and chemical extraction were considered by the investigators. Vitrification was considered to be the potential remedy of choice. This was based on its economically competitiveness, ability to convert hexavalent chromium to trivalent chromium that maintains a condition of lower toxicity, potential for creating end products with reuse value for highway construction, potential for volume reduction with no by-product as chromium is contained in glass, ability to distrust other trace organics and potential for applying this technology to other sites in New Jersey.

The cold-top vitrification technology with low cold top temperatures causes reduced emissions. There will be a cost saving as vitrified chromium contaminated soils will be used as sand and aggregate replacement material in both asphalt and structural concrete [2,3].

Processing Background

Vitrification has been successfully practiced in Japan and Europe for nearly 20 years with coal ash, municipal incinerator ash and wastewater treatment sludges.

This process produces environmentally safe byproduct aggregates for the construction industry, for use as road aggregate, for decorative home products, and for tennis court surfaces. The gas fired furnaces employed are usually capital intensive refractory-lined units with potential for generating high emissions. The cold top design minimizes heavy metal and volatile emissions during operation. Electric power is localized in the center of the furnace so that introduced feed material remains at low temperature, thereby acting as a barrier to evolving volatiles and particulates. A bag house and venturi scrubber are present to collect whatever particulates and acid gases, respectively, that do escape to the exhaust system.

## Nature of Glass Product

Modern glass technology has benefited from use of advanced x-ray absorption spectroscopic techniques (XAS) to obtain detailed information regarding the structural arrangement of ions in the glass structure. This includes their location in either network forming positions or in modifier sites, as well as the coordination number for surrounding oxygen atoms and the distance of closest approach.

Literature [4] shows that $CaCrO_4$, the predominant hexavalent chromium containing compound present in Jersey soil, begins to lose oxygen when heated above $800^oC$. Vitrification temperatures typically exceed $1400^oC$ so that trivalent chromium alone would be the likely species present after vitrification.

Transition metal ions similar to trivalent chromium, such as tetravalent titanium, have been shown to occupy network forming positions and to exhibit tetrahedral bonding coordination. This occurs at concentrations of less than 7 weight percentage when present in a silicate glass. Trivalent aluminum has a similar ionic radius (0.54A) to trivalent chromium (0.64A) and tetravalent titanium (0.68A) and also occupies network positions with tetrahedral coordination when present in glass Aluminum ion displays similar chemical behavior to the trivalent chromium ion in the corresponding oxide compound. Aluminum ion in silicate glass serves to raise the melting temperature and viscosity, stabilizes and strengthens the glass and lowers solubility for alkali metal components such as sodium ions. This leads to the conclusion that trivalent chromium ions present at concentrations of less than 7 weight percentage should also produce a stable glass resistant to the acid leaching of heavy metals.

The in situ soil will have a natural moisture content of between 10-35%. All soils will be dried to approximately 5 percent water to minimize steam generation during vitrification. This is important as cold top electric melting furnace virtually eliminates particulate and gaseous emissions during the vitrification step.

The mechanical strength of the resulting vitrified product is of paramount importance. Earlier studies in Japan [5] has demonstrated the importance of slow cooling during aggregate formation to produce aggregate with sufficient strength to meet the requirements of LA Abrasion test for top and substratum layers in black top roads.

472

The purpose of this study is to investigate the feasibility of using cold top ex-situ vitrification to vitrify chromium contaminated soils and to use vitrified chromium contaminated soils in highway and other construction projects. In order to ensure the highest possible strength for the aggregate produced, the molten vitrified product will be collected in graphite molds and cooled slowly from 1450°C to 1000°C to develop crystallinity. The resulting mass will be crushed to produce aggregates of particle size less than 2 inches. First the physical and chemical characteristics of the aggregate were tested. Then, the aggregate will be incorporated in structural and asphalt concrete mixes. The resulting mixes will be tested for strength, permeability, freeze-thaw durability, and heavy metal leachability.

## EXPERIMENTAL PROCEDURE AND TEST RESULTS

Ten soil samples were collected from nine chromium contaminated sites in Hudson Country, NJ. A second soil sample was taken from Diamond Shamrock site to resolve the difference inter-laboratory test results. The soil sample from the Liberty State Park site was taken from a point near a monitoring well. The area was noticeable by lack of vegetation. Nearly 5 gallons of soil was taken from the surface. Soil was brown in color with a reddish tint. The Cavern Point Road site was a marshy site with soil having the same appearance as Liberty State Park site. Both soils showed magnetic characteristics. Access to both sites was restricted by security fence. The Colony Diner site was an open access site and the sampling area was accentuated by lack of vegetation in an otherwise weeded area. The soil was dark brown in color with distinct green/yellow almost fluorescent color mixed throughout. It did not have any magnetic properties. The Diamond Shamrock site was capped with asphalt concrete and hence soil has to be obtained by boring as opposed to digging in other three cases. The soil from Diamond Shamrock site was dark brown in color without pieces of slag. That was the main difference between this soil and the other three soils. The soil from the Hackensack River Road site was obtained by boring on an access road leading to a bridge over the Hackensack River. The sampling point was 20 feet from the river. The soil was reddish brown with numerous pieces of slag. It was slightly magnetic. The Reed Mineral site is a paved industrial site off Central Avenue in Kearny. The soil was dark brown in color with black crystals resembling coal distributed throughout. The Roosevelt Drive-In site was capped with a HDPE liner. The HDPE liner was covered with several inches of Gravel. The soil was medium brown with red/yellow/green fluorescent color distributed. It had magnetic properties and numerous pieces of slag. The NJ Turnpike Bayview site was located under a bridge on the NJ Turnpike. The site was uncapped but the access to the site was restricted by a fence. The soil was reddish brown silty sand and it had magnetic characteristics. The Garfield Avenue site in Jersey City was capped with a HDPE liner and two inches of gravel. The initial screening for $Cr^{6+}$ showed low values. Therefore soil was taken from four locations and blended at the site. The second sample from

Diamond Shamrock site was taken on a rainy day and hence had a very high water content (see Table I). The date of soil sampling is indicated in the sample reference number.

All ten soil samples were stored in five gallon plastic buckets with waterproof lids. Each container had approximately 20 pounds of soil. An external chain of custody was generated to keep a track of this hazardous soil. The chain of custody was signed by the party relinquishing the custody of the sample as well as the party receiving the sample.

Hexavalent chromium analysis was performed by NJ certified laboratory within 48 hours. For this analysis 100 g of soil was hand delivered to PACE Laboratory, Edison, NJ. One pound of each soil was sent to Steven Institute of technology for the chemical analysis. One pound of each soil was crushed blended and used for physical tests at New Jersey Institute of Technology. A chain of custody form was generated for each recipient.

At NJIT following physical tests were performed; Water content (ASTM D2216), Liquid and Plastic limits (ASTM D4318), Grain Size Analysis mechanical and hydrometer (ASTM D421 & D422) and Specific Gravity (ASTM D854). Soil after specific gravity and hydrometer tests were air dried. All waste soils after physical tests were collected and stored to be used for in-situ vitrification, as they are hazardous waste.

## TABLE I- PHYSICAL SOIL TEST RESULTS

| Site | Ref. # | $G_s$ | W/C(%) | $D_{10}$(mm) | $C_u$ | $C_c$ |
|------|--------|-------|--------|--------------|-------|-------|
| Liberty State Park | LSP81794-S1 | 3.21 | 14.4 | 0.05 | 20 | 0.6 |
| Cavern Point Road | CPR81794-S2 | 3.12 | 25.4 | 0.04 | 21 | 0.8 |
| Colony Diner | CD81794-S3 | 2.72 | 29.6 | 0.06 | 10 | 1.2 |
| Diamond Shamrock | DS82594-S1 | 2.78 | 16.7 | 0.04 | 13 | 1.4 |
| Diamond Shamrock | DS120594-S1 | 2.83 | 56.2 | 0.06 | 24 | 0.7 |
| Hackensack River Rd | RR13182594-S1 | 3.02 | 14.7 | 0.15 | 21 | 0.1 |
| Reed Mineral | RM17682594-S3 | 2.90 | 6.1 | 0.08 | 25 | 1.6 |
| Roosevelt Drive-In | RDI82694-S1 | 2.76 | 22.0 | 0.11 | 7 | 1.1 |
| NJ Turnpike Bayview | TRP02082694-S2 | 2.76 | 32.5 | 0.01 | 40 | 0.0 |
| Garfield Avenue | GAR93094-S1 | 2.98 | 24.6 | 0.06 | 23 | 0.4 |

All ten soil samples were granular with zero liquid and plastic limits. The soils capped with asphalt had the lowest water content values (see Table I). The NJ Turnpike Bayview was a fine silty sand hence it had the highest water content. The large slag content resulted in lower water content value for soil from Liberty State park site. The other four soil samples had water content values between 20-30%. All nine soils had very high specific gravities indicating the presence of large quantities of heavy metals such as iron and chromium (c.f. specific gravities of typical soil vary from 2.65- 2.72). All nine soils were well graded as indicated by very high $C_u$ values.

Table II shows the hexavalent chromium concentration as determined from PACE laboratories. The hexavalent chromium concentration was determined by alkaline digestion (EPA method 3060A [6]) followed by colorimetric determination of chromium concentration (EPA method 7196A [6]). Table II also shows the chromium concentrations from TCLP, soft digestion and hydrofluoric digestion tests. The TCLP chromium concentration represents the quantity leached out from three grams crushed soil sample after it was treated with 60 ml of glacial acetic acid tumbled for 18 hours. The samples were subsequently acidified to a pH below 2 and refrigerated for two weeks before being analyzed using an Inductively Coupled Plasma (ICP) spectrometer. For the soft digestion test, one gram of soil was heated with nitric acid, hydrochloric acid and hydrogen peroxide (30%). Then the mixture was analyzed using ICP. This procedure is documented in EPA SW-846 procedure 3050-digestion of soils and sludges [6]. The soil pH was determined by measuring the mixture pH of five grams of soil and 5 ml of de-ionized water. The TCLP pH value is from the leachate from TCLP procedure after tumbling. This gives some measure of the neutralization capacity of the alkalines in soil.

Table II shows that higher the soil pH, the $Cr^{6+}$ results from the external laboratory are close to total Chromium concentration from soft digestion test. This also indicates that $Cr^{6+}$ can be easily removed from soil if the soil pH is above 11. A similar observation was made from coal tar contaminated soils in another research project of the first author. It can be observed from Table II that, if the total Chromium concentration in the soil is higher, then it had higher $Cr^{6+}$ concentration and hence higher pH.

## TABLE II- CHEMICAL TEST RESULTS OF SOILS

| Site | $Cr^{6+}$ (mg/kg) | Total Cr (mg/kg) Soft Digestion | Total Cr (mg/kg) HF Digestion | Cr (PPM) TCLP Leachate | Soil pH | TCLP pH |
|---|---|---|---|---|---|---|
| Colony Diner | 4800 | 5294 | 25573 | 68.6 | 11.5 | 7.9 |
| Roosevelt Drive-In | 4440 | 4600 | 20275 | 46.7 | 12.2 | 9.8 |
| Cavern Point Road | 29.2 | 1268 | 17738 | 5.81 | 9.0 | 6.4 |
| Liberty State Park | 1240 | 1544 | 16125 | 32.4 | 10.2 | 7.2 |
| NJ Turnpike Bayview | 29.2 | 544 | 12228 | 2.09 | 9.7 | 7.1 |
| Garfield Avenue | 246 | 1821 | 11729 | 8.67 | 9.2 | 6.9 |
| Diamond Shamrock | 61.7 | 950 | 8086 | 1.65 | 9.0 | 5.5 |
| Hackensack River Road | 19.7 | 587 | 2853 | 3.83 | 8.9 | 5.6 |
| Reed Mineral | B. D. L. | 455 | 1936 | 4.62 | 9.3 | 4.4 |

Table III shows the concentrations of important elements for vitrification process. Those concentrations were obtained from HF digestion tests. Table III shows that except for the soil from Liberty State Park site all other soils are deficient of Silica and Aluminum but high in Ferrous/Ferric ions. The high Fe concentration resulted from the original ore. It had high Cr and Fe concentrations.

## TABLE III- ELEMENT COMPOSITIONS OF SOILS FROM HF DIGESTION

| Site | Si % | Al % | Fe % | Mg % | Ca % | Na % | K % |
|------|------|------|------|------|------|------|-----|
| Colony Diner | 1.81 | 1.29 | 7.9 | 2.14 | 9.33 | 8.73 | 1.18 |
| Roosevelt Drive-In | 1.75 | 1.43 | 10.8 | 4.20 | 15.4 | 10.7 | 1.30 |
| Cavern Point Road | 0.06 | 1.44 | 14.2 | 5.52 | 4.60 | 4.51 | 2.62 |
| Liberty State Park | 12.6 | 1.44 | 13.9 | 5.18 | 9.41 | 4.2 | 2.45 |
| NJ Turnpike Bayview | 0.38 | 1.45 | 11.4 | 4.70 | 4.81 | 13.5 | 1.69 |
| Garfield Avenue | 0.77 | 1.47 | 8.27 | 4.36 | 4.89 | 4.98 | 0.68 |
| Diamond Shamrock | 0.22 | 1.36 | 35.0(?) | 1.56 | 2.85 | 6.35 | 0.85 |
| Hackensack River Road | 0.22 | 1.21 | 14.6 | 1.19 | 3.49 | 4.06 | 0.77 |
| Reed Mineral | 0.03(?) | 1.13 | 2.56 | 1.05 | 1.14 | 2.68 | 0.49 |

Vitrification

Since all ten soils were deficient of Silica (see Table III), the key ingredient for the formation of glass, it was augmented by adding sand 25% by total weight. The high pH of soils (see Table II) was another concern where high pH might prevent the reduction of $Cr^{6+}$ to $Cr^{3+}$ and the glass formation with Cr. To overcome this problem graphite 0.17% by total weight was added. To mimic the actual vitrification process where soil was dried to 5% by weight, each soil was oven dried and water 5% by total weight was added. A 2.5 pound mixture of each soil was shipped to Corning Glass Company in Corning NY.

At Corning each soil mixture was subjected to a melting temperature of 1590°C for three hours. The soil mixture was first placed inside a 600 cc silica crucible and was covered with a silica lid. The 600 cc covered silica crucible was placed inside a 1800 cc silica crucible to contain possible melt-through. After the three hour melt, samples were taken out and poured into a new 600 cc silica crucible to observe pouring characteristics and viscosity. This information is very important in actual vitrification where if the melt is highly viscous it will be difficult to remove the melt from the melting chamber. The glass was full of small bubbles but was a fluid (100 to 1000 poise). To mimic the actual vitrification process the following cooling sequence was adopted. The newly poured sample was transferred to a furnace set at 1350°C and kept for one hour. The sample was removed and poured into another crucible. At this temperature glass was lumpy and viscous. Freshly poured sample was transferred to another furnace set at 1250°C and held for one hour. Then the furnace temperature was lowered gradually to 1000°C over 80 minutes. Then the furnace was shut off to allow cooling to room temperature. Cooling from 1000°C to room temperature occurred overnight.

All four vitrified samples were dark gray in color and some areas had yellow-orange color. All four samples had a big void space at the center suggesting the difference in density of solid and liquid glass. The glass became solid at the outer most points first due to cooling and solidification progressed inward. This is causing the voids at the center. All four samples were slightly magnetic. The original soils were also slightly magnetic.

The RDI82694-S1 specimen when broken was uniform inside. The LSP81794-S1 soil mixture experienced a 6.3% weight loss. This would mainly due to the removal of moisture and carbon from the mixture. The added graphite was sufficient to create a reducing environment. The resulting low concentrations of Cr and Fe (see Table V) suggest perfect vitrification. When broken, the vitrified sample was not uniform. The CD81794-S3 soil mixture experienced a 17.2% weight loss. The vitrified sample surface was very shinny like a metal. When broken, the vitrified sample was uniform. The GAR93094-S1 soil mixture caused a melt-through to the outer crucible and could not measure the weight loss. It is estimated to be between 10% to 15%. The surface was slightly shinny. When broken, the vitrified sample was uniform. The low percentage of elements for both CD81794-S3 and GAR93094-S1 (see Table III) suggests the presence ions such as Carbonates, and Chromates. All the samples foamed very heavily during melting. The solidified samples had small voids throughout each sample. The heavy foaming and small void spaces throughout each sample were due to loss of volatiles. The vitrified sample from GAR93094-S1 site contained a small metal ball of diameter 7 cm at the bottom. This indicates some of the ferric and ferrous ions were reduced to iron by the presence of graphite.

At NJIT the following physical tests were performed on vitrified samples to determine the suitability as construction materials; Water content (ASTM D2216), Specific Gravity & Absorption (ASTM D127 & C128), Unit Weight & Porosity (ASTM C29), Grain Size Analysis mechanical (ASTM C136), LA Abrasion Test (ASTM C131 & C535), Sodium Sulfate Soundness (ASTM C88), & D422), Calorimetric Test (ASTM C40) and Friable Particle Content (ASTM C142). The test results are shown in Table VI. Please note that since laboratory vitrification did not produce sufficient quantity of the product for LA abrasion test was performed with one kilogram of material. This reduced weight was subjected to the LA abrasion test with the usual charge.

Since this modified LA abrasion test was a very severe test it produced high losses but still lower than the accepted value of 50% except the vitrified sample from Colony Diner site. The specific gravity values of all five samples were much higher than the original soil. With the addition of 25% sand, one should expect lower specific gravity values similar to those of the original soil. To understand the reason for higher $G_s$ values a small sample of soil from Roosevelt Drive-In site was heated to 500°C and performed the specific gravity test. That test resulted a 13.1% increase in $G_s$ value or a 13.1% weight loss. This value is comparable to the weight losses during vitrification. The loss of weight may be due to loss of crystalline water, $CO_2$ and $O_2$. Table IV also shows low Sodium Sulfate Soundness values indicating durable aggregates. As expected, all five samples had zero water content and very low Friable contents. The percentage absorption was low too.

## TABLE IV- PHYSICAL TEST RESULTS OF VITRIFIED SAMPLES

| Site | $G_s$ | % Absorption | Unit Wt. Kg/m$^3$ | Porosity (%) | % Porous | % Friable | % Na$_2$SO$_4$ Soundness | % Loss LA Abrasion |
|---|---|---|---|---|---|---|---|---|
| Roosevelt Drive-In | 2.98 | 0.8 | 1760 | 40.8 | 5.6 | 0.1 | 2.7 | 47.3 |
| Roosevelt Drive-In | 3.02 | 0.8 | 1758 | 41.6 | 5.3 | 0.3 | 2.3 | 47.3 |
| Liberty State Park | 3.13 | 1.0 | 1785 | 43.0 | 6.0 | 0.8 | 2.6 | 30.0 |
| Colony Diner | 2.97 | 1.3 | 1722 | 41.0 | 7.8 | 0.9 | 2.4 | 53.8 |
| Garfield Avenue | 3.00 | 1.0 | 1821 | 39.2 | 2.6 | 0.4 | 3.8 | 32.0 |

Table V shows the hexavalent chromium concentration as determined from PACE laboratories. It also shows the chromium and iron concentrations from TCLP, and soft digestion tests. The $Cr^{6+}$ concentrations as reported by PACE laboratory are below its detection limit. The Chromium concentrations in leachate from TCLP test except for the sample from Garfield site are of the order of several PPB. The total Chromium concentrations from soft digestion test are several orders of magnitude less than those of the original soil. These results indicate that the vitrification process worked well in immobilizing and bonding Chromium ions to the glass matrix. The iron concentrations from TCLP and soft digestion are given for reference as high iron concentrations were present in the original soil. Those iron concentrations are also a fraction of the iron concentrations found in original soil.

## TABLE V- CHEMICAL TEST RESULTS OF VITRIFIED SAMPLES

| Site | $Cr^{6+}$ (mg/kg) | Cr (PPM) TCLP Leachate | Total Cr (mg/kg) Soft Digestion | Fe (PPM) TCLP Leachate | Total Fe (mg/kg) Soft Digestion |
|---|---|---|---|---|---|
| Roosevelt Drive-In | B. D. L. | 0.015 | 0.049 | 7.55 | 10.1 |
| Roosevelt Drive-In | B. D. L. | 0.015 | 0.061 | 2.87 | 8.34 |
| Liberty State Park | B. D. L. | 0.041 | 76.00 | 9.18 | 44.9 |
| Colony Diner | B. D. L. | 0.012 | 32.60 | 7.41 | 27.4 |
| Garfield Avenue | B. D. L. | 0.284 | 9.650 | 2.70 | 51.5 |

## SUMMARY AND CONCLUSIONS

Ten chromium contaminated soil samples were obtained from sites in Hudson county, NJ to investigate the feasibility of vitrification and to used as construction materials. Each soil contained between 1000 PPM to 2.5% chromium with more than 20% as $Cr^{6+}$. All ten soils had high concentrations of iron but were deficient of silica and aluminum. Soil also had very high pH values and in-situ water contents. To augment the silica deficiency sand was added. To prevent detrimental effect of high pH to vitrification graphite was added to each soil. Five soils samples were vitrified in a commercial laboratory and slow cooled to room temperature. The vitrified product was free of $Cr^{6+}$ and had low digestible chromium contents. Vitrification seems to produce a stronger and durable aggregate with high specific gravity to be used as a construction material. This paper describes a research-in-progress. Results stated herein are of preliminary nature and may not accurately reflect the final results published at a later date.

## ACKNOWLEDGMENTS

Authors wish to acknowledge 1) the state of New Jersey department of environmental protection (NJDEP) for funding this research, 2) John Evenson of NJDEP, for providing test soils for tests, and 3) Thomas Tate of Geotech Development Co., King of Prussia, PA for providing valuable advice on Field vitrification. Although the research described in this paper has been funded by NJDEP, it has not been subjected to the Agency's review process for technical content, quality assurance/quality control (QA/QC) or administrative review. Therefore, it does not necessarily reflect the views of the agency. Mention of trade names or commercial products does not constitute endorsement or recommendation for use.

## REFERENCES

1. Environmental Science and Engineering Inc. 1989. "Risk Assessment for Chromium Sites in Hudson County, NJ" A report prepared to the NJDEP, April 1989.

2. Ezeldin, A. S. et al. 1992. "Stabilization and Solidification of Hydrocarbon Contaminated Soils in Concrete", *Journal of Contaminated Soils*, 1(1):61-79.

3. Meegoda, N, J. et al. 1993. "Use of Petroleum Contaminated Soils in Asphalt Concrete", Petroleum Contaminated Soils Volume V, Lewis Publisher, .

4. Clark, R. P. et al. 1979. "Thermo-analytical Investigation of CaCrO4" Thermochim Acta, 33:141-155.

5. Roy F. Weston. 1986. Report to Kawasaki Heavy Industries Ltd. on Cold Top Vitrification. Project # 2373-01-02-10.

6. EPA. 1990. "Toxicity Leaching Characteristics Procedure", Federal register, Vol. No. 261, Method 1311, EPA 530/SW-846, Cincinnati, Ohio.

# REMOVAL OF CYANIDES BY COMPLEXATION WITH FERROUS COMPOUNDS

C. Peter Varuntanya, D.Eng.Sc.
Project Manager

Walter Zabban, P.E., DEE
Technical Consultant

Chester Environmental
600 Clubhouse Drive
Moon Township, Pennsylvania 15108

## INTRODUCTION

Cyanide compounds are produced and used in metal plating, gold mining, photographic processing, and paint and chemical manufacturing. Cyanide is also generated in coke ovens and in potlining leachates from the smelting of aluminum. The cyanide from these sources eventually becomes an aqueous waste product. Ferricyanides ($[Fe(CN)_6]^{3-}$) and ferrocyanides ($[Fe(CN)_6]^{4-}$) are some of the iron cyanide complexes found in these waste streams.

Alkaline chlorination, an oxidation process with chlorine ($Cl_2$) or hypochlorite ($ClO^-$), is the most widely accepted method of cyanide treatment. However, removal of cyanide from wastewater to the extent required by the effluent limits imposed by Federal and State regulatory authorities is practically impossible, especially when the majority of the cyanide is present as an iron-cyanide complex. Chlorination at ambient temperature does not remove the iron-cyanide compounds. This form of cyanide is considered to be "unamenable" to treatment by chlorination. Furthermore, this process can convert part of the organic matter present into carcinogenic chlorinated organics.

Because of the disadvantages of the alkaline chlorination process, many alternative treatments have been investigated and utilized, including chemical oxidation by ultraviolet light (UV), hydrogen peroxide, and/or ozonated air; electrochemical oxidation; thermal hydrolysis; wet-air oxidation; ion exchange; and biological treatments. None of these processes are sufficiently effective to produce the desired results in treating certain complex cyanides, particularly ferri- and ferrocyanides.

One potential treatment method being further investigated uses ferrous ($Fe^{2+}$) compounds to react with free and complex cyanide ions and produce insoluble iron-cyanide complexes. However, sludges generated by this treatment method contain cyanide wastes which may be considered a hazardous waste by the U.S. Environmental Protection Agency (U.S. EPA). The studies reported in this paper demonstrate that ferrous ($Fe^{2+}$) precipitation can remove cyanide ions (both free and complex) to a concentration within the range of 1 to 2 mg/L.

The wastewaters utilized in these tests were collected from a coke plant facility. Synthetic cyanide solutions were used in the studies as well. Ferrous compounds used in the studies included commercial-grade ferrous sulfate, commercial-grade

480

ferrous chloride, and spent pickle liquor (containing ferrous ion). The desired effluent quality was successfully attained in the treatment of the above-mentioned wastewaters by using ferrous compounds as well as spent pickle liquor.

## BACKGROUND

Ferricyanide ($[Fe(CN)_6]^{3-}$) and ferrocyanide ($[Fe(CN)_6]^{4-}$) complex ions can bond with ferrous or ferric ions to form very insoluble complexed cyanide compounds. Iron ferrocyanides and iron ferricyanides are among not only the most stable complexes of the metallocyanides, but also the most insoluble. The stability constants for the ferrocyanide or ferricyanide ion are believed to be in excess of $10^{35}$, as compared to $10^{21}$ for the zinc cyanide complex ion and $10^{19}$ for the cadmium cyanide complex. The greater the constant, the smaller the concentration of the toxic free cyanide ($CN^-$) present under equilibrium conditions. Complex cyanides are less toxic than free cyanides.

For that reason, most regulatory agencies have chosen in the past to allow higher effluent concentration limits for total cyanide (which includes amenable cyanide and the cyanide contained in the iron cyanide complex ion) than for free cyanide (the ion from a simple cyanide compound). However, total cyanide effluent limitations imposed by regulatory agencies have been reduced considerably in the past few years.

As mentioned previously, the cyanide contained in the iron cyanide complex ion cannot be removed by alkaline chlorination at ambient temperature. One of the methods proposed for removal of complex cyanide uses ultraviolet (UV) light and/ or combinations of ozonated air and/or hydrogen peroxide (Merkel and Maziarz, 1984; Zaidi and Carey, 1984; Knorre and Griffiths, 1984; Gurol, et al., 1985; Varuntanya, et al., 1992). The UV radiation utilized is within the wavelength range of 250 and 350 nanometers (nm). The reaction time is usually in excess of 1 hour, depending on the initial concentration of total cyanide, the presence of other agents which will oxidize, the wattage on the UV light, and the design of the reactor, to name a few of the variables. The ultraviolet light accelerates the decomposition of the iron cyanide complex ion into a free cyanide ion and a simple iron salt.

Another method that has been used to destroy the iron cyanide complex is thermal hydrolysis in the alkaline range (Robuck and Luthy, 1989; Flaherty, 1992). The complex cyanide is decomposed to ammonia and a formate salt. The reaction takes place in the range of 450 to 500°F and under pressures of 600 to 850 psig. The necessary reaction time can vary between 0.5 to 2 hours, depending on a variety of factors. One problem with this treatment is that it generates ammonia, which also usually has strict effluent limits. The generation of ammonia can create a considerable problem unless facilities are available for ammonia removal in the effluent, such as a publicly owned treatment works (POTW) or a biological treatment plant operated by industry.

The complexation of cyanide with ferrous sulfate to produce insoluble iron cyanide compounds was used prior to 1945 primarily by the electroplating industry (Dodge and Reams, 1949). In this process, hydrated lime was used to maintain the pH at 10 or above. The reaction was apparently difficult to control, because the results showed a wide variation in removal efficiencies. This lack of process control is probably one of the reasons that alkaline chlorination was introduced. Alkaline chlorination almost guarantees complete removal of free cyanide as well as some of the major complex cyanide ions such as those of zinc, cadmium, copper, and silver, even though it does not remove the more stable iron cyanide complex ions (Dodge and Zabban, 1951; 1952).

An installation at which ferricyanide and ferrocyanide ions were removed on a commercial scale is discussed in a USSR article published in 1968 (Ozhiganov, et al.,

1968). Total cyanide was removed from scrubber wastewater generated by blast furnaces producing ferromanganese. The optimum dosage of ferrous sulfate used was 4 mg/L of $FeSO_4$ per mg of $CN^-$, which corresponds to 1.48 mg of $Fe^{2+}$ per mg of $CN^-$. The theoretical requirement is a ratio of 1.074 $Fe^{2+}/CN^-$ by weight to complex free cyanide into an assumed final product such as ferro-ferrocyanide $(Fe_2Fe(CN)_6)$, and a ratio of 0.72 $Fe^{2+}/CN^-$ by weight if cyanide is present as $[Fe(CN)_6]^{4-}$ (and forming the same ferro-ferrocyanide). Unfortunately, the paper did not contain information on the analytical procedure used for the determination of total cyanide. The trend of the results obtained as a function of pH, alkalinities, mg of $Fe^{2+}$ to $CN^-$ ratio is of value, even though there was no attempt to determine what percentage of the cyanide present was amenable to chlorination.

A similar study was conducted by Chester in 1972 (Zabban and Feather, 1973). The following case studies summarize the work performed by The Chester Engineers on the complexation of cyanide.

## CASE STUDY 1

In 1972, The Chester Engineers designed and operated a continuous treatment pilot plant to complex amenable cyanide with spent pickle liquor containing ferrous sulfate (Zabban and Feather, 1973). The plant, which was designed to treat 5 to 10 gpm of effluent from a flue dust thickener in a steel mill, consisted of two flash mix tanks in series; separate feed tanks to add the spent pickle liquor, caustic soda, and polymeric flocculant; and a clarifier. The pH value of the wastewater was maintained at 9 to 9.5 in the first flash mix tank by the manual addition of caustic soda solution. This pH adjustment was followed by treatment with the spent pickle liquor and polymeric flocculant, which were added to the second flash mix tank. The pH value of the effluent from the second flash mix tank varied between 8 and 9. The ratio of $Fe^{2+}/CN^-$ required for equivalent cyanide removal decreased with an increase in the cyanide concentration in the effluent. The applied $Fe^{2+}/CN^-$ weight ratio varied between 0.95 and 2.9 (Figure 1).

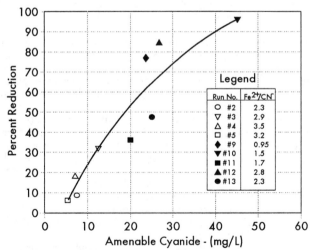

FIGURE 1 - PERCENT REDUCTION OF CYANIDE

The concentration of total solids in the sludge withdrawn from the pilot clarifier varied between 0.723 and 1.89 percent by weight, which is similar to that initially obtained from the precipitation of iron hydroxide floc. The sludge volume was in the range of 3 to 5 percent. Sludge vacuum filtration studies performed in the laboratory on a 0.0475 sq ft filter area, using Whatman No. 4 filter paper, resulted in a filtration rate which varied between 0.05 and 0.07 lb/hr sq ft. The solids in the filter cake averaged 19 percent, compared to 53 percent solids in the sludge from the flue dust thickener.

## CASE STUDY 2

In 1978, The Chester Engineers designed a cyanide complexation plant using ferrous sulfate heptahydrate ($FeSO_4 \cdot 7H_2O$) for the treatment of a 2,500 gpm wastewater flow from a proposed blast furnace blowdown recycle system (Prabhu and Helwick, 1978). The facility's National Pollutant Discharge Elimination System (NPDES) permit allowed an effluent concentration of 6 mg/L of total cyanide.

On-site laboratory studies were initially performed to precipitate cyanide by treatment with ferrous salts. The optimum pH was found to be 8.5. Flash mixing of the reagent for a few minutes, followed by two hours of quiescent settling, yielded the same results as mixing for two hours prior to settling. Most of the iron ferrocyanide settled in the first 15 minutes. Although a settling time of less than two hours was more than sufficient to meet the 6 mg/L total cyanide limit, the residual total cyanide concentration in the supernate was lowered from 1.6 mg/L after two hours to 0.48 mg/L after 18 hours of prolonged settling. Six pounds of iron (expressed as $Fe^{2+}$) as ferrous sulfate heptahydrate were added per pound of cyanide ($CN^-$). This treatment yielded less than 2.5 mg/L total cyanide in the effluent.

## DESCRIPTION OF THE INVESTIGATIONS

The wastewater was obtained from the effluent of an activated sludge treatment plant which receives liquid waste (ammonia still waste with cooling and dilution waters) from coke ovens in a steel mill. The treatment plant was designed for the removal of free cyanides, phenols, naphthalene, and benzene. Thiocyanates and ammonia are also removed, although these compounds are not regulated by the present NPDES permit for the facility. There is no evidence that ferrocyanides were removed in the activated sludge plant.

A commercially available $UV/O_3/H_2O_2$ pilot-scale system was initially used in this investigation to determine whether a physical/chemical treatment would satisfactorily remove the complex iron cyanide. The unit utilized ozonated air and hydrogen peroxide, and could be operated with or without ultraviolet light irradiation with a 254 nm wavelength. The feed to the unit was the effluent from the biological treatment plant. The flow varied from 0.5 to 10 gpm. The results (Varuntanya, et al., 1992), which are not reported here, indicated that the unit did not have sufficient capacity to remove the fluctuating, high concentrations of both cyanide and thiocyanate. Because the oxidation reaction rate of thiocyanate is much faster than that of cyanide (Layne, et al., 1984; Varuntanya, et al., 1992), the thiocyanate present in the waste stream was oxidized first. Since the removal of the organic compounds and total cyanide was of primary importance, the $UV/O_3/H_2O_2$ studies were discontinued and further investigations were initiated to examine complexation with ferrous salts.

## EXPERIMENTAL METHODS

The batch experiments were performed in the laboratory by adding a given amount of ferrous compounds. Ferrous sulfate, $FeSO_4 \cdot 7H_2O$; ferrous chloride, $FeCl_2 \cdot 4H_2O$; and spent pickle liquor containing iron (II) were utilized as a source of ferrous ion. These reagents were selected based upon their availability. All experiments were conducted with two wastewater streams from the subject coke plant facility: the bioplant effluent and the ammonia still effluent, as well as synthetic cyanide solutions.

In each experiment, the pH of the sample was adjusted to values ranging from 2.5 to 9.0. The reaction was promoted by continuous mixing with a magnetic stirrer for a pre-determined time period (30 minutes). In most cases, the solution was adjusted to pH 8.0 with sodium hydroxide and aerated with air. The aeration step was performed to oxidize unreacted ferrous ion ($Fe^{2+}$) present in the solution to ferric ion ($Fe^{3+}$), which readily precipitates as ferric hydroxide, $Fe_2O_3 \cdot nH_2O$. After the pH adjustment was completed, an anionic polymer was added and mixed with the suspension for an additional time period, up to 20 minutes. The treated solution was allowed to settle and was decanted. The sludge volume after settling was also determined.

## ANALYTICAL PROCEDURES

All samples were analyzed for total and amenable (free) cyanides by Standard Methods 4500-CN- C and 4500-CN- G, respectively. Amenable cyanide was determined as the difference of the total and unamenable (complex) cyanides. Amenable cyanide was also measured using Method 4500-CN- H. (This analytical method might not work as well when samples contain significant quantities of iron cyanides.) The concentration of iron not complexed with cyanide in the supernatant solution was determined with a flame atomic absorption spectrometer.

## RESULTS AND DISCUSSION

### EFFECT OF pH

The pH of each sample was adjusted to a desired level prior to the addition of ferrous compounds, as listed in Tables I to IV. The results of these experiments indicate that complexation reactions occurred and formed insoluble precipitates. The best removals of total cyanide were accomplished in the pH range of 6.0 to 7.0. At higher pH values, ferrous ion is less soluble, hence less available for complexing with free or soluble complex cyanide in the solution. Though ferrous ion is more soluble at low pH values, the results show that the complexing reaction might not be fast enough to remove soluble cyanides from the solution. Furthermore, it is not economical to adjust the pH of the solution to that level. Therefore, the initial pH adjustment for the complexation process is pivotal to the success of removing cyanide from the wastewater.

For final pH adjustment with an alkaline reagent, a neutral pH value of approximately 8.0 is desired. While mixing and aerating at this pH, ferrous ion is converted to ferric ion ($Fe^{3+}$). As ferric hydroxide ($Fe_2O_3 \cdot nH_2O$) is formed, the availability of ferrous ion to react with cyanides is diminished. Additionally, the ferric hydroxide improves the settling characteristics of the resulting sludge.

## TABLE I
### EFFECT OF pH AND FERROUS ION CONCENTRATION
### (FERROUS SULFATE)
### ON CYANIDE REMOVAL USING BIOPLANT EFFLUENT

| Solution | Initial Total CN$^-$ (mg/L) | pH Adjusted | Fe$^{2+}$/CN$^-$ (mg/mg) | Final pH | Final Total CN$^-$ (mg/L) | % Sludge Volume |
|---|---|---|---|---|---|---|
| A1 | 51 | 6.0 | 3.1 | 6.0 | 1.8 | 3 |
| A2 | 60 | 2.5 | 1.7 | 8.0 | 15 | 5 |
| A3 | 60 | 2.5 | 3.3 | 8.0 | 6.5 | 7 |
| A4 | 60 | 4.0 | 3.3 | 8.0 | 2.9 | 7.2 |
| A5 | 60 | 7.5 | 3.3 | 8.0 | 3.3 | 3 |
| A6 | 60 | 6.0 | 6.7 | 6.0 | 0.88 | 8 |
| A7 | 60 | 6.0 | 6.7 | 8.0 | 0.51 | 8 |
| A8 | 6.5 | 6.0 | 30.7 | 8.5 | 2.1 | - |
| A9 | 8.2 | 6.0 | 24.4 | 8.5 | 2.1 | - |


## TABLE II
### EFFECT OF pH AND FERROUS ION CONCENTRATION
### (FERROUS CHLORIDE)
### ON CYANIDE REMOVAL USING AMMONIA STILL WATER

| Solution | Initial Total CN$^-$ (mg/L) | pH Adjusted | Fe$^{2+}$/CN$^-$ (mg/mg) | Final pH | Final Total CN$^-$ (mg/L) | Final Total Fe (mg/L) |
|---|---|---|---|---|---|---|
| B1 | 35 | - | 1.4 | 9.2 | 35 | 37 |
| B2 | 35 | - | 2.9 | 9.0 | 23 | 17 |
| B3 | 35 | - | 4.3 | 8.8 | 25 | 8 |
| B4 | 35 | - | 4.3 | 6.0 | 2.0 | 33 |
| B5 | 17.8 | 6.0 | 2.4 | 6.0 | 1.6 | - |
| B6 | 17.8 | 7.0 | 2.4 | 7.0 | 2.2 | - |
| B7 | 17.8 | 7.0 | 4.8 | 7.0 | 1.8 | 0.8 |
| B8 | 17.8 | 8.0 | 2.4 | 8.0 | 12 | - |
| B9 | 17.8 | 9.0 | 2.4 | 9.0 | 16 | - |
| B10 | 17.8 | 7.0 | 2.4 | 8.0 | 3.3 | 2.0 |

## TABLE III
### EFFECT OF pH AND FERROUS ION CONCENTRATION
### (SPENT PICKLE LIQUOR)
### ON CYANIDE REMOVAL USING BIOPLANT EFFLUENT

| Solution | Initial Total CN⁻ (mg/L) | pH Adjusted | Fe²⁺/CN⁻ (mg/mg) | Final pH | Final Total CN⁻ (mg/L) | Final Total Fe (mg/L) |
|---|---|---|---|---|---|---|
| C1 | 15.9 | 7.0 | 9.4 | 8.0 | 3.2 | 1.4 |
| C2 | 15.9 | 7.0 | 18.9 | 8.0 | 1.5 | 14 |

## TABLE IV
### EFFECT OF pH AND FERROUS ION CONCENTRATION
### (FERROUS CHLORIDE)
### ON CYANIDE REMOVAL USING SYNTHETIC CYANIDE SOLUTION

| Solution | Initial Free CN⁻ (mg/L) | pH Adjusted | Fe²⁺/CN⁻ (mg/mg) | Final pH | Final TCN, UCN, ACN* (mg/L) | Final Total Fe (mg/L) |
|---|---|---|---|---|---|---|
| D1 | 5 | 6.0 | 6 | 6.0 | 1.7, <0.05, 1.7 | 1.8 |
| D2 | 50 | 6.0 | 6 | 6.0 | 8.5, <0.5, 8.5 | 1.5 |
| D3 | 5 | 6.0 | 6 | 8.0 | 2.7, <0.05, 2.7 | 2.0 |
| D4 | 50 | 6.0 | 6 | 8.0 | 21.7, 20.2, 1.5 | 19 |

Note:

    TCN = Total cyanide
    UCN = Unamenable cyanide
    ACN = Amenable cyanide

## EFFECT OF $Fe^{2+}/CN^-$ RATIO

The cyanide removal rates achieved by adding different dosages of ferrous ion to the wastewater solutions at the same pH value are compared in Table II (Solutions B6 and B7). As mentioned earlier, the stoichiometric ratio for the proposed two-stage complexing reactions is 1.074 mg $Fe^{2+}$ per mg of $CN^-$. Deviation from that ratio may be attributed to the reaction of residual ferrous with hydroxide ion.

$$Fe^{2+} + 6CN^- \rightarrow [Fe(CN)_6]^{4-} \text{ (soluble)} \tag{1}$$

$$2Fe^{2+} + [Fe(CN)_6]^{4-} \rightarrow Fe_2Fe(CN)_6 \text{ (insoluble)} \tag{2}$$

The results of the experiment (Solution A7, Table I) conducted at an initial pH of 6.0, with final adjustment to pH 8.0, show that a ferrous ion/cyanide ion ratio of 6.7 to 1 is required to obtain the best removal. The same procedure was repeated with synthetic solutions of simple cyanide (KCN) using a $Fe^{2+}/CN^-$ ratio of 6.0 (Table IV). The results obtained from this set of experiments did not duplicate the earlier results, which had demonstrated efficient removal of cyanide. As shown in Figure 2, the free cyanide ($CN^-$) concentration in the effluent after 30 minutes of reaction time was measured at 17 mg/L, and after 210 minutes, it was measured at 2.9 mg/L. The reason for the difference is that the wastewater collected at the coke plant (Tables I to III) and the synthetic solution (Table IV) contained different forms of cyanide: the former consisted of more complex cyanide (about 90 percent) while the latter contained all free cyanide ($CN^-$). As previously mentioned, the theoretical requirement is a ratio of 1.074 $Fe^{2+}/CN^-$ by weight to complex free cyanide, and a ratio of 0.72 $Fe^{2+}/CN^-$ by weight if cyanide is present as $[Fe(CN)_6]^{4-}$.

An additional experiment was performed utilizing 50 mg/L of soluble complex cyanide ($K_4Fe(CN)_6$) and the same $Fe^{2+}/CN^-$ ratio of 6.0. The total cyanide measured after 10 minutes of reaction time was 0.16 mg/L, and after 30 minutes, 0.091 mg/L (also shown in Figure 2). These results show that the soluble complex cyanide, $[Fe(CN)_6]^{4-}$, was complexed to the insoluble form within the first few minutes of the reaction. Therefore, the reaction between $Fe^{2+}$ and $[Fe(CN)_6]^{4-}$ is shown to be much more rapid than the reaction between $Fe^{2+}$ and $CN^-$.

## MECHANISMS OF CYANIDE COMPLEXATION

During the cyanide complexation process in the presence of hydroxide ion, two types of reactions should occur: the reactions that form soluble and insoluble cyanide complexes and those that form ferrous and/or ferric hydroxides. When iron ($Fe^{2+}$) is added to a solution containing simple cyanide ($CN^-$), a soluble complex cyanide such as ferrocyanide, $[Fe(CN)_6]^{4-}$, is produced. Another reaction between available $Fe^{2+}$ and soluble complexed cyanide, $[Fe(CN)_6]^{4-}$, occurs concurrently. The rate of the second reaction is believed to be more rapid than that of the first. It has been suggested that $[Fe(CN)_6]^{4-}$ has a great affinity for iron (Meeussen et al., 1992), and

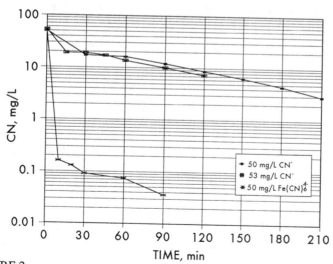

FIGURE 2 -
FIRST-ORDER PLOT FOR COMPLEXATION OF CYANIDE AT pH 6.0

that the possible products from this reaction are ferro-ferrocyanide [$Fe_4Fe(CN)_6$], ferri-ferrocyanide [$Fe_4(Fe(CN)_6)_3$], or even ferric ferro-ferricyanide [$Fe_7(FeFe(CN)_{12})_3$] (Williams, 1948). In Solution D4 (Table IV), the results indicate that sufficient $Fe^{2+}$ was not available to react with the remaining cyanides ($CN^-$ and/or [$Fe(CN)_6$]4-) because at the higher pH (8.0), $Fe^{2+}$ was oxidized by aeration to $Fe^{3+}$, which formed $Fe_2O_3 \cdot nH_2O$. The results further show that the remaining cyanides detected in this solution are in equilibrium at the concentrations of 20.2 mg/L (unamenable cyanide) and 1.5 mg/L (amenable cyanide).

## KINETICS OF CYANIDE COMPLEXATION

Kinetic experiments were performed at pH 6.0 with synthetic solutions containing free cyanide ($CN^-$) concentrations of 5 and 50 mg/L (Table IV). Figure 2 shows two similar curves obtained from duplicate experiments on the complexation of free cyanide. The results show typical first-order reaction kinetics in which there is a shift in the reaction rate within the first 60 minutes of the reaction. The shift confirms the theory that the two reactions occur concurrently, with the second, more rapid reaction continuing subsequently until the available $Fe^{2+}$ is depleted. These experiments did not attempt to identify the kinetic rates of the reactions.

## CONCLUSIONS

1.  Ferrous compounds were demonstrated to be effective in reducing cyanides in the wastewaters generated at a steel mill coke plant facility. The studies indicated that in order to achieve a concentration of 2 mg/L total cyanide in the effluent, an estimated 6 pounds of ferrous ion per pound of cyanide (as CN-) would be required.

2.  The proposed ferrous precipitation process includes: addition of sulfuric acid to reduce pH to 6.0; addition of a ferrous compound (or spent pickle liquor) to form soluble and insoluble iron cyanide complexes; readjustment of pH using sodium hydroxide to 8.0 with aeration, to oxidize any residual ferrous ions to the insoluble ferric state; flocculation with an anionic polymer; clarification; and sludge dewatering.

3.  The sludge volume generated from the ferrous precipitation process ranged between 4 and 6 percent by volume of the influent flow. The iron-cyanide complex produced was not solubilized at low pH values.

4.  The complexation of cyanide occurs through two reactions: one combines ferrous ion with free cyanide and the other involves the combination of ferrous ion with soluble complex cyanide.

## ACKNOWLEDGEMENTS

The authors wish to thank John Schrader, Jim Yarris, and Robert Helwick of Chester LabNet for their assistance in the experimental part of this work. They also thank Carla Robinson for editorial assistance in preparing this manuscript.

# REFERENCES

1.  Dodge, B. F., and Reams, D. C. (1949) Plating, 463-469, 512 (June). 571-4, 664 (July). 723-5, 728-31 (August).

2.  Dodge, B. F., and Zabban, W. (1951) Cyanide wastes: treatment with hypochlorites and removal of cyanates. Plating, **38**, 561.

3.  Flaherty, P. A. (1992) A non-chemical treatment process for the decomposition of metal cyanide complexes of metal cyanide complexes in wastewater. Paper presented at the 85th Air and Waste Management Association Annual Meeting, Paper No. 92-37.05.

4.  Gurol, M. R., Bremen, W. M., and Holden, T. E. (1985) Oxidation of cyanides in industrial wastewaters by ozone. Environmental Progress, **4**, 46.

5.  Knorre, H., and Griffiths, A. (1984) Cyanide detoxification with hydrogen peroxide using the Degussa process. Proceedings of the Conference on Cyanide and the Environment, Tucson, Arizona, Editor: Dirk Van Zyp, Vol. **2**, pp 519.

6.  Layne, M. E., Singer, P. C., and Lidwin, M. I. (1984) Ozonation of thiocyanate. Proceedings of the Conference on Cyanide and the Environment, Tucson, Arizona, Editor: Dirk Van Zyp, Vol. **2**, pp 433.

7.  Meeussen, J. C. L., et al. (1992) Dissolution behavior of iron cyanide (Prussian blue) in contaminated soils. Environ. Sci. Technol., **26**, 1832.

8.  Merkel, C. R., and Maziarz, E. F. (1984) The treatment of iron-complexed cyanides by ultraviolet light enhanced chemical oxidation. Aluminum Company of America Internal Report, Pittsburgh, PA.

9.  Ozhiganov, I. N., Regulenko, I. G., and Veselyi, A. G. (1968) Treatment to render harmless the cyanide-containing wastewater from blast furnace gas scrubbers. STAL **28** (2) (185-189), 169-173.

10. Prabhu, D., and Helwick, R. (1978) Treatability studies for the precipitation of cyanide in blowdown from blast furnace recirculation system. The Chester Engineers Internal Report, Pittsburgh, PA.

11. Robuck, S. J., and Luthy, R. G. (1989) Destruction of iron-complexed cyanide by alkaline chlorination. Water Sci. & Tech., **21**, 547.

12. Standard Methods for the Examination of Water and Wastewater, 18th Edition, 1992.

13. Varuntanya, C. P., Danicic, E. A., Feather, R. A., Hart, C. M., and Zabban, W. (1992) UV/Ozone/Hydrogen Peroxide pilot plant and source treatment studies. Chester Environmental Internal Report, Pittsburgh, PA.

14. Williams, H. E. (1948) Cyanogen Compounds: Their Chemistry, Detection, and Estimation, Edward Arnold & Co.

15. Zabban, W., and Feather, R. A. (1973) Cyanide complexation pilot plant studies on thickener effluent in blast furnace area. The Chester Engineers Internal Report, Pittsburgh, PA.

16. Zaidi, S. A., and Carey, J. (1984) Ultraviolet irradiation for removing iron cyanide from gold mill effluents. Proceedings of the Conference on Cyanide and the Environment, Tucson, Arizona, Editor: Dirk Van Zyp, Vol. **2**, pp 363.

# HYDROLYSIS OF FLUOROBORATE IN AN ELECTROPLATING WASTEWATER

DEBORAH M. WATKINS, P.E.
   Environmental Resources Management, Inc.
   855 Springdale Drive
   Exton, Pennsylvania 19341

PAUL J. USINOWICZ, Ph.D., P.E.
   Environmental Resources Management, Inc.
   855 Springdale Drive
   Exton, Pennsylvania 19341

## INTRODUCTION

Environmental Resources Management, Inc. (ERM) was retained by an electronic components manufacturing company to assist with National Pollution Discharge Elimination System (NPDES) permit compliance problems. Fluoride in the facility's wastewater treatment plant effluent consistently exceeded its NPDES permit limit of 40 mg/l. Further investigation revealed that the fluoride in the wastewater is present as fluoroborate. Standard calcium fluoride precipitation techniques, generally resulting in residual fluoride concentrations of less than 10 mg/l, are ineffective for the removal of the fluoride associated with the fluoroborate species.

## APPROACH TO MITIGATING FLUORIDE EXCURSIONS

A literature search was performed to identify fluoroborate removal technologies. The literature search uncovered methods to hydrolyze the fluoroborate and were successfully demonstrated through bench-scale testing. Raw wastewaters at the facility were analyzed for ionic fluoride, fluoroborate, and total fluoride as well as other indicators of treatment performance (i.e., aluminum, boron). Based on these data, a treatability test plan was developed and implemented. Using the results of the testing, modifications to the existing wastewater treatment system were designed to remove fluoride to below the facility's NPDES discharge limit. An operating manual was prepared that provides procedures, inspection recommendations, and a trouble-shooting guide to maintain NPDES compliance. This paper addresses various steps taken to achieve compliance with the facility's NPDES permit limit of 40 mg/l. Also, the facility's present compliance status with the NPDES fluoride limit is discussed.

## DESCRIPTION OF EXISTING WASTEWATER TREATMENT SYSTEM

The existing wastewater treatment system uses equalization, neutralization, precipitation, and clarification technologies. A schematic diagram of the existing wastewater treatment system is shown in Figure 1. Plant trenches empty into a scum tank which gravity flows to an equalization tank. "Killed" cyanide from cyanide alkaline chlorination pretreatment, filtrate from a rotary vacuum filter, and overflow from a sludge thickener also flow into the equalization tank. Sodium metabisulfite is introduced to the equalization tank via addition to the scum tank for dechlorination of the killed cyanide wastewater.

The equalization tank contents are pumped to a neutralization tank where lime slurry and alum are added for pH adjustment, coagulation, and precipitation of metals and fluoride at a pH of approximately 9.5. Spent acid and spent caustic from manufacturing can be added to the neutralization tank for pH control.

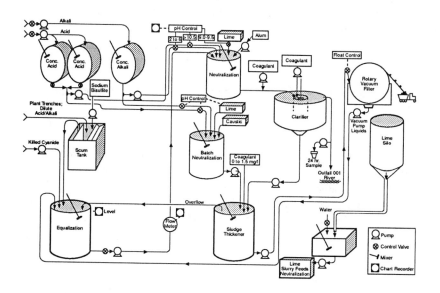

Figure 1 - Pre-improvement Wastewater Treatment System

The neutralization tank gravity discharges to a solids contact clarifier. Polymer is added to the clarifier to aid in the flocculation of suspended solids. The clarifier overflow is discharged to surface water. Sludge is pumped from the clarifier to a sludge thickener.

The sludge thickener receives clarifier under flow and the batch neutralization discharge. The sludge thickener under flow is pumped to a rotary vacuum filter for dewatering. The sludge thickener overflows to the equalization tank.

The rotary vacuum filter produces a sludge cake that is approximately 20% solids. The filtrate from the rotary vacuum filter is pumped to the equalization tank. The sludge cake is further dewatered in a sludge dryer to approximately 60% solids.

The batch neutralization process includes two concentrated acid storage tanks, a concentrated alkali storage tank, and a batch neutralization tank. Concentrated acid and alkali are pumped to the batch neutralization tank in suitable proportions to achieve a final pH of 8.5 to 9.0. Lime slurry can be added to the batch neutralization tank, if needed, to raise the pH.


## CHEMISTRY OF FLUORIDE PRECIPITATION AND FLUOROBORATE HYDROLYSIS

Fluorides can be effectively removed from wastewater via the precipitation of calcium fluoride using a calcium source such as lime or calcium chloride. Above pH 3.2 (i.e., above the $pK_a$ of hydrogen fluoride), hydrogen fluoride ionizes to ionic fluoride as follows:

$$HF + OH^- \Longleftrightarrow F^- + H_2O \qquad (1)$$

Addition of a calcium source results in precipitation of calcium fluoride:

$$2 F^- + Ca^{+2} \Longleftrightarrow CaF_2 (s) \qquad (2)$$

Fluoroborate, however, is a relatively soluble complex. Hydrolysis of fluoroborate to ionic fluoride is required prior to the precipitation of calcium fluoride. A literature search was conducted which uncovered two patents addressing the topic of fluoroborate hydrolysis [1,2]. These patents indicated that hydrolysis can be performed at an acidic pH using a calcium or aluminum catalyst, resulting in the cleaving of the boron-fluoride bond. The following reaction occurs:

$$\overset{\text{catalyst, } H^+}{3 BF_4^- + 12 H_2O \xrightarrow{\hspace{2cm}} 12 HF + 3 B(OH)_4^-} \qquad (3)$$

## ANALYSIS OF RAW WASTEWATER

Initially, samples of trench (dilute) wastewater, concentrated spent acid, and concentrated spent alkali were received from the wastewater treatment facility. The dilute wastewater was collected from the equalization tank and, because of its low total fluoride concentration, is believed to not have included batch dumps from the sludge thickener overflow. The concentrated acid and concentrated alkali were collected from the concentrated acid and concentrated alkali tanks, respectively.

Aliquots of dilute wastewater and concentrated acid were analyzed for ionic fluoride and total fluoride. Aluminum and boron analyses were also conducted so that the effects of alum addition and fluoroborate degradation could be determined in subsequent tests. The pH, ionic fluoride concentration, and fluoroborate concentration of the raw wastewater samples were measured. Analytical results are shown in Table I. Included in Table I are calculated values for total fluoride. These calculations assume that all complexed fluoride in the raw wastewaters is present as fluoroborate.

## TREATABILITY METHODOLOGY

Fluoroborate hydrolysis tests were performed on both the dilute wastewater and spent acid. First, each of the wastewater streams were characterized by measuring pH, fluoride, and fluoroborate concentrations. Following initial characterization, one liter of wastewater was poured into each of a series of 1-liter reactors. Concentrated sulfuric acid was used to adjust pH, as required. Calcium or aluminum was added at varying doses to the reactors. Reactors were mixed at 100 rpm using a gang stirrer. Fluoroborate measurements were performed at time

Table I - Analysis of Raw Wastewaters

| Constituent (mg/l) | Permit Limit (average) | Dilute Wastewater[a] | Concentrated Acid |
|---|---|---|---|
| pH (std. units) | 6 - 9 | 3.7 | 0.21 |
| Fluoride Ion | - | 4.6 | 3,085 |
| Total Fluoride | 40 | 20[b] | 9,175[b] |
| Fluoroborate | - | 17.2 | 7,000 |
| Aluminum | - | - | - |
| Boron | - | - | - |

[a]Dilute wastewater collected from equalization tank; batch dumps not believed to be present.
[b]Calculated.

493

intervals to monitor the kinetics of the reaction. Fluoride measurements were abandoned early in the experimentation because complexation of the fluoride with either calcium or aluminum prevented tracking of the fluoroborate hydrolysis to fluoride.

pH measurements were performed using a standard pH electrode and pH meter. Ion selective electrodes (ISEs) and a pH/ISE dual-input bench top meter were used to measure both fluoride and fluoroborate concentrations. A fluoride combination electrode was used for fluoride measurements. Fluoroborate and double-junction reference electrodes were used for fluoroborate measurements.

Fluoride calibration standards and wastewater samples were prepared by combining them with equal parts of Total Ionic Strength Adjuster and Buffer (TISAB). TISAB is a buffering solution that contains a chelating agent. Buffering ensures that the hydrogen fluoride ionizes to fluoride ion and eliminates interferences associated with high concentrations of hydroxyl ion. The chelating agent eliminates interferences associated with low to moderate concentrations of iron and aluminum.

Fluoroborate calibration standards and wastewater samples were prepared by combining them with 2 ml Ionic Strength Adjuster (ISA) for every 100 ml standard/sample. The ISA contains ammonium sulfate and is used to maintain a constant background ionic strength.


## HYDROLYSIS OF FLUOROBORATE IN DILUTE WASTEWATER

Raw wastewater characteristics of the dilute wastewater sample as measured by ERM were as follows:

| | |
|---|---|
| pH = | 3.7 |
| Fluoride = | 4.64 mg/l |
| Fluoroborate = | 17.2 mg/l |

One liter of dilute wastewater was poured into each of five reactors. The pHs in each of the reactors were adjusted with concentrated sulfuric acid to 1.0, 1.5, 2.0, 2.5, and 3.0, respectively. After pH adjustment, 300 mg/l of alum was added to each beaker. The beakers were mixed by a gang stirrer. Fluoroborate concentrations were measured at various time intervals (see Table II). The results are graphically illustrated in Figure 2. Fluoroborate measurements for the first two reactors, (pHs of 1.0 and 1.5) could not be measured because of the high acidity, which interfered with the test method.

Fluoroborate concentrations dropped from 17.2 to 11.0 in less than four hours for the reactors at pHs of 2.0, 2.5, and 3.0. The pH of the reactor had little effect on the kinetics of the hydrolysis reaction. Overall fluoroborate hydrolysis was 36% to 46% conversion to fluoride ion.

Table II - Hydrolysis of Fluoroborate in Dilute Wastewater[a]

| Time (hours) | Fluoroborate Concentration (mg/l) | | |
|---|---|---|---|
| | Beaker 3 (pH = 2.0) | Beaker 4 (pH = 2.5) | Beaker 5 (pH = 3.0) |
| 0 | 17.2 | 17.2 | 17.2 |
| 0.3 | 16.5 | 12.6 | 11.0 |
| 1.2 | 13.7 | 12.6 | 12.7 |
| 3.7 | 10.8 | 9.1 | 9.2 |

[a]Dilute wastewater collected from equalization tank; batch discharges believed to be not present.

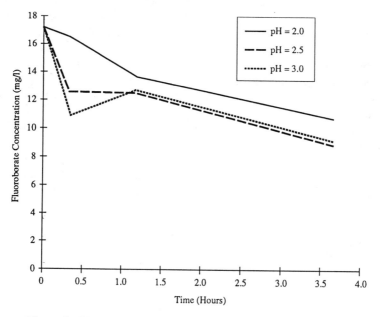

Figure 2 - Hydrolysis of Fluoroborate in Dilute Wastewater

# HYDROLYSIS OF FLUOROBORATE IN SPENT ACID

Because the majority of fluoroborate is present in the spent acid, fluoroborate hydrolysis of the spent acid prior to batch neutralization would be advantageous. The spent acid contains the most concentrated fluoroborate. Also, acid addition would not be necessary because of the acidity of the spent acid stream.

Raw spent acid characteristics were measured by ERM as follows:

| | |
|---|---|
| pH = | 0.21 |
| Fluoride = | 3,085 mg/l |
| Fluoroborate = | 7,000 mg/l |

Spent acid samples were diluted by a factor of 50 to reduce interferences for fluoride and fluoroborate measurement.

One liter of spent acid was poured into each of four reactors. Alum and calcium chloride were added to the reactors at the following concentrations:

| | | |
|---|---|---|
| Reactor 1 | Alum | 16 g/l |
| Reactor 2 | Alum | 32 g/l |
| Reactor 3 | Alum | 48 g/l |
| Reactor 4 | $CaCl_2$ | 27 g/l |

The reactors were mixed by a gang stirrer. Fluoroborate concentrations were measured at various intervals (see Table III). For Reactor 3, a fluoroborate concentration of 0 mg/l was indicated, but the calibration standards initially used did not bracket the concentrations in the diluted samples from Reactor 3. The minimum fluoroborate concentration detection limit for these samples is 500 mg/l, based on a 10 mg/l calibration standard and a dilution factor of 50. Recalibration was performed prior to further measurements using a 2% ISA solution as a standard.

The data are graphically illustrated in Figure 3. Fluoroborate concentrations dropped rapidly (within the first hour) for Reactors 1 through 3, after which the reaction appeared to be near completion. Fluoroborate hydrolysis was 55%, 88%, and >93% for Reactors 1, 2, and 3, respectively, within the first hour of reaction. The fluoroborate hydrolysis for Reactor 4 was 9% after the first hour of reaction.

Table III - Hydrolysis of Fluoroborate in Spent Acid

| Time (hours) | Fluoroborate Concentration (mg/l) | | | |
|---|---|---|---|---|
| | Reactor 1 (16 g/l alum) | Reactor 2 (32 g/l alum) | Reactor 3 (48 g/l alum) | Reactor 4 (27 g/l CaCl$_2$) |
| 0 | 7,000 | 7,000 | 7,000 | 7,000 |
| 0.7 | 3,165 | 870 | <500[a] | 6,400 |
| 1.3 | 3,060 | 885 | <500[a] | 6,650 |
| 1.9 | 3,180 | 855 | <500[a] | 5,800 |
| 2.6 | 2,840 | 755 | <500[a] | 6,100 |
| 5.3 | 2,520 | 745 | <500[a] | 5,750 |
| 8.8 | 2,775 | 745 | 73 | 4,505 |
| 22.3 | 2,515 | 760 | 92 | 5,650 |
| 28.2 | not measured | not measured | 412 | not measured |
| 46.8 | not measured | not measured | 108 | not measured |

[a]Fluoroborate concentration was measured at 0 mg/l due to the use of higher concentration calibration standards. The minimum fluoroborate concentration detection limit for these samples was 500 mg/l based on a 10 mg/l calibration standard and a dilution factor of 50.

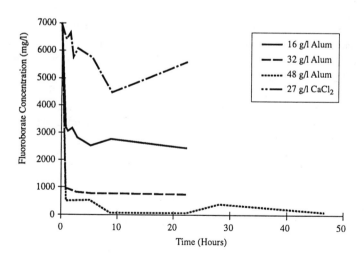

Figure 3 - Hydrolysis of Fluoroborate in Spent Acid

# RECOMMENDATIONS FOR MODIFICATION OF THE WASTEWATER TREATMENT SYSTEM

Based on the results of the treatability testing, the following modifications were recommended:

- Alum addition to the concentrated acid holding tanks for hydrolysis of the fluoroborate in these tanks.

- The use of two existing batch neutralization tanks which would alternate in function - one tank would be used for receiving and treating spent acid and alkali while the other tank would be used to meter treated batch wastewater to the continuous treatment operation.

- Alum addition to the scum tank for fluoroborate hydrolysis in the dilute wastewater.

- Clean acid addition, in lieu of spent acid addition, to the neutralization tank to eliminate the introduction of fluoroborate to the wastewater treatment system downstream of fluoroborate treatment.

- Calcium addition to the batch neutralization tanks to provide improved calcium fluoride precipitation and removal in the batch neutralization tanks.

The wastewater treatment system with recommended modifications is shown in Figure 4.


## PRESENT STATUS OF WASTEWATER TREATMENT FACILITY

Facility personnel are in the process of implementing the recommended modifications. Alum is currently being added to the scum tank. Clean acid is now used at the neutralization tank. The facility implemented these changes in 1994 and has been consistently meeting its NPDES discharge limit for fluoride. Future expansion of the facility's manufacturing operations will require the implementation of the remaining recommended modifications due to the expected increase of fluoroborate to the wastewater treatment system after the expansion.

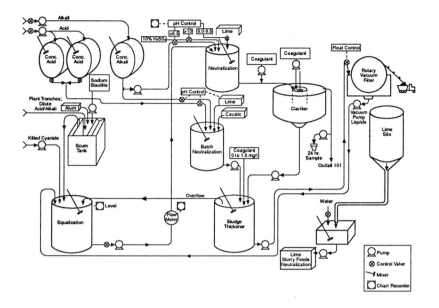

Figure 4 - Modified Wastewater Treatment System

## REFERENCES

1. Singh, Jaswant, U.S. Patent No. 3,959,132, "Methods for Removing Fluoborates from Aqueous Media", May 25, 1976.

2. Korenowski, Jerry et.al., U.S. Patent No. 4,008,162, "Waste Treatment of Fluoroborate Solutions", Feb. 15, 1977.

# Adsorption and Separation Processes

# A NEW DIMENSION IN PRECOAT TECHNOLOGY

JOHN R. SMITH
Graver Chemical Company
200 Lake Drive
Glasgow, DE 19702

## BACKGROUND

Adsorbents are widely used in many industries. Typically, granular adsorbents are used in columns, and powdered adsorbents are used in batch operations.

Based on adsorption theory, the smaller the particle size, the better the adsorption capacity and kinetics with poor hydraulics; and the larger the particle size, the better the hydraulics with poor kinetics and limited adsorption capacity. Maximizing both properties is impossible with conventional technologies. Current precoat filtration involves the addition of an inert, filterable material to a stream which is initially difficult to filter. Filter aids such as diatomaceous earth and Perlite are commonly used.

Conventional precoat filtration has no ability to remove soluble species. Because of this, most purification processes consist of two steps: the first being filtration followed by an adsorption step.

A very efficient (and very expensive) adsorption technique is the use of columns. Columns utilize the maximum capacity of the adsorbent and produce little solid waste. However, problems occur with the poisoning and fouling of the adsorbents and ion exchange resins. As adsorption capacities decrease with time, regeneration and energy costs increase. Also, the need for regeneration capabilities creates pollution problems from the waste chemical regenerants.

A less expensive and less efficient method (with powdered adsorbents) is the batch operation. The batch operation may be simple, but the capacity of the adsorbent is limited by the equilibrium constraints. This causes an inefficient use of the adsorbent. Also, the amount of solid waste generated from use of this system can be quite

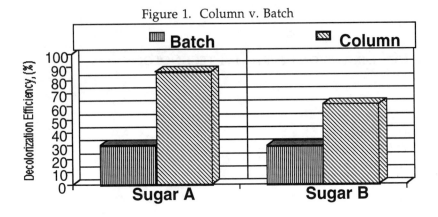

Figure 1. Column v. Batch

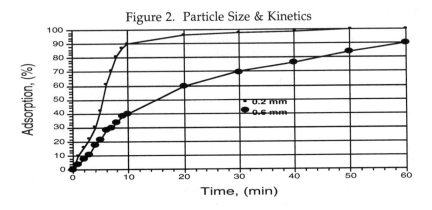

Figure 2. Particle Size & Kinetics

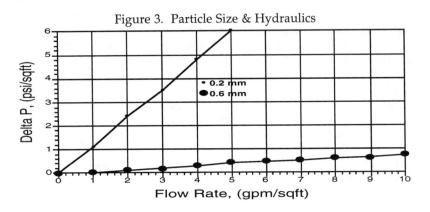

Figure 3. Particle Size & Hydraulics

large. Please refer to Figures 1-3 for a column vs. batch study and also for the effects of particle size on kinetics and on hydraulics.

Neither of these operations is ideal. Would it be possible to treat a stream containing suspended and dissolved solids in one simple step? The problem raised by such a question involves a kinetics/hydraulics dilemma. Is it possible to combine all the favorable properties of hydraulics, kinetics, and adsorption into one process? Better yet, how could fine-particle adsorbents such as activated carbons, zeolites, molecular sieves and powdered ion exchange resins possibly end up with good hydraulics while keeping their excellent kinetics and adsorption capabilities?

## INTRODUCTION

A group of adsorbent materials has been developed which combine all the favorable aspects of hydraulics, kinetics and adsorption. The adsorbent of choice is reduced to a small particle size (on purpose!) and immobilized onto a filter aid using a patented flocculation technology. See Figure 4.

This adsorptive matrix is easy to handle and is dust-free. Adsorption and filtration are combined into one unit operation with high permeability and excellent kinetics. Refer to Figures 5 and 6.

These new precoats are treated the same as conventional precoats. They are slurried and applied as a deep bed or to precoatable filters: candle, bag, leaf, sand, plate, tubular element, etc. Figure 7 shows that these new precoats are still efficient in removing suspended solids.

Figure 4.

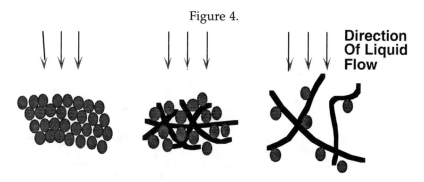

| Direction Of Liquid Flow |

**Powdered Resin Tight Bed Restricted Flow**  **Resin / Fiber Tight Bed Medium Flow**  **New Precoat Open Bed Good Flow**

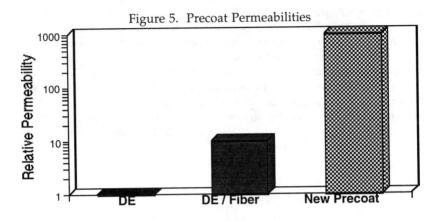

Figure 5. Precoat Permeabilities

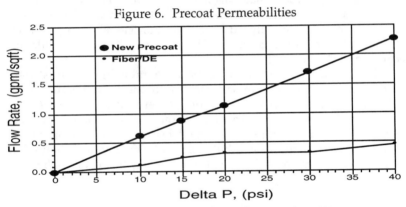

Figure 6. Precoat Permeabilities

Dosage = 0.25 lb/sqft, progressing cavity pump, in water

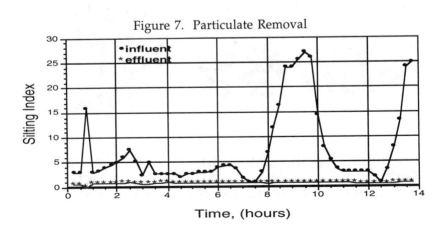

Figure 7. Particulate Removal

There are endless applications where dissolved and suspended solids are present and where both must be removed. Some of these include: (1) Treatment of metal finishing waste waters and ground waters, (2) Removal of odors and color, organics from radwaste water, toxic and noxious organics from potable water sources, trace acids, bases and salts from polar solvents, (3) Pretreatment for feeds of high purity water systems, (4) Polishing steam condensates, (5) Clarification of fluids containing traces of electrolytes and colloids, (6) Sugar/beverage processing, (7) Pharmaceutical processing, etc.

The increased concern over organic and inorganic contaminants in groundwater and wastewater streams prompted the study of the effectiveness of these precoats for the removal of halogenated hydrocarbons and trace metal contaminants.

## EXPERIMENTAL STUDY

Example 1: A water supply known to be contaminated with halogenated hydrocarbons was used to test the removal of organics with a formulation of these new precoats. Volatile organic analysis by GC/MS revealed the presence of the following compounds in the stream:

| | |
|---|---|
| Trichloroethene | 155.0 ppb |
| 1,1,1-Trichloroethane | 107.0 ppb |
| 1,1-Dichloroethene | 36.5 ppb |
| trans-1,2-Dichloroethene | 4.4 ppb |
| Tetrachloroethene | 3.5 ppb |

Leakage was monitored by GC/MS analysis. The precoat dosage was 0.30 dry lb/ft$^2$ and the stock feed was fed at a flow rate of 4.0 gpm/ft$^2$ to simulate normal field conditions.

Competition for sites on the adsorbent caused the efficiency of removal to vary from compound to compound. As can be seen from the data in Figures 8-10, run lengths depend on whether the endpoint is based on the removal of a single compound or on the removal of total organics.

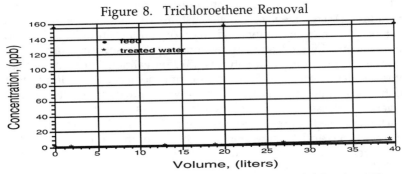

Figure 8. Trichloroethene Removal

Please Note: Data is based on results from a small-scale laboratory test.

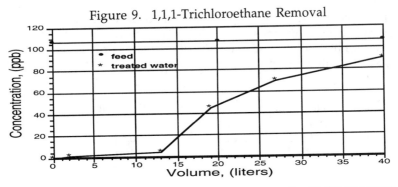

Figure 9. 1,1,1-Trichloroethane Removal

Please Note: Data is based on results from a small-scale laboratory test.

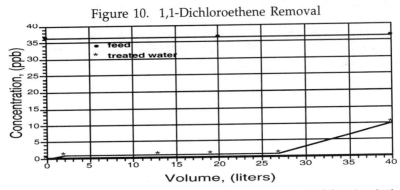

Figure 10. 1,1-Dichloroethene Removal

Please Note: Data is based on results from a small-scale laboratory test

Example 2: A water supply known to be contaminated with phenol was analyzed by GC/MS and was found to contain 10 ppb phenol. A specific formulation of these new precoats was applied to a Millipore pressure filter at a dosage of 0.30 lb/ft$^2$, and the contaminated water was fed through at a rate of 4.0 gpm/ft$^2$. Refer to Figure 11.

Example 3: Tap water spiked with metal salts of the cations for cadmium, chromium, copper, iron, manganese, zinc, lead and barium was used to test the removal of trace metals with yet another formulation of these new and improved precoats. Leakage was monitored by atomic absorption spectroscopy. Dosage applied to the Millipore pressure filter was 0.30 lb/ft$^2$, and the flow rate was 4.0 gpm/ft$^2$.

Based on a 10 ml precoat volume, the results after 1400 bed volumes of the feed have passed through the precoat were found. The detection limit for the metals was 0.05 ppm, and this value was used for non-detectable results. Therefore, the actual results may be better than shown in Figure 12.

Due to the competition for the adsorption sites, the removal of some of the metals would be more efficient in the absence of so much competition.

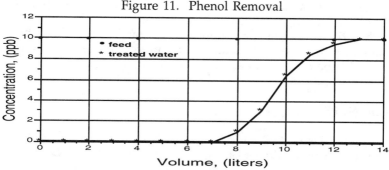

Figure 11. Phenol Removal

Please Note: Data is based on results from a small-scale laboratory test.

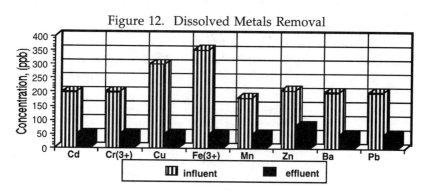

Figure 12. Dissolved Metals Removal

509

## CONCLUSIONS

This new dimension in precoating technology combines the favorable aspects of filtration with those of adsorption into one simple step while utilizing existing equipment. The performance of the adsorbents is enhanced to the point of maximum capacity thereby resulting in the elimination of a process step with an improvement in overall product quality.

The benefits of such a system include reduced adsorbent consumption, greater use of the adsorbent's capacity, simultaneous removal of soluble and insoluble contaminants, reduced waste generation, higher product quality, and lower cost than conventional processes.

Since the porosities of these new precoats are so high, the flow rates attained are very good, even in deep beds. While the precoat thickness rarely exceeds one inch, the process exhibits columnar properties.

These precoats can be custom formulated to meet special purification needs and can be tailored to provide specific performance characteristics of adsorption and filtration using a wide range of active surfaces. Current applications include many areas where low levels of contaminants must be removed from solution to polish product quality or to meet tight effluent discharge limits.

With current concerns for pollution, waste, and energy throughout the industry, the use of this innovative precoating technology has considerable potential in all areas of fluid treatment.

## REFERENCES

1.  Graver Chemical Product Bulletin No. 104, Ecosorb Product Selection Guide, Graver Chemical Company, Glasgow, DE, 1994.

2.  Graver Chemical Technical Bulletin No. GC-100, New Precoat Technology Combining Filtration, Ion Exchange and Adsorption, Graver Chemical Company, Glasgow, DE.

3.  Kunin, R., Wilber, G. Ion Exchange and Adsorbent Precoat Filters For Treating Toxic Wastes, Graver Chemical Company, Glasgow, DE, 1985.

# NOVEL CARBON AND CERAMIC RIGID MICROFILTRATION AND ULTRAFILTRATION MEMBRANES

JIM JABLONSKY
J & M Associates
1610 Hedgewood Road
Hatfield, PA 19440

Since filtration's inception, new and better materials for particulate separation in various hostile solution environments have been sought.

The past ten years have seen a movement from coarse woven natural fibers to advanced synthetic organic polymers with very consistent and reproducible pore sizes. This has allowed many industries new avenues of filtration with very specific cut off points and separation by various molecular weights. (Slide # 1)

With finer and finer filtration applications, the old nemesis "clogging" began to play a larger part in limiting the role these new membranes could attain.

Recently, tangential flow was introduced to eliminate clogging and maintain higher filtration rates (flux). This is a great advancement over classical dead end filtering which experiences reduced flux rates and quicker clogging as finer filtration parameters are implemented.

However, tangential flow cannot help the systemic shortcoming of many of synthetic polymer membranes. Many polymer membranes are sensitive to temperature and cannot be run in applications needing above 140° to 150° F. Strong oxidizing agents such as sodium hypochlorite and ozone can greatly compromise the polymers, causing premature failures. Extreme pH, either very low or very high, can cause failures as well as certain solvents. Low physical and mechanical resistances, coupled with limited back flushing capabilities, as well as aggressive cleaning chemicals and methods has greatly increased the cost of using these membranes.

The need for a high tech membrane precipitated the development of rigid carbon and ceramic micro and ultra filtration membranes. Their structure is rigid, symmetrical, self-supporting, and free of supporting elements.

Two commercially available types are the mineral based support matrix (carbon)[slide #2], and the monolithic $Al^2O^2$ - $TiO^2$ (ceramic). Both support structures make use of $TiO^2$ or $ZrO^2$ as the membrane layer.[Slide 3 & 4]

Detailed explanation of manufacturing processes for the support matrixes can be found in U.S. Patent No. 3,790,654 and monolithic ceramic U.F. such as is being sold in the U.S. today can be found in U.S. Patent No. 4,069,157.

The void space resulting from the porous nature of the support material represents approximately 30% to 60% of the total structure and helps with flux production. [Slide 5].

Ceramics address and surpass most of the shortcomings of the organic synthetic polymers. The following are some of their advantages:

- Unequaled performance in micro and ultra filtration
- Very high filtration flux
- Low operating cost
- Very high physical, mechanical, and thermal resistance
- Steam sterilization capability
- High temperature application (350°C)
- Resistant to pH extreme (0 to 14)
- Unaffected by chemicals (except hydrofluoric acid)
- Unaffected by solvents
- Back flushing capability
- Resistant to radiation
- Long membrane life
- Smooth cylindrical format permits leak free sealing
- Resistant to strong oxidizers
- Resists clogging and fouling even up to 50% free oil
- FDA approval
- Not affected by very aggressive cleaners

These carbon and ceramic filtration elements are packaged in bundles to increase effective surface area thereby increasing flowrate and increasing flux.[Slide 6 & 7] Not only element increases are possible, but standard bundle packages are available , thereby simplifying engineering and design costs.

Carbon Micro and Ultra Filtration

| Length | 1,200 mm |
|---|---|
| Diameter | 6 mm |
| Channels | 1 |
| Bursting press | 40 hours |
| Service press | 15 hours |

Ceramic Micro and Ultra Filtration

| Lengths | 856 mm |
|---|---|
| | 1178 mm |
| Diameter | 20 mm |
| | 25 mm |
| Channels | 7 and 19 |
| Internal Channels | 2.5 - 3.5- and 6 mm |
| Bursting pressure | 50 bars |
| Service presses | 10 bars |
| Microfiltration Pore Diameter | 0.10, 0.2, and 0.45 |

CARBON

| No. of Tubes | Membrane Area m$^2$ | Circulation Rate m$^3$/h at 1 m/s |
|---|---|---|
| 1 | 0.025 | 0.10 |
| 3 | 0.068 | 0.30 |
| 7 | 0.16 | 0.70 |
| 37 | 0.84 | 3.70 |
| 55 | 1.25 | 5.50 |
| 151 | 3.40 | 15.10 |
| 252 | 5.70 | 25.20 |

| Ceramic No. of Tubes* | Membrane area m$^2$ | Circulation Rate m$^3$/h at 1 m/s |
|---|---|---|
| 1 | .155 | 0.71 |
| 7 | 1.08 | 4.97 |
| 19 | 2.95 | 13.49 |
| 37 | 5.73 | 26.27 |
| 99 | 15.30 | 70.29 |

* 1 tube has 7 channels
100 l/h = 1.0 m/s

In commercial application, I have found that carbon membranes typically operate at .5 to 4 bar pressure with recirculating rates of a low (30 liters per hour) to a high (500 l/h), depending on product to be processed. Ceramic typically runs at 2 to 8 bars.

It should be noted that these flows are very dependent on solution viscosity, total solids, particulate surface charge, and temperature. Flux permeate rates are also dependent on these same solution characteristics, and should be lab tested to determine the best recirculation flows and pressures to optimize production.

When one is designing a system for a particular solution, temperature can be one of the most important factors that affect permeate production and reduce total system cost. A rule of thumb states that for every degree of increase in temperature there is an approximate 2% increase in permeate flux.

If solutions are not temperature sensitive, then heating the solutions can actually enable one to use physically smaller surface areas to attain the same flux production. Since these membranes are fairly expensive, a considerable saving can be gained by use of elevated temperatures in processing your solution separations.

The system can be configured to a simple batch treatment, feed and bleed batch, or simple continuous system. [Slides 8, 9, and 10]

Batch systems typically are less expensive due to the fact that offline 24 hour process days can be employed, allowing use of smaller systems and costing less.

Feed and bleed batch systems allow "S.P.C." procedures to be applied to manufacturing processes, increasing quality and decreasing scrap as well as reducing unit production costs.

For the most critical applications, a continuous system can be applied, yielding the highest consistent level of quality.

When one faces real world recycling, waste water treatment and industrial processes-very often with high dirt load, oily and emulsified oil in water, and greatly fluctuating consistency - the dynamic or tangential micro/ultra filtration has been far more advantageous to use.

Such heavily contaminated liquids quickly form a thick filter cake that reduces the filtration efficiency to a very low point. Pressure increases show improvement for only a very short time, until almost no flow is seen.

A superimposed tangential flow (TF) during filtration across the micro/ultra membrane removes a large portion of the filter cake. Depending on the velocity of the tangential flow and the firmness of the filter cake, the membrane can be partially or almost entirely cleaned. The thickness of the filter cake decreases, and the flux of the ceramic membrane increases. This efficiency improvement, however, is made possible by an additional input of flow energy by means of an appropriate high velocity pumping system. There is higher initial investment cost with "TF" ceramic filter systems compared with dead end or cartridge filtration. The advantages, however, by far outweigh the initial higher expenses.

Once again, the particles separated from the liquid by ceramic rigid membranes are continuously removed by tangential flow and turbulent action at the oxide membrane surface.

Under even ideal conditions, a thin layer of filter cake will be formed on the oxide wall of the membrane. This efficiency reducing "caking or layering" can further be removed by incorporating intermittent back flow (IBF) process into the micro/ ultra filtration system.

Depending on the separation problems encountered, the clean permeate at certain predetermined intervals is pressed back through the membrane for a few seconds in the opposite direction without interrupting the recirculation.

In this way, more of the cake or layer is separated from the oxide membrane surface and washed away, thus restoring original efficiency of the separation process, and sustaining high flow flux rates indefinitely.

It is only the combination of "TF" principle linked with the "IBF" process that will give these high filtration rates over prolonged periods of time, even for suspensions having high solid content and tenacious emulsified oil suspensions.

New studies show promise of further reducing fouling and increasing flow flux on ceramic membranes by altering surface charges. "Surface charge and ion sorption properties are major contributory factors to fouling by polar or ionic substrates, since surface sorption is often the first step in fouling leading to pore blocking within the membrane structure. By altering the surface charge density and changing its polarity, the opportunity exists for minimizing fouling and optimizing separation processes. A fundamental understanding of the surface charge and sorptive properties within the pores of the active layers of ceramic membranes is therefore essential to their efficient use" [1].

A problem waste stream at an indigo dyeing plant was resolved by using a two stage U.F. elevated temperature system. [Slide #11]

Many uses for this technology have emerged and the following are but a few of the major area applications.

I       ENVIRONMENTAL PROTECTION

- Oily effluents from oil
- Industrial waste water
- Extraction of heavy metals
- Municipal waste water
- Bioreactors with membranes

- Radioactive effluent decontamination

## II PROCESS OPTIMIZATION AND RECYCLING

### Chemical and Mechanical Industries

- Recycling of aqueous cleaner (hot and cold)
- Recycling of catalysts
- Recycling of paints, solvents
- Recycling of machine cutting coolants
- Recycling of machine cutting oils

### Paper Industry

- Black liquors, white water, drinking
- Pulp fiber clarification

### Textile Industry

- Dye recovery
- Reduction of T.D.S. in rinse streams

### Petroleum Industry

- Separation of hydrocarbons
- Track tank cleaner recycling

### Biotechnology

- Separation and concentration after fermentation

### Food Industry

- Recycling of strong alkaline cleaners
- Recycling of brine solutions
- Separation of bacteria from rinse streams
- Reduction of fats ans oil in waste streams

### Printing

- Separation of inks and paper dust from fountain streams
- Recycling rinse streams

- Process water recycling
- Condensate treatment
- Ultrapure and pyrogen free water
- Clarification of natural water, lakes, rivers, spring water, drilling, and bore hole water.
- Cooling tower water recycling

As we approach the "zero discharge era", ceramic UF/ micro filtration will play a pivotal part in helping us to meet even stricter environmental waste discharge limits.

Ceramic filtration has been able to not only cut costs for companies, but also to start to reduce process expenses and add profit to the bottom line by it's unique ability to recycle chemicals and waste water.

Waste treatment costs are cut dramatically, and better process systems are reducing process cost, enabling industries to build a better product at a lower cost while being environmentally conscious.

REFERENCE

1. Gallagher, Stephen , Paterson, Russell, Etienne, Jocelyn , Larbot Andre, and Cot Louis. 1992 "Surface Charge and Ion Sorption Properties Influencing the Fouling and Flow Characteristics of Ceramic Membranes" *Ion Exchange Advances - Proceedings of IEX '92*, pp.318, 319.

# VAPOR PHASE ADSORPTION-DESORPTION OF 1,1,1-TRICHLOROETHANE ON DRY SOIL

NILUFER H. DURAL
Department of Chemical Engineering
Cleveland State University
Cleveland, Ohio 44115

CHAO-HSI CHEN
Department of Chemical Engineering
University of Missouri-Columbia
Columbia, Missouri 65211

## ABSTRACT

Vapor phase adsorption of 1,1,1-trichloroethane (TCA) on dry soils was studied at 288, 293, and 298 K. Using a gravimetric adsorption apparatus, adsorption/desorption isotherms of TCA were generated on two representative soil samples with different physical/chemical characteristics. Influence of temperature and soil properties were investigated. Isosteric heats of adsorption were calculated and heat curves were established. The experimental data were correlated by the Polanyi Potential, the BET, and the GAB models.

Equilibrium isotherms of TCA on both soils were Type II, characterizing vapor condensation to form multilayers, and they exhibited hysteresis upon desorption. A positive correlation between the soil's specific surface area and its sorption capacity was observed. Clay content and pore size were also dominating factors. Thermal data showed that the adsorption of TCA vapor on soil was primarily due to physical forces and both samples exhibited energetically heterogeneous surfaces. Results followed the Potential Theory satisfactorily and led to a single temperature-independent characteristic curve for each soil-TCA pair. The BET equation gave accurate correlations for up to 40% of the saturation pressure, while the GAB equation provided superior correlation of the data for the entire relative pressure range.

## INTRODUCTION

Chlorinated hydrocarbons are recognized as a major class of subsurface

518

contaminants although they are widely used as solvents and degreasing agents by numerous industries and military. It is estimated that of the chlorinated solvents that have been used as degrasers across the U.S., 60% has been 1,1,1-trichloroethane (TCA) which is also known as methyl chloroform [1]. In addition to being an ozone depleting chemical, spent TCA is a listed hazardous waste and it has been designated as a priority pollutant contaminating soil and ground water in several regions.

The transport, fate and removal of organic pollutants in soils are highly dependent upon their sorptive behavior. The focus of preponderance of research on halogenated organic compounds has been on the sorption from the aqueous or nonaqueous liquid phase [2-4]. Studies including sorption from the gas phase are generally lacking. Many chlorinated solvents, including 1,1,1-trichloroethane, are volatile compounds. Unlike the saturated zone, the unsaturated zone may vary from a completely dry content to a saturated content for moisture. Generally, soils at the top layers of the ground are completely dry because of disclosure to higher temperatures and wind. Organic vapors may diffuse from subsurface to surface or even traces of organic vapor in the air may be adsorbed. Hence, for volatile organic compounds, gas phase sorption can be significant in the unsaturated zone, especially for dry soils.

Earlier research suggests that the temperature, and the physical/chemical properties of the soil system must be given serious consideration [2,5-10]. Change in soil temperature, due to climatic changes and depth, causes varying soil sorption capacity for a specific pollutant. Furthermore, depending on surface properties (specific area and pore size) and chemical composition, the sorption characteristics of different soils may vary widely. Hence, an accurate analysis of the adsorption process for a pollutant-soil pair should include the impact of these factors.

The present work was undertaken with the goal to analyze the vapor phase adsorption/desorption of 1,1,1-trichloroethane on dry soils. Equilibrium isotherms were determined on two representative soil samples with different physical/chemical properties and at three different temperatures in order to examine and quantify the factors involved. This, in turn would provide information in regard to pollutant's migration in and removal from the soil.


## EXPERIMENTAL SECTION

1,1,1-trichloroethane was supplied by the Aldrich Co with a purity of 99+%. The soil samples from Visalia-California (1.7% organic matter, 45.1% sand, 35.2% silt, 21.7% clay) and Times Beach-Missouri (2.4% organic matter, 11.4% sand, 52.7% silt, 33.4% clay) were used as adsorbents. Prior to their use, the samples were further characterized with respect to specific surface area and pore size by using an automated BET sorptometer (Porous Materials, Inc.) and nitrogen gas at liquid nitrogen temperature (-195.8°C). The results of the physical characterization are presented in Table I.

The adsorption and desorption data were measured gravimetrically using a

Table I - Physical Characteristics of Soil Samples

|  | Missouri Soil | California Soil |
|---|---|---|
| Specific Surface Area, $m^2/g$ | 44.1444 | 25.3263 |
| Average Pore Diameter, A° | 17.6901 | 15.9946 |
| Total Pore Volume, $cm^3/g$ | 0.0195 | 0.0101 |
| Median Pore Diameter, A° | 24.4620 | 17.6740 |

Cahn-2000 electrobalance (CAHN Instruments), with a sensitivity of 0.1 micrograms, mounted in a glass vacuum chamber assembly. The adsorption apparatus was assembled to contain the Cahn-2000 electrobalance, the pressure gauges, the temperature controller, and the vacuum system. The electrobalance was placed in a vacuum bottle and the assembly was equipped with hangdown tubes for sample and the counterweight pails. The sample weight was monitored with a stripped chart recorder connected to the electrobalance control unit. The sample hangdown tube was wrapped by a copper circulation coil connected to a refrigerated/heated constant temperature circulator with a temperature control within $\pm 0.02°C$ of set point. The sorption temperature was monitored by a thermocouple connected to a digital thermometer with an accuracy of $\pm 0.1°C$. The vacuum system consisted of a mechanical pump for roughing, a diffusion pump for high vacuum, and a dry ice sorption trap. A vacuum of approximately $4.6 \times 10^{-3}$ mmHg was obtained and the leak rate of the whole system was 0.002 mmHg/hr. In order to prevent the 1,1,1-trichloroethane vapor being discharged to the outdoors from the vent, an activated carbon trap was set up between the dry ice sorption trap and the vacuum pump. Two convectron gauges were used to monitor the pressure during evacuation. Equilibrium pressures were measured with a Wallace and Tiernan absolute pressure gauge.

After the electrobalance was zeroed and calibrated, the soil sample was placed on the hangdown pan suspended from the electrobalance. The regeneration was carried out by evacuating the system and applying heat at 343 K until a constant sample weight was obtained. Typically, this procedure needed 10 hours or longer.

The soil sample was cooled to the adsorption temperature, and the 1,1,1-trichloroethane vapor was introduced into the system. The weight change of the sample was monitored by the electrobalance and a recorder. After equilibrium was reached, as indicated by a constant sample weight, the system pressure and weight of the sample were recorded. Then, more TCA vapor was introduced into the system, and the procedure was repeated for a new system pressure. Equilibrium isotherm data were taken from a pressure of 3 mmHg approximately up to 90% of the saturation pressure. Following adsorption, desorption measurements were made by reducing the system pressure in steps.

## RESULTS AND DISCUSSION

Equilibrium isotherms were obtained for TCA vapor on two representative soil samples at 288, 293, and 298 K. These temperatures can represent the typical dry soil temperature for the most part of the year in several regions. For both TCA-soil pairs examined, the isotherms were Type II, according to Brunauer's classification [10], characterizing the formation of multiple layers of adsorbate molecules on the solid particle surface. In Figure 1, adsorption/desorption isotherms of TCA on Missouri soil at 298 K is illustrated as a typical example. These isotherms are plotted in the standard manner, namely the amount of TCA vapor adsorbed per gram of dry soil (M) as a function of the relative pressure, $P_r(=P/P_0)$, or the equilibrium pressure, P. The sharp rise of the isotherm at high values of the relative pressure (approximately $P_r > 0.4$) can be attributed to a multilayer adsorption (i.e. an induced vapor condensation) on the surface of the soil sample [3]. The curvatures at low relative pressure ($P_r \leq 0.4$), referred to as monolayer region, are scarcely visible, which can be ascribed to reflect some selective adsorption.

Considerable hysteresis effects, associated with capillary condensation, were observed upon desorption. As expected, the desorption curve was always above the adsorption curve. This can be attributed to the pore geometry in that the surface curvature in contact with the vapor at a given value of $P_r$ is different during adsorption and desorption. As pointed out by Satterfield [11] and many others, ink-bottle shaped pore structure leads to hysteresis. Accordingly, the value of $P_r$ at which condensation occurs is determined by a larger effective radius of curvature (body of bottle), while $P_r$ for evaporation from a filled pore is determined by a smaller effective radius of curvature (neck). Thus, the difference in the adsorption and desorption isotherms obtained in the present

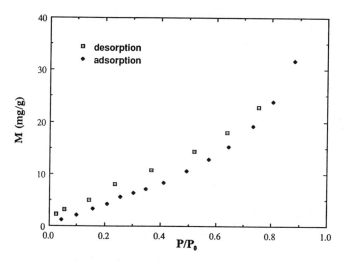

Figure 1 - Adsorption Desorption of TCA on Missouri Soil at 298 K.

work are indication of capillary condensation and ink-bottle type pore structure of the soils.

The adsorption isotherms of TCA on two soils were qualitatively similar, but the equilibrium uptakes corresponding to the same equilibrium pressure were different. In Figure 2, adsorption of TCA at 298 K on different soils is compared. At other temperatures, similar behavior was observed. In all cases, Missouri soil adsorbed more TCA vapor. The different uptake capacities of the soils can be attributed to both the chemical composition and the physical structure of the soil.

Since adsorption is a surface phenomenon, the surface area is of the most direct significance. In the present study, the specific surface areas of the soil samples were determined, independently, by using a BET sorptometer surface analyzer with nitrogen gas at liquid nitrogen temperature as the adsorbate. As can be seen from the results of the surface area analyses (Table 1), the Missouri soil has larger specific surface area than the California soil does. Furthermore, it has larger pore diameter and total pore volume. Therefore, the Missouri soil should have higher adsorption capacity than the California soil does. This conclusion is supported by the entire data set.

For the adsorption of organic compounds by soil, the mineral content and the organic matter provide most of the surface area. However, it has been postulated that [9], the mineral matter, especially clay content, dominates the adsorption of organic compounds and organic matter has secondary influence on adsorption by dry soils. The present results are in agreement with this claim. Missouri soil with 33.4% clay content adsorbs more than California soil with only 21.7% clay content. It should also be noted that the % increase in the clay content of the soils (from 33.4% to 21.7%: 35%) is very close to the % increase in their specific surface areas (from 44.14% to 25.33%: 42%).

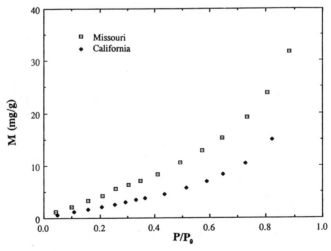

Figure 2- Adsorption Isotherms of TCA on Missouri and California Soil at 298K

The isotherms were measured at 288, 293, and 298 K to investigate the influence of temperature and provide data that could be used to evaluate the isosteric heat of adsorption. Since physical adsorption is an exothermic process, increasing the temperature resulted in a downward shift of isotherms (Figure 3).

The heat of adsorption provides a direct measure of the strength of the binding forces between the adsorbate (TCA) molecule and the adsorbent (soil) surface. If the amount of vapor sorption at different temperatures is known, the isosteric heat of adsorption, $\Delta H_{ads}$, can be calculated in the usual manner from the Clasius-Clapeyron equation:

$$\Delta H_{ads} = R[\partial \ln P / \partial (1/T)]_M \qquad (1)$$

Accordingly, a semilogarithmic plot of equilibrium pressure versus reciprocal absolute temperature at constant adsorbent loading should give a straight line with a slope of $\Delta H_{ads}/R$. Figure 4 shows the isosteric heats of adsorption for the two soils, plotted as a function of the amount of TCA vapor adsorbed. Such plots, commonly referred to as 'heat curves' are useful in determining the thermodynamic properties of the system and in characterizing the adsorbent surface.

The shapes of the curves given in Figure 4 clearly imply that the soil samples studied have energetically heterogeneous surfaces [12,13]. Furthermore, the heats of adsorption shown in the figures are of the same order of magnitude as the heat of condensation of the adsorbate, which confirms that the adsorption of TCA vapor on soils is basically physical adsorption. In the monolayer region, the Missouri soil binds the adsorbate much stronger than the California soil does. This is primarily due to its higher clay content and indicates higher energy requirements for decontamination via desorption techniques.

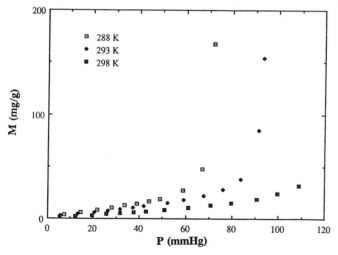

Figure 3-Influence of Temperature on TCA Adsorption (Missouri Soil).

The experimental data were correlated by the BET, the GAB and the Polanyi Potential multilayer isotherm models for future simulations. Probably the most important model for multilayer adsorption of vapors is the BET equation [10,14,15]. It is essentially an extension of the Langmuir isotherm, but it accounts for multilayer adsorption by assuming that each molecule in the first adsorbed layer provides one site for the second and subsequent layer. The molecules in the second and subsequent layers are considered to behave as saturated liquid, while the heat of adsorption for the first layer of molecules is different. The resulting equation for the BET equilibrium model is

$$M \, / \, M_m = CP_r \, / \, \{(1-P_r)(1-P_r+CP_r)\} \qquad (2)$$

where $P_r$ is the relative pressure, $M_m$ is the monolayer capacity, and C is a constant related to the net heat of adsorption.

In independent studies, Anderson [16], De Boer [17] and Guggenheim [18] improved the BET model by assuming that the heat of adsorption of the second to about to ninth layers differs from the heat of liquefaction by a constant amount, and that the heat of adsorption is equal to the heat of liquefaction in the layers following these. The final equation has the following form:

$$M \, / \, M_m = CkP_r \, / \, \{(1-kP_r)(1-kP_r+CkP_r)\} \qquad (3)$$

where $M_m$ and C are the same as defined in the BET equation, and the additional parameter k represents the difference between the heat of adsorption of the multilayers and the heat of liquefaction.

The results of the data correlation using the BET and GAB models, including the average percentage deviations of the predicted isotherms from the experimental ones, are presented in Table II.

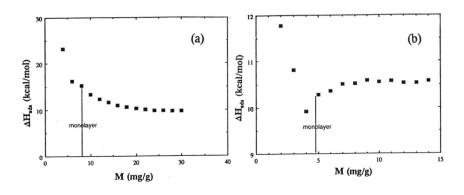

Figure 4- Heat of Adsorption of TCA. (a) Missouri Soil; (b) California Soil

Table II- Model Predictions for 1,1,1-Trichloroethane Adsorption on Soil

| Soil Type | T,K | BET | | | GAB | | | |
|-----------|-----|-------|-------|--------------------|-------|-------|------|--------------------------|
| | | $M_m$ | C | Error,% (Pr<0.4) | $M_m$ | C | k | Error,% (entire set) |
| California | 288 | 3.31 | 6.63 | 4.90 | 5.35 | 2.25 | 0.96 | 4.22 |
| | 293 | 3.19 | 4.40 | 4.98 | 5.87 | 1.58 | 0.94 | 3.36 |
| | 298 | 3.75 | 3.14 | 2.84 | 4.08 | 3.19 | 0.91 | 2.14 |
| Missouri | 288 | 8.09 | 8.08 | 1.23 | 7.83 | 10.47 | 0.97 | 2.84 |
| | 293 | 7.74 | 8.05 | 4.83 | 9.05 | 6.35 | 0.91 | 3.18 |
| | 298 | 7.66 | 3.104 | 2.02 | 8.61 | 3.41 | 0.85 | 1.46 |

BET model predicted the experimental data excellently within the deviation range of 1.23 to 4.98%, for the relative pressure less than 0.4 (monolayer region). In the multilayer region, however, the average absolute percentage deviation was greatly increased (always greater than 10%) and even exceeded 100% in many cases. This can be attributed to the oversimplifying assumptions of the BET model concerning the multilayer region. On the other hand, GAB model provided much better overall predictions than did the BET model did. The average absolute percentage deviations for the entire relative pressure range were in a range varying from 1.48 to 4.22 %. Model parameters for both BET and GAB equations did not show a definite trend with increasing temperature.

The Potential Theory of Polanyi [14] has been applied successfully for monolayer and multilayer adsorption of gases and vapors on both porous and microporous adsorbents of commercial use. However, no application regarding sorption on soils has been located in the existing literature. The potential theory assumes that a potential field exists at the surface of the adsorbent and it exerts strong long-range attractive forces on the surrounding gas or vapor. The adsorption potential, $\epsilon$, which is related to the difference in free energy between the adsorbed phase and the saturated liquid sorbate at the same temperature, may be calculated directly from the ratio of the equilibrium pressure and the saturation vapor pressure:

$$\epsilon = -RT \ln(f/f_0) = -RT \ln(P/P_0) \qquad (4)$$

where $f_0$ and $P_0$ refer to the saturation fugacity and pressure for the liquid sorbate and f and P are the corresponding equilibrium quantities for the adsorbed phase. A unique temperature independent relation between the adsorption potential and the volume of adsorbed species exists for a given adsorbate-adsorbent system. In other words, a plot of the volume adsorbed vs. the adsorption potential should yield a temperature independent curve which is referred to as the characteristic curve. The characteristic curve provides an excellent means of summarizing equilibrium data over a wide range of temperatures. Provided that the system follows the Potential Theory, this property can be very useful for forecasting the adsorption capacities for contamination of soils in different seasons.

The characteristic curves of TCA on Missouri and California soils, plotted as $cm^3$ TCA adsorbed per gram of soil vs $\epsilon/R$ (i.e; $Tln(P_o/P)$), are given in Figure 5. As shown, the data for all temperatures lie close to a single curve as required by the Polanyi theory. Consequently, they can be used to predict the sorption capacities at other temperatures by simple extrapolation or interpolation.

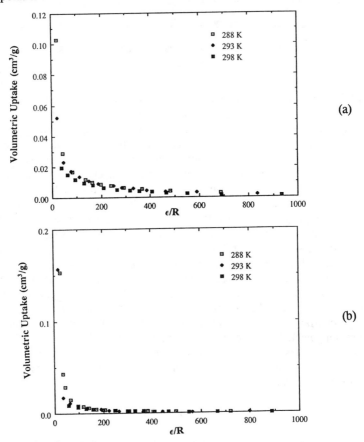

Figure 5-Characteristic Curve for 1,1,1-Trichloroethane. (a) Missouri Soil; (b) California Soil

# REFERENCES

1. Wolf, K. An Analysis of Alternatives to Ozone Depleting Solvents in Cleaning Applications, *Pollution Prevention Rev.*, 2 (1993) 1.
2. Ong, S. K. and L. W. Lion, Effects of Soil Properties and Moisture on the Sorption of Trichloroethylene Vapor, *Water Res.*, 25 (1991) 29-36.
3. Chiou, C. T., D. E. Kile, and R. L. Malcolm, Sorption of Vapors of Some Organic Liquids on Soil Humic Acid and its Relation to Partitioning of Organic Compounds in Soil Organic Matter, *Environ. Sci. Tech.*, 22 (1988) 298-303.
4. Poe, S. H., K. T. Valsaraj, L. J. Thibodeaux, and C. Springer, Equilibrium Vapor Phase Adsorption of Volatile Organic Chemicals on Dry Soils, *J. Hazardous Mater.*, 19 (1988) 17-32.
5. Chiou, T. C., P. E. Porter, D. W. Schmedding, Partition Equilibria of Nonionic Organic Compounds Between Soil Organic Matter and Water, *Environ. Sci. Tech.*, 17 (1983) 227.
6. Smith, J. A., C. T. Chiou, J. A. Kammer and D. E. Kile, Effect of Soil Moisture on the Sorption of Trichloroethylene Vapor to Vadose Zone Soil at Picatinny Arsenal, New Jersey, *Environ. Sci. Tech.*, 24 (1990) 676-683.
7. Koo, J. K., D. W. Ahn, S. P. Yoon, and D. H. Kim, Effects of Water Content and Temperature on Equilibrium Distribution of Organic Pollutants in Unsaturated Soil, *Water, Air, and Soil Pollut.*, 53 (1990) 267-277.
8. Alben, K. T., E. Shpirt, and J. H. Kaczmarczyk, Temperature Dependence of Trihalomethane Adsorption on Activated Carbon: Implications for Systems with Seasonal Variations with Temperature and Concentration. *Environ. Sci. Tech.*, 22 (1988) 22.
9. Chiou, C. T., and T. D. Shoup, Soil Sorption of Organic Vapors and Effects of Humidity on Sorptive Mechanism and Capacity, *Environ. Sci. Tech.*, 19 (1985) 1196-1200.
10. Brunauer, S., P. H. Emmett, and E. Teller, Adsorption of Gases in Multimolecular Layers, *J. Am. Chem. Soc.*, 60 (1938) 309.
11. Satterfield, C. N. Heterogeneous Catalysis in Practice, McGraw Hill, New York, 1980.
12. Ruthven, D. M. Principals of Adsorption and Adsorption Processes, John Wiley, New York, 1984.
13. Young, D. M. and A. D. Crowell, Physical Adsorption of Gases, Butterworths, London, 1962.
14. Hines, A. L. and R. N. Maddox, Mass Transfer-Fundamentals and Applications, Printice Hall, New Jersey, 1985.
15. Hill, T. L. Statistical Mechanics, McGraw Hill, New York, 1960.
16. Anderson, R. B. Modification of Brunauer, Emmet and Teller Equation, *J. Am. Chem. Soc.*, 68 (1946) 686-691.
17. De Boer, J. H. The Dynamical Character of Adsorption, Clarendon Press, Oxford, 1953.
18. Guggenheim, E. A. Applications of Statistical Mechanics, Clarendon Press, Oxford, 1966.

# COMPETITIVE ADSORPTION OF CHLORINATED ALIPHATIC HYDROCARBONS FROM AQUEOUS MIXTURES ONTO SOIL

NILUFER H. DURAL
Department of Chemical Engineering
Cleveland State University
1960 East 24th Street, SH 455
Cleveland, Ohio 44115

DONGLIN PENG
Department of Chemical Engineering
Cleveland State University
1960 East 24th Street, SH 455
Cleveland, Ohio 44115

## ABSTRACT

Competitive adsorption of carbon tetrachloride, chloroform and 1,1,1-trichlorethane from aqueous mixture onto soil was investigated. The experimental data were measured through batch equilibration studies in conjunction with GC/MS analysis. Single component and multicomponent sorption isotherms were generated on three representative soil samples with different properties. Competitive effects were analyzed by comparing single component partitioning to binary and ternary solute partitioning, and were quantified in terms of selectivity of an individual compound with respect to another constituent in the mixture.

Sorption linearity was observed in all cases, even in the vicinity of saturation, indicating constant retardation during the subsurface transport. Sorption of the same compound on different soils was closely related to the soil's organic matter content. The order of sorption uptake by different soils was also consistent with the order of clay percentage and the specific surface area. It was concluded that the sorption process was governed by, both, partitioning of the compound on the soil organic matter and accumulation on the inorganic surface. Competitive sorption from multisolute solutions was within a selectivity range of 1.17-2.22. While the contaminant properties dominated the order of competition, soil properties determined its quantification. An opposite relationship between the organic matter content and the selectivity for the compound with higher polarity was observed.

## INTRODUCTION

Halogenated organic solvents are widely used as degreasing/cleaning agents as well as in production of polymers, fertilizers, pesticides and pharmaceuticals. Many of these chemicals, including carbon tetrachloride, chloroform, and 1,1,1-trichloroethane, are aliphatic chlorinated organic compounds which have been identified as priority substances by the U.S.E.P.A. in several regions. To assess the potential for them reaching groundwater, the kind of physical and chemical processes that occur in the subsurface must be understood.

Sorption processes, in combination with physicochemical properties, play a major role in determining the mobility and the fate of a pollutant in soil. More precisely, soil sorption determines to what extent a compound will vaporize from soil surface or will be transported by water or any other solvent to groundwater. The main focus of research on soil sorption has been on single solute behavior. Studies including multisolute behavior are limited [1-6]. Contaminating solutes are often released to the subsurface environment as multicomponent mixtures rather than a single solute waste stream. The sorption behavior of a single compound may be changed when other contaminants present in the media compete for the available sorption sites, and previously adsorbed chemicals can be remobilized by an increase in the concentration of competing compounds that are strongly adsorbed. Thus, any realistic consideration of chemical fate of a given pollutant requires information on the sorption characteristics of individual compound as modified by other pollutants in the mixture.

The objective of the work reported herein was to examine the multisolute sorption equilibrium of three chlorinated priority compounds, namely chloroform, carbon tetrachloride and 1,1,1-trichloroethane, on three soil samples with different chemical/physical properties. This objective was achieved by conducting single, binary, and ternary solute batch sorption studies. The results are expected to provide insight to the interactions between multisolute systems and soil, which can be used in risk assessment and contaminant removal studies.

## EXPERIMENTAL SECTION

Chloroform and carbon tetrachloride were obtained from Aldrich Co. and had purities of 99.8% and 99.9+%, respectively. 1,1,1-trichloroethane with 99.8% purity was supplied by Fisher Scientific Co. Three representative soil samples obtained from Times-Beach-Missouri, Visalia-California and Eglin-Florida were used as adsorbents. Soil compositions are shown in Table I. Prior to their use, the samples were further characterized with respect to specific surface area and pore size by using an automated BET sorptometer-surface analyzer and nitrogen gas at liquid nitrogen temperature (-195.8°C). Physical characteristics of the soils are presented in Table II. Aqueous solutions were prepared with double distilled water.

Experimental methods of study for adsorption from solutions are based on

Table I - Soil Compositions

| Origin | Sand, % | Silt, % | Clay, % | OM* | pH |
|--------|---------|---------|---------|-----|-----|
| Missouri | 11.4 | 52.7 | 33.4 | 2.4 | 6.9 |
| California | 45.1 | 35.2 | 21.7 | 1.7 | 8.1 |
| Florida | 91.7 | 6.3 | 2.0 | 1.6 | 4.7 |

* OM: Organic Matter

the determination of the concentration of a solute in a solution before and after this solution is brought into contact with the adsorbent. While adsorption equilibrium is being established, the solution is vigorously shaken to ensure uniformity. After the equilibrium is established, the sample is separated or filtered before the final concentration is measured. In the present study, adsorption of chlorinated hydrocarbons by soils was estimated from the change in solute concentration on exposure of the solution to soil as outlined below.

Prior to the experiments the soil samples were oven dried and kept in a desiccator. 1,1,1-trichloroethane, chloroform and carbon tetrachloride were dissolved in distilled water to form aqueus solutions of various concentrations (1-15 mg/L). 40 ml amber glass serum bottles with teflon lined septa were used

Table II - Physical Characteristics of the Soil Samples

| | Missouri Soil | California Soil | Florida Soil |
|--------|---------------|-----------------|--------------|
| Specific Surface Area, $m^2/g$ | 44.1444 | 25.3263 | 23.4772 |
| Average Pore Diameter, A° | 17.6901 | 15.9946 | 13.2829 |
| Total Pore Volume, $m^3/g$ | 0.0195 | 0.0101 | 0.0089 |
| Median Pore Diameter, A° | 24.4620 | 17.6741 | 14.0771 |

as experiment containers. A soil:solution ratio of about 1:7 was choosen to provide optimum contact. In a typical experiment, approximately 5 grams of soil sample and 35 ml of solution was added to the serum bottle. Special attention was paid to eliminate the headspace. Initially, a wide initial concentration range, up to 90% of solubility was tested on Missouri soil. It was observed that the isotherms were linear, showing no indication of curvature even at concentrations approaching saturation. Therefore, the equilibrium uptake values for the rest of the experimental studies were determined using a lower concentration range with the intend to increase the reliability of the distribution coefficients obtained.

Each measurement was carried out at constant temperature (20°C), until the equilibrium is reached, in a reciprocating shaker bath fitted with a thermostat. The system was shaken at a rate which ensured that all soils were mixed well with the solution during the experiment. The equilibrium time was found from the kinetic batch studies, which was approximately 40 hours. To ensure complete equilibration, 48 hours shaking period was employed.

At the end of the shaking period, the samples were directly centrifuged in serum bottles in order to prevent volatilization. The supernatant was removed from the bottle by using a syringe and was taken for concentration measurement. The equilibrium (final) concentrations were measured with a gas chromatograph/mass spectrometer (Perkin-Elmer-Qmass 910). Measurements were carried out in duplicate.

In calculating the adsorption equilibrium uptake, it was assumed that the solution was uniformly distributed throughout the pores of the adsorbent and the supernatant, and that any decrease in the concentration of chlorinated hydrocarbons in the mixture was due to the adsorption by soil. Since the environmental concentration levels of the contaminant solutes are usually in ppm range, as in the present work, the total solution volume was assumed to remain constant. In order to check the validity of this assumption, volume of the supernatant was measured for a few samples. No noticable change was observed within the accuracy provided by the experimental instrumentation.


## RESULTS AND DISCUSSION

The adsorption equilibrium isotherms describing binding tendencies of chloroform, carbon tetrachloride and 1,1,1-trichloroethane on three well characterized soil samples were measured at 20°C. For all chlorinated hydrocarbons examined, the isotherms were essentially linear, showing no indication of curvature at even concentrations approaching saturation. This implies constant retardation factor for subsurface transport.

The results of single solute isotherm studies are presented in Table III for each solute-soil pair. The Henry's isotherm constants (K) were calculated in the standard manner, namely the ratio of the amount of contaminant adsorbed in micrograms per gram of soil to the equilibrium concentration in parts per million (mg/L). As can be observed from Table III, all soils adsorbed appreciable quantities of chloroform in comparison to 1,1,1-trichloroethane and

531

Table III - Single Solute Linear Isotherm Constants (K)

| Soil Sample | Compound | K | Standard Error |
|---|---|---|---|
| Missouri | $CHCl_3$ | 2.133 | 0.034 |
| | $C_2H_3Cl_3$ | 1.804 | 0.040 |
| | $CCl_4$ | 1.695 | 0.039 |
| California | $CHCl_3$ | 1.941 | 0.078 |
| | $C_2H_3Cl_3$ | 1.592 | 0.053 |
| | $CCl_4$ | 1.483 | 0.061 |
| Florida | $CHCl_3$ | 1.763 | 0.027 |
| | $C_2H_3Cl_3$ | 1.338 | 0.029 |
| | $CCl_4$ | 1.123 | 0.012 |

carbon tetrachloride; while carbon tetrachloride was adsorbed in the least amount. In contrast to the claims for strong impact of solubility on sorption [7], the Henry's isotherm constants were not inversely correlated with the corresponding solubilities. This can be related to the extent of the forces involved in the binding process, which demonstrates that chloroform interacts significantly with the binding sites on the soil samples. Consequently, the strong positive interaction between the surface and the adsorbate overcomes even a fairly strong solute-solvent interaction. In a previous investigation [8], in which vapor phase adsorption of the same chlorinated hydrocarbon-soil pairs was studied, similar trends were observed and the trend was in agreement with the heat of adsorption values calculated. Hence, although the isotherms obtained in the present work are linear in shape, the sorption was primarily as a result of adsorption (soil-solute interaction) forces rather than partitioning.

The isotherms of a specific solute on three soils were qualitatively similar but equilibrium uptakes corresponding to the same equilibrium concentration were different. In all cases, different soils adsorbed the same solute with the following order: Missouri > California > Florida. The different uptake capacities of the soils can be attributed to both chemical composition and physical structure of the soil (Figures 1 and 2).

Earlier studies imply that the organic matter present in the soil is an important factor in the adsorption process [7]. As can be observed from Tables I and III, the trend of adsorption capacities of the soils is the same as the trend of their organic matter content.

As the clay percentage increases or the fraction of organic matter becomes

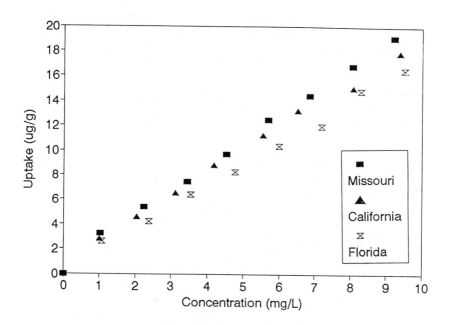

Figure 1 Adsorption Isotherms of Chloroform on Different Soils

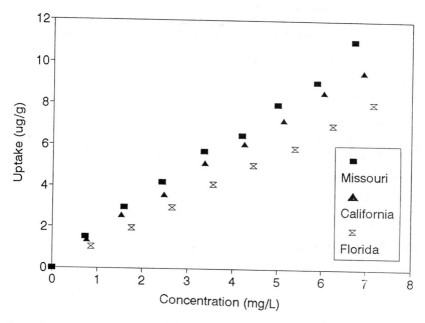

Figure 2 Adsorption Isotherms of Carbon Tetrachloride on Different Soils

small, adsorption to inorganic mineral surface becomes significant. The mineral content and the organic matter provide most of the surface area available for adsorption. Since adsorption is a surface phenomena, the uptake capacity of a specific soil depends on its specific surface area and pore size as well as the size of the adsorbate molecule. The specific surface areas, measured by BET sorptometer (Table II) exhibited the following trend: Missouri > California > Florida. Therefore the order of adsorption uptake, for all pollutants, was expected to follow the same trend. This is exactly reflected in Table III. Furthermore, this trend is consistent with the clay percentage of the samples, indicating that higher retardation factors are to be expected in soils made up with mostly clays. On the other hand, pore size and its distribution are of importance in adsorption process. As shown in Table II, the average pore diameter of Missouri soil is the largest one and that of Florida soil is the smallest. Thus, the wide pore spaces provided by Missouri soil permit the molecules of the adsorbate to penetrate and proceed the adsorption.

Competitive sorption studies were carried out in two steps. First, the sorption isotherms of $CHCl_3$-$CCl_4$, $CHCl_3$-$C_2CH_3Cl_3$, and $C_2CH_3Cl_3$-$CCl_4$ binary pairs on three soils were generated. The results were evaluated in terms of binary distribution coefficients ($K_b$), relative distribution coefficients ($K_b/K$) and selectivities. Then, ternary solute isotherms ($CHCl_3$, $C_2CH_3Cl_3$, $CCl_4$) were measured and analyzed in terms of ternary distribution coefficients ($K_T$), relative distribution coefficients ($K_T/K$) and selectivity of an individual compound with respect to another constituent in the mixture. The intent was to observe the influence of the additional species on the previously studied binary pairs.

The binary distribution coefficients ($K_b$), presented in Tables IV through VI, are significantly lower than the single solute coefficients ($K$). However, they followed a similar trend on a specific soil: $CHCl_3 > C_2CH_3Cl_3 > CCl_4$. The

Table IV - Binary Solute Distribution Coefficients ($K_b$) of $CHCl_3$ and $CCl_4$

| Soil Sample | Compound | $K_b$ | Standard Error | $K_b/K$ |
|---|---|---|---|---|
| Missouri | $CHCl_3$ | 1.179 | 0.030 | 0.55 |
| | $CCl_4$ | 0.813 | 0.017 | 0.48 |
| California | $CHCl_3$ | 1.128 | 0.028 | 0.58 |
| | $CCl_4$ | 0.734 | 0.011 | 0.49 |
| Florida | $CHCl_3$ | 0.844 | 0.018 | 0.47 |
| | $CCl_4$ | 0.410 | 0.008 | 0.36 |

Table V - Binary Solute Distribution Coefficients ($K_b$) of $CHCl_3$ and $C_2H_3Cl_3$

| Soil Sample | Compound | $K_b$ | Standard Error | $K_b/K$ |
|---|---|---|---|---|
| Missouri | $CHCl_3$ | 1.303 | 0.047 | 0.61 |
| | $C_2H_3Cl_3$ | 0.847 | 0.026 | 0.47 |
| California | $CHCl_3$ | 1.252 | 0.037 | 0.64 |
| | $C_2H_3Cl_3$ | 0.789 | 0.020 | 0.44 |
| Florida | $CHCl_3$ | 1.205 | 0.035 | 0.68 |
| | $C_2H_3Cl_3$ | 0.542 | 0.019 | 0.40 |

magnitude of decrease in $K_b$, which can be deducted from the relative distribution coefficients of Tables IV, V and VI, is different for each compound with the following order: $CHCl_3 < C_2CH_3Cl_3 < CCl_4$. This order is inversely related to their polarity, which suggests that different molecules are inclined to compete for the same site primarily because of their different polarity.

The multisolute distribution coefficients ($K_T$) are presented in Table VII. As expected, $K_T$ values are substantially lower than the $K_b$ values. However, they followed the same trend, both, for their magnitude and for the decrease in their

Table VI - Binary Solute Distribution Coefficients ($K_b$) of $CCl_4$ and $C_2H_3Cl_3$

| Soil Sample | Compound | $K_b$ | Standard Error | $K_b/K$ |
|---|---|---|---|---|
| Missouri | $CCl_4$ | 0.914 | 0.016 | 0.54 |
| | $C_2H_3Cl_3$ | 1.137 | 0.026 | 0.63 |
| California | $CCl_4$ | 0.858 | 0.009 | 0.57 |
| | $C_2H_3Cl_3$ | 1.075 | 0.023 | 0.67 |
| Florida | $CCl_4$ | 0.710 | 0.024 | 0.62 |
| | $C_2H_3Cl_3$ | 0.938 | 0.014 | 0.70 |

Table VII - Ternary Solute Distribution Coefficients ($K_T$)

| Soil Sample | Compound | $K_T$ | Standard Error | $K_T/K$ |
|---|---|---|---|---|
| Missouri | $CHCl_3$ | 0.920 | 0.024 | 0.43 |
| | $C_2H_3Cl_3$ | 0.710 | 0.020 | 0.39 |
| | $CCl_4$ | 0.607 | 0.012 | 0.36 |
| California | $CHCl_3$ | 0.555 | 0.014 | 0.29 |
| | $C_2H_3Cl_3$ | 0.401 | 0.011 | 0.25 |
| | $CCl_4$ | 0.334 | 0.012 | 0.22 |
| Florida | $CHCl_3$ | 0.490 | 0.008 | 0.28 |
| | $C_2H_3Cl_3$ | 0.317 | 0.007 | 0.23 |
| | $CCl_4$ | 0.242 | 0.005 | 0.21 |

magnitudes. These results confirm that the polarity of the compound, rather than its aqueous solubility, dominates the competitive adsorption.

The selectivity information is presented in Table VIII. Accordingly, chlorinated aliphatic hydrocarbons compete moderately for the sites available for adsorption on soil, within a selectivity range from 1.17 to 2.22. While the adsorbate properties determine the nature of the competition, soil properties determine its quantification. Although all selectivities from the ternary solute mixture are in accordance with the polarity of the compounds, selectivities of $CHCl_3$-$C_2H_3Cl_3$ and $CHCl_3$-$CCl_4$ from the binary solute mixtures display slightly different trend. Since carbon tetrachloride is nonpolar, chloroform would have been expected to have higher selectivity with respect to carbon tetrachloride (as in the case of ternary mixture). This behavior can be attributed to the aqueous solubilities of the compounds. The solubility of chloroform is 8.3 times that of carbon tetrachloride while it is only 2.22 times that of 1,1,1-trichloroethane. The great difference in the solubilities result in a higher tendency for chloroform to stay in the aqueous pfase for the $CHCl_3$-$CCl_4$ pair with respect to other cases. Nevertheless, the selectivity in favor of chloroform (1.45) implies the dominance of polarity effect.

Further examination of Table VIII shows that selectivity of a compound from the same mixture is different on different soils. In all cases, the selectivity for the compound with higher polarity increased on different soils with the following order: Missouri < California < Florida. This trend is opposite to the organic matter content of the soils, suggesting that inorganic (mineral) constituents of the soil is more selective to the polar compounds. Thus, in soils with low organic matter, higher competition among the organic contaminants should be expected. However, total sorption capacity will tend to be low. These findings support Wang et al's [9] claim related to sorption of toxic organic compounds on wastewater solids. Accordingly, the degree of competition depends on whether partitioning or adsorption dominates the sorption. For compounds whose sorption uptake is primarily that of adsorption, the competition will be significant; for compounds whose sorption uptake is primarily that of partitioning, the competition effect will be negligible.

Table VIII - Selectivity (S) Information

| Soil Type | B/T | $S_{1-2}$ | $S_{1-3}$ | $S_{2-3}$ |
|-----------|-----|-----------|-----------|-----------|
| Missouri | B | 1.54 | 1.45 | 1.24 |
| | T | 1.30 | 1.52 | 1.24 |
| California | B | 1.58 | 1.54 | 1.25 |
| | T | 1.38 | 1.66 | 1.20 |
| Florida | B | 2.22 | 2.05 | 1.32 |
| | T | 1.54 | 2.07 | 1.31 |

$1 = CHCl_3$      B= in binary mixture
$2 = C2CH_3Cl_3$      T= in ternary mixture
$3 = CCl_4$

**REFERENCES**

1. Goldberg, S. 1985. *Soil Sci. Soc. Am. J.*, **49**, 851.
2. Murali, V. 1983. *Soil Sci.*, **136**, 279.
3. Zachara, J. M. 1987. *Environ. Sci. Technol.*, **21**, 397.
4. Trania, S. J. 1991. *J. Contam. Hydrol.*, **7**, 237.
5. Lee, J., J. R. Crum, and S. A. Boyd. 1989. *Environ. Sci. Technol.*, **23**, 1365.
6. Wood, A. L., D. C. Bouchard, M. L. Brusseau, and S. C. Rao. 1990. *Chemosphere*, **21**, 575.
7. Chiou, T. C., J. K. Peters, and H. V. Freed. 1979. *Science*, **206**, 831.
8. Chen, C. H. "Adsorption of Chlorinated Hydrocarbon Vapors on Soils". 1993. M. S. Thesis. University of Missouri-Columbia.
9. Wang, L., and R. Govind. 1993. *Environ. Sci. Tech.*, **27**, 152.

# NEURAL NETWORKS: CAN THEY PREDICT THE PERFORMANCE OF ADSORPTION COLUMNS?

**YACOUB M. NAJJAR and IMAD A. BASHEER**
Department of Civil Engineering
Kansas State University
Manhattan, KS 66506

## ABSTRACT

Removal of contaminants from water by adsorption onto activated carbon is an effective unit operation in the water industry. In the present study, neural networks (NNs) are used to determine the performance of activated carbon adsorption columns based on parameters pertaining to both the hydrodynamic conditions of the column and the physical properties of the contaminant. Two examples representing two different classes of data quality are considered. The data of the first example pertain to pilot and large-scale adsorption columns operating to remove a number of volatile organic compounds (VOCs) from groundwater. The other example deals with synthetic data extracted from a numerically-solved mathematical model of the column adsorption process. Because of the uncertainty and the high variability in the large-scale data, the developed NN was found to be relatively less capable of generalizing the pattern involved as compared to the excellent generalization observed in the case of the synthetic data.

## INTRODUCTION

The public awareness of the dangers associated with water contamination has opened new horizons for developing effective unit process operations for treating waters. Adsorption on granular activated carbon (GAC) is recognized as an effective treatment technology for removal of a wide spectrum of contaminants from water and wastewater [1]. Commonly, adsorption is operated in fixed beds where the GAC is held stationary in the column and the contaminated water is allowed to pass (in downflow or upflow mode) through the bed. Through its passage, the contaminants in the water are adsorbed on the GAC. For long passage times, the adsorbent closer to the inlet of the fluid

538

becomes saturated with the contaminant. This gives rise to a concentration front that divides the bed into a contaminated region and another intact one. This concentration front (also called the mass transfer zone) continues to travel along the bed as more influent is being pushed into the bed. At the time the leading edge of the concentration front has reached the exit end of the bed, breakthrough is said to have been attained and the first emergence of the contaminant has just taken place. The breakthrough curve represents the history of effluent concentration along the course of operation. The point at which breakthrough occurs is called the breakthrough point. The location of the breakthrough point and the shape of the breakthrough curve (usually S-shaped) are influenced by many parameters pertaining to the GAC, the adsorbate (contaminant), the column operating conditions, and the column dimensions. For the design of fixed-bed adsorbers, the dynamic response of the adsorption process is considered as an essential part for the design. Since pilot and laboratory-scale fixed-bed adsorption experiments on a specific target compound tend to be lengthy and expensive, predictive mathematical models that describe the adsorption behavior of fixed beds have been extensively investigated to replace experimental testing. For the operators of adsorption columns, prediction of the breakthrough time can be an asset for identifying, long ahead of time, the point at which the operation is to be terminated and the column is to be replaced. In the present study, prediction of the breakthrough point as well as the entire breakthrough curve will be investigated using NNs that utilize data relevant to adsorption process. Since the quality of input data has a great influence on the prediction accuracy of the developed NNs, prediction accuracy is compared with data obtained from pilot-scale column runs (VOC adsorption) and from a mathematically-based model outputs (phenol adsorption).

## NEURAL NETWORKS

A neural network (NN) is a highly interconnected network of many simple processing elements (called neurons) that can perform massively parallel computation for data processing and knowledge representation. Such a network is capable of learning the pattern associated with a large body of data that describe a physical phenomenon. It is usually constructed of three or more different layers each containing a number of neurons. The input layer contains the input variables that are speculated to affect the model output while the output layer contains the desired output variables. The hidden layer(s), is placed between the input and output layers to enable the network deal with highly nonlinear and complex problems. The neurons are interconnected to each other through connection weights that represent the relative strength of an input neuron in contributing to the output neuron(s). These connection strengths are incrementally changed through supervised learning process every time the network is presented with a new example. Eventually, the network converges to a set of connection values that enable prediction of the output(s) for all training sets at the desired pre-assigned degree of accuracy. The required number of nodes in a hidden layer is generally determined by a-trial-and-error procedure. This can be performed by varying the number of hidden nodes until the network becomes able, to a certain level of accuracy, to learn the pattern involved in the training data sets. It is obvious that this step can only be implemented after selecting the training sets which the NN is to be presented with in order for the network to identify the patterns involved. The training sets should represent a wide variety of all possible features and sub-features that the network is required to learn. There are several NN paradigms available in the literature [2]. The type of network to be used can be selected depending on the type of problem under investigation. In most civil engineering applications, the supervised learning error-backpropagation neural network (BPNN) algorithm is found to be successful in capturing patterns embodied in a specific set of data. A computer program for the BPNN algorithm was written for this purpose. The algorithm of the BPNN adopted in this study can be found in [3].

## MODELING ADSORPTION OF PILOT-STUDY COLUMNS

Major sources of organic compounds traced in groundwater are commonly linked with the end-products of nearby manufacturing industrial plants. Most frequently occurring organic contaminants in groundwater are VOCs. By 1983, 16 VOCs were found to be the chief pollutants of groundwater and can impart health hazards even at very low concentration [4]. Purification of drinking water by adsorption on GAC has proved to be a cost-effective treatment technology for a large number of VOCs [5].

### The VOC Column Adsorption Database

For a specific adsorbent, the adsorbate-adsorbent interaction is simply reflected by the adsorption isotherm constants. Hence, the Freundlich isotherm constants, K and 1/n [6] are essential to determine the breakthrough time. A large data set containing adsorption of a wide variety of VOCs from water onto GAC under a large number of operating conditions is rarely found in the literature. A summary of 40 adsorption laboratory and pilot column studies conducted on 10 different VOCs are reported by [4]. Due to the high variation in the influent concentration and other operating conditions the value of the breakthrough time was not clearly identified for 12 of the 40 cases reported by the original report of [4]. Consequently, average values were adopted for the breakthrough time to replace these cases where ranges of breakthrough time are only reported. The values of K and 1/n for all the VOCs studied are obtained from [6]. Table I is the revised database to be used in developing the NN for estimating the adsorption breakthrough time of the VOCs considered in the present study. As can be noted from Table I, the final database includes 7 different VOCs and 34 data sets. The empty bed contact time (EBCT) is a combined measure of both the volume of the GAC bed, V, and the influent flowrate, Q, given by EBCT=V/Q. Since the bed depth, H, is also given, the

# TABLE I-DATA USED TO DEVELOP THE VOC NEURAL NETWORK

| Volatile Organic Compound | K | 1/n | Co (ug/L) | H (m) | EBCT (min) | BV |
|---|---|---|---|---|---|---|
| Trichloroethylene | 28.00 | 0.62 | 177 | 0.8 | 9.0 | 20160 |
| | 28.00 | 0.62 | 4 | 0.8 | 8.5 | 92120 |
| | 28.00 | 0.62 | 3 | 0.8 | 18.0 | 32500 |
| | 28.00 | 0.62 | 0.4 | 0.8 | 18.0 | 32500 |
| Tetrachloroethylene | 51.00 | 0.60 | 1400 | 0.8 | 9.0 | 12300 |
| | 51.00 | 0.60 | 94 | 0.8 | 18.0 | 32500 |
| | 51.00 | 0.60 | 9 | 0.8 | 18.0 | 32500 |
| | 51.00 | 0.60 | 4 | 0.8 | 8.5 | 92120 |
| | 51.00 | 0.60 | 1 | 0.8 | 9.0 | 20160 |
| 1,1,1-Trichloroethane | 2.50 | 0.30 | 100 | 0.6 | 7.5 | 1300 |
| | 2.50 | 0.30 | 100 | 1.2 | 15.0 | 2700 |
| | 2.50 | 0.30 | 100 | 1.8 | 22.5 | 3900 |
| | 2.50 | 0.30 | 237 | 0.8 | 18.0 | 15700 |
| | 2.50 | 0.30 | 23 | 0.8 | 18.0 | 32500 |
| | 2.50 | 0.30 | 38 | 0.8 | 8.5 | 11800 |
| | 2.50 | 0.30 | 1 | 0.8 | 9.0 | 16400 |
| Carbon Tetrachloride | 11.10 | 0.83 | 12 | 0.8 | 10.0 | 14000 |
| | 11.10 | 0.83 | 12 | 0.8 | 5.0 | 6050 |
| Cis-1,2-Dichloroethylene | 6.50 | 0.70 | 18 | 0.8 | 6.0 | 4100 |
| | 6.50 | 0.70 | 18 | 1.5 | 12.0 | 7100 |
| | 6.50 | 0.70 | 18 | 2.3 | 18.0 | 8100 |
| | 6.50 | 0.70 | 6 | 0.8 | 9.0 | 14200 |
| | 6.50 | 0.70 | 2 | 0.8 | 8.5 | 29600 |
| 1,2-Dichloroethane | 3.57 | 0.83 | 8 | 0.8 | 20.0 | 1700 |
| | 3.57 | 0.83 | 1.4 | 0.8 | 18.0 | 17400 |
| | 3.57 | 0.83 | 0.8 | 0.8 | 18.0 | 32500 |
| | 3.57 | 0.83 | 2 | 0.9 | 11.0 | 3300 |
| | 3.57 | 0.83 | 2 | 1.8 | 22.0 | 3400 |
| | 3.57 | 0.83 | 2 | 2.7 | 33.0 | 4150 |
| | 3.57 | 0.83 | 2 | 3.6 | 44.0 | 7000 |
| | 3.57 | 0.83 | 2 | 0.8 | 17.5 | 2500 |
| | 3.57 | 0.83 | 2 | 0.7 | 17.0 | 2500 |
| 1,1-Dichloroethylene | 4.90 | 0.50 | 122 | 0.8 | 18.0 | 22400 |
| | 4.90 | 0.50 | 4 | 0.8 | 18.0 | 33600 |

EBCT also implies the surface area of the column adsorber or alternatively the flow velocity which can be determined as $v = H/EBCT$. The breakthrough time, $T_B$ is given in Table I in units of number of bed volumes (BV) passed through the bed at time of breakthrough and can be calculated from $BV = QT_B/V$. Therefore, the time to breakthrough can be determined as $T_B = BV*(EBCT)$.

## Developing the VOC Neural Network

For the present adsorption data, the input layer of the NN should contain five input variable representing both the column-operating parameters and the adsorbate/adsorbent characteristics parameters. The former category includes the influent concentration $C_o$ ($\mu g/L$), the depth of the GAC bed, H (m), and the

EBCT (min). The adsorbate/adsorbent interaction category includes the Freundlich isotherms constants, K and 1/n. On the other hand, the output layer includes the number of bed volumes (BV) that has passed through the bed till the breakthrough time is achieved. The optimal number of hidden nodes is determined by varying the number of hidden nodes in a network and calculating the corresponding coefficient of determination at the end of each training phase (using only 100,000 training cycles). The NN with the maximum coefficient of determination ($r^2 = 0.800$) was found to be the one with two hidden nodes. A schematic architecture of the developed NN is shown in Fig. 1. Although the obtained coefficient of determination may not be considered high enough, it may be reasonably regarded sufficient in the present study to illustrate the overall generalization for the given poor-quality data. A comparison between the NN-predicted BV and actual BV for the 34 cases of column adsorption is shown in Fig. 2. As can be seen from the plot, there is a number of cases where the points have deviated substantially from the line of equality. The insufficient level of generalization obtained in modeling the data given in Table I may imply that the data contains errors especially in the value of breakthrough BV.

## MATHEMATICAL MODEL OUTPUT-BASED NEURAL NETWORK

### The HSDM

As can be noted from the cases of VOC adsorption, the generalization and prediction accuracy of the NN significantly depend on the quality of the training sets (or the input-output patterns) used. In order to test the ability of NNs to obtain generalization for adsorption data of high quality, a mathematical model is utilized herein to generate the systematic data. The HSDM (Homogenous Surface Diffusion Model) has been used by many investigators and found to successfully simulate fixed-bed adsorption breakthrough curves for a large

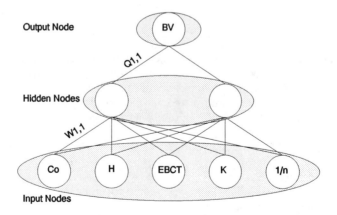

Figure 1 A Schematic of the Developed VOC Neural Network.

542

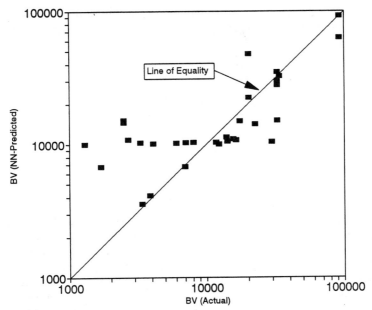

Figure 2 Comparison Between Actual and VOC NN-Predicted BV.

number of adsorbate-adsorbent systems [7,8]. Therefore, to investigate the feasibility of utilizing NN techniques in modeling the adsorption behavior, the HSDM is used to generate the breakthrough point as well as the entire breakthrough curves for various desired sets of operating conditions.

Generally, the HSDM represents a numerical simulation of a set of partial differential equations that describe the mass balance of adsorbate in both the entire bed and the adsorbent particle, linked with an adsorption equilibrium expression. Depending on the assumptions and simplifications involved in developing the model, and the type of the initial and boundary conditions imposed, the model in most cases may not have an analytical solution and it is most likely that numerical techniques are needed for solution. The assumptions and the various differential equations that describe the HSDM are discussed elsewhere [7]. The various governing differential equations may be stated as:

(1) The mass balance of the adsorbate in the entire bed and the corresponding initial and boundary conditions is expressed as:

$$\frac{\partial C}{\partial t} = [-v\frac{\partial C}{\partial z} - \frac{3(1-\epsilon_b)}{R_p\epsilon_b}] k_f (C-C_s)$$
$$@t=0 ; 0 \leq z \leq L : C=0$$
$$@t \geq 0 ; z=0 : C=C_o$$

(1)

543

(2)  The Rate of accumulation of adsorbate on surface of adsorbent particle can be described by the following partial differential equation and initial and boundary conditions [8]:

$$\frac{\partial q}{\partial t} = \frac{D_s}{r^2} \frac{\partial}{\partial r} (r^2 \frac{\partial q}{\partial r})$$

$$@t=0 \; ; 0 \leq r \leq R_p : q=0$$

$$@t \geq 0 \; ; r=0 : \frac{\partial q}{\partial r} = 0$$

$$@t \geq 0 \; ; r=R_p : R_p^2 k_f (C-C_s) / \rho = \frac{\partial}{\partial t} \int_0^{R_p} q r^2 dr$$

$$@r=R_p \; ; C_s = f(q_s)$$

(2)

where r is the radial distance within the spherical adsorbent particle, $R_p$ is the radius of the particle, z is the axial distance along the column, L is the length of the column, v is the superstitial fluid velocity in the bed, $\epsilon_b$ is the bed porosity, q is the surface adsorbate concentration, $D_s$ is the solid phase diffusion coefficient, C is the concentration of adsorbate in the liquid phase, $C_0$ is the influent concentration, $C_s$ and $q_s$ are the adsorbate liquid and solid concentration at the external surface of the particle, respectively, $\rho$ is the apparent density of the particle, $k_f$ is the mass transfer coefficient, and t is the time since the beginning of operation. The equation $C_s = f(q_s)$ represents a general expression for the adsorption isotherm. The present model equations are numerically solved by [7] using the orthogonal collocation method. It is to be mentioned herein that the solid phase diffusion coefficient, $D_s$, has to be determined from batch experiments. On the other hand, the mass transfer coefficient, $k_f$ can also be determined from batch experiments as well as from correlations such as those described by [9]. The present model (HSDM) is used in this research to generate the fixed-bed adsorption database.

**Defining the Problem Domain**

The problem used herein to generate the adsorption breakthrough data involves treating, by GAC adsorption, water contaminated with phenol. The various fixed input parameters and the assumption used in running the HSDM are (i) the solid phase diffusion coefficient, $D_s$, of phenol on GAC is assumed to be constant regardless of the adsorbent particle size. A value of $D_s = 3.5 \times 10^{-8}$ cm/sec [8] is used in all runs, (ii) an apparent density of GAC = 0.68 and a constant bed porosity (voidage) of 0.40 are assumed constant for all runs, (iii) the three-parameter isotherm expression $q = AC/(1+BC^\beta)$, where A, B, and $\beta$ are constants, is selected to represent the equilibrium. For phenol adsorption from aqueous solutions at 25°C and a pH of 7, the values of A, B, and $\beta$ are taken as 15.11, 7.547, and 0.8685, respectively [8], and (iv) uniform adsorbent particle size, constant influent flowrate, and constant influent concentration of phenol are assumed to be prevailing conditions in all tests. Based on these assumptions, seven input variables are found to be essential to run the HSDM.

These variables are: the influent concentration, $C_0$, the influent flowrate, Q, the weight of GAC in the bed, W, the diameter of the cylindrical adsorber, D, the length of the bed, L, the adsorbent particle diameter, $d_p$, and the external mass transfer coefficient, $k_f$. However, the length of the bed could be related to both the weight of GAC and the adsorber diameter using the equation [L=2.45 W/A] where A is the cross sectional area of column in $cm^2$, and W in grams (g). Moreover, the external mass transfer coefficient could also be related to the cross sectional area of the adsorber, the flowrate, and the particle diameter using the correlation of [9]. This reduces the number of independent variables to five; namely Co, Q, A, W, and $d_p$. To further reduce the number of combinations involved, all runs were made at constant flowrate of Q=500 $cm^3$/min and a column diameter of 2" (5.08 cm). This in turn reduces the input parameters to $C_0$, W, and $d_p$.

## Generating Data for Training

To create the database, several values of the various input parameters are selected within certain ranges that represent values practically encountered in laboratory adsorption studies. Consequently, the HSDM program is run to obtain the corresponding predicted breakthrough curves. The data for the different runs are designed such that they simulate only laboratory fixed-bed adsorption experiments. Three particle sizes; represented by their geometric mean particle radii ($R_p$= 0.077, 0.055, and 0.027 cm) correspond to the U.S. mesh sizes (retained/passing):#12/#14, #16/#18, and #30/#35, respectively are used for the various runs. Moreover, the infleunt concentration was varied between 50 and 1000 mg/L, while the weight of GAC in each column was varied between 100 and 1000 g. For each run, a breakthrough curve is obtained which reflects the variation of effluent phenol concentration with the operation time. To fit the output nodes in the NN, the output breakthrough curve was discretized to a number of points equivalent to the number of nodes in the output layer. In this paper, the breakthrough curve is assumed to be represented by six (C,t) pairs which make a six-output NN. The six concentration levels considered are those corresponding to the non-dimensional parameter (C/Co) values of 1, 5, 25, 50, 75, and 95 percent. Therefore, the outputs are the times corresponding to these concentration which are denoted as $t_1$, $t_5$, $t_{25}$, $t_{50}$, $t_{75}$, and $t_{95}$. The two points at 1 and 5 percent are used to capture the breakthrough point which depicts an important element in designing the useful life of fixed-bed adsorbers. The six-point method proposed here is assumed to be sufficient to sketch an approximate yet a close enough breakthrough curve.

## Developing the Phenol Neural Network

A total of 30 training sets are used to train several networks varying with their number of hidden nodes. Five testing sets are used to test the prediction accuracy of the developed NNs. In this case, neural network predictions are compared with their corresponding exact values obtained from the HSDM. The

545

mean of the absolute values of the relative error between the NN and HSDM values of the six outputs and for the five testing sets is used for the comparison. An optimum architecture at five hidden nodes produced an average relative error of only 2.66%. The phenol NN is schematically shown in Fig. 3. This network is selected as the appropriate network for prediction of the breakthrough curves

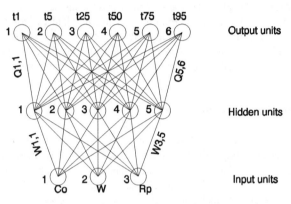

Figure 3 A Schematic of the Developed Phenol Neural Network.

for the problem under investigation. A comparison between breakthrough curves predicted by the 5-hidden nodes-NN and the corresponding HSDM curves for the five testing sets is depicted in Fig. 4. As can be seen from Fig. 4, the NN-predicted breakthrough curves are very close to those simulated by the HSDM.

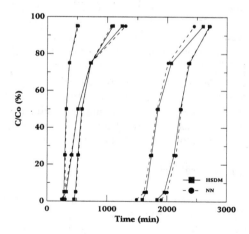

Figure 4 Comparison Between HSDM and Phenol-NN Prediction of BV.

## CONCLUSIONS

Neural networks have been applied to a limited number of environmental engineering problems. In this paper, the feasibility of NN in developing a tool for predicting adsorption breakthrough curves for common contaminants in water and wastewaters is investigated. Prediction of the breakthrough point in adsorption columns is an essential element that influences controlling the water treatment unit operation. Like all regression and curve fitting techniques, NNs are found to be affected by the quality of the data (i.e. training sets). The plot that shows the prediction versus experimental data for the VOC adsorption (Fig. 2) may indicate the accuracy of the developed NN to predict breakthrough point for cases where the number of bed volumes to breakthrough is expected to be high. In other words, the developed VOC-NN can best be used to predict breakthrough times for cases with low $C_0$, small EBCT, and/or large H. Conversely, when the data are of high certainty, as in the case of phenol adsorption, the NN did not only predict the breakthrough point with a high accuracy, but the entire breakthrough curve as well. It is worthy to mention herein that replacement of the HSDM with a NN can be efficient in reducing computer time as well as performing the inverse problem without the need to reformulate the governing equations.

## REFERENCES

1. Clark, R.M. and J.Q. Adams. 1991." Evaluation of BAT for VOCs in Drinking Water," *Journal of Env. Engrg.*, ASCE, 117(2):247-268.
2. Simpson, P.K. 1990. *Artificial Neural Systems: Foundations, Paradigms, Applications, and Implementations*, Pergamon Press, New York, NY.
3. Basheer, I.A. and Y.M. Najjar. 1995. "Designing and Analyzing Fixed-Bed Adsorption Systems With Artificial Neural Networks," Accepted for Publication in the *Journal of Env. Systems*.
4. Love, O.T., R.J. Miltner, R.G. Eilers, and C.A. Fronk-Leist. 1983. *Treatment of Volatile Organic Compounds in Drinking Waters*, EPA-600/8-83-019.
5. ESE, Inc. 1986. *Removal of Volatile Organic Chemicals from Potable Water: Technologies and Costs*, NOYES Data Corp., Park Ridge, NJ.
6. Dobbs, R.A. and J.M. Cohen. 1980. *Carbon Adsorption Isotherms for toxic Organics*, EPA-600/880-023.
7. Thacker, W.E., V.L. Snoeyink, and J.C. Crittenden. 1981. *Modeling of Activated carbon and Coal Gasification Char Adsorbents in Single-Solute and Bisolute systems*, Water Resources Report, No.161, Univ. of Illinois, Ill.
8. Kunjupalu, T. 1986. *Effect of particle Size Distribution on Activated Carbon Adsorption*, M.S. Thesis, Kansas State University, Kansas.
9. Dwiwedi, P.N. and S.N. Upadhyay. 1977." particle-Fluid mass Transfer in Fixed and Fluidized Bed," *Ind. Eng. Chem. Proc. Des. and Dev.*, 16: 157-165.

# CATEGORY VIII:
# *Heavy Metals Removal*

# REMEDIATION OF METAL/ORGANIC CONTAMINATED SOILS BY COMBINED ACID EXTRACTION AND SURFACTANT WASHING

JOHN E. VAN BENSCHOTEN
Associate Professor
State University of New York at Buffalo
Civil Engineering, Buffalo, NY 14260

MICHAEL E. RYAN
Associate Professor
State University of New York at Buffalo
Chemical Engineering, Buffalo, NY 14260

CHIH HUANG
TARA C. HEALY
PAUL J. BRANDL
Graduate Students
State University of New York at Buffalo
Civil Engineering, Buffalo, NY 14260

## INTRODUCTION

The occurrence of metal and hydrophobic organic contaminants (HOCs) at hazardous waste sites is common. In New York State in 1993, there were 934 inactive hazardous waste disposal sites [1]. The most frequently found metals at these contaminated sites included Cr, Cd, Pb, Hg, and As [2]. In a survey by the Department of Energy (DOE), it was found that at 11 disposal sites (out of 91) and 9 facilities (out of 18), the major binary contaminants were metals and hydrocarbons (i.e., HOCs) [3]. The most commonly reported metals in soil/sediment were Cu, Cr, Zn, Hg, and Pb. Due to the differences of chemical and physical properties between metals and organic contaminants, mixed wastes are particularly difficult to remediate. For organic pollutants in soils, several techniques can be used for remediation, including thermal treatment, steam and air stripping, and chemical oxidation. In contrast, there are few treatment techniques available for the remediation of metal-laden soils. Metals can be removed either by flotation or soil washing [4]. The chemical composition of the

551

fluid used for soil washing depends on the contaminants to be removed. For metals, the washing medium may consist of acids, where the acidic solution aids in the solubilization and desorption of metals from soils [5]. In aqueous systems, HOC removal may be aided by surfactants where the surfactant micelles enhance the solubility of HOCs [6]. It was hypothesized that a combination of acid extraction and surfactant washing might enhance the combined extraction of both HOCs and metals in contaminated soils. However, there is little information concerning the feasibility of using surfactants under acidic conditions to treat co-contaminated soils. This study is expected to: (1) provide greater knowledge concerning the interactions among surfactants, HOCs, metals, and soils; (2) identify the factors which affect these interactions; and (3) identify the feasibility of remediating metal/organic contaminated soils by acid extraction and surfactant washing.

## OBJECTIVES

The specific objectives of this research are to: (1) determine the solubilization of two polycyclic aromatic compounds (PAHs) (naphthalene and pyrene) using several surfactants at low pH conditions; (2) determine the losses of candidate surfactants due to precipitation or adsorption to a test soil as a function of pH; and (3) evaluate the performance of surfactants under acidic conditions for removal of lead and PAH compounds from a contaminated soil. In this paper, experimental results related to the first two objectives are presented.

## BACKGROUND

### Removal of Metals by Acid Extraction

The ability to extract metals under acidic conditions is well known. Standard digestion procedures [7] and methods such as sequential extraction techniques [8] are applied under acidic conditions to solubilize metals. Metals in soil systems can exist at several interfacial regions in these heterogeneous systems including the lattice of crystalline minerals, the interlayer positions of clay minerals, adsorption sites on mineral surfaces, adsorption sites on hydrous Fe and Mn oxides and hydroxides, and complexation sites on natural organic matter [9]. For all of these types of binding sites and for nearly all metals, equilibrium between the solution and solid phase is pH dependent, with increasing metal solubility as pH decreases. A recent study of lead contaminated soils showed that soil washing by acidic extractants met TCLP limits for all soils and lead clean-up goals for five of the seven soils tested [10].

A surfactant molecule is amphiphilic, having two distinct structural moieties, one polar and the other nonpolar. The surfactant concentration at which monomers begin to aggregate is called the critical micelle concentration (CMC). Typically, the polar or ionic portions of the micelle are oriented toward the aqueous phase, while the nonpolar tails of the molecules are shielded away from the water molecules. Thus, the central region of the micelle consists of a distinct, nonpolar, hydrocarbon core [11]. At surfactant concentrations greater than the CMC, additional surfactant is incorporated into the bulk solution through micelles which increase the available micellar solubilization volume [12]. However, the aggregation number, defined as the average number of monomers per micelle, essentially remains constant.

Enhanced solubility of HOCs in surfactant solutions is due to partitioning of the HOC between the water and micellar phases. The effectiveness of a particular surfactant is described by the molar solubilization ratio (MSR), which is defined as the number of moles of organic compound solubilized per mole of surfactant added to solution [13]. In the presence of a separate phase, or an excess of the HOC, the MSR is determined from the slope of a plot of solubilizate concentration versus surfactant concentration [14].

Another perspective in quantifying surfactant solubilization consists of characterizing the partitioning of the solubilizate between the micellar phase and the aqueous phase of the solution. The micellar-phase / aqueous-phase partition coefficient, $K_m$, is defined as the molar ratio of the amount of hydrophobic organic compound in the micellar phase ($X_m$) to the amount in the aqueous phase ($X_A$) [14]. The value of $X_m$ may be calculated from the MSR value. The value of $X_A$ may be approximated for dilute solutions by multiplying the solubility of the organic at the CMC by the molar volume of water (18 $cm^3$/mol). The value of $K_m$ is dependent upon surfactant chemistry, solubilizate chemistry, and temperature [14]. However, $K_m$ is independent of saturation or the presence of a separate phase of organic contaminant. Thus, the value of $K_m$ can be applied to solutions with or without soil.

Surfactant-enhanced solubilization of HOCs typically has been applied to pump and treat systems for in-situ groundwater remediation [14-17] [20]. However, this research focuses on ex-situ treatment of PAH contaminated soil under acidic conditions. Information concerning the effect of pH on surfactant-enhanced solubilization is scarce due to the dependence on specific surfactant chemistry [20]. Thus, surfactant solubilization experiments were performed under acidic conditions as well as at pH 8.3 to determine if the acidic environment affects solubilization. Anionic and nonionic surfactants were selected for testing.

*Surfactant Sorption*

The loss of surfactant by sorption onto soil or aquifer sediment will adversely affect the economics of surfactant-enhanced remediation. When surfactant is added to a soil/water system, sorption to the soil occurs until an equilibrium is reached with the level of surfactant in the aqueous phase. Nonionic surfactants and sulfonated anionic surfactants have been shown to reach sorption saturation at or near the CMC [15][16]. Once sorption saturation is achieved, any further surfactant added to the system will be incorporated into micelles. As a result of sorption, the amount of surfactant that must be added for the solution concentration to reach the CMC is larger than the amount required in a soil-free system [17]. Thus, it is advantageous to minimize the sorption of surfactant thereby making more surfactant available for micellar solubilization of HOCs.

Disulfonated anionic surfactants have been shown to sorb less than ethoxylated nonionic surfactants to subsurface media [16]. This may be due to electrostatic repulsion between the sulfonate groups and the negative charges found on silica surfaces under natural conditions. However, sorption characteristics at low pH have not been investigated. Although disulfonated anionic surfactants exhibit more favorable sorption characteristics, nonionic surfactants are known to provide better solubilization of HOCs [14].

## SYSTEM CONCEPTUALIZATION

Based on the foregoing discussion, a conceptualization of the soil, surfactant, metal, and HOC system is shown in Figure 1. For this type of soil remediation process to be successful, metal desorption/dissolution and HOC solubilization must be maximized and the surfactant sorption must be minimized. Surfactant sorption results in added cost as well as potentially poorer HOC solubilization due to an increase in the organic carbon content of the soil. The mechanisms of interaction between metals in the aqueous phase and surfactant molecules as well as the interactions between sorbed metals and surfactants are not clearly understood. For anionic surfactants, interactions between metal cations and the anionic surfactant moieties may be possible.

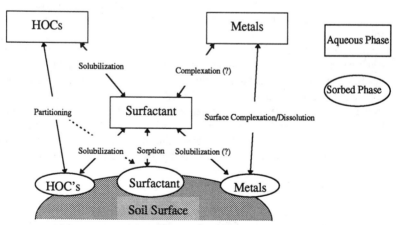

Figure 1. Conceptualization of HOC/Metal/Surfactant/Soil System.

## MATERIALS AND METHODS

Two polycyclic aromatic hydrocarbons (naphthalene and pyrene) were evaluated in this study. These compounds were selected because they are common soil and sediment contaminants, exhibit distinct differences in hydrophobicity, and have been previously studied for removal enhancement using surfactants. The hydrocarbons (Fluka) were 97%+ pure and used as received.

Two straight-chain diphenyl oxide disulfonates (DPDSs) were selected as the sulfonated anionic surfactants: Dowfax C10L and Dowfax 8390 (C16). The surfactants were characterized and donated by Dow Chemical, and were used as received. Sulfonated surfactants were selected for study due to their low sorption onto soil, stability in a low pH environment, FDA food additive approval, and biodegradability (SCAS testing) [18].

A phosphate ester blend, Rexophos 25/97, was selected as the model weak-acid anionic surfactant. The Rexophos 25/97 surfactant consists of a nonionic surfactant base which is "phosphated" by mixing with phosphoric acid. The Rexophos 25/97 (Huntsman Corporation) was 99%+ pure active ingredient and used as received.

A secondary alcohol ethoxylate (Tergitol 15-S-9) and an alkylphenol ethoxylated surfactant (Triton X-100) were selected as model ethoxylated nonionic surfactants. Ethoxylated nonionic surfactants were selected for their high solubilization efficiency of PAHs. The Tergitol 15-S-9 and the Triton X-100 (Union Carbide) were 99%+ pure active ingredient and used as received.

The soil employed for the experiments was Minoa Very Fine Sandy Loam which consists of 93.6% fine sand, 4.1% median to coarse sand and 2.4% silt/clay [19], and was determined to have an organic carbon content of 1.19% ± 0.45%. The soil was sieved with a No. 10 (2 mm) sieve before use to remove larger material.

*Solubilization Experiments*

Solubilization experiments were conducted at pH 8.3 and pH 1. For pH 8.3, sodium bicarbonate was used as a buffer (99.9% pure, Sigma Chemical) and used as received. Hydrochloric acid (NF/FCC grade, Fisher Scientific) was used with the anionic surfactants and with the Triton X-100. Initially, reagent grade, 85% phosphoric acid was used with the nonionic surfactants to assure stability in the low pH environment. Both the sodium bicarbonate and the acids were diluted with nanopure water.

For each hydrocarbon surfactant pair, experiments were conducted to determine the concentration of hydrocarbon solubilized at neutral and acidic pH values. All experiments were carried out at room temperature. Stock surfactant solutions were created and dilutions were made from the stock solutions to obtain a range of surfactant concentrations. Diluted samples were placed in 40 mL amber glass vials with open-top screw caps and Teflon-backed septa. To each vial, either 40 mg of naphthalene or 20 mg of pyrene was added. Samples were sealed and then mixed on an end-over-end tumbler (Browning, VPE-110) for 24 hours to allow for equilibration. After equilibration, the samples were filtered with a 0.2 micron Teflon filter (Millipore) to separate the solid-phase PAH: dissolved PAH compounds in the aqueous pseudophase and solubilized PAHs in the micellar pseudophase passed through the filter. Filtered samples were analyzed for PAH and surfactant concentrations using HPLC.

*Sorption Experiments*

The sorption experiments were conducted using a batch equilibrium method to create isotherms for the Dowfax C10L and Triton X-100 surfactants. The experiments were conducted at pH 6 and pH 1. All surfactant solutions were created by weighing surfactant stock solution on a high-precision balance (Mettler AE 100) and then diluting to the appropriate concentrations with nanopure water. pH adjustment was made using hydrochloric acid. Twenty-five mL of surfactant solution were added to 8.33 grams of Minoa soil in 40 mL amber glass vials. The vials were then placed in an end-over-end tumbler (Browning, VPE-110) for 24 hours allowing the surfactant to reach equilibrium with the soil. The supernatant was then decanted from the soil and centrifuged at 5000 RPM for 20 minutes (Damon/IEC HT). The samples were then passed through 0.22 micron cellulose syringe filters (Millipore) to remove any remaining particles and then analyzed for surfactant concentration using HPLC.

*Analytical Methods*

Surfactant and PAH concentrations were measured using HPLC on a Shimadzu System which included a system controller, a pump, and a UV-VIS spectrophotometric detector. An Envirosepp C-18 reverse phase column (250 x 4.6 mm) (Phenomenex) was utilized for separation in conjunction with replaceable column guards (Alltech).

The mobile phase, 70% acetonitrile and 30% nanopure water, remained constant for all measurements and isocratic conditions were utilized. However, the flow rate was decreased from 1.5 mL/min to 1 mL/min for the surfactants to improve peak resolution. The mobile phase was filtered and degassed by a special filtration apparatus prior to use. Likewise, nanopure water, used for flushing and cleaning of the system, was filtered and degassed.

Fifty microliters of sample were injected using a 50 uL SGE flat-tipped syringe (Fisher Chemical). Twenty microliters were separated and analyzed. Between sample runs, the syringe was cleaned by rinsing 5-10 times with methanol and then rinsing with nanopure water. Standard solutions were used to create calibration curves for the hydrocarbons and the surfactants. Dilutions of filtered sample were made so as to remain in the linear range of the calibration curve.

## RESULTS

Table I. Comparison of MSR and Log $K_m$ values for PAHs

| Surfactant Name | Surfactant Type | PAH | pH | MSR | Log $K_m$ |
|---|---|---|---|---|---|
| Dowfax C10L | Anionic-Sulfonate | Naphthalene | 1.5 | 0.067 | 4.23 |
| Dowfax C10L | Anionic-Sulfonate | Naphthalene | 8.3 | 0.073 | 4.21 |
| Dowfax C10L | Anionic-Sulfonate | Pyrene | 1.5 | 0.009 | 5.67 |
| Dowfax C10L | Anionic-Sulfonate | Pyrene | 8.3 | 0.008 | 5.61 |
| Dowfax 8390 (C16) | Anionic-Sulfonate | Naphthalene | 1.5 | 0.146 | 4.52 |
| Dowfax 8390 (C16) | Anionic-Sulfonate | Naphthalene | 8.3 | 0.178 | 4.55 |
| Dowfax 8390 (C16) | Anionic-Sulfonate | Pyrene | 1.5 | 0.018 | 5.97 |
| Dowfax 8390 (C16) | Anionic-Sulfonate | Pyrene | 8.3 | 0.016 | 5.92 |
| Rexophos 25/97 | Anionic Phosphate Ester | Naphthalene | 2 | 0.376 | 4.89 |
| Rexophos 25/97 | Anionic Phosphate Ester | Naphthalene | 8.3 | 0.320 | 4.79 |
| Tergitol 15-S-9 | Nonionic | Naphthalene | 1* | 0.188 | 4.56 |
| Tergitol 15-S-9 | Nonionic | Naphthalene | 8.3 | 0.201 | 4.74 |
| Triton X-100 | Nonionic | Naphthalene | 1* | 0.155 | 4.48 |
| Triton X-100 | Nonionic | Naphthalene | 8.3 | 0.225 | 4.77 |
| Triton X-100 | Nonionic | Naphthalene | 1 | 0.255 | 4.73 |

* Acidic solution made from phosphoric acid, all other low pH values used a hydrochloric acid solution

Solubilization results for naphthalene and pyrene are shown in Table 1. For the sulfonated surfactants, relatively low MSR values were obtained, especially for the C10L. Thus, the Dowfax surfactants are not as good as the other surfactants for solubilizing naphthalene and pyrene. The MSR and Log $K_m$ values are in agreement with results found in the literature for naphthalene and pyrene [16]. Over the pH range studied, the MSR and log $K_m$ values calculated are not significantly different for both naphthalene and pyrene. Thus, solubilization of these compounds is not significantly affected at low pH.

For both sulfonated surfactants, the solubilization trends of naphthalene and pyrene are similar at the two pHs investigated. However, the MSR for naphthalene is an order of magnitude greater than pyrene. Pyrene is a larger molecule and for PAHs, the extent of solubilization appears to decrease with an increase in molecular size [17]. The $K_m$ value for pyrene is over an order of magnitude greater than for naphthalene. Although the $K_m$ value is a function of the MSR, contaminant hydrophobicity has a greater effect on partitioning. Pyrene is more hydrophobic than naphthalene (Log $K_{ow}$ values 5.2 and 3.4, respectively). Since, for the sulfonated surfactants, the solubilization trends of naphthalene and pyrene were similar, pyrene was not investigated using the other surfactants.

The Rexophos 25/97 solubilization run was at pH 2 and not pH 1, because of surfactant precipitation at pH<2. Relatively high MSR values were obtained for the Rexophos: an indication that it is efficient for solubilizing naphthalene. Between pH 2 and pH 8.3, solubilization effects were statistically different: the MSR value for Rexophos at pH 2 was greater than the value at pH 8.3. Data correlating pH and solubilization are scarce. For ionic surfactants, small increases in the electrolyte concentration can either increase or show no effect on solubilization efficiencies [11][13].

Nonionic surfactants possess advantages including a significantly lower sensitivity to the presence of electrolytes in the system and a lessened effect of solution pH [20]. The nonionic surfactants exhibited average to high MSR values, thus the surfactants were good for solubilizing naphthalene. Tergitol solubilization of naphthalene is not significantly affected by pH changes. Likewise, Triton X exhibited slight MSR differences between pH 8.3 and the pH 1 value using HCl (0.225 vs. 0.255). The MSR and $K_m$ values for the neutral pH value agree with values reported in the literature [16]. Therefore, no significant change in naphthalene solubilization is evident for acidic pH conditions using HCl. However, the MSR value obtained at pH 1 using a phosphoric acid solution was significantly less than the neutral pH MSR value.

Figure 2 shows the results of the sorption experiments. At both pH values, the Triton X-100 increased with equilibrium aqueous surfactant concentration until a point at or near the CMC (0.12 g/L [15]). Once the CMC was attained, the sorptive capacity ($q_{max}$) of the soil was reached and the isotherms appear to reach a plateau. The same behavior was observed for the C10L at pH 6 but a clear plateau was not reached at pH 1 (CMC = 3 g/L [16]).

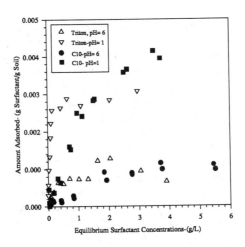

Figure 2. Effect of pH on Sorption of Surfactants.

The soil showed an increase in $q_{max}$ for both the C10L and the Triton X-100 with decreasing pH. The value of $q_{max}$ for the C10L at pH 6 was determined to be about 0.001 (g/g) which is slightly higher than the value of 0.00075 (g/g) found for the Triton X-100. At pH=1, $q_{max}$ increased by a factor of about 5 for the C10L and it increased by about a factor of 4 for the Triton X-100.

## SUMMARY

A decrease in pH from 8.3 to 1 showed no significant effects on solubilization of naphthalene and pyrene. Thus, the solubilization of these contaminants should not be inhibited by the use of surfactants at low pH. However, sorption of C10L and Triton X-100 increased significantly with a decrease in pH. Additional testing will involve the use of contaminated soils to test the feasibility of surfactant-based soil remediation at low pH.

## REFERENCES

[1] DEC. October, 1993. "Quarterly Status Report of Inactive Hazardous Waste Disposal Sites," Department of Environmental Conservation, Albany, NY.

[2] NYSDOH, 1989. Unpublished list of chemicals at hazardous waste sites, New York State Department of Health.

[3] Riley, R.G., and J.M Zachara. April 1992. "Chemical Contaminants on DOE Lands and Selection of Contaminant Mixtures for Subsurface Research," DOE/ER-0547T, Department of Energy.

[4] Peters, R.W. and L. Shem. 1992. "Use of Chelating Agents for Remediation of Heavy Metal Contaminated Soil," in *Environmental Remediation-Removing Organic and Metal Ion Pollutants,* G.F. Vandergrift, D.T. Reed, and I.R. Tasker, eds. Washington, DC: ACS Symposium Series 509, American Chemical Society, p. 71.

[5] Manahan, S.E. 1990. *Hazardous Waste Chemistry, Toxicology, and Treatment,* Chelsea, MI: Lewis Publishers.

[6] Sun, S.B., W.P. Inskeep, and S.A. Boyd. April 1995. "Sorption of Nonionic Organic Compounds in Soil-Water Systems Containing a Micelle-Forming Surfactant," *Environ. Sci. Technol.,* 29(4): 903.

[7] EPA. 1986. "Test Methods for Evaluating Solid Waste," 3rd. ed., Office of Solid Waste and Emergency Response, Environmental Protection Agency, Washington, D.C.

[8] Tessier, A. et al. 1979. "Sequential Extraction Procedure for the Speciation of Particulate Trace Metals," *Analytical Chemistry*, 51(7), p. 844.

[9] Engler, T. et al. 1977. "A Practical Selective Extraction Procedure for Sediment Chemistry," *Chemistry of Marine Sediments*, T.F. Yen, ed. Ann Arbor, MI: Ann Arbor Science.

[10] Young, W. H., M. E. Buck, J. E. Van Benschoten and M. R. Matsumoto. May, 1994. *Proceedings of the 49th Industrial Waste Conferene, Purdue University,* Ann Arbor, MI: Lewis Publishers.

[11] Rosen, M.J. 1989. *Surfactants and Interfacial Phenomena*, 2nd ed.; New York: John Wiley and Sons.

[12] Clarke, A.N., R.D. Mutch, D.J. Wilson, and K.H. Oma. 1992. "Design and Implementation of Pilot Scale Surfactant Washing/Flushing Technologies Including Surfactant Reuse," *Wat. Sci. Tech.*, 26:127-135.

[13] Attwood, D., and A.T. Florence. 1983. *Surfactant Systems*, London: Chapman and Hall, Ltd.

[14] Edwards, D., R. Luthy, and Z. Liu. 1991. "Solubilization of Polycyclic Aromatic Compounds in Micellar Nonionic Surfactant Solutions," *Environ. Sci. Technol.*, 25: 125-133.

[15] Liu, Z., D. Edwards, and R. Luthy. 1992. "Sorption of Nonionic Surfactants onto Soil," *Wat. Res.*, 26(10):1337-1345.

[16] Rouse, J., 1993. "Minimizing Surfactant Losses Using Twin-Head Anionic Surfactants Substance Remediation," *Environ. Sci. Technol.* 27: 2072-2078.

[17] Edwards, D., and R. Luthy. 1990. "Nonionic surfactant solubilization of polycyclic aromatic hydrocarbons in aqueous and soil/water systems," *National Conference on Environmental Engineering*, New York: ASCE, 286-289.

[18] Dow Chemical Company. Undated. *Dowfax Anionic Surfactants*, Midland, MI.

[19] Garland, M., Soil Flushing of Lead-Contaminated Soils, MS Thesis, State University of New York at Buffalo, 1994.

[20] Myers, D. 1992. *Surfactant Science and Technology*, 2nd ed.; New York: VCH Publishers.

# SOLID-PHASE HEAVY-METAL SEPARATION UNDER UNFAVORABLE BACKGROUND CONDITIONS BY COMPOSITE MEMBRANES

SUKALYAN SENGUPTA
Assistant Professor, Dept. of Civil & Environmental Engineering
University of Massachusetts Dartmouth
North Dartmouth, MA 02747-2300

ARUP K. SENGUPTA
Professor, Dept. of Civil & Environmental Engineering
Lehigh University
Bethlehem, PA 18015

## Introduction

Disposal of sludges or treatment of soil contaminated with minor fraction (often less than 5%) of heavy metals in the solid phase in an otherwise innocuous background is a widespread problem. This is due to the fact that the heavy metals present cause the sludge/soil to be designated as a "hazardous waste", thus requiring extensive treatment before ultimate disposal. Selective and targeted removal of the heavy metals from the background solid phase would constitute an efficient treatment process as it would be able to reduce the volume of hazardous sludge considerably and also may make it possible for the heavy metals to be concentrated and recycled/reused.

A new class of sorptive/desorptive ion-exchange composite membranes available commercially is extremely suitable for heavy metal decontamination from sludges/slurries[1,2]. In this material, fine spherical beads (<100 $\mu$ in dia) of heavy-metal selective chelating ion-exchangers are physically enmeshed or trapped in thin sheets ($\approx$0.5 mm thick) of highly porous polytetrafluoroethylene (PTFE), as shown in the electron microphotograph (Figure 1). These composite membranes, because of their thin-sheet like physical configuration, can be easily introduced into and withdrawn from any reactor containing sludge/slurry and the target solutes can be adsorbed onto the microbeads. These membranes are not fouled by high concentration of suspended solids but retain the retain the original properties of the chelating exchangers even after use for a number of cycles. This paper explores the efficacy of the composite membrane for heavy metal decontamination under unfavorable conditions, namely

1) Heavy metal presence amidst a background of high buffer capacity; the

background solid phase compounds can interact chemically with the heavy metal dissolution. This will create an alkaline condition with low aqueous phase heavy metal concentration compounded by the fact that the buffer capacity contributing cation would be present at an aqueous phase concentration orders of magnitude higher than the heavy metal cation.

2) Heavy metal cations associated with functional sites of the background material (primarily soil) by adsorption or ion-exchange. Removal of the heavy metal entails a two-step process:
    a) desorption of the heavy metal from adsorption/ion-exchange sites,
    b) uptake of heavy metal cations by the composite membrane.
The process is shown schematically in Figure 2.

3) Heavy metal present in a solid phase containing humic/fulvic material. are good complexing agents , thus the heavy metal cation will be complexed with H/F functional groups. As with 2), a decontamination process would require splitting of the heavy metal complex with H/F material followed by removal of the heavy metal cation from the aqueous phase.

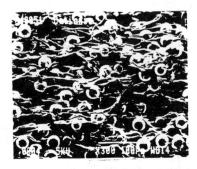

Figure 1. Scanning Electron Microphotograph of the composite membrane.

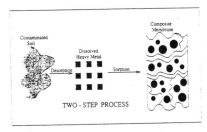

TWO - STEP PROCESS

Figure 2. Conceptual diagram of heavy metal removal from ion-exchange/adsorption sites of the solid phase.

## The Proposed Decontamination Process

Since this study is aimed at removing heavy-metals, the microbeads chosen were polymeric ion-exchangers containing covalently attached chelating iminodiacetate (IDA) moiety. This functional group was chosen since it has high selectivity towards dissolved heavy-metals over alkali metal or alkaline earth metal cations, which are likely to be present as background materials. Table 1 provides the salient properties of the composite IDA membrane, from which it can be seen that the chelating microbeads constitute up to 90% of the composite membrane by mass. Various characterization studies of the composite membrane, e.g., water uptake in different counterion forms, titration, hysteresis, and equilibrium heavy-metal uptake isotherm, etc. proved[1,3] that the sorption and regeneration behaviors of the composite membranes are essentially the same as the parent chelating microbeads (IDA in this case).

## Table 1
### Properties of the Composite Chelex Membrane

| Composition | 90% Chelex chelating resin, 10% teflon |
|---|---|
| Pore size (nominal) | 0.4 µm |
| Nominal capacity | 3.2 meq/gm dry membrane |
| Membrane thickness | 0.4-0.6 mm |
| Ionic form (as supplied) | Sodium |
| Resin matrix | Styrene-divinylbenzene |
| Functional group | Iminodiacetate |
| pH stability | 1-14 |
| Temperature operating range | 0-75 $^\circ$C |
| Chemical stability | methanol; 1N NaOH; 1N $H_2SO_4$ |
| Commercial availability | Bio-Rad Inc., California |

For a sludge containing small amount (<5%) of heavy-metal, say as $Me(OH)_2(s)$, the composite membrane can be used in a cyclic two-step process as follows:

Sorption Step:

The composite membrane sheet, when introduced into a vigorously stirred sludge reactor, selectively removes dissolved heavy-metals from the aqueous phase in preference to other non-toxic cations such as $Na^+$, $Ca^{2+}$, $Mg^{2+}$, etc. The reaction can be written as follows:

$$\text{Memb. Uptake--} \quad \overline{RN(CH_2COOH)_2} + Me^{2+} \rightleftharpoons \overline{RN(CH_2COO^-)_2Me^{2+}} + 2H^+ \quad (1)$$

The overbar represents the membrane phase, in which the polymer matrix R is covalently attached to the chelating iminodiacetate functional group.

## Desorption Step:

The second step of the process entails withdrawing the composite membrane sheet from the sludge and introducing it into a gently stirred tank containing 2-5% v/v $H_2SO_4$ or any other mineral acid. In this step, the exchanger microbeads in the membrane are efficiently regenerated in hydrogen form (because IDA is a weak acid and thus prefers hydrogen ion the most) according to the following reaction:

$$Regen. -- \quad \overline{RN(CH_2COO^-)_2Me^{2+}} + H_2SO_4 \rightarrow \overline{RN(CH_2COOH)_2} + Me^{2+} + SO_4^{2-} \quad (2)$$

The regenerated membrane is subsequently withdrawn and reintroduced into the sludge reactor as a continuation of the cyclic process.

## Heavy-Metal Precipitate in a Buffered Sludge at Alkaline pH

Sludges/slurries may contain a heavy-metal precipitate swamped in a solid phase with high buffer capacity (e.g., calcite) and organic substances capable of forming strong complexes with toxic metals. Due to high buffer capacity of the solid phase, pH of the sludge will be alkaline, thereby reducing the aqueous phase free heavy-metal cation concentration, $\{Me^{2+}\}$. This will significantly reduce the uptake capacity of $Me^{2+}$ by the composite membrane, thus making direct application of the above process inefficient. Also, it would be impractical to add acid to the reactor to lower the pH. Our approach was to introduce an aqueous phase ligand into the sludge reactor to selectively increase aqueous phase heavy-metal concentration. However, the chelating functionality of the composite membrane microbead would have stronger preference for the heavy-metal as compared to the aqueous phase ligand and so would split the aqueous phase heavy-metal - ligand bond to attract the heavy-metal cation to its ion-exchange sites. A cyclic recovery process was tried for a sludge phase containing sand, calcite, and copper oxide. Two aqueous phase ligands were chosen, oxalate and citrate. The pH of the sludge reactor was maintained at 9.0. It can be seen from Figure 3 that the cyclic process was quite effective for selective separation of copper from a very unfavorable sludge composition. Explanation of oxalate, citrate, copper, and calcium uptake is provided by the authors[1,2].

## Heavy-Metals Extraction from Ion-Exchange Sites of Soil

If the heavy-metal contaminants are bound to the ion-exchange sites of soil, the composite membrane may be used to remove them selectively in a single reactor. In order to confirm the same, analytical grade bentonite (-200 mesh from J. T. Baker Co., Pennsylvania) was chosen as the soil type and was loaded with copper by equilibrating with an aqueous phase containing 400 mg/L copper concentration added as copper nitrate and at pH = 5.5. By mass balance, the total exchange capacity was found to be approximately 50 *mequiv./100 g* of dry bentonite (magnesium alumino silicate). Copper- loaded bentonite was then introduced into a plastic container containing 200 *mL* of 500 *mg/L* Na⁺ (added as NaCl) solution. Copper-loaded bentonite was the only solid phase present in the slurry and the solids loading was 2.2%. A strip of composite membrane (weighing 0.112 *g*) was introduced into the container and the two-step process (i.e. sorption and desorption) was run, the results of which are presented in Figure 4. It can be seen that >60% of Cu(II) removal was achieved in less than 30 cycles. Copper concentration in the regenerant was used to compute the percentage copper recovery. The mechanism of Cu(II) removal is discussed by the authors in open literature[1,2].

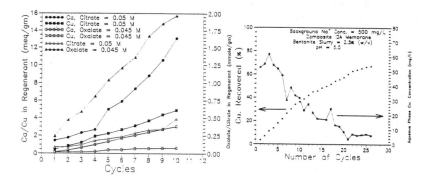

Figure 3. Cumulative Cu(II), Ca(II), oxalate, and citrate recovery with cycles at alkaline pH.

Figure 4. Cu(II) recovery from the ion-exchange sites of bentonite clay during the cyclic process.

## Heavy Metals Removal from Humic Acid Sludge

Humic substances are present in all natural aquatic and soil environments; thus it is important to study the fate and transport of heavy metals in the presence of humic substances.  It is generally agreed that humic substances contain carboxylic and phenolic groups which act as ligands in the presence of heavy metals to give rise to metal-humic complex.  Humic acid extract (HAE), sodium salt [1415-93-6]; supplied by Aldrich Chemical Company was taken to be the reference humic acid sample.  Heavy metal capacity of the HAE was determined by the procedure described by Sengupta[2] and the following data was obtained:

At pH = 4.05, pCu = 2.544, vCu = 3.22 ($10^{-3.22}$ moles of Cu/$g$ of original HAE)

Experiments were conducted to find out copper removal capacity of a humic acid sludge.  Copper-loaded humic acid sludge was taken as the contaminated sample and cyclic extraction test was performed on this sludge at a constant pH of 3.0.  A strip of the composite membrane in H$^+$ form weighing 0.5496 $g$ was added to the reactor in each exhaustion cycle, which lasted for 4 hours.  At the end of each exhaustion cycle, the membrane was taken out, rinsed with tap water and introduced into the regeneration bottle which had initially 200 mL. of 5% $H_2SO_4$.  Filtered samples from each cycle (exhaustion and regeneration) were collected and analyzed for Cu(II) and Dissolved Organic Carbon (DOC).  It can be seen from Figure 5 that after 11 cycles, approximately 44% of Cu(II) was transferred from the humic acid sludge reactor to the regenerant solution.  Explanation of the mechanism of Cu(II) transfer is provided by Sengupta[2].

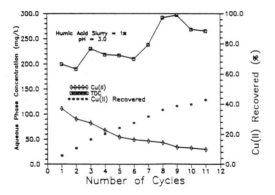

Figure 5.  Cu(II) recovery from humic acid-extract sludge in a cyclic process.

## Conclusions

The composite sorptive/desorptive ion-exchange membrane shows promise in selective sorption of heavy metals from complex sludges, e.g. from highly buffered alkaline sludges, from soils with cation-exchange capacity, from humic/fulvic acid containing soil, etc. Its tough physical texture allows it to be used in vigorously stirred slurry reactors containing high percentage of suspended solids without any tearing/rupture or membrane fouling by particulate matter. At the same time, the entrapped chelating ion-exchanger microbodies which comprise up to 90% of the material by mass, are uniquely suited to selectively sorb heavy metals from the aqueous phase under severe competition from such common background ions as $Ca^{2+}$ and $Na^+$. The composite membrane can be used in a simple two-step process, sorption and desorption, allowing heavy metals to be selectively recovered and concentrated in an acid solution, which may be suitable for recycle/reuse in any process.

## References

1. Sengupta, S., and Sengupta, A. K., "Characterizing a New Class of Sorptive/Desorptive Ion Exchange Membranes for Decontamination of Heavy-Metal-Laden Sludges", *Env. Sci. & Tech.*, Vol. 27, pp. 2133-2140, 1993.

2. Sengupta, S., "A New Separation and Decontamination Technique for Heavy Metal-Laden Sludges Using Sorptive/Desorptive Ion-Exchange Membranes", *Ph. D. Dissertation*, Lehigh University, Bethlehem, PA, Nov. 1993.

3. Sengupta, S., and Sengupta, A. K., "Treatment of Heavy-Metal-Laden Sludges by a New Class of Sorptive/Desorptive Membranes", *Membrane Technology Reviews*, pp. 1-18, 1994.

# THE REMOVAL OF HEXAVALENT CHROMIUM FROM WATER BY FERROUS SULFATE

CHE-JEN JERRY LIN, Graduate Student

P. AARNE VESILIND, Professor

Department of Civil and Environmental Engineering
Duke University
Durham, NC 27708-0287

## ABSTRACT

The redox reaction of hexavalent chromium and ferrous sulfate is investigated in this study. Hexavalent chromium, a highly toxic and mobile anion, could exist in raw water used as a public water supply due to the industrial chromium contamination of natural water or due to natural oxidation of trivalent chromium. Ferrous sulfate is one of the widely used coagulants in water treatment plants and has good reducing ability. Because of its reducing capacity, ferrous sulfate can be applied to remove hexavalent chromium from water.

The required contact time to reach equilibrium, the effectiveness of Cr(VI) reduction at different initial pH, and the required ferrous sulfate dosage for complete reduction are investigated. The redox reaction can be completed within 10 minutes, allowing 30 mg/L of hexavalent chromium to react with stoichiometric dosage of ferrous sulfate in deionized water, regardless of the initial pH. The pH of the solution is depressed during the progress of the reaction due to the hydrolysis of the produced Fe(III) and Cr(III) ions from the reaction. Dissolved oxygen in water is found to interfere with the redox reaction by consuming ferrous ions when the initial pH of solutions is high. In deionized water, complete Cr(VI) reduction can be achieved by applying excess ferrous sulfate under the condition of this study. It is also achievable when the raw water from Durham Water Treatment Plant is used as the reaction medium, without additional dosage of ferrous sulfate. Based on the results, simultaneous removal of hexavalent chromium in water treatment by applying ferrous sulfate as the coagulant is theoretically feasible.

Key words: Cr(VI) reduction, water treatment, ferrous sulfate.

# INTRODUCTION

Chromium is a metal widely used in metallurgical, refractory and chemicals manufacturing industries for over a hundred years. However, the extensive use of chromium has caused chromium contamination of the environment, especially in natural water of highly industrialized areas [1-7]. The contamination is mostly caused by the discharge of industrial wastewater or the disposal of industrial waste sludge. Another pathway is the land application of municipal sewage sludge containing insoluble chromium compounds due to industrial chromium discharge to municipal sewage systems.

Chromium exists in the aquatic environment as two stable oxidation states: trivalent [Cr(III)] and hexavalent [Cr(VI)] forms. Trivalent chromium is considered a nutrient for humans and many living organisms. It is relatively immobile in natural water because of the poor solubility of Cr(III) compounds. It is found that trivalent chromium can be oxidized to hexavalent forms under natural conditions through a series of oxidation-reduction mechanisms in soils and natural water [8]. Hexavalent chromium is toxic and has been proven to be carcinogenic [9]. Its mobility is high since it exists in water as soluble anions ($CrO_4^{2-}$ or $Cr_2O_7^{2-}$). It is reported that hexavalent chromium will travel in the water course until it ends up in the sediment [10]. Due to the toxicity and carcinogenicity at its hexavalent state, EPA has set the drinking water limit for chromium at 0.10 mg/L. As a result, removing chromium from water to meet the drinking water standards becomes necessary if it occurs in the water source of a public water supply.

In industrial wastewater treatment, the removal of hexavalent chromium involves a two-step process: the reduction of hexavalent chromium under acidic condition (usually at pH 2-3), and the hydroxyl precipitation of trivalent chromium at pH 8-10 [11]. The precipitates then can be removed by conventional solid-liquid separation processes. Reducing agents most commonly employed are gaseous sulfur dioxide, sodium sulfite, sodium bisulfite, and ferrous sulfate.

Cr(VI) reduction by using the above reducing agents are almost spontaneous at strongly acidic conditions, due to the instability of Cr(VI) species in acidic environment. The rate of the redox reaction has been reported to decrease with the increase of pH for the S(IV) reducing agents [12]. However, the effects of pH on the rate of chromium reduction by ferrous ions have been questioned. It is reported that a pH less than 3 is required for reasonably rapid reduction by ferrous sulfate [13,14]. On the other hand, the rate of Cr(VI) reduction by ferrous ion at higher pH has also been proven to be relatively rapid. A study on the treatment of high Cr(VI) content waste suggested that the use of ferrous ammonium sulfate [$Fe(NH_4)_2(SO_4)_2$] for Cr(VI) reduction is sufficiently rapid at neutral or even alkaline pH range [12]. Another study on ISP (insoluble sulfide precipitation) process also demonstrated that ferrous reduction of hexavalent chromium is sufficiently rapid at neutral to alkaline pH (7-10) for the reaction to be carried out in line [15].

Among the four reducing agents mentioned, ferrous sulfate ($FeSO_4$) is less favorable because of its higher required dosage and the large amount of ferric hydroxide sludge generated during the treatment. However, the application of ferrous sulfate in treating chromium-bearing raw water seems to be an ideal alternative. Since the chromium content in the contaminated water is relatively low compared to that of industrial waste effluents, no high dosage of reducing agent is required. In addition, ferric ions have been proven to be effective in precipitating the resulting Cr(III) ions [16,17]. They can also serve as an coagulant to settle the colloidal particles in water treatment.

The objective of this research is to study the redox reaction of hexavalent chromium and ferrous sulfate under various reaction conditions. Goals are set to investigate the required contact time to reach equilibrium, the effectiveness of reduction at different initial pH, and the required dosage for complete Cr(VI) reduction. Since ferrous sulfate has been used as a coagulant in drinking water treatment, conditions assimilating those into a conventional water treatment plant are especially of interest. If ferrous sulfate is effective in reducing hexavalent chromium under the conditions of water treatment and Cr(III) precipitates can be retained in the in the liquid-solid separation processes, simultaneous removal of hexavalent chromium by using ferrous sulfate as the coagulant in water treatment may be possible.

## MATERIALS AND METHODS

Batch reactions are performed at room temperature (22-25 °C) and the residual Cr(VI) concentration after reaction is determined. The experimental procedure mainly consists of four steps: pH adjustment, reduction of hexavalent chromium, filtration and Cr(VI) analysis.

Solution pH is adjusted by dropping previously prepared sodium hydroxide or nitric acid solution to a desired value. The pH and temperature of solutions are measure by Cole Parmer 59002-12 pH/mV/temperature meter. Reagent-grade potassium dichromate ($K_2Cr_2O_7$) is used as the synthetic Cr(VI) contamination. The reducing agent is prepared by dissolving reagent-grade ferrous sulfate containing seven moles of water ($FeSO_4 \cdot 7H_2O$) in deionized water. This ferrous sulfate solution is made freshly and injected into Cr(VI) solution under stirring to allow reaction to proceed for 20 minutes. The reaction medium is deionized water, except the reactions run in the raw water from the Durham Water Treatment Plant. Vacuum filtration by using 0.45 μm pore size cellulose nitrate membrane is employed to remove the precipitates from the reacted solutions. The filtrate is then taken for Cr(VI) analysis by the DPC (diphenylcarbazide) colorimetric method developed by Bartlett and James [10]. The DPC reagent is prepared by adding 120 ml of 85 % phosphoric acid, diluted with 280 ml distilled water, to 0.38 gram of s-diphenylcarbazide previously dissolved in 100 ml of 95 % ethanol. DPC reagent must be stored in brown bottle in the dark at 4 °C. The colorimeter used is Hach Type DR/2000 Direct Reading Spectrophotometer.

The reactions also proceed in an oxygen-free solution, achieved by nitrogen stripping. YSI Model 57 Dissolved Oxygen Meter is applied to detect the dissolved oxygen level. The dissolved oxygen in solutions can approach zero after 5-minute of nitrogen stripping.

## RESULTS AND DISCUSSION

**The redox reaction of ferrous sulfate and hexavalent chromium.** In this study, the redox reaction proceeds at initial pH from 2.5 to 11.0 without buffering the solutions. The predominant Cr(VI) species is $HCrO_4^-$ at pH 1-6 while it is $CrO_4^{2-}$ for pH greater than 6 [18], and the redox reaction is expressed as equation (1) and (2) for different pH ranges. The produced Fe(III) and Cr(III) ions will continue to react with water through a series of hydrolysis mechanisms and form various hydroxyl compounds. The stoichiometric ratio of Fe(II) to Cr(VI) is 3 by moles and 3.22 by weight.

$$HCrO_4^- + 3\,Fe^{2+} + 7\,H^+ \rightarrow Cr^{3+} + 4\,H_2O + 3\,Fe^{3+} \quad \text{for pH 1-6} \quad (1)$$
$$CrO_4^{2-} + 3\,Fe^{2+} + 8\,H^+ \rightarrow Cr^{3+} + 4\,H_2O + 3\,Fe^{3+} \quad \text{for pH} > 6 \quad (2)$$

**Required contact time.** To test required contact time for the reaction, stoichiometric dosage of ferrous sulfate (100% dosage) is used to react with 30 mg/L Cr(VI) at initial pH of 2.5, 5.0, 7.5 and 10.0 for 100 minutes, and the solution pH is monitored. The use of 30 mg/L Cr(VI) is to represent the worst case of chromium contamination in natural water since the Cr(VI) level in the contaminated water mostly ranges from 0 to 30 mg/L [1-7]. It is observed that the redox reaction can almost go to completion within 10 minutes, regardless of the initial pH of solutions. The pH is depressed to below 4 within 5 minutes except the reaction proceeded at initial pH 2.5 in which the pH rises to 2.9 in first 5 minutes then continues to drop gently to 2.6. The residual Cr(VI) concentration in solutions after 10-minute reaction is all below 0.8 mg/L and almost keeps constant for the rest of the reaction. The drastic pH drop is caused by the very rapid hydrolysis of the produced Fe(III) and Cr(III) ions from the redox reaction, which releases a great deal of hydrogen ions. The first rise of pH for the reaction with initial pH 2.5 probably results from the redox reaction itself which consumes hydrogen ions.

**Effects of initial pH on Cr(VI) reduction.** Although the Cr(VI) level does not change after a 10-minute reaction, all the experiments afterwards still use 20-minute contact time to ensure the redox reaction to reach equilibrium. To see the effects, various concentrations of hexavalent chromium are allowed to react with the respective stoichiometric dosage of ferrous ions at different initial pH. Figure-I shows the final Cr(VI) level after a 20-minute reduction at different initial pH. As seen in the figure, at the initial pH from 5 to 9, the residual Cr(VI) concentration in solutions is low (all below 0.4 mg/L) and does not fluctuate much for each initial Cr(VI) concentration. The results indicate that good Cr(VI) reduction can be obtained within this pH range.

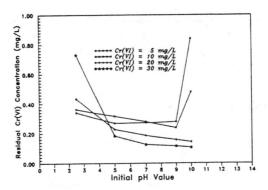

Figure-I Residual Cr(VI) concentration after 20-minute
reduction at different initial pH

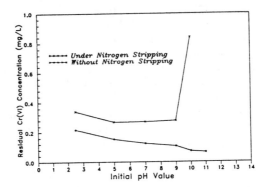

Figure-II Effects of dissolved oxygen on Cr(VI) reduction
for 5 mg/L initial Cr(VI) concentration

On the other hand, there are also some unexpected results shown in Figure-I. First of all, the final Cr(VI) for the 5 and 10 mg/L initial Cr(VI) concentration increases sharply at initial pH 10. But this does not occur for those reactions using higher Cr(VI) concentrations (20 and 30 mg/L). Secondly, assuming the reaction reaches equilibrium after 20 minutes, the residual Cr(VI) concentration between initial pH 5 and 9 should be higher for those reactions using higher initial Cr(VI) concentration due to the greater amount of $Fe^{3+}$ and $Cr^{3+}$ present in the solutions. But the findings, as shown in Figure-I, contradict the expected results.

Ferrous ions can also be oxidized by dissolved oxygen in water, and the rate of this oxygenation reaction rises with the increase in pH [19]. Therefore an assumption is made that *the dissolved oxygen is responsible for the abnormal consequence in Figure-I* since it could compete with hexavalent chromium to oxidized Fe(II), especially when the pH is high. To verify this assumption, tests

572

were conducted in an oxygen-free medium from initial pH 2.5 to 11 to see the effect of dissolved oxygen.

The reaction in oxygen-free medium is achieved by nitrogen stripping. Figure-II shows the effect of oxygen-free medium (under nitrogen stripping) on the reaction with 5 mg/L initial Cr(VI). According to the figure, the residual Cr(VI) concentration is reduced as expected. The difference in the residual Cr(VI) concentrations between the two curves in Figure-II also increases with the increasing pH. These results indicate that the dissolved oxygen does become involved in the redox reaction, particularly at high pH.

From the results above, the causes for the abnormal findings can be explained. At pH 10, the high residual Cr(VI) concentration for the reactions with low initial Cr(VI) concentrations (5 and 10 mg/L) is caused by the dissolved oxygen that "consumes" a part of ferrous sulfate injected into the solutions. While, for the reactions with high initial Cr(VI) concentration (20 and 30 mg/L), the pH of the solutions can be reduced to a level where the dissolved oxygen does not compete with hexavalent chromium to consume ferrous ions at the very beginning of the reactions, due to the hydrolysis of large amount of $Fe^{3+}$ and $Cr^{3+}$. The higher levels of residual Cr(VI) concentration at pH 5-9 for the reactions with lower initial Cr(VI) concentration can also be explained by the partial consumption of ferrous ions by the dissolved oxygen in water.

**Required dosage of ferrous sulfate for Cr(VI) reduction.** The required dosage of the reducing agent for complete Cr(VI) reduction was investigated. Hexavalent chromium is considered "completely" reduced when the residual Cr(VI) level is below the detection limit of DPC test (less than 0.01 mg/L). Since dissolved oxygen could interfere with Cr(VI) reduction, the dosage tests performed start from 100 % (stoichiometric) dosage and the ferrous sulfate dosage is increased gradually until the final Cr(VI) level after reaction is below the detection limit. The reactions proceed without prior adjustment of pH since this would be the case in practical application.

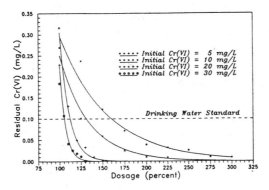

Figure-III Effects of ferrous sulfate dosage on Cr(VI)
reduction for various initial Cr(VI) concentration

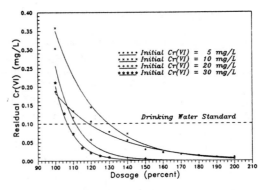

Figure-IV Effects of ferrous sulfate dosage on Cr(VI)
reduction when using the raw water as reaction
medium

Figure-III shows the dosage effects when deionized water is used as the reaction medium. As seen in Figure-III, the residual Cr(VI) concentration decreases with the increasing ferrous sulfate dosage. The final pH is found to be between 3 and 4 since deionized water does not have any buffering capacity. By using an exponential curve to fit the data, it is calculated that 20-30 mg/L of excess ferrous ion is required for complete Cr(VI) reduction. The excess dosage of ferrous ions for various initial Cr(VI) concentration is tabulated in Table-I.

To see if the redox reaction can proceed well in natural water, the same dosage test is run by using raw water from the Durham Water Treatment Plant. Table-II lists some properties of this raw water.

Shown in Figure-IV and Figure-V are the dosage effects and the final pH values when the raw water is used. As seen in Figure-IV, the results do not show major differences compared to those reactions run in deionized water except the reaction initially using 5 mg/L Cr(VI), which has a better Cr(VI) removal than expected. This phenomenon could result from the adsorption of the residual Cr(VI) by the metal hydroxide precipitates and flocculated colloidal particles in water.

### Table-I Excess Dosage of Ferrous Ion for Complete Cr(VI) Reduction

| Initial Cr(VI) Concentration | Overdosage of Ferrous Ions % of stoichiometry | mg/L |
|---|---|---|
| 5 mg/L | 168 % | 27.06 |
| 10 mg/L | 90 % | 28.99 |
| 20 mg/L | 38 % | 24.49 |
| 30 mg/L | 24 % | 23.20 |

### Table-II Properties of the Raw Water*

| Property | Value |
|---|---|
| pH | 6.7-6.8 |
| Alkalinity | 16-19 meq/L |
| Total Hardness | 22-23 mg/L |
| Specific Conductance | 76 |
| Dissolved Oxygen | 6.3-6.8 mg/L |
| Total Iron | 0.58 mg/L |
| Total Manganese | 0.16 mg/L |

* Data from Durham Water Treatment Plant.

As shown in Figure-V, the pH does not drop much for the reaction with 5 mg/L initial Cr(VI), which causes most of the generated Fe(III) and Cr(III) ions to form $Fe(OH)_3$ and $Cr(OH)_3$ precipitates in the solution. During the formation of the metal hydroxyl precipitates, the colloidal particles in the raw water are flocculated and larger suspended particles are formed (sweep flocs). These suspended particles could adsorb a part of the residual Cr(VI) in the solution and lead to the lower residual Cr(VI) concentration. But the reactions with higher Cr(VI) concentrations will reduce the pH down to 3 to 4 due to the hydrolysis reactions caused by the large amount of $Fe^{3+}$ and $Cr^{3+}$ produced, in which case most Fe(III) and Cr(III) ions exist as soluble hydrolyzed species. Because little sweep flocs is formed under this condition, almost no adsorption of hexavalent chromium occurs.

Figure-VI shows the comparison of the calculated required dosage of ferrous sulfate for complete Cr(VI) reduction when different reaction media are used. It is shown that no additional dosage of ferrous sulfate is required when the raw water is used as the reaction medium.

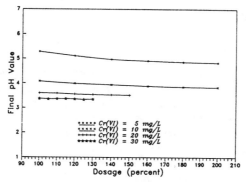

Figure-V Final pH value after reaction when using the raw
water as the reaction medium

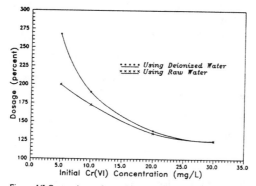

Figure-VI Comparison of the calculated required dosage of
ferrous sulfate for complete Cr(VI) reduction
when different reaction media are used

In summary, the simultaneous removal of hexavalent chromium by ferrous sulfate as the coagulant in water treatment is theoretically feasible. Because Cr(VI) reduction is rapid enough to be carried out in a normal pH range, and complete Cr(VI) reduction is achievable by applying excess dosage of ferrous sulfate. In addition, the produced ferric ions can serve as a promoter to precipitate the colloidal particles and Cr(III) hydroxide from water. The precipitates then can be remove by filtration in a water treatment plant.

The practical application of this technique in water requires further consideration since there are at least two more factors that could affect the redox reaction of ferrous sulfate and hexavalent chromium. One is the carbonate content in natural water, which could cause the precipitation of the injected ferrous ions. The other is the effect of temperature, which could change the redox dynamics of the reaction. The effects of reaction temperature and carbonate content in natural water on the redox reaction need to be investigated in the future.

## CONCLUSIONS

Based on the experimental results, the following conclusions are drawn:

(1) The redox reaction in which 30 mg/L of hexavalent chromium reacts with stoichiometric dosage of ferrous sulfate, regardless of the initial pH, is completed in less than 10 minutes. There is a drastic pH drop during the progress of the reaction, which is caused by the hydrolysis of the produced Cr(III) and Fe(III) ions.

(2) The initial pH of water does not seem to affect Cr(VI) reduction within the pH range from 5 to 9. However, when the pH is high, dissolved oxygen in water will interfere with the redox reaction by the consumption of ferrous ions.

(4) The required dosage of ferrous sulfate to completely reduce hexavalent chromium will exceed the stoichiometric value due to the ferrous oxygenation reaction in water. For the experiments conducted in deionized water, 20-30 mg/L of excess ferrous ion is required for complete Cr(VI) reduction.

(5) The reactions run in the raw water from the Durham Water Treatment Plant show that no additional dosage of ferrous sulfate is required to reduce hexavalent chromium when a natural water, instead of deionized water, is used as the reaction medium.

## REFERENCE

1. Perlmutter, N.M. and M. Lieber. 1970. *Dispersal of Plating Waste and Sewage Contaminants in Ground Water and Surface Water, South Farmingdale*

*Massapequa Area, Nassau County, New York*. U.S. Geological Survey Water Supply Paper 1879-G, pp. 67.

2. Pinder, G.F. 1973. "A Galerkin-finite Element Simulation of Groundwater Contamination on Long Island, New York," *Water Resource Res.*, 9, pp. 1657-1669.

3. Grove, D.B. 1985. "Hexavalent Chromium Contamination of an Alluvial Aquifer near Telluride, Colorado," *2rd Annual U.S.G.S. Conference on Hazardous Waste Disposal*. Cape Cod, October. 1985, Extended Abstract.

4. Grove, D.B. and K.G. Stollenwerk 1985. "Modeling the Rate-controlled Sorption of Hexavalent Chromium," *Water Resource Res.*, 21, pp. 1703-1709.

5. Zotter, K. and I. Licsko. 1991. "Removal of Chromium(VI) and Other Heavy Metals from Groundwater in Neutral and Alkaline Media," *Wat. Sci. Tech.*, 26, pp. 207-216.

6. Philipot, J.M., F. Chaffange, and J. Sibony. 1985. "Hexavalent Chromium Removal from Drinking Water," *Wat. Sci. Tech.*, 17, pp. 1121-1128.

7. Forstner, U. and G.T.W. Wittmann. 1981. *Metal Pollution in the Aquatic Environment*, Springer-Verlag Berlin Heideberg, New York, pp. 30-61.

8. Bartlett, R.J. 1991. "Chromium Cycling in Soils and Water: Links, Gaps, and Methods." *Environmental Health Perspectives*, 92, pp. 17-24.

9. Yassi, A. and E. Nieboer. 1988. "Carcinogenicity of Chromium Compounds," In Nriagu, J.O., Nieboer, E. Ed., *Chromium in Natural and Human Environments*, John Wiley & Sons, Inc., pp. 81-104.

10. Merian, E. 1991. *Metals and Their Compounds in the Environment*, VCH Publishers, Inc., New York.

11. U.S. EPA. 1973. "Waste Treatment: Upgrading Metal-finishing Facilities to Reduce Pollution," *EPA Technol. Transfer Seminar Publ.*, July, 1973.

12. Campbell, H.J. Jr., N.C. Scrivner, K. Bartzer, and R.F. White. 1977. "Evaluation of Chromium Removal from a Highly Variable Wastewater Stream," In *Proc. 15th Ind. Waste Conf.*. Purdue Univ., pp. 1-8.

13. Kunz R.G., T.C. Hess, A.F. Yen, and A.A. Arseneaux. 1980. "Kinetic Model for Chromate Reduction in Cooling Tower Blowdown," *Journal of Wat. Pollut. Control Fed.*, 52, pp. 2327-2339.

14. Jacobs, J.H. 1992. "Treatment and Stabilization of a Hexavalent Chromium Containing Waste Material," *Environmental Progress*, 11, pp. 123-126.

15. Higgin, T.E. and B.R. Marshall. 1985. "Combined Treatment of Hexavalent Chromium with Other Heavy Metals at Alkaline pH," In *Toxic and Hazardous Waste: Proc. 27th Mid-Atlantic Indus. Waste Conf.*, Technomic Publishing Co., pp. 432-443.

16. Schroeder D.C. and G.F. Lee. 1975. "Potential Transformation of Chromium in Natural Waters," *Water, Air and Soil Pollution*, 4, pp. 355-365.

17. Thomas, M.J. and T.L. Theis. 1976. "Effects of Selected Ions on the Removal of Chrome(III) Hydroxide," *Journal of Water Pollution Control Federation*, 48, pp. 2032-2045.

18. Cotton, F.A. and G. Wilkinson. 1980. *Advanced Inorganic Chemistry. A Comprehensive Text*, 4th ed., John Wiley & Sons. New York.

19. Stumm W. and J.J. Morgan. 1981. *Aquatic Chemistry - An Introduction Emphasizing Chemical Equilibrium in Natural Waters*. 2nd Ed., John Wiley and Sons. New York.

577

# SELECTIVE REMOVAL OF METAL-LIGAND COMPLEXES USING SYNTHETIC SORBENTS

ARTHUR KNEY and ARUP SENGUPTA
Lehigh University
Civil & Environmental Engineering Department
13 East Packer, Fritz Lab
Bethlehem, PA 10815

## INTRODUCTION

When dissolved heavy metals are accompanied by ligands (organic and inorganic) in a wastewater stream, conventional removal methods, such as precipitation, become hindered. Heavy metals remain in solution due to complexation with ligands. Non-conventional methods such as selective ion-exchange also do not work because of reduced selectivity when strong ligands are present in a system.

As industry forges ahead improving purification processes used in the metal plating and electronics sectors, the removal of metals in the presence of strong ligands becomes more of a problem. Heavy metals that would have normally precipitated out, using past purification methods, are remaining in solution because of complexation with heavy metals. Trace levels are approaching or surpassing, the Environmental Protection Agencies (EPA) maximum concentration limits (MCL).

Methods such as hydroxide precipitation [1] and anion exchange [2] procedures have been redefined to deal with the complexed metals. Both of these methods work but with limitations; the hydroxide method must be within the proper pH range, the anion exchange method does not have much flexibility with respect to the various species present (metal (II) ions, ligands and metals-ligand complexes).

The primary objective of this project is to characterize some new and tailored polymeric sorbents for the selective sorption of anionic metal-ligand complexes. The **basic concept of the process** we are developing is to tailor a resin in such a way that it will have a high affinity for an array of predefined target ions (metal (II) ions, ligands and metal-ligand complexes) present in trace concentrations, all at the same time.

## PREVIOUS WORK

Previous work involving the binding of metal (II) ions onto chelating exchangers with nitrogen donor atoms was preformed by Sengupta, Zhu, and Hauze [3].

Sengupta, Zhu, Hauze, and Ramana have shown that the DOW 2N and DOW 3N resins can be easily loaded with metal (II) ions. With this modification it has been proven, by Ramana and Sengupta [4], that anionic ligands can be removed selectively from a wastewater stream. Therefore, previous work indicates the removal of both cations and anions is possible using these specialty chelating polymers tailored in different ways. Through the cooperative efforts of these two teams we have been able to recognize the potential use of such a resin.

Work, done by Dudzinska and Clifford [2], has shown that lead-EDTA complexes can be removed using anion exchangers. The viability of this process is greatly reduced when high amounts of sulfates and chlorides are present.

Building on the previous work, we have found that by loading DOW 3N with copper we can selectively remove predefined target ions in the presence of high concentrations of sulfate and chloride, as represented by Figure 1, using a process we have coined, "The SMART Process".

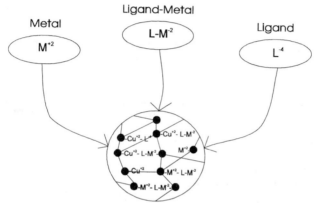

Figure 1 - Metal (II), Ligand, Ligand-Metal Sorption Process

## MATERIALS AND PROCEDURE

RESINS

| Resin Trade Name | Donor Atoms | Matrix | Acid-Base Characteristic | Manufacturer |
|---|---|---|---|---|
| IRC 718 | 1 Nitrogen 2 Oxygen | Polystyrene | weak base & weak acid | Rohm and Hass |
| DOW 2N (XFS 43084) | 2 Nitrogen | Polystyrene | weak base | Dow Chemical |
| DOW 3N (XFS 4195) | 3 Nitrogen | Polystyrene | weak base | Dow Chemical |

Table I - Resin Characteristics

Defined in Table 1, and represented in Figure 2, are the three types of resins we used in this project to study removals of metal-ligand complexes. The resins are

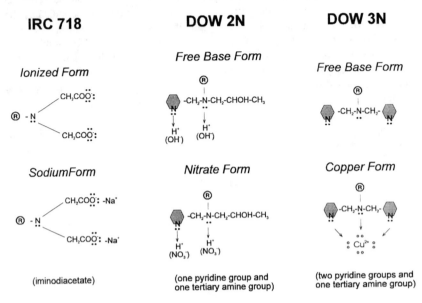

Figure 2 - Functionality of the Resins

obtained in spherical bead forms varying from 0.3 to 0.8 mm. The IRC 718 is conditioned using the standard procedure of cyclic exhaustion with 1 N hydrochloric acid and 1 N sodium hydroxide. The DOW 2N and DOW 3N are used in three different forms:

- Nitrate
- Free Base
- Copper

By passing 2% $HNO_3$ through a column the pyridine and amine groups becomes protonated with the nitrate as the counter ion (nitrate form). The resin is then rinsed with deionized (DI) water, removed from the column and air dried for 72 hours.

By passing 1 N sodium hydroxide through the column the nitrate counter ion is replaced by the hydroxyl ion (free base form). The free base form can be represented in two different ways, as illustrated in Figure 2. DOW 2N shows the free base form with a hydrogen and hydroxyl ion accompanying each nitrogen group. DOW 3N is displayed with two donor electrons at each nitrogen site. The latter is a more effective representation. Once cyclic exhaustion has been reached the resin is rinsed with DI water, removed from the column and air dried for 72 hours.

Experimentally determined capacities of the DOW 2N and DOW 3N in nitrate form are 1.72 and 1.73 mmol/g of resin respectively [3]. Using the 1.73 mmol/g of resin for the DOW 3N, for example, the appropriate loading of copper, or other metal such as zinc or cobalt, was determined (100%, 80%, etc.). To achieve an even loading distribution of the metal at various loading scenarios, a batch reactor arrangement is used. The batch reactor consists of a 125 mL plastic bottle filled with the proper concentration of aqueous samples and measured mass of resin. The bottle containing the aqueous solution and resin are placed in a rotating chamber to maintain a constant stirring action. The contents of the bottle are allowed to come to an equilibrium over an average period of five days. When loading the metal (II) it may be necessary to maintain a pH of around four or less, so not to exceed the solubility product of the particular metal. Once equilibrium has been reached the resin beads are removed from the bottle and allowed to dry for a period of 72 hours.

Regeneration is performed by passing 2 to 4% $NH_4OH$ through an exhausted column until the regeneration process is complete. The resin, after regeneration, would be considered in free base form. Equation 1 represents the reaction that takes place during the regeneration process.

$$R\text{-}Cu\text{-}EDTA\text{-}Cu + 2NH_4OH \leftrightarrow R\text{-}2H\text{-}2OH + 2Cu(NH_3)^{2+} + EDTA^{4-} \quad (1)$$

## CHEMICALS

All the standards and aqueous solutions were prepared using analytical grade chemicals; $Cu(NO_3)_2 \cdot 2\frac{1}{2}H_2O$, $Zn(NO_3)_2 \cdot 6H_2O$, $NaHCO_3$, $NaC_{10}H_{14}O_8N_2 \cdot 2H_2O_3$, and $C_6H_5O_7$ (Fisher Chemical). Also used were analytical grade $CoCl_2 \cdot H_2O$, $(HO_2^{14}CCH_2)_2NCH_2CH_2N(CH_2^{14}CO_2H)_2$, and $N(CH_2COOH)_2$ (Sigma Chemical).

| Element | M. M. (mg/mmol) | Mass (mg/L) | Moles (mole/L) | Equivalent (meqv/L) | Activity (A) |
|---------|-----------------|-------------|----------------|---------------------|--------------|
| $Zn^{+2}$ | 65.38 | 0.25 | 3.82E-6 | 7.65E-6 | 2.86E-6 |
| $C_{10}H_{14}O_8N_3^{-2}$ | 290.24 | 2.21 | 7.62E-6 | 1.52E-5 | - |
| $HCO_3^{-1}$ | 61 | 260.36 | 4.27E-3 | 4.27E-3 | 3.98E-3 |
| $Na^{+1}$ | 23 | 100.7 | 4.38E-3 | 4.38E-3 | 4.08E-3 |
| $NO_3^{-1}$ | 62 | 1.92 | 3.06E-5 | 3.06E-5 | 2.84E-3 |

Table II - Typical 1:2 (metal:ligand) Influent Concentration
Ionic Strength ($\mu$) = 4.36E-3 M

## SOFTWARE

To assist in the determination of the various species present in solution a software program called, "MINEQL$^+$" (version 3.01), was used [5]. This is a chemical equilibrium program for personal computers that includes many stability constants.

## COLUMN RUNS, BATCH TEST, AND METHODS OF ANALYSIS

Experimental details are similar to the ones used earlier studies [4], and are not being provided for the sake of brevity.

## RESULTS AND DISCUSSION

By running various batch tests we were able to characterize the effects of ligands in the presence of heavy metals, when using a typical cation exchanger (IRC 718). Figure 3 represents a batch test with three ligands; citrate, nitrilotriacetate (NTA), and ethylenediaminetetraacetate (EDTA). The binding effect of the stronger ligand becomes very apparent in this separation process. Of the three organic ligands, citrate is the weakest and EDTA the strongest.

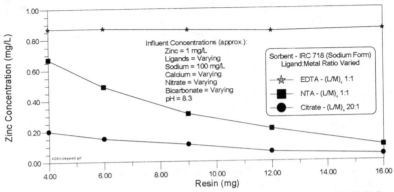

Figure 3 - Comparison of Ligand-Metal Sorption Capacity Using IRC 718

Lewis acid-base properties (the ability of a metal to accept electrons and a ligand to donate them) play key role in the chelating ability of a resin and the degree of interaction it has with its surrounding environment. Note, the interaction of a metal, ligand, and hydrogen can be predicted using stability constants (Table III).

| Ligand | Zinc ($Zn^{2+}$) | Copper ($Cu^{2+}$) | Cobalt ($Co^{2+}$) |
|---|---|---|---|
| Citrate - $ML^{-1}$ | 6.1 | 7.2 | 6.3 |
| NTA - $ML^{-2}$ | 12.0 | 14.2 | 11.7 |
| EDTA - $ML^{-2}$ | 18.2 | 20.5 | 18.1 |
| $HCO_3^{-1}$ - $ML^{+1}$ | 12.4 | 13.0 | - |
| $CO_3^{-2}$ - $ML2^{-2}$ | 9.6 | 9.8 | - |
| IDA* - ML | 8.3 | 11.5 | 7.9 |
| Di-2-picolyamine** - ML | 7.6 | - | - |

* Iminodiacetate (IDA) is the functional group on the IRC 718. This stability constant represents metal-ligand equilibrium in solution. It will **NOT** be the same as the equilibrium between the functional group on the resin and the metal, it will be somewhat lower [3].
** Di-2-picolyamine is the functional group on the DOW 3N. This stability constant represents the metal-ligand equilibrium in solution. It will **NOT** be the same as the equilibrium between the functional group on the resin and the metal, it will be somewhat lower [3].

Table III - Stability Constants (log $K_f$) [5]

Figure 3 also shows that the EDTA keeps the zinc in solution at about the same molar concentration as the total zinc. The curves, representing NTA and citrate, show a decreasing amount of zinc with respect to the mass of the resin present. NTA, a stronger ligand than citrate, maintains a higher concentration of zinc in solution. Figure 3 displays graphicly the problem that exists today with both organics and inorganics present in wastewater streams; how do we deal with the metal-ligand complexation that takes place? Accepted methods of separation prove to be ineffective in the presence of ligands!

Characterization of the relationship between the strongest ligand (EDTA), the metal ($Zn^{2+}$), and the resin (IRC 718) with respect to pH is extremely important. Figure 4 shows this relationship.

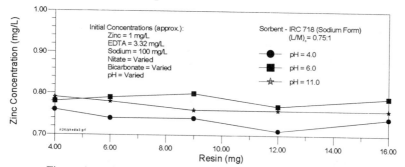

Figure 4 - Affects of pH With Respect to IRC-718 and EDTA

Since EDTA is a weak acid, hydrogen ion competition increases as the pH decreased. This behavior would explain why the zinc concentration is less at a pH of 6 versus a pH of 4.

As the pH approaches 11 the concentration of the zinc proves to be the lowest of the three pH values considered. This can be explained by analyzing the speciation of the system we are dealing with. A titration curve, similar to Figure 8, can be used to determine the speciation at various values of pH. Table IV lists the predominant species of this system at different pH values.

| pH < 3 | 3 < pH < 8 | pH > 8 |
|---|---|---|
| $H_5L^{+1}$, $Zn^{2+}$ | $ZnL^{2-}$ | $L^{4-}$, $Zn(CO_3)_2^{2-}$ |

Table IV - Predominant Species, Metal:Ligand Ratio $(1:2)_o$

At a pH of 11 the ligand complexed with the zinc would be $Zn(CO_3)_2^{2-}$, not EDTA as might be expected. Because of the carbonate complexation, some degree of precipitation might be expected. This precipitation explains the depletion of zinc and why the concentration is not the highest at a pH of 11.

DOW 3N was examined with respect to four different scenarios using the batch method;

- 100% Nitrate form

- 50% Copper (II) form - 50% Free Base form
- 80% Copper (II) form - 20% Free Base form
- 100% Copper (II) form

The idea behind the varied loading of the resin is as follows: by leaving some of the functional sites loaded (metal (II)) and others unloaded (free base form) the ligand and metal-ligand complexes could bind at the loaded sites (acting as an anion exchanger) and the unloaded sites would be able to chelate with any metal (II) in solution. Figure 5 shows the results of this batch test.

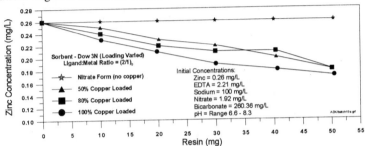

Figure 5 - Varied Loading of DOW 3N

The 100% loaded resin was working exceptionally well, the 80% a bit less, and the nitrate form showing no zinc (II) removal. We had expected that the 50% or 80% might show a greater capacity than the 100% copper loaded. Column runs, done at a later date, helped to answer the reason for this. This process causes an exceptional amount of copper bleeding. Zinc (II) competes for the same sites as the desorbed copper (II), therefore less removal of the zinc (II). This phenomenon is still under investigation.

Initial column runs were done to verify the result from the batch tests. Figure 6 shows the results from these runs. DOW 3N displays the best results. IRC 718, in both sodium and copper form, along with DOW 3N in free base form show little or no sorptive capacity with respect to zinc in the present twice the molar concentration of EDTA.

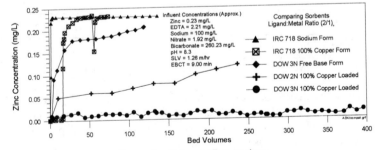

Figure 6 - Column Comparison

At this point we began to take a closer look at what was happening in the system using the 100% copper loaded resin. We found that the pH dropped to between 3 and 4 for the first 100 to 200 bed volumes and then rose back to the influent concentration of 8.3, when using the typical 1:2 metal-ligand influent concentration defined earlier (Table II). The pH drop can be attributed to the copper (II) loaded functional groups acting as anion exchange sites and adsorbing the hydroxyl ions. The rising pH is an indication hydroxyl ions are being chromatographicly eluded, most likely replaced by bicarbonate anions and metal-ligand complexes present in solution.

We also found that the amount of copper present in solution was much higher than the concentration of zinc present due to the bleeding taking place. This meant that copper (II) was in competition with the zinc (II) for sites on the resin. Also, copper (II) complexation with the EDTA was favored since its stability constant is higher than zinc-EDTA (Table III). Because of the copper bleeding the species in solution, inside the column, were not what we expected them to be.

Using MINEQL we redefined the species present in the system. For this particular column run the influent metal-ligand ratio was 1:1, the remaining constituents were unchanged (Table II). But, when the influent enters the column the interaction of the copper (II) must be accounted for. We choose to account for only the copper (II) that could bind with the EDTA, i.e., one part. Therefore, the metal-ligand ratio in solution would now be 2:1 where the fraction of metal is represented in an equal 1:1 ratio ($Cu^{2+}:Zn^{2+}$). This system just described is what the following set of titration curves represent (Figure 7).

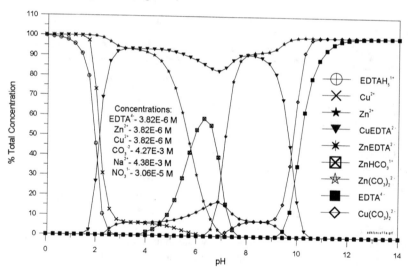

Figure 7 - Titration Curve, % of Predominant Species
System - EDTA/Zinc/Carbonate, Metal:Ligand Ratio $(2:1)_o$

Table V lists the predominant species at different pH values.

| pH < 3 | 3 < pH < 8 | pH > 8 |
|--------|-----------|--------|
| $H_5L^{+1}$, $Zn^{2+}$, $Cu2+$ | $CuL^{2-}$, $Zn^{2+}$, $ZnHCO_3^+$, $Zn(CO_3)_2^{2-}$ | $L4-$, $Zn(CO_3)_2^{2-}$, $Cu(CO_3)_2^{2-}$ |

Table V - Predominant Species, Metal-Ligand Ratio $(1:2)_o$

The experimental results of this system are displayed in Figure 8a. The metal-ligand ratio in the influent is 1:1. It must be kept in mind, when the influent enters the column the initial relationship changes due to the copper that is desorbed. This is the system the titration curve is representing.

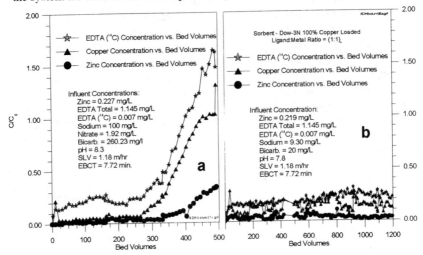

Figure 8 - Column Run, 100% Copper Loaded Resin
Metal:Ligand Ratio $(1:1)_o$, Comparing Bicarbonate Competition

The removal of the target ions is successfully accomplished as represented in Figure 8a. Using the titration curves we can predetermine the species present and specifically target the key ions at various values of pH.

By decreasing the concentration of bicarbonate in the system the capacity of the bed increases accordingly due to the decreased concentration of the competing ligand (Figure 8b). Prior work with DOW 2N copper loaded [6], has proven that in the presence of sulfates and chlorides, weak ligands, the competing effect would be very small to nonexistent with respect to the same concentrations of bicarbonate used.

## CLOSING REMARKS

The selective removal of metal (II) ions, ligands, and metal-ligand complexes can be accomplished successfully with one resin, DOW 3N copper loaded. By modifying the functional sites of DOW 3N we can improve the selective sorption affinity of the resin. With the use of titration curves to help determine the speciation at any particular pH it appears possible to predict what species will be adsorbed and possibly model the behavior of the chelating exchanger. The "SMART Process" is proving to be a promising and cost effective method for dealing with the issue of complexed heavy metals. Research is still in progress concerning this process.

## ACKNOWLEDGMENT

This study received financial assistance from the Environmental Protection Agency through grant number R-8192298.

## REFERENCES

1. Tunay, O., and N. I. Kabdasli. 1994. "Hydroxide Precipitation of Complexed Metals," *Water Resources*, 28(10):2117-2124

2. Dudzinska, Marzenna R., and Dennis A. Clifford. 1991. "Anion Exchange Studies of Lead-EDTA Complexes," *Reactive Polymers*, pp. 71-80

3. Sengupta, Arup K., Yuewei Zhu, and Diane Hauze. 1991. "Metal (II) Ion Binding onto Chelating Exchangers with Nitrogen Donor Atoms: Some New Observations and Related Implications," *Environmental Science & Technology*, 25(3):481-488.

4. Ramana, Anuradha, and Arup K. Sengupta. 1992. "Removing Selenium(IV) and Arsenic(V) Oxyanions with Tailored Chelating Polymers," *Journal of Environmental Engineering*, 118(5):755-775.

5. Schecher, William D., and Drew C. McAvoy. 1994. MINEQL$^+$, *Environmental Research Software,* software program

6. Sengupta, Arup K., and Yuewei Zhu. 1994. "Selective and Reversible Ligands Sorption through a Novel Regeneration Scheme," *I & EC Research*, 33(2):382-386

# FLY ASH ENHANCED METAL REMOVAL PROCESS

SUJITHKUMAR NONAVINAKERE
Plexus Scientific Corporation
980, Awald Drive, Suite 202
Annapolis, MD 21403

BRIAN E. REED
P O Box 6101
Department of Civil Engineering
West Virginia University
Morgantown, WV 26505-6101

## INTRODUCTION

Coal fly ash, an industrial solid waste by-product, is produced in enormous quantities by thermal power plants. As of 1990, 78.9 million metric tons of waste, including fly ash, bottom ash, boiler slag, and flue gas desulfurization sludge was produced in the United States (American Coal Ash Association, 1992). Constructive utilization of coal fly ash was about 25.4 percent in 1990. Fly ash is used in structural fills, road base, asphalt fillers, snow and ice control, blasting grit/roofing granules, grouting, mining applications, and in preparing bricks, Portland cement, puzzolona *etc*. In addition to these used, it has been demonstrated that fly ash can be used to remove trace amounts of heavy metals from aqueous waste streams (Asit *et al.*, 1987; Weng, 1990).

Heavy metals present in industrial waste streams are toxic to humans, and can have a long-term adverse environmental impact. Anthropogenic sources of heavy metals include wastewater discharges from electroplating and metal processing, storage battery manufacturing, pigment manufacturing, tanning industries, leachate from landfills, and land application of heavy metal-laden sludge. Due to stringent wastewater discharge standards set by state and federal regulatory agencies, it is imperative to find appropriate technologies for their removal from waste streams. The metals of immediate concern are cadmium, nickel, lead, chromium, copper, mercury, and zinc because of their extreme toxicity to humans and being listed as priority pollutants by the EPA.

Removal of heavy metals by adsorption at solid-solution interface is a feasible means of treating metal-bearing waste streams. Various materials such as activated carbon, metal oxides, and ion exchange resins have been used to remove trace metals from industrial and drinking water streams. Fly ash, being readily available and relatively inexpensive may provide an economically viable alternative for the removal of heavy metals.

An example of a industrial metal-bearing waste stream is electroless nickel (EN) plating waste. Presently, (EN) plating is gaining popularity due to better plating characteristics. Typical spent electroless nickel (EN) baths contain high concentrations of nickel (>6000 mg/L), hypophosphite, phosphite, and/or any of the organic chelating agents such as citric, tartaric, lactic, succinic, and glycolic acids. The most popular method for treating EN plating waste is by conventional alkaline-neutralization process. Hydrated lime is added to the waste stream to precipitate nickel ions as nickel hydroxide. The resulting sludge is then disposed off in landfills. The presence of complexing agents can decrease the treatment efficiency. Also, the poor settling characteristics of nickel hydroxide precipitates produce large volumes of nickel sludge requiring further treatment and disposal. Therefore, there exists a need to find waste-treatment methods that decrease the volume of metallic sludge.

The addition of a solid, having an adsorptive affinity for heavy metals, could increase the efficiency of precipitation process by adsorption and surface precipitation of the metal. The acid-base characteristics of fly ashes makes their use in the treatment of metal-bearing waste streams and the stabilization of metal sludges especially attractive. Fly ash can be used as a "coagulant" to aid the settling of nickel floc in precipitation treatment of spent EN plating baths (Reed, 1991).

The primary objective of the study was to evaluate the effectiveness of fly ashes from local thermal power plants in the removal of cadmium, nickel, chromium, lead, and copper from aqueous waste streams. Physical and chemical characteristics of fly ashes were determined, batch isotherm studies were conducted. A practical application of using fly ash in treating spent electroless nickel (EN) plating baths by modified conventional precipitation or solid enhanced metal removal process (SEMR) was investigated. In addition to nickel the EN baths also contains complexing agents such as ammonium citrate and succinic acid reducing agents such as phosphite and hypophosphite. SEMR experiments were conducted at different pHs, fly ash type and concentrations, and settling times.


## BACKGROUND STUDY

### Electroless Nickel Plating Waste

The life of electroless nickel (EN) plating bath is relatively short in comparison to other electroplating baths. Usually EN baths must be "dumped" after 3 or 4 turnovers. Spent EN bath typically contains 5,000 - 7,000 mg/L nickel (Parker, 1982). A typical effluent standard for EN industries is 10 mg/L of Ni (40 CFR, Part 413). In addition to nickel, EN bath contains large amounts of orthophosphates and organic complexants, that further complicates the treatment process. Treatment processes used in the treatment of EN spent bath include hydroxide precipitation, ion exchange, reverse osmosis, and adsorption. Hydroxide precipitation has remained the most popular and economical treatment alternative for treating EN spent baths. It is difficult to meet the stringent effluent discharge limits with precipitation treatment alone, and other downstream treatment methods such as activated carbon adsorption or ion exchange may be required.

Numerous researchers have demonstrated that spent EN baths can be successfully treated with caustic and/or lime precipitation in conjunction with oxidation-reduction treatment. Parker (1982) used dimethyl or diethyl-thio-carbamate (DTC) to precipitate the nickel. DTC is a special precipitant that produce nickel compounds with extremely low solubility. Nickel concentrations were reduced from 4,000 - 12,000 mg/L to 5 - 70 mg/L. Ying and Bonk (1987) reported that caustic soda treatment was effective in removing nickel, as nickel hydroxide, from spent EN bath containing acetic acid, but was ineffective in removing nickel from spent EN baths containing citrate as a complexant. Brooks (1989) indicated that treating electroless nickel waste by aeration followed by hydrogen peroxide oxidation and finally precipitating nickel with oxalic acid reduced nickel concentration by 99.7% in a EN bath containing 2,880 mg/L of Ni. Reed (1991) demonstrated that addition of activated carbon along with pH adjustment increased removal of nickel compared to pH adjustment alone. A schematic of the process, refereed to as the solid enhanced metal removal (SEMR) process is presented below.

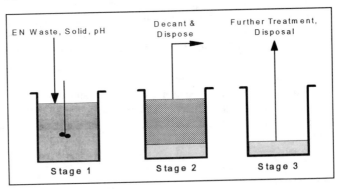

**Figure 1. Schematic of PACMS process (Reed, 1991)**

The addition of a properly selected solid can decrease the amount of aqueous nickel and improve the settling characteristics of the nickel sludge. The latter results in a reduction in the volume of sludge requiring further treatment and disposal. The demerits of the SEMR when fly ash is added as the enhancing solid is the chance that trace metals, originally bound to the fly ash may desorb into the effluent being discharged.

## MATERIALS AND METHODS

### Fly Ash

Fly ashes were obtained from Ft.Martin, Ft.Albright, and Rivesville thermal power plants in and around West Virginia. The fly ashes were air dried and pulverized to maintain a uniform particle size in subsequent analyses.

### Stock Metal Solutions

Cadmium, nickel, lead, chromium, and copper were chosen as the study heavy metals. Stock solution for each metal (500 ppm) was made up by adding appropriate amount of reagent grade nitrate salt of the desired metal into one liter of deionized water.

### Batch Adsorption Studies

Adsorption isotherms were conducted over a range of pH values (3 - 11) and varying metal and fly ash concentration. Metal concentrations of approximately 10 and 50 mg/L and fly ash concentration of 5 and 10 g/L were investigated. Background electrolytes were not used. Isotherms were also conducted without fly ash to determine the removal that would occur via solution precipitation. Fly ash was hydrated for a period of 12 hours before being used in batch adsorption experiments. Predetermined volumes of metal and hydrated fly ash solutions were added to a 500 ml volumetric flask such that upon dilution 500 ml the desired metal and fly ash concentration would be achieved. Aliquots of 50 ml were then placed in 75 ml Nalgene bottles and varying amounts of either NaOH (0.1 or 1.0 N) or $HNO_3$ (0.1 or 1.0 N) were added to individual samples to obtain desired pH. The samples were sealed and placed on a mechanical shaker and agitated for 24 hours. At the end of 24 hour period the samples were removed, the pH measured, slurries were filtered, and the acidified filtrate analyzed for respective metal.

All samples were filtered through 0.45 μm cellulose nitrate filters. Metal content was determined by flame atomic adsorption spectrophotometry using a Perkin Elmer 3000 Atomic Absorption Spectrophotometer. Wavelengths 231.1, 228.8, 217.0, 357.9 and 324.8 nm were used for nickel, cadmium, lead, chromium, and copper, respectively. If necessary samples were diluted with deionized water prior to metal analysis. An Orion Research combination pH electrode (Model #810200) was used for pH measurements. An approximate temperature of 22° C was used for all experiments. Analytical grade reagents of NaOH and $HNO_3$ were used to prepare the acid and base stock solutions. NaOH was stored in a reservoir with a soda lime scrubber to prevent atmospheric $CO_2$ contamination. NaOH was standardized using Method 2310

B.3(c) of Standard Methods (1989). $HNO_3$ was standardized by titrating a known volume of $HNO_3$ with NaOH of known normality to pH 7.

## Solid Enhanced Metal Removal Process (SEMR)

Experiments were carried out to determine the extent of nickel removal from two synthetic nickel-organic acid-hypophosphite/phosphite spent plating baths after the addition of fly ash as a "coagulant" aid. Composition of the synthetic spent EN plating baths are given in Table 1.

The effect of addition of three different fly ashes on caustic precipitation treatment of EN spent bath was studied. Settling of of 2.5 hrs. and 18 hrs. were employed. Fly ash concentrations of 10 g/L and 50 g/L were used. Experiments were also conducted without using fly ash to assess the difference between the ash and no-ash systems. Aliquots of 50 ml of the spent baths were placed in 80 ml glass beakers to which appropriate amount of fly ash was added to obtain a desired final fly ash concentration. Required amounts of 6N NaOH were added to each beaker using a 50 ml serological pipette. Following the addition of NaOH the beakers were covered with parafilm, vigorously shaken to obtain a complete mix, and were then gently mixed for a period of one hour using a mechanical shaker. The beakers were then placed on an observation table for settling. After 2.5 or 18 hours the sludge height in each beaker was measured. The supernatant was decanted into test tubes and acidified and analyzed for Ni.

**Table 1. Typical Spent Electroless Nickel Plating Bath Composition**

| Constituents | Succinic Acid Bath (pH = 4.13) | Ammonium Citrate Bath (pH = 5.31) |
|---|---|---|
| Nickel Ni | 5870 mg/L | 5870 mg/L |
| Sodium hypophosphite $NaH_2PO_2.H_2O$ | 7.95 g/L | 7.95 g/L |
| Sodium phosphite $NaH_2PO_3.H_2O$ | 31.18 g/L | 31.18 g/L |
| Succinic Acid $HOOCCH_2CH_2COOH$ | 35.43 g/L | --- |
| Ammonium Citrate $HOOCOHC(CH_2COONH_4)_2$ | --- | 67.86 g/L |
| Sodium Hydroxide NaOH | To pH | To pH |

# RESULTS AND DISCUSSION

## Batch Adsorption Isotherms

The effect of pH on the adsorption of cadmium, nickel, lead, copper, and chromium on fly ash were studied by varying the pH between 3 and 12. Adsorption isotherms were conducted at two metal concentration of 10 mg/L and 50 mg/L. For all metal-fly ash experiments, metal removal increased with increasing pH. No-ash systems were investigated for all metals to determine if the metals were being removed by precipitation alone. Chromium(III) exhibited strong amphoteric characteristics. In the absence of fly ash, the concentration of chromium(III) rapidly decreased to less than 1 percent at pH ~ 6.0 and gradually increased at pH ~ 10 and above. However, in the presence of all three fly ashes chromium(III) was completely removed from the solution at pH ~ 6.0 and above.

Similar results were obtained for all other metals and fly ashes. Due to space constraint all the results cannot be reported in this format. Thus, metal removal at two pH units are presented and discussed in the following paragraphs. The pH units were chosen on the basis of precipitation pH for each metal at different concentration, as obtained from the no-ash systems.

In Figure 2, lead removals at pH values 4.5 and 5.5 are presented. At lower lead concentration (10 mg/L), the removal increased with increasing fly ash concentrations. Same trend was observed for 50 mg/L lead. In the no-ash systems, only 5 percent of lead was removed in 10 mg/L systems and 15 percent of lead was removed in 50 mg/L systems at pH 4.5 and 5.5. Increased removal was observed at all concentrations of fly ash and lead as the pH was increased. Ft. Albright fly ash performed better when compared to other fly ashes.

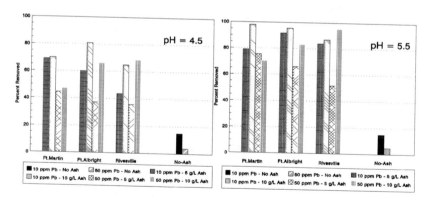

## Figure 2. Removal of Lead at pH 4.5 and 5.5

In Figure 3, nickel removals at pH values 5.0 and 7.0 are presented. The removal increased as the fly ash concentration was increased. The removal decreased as the nickel concentration was increased at all fly ash concentrations except for Rivesville fly ash at 5 g/L and Ft. Albright fly ash at 10 g/L with a nickel concentration of 50 mg/l. Ft. Martin fly ash performed better at pH 5.0 and maximum nickel removal was obtained by Ft. Albright fly ash at pH 7.0. Zero removal was observed in the no-ash systems. Removals at pH 7.0 were higher than the removals at pH 5.0 for all systems.

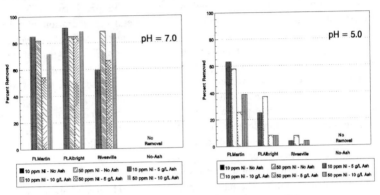

**Figure 3. Removal of Nickel at pH 5.0 and 7.0**

In Figure 4, copper removal at pH value of 6.0 is presented. Removal of copper increased as the fly ash concentration was increased, except for Ft. Martin and Ft. Albright fly ash at 10 g/L fly ash concentration and 50 mg/L Cu. In no-ash systems, 18.6 percent Cu precipitated in 10 mg/L systems and 5.3 percent of Cu precipitated in 50 mg/L systems. Maximum removal was obtained by Rivesville fly ash in 10 mg/L Cu systems.

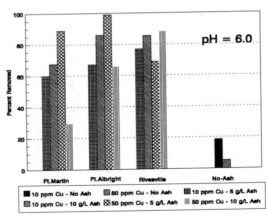

**Figure 4. Removal of Copper at pH 6.0**

In Figure 5, chromium removal at pH value of 4.5 is presented. Chromium precipitated completely around pH 5.0. Hence, pH 4.5 was selected as the optimum value as the precipitation of Cr did not exceed 17 percent for both 10 mg/L and 50 mg/L Cr. Removal of Cr decreased with the increase in metal concentration. Ft. Martin fly ash performed better in at all concentrations of fly ash and Cr.

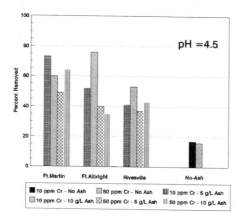

**Figure 5. Removal of Chromium at pH 4.5**

In Figure 6, cadmium removals at pH values of 6.0 and 7.5 are presented. Removal of cadmium increased with increasing pH. In the no-ash systems, removal via precipitation did not exceed 10 mg/L and no removal was obtained at 50 mg/L. Cadmium removal increased as the fly ash concentration was increased from 5 g/L to 10 g/L. Removal decreased with increasing cadmium concentration. Ft. Albright performed better than Ft. Martin and Rivesville fly ashes.

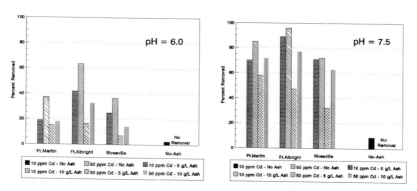

**Figure 6. Removal of Cadmium at pH 6.0 and 7.5**

Primary metal removal mechanism by fly ash is attributed to surface precipitation. In general, all three fly ashes removed significant amounts of metal ions from solution prior to the onset of solution precipitation. Farley *et al.*, (1985) reported that surface precipitation of a metal can occur prior to precipitation in the bulk solution. High oxide contents are deemed responsible for the adsorption and/or surface precipitation of metal ions from solution by fly ash. Metal ions are possibly precipitated on the surface of either alumina, silica, ferric or a combination of these and other oxides. For a given surface loading, the adsorption of study metals on to fly ash was in the following order: $Cr^{3+} > Pb^{2+} > Ni^{2+} > Cd^{2+} > Cu^{2+}$.

### EN Spent Bath Treatment by SEMR

Conventional alkaline precipitation treatment of spent EN bath yields large volumes of nickel floc having poor settling characteristics. The use of fly ash as a coagulant in aiding metal removal and improvement of settling characteristics were evaluated. Two different concentrations of fly ash, 10 g/L and 50 g/L and effect of settling time (2.5 hours and 18 hours) were examined.

Base titration curves for the spent EN baths are presented in Figure 7. The compositions of two synthetic spent EN baths were presented earlier. The titration curves were used in estimating the amount of NaOH required to reach a certain pH for each individual bath. Ammonium citrate bath had a lower initial pH (2.42) in comparison to succinic acid bath (pH ~ 3.85) and much higher buffering capacity for the range of the titration study.

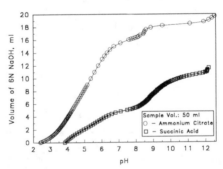

**Figure 7. Titration Curves for 50 ml of Spent EN Plating Baths**

Results from the SEMR experiments using the succinic acid bath are presented in Table 3. The removals at different pH values (9.3,10.11.3 and 12) in were obtained by linear interpolation assuming that the relationship between nickel removal and pH is linear. Some discrepancies in removal calculations may have occurred if this relationship was not linear. Twenty four experiments were conducted at each settling time of 2.5 and 18 hours. For almost all cases, the addition of fly ash significantly improved removal of nickel in comparison to no-ash systems. The beneficial affect of fly ash was very much apparent at pH 9.3 when compared to no-ash systems. At pH values above 10 no improvement was obtained in nickel removal. Ft.Martin and Rivesville performed basically the same when compared to Ft. Albright fly ash. Increasing the settling time resulted in increased nickel removal in some cases. Based on sludge height observations, minimum sludge was produced in Ft. Albright fly ash systems after 18 hours settling in comparison to other two fly ashes. This may be due to surface precipitation, as Ft. Albright fly ash has the largest surface area in comparison to Ft. Martin and Rivesville fly ashes. The increase in fly ash concentration resulted in increased removal for 14 out of 24 experiments at 2.5 and 18 hours settling times.

**Table 3. Results from SEMR Treatment of Succinic Acid Bath**

| Ash Type | pH | Effluent Nickel Concentrations* | | | |
|---|---|---|---|---|---|
| | | 2.5 hours | | 18 hours | |
| | | 10 g/L | 50 g/L | 10 g/L | 50 g/L |
| Ft. Martin | 9.3 | 37.65 | 17.89 | 27.32 | 22.28 |
| | 10.0 | 7.6 | 3.96 | 4.71 | 5.71 |
| | 11.3 | 2.72 | 2.32 | 1.97 | 2.97 |
| | 12.0 | 3.20 | 2.55 | 1.83 | 3.32 |
| Ft. Albright | 9.3 | 86.84 | 82.69 | 104.53 | 96.60 |
| | 10.0 | 25.41 | 18.42 | 38.96 | 39.77 |
| | 11.3 | 2.49 | 2.44 | 2.17 | 1.85 |
| | 12.0 | 2.05 | 2.30 | 1.70 | 1.60 |
| Rivesville | 9.3 | 41.67 | 14.40 | 22.26 | 44.66 |
| | 10.0 | 6.45 | 7.16 | 14.33 | 22.57 |
| | 11.3 | 1.18 | 1.20 | 2.21 | 1.82 |
| | 12.0 | 1.12 | 1.13 | 2.13 | 1.63 |
| No Ash | 9.3 | 106.80 | | 95.40 | |
| | 10.0 | 81.20 | | 14.90 | |
| | 11.3 | 18.30 | | 10.60 | |
| | 12.0 | 23.10 | | 10.60 | |

*Initial concentration of Ni in the EN spent bath = 5532 mg/L.

Results for ammonium citrate bath are present in Table 4. There was little (<3%) removal of nickel from solution by hydroxide precipitation. Improved nickel removal was observed in 17 out of 24 experiments with increasing fly ash concentration. Ft. Albright and Rivesville fly ashes performed better at pH 12.0 with a fly ash concentration of 50 g/L and 18 hours settling time. Increase in settling time increased the removal in 21 out of 24 experiments. Settling of fly ash-metal floc did not occur, hence it was difficult to visibly assess the affect of fly ash on settling characteristics of the nickel sludge.

## Table 4. Results from SEMR Treatment of Ammonium Citrate Bath

| Ash Type | pH | Effluent Nickel Concentrations* | | | |
| | | 2.5 hours | | 18 hours | |
| | | 10 g/L | 50 g/L | 10 g/L | 50 g/L |
|---|---|---|---|---|---|
| Ft. Martin | 9.3 | 5388 | 5293 | 5155 | 5219 |
| | 10.0 | 5291 | 5317 | 4723 | 4987 |
| | 11.3 | 4996 | 5334 | 3888 | 3549 |
| | 12.0 | 4616 | 5556 | 3684 | 3232 |
| Ft. Albright | 9.3 | 5146 | 4847 | 4946 | 4875 |
| | 10.0 | 5079 | 4777 | 5012 | 4460 |
| | 11.3 | 5017 | 4693 | 4015 | 3107 |
| | 12.0 | 4842 | 3067 | 2942 | 2749 |
| Rivesville | 9.3 | 4981 | 5118 | 5399 | 4905 |
| | 10.0 | 4908 | 5124 | 5217 | 3914 |
| | 11.3 | 4874 | 4686 | 4675 | 2391 |
| | 12.0 | 4951 | 3359 | 4387 | 2489 |
| No Ash | 9.3 | 5209 | | 5827 | |
| | 10.0 | 5050 | | 5065 | |
| | 11.3 | 5071 | | 5560 | |
| | 12.0 | 5099 | | 5307 | |

*Initial concentration of Ni in the EN spent bath = 5395 mg/L*

## SUMMARY

Batch adsorption isotherm experiments have demonstrated that fly ash can be used to remove trace amounts of heavy metals from waste water streams. Adsorption of various metals on to all three fly ashes at a given pH was in the following order: $Cr^{3+} > Pb^{2+} > Ni^{2+} > Cd^{2+} > Cu^{2+}$. Present concerns about fly ash is its leachable trace metal content But its easy availability and low cost may make it a desirable adsorbent in treating heavy metal waste streams. One use of fly ash may be as a secondary or tertiary treatment before waste effluent is subjected to conventional treatment such activated carbon adsorption, ion exchange or reverse osmosis.

Addition of fly ash to conventional precipitation treatment of succinic acid EN plating baths will result in reduced nickel sludge production. Ft. Martin and Rivesville fly ashes performed basically the same. Increase in fly ash concentration will result in increased nickel removal. Settling times may or may not affect the removal of nickel. Fly ash did not aid the metal floc formation in the case of ammonium citrate spent EN baths. Removal obtained in ammonium citrate spent EN baths were not comparable to the removal obtained in succinic acid spent EN baths. In general, fly ash can be used in reducing nickel concentration in succinic acid spent EN baths.

## ACKNOWLEDGMENTS

The study was funded by National Research Center for Coal and Energy, Morgantown, WV.

## REFERENCES

American Coal Ash Association, Inc., *1990 Coal Combustion By-product-Production and Consumption*, Revision No.1, June 15, 1992, Washington D.C.

Brooks, C.S. (1989). " Metal Recovery from Electroless Plating Wastes", *Metal Finishing*, May 1989, 33-36.

Farley, K.J., Dzombak, D.A., and Morel, F.M.M. (1985). "A Surface Precipitation Model for the Sorption of Cations on Metal Oxides", *Journal of Colloid and Interface Science*, 106(1),226-242.

Lorenz, K. (1954). " Secondary Treatment of Power Plant Phenol Waste with Fly Ash and Cinders", *Gesundh. Ing.*, Vol.75, 189.

Parker, K. (1982). " The Waste Treatment of Spent Electroless Nickel Baths", *The 1st AES Electroless Plating Symposium*, March 23-24, 1982, St.Louis, MO.

Reed, B.E. (1991). " Treatment of Metal-Bearing Waste streams Using the Powdered Activated Carbon Metal Sorption Process", *1991 Water Pollution Control Federation Annual Conference*, Toronto, Canada, October 6-10, 1991.

Schure, M.R., Soltys, P.A., Natusch, D.F.S., and Mauney, T. (1985). "Surface Area and Porosity of Coal Fly Ash", *Environ. Sci. Techno.*, Vol.19, No.1, 82-86.

*Standard Methods for the Examination of Water and Wastewater.* (1989), 20th Edition, American public Health Association, Washington, D.C.

Weng, C.H. (1990). *Removal of Heavy Metals by Fly Ash*, thesis presented to University of Delaware in partial fulfillment of the requirements for the degree of Master of Civil Engineering.

Ying, W. and Bonk, R.R, (1987). " Removal of Nickel and Phosphorus from Electroless Nickel Plating Baths", *Metal Finishing*, December 1987, 23-31.

# ADSORPTION OF ZINC ON MAGNETITE PELLETS

**DANIEL A. CARGNEL, P.E.**
  RUST Environment & Infrastructure, Inc.
  Mechanicsburg, PA

**CHARLES A. COLE, Ph.D.**
  Associate Chair, Environmental Pollution Control Program
  Penn State Harrisburg

## INTRODUCTION

Zinc is a common contaminant in wastewater from electroplating, metal finishing, and many other industrial processes. Zinc can interfere with municipal wastewater treatment processes at concentrations as low as 0.08 mg/l and can be toxic to aquatic life at less than 0.10 mg/l [1]. Facilities that discharge industrial wastewater to surface water or to municipal wastewater treatment facilities have found themselves under increasing pressure to reduce the concentration of zinc and other metals in the effluent from their treatment systems. The most frequently employed method for the removal of metals from industrial wastewater is hydroxide precipitation. This process involves the addition of lime or sodium hydroxide to raise the pH to approximately 8 to 11, coagulation/flocculation and clarification to remove the precipitated solids, and the addition of sulfuric acid to neutralize the wastewater prior to discharge. Filtration is often necessary as a polishing step. Sludge storage and dewatering facilities are also required. There are several limitations to hydroxide precipitation, including:

◊ The relatively high minimum solubility of many metal hydroxides. An effective minimum solubility of greater than 0.10 mg/l has been reported for zinc hydroxide [2].
◊ Difficulties associated with the simultaneous removal of two or more metals. The pH of minimum solubility varies for the different metal hydroxides.
◊ Inability of the process to remove complexed metals.

◊   Labor intensive nature of the chemical feed and sludge handling systems associated with the process.
◊   Large volume of sludge generated by the process.

These limitations have prompted research into other methods of metals removal including many adsorption processes. Adsorption processes have the potential to overcome many of the limitations of hydroxide precipitation.

Adsorption is the accumulation of a substance on the surface of an adsorbent solid. Adsorption may be classified as either physical adsorption or chemical adsorption. Chemical adsorption is rarely used in water or wastewater treatment [3]. Physical adsorption is due primarily to electrostatic forces [4] and is a reversible occurrence. When the molecular forces of attraction between the solute and the adsorbent are greater than the forces between the solute and the solvent, the solute will be adsorbed on the surface of the adsorbent [3].

As the adsorption process proceeds, the adsorbed solute tends to desorb into the solvent. Equal amounts of solute are eventually being adsorbed and desorbed simultaneously. The rates of adsorption and desorption will eventually reach equal levels, a state called adsorption equilibrium [5]. The quantity of solute adsorbed per unit weight of adsorbent usually increases with an increase in solute concentration. A graphic presentation of the amount of solute adsorbed per unit weight of adsorbent as a function of the equilibrium concentration of the solute in the solution is called an adsorption isotherm [5]. Several adsorption isotherm models are available, of which Langmuir's and Freundlich's are the most commonly used. The Langmuir model is based on the assumption that the energy of adsorption is the same for all surface sites and is not dependent of the degree of coverage. In reality, the energy of adsorption may vary from site to site because most surfaces are heterogeneous [4]. The Freundlich isotherm attempts to account for this heterogeneity. The Freundlich isotherm is widely used to model adsorption data empirically, even when there is no basis for it's assumptions [4].

The contact time required for adsorption equilibrium to be achieved can be determined for batch systems by adding a known quantity of adsorbent to a given volume of water and mixing the solution. Samples are removed at various time intervals and the residual concentration of the solute determined. Equilibrium is achieved at the contact time when no significant change in residual concentration is observed with increased time. An isotherm can be constructed by varying the adsorbent dosage, mixing until equilibrium is achieved, and determining the equilibrium solute concentration [5]. This data can then be plotted according to one of the available isotherm models.

Some of the most promising of the adsorbents currently under investigation for the removal of metals from water are iron oxides. The adsorption of trace metals onto freshly precipitated iron oxides has been demonstrated [6,7]. Problems with this process include the high cost of preparing the fresh adsorbent, the resulting large volume of sludge, and difficulty in separating the solids from the liquid phase. To address these, Benjamin and Edwards [8] developed iron oxide coated filter sand and demonstrated effective metals

601

removal using a packed column. Schultz *et al.* [9] examined the adsorption and desorption of metal onto iron oxide. These studies evaluated freshly precipitated ferric oxide and the possibility of re-using the adsorbent. Chen *et al.* [10] investigated the use of ferric oxide coated magnetite for adsorption of hexavalent chromium and zinc. The feasibility of re-using the adsorbent was again evaluated. Magnetite was used as a base material because it allowed the adsorbent to be magnetically removed from solution. Magnetite was described as an inefficient adsorbent by itself because of a small specific surface area (low binding site density). The results indicated that the ferric oxide coated magnetite had nearly twice the hexavalent chromium adsorption capacity as the uncoated magnetite, but the desorption efficiency was found to decline significantly after only one adsorption/desorption cycle. The authors also discussed the advantages of adsorption over precipitation, including the fact that adsorption is unaffected, and possibility even enhanced, by the presence of complexing agents.

The use of magnetite as a coagulant aid has also been investigated [11]. The magnetite assisted coagulation by adsorption of the precipitated solids. Floc growth occurred very rapidly in comparison to conventional methods, and the solids could be removed efficiently by magnetic separation.

This paper presents the results of work which is intended to be the first step in an evaluation of the use of concentrated and pelletized magnetite for the adsorption of metals from industrial wastewater. The magnetite used is a cold carbon bonded material which is formulated for the steel industry as a complete product ready for feed to the furnaces. The specific objective of this work was to determine the zinc adsorption capacity of the prepared magnetite pellets through batch tests that were designed to allow the development of an adsorption isotherm. Future work would explore the potential for use of the spent adsorbent in the steel making process, thereby allowing the recovered metals to be recycled into steel products, while avoiding spent adsorbent disposal costs. Although not evaluated in this study, an additional advantage of the use of magnetite as an adsorbent is that it can be magnetically separated from the wastewater.

Zinc was selected because it is commonly found in electroplating, metal finishing, and other industrial wastewaters. Also, the relatively high minimum solubility and amphoteric nature of zinc hydroxide often complicates the removal of zinc to low levels from wastewater containing multiple metals.

## EXPERIMENTAL METHODS

### Adsorbent

The magnetite used in this study was manufactured and supplied by Pellestar, Ltd. The raw iron ore was from the Cleveland Cliffs Iron Company's Empire Mine. The sample received from the manufacturer was in the form of round pellets with a diameter of approximately twelve millimeters. The pellets consisted of a blend of magnetite ($Fe_3O_4$), carbon, limestone, and a small amount of silica. The pellets have a total iron content of approximately 55 to 60 percent [12]. Laboratory analyses of the pellets were not available from the manufacturer.

Preliminary testing was performed during the development of the experimental method using the full size pellets. Estimates performed using this preliminary data indicated limited adsorption capacity for the full size pellets. Because the pellets had a very low surface area to mass ratio, it was estimated that much better adsorption capacity could be achieved if the pellets were broken into smaller pieces. The pellets were broken into pieces averaging approximately six millimeters in diameter and weighing approximately 0.35 grams. The pellets were then rinsed with deionized water to remove fines and dried in an oven at 100 °C.

It is estimated that the surface area to mass ratio of the pellets was increased by approximately two and one-half times when the pellets were broken. Data generated in the adsorption trials discussed below show an adsorption capacity of more than eight times that suggested for the full size pellets during the preliminary testing

## Adsorption Trials

All adsorption experiments were conducted using synthetic wastewater made by the addition of a stock zinc standard solution (zinc metal solution, 0.3 M in nitric acid) to deionized water. The synthetic wastewater had a pH of approximately 7.8. Samples of the synthetic wastewater samples were analyzed for zinc by flame atomic absorption spectroscopy using a Perkin Elmer Model 360. The manufacturer's recommended procedures were followed for all analyses. Manufacturer's literature reports a lower detection limit of 0.018 mg/l for zinc. Dilutions with deionized water were performed as required for the analysis of each individual sample. The instrument was calibrated before the samples were analyzed for each trial.

The experiments were conducted by placing a predetermined amount of magnetite pellets in the middle of a two-liter polyethylene bottle filled with glass marbles. Exactly 840 mils of synthetic wastewater was then added to the bottle and the bottle was immediately placed in a tumbler. A volume of headspace remained above the water in the bottle, allowing the water to flow back and forth through the magnetite. Refer to Figure I for a diagram of the experimental setup.

A sample of the raw synthetic wastewater was retained for analysis and five milliliter samples were withdrawn from the bottle at 20 or 30 minute intervals during tumbling. No more than five percent of the total volume of sample was withdrawn for analysis during each trial.

The marbles were used to keep the pellets from being broken during tumbling, although magnetite fines were increasingly evident in the samples as the contact time progressed. For this reason, each sample was filtered using a 0.45 μm syringe filter before analysis for zinc. Three trials were performed using different magnetite dosages and initial zinc concentrations so that an adsorption isotherm could be developed.

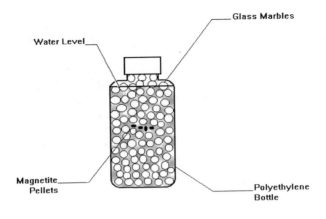

Water Level

Glass Marbles

Magnetite
Pellets

Polyethylene
Bottle

**Figure I - Experimental Setup**

## Control Trial

A control trial was also performed using the experimental setup described above, except no magnetite was added to the bottle and only one sample was collected after 180 minutes of tumbling. This trial was intended to determine if the polyethylene bottle or glass marbles were contributing or adsorbing zinc from the synthetic wastewater.

## RESULTS

### Adsorption Trials

The three adsorption trials were designed to establish the contact time required before equilibrium is achieved and to provide three equilibrium data points for the development of an adsorption isotherm. The magnetite and initial zinc concentrations were varied to achieve the range of equilibrium zinc concentrations which are required for an adsorption isotherm to be constructed.

The trials were terminated when no decline in zinc concentration occurred for two consecutive data points. This point was established as the equilibrium concentration and required contact time. Data for the three adsorption trials are summarized on Table I and presented graphically in Figures II through IV. The initial zinc and magnetite pellet concentrations are shown with the figures.

The data show a generally declining rate of adsorption, with equilibrium achieved after approximately 180 to 210 minutes of contact time. Trials No. 1 and 2 show a slight increase in zinc concentration from the initial sample (before being placed in the container) to the first sample taken during tumbling (after 20 and 30 minutes of contact time, respectively).

# Table I - Adsorption Results

| Contact Time (minutes) | Trial No. 1 Zn Conc. (mg/l) | Trial No.2 Zn Conc. (mg/l) | Trial No. 3 Zn Conc. (mg/l) |
|---|---|---|---|
| 0 | 8.15 | 2.79 | 3.15 |
| 20 | - | 3.09 | - |
| 30 | 9.10 | - | 2.38 |
| 40 | - | 2.70 | - |
| 60 | 7.26 | 2.49 | 0.87 |
| 90 | 4.73 | 1.95 | 1.41 |
| 120 | 4.84 | 1.62 | 0.92 |
| 150 | 3.85 | 1.02 | 0.22 |
| 180 | 3.41 | 0.99 | 0.04 |
| 210 | 2.70 | - | 0.04 |
| 240 | 2.75 | - | - |

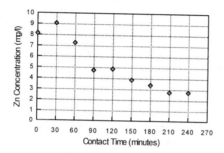

**Figure II - Adsorption Trial No. 1 Results
Adsorbent Dosage  21.4 g/l
Initial Zn Concentration  8.2 mg/l**

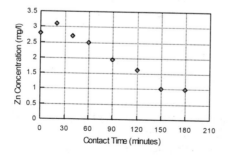

**Figure III - Adsorption Trial No. 2 Results
Adsorbent Dosage  10.7 g/l
Initial Zn Concentration  2.8 mg/l**

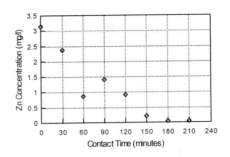

**Figure IV - Adsorption Trial No. 3 Results**
**Adsorbent Dosage 21.4 g/l**
**Initial Zn Concentration 3.2 mg/l**

Equilibrium data from Trial No. 3, shown in Figure IV, indicates that zinc can be removed to levels below the effective solubility limit of zinc hydroxide (approximately 0.10 mg/l at a pH of approximately 9.0). The ability of the process to remove zinc to very low levels could not, however, be fully evaluated because of the detection limit of the analytical instrument employed.

**Control Trial**

The initial zinc concentration of the synthetic wastewater in the control trial (before being placed in the container) was 3.80 mg/l. After tumbling for 180 minutes, the sample was found to contain zinc at a concentration of 4.00 mg/l. This slight increase in zinc concentration is in general agreement with the data from Trials No. 1 and 2 and may indicate a source of zinc contamination in the experimental set-up.

**Adsorption Isotherm**

The adsorption isotherm for zinc on magnetite pellets is shown on Figure V. The experimental data was fitted to a linearized Freundlich equation, in the form:

$$Log(X/M) = Log\ K + 1/n\ Log\ C_e$$

Where:

$X/M$ = zinc adsorbed per unit weight of magnetite pellets (mg/g)
$C_e$ = equilibrium concentration of zinc in solution (mg/l)
$K$ = adsorption capacity constant
$1/n$ = adsorption intensity constant

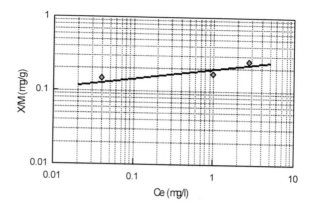

**Figure V - Freundlich Adsorption Isotherm
Zinc on Magnetite Pellets**

This equation can be used to calculate the amount of magnetite pellets required to reduce any initial concentration to a predetermined final concentration by substituting $(C_o - C_e)$ for X, where $C_o$ is the initial concentration.

The relatively shallow slope of the isotherm ($1/n = 0.11$) indicates that adsorption capacity is only slightly reduced at lower equilibrium concentrations. The characteristic of the isotherm indicates that magnetite pellets could be used effectively for the removal of metals to relatively low final concentrations.

## DISCUSSION AND CONCLUSIONS

The data presented in this paper indicate that the prepared magnetite pellets have appreciable capacity for the adsorption of zinc from wastewater. The minutes was found to be required. A Freundlich adsorption isotherm was used to model the adsorption of zinc on magnetite pellets. The Freundlich isotherm constants were found to be :

$$K = 0.20$$
$$1/n = 0.11$$

for X/M units of mg/g and $C_e$ units of mg/l.

As suggested by preliminary testing, the size of the pellets has a significant impact on the adsorption capacity of the material. The smallest available size pellet or flake should be used for future testing.

An advantage of this process is that no chemical feed systems are required and no sludge is generated. Operation and control requirements of packed bed type adsorption processes are typically minimal, an important consideration for many industrial wastewater treatment system operators. It is expected that the spent pellets could be used in the steel manufacturing process, allowing recycle of the adsorbed metals and eliminating the cost of spent adsorbent disposal. The primary cost associated with the process would be shipment of the material to and from the treatment facility. The proximity of the wastewater treatment facility to an ultimate end user of the material (steel mill) could significantly impact the cost of this process. For this reason, the process may be particularly applicable to the treatment of metals bearing wastewater generated at steel mills and associated facilities. The cost associated with the process in this type of application would be minimal. Depending on the economics of shipping the material, other potential applications include the treatment of low or intermittent wastewater flows or for the polishing of effluent from existing hydroxide precipitation systems.

An example of a polishing application would be the removal of zinc from a 50,000 gallon per day industrial wastewater stream which has been treated by hydroxide precipitation to an effluent concentration of 0.20 mg/l. Using the adsorption isotherm developed, approximately 430 pounds per day of magnetite pellets would be required to further reduce the concentration in the effluent to 0.05 mg/l. In this application, a 20-ton shipment of pellets would be required approximately once every three months. For comparison, if the wastewater had an initial concentration of 2.0 mg/l and the hydroxide precipitation system was eliminated, approximately 2.8 tons per day would be required to produce an effluent with a Zn concentration of 0.05 mg/l. In this case, a 20-ton shipment would be required approximately once every week.

It is important to note that the amount of magnetite pellets required under the above described scenarios could be reduced significantly if a smaller pellet or flaked product were used.

## REFERENCES

1. USEPA. 1993. *Procedures for Determining Water Quality Based Effluent Limits in Order to Calculate Local Limits.*

2. Eckenfelder, W.W. Jr. 1989 *Industrial Water Pollution Control,* McGraw-Hill.

3. Montgomery, J.M., Consulting Engineers. 1985. *WaterTreatment Principles and Design,* John Wiley and Sons.

4. Reynolds, Tom D. 1982 *Unit Operations and Processes in Environmental Engineering,* Brooks/Cole Engineering Division, Wadsworth, Inc.

5. Faust, Samuel D. 1987. *Adsorption Processes for Water Treatment,* Butterworth.

6. Benjamin, M.M. 1983. "Adsorption and Surface Precipitation of Metals on Amorphous Iron Oxyhydroxide", *Environ. Sci. Technol.,* 17: 686.

7.  Swallow, K.C., D.N. Hume, and F.M.M. Morel. 1980. "Sorption of Copper and Lead by Hydrous Ferric Oxide", *Environ. Sci. Technol.*, 11: 1326.

8.  Benjamin, M.M. and M. Edwards. 1989 "Adsorptive Filtration Using Coated Sand: A New Approach to Treatment of Metal Bearing Wastes", *JWPCF*, 61: 1523.

9.  Schultz, M.F., *et al.* 1987 "Adsorption and Desorption of Metals on Ferrihydrite: Reversibility of the Reaction and Sorption Properties of the Regenerated Solid", *Environ. Sci. Technol.*, 21: 863.

10. Chen, W.Y., P.R. Anderson, and T.M. Holsen. 1991 "Recovery and Recycle of Metals from Wastewater with a Magnetite-Based Adsorption Process", *JWPCF*, 63: 658.

11. Akyel, G., E. Booker, E. Cooney, A. Priestley. 1994 "Rapid Sewage Clarification Using Magnetite Particles", *Waterworld Review*, Sept/Oct.

12. Bal, Dennis, Engineer, Pellestar, Ltd. 1992 Ann Arbor, Michigan, Personal Communication.

# THE USE OF RESIDUAL MATERIALS FOR THE ADSORPTION OF LEAD IN WASTE DISPOSAL FACILITY SUBGRADES

## TIMOTHY M. LAUMAKIS
Project Engineer
CH2M Hill
1700 Market Street, Suite 1600
Philadelphia, PA 19103

## JOSEPH P. MARTIN
Associate Professor, Civil Engineering Dept.
Drexel University
Philadelphia, PA 19104

## SIBEL PAMUKCU
Associate Professor, Civil Engineering Dept.
Lehigh University
Bethlehem, PA 18015

## SHI-CHIEH J. CHENG
Professor, Civil Engineering Dept.
Drexel University
Philadelphia, PA 19104

## INTRODUCTION

To prevent groundwater contamination from landfills, mining facilities, and transfer stations, the customary focus has been on hydraulic containment through the use of liners and collection systems. However, it is not the seepage that is the concern, but the solutes dissolved in it. Hence, the more recent emphasis is on attenuation of contaminants by filtration, sorption, neutralization, precipitation, ion exchange and other mechanisms. An opportunity exists to improve confidence in liquid and contaminant migration control when it is necessary to use control fill material to adjust the elevation and alignment of site subgrades.

Borrow soils have generally been selected on the basis of mechanical properties that assure dimensional stability, and more recently, low permeability. As shown in Figure 1, large volumes of fill may be required. Consequently, the use of soil-textured industrial by-products such as mine tailings or fly ash can be an economically attractive alternative, since such materials are available in large quantities and must be disposed of in some

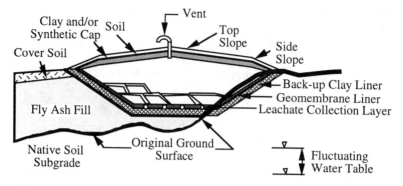

**Figure 1. Typical Landfill Cross-Section and Subgrade Alteration**

manner anyway. Often, these residual materials are products of chemical, thermal or physical processing. In this regard, there is always a concern with contaminant leaching. Howeverer, it is also possible that these materials may exhibit retardation characteristics.

## STUDY OVERVIEW

This study investigates the properties of an Appalachian bituminous fly ash and clean mine spoil as potential environmental facility subgrades. The concentration is on the ability of these materials to limit impacts of an acidic leachate with a high lead content, i.e. restrict seepage rate, neutralize the effluent, and most importantly, retard lead migration.

The work reported herein includes a first flush extraction of the leachable lead content present in both materials. The focus is then on batch sorption tests to establish a baseline distribution coefficient ($K_d$) and column permeation tests to verify the batch test results for each material. The lead influent was derived from leaching battery casings with synthetic rainfall and distilled water using the Toxicity Characteristic Leaching Procedure (TCLP) test techniques and apparatus.

Using the $K_d$ values obtained from the batch tests, "breakthrough" of the lead was predicted for column test samples. Also, during column permeation, effluent pH and specific conductivity, as well as hydraulic conductivity (permeability) were monitored for each sample. The latter was studied to determine if the leachate had any adverse effect on the performance of the residual soil as a liner material.

## COAL FLY ASH STUDIES

Fly ash is derived from electric power generation. It includes both the noncombustible mineral matter in the source coal, (silicon, aluminum, iron oxides), and residual carbon. The texture is usually silt-sized, forming an artificial soil in the same sense as mine tailings, blast-furnace slag and other by-products from the processing of geologic materials. Power plants produce millions of tons of fly ash annually, of which only a fraction is productively employed, with the balance being disposed of in impoundments and landfills.

Type-F ashes are generally derived from the bituminous coal that predominates in the eastern United States. The mechanical properties are based on friction between discrete particles that form a single-grained structure, as opposed to the self-cementing type-C ashes which tend to form a cohesive mass. McLaren and DiGoia (1987) showed that the geotechnical properties of type-F fly ash can be determined using traditional methods. Fly ash is employed in site development as common fill and as a component in concrete mixtures. Environmental applications include a pozzolanic cement component in waste stabilization (Martin et al. 1987), and cover material in refuse disposal sites (Mamane & Gottlieb 1992). Another area of fly ash usage is for treatment of waters contaminated with trace metals. Sharma et al. (1990), investigated fly ash treated with hydrochloric acid as an ion exchanger for the removal of heavy metals. The combination of these uses shows promise for constructing low permeability, dimensionally stable and sorbing subgrades.

## GEOTECHNICAL AND GEOCHEMICAL PROPERTIES

### Fly Ash

Table 1 shows that silicon and aluminum are the primary chemical constituents of the fly ash. Calcium was only a minor component, but the iron oxide content was relatively high at 14%. The loss on ignition (LOI) test indicated a total carbon content of 2.3% by weight. A soil pH of 4.2 was obtained with both water and calcium chloride solutions. A cation exchange capacity of 131.8 meq per 100g of soil was measured. This value is similar to that found for clay materials.

Fly ash is typically a relatively uniform, silt-sized material. The process of condensation in the stack generally produces spherical particles, some of which may be hollow. The fly ash specific gravity ($G_s$) is 2.45, which is slightly less than that of most natural soils. This fly ash contains some fine sand size particles, but is 62% (by weight) finer than the #200 sieve (0.075mm) with a mean grain diameter of 0.012mm.

The standard Proctor test showed a maximum dry unit weight of 15.6 kN/m$^3$ (99.5 pcf) and an optimum moisture content (OMC) of 16%. Even though the material is frictional and single-grained, the sensitivity of the unit weight to compaction moisture implies that the pore fabric or structure can vary. Permeability was determined using the falling head test method. Samples were prepared at four water contents bracketing the OMC. The results ranged from 1.70e-4 to 5.79e-5 cm/sec with increasing compaction moisture content. Compaction on the wet side of OMC apparently causes the finer fraction of fly ash to fill the pores between larger diameter particles, but was difficult to perform due to decreased workability.

| mineral | $SiO_2$ | CaO | $Al_2O_3$ | $Fe_2O_3$ | MgO | $SO_3$ | $K_2O$ | $Na_2O$ |
|---------|---------|-----|-----------|-----------|-----|--------|--------|---------|
| % | 49 | 3 | 30 | 14 | 1 | 0.5 | <1 | <1 |

**Table 1. Chemical Composition of Test Fly Ash**

## Clean Mine Spoils

The mine spoils are composed primarily of weathered sandstone and shale with a total carbon (LOI) and free iron oxide content of 6.8% and <1%, respectively, by weight.

A soil pH of 4.2 was also obtained with both water and calcium chloride solutions, while the cation exchange capacity was 131.8 meq per 100g of soil. Consequently, by these conventional geochemical index tests, the fly ash and mine spoils appear to be similar.

The mine spoil is a nonplastic, silty sand material with a specific gravity ($G_s$) of 2.69, which is typical for most natural soils. The material was of angular texture, and often a chip shape. Investigation was limited to samples "scalped" of gravel-sized particles, and finer than a #4 sieve (0.25"). The shale fragments did not slake, soften or disintegrate upon soaking. The mine spoils included 16% (by weight) of the material finer than the #200 sieve (0.075mm) with a mean grain diameter of 0.05mm.

The standard Proctor test showed a maximum dry unit weight of 18.2 kN/m$^3$ (115.6 pcf) and an optimum moisture content (OMC) of 12%. Permeability was determined using the falling head test method. Samples were prepared at three water contents bracketing the OMC. The results ranged from 1.30e-4 to 1.0e-6 cm/sec with increasing compaction moisture content. Compaction on the wet side of the OMC did not decrease soil workability.

## BASELINE LEACHING CHARACTERIZATION

Concentrations of trace metals in fly ash and mine spoils are highly variable. To establish a baseline leachable lead concentration, batch extraction tests were performed with a 20:1 liquid to solid ratio. Soil samples were placed into eight-2 liter polyethylene jars. Four were filled with 1400ml of distilled water (DW) of pH = 6.0 and four with synthetic rain (SR) having pH = 4.0. The synthetic rain contained $2.95 \times 10^{-5}$ molar $H_2SO_4$, $3.2 \times 10^{-5}$ molar $HNO_3$, and $1.2 \times 10^{-5}$ molar HCl. The jars were mounted on a rotary agitator and rotated for 18 hours. Decanted aqueous samples were filtered through cellulose nitrate filters with 0.45 $um$ pore openings in an Analytical Test Equipment Pressure Filter. Effluent was collected in 125 ml polyethylene (PE) bottles, with pH and conductivity measured before being fixed with a 5% $HNO_3$ solution and chilled. Lead content in the filtered samples was measured by atomic adsorption spectrophotometer.

The lead concentration in the fly ash effluent was nondetect in the distilled water extraction, and 1.704 ppm in the synthetic rain solvent extract. The mine spoils effluent had a leachable lead content of 0.187 ppm and 1.730 ppm in the distilled water and synthetic rain extract, respectively.

## LEACHATE GENERATION

Two standard influents were generated to assess the sorption capacity of the fly ash and mine spoil by leaching battery casings from a lead recovery facility. 70 grams of battery casings were placed into each of twenty two-2 liter polyethylene jars. Sixteen of them were filled with synthetic rain (SR) and the rest with distilled water (DW) at a 20:1 liquid to solid ratio and rotated for 18 hours.

| Effluent | Pb conc. (ppm) | pH | Conductivity (mS) |
|----------|----------------|-----|-------------------|
| CSRE-4A | 36.97 | 3.2 | 0.97 |
| CSRE-4A2 | 12.84 | 3.2 | 0.68 |
| CSRE-4A3 | 8.95 | 5.3 | 0.08 |

**Table 2. Lead Effluent Characteristics**

Following agitation, samples were filtered as described previously, with pH, conductivity and lead content recorded for all samples. The SR and DW sample extraction solutions were blended in a 15 gallon High Density Polyethylene (HDPE) tank. This contaminated solution was used as influent leachate for the column permeation and batch sorption tests. The pH, conductivity, and lead content of the three sets ultimately used are listed in Table 2.

## BATCH SORPTION

Batch sorption tests, like the extraction tests, produce what is generally thought of as a long-term equilibrium condition where the source leachate and solid surfaces are fully in contact for an extended period. The initial assumption is that the ratio between the equilibrium concentrations of the contaminant dissolved in solution and sorbed on the solid is not a constant over a wide range of concentrations; therefore, multiple tests are required to develop an adsorption relationship. Using the rotary agitation apparatus described earlier, the major variables are the liquid:solid ratio and the agitation time. One method employs keeping the liquid:solid ratio constant by successively diluting the leachate, while another approach maintains a constant leachate concentration, and varies the liquid:solid ratio i.e. adsorbent weight. The former method was employed, giving direct data on the effects of infiltration dilution on leachate strength.

In the mine spoil batch tests, the composite leachate CSRE-4A, (Table 2) was the base source liquid with a 36.97 mg/l lead content. Distilled water was used to generate 75%, 50%, and 25% diluted lead solutions. The original composite will be referred to as CSRE-4A(100%), while the new dilutions were labeled CSRE-4A(75%), CSRE-4A(50%), CSRE-4A(25%). Similarly, the leachate CSRE-4A2 was diluted in the same manner and used in the batch tests for the fly ash. The use of the stronger leachate with the mine spoil modeled an actual field condition.

## BATCH SORPTION RESULTS

Table 3 and 4 show the before and after results used to determine the distribution coefficient, $K_d$, for the fly ash and mine spoils, respectively. It was assumed that the change in lead content in the solvent represented material sorbed onto the soil surfaces. Table 5 illustrates the pH and conductivity results for the fly ash. It can be seen that, the distilled water dilutions of CSRE-4A2 only marginally effected the pH of the test liquid, such that pH was essentially constant. The fly ash showed a high capacity to buffer these aggressive leachates, with all four batches producing an effluent with a pH over 6.

In accomplishing this neutralization, however, the ion concentration in the effluents increased markedly, as indicated by the specific conductance results. The most important result is the decrease in dissolved lead content, approaching

614

| Effluent | Co (mg / l) | Ceq (mg / l) | Adsorbent (g) | Mass sorbed (ug/g) |
|---|---|---|---|---|
| CSRE-4A2 (100%) | 12.84 | 0.572 | 50.00 | 250 |
| CSRE-4A2 (75%) | 10.41 | 0.497 | 50.00 | 200 |
| CSRE-4A2 (50%) | 4.78 | 0.347 | 50.00 | 90 |
| CSRE-4A2 (25%) | 2.63 | 0.347 | 50.00 | 50 |

Table 3. Fly Ash Batch Sorption Results

| Effluent | Co (mg / l) | Ceq (mg / l) | Adsorbent (g) | Mass sorbed (ug/g) |
|---|---|---|---|---|
| CSRE-4A (100%) | 36.97 | 22.28 | 50.00 | 294 |
| CSRE-4A (75%) | 28.30 | 15.21 | 50.00 | 262 |
| CSRE-4A (50%) | 17.21 | 6.65 | 50.00 | 211 |
| CSRE-4A (25%) | 8.57 | 2.82 | 50.00 | 115 |

Table 4. Mine Spoil Batch Sorption Results

| Effluent | Pre-Agitation | | Post-Agitation | |
|---|---|---|---|---|
| | pH | Cond. (mS) | pH | Cond. (mS) |
| CSRE-4A2 (100%) | 3.22 | 0.68 | 6.69 | 1.20 |
| CSRE-4A2 (75%) | 3.29 | 0.50 | 6.30 | 1.22 |
| CSRE-4A2 (50%) | 3.45 | 0.31 | 6.48 | 1.19 |
| CSRE-4A2 (25%) | 3.72 | 0.12 | 6.39 | 1.23 |

Table 5. Fly Ash Batch Test Effluent Characteristics

| Effluent | Pre-Agitation | | Post-Agitation | |
|---|---|---|---|---|
| | pH | Cond. (mS) | pH | Cond. (mS) |
| CSRE-4A2 (100%) | 2.9 | 0.87 | 3.0 | 0.68 |
| CSRE-4A2 (75%) | 2.9 | 0.63 | 1.8 | 8.13 |
| CSRE-4A2 (50%) | 3.0 | 0.44 | 2.6 | 1.21 |
| CSRE-4A2 (25%) | 3.2 | 0.16 | 3.3 | 0.23 |

Table 6. Mine Spoil Batch Test Effluent Characteristics

two orders of magnitude. The effluent lead concentrations that was obtained with the synthetic rain leaching of the fly ash was lower. Evidently, other constituents in the battery casings leachate produced conditions that reduce lead solubility and increased pH.

In contrast, the mine spoils only adsorbed a fraction of the lead from solution. The ability of the mine spoils to control the impacts of an acidic leachate were marginal as seen by the pH results listed in Table 6. In addition, the specific conductivity varied slightly from the source leachate, except for the sample using CSRE-4A2 (75%) as the batching solution.

To make a more generic characterization of sorption, the two most common adsorption equations, created by H. Freundlich (1909) and Langmuir (1918), are often employed to represent laboratory data. It should be noted that these adsorption equations may not properly represent every solid-liquid system (Kinniburgh 1986). The batch test results shown in Table 3 are plotted on Figure 2. The plot itself is unusual, being of increasing slope, indicating an infinite amount of sorption sites. The basis for the sorption is not completely known at this time. There is substantial amorphous iron content, low but finite carbon content, a significant cation exchange capacity, along with a high neutralization capacity. The property, or combinations thereof, that dominates sorption is not clear.

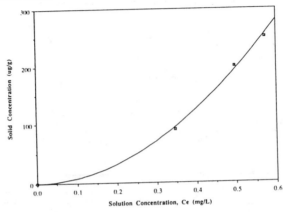

**Figure 2. Fly Ash Adsorption Isotherm**

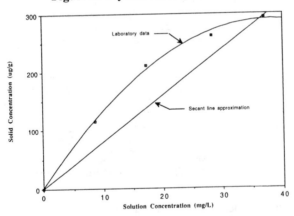

**Figure 3. Mine Spoil Adsorption Isotherm**

The adsorption isotherm for the mine spoil is shown in Figure 3. The slope of this curve does appear to level off, indicating the presence of only a fixed number of available sorption sites. This is evident from the threefold decrease in lead concentration at higher leachate dilutions for the fly ash (Table 3), compared to only a 30% decrease with the "stronger" leachate for the mine spoil. Consequently, an empirical fit was used in determining partitioning coefficients for both soils.

A Freundlich and Langmuir nonlinear regression analysis was performed using Microsoft Excel 4.0, an integrated spreadsheet with a built-in macro solver that functions as a general purpose optimization device, for the Apple Macintosh. The analysis indicated that the Freundlich equation best represented the laboratory data for both soils. The results yielded $K_f$ values of 795.88 (ug$^{(1-1/n)}$mL$^{1/n}$/g) and 122.85 (ug$^{(1-1/n)}$mL$^{1/n}$/g) for the fly ash and mine spoils, respectively. The values for the constant $n$ were 0.487 and 3.57 for the fly ash and mine spoils, respectively.

## COLUMN SORPTION

The fitted parameters, $K_f$ and $n$, obtained from the batch tests were used to calculated the corresponding weighted mean retardation factor ($R_{avg}$) for plume movement. Table 7 lists the predicted number of pore volumes (T) required to achieve throughput of lead for the fly ash and mine spoil column permeation test samples.

It can be seen that a significant number of pore volumes are required for the fly ash to begin to reach the theoretical values. However, the number of pore volumes required for the mine spoil column samples is approximately three orders of magnitude less than for the fly ash samples. Nevertheless, column tests were performed to confirm this and to ensure there was no short circuiting of lead through the samples.

Fly ash samples were prepared at 13%, 16%, and 19% water contents in 2.5" diameter plexiglass molds (permeameters) with HDPE fittings. Earlier tests indicated a sensitivity of permeability to compaction conditions, so it was assumed that contaminant-sorption site contact would be affected by the pore channel structure as well. The compaction procedures were modeled after those developed by Lafleur et al. (1960). Two mine spoil column samples were prepared at the same water content, but varing compaction energies. This resulted in a characteristic dry density and porosity for each sample as shown in Table 7.

Fly ash samples were prepared by varying both the moisture content and compaction energy. The applied compaction energy varied over an order of magnitude, but the resultant unit weights and porosities for the fly ash samples, listed in Table 5, did not vary nearly as much. The loosest sample was compacted at the drier end, and had a porosity of 0. 407. The densest sample was at the highest compaction moisture content and compaction energy, producing a porosity of 0.353.

| Sample | w.c. (%) | $\gamma_b$(g/ml) | n | PV (ml) | $K_f$ | n | $R_{avg}$ (T) |
|---|---|---|---|---|---|---|---|
| **Fly Ash** | | | | | | | |
| A | 14.3 | 1.452 | 0.407 | 85.0 | 795.88 | 0.486 | 1.27e5 |
| B | 16.6 | 1.458 | 0.404 | 87.6 | 795.88 | 0.486 | 1.29e5 |
| C | 16.6 | 1.532 | 0.374 | 77.0 | 795.88 | 0.486 | 1.46e5 |
| D | 18.8 | 1.584 | 0.353 | 27.6 | 795.88 | 0.486 | 1.60e5 |
| **Mine** | | | | | | | |
| **Spoil** | | | | | | | |
| 1 | 12 | 1.706 | 0.365 | | 122.85 | 3.567 | 119 |
| 2 | 12 | 1.786 | 0.335 | | 122.85 | 3.567 | 136 |

Table 7. Column Leach Sample Characteristics

It can be seen from Table 7 that for the same compaction energies, the dry density was greater and corresponding porosity smaller for the mine spoil as compared with the fly ash samples. Therefore, the increase in density appears to be due in large part to the nature of the material and not the compaction technique.

The permeameters were then connected to a falling head apparatus and saturated with distilled water . All samples were first permeated with distilled water up to a throughput of five pore volumes (pv). Then, the distilled water was evacuated from the permeameter reservoir using a pipette and then refilled with the lead solution CSRE-4A(100%) in the fly ash columns and CSRE-4A3(100%) in the mine spoil columns. Based on the permeabilities determined from the distilled water permeation, a hydraulic gradient was established for each sample so that roughly 0.5 pv were permeated daily , yielding an average hydraulic detention time of two days. The effluent pH, conductivity, and lead concentration of each sample were recorded after each pore volume.

## COLUMN RESULTS

Column tests describe attenuation processes by simulating field conditions. Often, column tests show decreased hydraulic conductivities over time due to the formation of precipitates in voids or transformations within the soil. There could also be internal instability, such that the fine particles migrated and were captured on and clogged the outflow filter.

In this study there was a steady increase in permeability for all fly ash samples except sample D, which is the densest. Figure 4 shows an increase in permeability that is probably a result of the fly ash particles entering into solution, thereby, forming larger pore openings. The gradual trend of decreasing permeability in sample D could be in part due to the high unit weight (i.e. low porosity) and the adsorption of the larger lead precipitates into these smaller areas. Also, algae growth on the soil surface of all fly ash samples was noticed near the fortieth pore volume. In general, however, the permeability can be approximated as steady.

While Figure 5 shows a more theoretical curve with a gradual decrease in permeability and only a slight increase around pore volume thirty. This increase could also be due to the mine spoil particles entering into solution thereby, creating larger pore channels.

The pH and conductivity were monitored. Figure 6 shows the ability of the fly ash to buffer the acidic lead solution (pH = 3.2) to three to four pH units higher than the influent. At very high throughputs, this buffering capacity started to decline. In contrast, Figure 7 shows an initial decrease in the mine spoil effluent pH using distilled water as the permeant. Upon permeation with the CSRE-4A3(100%), the mine spoil samples were only able to slightly control the impacts of the acidic lead leachate.

To date, having permeated 62 pore volumes of CSRE-4A(100%, 36.97 mg/l) through the fly ash and 32 pore volumes of CSRE-4A3(100%) through the mine spoil there has been no measurable break through of lead in any column sample. Currently, the mass of lead adsorbed by each fly ash and mine spoil sample is approximately 195 mg and 19 mg, respectively. These results indicate the high sorptive capacity of each material, which was initially predicted by the batch tests.

618

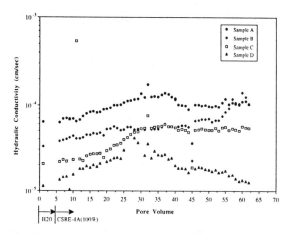

**Figure 4. Permeability Results for Fly Ash Column Leach Samples**

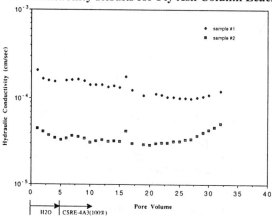

**Figure 5. Permeability Results for Mine Spoil Column Leach Samples**

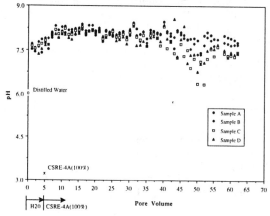

**Figure 6. pH Results for Fly Ash Column Leach Samples**

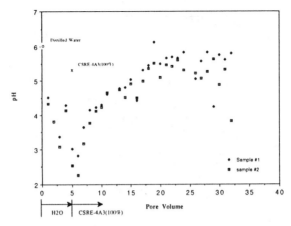

**Figure 7. pH Results for Mine Spoil Column Leach Samples**

## CONCLUSIONS

The premise of this study was to analyze the performance of industrial by-products as environmental facility subgrades by examining their ability to control the impact of an acidic leachate with a high lead content by restricting its seepage rate, neutralizing the effluent, and most importantly, attenuating lead migration.

This was investigated by performing batch and column leach testing on a type-F fly ash and a ground sandstone mine spoils. The results of the batch tests indicated the ability of the fly ash to buffer an acidic solution, while the mine spoils had only a minimal buffering capacity. The adsorption data was used in determining the partitioning coefficient ($K_p$) and retardation factors. Then the theoretical number of pore volumes necessary to obtain breakthrough of lead in column samples was determined.

The column leach results indicated that dissolution of the fly ash began to occur in the less dense samples. This would affect the ability of the soil to retard contaminants and favors compacting subgrades dense and wet of optimum to assure stability of the material. The larger void volume would appear to limit the contaminant-soil interface and corresponding adsorption and buffering capacity of the soil. However, even in the less dense fly ash samples, no lead breakthrough was seen. The mine spoils, were physically and hydraulically stable, while the permeability and sorption behavior appeared to follow a more expected pattern than the fly ash.

## REFERENCES

Freundlich, H. 1909. Kapillarchemie:eine Darstellung der Chemie der Kolloide und Verwandter Gebiete, Leipzig, Akademische Verlagsgesellschaft, p. 591.

Kinniburgh, D.G. 1986. "General Purpose Adsorption Isotherms," *Environ. Science and Technol.*, Vol. 20, 895-904.

Lafleur, J.D., Davidson, D.T., Katti, R.J., and Gurland, J. (1960). " Relationship Between the California Bearing Ratio and Iowa Bearing Value," *Methods for Testing Engineering Soils*, Iowa State University, Ames, Iowa, U.S.A.

Langmuir, I. 1918. "The Adsorption of Gases on Plane Surfaces of Glass Mica and Platinum," *J. of the Amer. Chem. Society*, Vol. 40, 136-1403.

Martin, J.P., Felser, A.J., and Vankeuren, E.L. 1987. "Hydrocarbon Refining Waste Stabilization for Landfills," *Geotechnical Practice for Waste Disposal 87:* Geotechnical Special Publication No. 13, R.D. Woods, ed., 668-682.

Mamane, Y., and Gottlieb, J. 1992. "Utilization of Coal Ash as a Cover Material in a Municipal Refuse Disposal Site," *Environmental Geotechnology*, Usmen and Acar, eds., 429-435.

McLaren, R.J., and DiGioia, A.M. Jr. 1987. "The Typical Engineering Properties of Fly Ash," *Geotechnical Practice for Waste Disposal '87:* Geotechnical Special Publication No. 13, R.D. Woods, ed., 683-697.

Sharma, R.K., Kumar, S., De, A.K., and Ray, P.K. 1990. "Use of Fly Ash as an Ion Exchanger in Water Filtration Studies for the Removal of Heavy Metals," J. *Environ. Sci. Health*, A25(6), 637-651.

# TREATMENT OF ORGANIC-HEAVY METAL WASTEWATERS USING GRANULAR ACTIVATED CARBON (GAC) COLUMNS

BRIAN E. REED
MAQBUL JAMIL
PATRICK CARRIERE
Department of Civil and Environmental Engineering
West Virginia University
Morgantown, WV 26506-6103
BOB THOMAS
Norit Americas, Inc.
1050 Crown Pointe Parkway, Suite 1500
Atlanta, GA 30338

## INTRODUCTION

In the past, granular activated carbon (GAC) columns have been used to remove trace amounts of organic compounds from various aqueous waste streams. If heavy metals were present, other processes such as precipitation, reverse osmosis, or ion exchange would be added to the treatment flow sheet. Significant savings could result if organic-heavy metal wastewaters could be successfully treated using a **single** process. Researchers at West Virginia University in conjunction with Norit Americas, Inc. have recently developed a GAC column process that removes heavy metals from the aqueous phase using GAC columns (Reed *et al.* 1994, Reed and Arunachalam 1994, Reed 1995, Reed *et al.* 1995). Because the ability of using GAC columns to remove organic contaminants is well known, the next logical step is to demonstrate the feasibility of the GAC process to simultaneously remove organic and inorganic contaminants. In this study, two synthetic wastewaters, 1 mg/L Pb-10 mg/L phenol and 1 mg/L Pb-1 mg/L TCE, were treated by GAC columns containing the Norit Americas, Inc. carbon Hydrodarco (HD) 4000.

## BACKGROUND

Solution pH has been identified as the variable governing metal adsorption onto hydrous solids (Stumm and Morgan 1981). For most cationic heavy metals, adsorption increases with increasing pH. The opposite is observed when the metal exists as an anion (*e.g.*, $Cr_2O_7^{-2}$). The fraction of metal ions removed increases from zero to one over a relatively narrow pH range (referred to as the "pH-adsorption edge"). The pH affects the status of the outer hydration sheaths of the metal ion, metal solution speciation and solubility, and the behavior of the carbon surface groups. Reed and others have documented the effect of pH on metal removal by PAC and GAC columns (Corapcioglu and Huang 1987, Huang and Ostovic 1978, Reed and Matsumoto 1993, Reed *et al.* 1994, Reed and Arunachalam 1994, Reed 1995, Reed *et al.* 1995). Solution pH can also affect organic contaminant removal by activated carbon. An example of this are organics that behave as a weak acid-base (*e.g.*, phenol). The ionic form (or the disassociated form) may not be as adsorbable as the neutral species.

The most reliable method for determining if GAC columns are a feasible treatment method for a particular wastewater is to conduct column breakthrough studies. Wastewater is introduced to the column and effluent contaminant(s) concentration is monitored with time (or volume of waste treated). Effluent concentration versus the volume of wastewater treated is referred to as a breakthrough curve. When the effluent contaminant concentration exceeds a predetermined value, breakthrough has occurred. The breakthrough concentration is determined by regulatory or down-line treatment concerns. When the effluent concentration is equal to the influent concentration, the column is said to be exhausted. Operationally, exhaustion is often taken as $C_{eff} = 0.95C_{inf}$. Process performance parameters include: Bed volumes of wastewater treated at breakthrough ($V_b$) and exhaustion ($V_{exh}$); surface loading (mg contaminant/g carbon) at breakthrough ($X/M_b$) and exhaustion ($X/M_{exh}$); carbon usage rate (g carbon/L of wastewater treated at breakthrough); and the degree of column utilization ($X/M_b / X/M_{exh}$).

## EXPERIMENTAL DESIGN

GAC columns containing Hydrodarco (HD) 4000, a lignite based carbon from Norit Americas, Inc., were used to treat two synthetic wastewaters containing either Pb-phenol or Pb-TCE. In Table 1, the experimental conditions used for both wastewaters are presented. Columns were carefully packed with washed HD4000 to minimize the introduction of air bubbles to the column. The columns were then pretreated with an acid/base rinsing procedure. For the phenol experiments, the columns were rinsed with 10 bed volumes (BV) of 0.1 N HCl followed by 10 BV of 0.1 N NaOH. For the TCE experiments, the columns were rinsed with 10 bed volumes (BV) of either 0.1 N or 1 N acetic acid (HAc) followed by 10 BV of 0.1 N NaOH. Wastewater was then introduced into the column and the effluent Pb, pH, phenol or TCE concentrations were measured periodically. Following column exhaustion with respect to Pb, columns were regenerated using the pretreatment procedure described earlier. Pb concentration and volume of all solutions (effluent samples, bulk effluent, regeneration solutions) were determined so that a Pb mass balance and Pb regeneration efficiencies could be calculated. Concentrations of phenol and TCE were also determined but since thermal regeneration is normally practiced with organic contaminants, a mass balance and regeneration efficiencies were not calculated.

Breakthrough and exhaustion are defined as $C_{eff} = 0.03C_{inf}$ and $C_{eff} = 0.95C_{inf}$, respectively. Following regeneration, the column was placed on-line and the procedure repeated. Seven runs were conducted for Pb-phenol and two runs were conducted for Pb-TCE. A third Pb-TCE run is ongoing and a fourth run is planned.

Table 1. Summary of Experimental Conditions

| EXPERIMENT CONDITION | VALUE |
|---|---|
| **Column Parameter** | |
| Empty Bed Contact Time (EBCT), min | 12.75 |
| Hydraulic Loading Rate (HLR), gpm/ft$^2$ | 2.0 |
| Amount of Carbon, g | 450-490 |
| Liter/Bed Volume | 1.18 |
| Flow Direction | Up-Flow |
| **Wastewater** | |
| Influent pH | 5.4 |
| Pb, mg/L | 1.0 |
| Phenol, mg/L | 10 |
| TCE, mg/L | 1.0 |
| Ionic Strength (as NaNO$_3$), N | 0.01 |

## RESULTS

### Pb-Phenol Wastewater

In Figure 1, Pb and phenol breakthrough curves and effluent pH curves are presented. In Table 2, BVs treated at Pb breakthrough ($V_b$) and exhaustion ($V_{exh}$), Pb surface loadings at exhaustion, and Pb desorption efficiencies are presented. The column was regenerated using 0.1 N HCl - 0.1 N NaOH rinses at Pb exhaustion or when the column clogged (Run 2 and 6).

Table 2. Summary of Pb Results for Pb-Phenol Experiments

| Run # | $V_b{}^1$ (BV) | $V_{exh}{}^2$ (BV) | X/M$_{exh}$ (mg Pb/g carbon) | Pb Desorption Efficiency (%) |
|---|---|---|---|---|
| 1 | 870 | 1270 | 2.59 | 74 |
| 2 | 850 | 850[3] | 1.93[3] | 90 |
| 3 | 835 | 1160 | 2.35 | 77 |
| 4 | 840 | 1070 | 2.14 | 89 |
| 5 | 650 | 870 | 1.77 | 83 |
| 6 | 770 | 870[3] | 1.89[3] | 97 |
| 7 | 850 | 1100 | 2.36 | 92 |

[1]Breakthrough (C = 0.03C$_o$). [2]Exhaustion (C = 0.95). [3]Column clogged before exhaustion, results represent situation at end of column run.

In Run 1, Pb breakthrough and exhaustion occurred at 870 and 1270 BVs, respectively. At the time of Pb exhaustion (1270 BVs), phenol breakthrough had yet to occur. After regeneration, the pH of the column was high ($\approx$ 12) and a portion of the carbon-bound phenol was desorbed (note the spike in phenol concentration immediately following regeneration). Above pH = 10.5, phenol exists as $C_6H_5O^-$. Obviously, the ionic form of phenol is desorbable at high pH values. As the column pH decreased, the concentration of phenol in the effluent dropped to below the detection limit (0.1 mg/L).

In Run 2, Pb breakthrough occurred at 850 BVs. At this point pressure in the column began to buildup and the run was stopped and the column was regenerated. As in Run 1, the pH of the column was high ($\approx$ 12) after regeneration and a portion of the carbon-bound phenol was desorbed. As the column pH decreased, the concentration of phenol in the effluent dropped to below the detection limit (0.1 mg/L).

During Run 3, breakthrough of phenol ($C = 0.03C_0 = 0.3$ mg/L) occurred after 2700 BVs. The phenol surface loading at this point was 54.2 mg phenol/g carbon. Pb breakthrough and exhaustion for Run 3 occurred at 835 and 1160 BVs (relative to the start of Run 3), respectively. The column was not exhausted for phenol at this point but at end of Run 3, 3500 BVs of phenol wastewater had been treated. The effluent phenol concentration and phenol surface loading at this point was 6.60 mg/L and 71.7 mg phenol/g carbon, respectively.

Column behavior for Runs 4 through 7 with respect to Pb removal were similar to that observed in earlier runs. In Run 6, pressure buildup occurred and the column was regenerated prior to Pb exhaustion. Column performance after regeneration of the clogged column (Runs 2 and 6) was not adversely affected. Clogging of GAC columns during full-scale operation with a subsequent buildup of column headloss is common. In full-scale operation, columns will often be taken off-line when a maximum head loss has been reached. Thus, column clogging should not prevent the use of GAC columns for the removal of inorganic and organic contaminants in the field.

Pb desorption efficiencies ranged from 74 to 97 percent with a mean value of 86 percent. Incomplete lead removal during regeneration did not significantly decrease Pb removal in the subsequent run. The slight decrease in Pb removal observed in Runs 5 and 6 was most likely caused by the presence of large amounts of phenol in the effluent.

After Regeneration 3, a portion of the phenol adsorption capacity had been reclaimed via the base rinse but the effluent phenol concentration was always greater than the breakthrough concentration ($0.03C_0$). In an actual treatment system, the column at this point would be taken off-line and regenerated for phenol. Regeneration of GAC columns loaded with organics is typically accomplished using thermal methods, however, caustic regeneration is sometimes used. In Regeneration 5, 1 N NaOH was used in place of 0.1 N NaOH to ascertain if better phenol regeneration could be accomplished using a more concentrated caustic solution. In Table 3, the average concentration of phenol in the base regenerant, the value of the phenol "spike" in column effluent immediately following regeneration, and the volume of column effluent immediately following column regeneration that had a phenol concentration greater than 0.3 mg/L (breakthrough concentration) are provided for each regeneration.

Table 3. Phenol Concentrations in Various Solutions

| Run # | Avg. Phenol Conc. in Base Regenerant (mg/L) | Phenol Spike (mg/L) | BVs With Phenol Conc. > 0.3 mg/L |
|---|---|---|---|
| 1 | 136.5 | 24.9 | 75 |
| 2 | 192.9 | 15.3 | 82 |
| 3 | 351.1 | 24.9 | --[1] |
| 4 | 367.4 | 304 | --[1] |
| 5[2] | 587.0 | 344 | --[1] |
| 6 | 412.5 | 24.1 | --[1] |
| 7[3] | 428.4 | NA | NA |

NA: Not Applicable  [1]All Column Effluent > 0.3 mg/L
[2]1 N NaOH Used To Regenerate Column In Place Of 0.1 N NaOH.
[3]Run #8 Was Not Conducted But Column Was Regenerated At The End Of Run 7.

The amount of phenol in the base regenerant and in the phenol spike was largest for Regeneration 5 where 1 N NaOH was used in place of 0.1 N NaOH. In addition, more phenol was removed subsequent to Regeneration 5 (Run 6, see Figure 1) compared to Runs 4 and 5. In Regeneration 6, 0.1 N NaOH was used, and the phenol spike was less than that observed in Regeneration 5 and less phenol was removed in Run 7 compared to Run 6. It is apparent increasing the strength of the base rinse increased phenol desorption and phenol removal during the subsequent treatment run. Based on these observations, a hot caustic rinse could reclaim a significant amount of the carbon's phenol removal capacity thus, increasing the number of BVs of wastewater treated for phenol removal. This is a topic for future research.

The carbon usage rate (CUR), based on Pb removal, ranged from 0.500 to 0.668 g carbon/L of wastewater treated and averaged 0.541 g/L. The degree of column utilization (DoCU), based on Pb removal, ranged from 78 to 91 percent with a mean value of 84 percent. For the runs in which the column clogged (Runs 2 and 6), calculation of the DoCU was not possible because the column did not reach exhaustion. The CUR based on phenol can only be calculated for the first phenol breakthrough that occurred at 2700 BVs. At this point the CUR for phenol was 0.185 g/L The degree of column utilization with respect to phenol could not be calculated because the column was never exhausted for phenol.

## Pb-TCE Wastewater

In Figures 2 and 3, influent and effluent Pb, TCE, and pH are presented for Runs 1 and 2, respectively. In Table 4, a summary of column performance parameters are presented for the data presented in Figures 2 and 3. Performance decreased slightly from Run 1 to Run 2. However, based on preliminary results from Run 3 (Run 3 was ongoing at the time this paper was written), column performance improved from Run 2 to Run 3. Column performance parameters based on Pb removal were similar to those observed in the Pb-phenol experiments. Unlike phenol, TCE, which does not ionize, did not desorb after the acid-base regeneration procedure. Except for one point in Run 1, TCE concentration was not detectable in the effluent. A total of 1930 BVs (2275 L) of TCE wastewater were treated for the two runs resulting in a commutative TCE surface loading of 5.05 mg/g. As was mentioned previously, the Pb-TCE experiment was ongoing at the time this paper was written and results from Runs 3 and 4 will be presented at the conference.

**Table 2. Summary of Results for Pb-TCE Experiments**

| Parameter | Run 1 | Run 2 |
|-----------|-------|-------|
| $V_b$ | 750 | 625 |
| $V_{exh}$ | 1080 | 850 |
| $X/M_b$ | 1.92 | 1.60 |
| $X/M_{exh}$ | 2.36 | 1.84 |
| CUR, g/L | 0.510 | 0.606 |
| DoCU | 81 | 86 |
| Pb Desorption Efficiencies, % | 100 | 93 |
| TCE X/M[1] | 2.83 | 2.22 (5.05)[2] |

[1]X/M (mg TCE/g carbon) for individual run
[2]Total X/M (mg TCE/g carbon) at end of run 2

## SUMMARY

Two synthetic wastewaters containing 1 mg/L Pb and either 10 mg/L phenol or 1 mg/L Pb were treated using granular activated carbon (GAC) columns. Several process cycles (wastewater treatment-regeneration) for each wastewater were conducted. The study carbon, Hydrodarco (HD) 4000, was a lignite based carbon manufactured by Norit Americas, Inc. Regardless of the organic contaminant, column performance parameters based on Pb removal were relatively constant: bed volumes at breakthrough ranged from 625 to 870, X/M at exhaustion ranged from 1.60 to 2.59 mg Pb/g carbon, carbon usage rate ranged from 0.500 to 0.668 g carbon/L of wastewater treated, the degree of column utilization ranged from 78 to 91 percent, and the Pb desorption efficiencies ranged from 74 to 100 percent.

Phenol desorbed during the base rinse portion of the regeneration procedure as well as during the first several bed volumes of the subsequent treatment run. The high column pH was due to the base conditioning step that follows the acid desorption step of the regeneration procedure. Phenol ionizes at high pH and the ionic form of phenol is not as adsorbable as the nonionic form. TCE did not desorb to any great extent during the regeneration procedure. After 2275 L of Pb-TCE wastewater, TCE was not present in the column effluent (*i.e.*, TCE had not broken through). The TCE surface loading at this point was 5.05 mg TCE/g carbon. Simultaneous removal of organics and heavy metals is feasible provided that the organics contaminants do not desorb at the extreme pHs experienced during regeneration for heavy metals. If desorption does occur, the portion of the column effluent with an acceptable concentration of organics can be recycled to the column influent.

## REFERENCES

Corapcioglu, M.O. and Huang, C.P. (1987). The Adsorption of Heavy Metals onto Hydrous Activated Carbon. **Wat. Res.** 9 (9), 1031-1044.

Huang C.P. and Ostovic, F.B. (1978) Removal of Cadmium (II) by Activated Carbon Adsorption. **J. Envir. Engrg Div. ASCE,** 104(5), 863-878.

Reed B.E. and Matsumoto, M.R. (1993) Modeling Cadmium Adsorption in Single and Binary Adsorbent (Powdered Activated Carbon) Systems. **Journal of Environmental Engineering.** Vol. 119, No. 2, pp 332-348.

Reed B.E., Arunachalam, S. and Thomas, B. (1994a) Removal of Lead and Cadmium from Aqueous Waste Streams Using Granular Activated Carbon (GAC) Columns". **Environmental Progress,** Vol. 13, No. 1, pp 60-64.

Reed, B.E. and Arunachalam, S. (1994). Use of Granular Activated Carbon Columns for Lead Removal. **Journal of Environmental Engineering.** Vol. 120, No. 2, pp 416-436.

Reed B.E. (1995) Identification of Removal Mechanisms for Lead in Granular Activated Carbon (GAC) Columns. **Sep. Sci. and Technol.** 30(1), pp 101-116.

Reed, B.E., Robertson, J. and Jamil, M. (1995). Regeneration of Granular Activated Carbon (GAC) Columns used for Removal of Lead. Accepted for publication in **Journal of Environmental Engineering.**

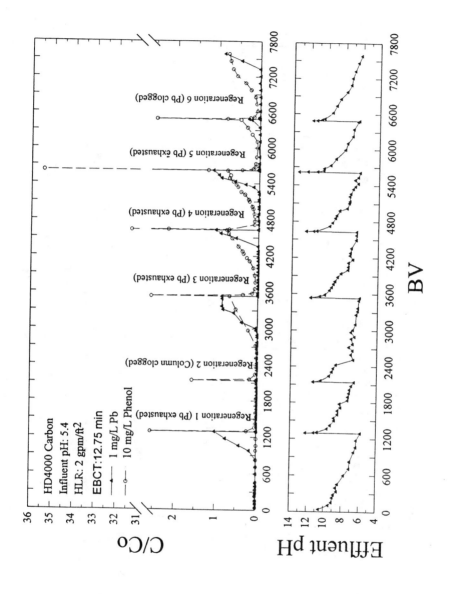

Figure 1. Effluent Pb, Phenol, and pH versus Bed Volumes of Wastewater Treated.

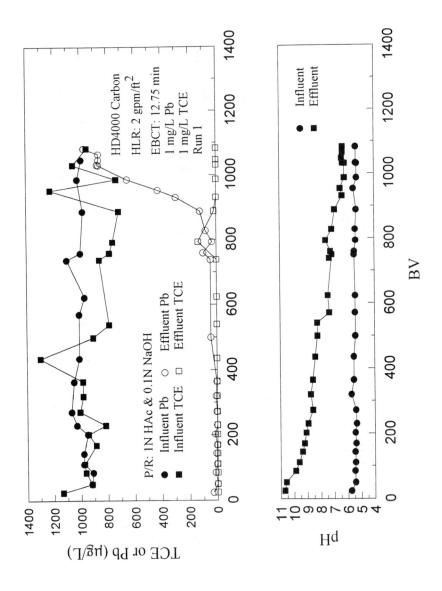

Figure 2. Run 1 Effluent Pb, TCE, and pH versus Bed Volumes of Wastewater Treated.

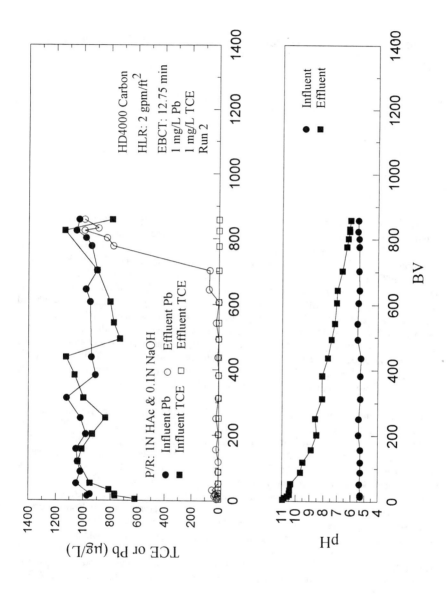

Figure 3. Run 2 Effluent Pb, TCE, and pH versus Bed Volumes of Wastewater Treated.

# "*REDOX* MANIPULATION TO ENHANCE CHELANT EXTRACTION OF HEAVY METALS FROM CONTAMINATED SOILS"

ROBERT W. PETERS*, WEN LI, GREGORY MILLER, MICHAEL D. BREWSTER[+],
TERRI L. PATTON, AND LOUIS E. MARTINO
Argonne National Laboratory
9700 South Cass Avenue
Argonne, Illinois 60439

## INTRODUCTION

J-Field on the Gunpowder Neck Peninsula at the Aberdeen Proving Ground has contamination resulting from past field activities at the facility. Disposal operations conducted in this region include: toxic burning pits, white phosphorus burning pits, riot control burning pits, South Beach Demolition Ground, South Beach trench, Robins Point Demolition Ground, Robins Point Tower Site, and the Ruins Site. J-Field is relatively flat, with a maximum relief of about 10 ft. The ground surface slopes gently toward marshy areas or toward Chesapeake Bay and on-site surface water. The Toxic Burning Pits (TBP) area is bounded to the northeast by marsh and to the south and southeast by woods and marsh. Because the ground surface elevation is highest in the northwestern portion of the TBP area, surface water drainage probably flows toward the south-southeast into the marsh area. The direction of groundwater flow in the surficial aquifer is also probably toward the marsh. Sieve analyses indicated that the surface soil (the upper 6-inches in the TBP area is mainly silty sand (with a silt/clay fraction of up to 47.1%) and sandy silt (with a silt/clay fraction of up to 64.4%) with an organic content ranging from 1.4% to 10.5%.

The Toxic Burning Pits were used to dispose of munitions, explosives, nerve and chemical agents, mustard gas, liquid smoke, chlorinated solvents, and radioactive chemicals. Three of the burning pits were used to dispose of methylphosphonothioic acid (VX), dichlorodiethyl sulfide (mustard gas), and the primary components of liquid smoke -- titanium tetrachloride (FM) and sulfur trioxide/chlorosulfonic acid (FS). In addition, fuel was used to ignite materials placed in the pits. From soil gas measurements collected between 1987 and 1992 by the U.S. Geological Survey, contaminants identified included trichloroethylene, tetrachloroethylene, dichloroethylene, trichloroethane, phthalates, combined hydrocarbons, simple aromatics, and heavy aromatic hydrocarbons. The surface soil contain elevated levels of metals, especially lead (up to 2.6% in places), mercury (up to 10 mg/kg), and cadmium (up to 16.6 mg/kg). Other contaminants include PCBs (up to 143,000 µg/kg), 1,1,2,2-tetrachloroethane (up to 3,270,000 µg/kg, 1,1,2-trichloroethane (up to 8,500 µg/kg), tetrachloroethylene (up to 25,700 µg/kg), and trichloroethene (up to 263,000 µg/kg). The highest level of organic compounds were found at a depth of 6 ft.

[+]Formerly with Argonne National Laboratory; currently working for CSK Technical, Inc., Tonawanda, NY.
*Correspondence should be directed to this author. References in this paper to commercial products do not constitute or imply endorsement of these products by the authors or the Laboratory. The viewpoints expressed here are not necessarily those of the Laboratory or its sponsors.

Soil samples collected in 1983 indicated the presence of lead, zinc, nitrate, and petroleum hydrocarbons in each of the samples, and mercury and cadmium in one of the samples. Lead has been detected both by xray fluorescence (XRF) data and atomic absorption data, ranging from non-detectable to 8% lead. The whole area is probably above the 500 mg/kg Pb preliminary clean-up level.

Argonne National Laboratory is conducting a Focused Feasibility Study (FFS) to identify and evaluate remedial technologies that may be implemented to address contamination at the Toxic Burning Pits (TBP) area at J-Field, Aberdeen Proving Ground, Maryland. As part of the FFS, several treatability studies were performed to collect additional data on potential technologies and determine their applicability on conditions specific to the TBP area. This paper summarizes the results soil washing and enhancements to soil washing portions of the study. On the basis of their concentrations in the untreated TBP soils and degree of environmental impact, the primary heavy metals of concern for the FFS were copper, lead, and zinc.

## PROCEDURES AND EQUIPMENT

Research relating to the soil washing task performed in this study are described below.

### Physical and Chemical Soil Characterization

Soil samples (background, representative, and worst-case) used in the study were composited over a depth interval of 4 ft. Before being shipped to Argonne, all samples were screened and found to be free of agent materials. Physical and chemical analyses were performed to characterize the soil. Characterization analyses included total extractable metals, toxicity characteristic leaching procedure (TCLP) for metals only, cation exchange capacity, soil pH, metal speciation via sequential extraction, moisture content, color, bulk density, soil texture, and particle size analysis.

The soils were analyzed for total extractable metals, Toxicity Characteristic Leaching Procedure (TCLP) metals, and metal speciation. The sequential extraction procedure (developed by the U.S. Army Corps of Engineers) used to identify the metal speciations is described in Table 1.

Table 1. Heavy Metal Speciation Procedure by Sequential Extraction

| Classification | Extraction Reagent and Conditions |
|---|---|
| Exchangeable | 1 M $MgCl_2$, 60 min |
| Carbonate | 1 M acetate buffer, pH~5, 5 h |
| Reducible Oxides | 0.04 M hydroxyamine hydrochloride in 25% acetic acid, 96°C for 6 h |
| Organically-Bound | 30% $H_2O_2$ and 0.02M $HNO_3$, 85°C for 2 h, followed by extraction with 3.2 M ammonium acetate in 20% $HNO_3$ |
| Residual | Concentrated $HNO_3$, 95°C for 8 h |

## Soil Washing

### Batch Shaker Tests

Soils contaminated with heavy metals, primarily lead, were subjected to a series of batch-shaker flask experiments to identify the chelating agents and surfactants that show promise in mobilizing lead and other metals from the TBP soils. Chelating and mobilizing agents evaluated included ethylenediamine-tetraacetic acid (EDTA), citric acid, Citranox, gluconic acid, phosphoric acid, oxalic acid, nitrilotriacetic acid (NTA), and ammonium acetate, in addition to pH-adjusted water. Soil washing experiments were performed by first placing nominal 5-g portions of TBP soils in plastic shaker containers. To these containers, 45-mL of extractant solution (0.01M, 0.05M, or 0.1M) were added. Contact time was maintained at 3 h. This time requirement was determined from a previous study to be adequate for equilibrium conditions to be achieved [Peters and Shem, 1992]. Following this agitation, the samples were centrifuged in plastic Nalgene centrifuge tubes equipped with snap-on caps, filtered using No. 42 Whatman filter paper, and stored in glass vials maintained at pH<2 (prepped using ultrapure $HNO_3$) to await atomic absorption spectrophotometry (AAS) analysis. At least 10-mL of sample was collected for the AAS analysis. The filtrates collected were analyzed for copper, lead, and zinc by AAS. The analyses were performed in accordance with the procedures described in Standard Methods [1992].

Data collected in these studies included the following: operating temperature, extractant type and concentration, heavy metals concentration on the soil before treatment (and after treatment as determined by calculation), heavy metals concentration in the extract solution after treatment, pH of the solution before and after treatment, and batch shaking time.

## Enhancements to Soil Washing/Soil Flushing

These sets of experiments were aimed at improving the performance of soil washing/soil flushing by pretreating the soils before performing soil washing/soil flushing operations. The use of sonication and REDOX manipulation to increase the removal of heavy metals from the Toxic Burning Pits was investigated. Sonication involves the application of high-energy sound waves to degrade organic pollutants and enhance the removal of heavy metals from the soils. A laboratory scale apparatus (Sonics & Materials, VC 600) was used for the sonication treatment. Variables investigated included input power, operation temperature, pH, and addition of chemical enhancements.

The sonication-enhanced soil washing experiments were performed by first placing nominal 5-g portions of TBP soils in 50-mL plastic centrifuge tubes. After 25-mL of deionized water was added to each centrifuge tube, the samples were subjected to sonication for 10 min. Then, the lids to the centrifuge tubes were replaced, and the tubes were centrifuged to separate the solid and liquid phases. Aliquots (5 mL) were collected and analyzed for metals by atomic absorption spectrophotometry. To the solutions remaining in the centrifuge tubes, 25-mL of the 0.05M chelants (citric acid or EDTA) was added. The extractant solutions were pH-adjusted (pH~5 or pH~9) before being added to the sonication-treated TBP soils. Standard batch-shaker soil washing tests were then performed on the samples to determine the effect of sonication on heavy metal extraction by soil washing.

REDOX manipulation can provide conditions that maximize the solubilities of contaminants and promote their removal. Reducing agents studied included sodium borohydride, sodium metabisulfite, and thiourea dioxide. For some soils, oxidizing agents may enhance metals removal by degrading organometallic complexes and releasing the metals that have an affinity for natural organic matter. Hydrogen peroxide, sodium percarbonate, sodium hypochlorite, and potassium permanganate were the oxidants evaluated for REDOX manipulation. The oxidizing and reducing agents used were chosen on the basis of their REDOX characteristics, operating conditions (e.g., pH, concentration, etc.), ionic content, and availability.

The oxidizing and reducing agents used for REDOX modification were screened by adding 45-mL aliquots of 1000-ppm solutions of each reagent to nominal 5-g portions of the representative TBP soils. The soil samples were then processed in a manner similar to the batch shaker flask soil washing method. Variables monitored to determine the effect of REDOX manipulation included pH, ORP, and metals removal efficiencies. As the result of the screening tests, sodium borohydride (highest change in ORP), sodium metabisulfite (most common and versatile of the reducing agents studies), and sodium percarbonate (highest lead removal of the oxidants studied) were used to further study the enhanced heavy metal extraction by treating the TBP soils with the REDOX modifiers before performing the chelant extraction procedures.

REDOX modification was combined with chelant extraction to extract copper, lead, and zinc from the representative TBP soil sample. Aliquots (45-mL) of the 1000-ppm solutions of the oxidizing and reducing agents were combined with nominal 5-g portions of the TBP soils. After the required contact time, solid/liquid separation was performed by vacuum filtration. To the residual soils, 45-mL of the 0.05M chelant (EDTA or citric acid) solutions were added. The chelant extraction step was done according to the soil washing procedure described earlier. Variables measured in the intermediate (following REDOX manipulation) and final (after chelant extraction) samples to quantify and explain the combined REDOX modification/chelant extraction approach included pH, oxidation/reduction potential (ORP), and copper, lead and zinc concentrations.

## RESULTS AND DISCUSSION

### SOIL CHARACTERIZATION RESULTS

The soils (background, representative, and worst case) all are generally brownish in color, have a low cation exchange capacity (1.2 - 4.0 meq/100-g), are slightly alkaline in nature (soil pH in the range of 7.5 to 8.4), have a moderate volatile solids content (2.5 - 8.8%), and have a sandy loam soil texture. The particle size distribution characteristics of the soils determined from hydrometer tests are approximately 60% sand, 30% silt, and 10% clay. The results for the individual soil analyses are summarized below.

| Sample No. | Color | Soil pH | Moisture Content, (%) | Cation Exchange Capacity, (meq/100-g) | Volatile Solids, (%) | Particle Size Distribution, (wt%) | | | Soil Texture |
|---|---|---|---|---|---|---|---|---|---|
| | | | | | | Sand | Silt | Clay | |
| REP TBP1 | Brown | 7.786 | 24.3315 | 1.17 | 8.76 | 59.2 | 30.8 | 10.0 | Sandy Loam |
| REP TBP1-dup | | 7.801 | 24.8182 | N.S. | | | | | |
| REP TBP2 | | 7.793 | 25.0108 | 1.32 | | | | | |
| REP TBP2-dup | | 7.818 | 24.0818 | N.S. | | | | | |
| REP TBP3 | | N.A. | N.A. | 4.03 | | | | | |
| WC TBP1 | Light Olive Brown | 7.507 | 18.4903 | 1.83 | 7.33 | 61.2 | 29.8 | 9.0 | Sandy Loam |
| WC TBP1-dup | | 7.526 | 18.1375 | N.S. | | 61.2 | 29.8 | 9.0 | Sandy Loam |
| WC TBP2 | | 7.559 | 19.2314 | 1.67 | | | | | |
| WC TBP2-dup | | 7.532 | 19.4814 | N.S. | | | | | |
| WC TBP3 | | N.A. | N.A. | 2.10 | | | | | |

N.S.: No sample
N.A.: No analysis

635

The total extractable metal characteristics of these soils are summarized below for the metals Cd, Cr, Cu, Fe, Hg, Mn, Ni, Pb, Zn, and As:

Table 2.  Soil Characterization Results for the Aberdeen Proving Ground J-Field Soils

| Heavy Metal | Total Extractable Metals, (mg/kg) | | |
|---|---|---|---|
| | Worst Case | Representative | Background |
| Cd | 7.4 | 6.6 | 2.2 |
| Cr | 238.7 | 311.7 | 38.7 |
| Cu | 1241.3 | 1533.2 | 88.4 |
| Fe | 39,858 | 48,312 | 10,913 |
| Hg | 1.52 | 1.52 | 1.39 |
| Mn | 203.5 | 286.3 | 92.7 |
| Ni | 27.7 | 35.7 | 4.0 |
| Pb | 21,560 | 15,294 | 56.9 |
| Zn | 3729.0 | 3677.0 | 64.7 |
| As | 17.8 | 21.8 | 9.5 |

These results indicate that the soils contain appreciable Cu, Pb, Zn, and Fe concentrations, and moderate concentrations of Cr and Mn, with minor concentrations of Cd, Hg, Ni, and As.

Sequential extractions were performed on the "as-received" soils (worst case and representative) to determine the speciation of the metal forms.  This technique speciates the heavy metal distribution into an easily extractable (exchangeable) form, carbonates, reducible oxides, organically bound, and residual forms.  Sequential extractions were performed on the worst case and representative bulk soil samples; in addition, sequential extractions were performed on the sand and silt/clay fractions of the worst case soil.  The heavy metals analyzed were: Cd, Cu, Cr, Pb, Zn, Mn, Fe, and Ca.  Most of the metals are amenable to a soil washing technique (i.e., Exchangeable + Carbonate + Reducible Oxides).  The metals Cu, Pb, Zn, and Cr have greater than 70% of their distribution in forms amenable to soil washing techniques, while Cd, Mn, and Fe are somewhat less amenable to soil washing using chelant extraction.

The heavy metals are distributed throughout the soil matrix and are not concentrated in a single soil fraction (such as the clay fraction).  As a consequence, pretreatments such as hydrocycloning would not be effective in an effort of reducing the amount of soil to be treated, since the heavy metals are found in all three soil fractions.  The soils are predominantly sand and silt in content; therefore, soil washing and soil flushing techniques should be effective in treating these soils, particularly given that greater than 70% of the metals of concern are present in forms that are conducive to soil washing techniques.

## SOIL FEASIBILITY/TREATABILITY RESULTS

### Soil Washing

#### Batch Soil Washing Experiments

Soils contaminated with heavy metals (primarily lead) were subjected to a series of batch-shaker flask experiments to identify the chelating agents and surfactants that showed promise in mobilizing lead and other heavy metals from the TBP soils. The chelating and mobilizing agents investigated included EDTA, citric acid, Citranox, gluconic acid, phosphoric acid, oxalic acid, NTA, and ammonium acetate, in addition to pH-adjusted water. These chelating agents have been used in other studies as means to solubilize, sequester, and extract heavy metals into solution. The contact time in all the batch shaker flask experiments was maintained at 3 h.

It is important to realize that all eight heavy metals are being extracted simultaneously. Further, some of the concentrations of heavy metals appear to plateau out (i.e., become saturated). It should also be pointed out that the soil contains an appreciable amount of iron; iron is not a serious contaminant of concern, due to its common presence in many soils and groundwaters. Therefore, an important property of the optimum chelating agent is one that minimizes the extraction of iron into solution, while maximizing the extraction of the other heavy metals of concern.

Table 3 summarizes the range and mean extraction efficiencies for the worst-case and representative TBP soils for the eight heavy metals and nine extracting agents used. Note that the data contained in this table neglects the chelating agent concentration used and pH-effects associated with the extractions. As examples, for the representative soil, the overall removal efficiency of zinc was 11.450% for gluconate; at pH~4, the overall removal efficiency was ~36.798%, as compared to 3.001% for pH in the range of 7 to 9. Similarly, for the representative soil, the overall removal efficiency of zinc was 14.616% for ammonium acetate; at pH~4, the overall removal efficiency was ~30.928%, as compared to 1.977% for pH in the range of 7 to 9.

Table 3. Comparison of Heavy Metal Removal Efficiencies Using Different Chelating Agents

| Soil Sample | Heavy Metal | Statistic | % Heavy Metal Removal | | | | | | | | |
|---|---|---|---|---|---|---|---|---|---|---|---|
| | | | EDTA | Oxalate | Citrate | Citranox | Gluconate | H3PO4 | NH4-Ac | NTA | pH-Adjusted H2O |
| WC | Cd | Range | 6.262-89.252 | 6.044-18.137 | 7.743-~100 | 3.687-20.502 | 2.418-26.114 | 2.439-9.869 | 2.322-29.917 | 2.431-42.717 | 5.912-7.342 |
| | | Mean | 25.811 | 8.843 | 40.007 | 10.402 | 9.352 | 5.235 | 23.189 | 23.189 | 6.381 |
| | Cu | Range | 24.124-62.996 | 1.477-32.114 | 11.364-35.929 | 2.666-22.545 | 0.438-22.867 | 0.285-2.186 | 0.142-16.487 | 13.503-45.466 | 0.077-0.225 |
| | | Mean | 40.574 | 15.339 | 23.717 | 10.351 | 7.169 | 1.306 | 2.615 | 29.375 | 0.119 |
| | Pb | Range | 23.494-88.123 | 0.051-0.987 | 1.599-11.734 | 0.847-3.701 | 0.385-14.704 | 0.006-0.099 | 0.042-17.297 | 11.469-92.988 | 0.006-0.054 |
| | | Mean | 31.910 | 0.390 | 5.261 | 1.866 | 5.319 | 0.029 | 2.379 | 35.530 | 0.022 |
| | Zn | Range | 18.270-57.892 | 0.740-11.454 | 3.163-84.480 | 1.568-35.137 | 0.904-41.412 | 0.472-21.005 | 0.919-49.586 | 15.866-53.134 | 0.272-4.879 |
| | | Mean | 36.133 | 4.983 | 21.880 | 14.683 | 10.867 | 5.350 | 11.634 | 30.543 | 1.562 |
| | Fe | Range | 0.329-3.298 | 0.020-7.337 | 0.177-2.184 | 0.052-1.188 | 0.042-0.989 | 0.005-0.089 | 0.001-0.015 | 0.193-3.622 | 0.006-0.013 |
| | | Mean | 1.619 | 1.511 | 0.784 | 0.455 | 0.214 | 0.018 | 0.0066 | 1.076 | 0.0085 |
| | Cr | Range | 0.375-7.661 | 0.270-9.201 | 0.319-8.161 | 0.189-1.964 | 0.386-6.763 | 0.371-0.526 | 0.110-0.846 | 0.111-5.735 | 0.183-0.189 |
| | | Mean | 3.437 | 2.081 | 2.182 | 0.727 | 1.523 | 0.433 | 0.180 | 2.186 | 0.187 |
| | As | Range | 1.152-2.875 | 2.322-9.802 | 1.978-9.549 | 0.185-2.249 | 0.392-1.306 | 0.539-1.492 | 0.084-0.257 | 0.129-0.880 | 0.245-2.273 |
| | | Mean | 1.684 | 4.837 | 4.892 | 0.706 | 0.587 | 0.869 | 0.120 | 0.391 | 1.212 |
| | Hg | Range | 2.322-~100 | 2.080-42.922 | 1.191-90.908 | 1.170-31.422 | 0.704-18.691 | 1.723-42.059 | 1.127-26.577 | 1.330-25.423 | 1.187-5.912 |
| | | Mean | 31.611 | 14.245 | 18.245 | 8.364 | 5.134 | 7.831 | 3.360 | 7.716 | 2.649 |
| REP | Cd | Range | 6.796-41.803 | 8.855-14.012 | 8.471-~100 | 2.703-35.428 | 2.701-33.548 | 2.701-11.018 | 2.826-25.530 | 4.090-40.284 | 8.596-7.192 |
| | | Mean | 23.805 | 8.859 | 40.735 | 10.959 | 12.071 | 5.928 | 9.466 | 24.729 | 6.796 |
| | Cu | Range | 26.463-81.847 | 1.845-26.342 | 11.563-68.823 | 2.169-17.181 | 0.385-25.484 | 0.244-1.594 | 0.114-9.974 | 13.417-41.286 | 0.099-0.178 |
| | | Mean | 45.548 | 12.001 | 23.949 | 7.020 | 7.448 | 0.915 | 2.071 | 27.431 | 0.138 |
| | Pb | Range | 20.421-~100 | 0.104-1.633 | 1.058-12.426 | 0.333-2.181 | 0.096-12.547 | 0.011-0.922 | 0.025-8.868 | 12.389-77.988 | 0.025-0.048 |
| | | Mean | 37.804 | 0.498 | 3.828 | 0.989 | 2.452 | 0.105 | 1.354 | 34.525 | 0.033 |
| | Zn | Range | 20.489-92.893 | 0.793-12.789 | 3.637-42.015 | 1.486-45.481 | 1.360-44.006 | 0.273-9.202 | 0.611-48.240 | 2.375-50.386 | 0.420-1.327 |
| | | Mean | 36.752 | 4.832 | 16.535 | 14.194 | 11.45 | 3.731 | 9.215 | 24.570 | 0.688 |
| | Fe | Range | 0.616-8.156 | 0.040-8.674 | 0.349-3.260 | 0.030-1.480 | 0.093-0.928 | 0.003-0.053 | 0.001-0.078 | 0.001-0.079 | 0.001-0.013 |
| | | Mean | 1.743 | 1.288 | 0.494 | 0.279 | 0.0136 | 0.0207 | 0.0207 | 0.0065 | 0.165 |
| | Cr | Range | 0.729-8.888 | 0.232-6.338 | 0.382-4.678 | 0.143-1.500 | 0.281-6.866 | 0.286-2.825 | 0.084-0.574 | 0.322-3.621 | 0.139-0.223 |
| | | Mean | 3.585 | 1.820 | 1.718 | 0.567 | 1.423 | 0.515 | 0.133 | 2.100 | 0.165 |
| | As | Range | 0.292-4.113 | 1.188-9.057 | 1.719-5.585 | 0.220-1.541 | 0.080-3.043 | 0.368-0.714 | 0.057-0.159 | 0.145-0.650 | 0.159-1.653 |
| | | Mean | 1.204 | 3.500 | 3.414 | 0.616 | 0.757 | 0.569 | 0.087 | 0.385 | 0.5645 |
| | Hg | Range | 1.150-90.334 | 1.869-~100 | 1.226-36.717 | 1.186-31.088 | 0.878-10.333 | 1.912-16.539 | 1.141-12.471 | 1.201-~100 | 1.142-1.249 |
| | | Mean | 20.488 | 32.614 | 13.351 | 11.294 | 1.874 | 5.249 | 3.942 | 23.280 | 1.1805 |

To summarize, of the chelating agents investigated, EDTA and citric acid appear to offer the greatest potential as chelating agents to use in soil washing Aberdeen Proving Ground soils. NTA is also a very effective chelant; however, it is a Class II carcinogen, and as such would probably not be used in remediating the site. The other chelating agents studied (gluconate, oxalate, Citranox, ammonium acetate, and phosphoric acid, along with pH-adjusted water) were generally ineffective in mobilizing the heavy metals from the soils. It is particularly interesting to note that phosphoric acid was generally one of the least effective extractants used in this study, despite being a strong acid.

## Sequential Batch Chelant Extraction Studies

Due to the observation from both the batch shaker test study and the columnar chelant soil washing study that the solutions became nearly saturated, several batch experiments were performed in which the soil was repeated subjected to chelant extraction, followed by washing with deionized water. A total of six cycles of operation were performed to monitor the extraction of the three primary heavy metals of concern (copper, lead, and zinc) as a function of the number of extractions performed.

In order to compare the results of the metal speciations via sequential extractions, six stage batch extractions were performed using the worst-case soil. In addition, TCLP tests were performed on the untreated soil and on the soils after the 1st-, 3rd-, and 5th-stage extractions, respectively. The results, describing the concentrations of heavy metals remaining in the soil, removal efficiency of the heavy metals, and TCLP vs. number of stage extractions for lead, copper, and zinc, are presented in Table 4.

Table 4.  Multi-Stage Batch Extractions with EDTA for Pb, Cu, and Zn on the TBP Worst-Case Soil

| Contaminant Concentration | Pb | Cu | Zn |
|---|---|---|---|
| Untreated Soil-- | | | |
| Total Extractable Metals, (mg/kg): | 21,560 | 1241.3 | 3729.0 |
| Exchangeable + Carbonates, (%) | 57.80 | 44.93 | 54.92 |
| Exchangeable + Carbonates + Reduc. Oxides, (%) | 81.71 | 87.91 | 89.18 |
| Organic + Residual, (%) | 18.28 | 12.09 | 10.82 |
| TCLP (0), (mg/L) | 340.91 | 5.71 | 56.07 |
| *After 1st Washing*: | | | |
| Metal Concentration Remaining in Soil, (mg/kg) | 13000 | 668.89 | 1365.15 |
| Heavy Metal Removal, (%) | 49.94 | 54.63 | 63.39 |
| TCLP (1), (mg/L) | 30.39 | 2.95 | 6.38 |
| *After 2nd Washing*: | | | |
| Metal Concentration Remaining in Soil, (mg/kg) | 10137 | 390.45 | 737.50 |
| Heavy Metal Removal, (%) | 60.96 | 73.52 | 79.33 |
| TCLP (2), (mg/L) | NA | NA | NA |

| | | | |
|---|---|---|---|
| *After 3rd Washing:* | | | |
| Metal Concentration Remaining in Soil, (mg/kg) | 8063.2 | 264.37 | 489.04 |
| Heavy Metal Removal, (%) | 68.95 | 82.07 | 86.46 |
| TCLP (3), (mg/L) | 29.31 | 0.32 | 1.31 |
| *After 4th Washing:* | | | |
| Metal Concentration Remaining in Soil, (mg/kg) | 7327.5 | 209.11 | 386.77 |
| Heavy Metal Removal, (%) | 71.78 | 85.82 | 89.41 |
| TCLP (4), (mg/L) | NA | NA | NA |
| *After 5th Washing:* | | | |
| Metal Concentration Remaining in Soil, (mg/kg) | 3383.5 | 112.68 | 208.36 |
| Heavy Metal Removal, (%) | 86.97 | 92.36 | 93.95 |
| TCLP (5), (mg/L) | 1.56 | 0.14 | 0.49 |
| *After 6th Washing:* | | | |
| Metal Concentration Remaining in Soil, (mg/kg) | 297.18 | 15.85 | 74.02 |
| Heavy Metal Removal, (%) | 98.86 | 98.92 | 97.20 |
| TCLP (6), (mg/L) | NA | NA | NA |

The results show that the heavy metals, Cu and Zn, present as exchangeable and carbonate fractions, are completely extracted in the first extraction stage, whereas these same fraction for Pb were not extracted until after the second stage of extraction. Removal of Pb, Cu, and Zn present are exchangeable, carbonates, and reducible oxides occurred between the fourth- and fifth-stage extractions. Also between these two extraction stages, the Pb TCLP passed the EPA limit for lead of 5.0 mg/L. The corresponding Pb removal at this point was 87.0%, and the residual concentration of Pb remaining in the soil was about 3,400 mg/kg, well above the EPA Total Extractable Metal Limit for Pb of 500 mg/kg. However, by treating with a sixth EDTA extraction stage (operated at pH~9), the residual lead concentration was reduced to about 300 mg/kg (thereby passing the EPA Total Extractable Metal Limit). After the sixth stage of treatment, the residual concentrations of Pb, Cu, and Zn in the soil were approximately 300, 16, and 75 mg/kg, respectively. The overall removals of copper, lead, and zinc from the multiple-stage soil washing were 98.9%, 98.9%, and 97.2%, respectively, using EDTA as the chelant. Note during the conduct of these experiments, the concentration and operating conditions for the extractions were not necessarily optimized. If the conditions had been optimized, it is the belief of this researcher that the TCLP and residual heavy metal concentrations could probably be met within three or four extractions. The above results, however, show that it is very possible to treat the J-Field contaminated soils using a soil washing technique; the treated soil can meet EPA's TCLP and Total Extractable Metal Limits.

## Enhancements to Soil Washing/Soil Flushing (Sonication and REDOX Manipulation)

### *Sonication*:

The use of sonication to enhance the extraction of copper, lead, and zinc from the TBP soils was investigated. The results of the treatability studies indicated that the use of sonication to pretreat the TBP soils before soil washing with chelants had very little effect on heavy metal extraction efficiencies. Although sonication may have mobilized the heavy metals from the soil matrix, it is likely that the metals readsorbed back onto the soil matrix. It was concluded that the application of high-energy sound waves (sonication) was not a viable technique for increasing the chelant extraction rate of copper, lead, and zinc from the TBP soils.

*REDOX Manipulation*:

Initial screening experiments were performed investigating sodium borohydride, sodium metabisulfite, thiourea dioxide, hydrogen peroxide, sodium percarbonate, sodium hypochlorite, and potassium permanganate, for their effectiveness in solubilizing contaminants from the soil matrix. The results of these screening tests identified the following REDOX agents to pursue in further studies: sodium borohydride (highest change in ORP), sodium metabisulfite (most common and versatile of the reducing agents studied), and sodium percarbonate (highest lead removal of the oxidants studied).

Results presented in Table 5 below indicate that lead and copper removal by chelant extraction with EDTA and citric acid is minimally affected by pretreatment with sodium borohydride, sodium metabisulfite, and sodium percarbonate. Zinc removal by the stronger chelant (EDTA) was slightly increased by each REDOX agent studied. The reagents used for REDOX manipulation significantly improved the performance of citric acid for removing zinc from the worst-case TBP soils.

Table 5. Effect of REDOX Manipulation

| REDOX Agent | Chelant | Removal Increase, (%) | | |
|---|---|---|---|---|
| | | Copper | Lead | Zinc |
| Sodium Borohydride | EDTA | 0 | 1.4 | 5.7 |
| Sodium Metabisulfite | EDTA | 6.3 | 0 | 13.7 |
| Sodium Percarbonate | EDTA | 2.7 | 0 | 4.0 |
| Sodium Borohydride | Citric Acid | 0 | 3.5 | 27.0 |
| Sodium Metabisulfite | Citric Acid | 0.6 | 0 | 20.5 |
| Sodium Percarbonate | Citric Acid | 0 | 2.6 | 24.8 |

Figures 1a, 1b, and 1c summarize the results of soil washing/soil flushing (i.e., EDTA and citric acid) and enhancement to soil washing/soil flushing portions of this study. The results indicate that EDTA is much more effective than citric acid for removing copper, lead, and zinc from the worst-case TBP soils. For chelant extraction with EDTA, the removal efficiencies of copper, lead and zinc tend to plateau at values comparable to the forms present as exchangeable and carbonate species. The results in Figure 1c indicate that REDOX manipulation combined with chelant extraction with citric acid can be used to achieve zinc removal efficiencies comparable to those of EDTA. Depending on the method used to treat the heavy metal-containing extraction solutions, it may be desirable to use REDOX manipulation and mild chelation in place of EDTA. Because it is more difficult to remove heavy metals from extraction solution containing EDTA, the citrate-containing effluent will be easier to treat by conventional wastewater treatment technologies. Treatment of the citrate-containing effluent may result in citrate recovery and reuse.

The results from the REDOX manipulation followed by chelant extraction are summarized below:

*Copper*:    EDTA: Metabisulfite > Percarbonate > Borohydride
              Citrate: Metabisulfite > Percarbonate ~ Borohydride

*Lead*:    EDTA: Borohydride > Metabisulfite ~ Percarbonate
              Citrate: Borohydride > Percarbonate > Metabisulfide

*Zinc*:    EDTA: Metabisulfite >> Borohydride > Percarbonate
              Citrate: Metabisulfite > Borohydride >> Percarbonate

*Overall*:    EDTA: Metabisulfite >> Borohydride ~ Percarbonate
              Citrate: Metabisulfite ~ Borohydride ~ Percarbonate

## SUMMARY AND CONCLUSIONS

Characterization of the worst-case and representative soils from Aberdeen Proving Ground's J-Field indicated that the soils were generally brownish in color, have a low cation exchange capacity (1.4-4.0 meq/100-g), are slightly alkaline in nature (soil pH in the range of 7.5 to 8.4), have a moderate volatile solids content (2.5% to 8.8%), and have a sandy loam soil texture. The particle size distribution characteristics of the soils determined from hydrometer tests are approximately 60% sand, 30% silt, and 10% clay.

Sequential extractions were performed on the "as-received" soils (worst case and representative) to determine the speciation of the metal forms. The technique speciates the heavy metal distribution into an easily extractable (exchangeable) form, carbonates, reducible oxides, organically-bound, and residual forms. The results indicated that most of the metals are in forms that are amenable to soil washing (i.e., exchangeable + carbonate + reducible oxides). The metals Cu, Pb, Zn, and Cr have greater than 70% of their distribution in forms amenable to soil washing techniques, while Cd, Mn, and Fe are somewhat less amenable to soil washing using chelant extraction. However, the concentrations of Cd and Mn are low in the contaminated soil.

From the batch chelant extraction studies, EDTA, citric acid, and NTA were all effective in removing copper, lead, and zinc from the J-Field soils. Due to NTA being a Class II carcinogen, it is not recommended for use in remediating contaminated soils. EDTA and citric acid appear to offer the greatest potential as chelating agents to use in soil washing the Aberdeen Proving Ground soils. The other chelating agents studied (gluconate, oxalate, Citranox, ammonium acetate, and phosphoric acid, along with pH-adjusted water) were generally ineffective in mobilizing the heavy metals from the soils. The chelant solution remove the heavy metals (Cd, Cu, Pb, Zn, Fe, Cr, As, and Hg) simultaneously.

Sonication was ineffective in enhancing the heavy metal extraction efficiencies associated with chelant extraction. Although sonication may have mobilized the heavy metals from the soil matrix, it is likely that the metals readsorbed back onto the soil matrix during the solid/liquid separation phase for analysis.

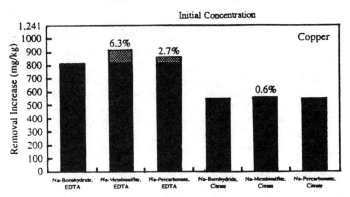

Figure 11a.   Copper Removal by REDOX Manipulation and Chelant Extraction.

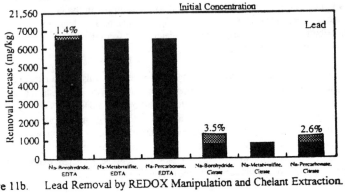

Figure 11b.   Lead Removal by REDOX Manipulation and Chelant Extraction.

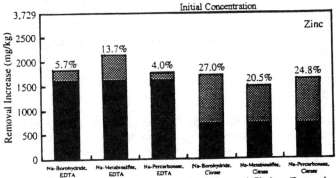

Figure 11c.   Zinc Removal by REDOX Manipulation and Chelant Extraction.

REDOX manipulation offers potential to enhance the removal of heavy metals associated with chelant extraction. Of the oxidizing and reducing agents studied, sodium borohydride, sodium metabisulfite, and sodium percarbonate enhanced removal of copper, lead, and zinc during screening experiments. Due to the ability to enhance the oxidation/reduction potential (ORP), sodium borohydride was selected for further study and was used in conjunction with the soil flooding experiments. Enhanced removal of copper, lead, and zinc was observed in these soil column flooding experiments for the EDTA extraction system.

Using a multiple-stage batch extraction, the soil was successfully treated passing both the TCLP and EPA Total Extractable Metal Limit. The final residual Pb concentration was about 300 mg/kg, with a corresponding TCLP of 1.5 mg/L. Removal of the exchangeable and carbonate fractions for Cu and Zn was achieved during the first extraction stage, whereas it required two extraction stages for the same fractions for Pb. Removal of Pb, Cu, and Zn present as exchangeable, carbonates, and reducible oxides occurred between the fourth- and fifth-stage extractions. The overall removal of copper, lead, and zinc from the multiple-stage washing were 98.9%, 98.9%, and 97.2%, respectively. The concentration and operating conditions for the soil washing extractions were not necessarily optimized. If the conditions had been optimized and using a more representative Pb concentration (~12,000 mg/kg), it is likely that the TCLP and residual heavy metal soil concentrations could be achieved within 2-3 extractions. The results indicate that the J-Field contaminated soils can be successfully treated using a soil washing technique.

## ACKNOWLEDGMENTS

This work is funded by the U.S. Department of Defense, U.S. Army, Directorate of Safety, Health, and Environment, through the Environmental Assessment Division (EAD) of Argonne National Laboratory. The authors express their appreciation to John D. Taylor and Laura R, Skubal in the Energy Systems Division at ANL for their contributions to the outstanding analytical support.

## REFERENCES CITED

American Public Health Association, Greenberg, A.E., L.S. Clesceri, and A.D. Eaton, Eds., (1992). *Standard Methods for the Examination of Water and Wastewater*, 18th Ed., American Public Health Association, Washington, D.C.

Peters, R.W., and L. Shem, (1992). "Use of Chelating Agents for Remediation of Heavy Metal Contaminated Soil", *ACS Sympos. Series 509 on Environmental Remediation: Removing Organic and Metal Ion Pollutants*, G.F. Vandegrift, D.T. Reed, and I.R. Tasker, Eds., *509*: 70-84, American Chemical Society, Washington, D.C.

# POLLUTION POTENTIAL OF MURGUL COPPER PLANT SOLID WASTES

RÜSTEM GÜL
   Atatürk University, Eng. Faculty
   25240, ERZURUM, TURKEY

AHMET YARTAŞI

AHMET EKMEKYAPAR

ABDURRAHMAN SERT

## INTRODUCTION

Heavy metals, which is of important pollutants, penetrate into food chains ant reach at the amounts of toxic level, and play a negative role on inhabitants (1-2).Various methods have been employed and different adsorbants have been used for the removal of heavy metals from water (3-4).

In the present study, the pollution of heavy metals to the environment by MBI solid wastes was investigated. For this reason, the amount of metal passing into water from the solid was investigated using column tests for different solid waste-to-water ratios at atmospheric conditions, and the capacity of the region soil for remaing these metals was studing. In addition, the amount of metal leaching into enviroment was observed employing an experimental set, at atmospheric conditions.

## MATERIAL AND METHOD

The solid waste used in the study was provided from MBI. This waste in a particle size of fine sand was used in experiments without any pretreatment. The soil was provided from the region of Erzurum-Turkey. The soil sample was sieved by 2 mm sieve  (6). The Leakage experiments were carried out with the columns of 50 and 100 mm in diameter and 500 mm in height at Laboratory  conditions and with the column of 300 mm in diameter and 200 mm in height at atmospheric conditions .

The experiments were performed in two stage. In the first stage, snow water was regularly passed through the solid wastes in the columns for six months. The amounts of metal passing into water and the capacity of the adsorbant soil for the removal of metals were determined once every 15 days. In the second stage, the amount of heavy metals leached into enviroment was determined for a period of one year at atmospheric conditions. The amounts of solid waste and of water for Laboratory conditions were given in Table I.

The chemical analysis of the waste and the soil were carried out dissolving in the acid solution of $1HNO_3 + 3HCl$ and evaporating it until dry and then again dissolving with HCl solutioon (6).The metals In to the water Leaked through the waste in the columns were directly analysed by atomic absorption spectrophotometer (AAS). pH measurement were also performed.

It was determined that the solid waste samples contain 21 % $SiO_2$, 1.80 % S and 73

% $Fe_2 O_3$ , and the soil samples 44 % $Si O_2$ , 25 % $Al_2 O_3$ and 18 % $Fe_2 O_3$ in weight. The heavy metal content of solid waste and soil is given in Table II.

The solid waste contains considerable amounts of heavy metals of Fe, Zn, Al, Cu and Mg and soil Al, Fe, Mg and pb in increasing order as seen in the Table II. Moreover, the amount of Al, Mg, Mn and Ni in the soil sample are much more than in the solid waste sample. As the results obtained was examined, it can be seen that MBI solid waste contains a lot of soluble metal ions. In the experiments for columns 1,2 and 3, the amount of solid waste was kept constant while the amount of water was different for each column.


TABLE I. THE AMOUNTS OF MATERIAL IN COLUMNS


|  | Column number | | | | | | | |
|---|---|---|---|---|---|---|---|---|
| Materials | 1 | 2 | 3 | 4 | 5 | 6 | 7 | 8 |
| Solid Waste (g) | 250 | 250 | 250 | 500 | 1000 | 2000 | ---- | --- |
| Soil (g) | --- | --- | --- | --- | --- | --- | 200 | 200 |
| Snow Water (ml) | 250 | 500 | 1000 | 1000 | 1000 | 1000 | 300 | --- |
| The Solution Contaning Heavy Metal (ml) | --- | --- | --- | --- | --- | --- | --- | 300 |
| Solid Waste/Water | 1 | 1/2 | 1/4 | 1/2 | 1 | 2 | --- | --- |

TABLE II . THE HEAVY METAL CONTENT OF SAMPLE (mg/kg)

| Metals | Solid Waste | Soil |
|--------|-------------|------|
| Fe | 40,610 | 25,390 |
| Zn | 28,730 | 473 |
| Mg | 3,260 | 4,489 |
| Ni | 86 | 137 |
| Mn | 121 | 1,290 |
| Co | 291 | 202 |
| Al | 13,120 | 40,740 |
| Cd | 57 | 32 |
| Cu | 5,780 | 136 |
| Cr | 843 | 296 |
| Pb | 1,504 | 534 |

## RESULT AND DISCUSSION

During the experiments, it was observed that the most soluble metal was Mg and then Zn; Ni, Al, Fe, Cr, Co, Cd, Mn, Cu and Pb in increasing order in the columns 3 (Solid/Water : 1/4).

In the columns 1 (Solid/Water:1/1) and 2 (Solid/Water: 1/2) the total amount of Zn is the most soluble metal instead of Mg and the order of the other metals is almost the same as in the column 3. It was determined that the highest amount of metal was leaked into water from the column 3 among three column, for the same amount of solid waste (250 g) and then columns 2 and 1, in increasing order. In the experiments for columns 4,5 and 6, the amount of water was kept constant. In these serious of experiments, in the 6th column, the most soluble metal was Zn and then Mg,Al, Ni, Fe, Co, Mn, Cd, Cu, Pb and Cr in increasing order. In the 4th and 5th columns, the dissolved amount of Fe is higher than that of Al while the dissolved of other metals main tained almost the same order. For these three columns, the amount of total metal passed into the water of 1000 ml is the highest for the 6th column and then the 5th and 4th columns follow. This can be explained by the fact that the amount of the solid waste in the 6th column was much more than those in the 4th and 5th columns. The solubility of the metals in the solid waste as the amount of solid waste was increased keeping the amount of water constant, was found to be lower than for the case as the amount of water was increased keeping the amount of solid constant.

When keeping constant the ratio of solid waste to-water, but increasing their amount fourfold, it can be seen that te amount of metal dissolving into water increased more than twice as the results of columns 2 and 4 was compared. This shows the necessity of increasing the amounts of both solid waste and water simultaneously to reach at an optimum solubility. The change in the metal amounts dissolving into water with the ratios of solid waste -to-water are shown in Figure 1. Some metals show an increase in their dissolved amounts while some show a decrease in their dissolved amounts for the analysis of the samples taken. It was determined that pH of solutions ranged in 6-8.

It can be seen that the amount of metals dissolved into water from the solid waste of 1000 g at atmospheric conditions are appraximately five time that from the same amount of the solid waste at laboratory conditions. The big difference of dissolved metal amount can be attributed to the appeciable temperature difference and longer contact time of water with the solid. From these results, it is understood that appreciable amounts of metal in the solid waste can be dissolved into enviroment at atmospheric conditions. As the time passes, the amounts of metals dissolving into water decreases since the amount of metal goes to a minimum with the contact of the solid with water for a certain period. The metal amounts dissolved into water at atmospheric and laboratory conditions are compared in Figure 2a. The adsorbtion capacity of the soil used changes between 22-100 percent. Cu and Cr are completly adsorbed by the soil while Fe, Zn, Mg, Ni, Mn, Co, Al and Cd are adsorbed in the ratios of 34,33,22,35,31,60,41 and 45 %, respectively. This different adsorbed amount can be explained by the structure and the inital amount of the metals, and the structure of the soil.The complete adsorption of Cu and Cr in the column can be attributed to the their lower initial amounts. The least adsorption of Mg can be due to its higher inital concentration since the excessive amount of the adsorption capacity of soil passes into water. Moreover, the fact that some metals form some compenents with some material in the soil can be a reason of different adsorbing level (7).

In addition to the adsorption mechanism, precipitation and ion exchange can also play role in the trapping of metal (8). Different behaviours of the metals in the soil can be a reason of different adsorbing capacity by the soil (9). In the adsorbtion of all metals except Mn, there is a decrease in the adsorbed amounts, especially after 9th analysis. The adsorbed amounts of the most adsorbed metal, Co, after Cr and Cu more goes to none after 10th measurement, Al after 7th measurement. The adsorbed amounts of Fe, Zn and Ni decreases appreciably after 7th , 9th and 6th  measurements, respectively. It Can be seen that during the experiments the most adsorbed metal is Zn (5.11 mg) and Mg (3.39 mg), Fe (1,66 mg), Al (1.331 mg), Ni (0,735 mg), Co (0,364 mg), Cu (0,147 mg), Cd (0,139 mg) and Cr (0,018 mg) follow it in increasing order. The initial amounts of metals are compared with the adsorbed a mount by soil and dissolved into waste in Figure 2b.

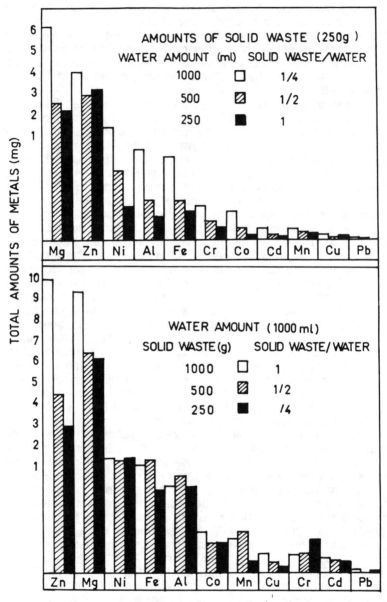

Figure 1. Amounts of metal dissolved into water depend on solid-to-water ratio.

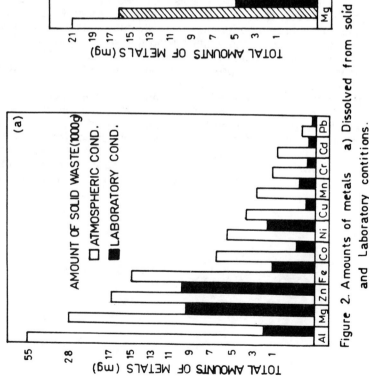

Figure 2. Amounts of metals    a) Dissolved from solid waste   b) Adsorbed on to soil at Atmospheric and Laboratory contitions.

649

# CONCLUSIONS

1)The dissolution level of MBI solid wastes are different as they are watered in laboratory and are subjected to the atmospheric conditions. It was determined that the metal amount dissolved into water from1000 g solid waste at otmospheric conditions is much more than at laboratory conditions. Generally, it was observed that, at atmospheric conditions, the dissolved metal decreases with time. This shows that metal in solid waste can reach at a minimum after a certain period.

2)In addition to the atmospheric conditions, it was observed tha the ratio of solid waste-to-water is also efficient for dissolution metals in the waste and that the dissolved amount of metals increased with increasing water.

3)It was determined that the region soil can adsorb 22-100 percent of the dissolved metals. Using these experimental results, to reduce metal content of a solid waste, to a certain level, the necessary soil amount and the dissoluble amount of metal can be predicted

4)It was determined that the concentration of Fe and Al in all solutions and those of the other metals in a few solution are higher than the standarts for drinking water, but under the limit of allowable discharge to the environment. It was detected that the concentration of Cd is over the allowable limit for drinking water and for discharge to the enviroment. It is known that the metals dissolved into water above the allowable limits can directly pollute the environment, and the metals under the limits can also be harmful indirectly due to the accumulation in the living things using the water mixing this waste water. For this reason the solid waste from MBI must be under control.

5)Under the light of the present experimental conditions, to avoid the negative effect of MBI solid wastes to the water resources, and to take the waste under control, this waste must not be left around water resources and to the field which is underground water not deep enough. The waste can be left to the enviroment after extracting its metal content by water in pools. The metals dissolved from can be adsorbed by soil. Keeping the metal contents of the polluted water under the allowable discharge limits by avoiding the contact of the waste by rain and snow can also reduce apperciably the enviromental pollution.

# REFERENCES

1.Salihoğlu,I.1978.Su Kirlenmesi ve Denetimi, DSİ, pp.1-5.

2.Srinivasan, B.A.M. 1988. Metal Pollution in the Estuarina Regions, Çevre '88 Environment, I, İzmir, Srinivasan,pp. 1-11.

3.Kompelman, M.H., J. Dillard. 1977. A study of the Adsorption of Ni (II) and Cu (II) by Clay Minerals, Clays-Clay Miner, 25 (6),457-462.

4.Bowers A.R., C.P. Huang. 1981. Activated Carbon Processes for the Treatment of Chromium (VI) Containing İndustrial Waste Waters, Water Sci.Technd., 13 (1) , 629-650.

5. Hansen, A., R.E. Weeks. 1983. Evulation of on-site soil for use as an Impoundment liner, Hazardous and Industrial Solid Waste Testing, Second Symposium, pp. 231-245.

6.Gündüz, T. 1984. Kantitatif Analiz Laboratuvar Kitabı, Ankara Üniversitesi Fen Fakültesi Yayınları, Ankara pp.224-225.

7.Ray, C., P.C. Chan. 1986. Heavy Metal in Landfiil Leachate, Int, J, Environ, Studies, 27, 225-237.

8.Doğan, L. 1981. Hidrojeolojide Su Kimyası, DSİ, pp.131-145

9.Bruggenwent, M.G.M.1979. Komphorst, Survey of Experimental İnformation on Cation Exchange in Soil Systems, Chapter 5 in Soil Chemistry. B.Physico-Chemical Models (ed.G.H.Bolt), Elsevier, Amsterdam, Oxfort, New york.

# CATEGORY IX:
# *Monitoring and Assessment*

# AOX IN SEWER SLIME -IDENTIFICATION OF INDUSTRIAL WASTEWATER DISCHARGES INTO PUBLIC SEWERS

ERNST ANTUSCH
Universität Karlsruhe
Institut für Siedlungswasserwirtschaft
D-76128 Karlsruhe, GERMANY

CHRISTIAN RIPP
Same address

HERMANN H. HAHN
Same address

## 1. INTRODUCTION

The successful protection of receiving waters is the most important aim of sewage and wastewater-treatment. Whereas the elimination of visible pollutants like suspended or non-settleable solids or oxygen-consuming compounds previously has been the main emphasis in wastewater-treatment, the current problems are the accumulation of micropollutants, like heavy-metals and the large number of hazardous organic substances. The discharge of such pollutants can disturb or stop physico- and biochemical treatment. A typical example is the heavy metals, which inhibit biochemical degradation if they exceed a certain concentration. Trace organic compounds have similar effects. Both groups of micropollutants complicate the utilization of sewage sludge for agricultural purposes. The environment suffers from the discharge of organic or inorganic anthropogenic pollutants by nutrient enrichment of the food-chain.

To solve these problems we should have a closer look at the sources of emission. In many countries, a lot of different regulations exist that follow the "polluter-pays-principle". Prevention is always better than treatment or deposition of sewage sludge. Therefore, the search for the polluters, regardless if they are aware of their pollutant discharge or not, is the main job.

Voluntary information from the polluter is not dependable, because company loyalty still has precedence over environmental consciousness in the mind of most employees. Local monitoring by regulary analyzing water samples is not practical.

655

## AOX - Adsorbable organic halogenated compounds

In this study, we present the measurements of halogenated organic compounds in sewer slimes. Many of the halogenated organic substances are of anthropogenic origin, and, although only some are hazardous, their emission into the natural environment should be avoided. Therefore, the summary parameter AOX has become one of the most important criteria for regulating industrial wastewater discharge in German water quality legislation.

The discharge limits have a preventative character, as there is no quantitative relation between the concentration of AOX and its toxicity [7,11]. If an exceeding value is found in the sewer system, one should look for single components to indicate or to exclude toxic substances. We used this method to determine total organic halides as Chloride by active carbon adsorption and microcoulometric-titration detection, as regulated in DIN 38 404, part 14 (DIN: Deutsches Institut für Normung, comparable to ASTM regulation) [12]. All samples had been run in duplicate and the reliable limit of sensitivity under these conditions was 5 µg/L.

In the following sections, the "sewer-slime-method" is explained as a useful tool for localization and identification of indirect discharges, developed and successfully tested at the University of Karlsruhe, Institute of Aquatic Environmental Engineering.

## 2. ACCUMULATION OF HAZARDOUS SUBSTANCES IN BIOFILMS

### 2.1. SEWER SLIME, THE BIOFILM WITH "MEMORY-EFFECT"

Sewer slime is the biofilm which results from the fat depositions and microorganisms of wastewater. It is mostly composed of organic matter, but it also contains an inorganic sediment component as well.

Because of bio- and physicochemical sorption-processes, heavy metals accumulate in extracellular polymeric substances (so called EPS), like anionic saccharides, proteins and lipoproteins, which are built up by the microorganisms of the sewer slime [3,4].

The accumulation-factors for heavy metals in sewer slime, shown in Table 1, have been investigated in our Institute by *Gutekunst and Hahn* [6].

**Table 1.** Accumulation factors of some heavy metals in sewer slime [6].

|          | Cd     | Cu     | Ni    | Zn     |
|----------|--------|--------|-------|--------|
| $K_P{}^*$ | 15,800 | 25,600 | 1,250 | 10,000 |

$^*$ $K_P = \dfrac{Q}{[me]}$    [l/kg dry weight]

with     Q     = load in sewer slime [mg/kg dryweight]
and     [me]     = concentration of heavy metal in the solution [mg/L]

656

## 2.2 SORPTION OF ORGANIC COMPOUNDS ONTO BIOFILMS

A high bioaccumulation of pesticides and herbicides was found in the microorganisms of activated sludge by *Paris* and *Urey* and has been well known for nearly 20 years [8,10]. The accumulation factors of other organic substances have been investigated in activated sludge by *Freitag et al.*(Table 2) [5].

Table 2. Accumulation factors of organic substances in activated sludge [5].

Bioaccumulation factor: $BF_n = \dfrac{\text{concentration of chemicals in sludge } [\mu g / g]}{\text{final conc. of chemicals in water } [\mu g / g]}$

n = 5 days

| | | | |
|---|---|---|---|
| Dibenz(a,h)antracene | 42,800 | 1,2,4-Trichlorobenzene | 1,400 |
| Hexachlorobenzene | 35,000 | 3-Cresol | 1,100 |
| 2,5,4'-Trichlorbiphenyl | 32,000 | Pentachlorophenol | 1,100 |
| 2,4,6,2',4'-Pentachlorobiphenyl | 27,800 | ß-Hexachlorocyclo hexane | 1,200 |
| Chlorhexidine | 26,700 | Naphthalene | 1,000 |
| Benz(a)anthracene | 24,400 | Trichloroethylene | 990 |
| Perylene | 22,900 | Phenanthrene | 930 |
| Aldrin | 18,000 | 2,4,6-Trichloroaniline | 870 |
| Dieldrin | 17,600 | 1,4-Dichlorobenzene | 560 |
| Pentachlorobenzene | 14,300 | Aniline | 500 |
| DDT | 14,000 | Carbon tetrachloride | 480 |
| Benzo(a)pyrene | 10,000 | 2,4-Dichlorophenol | 340 |
| 2,4'-Dichlorobiphenyl | 9,800 | 2,4-Dichloro-nitrobenzene | 310 |
| Anthracene | 6,700 | Coumaphos | 290 |
| 2,4,6,2'-Tetrachlorobiphenyl | 6,500 | 4-Chloroaniline | 280 |
| 2,2'-Dichlorobiphenyl | 6,300 | Maneb | 250 |
| Quintocene | 4,500 | 4-Chlorobenzoic acid Dodecylbenzene | 170 |
| 3,3'-Dichlorobenzidine | 3,100 | Chlorferon | 170 |
| Phtalic acid bis-(2-ethylhexyl)ester | 3,000 | Zineb | 130 |
| Hexachlorocyclo-pentadiene | 2,400 | 2,6-Dichloro-benzonitrile | 90 |
| Phenol | 2,200 | 2,4,6-Trichlorophenol | 60 |
| Toluene | 1,900 | Atrazine | 40 |
| Chlorobenzene | 1,700 | 2,6-Dichlorbenzamide | 30 |
| Benzene | 1,700 | 4-Nitrophenol | 30 |
| Bromebenzene | 1,500 | 2,4-Dichlorobenz. acid | 10 |

These results lead us to the conclusion that hazardous <u>organic</u> substances accumulate as well in sewer slime.

After all, the accumulation in the sewer slime has a similar effect as the solid phase extraction and leads in the same way to better analytical detection. Therefore the sewer slime "remembers" every pollutant for about 2 months until it gets flushed away by higher flow velocities. This "memory effect" has another big advantage compared to a random-test of water samples: the evidence of pollution discharges are stored in the sewer slime.

# 3. EXPERIMENTAL INVESTIGATIONS AND RESULTS

## 3.1 LABORATORY EXPERIMENTS

The experiments of sorption-kinetics were made in "jar-tests" in our laboratory to determine and to compare the maximum load of organic substances with different polarities in sewer slime. To investigate the adsorption-velocity, equal amounts of sewer slime were stirred continously in different flasks with solutions of chlorophenols and phenantrene. The adsorption process was interrupted at different times. The suspensions were filtered, and the solid and the liquid phase were analyzed as well. Chlorophenols were measured as AOX [13], phenantrene was determined by HPLC combined with fluorenscence detection. The desorption was investigated by exchanging the solution after adsorption with deionized water and then continous stirring for several hours. At certain times we took samples of the water phase for analyzing the resulting concentration.

In the following graphs, we present the results of our sorption experiments for two different classes of organic compounds, phenantrene as a congener of PAH and two chlorophenols, dichlorophenol (DCP) and trichlorophenol (TCP), in sewer slime with comparable fat and organic carbon content.

In Figure 1, it is evident that the kinetics of adsorption of Phenantrene in sewer slime is nearly complete after several minutes, but the desorption needs more than 1 hour.

**Figure 1.**    Adsorption and desorption kinetics of phenanthrene onto sewer slime

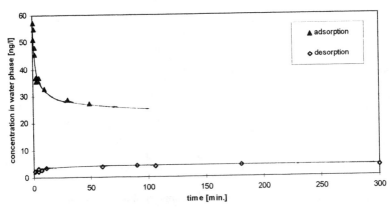

Furthermore it can be observed, that more than 50 % of the original concentration of phenantrene was adsorbed (60 ng/L down to 27 ng/L), while only 4.5 ng/L could be found in solution after 5 hours of desorption. That underlines the high accumulation potential of the sewer slime for the nonpolar phenantrene.

The desorption of the investigated chlorophenols, shown in Figure 2 with the example of DCP, occurs much faster than the desorption of PAH. After 15 minutes, 70 - 80 % of Dichlorophenol has already desorbed.

**Figure 2.** Adsorption and desorption kinetics of Dichlorophenol onto sewer slime (13 % fat) at pH 7.

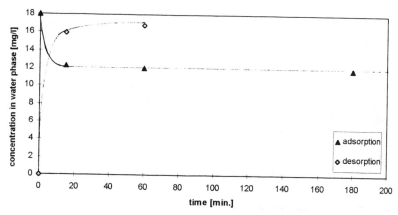

In Figure 3, the sorption of Di- and Trichlorophenol in sewer slime with 13 % fat and 1.3 % fat is shown to be dependent on the pH-value of the solution. The initial concentration of DCP and TCP were 18 µg/ml and 14.8 µg/ml, respectively. At the ordinate, the concentration in sewer slime is in equilibrium [µg/mg dry weight].

**Figure 3.** pH dependence of the sorption equilibrium of Di- and Trichlorophenol onto different sewer slimes; AOX initial concentration 8µg/ml.

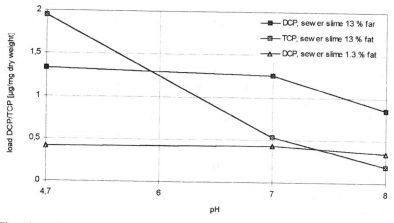

The adsorption of TCP at the pH-value of 4.7 is much stronger than for DCP. With higher pH-values, the sorption attitudes of TCP and DCP change.

Here, DCP shows a better sorption to sewer slime than TCP. The pH-influence on the sorption of DCP onto sewer slime with 1.3 % fat is very small. Overall, lower pH-values promote the sorption of these chlorophenols. Higher pH-values have a negative influence on the adsorption, because of the decreased adsorption of the polar phenolate-ion. The pH-dependence of TCP is much stronger than that of DCP.

Furthermore, our experiments showed that the fat content of the sewer slime has a strong influence on the chlorophenol sorption. The rate of sorption can be estimated by the octanol/water coefficients of the organic compounds and the content of fat and organic carbon in the sewer slime.

In contrast to the chlorophenols, many hazardous halogenated organic compounds are non-polar (eg. PCB, DDT, PCDD, HCB), and their behavior in sewage should be comparable to PAH, i.e. they should show a high adsorption tendency to the sewer slime. Therefore our field experiments focused on the summary parameter AOX.


## 3.2 FIELD EXPERIMENTS

**Developing a pragmatic method for identification of Indirect discharges of industrial wastewater into public sewers**

The first step of every localization of indirect discharges is the sewage treatment plant. If high values of AOX in the sewage sludge are found (In Germany: exceeding 500 mg/kg dryweight), it is necessary to locate the source of emission. As this can be extremely difficult, the sewer slime method is a very useful tool.

The next step is to take sewer slime samples in the single main arms of the sewage system entering the sewage treatment plant. Comparing analyses of those samples leads to the first piece of information about the origin of the contamination: the highest load indicates the direction from which the discharge is expected.

To locate the exact source of emission of AOX, the third step is the evaluation of city maps showing the sewerage system and, if available, cadastres for indirect discharge. In this way, you can focus on the industrial areas and sites, which are the most probable sources of AOX-emission. A further criteria is knowledge about the typical wastewater components of certain industries.

A subsequent sample taking and analysis at important outlets of the evaluated areas in the sewer system leads to a more detailed view of the paths of the contamination. Finally, the problematic discharge can be found when analyzing the sewer slime samples along the single streets of the suspected area.
Figure 5 shows the map of an industrial zone of approximately 0.25 km$^2$, which was determined to be the main source of the AOX-problem of a city.

660

**Figure 5.** Localization of AOX-Discharge

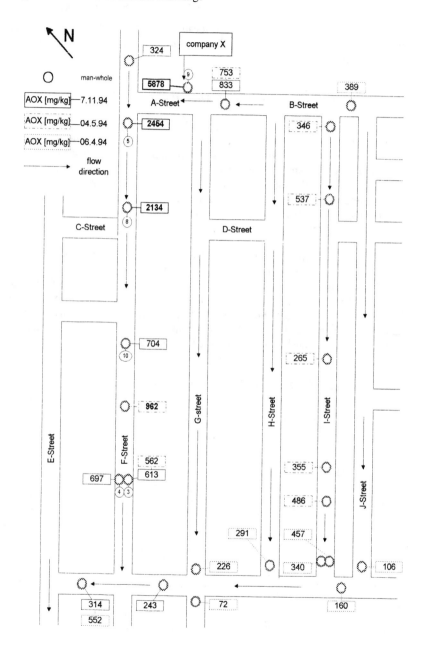

Samples were taken several times. We found significantly increasing AOX-loads in the sewer slime along "F-street" from 962 up to 5,878 mg/kg. This high AOX-value established the discharge at point 9.

On this analytical base, you are able to induce further administrative activities.

This method was successfully tested to localize and identify discharges of other non polar-organic substances as dioxines performed in exactly the same way by *Rieger and Ballschmiter* [9].

## CONCLUSIONS AND RECOMMENDATIONS

In this paper, we presented a method, which is suitable to identify and localize indirect discharges of AOX in municipal sewer systems. Investigation of the sorption behavior of two different classes of substances showed that the sorption of chlorophenols is extremely dependent on the pH-value of the water phase.

Parameters like AOX and non-polar substances like PCB and dioxines were accumulated and can be analyzed in sewer slime. Therefore, we can use the sewer slime method not only for heavy metals, but for those organic pollutants as well to identify indirect discharges.

Futhermore, we found other substances in our current gas chromatographic investigations with mass spectrometric detection, which may become important for ecological assessment in the future.

Additonal substances that we found accumulating in sewer slime are for example: 4-nonylphenols, degradation products of surfactants with negative hormonal effects on aquatical animals [2], phtalates (DEHP), halogenated antimicrobial compounds (Triclosane, Chlorophene) and the synthetic fragances musk ketone and musk xylene (the last two substances not in the extract of Figure 6) [1].

**Figure 6.** Capillary GC/MS-scan of a cleaned sewer slime extract [1].

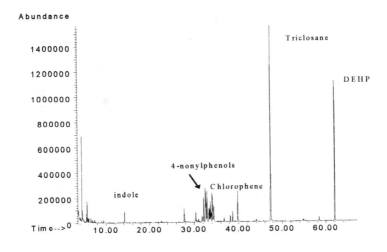

## Aknowledgements

The authors thank the Bundesministerium für Bildung und Forschung and the Oswald-Schulze-Stiftung for the financial support for research and this presentation.

## LITERATURE

[1] **Antusch, E., Sauer, J., Ripp, C.**: „Untersuchungen zur Identifizierung organischer Schadstoffe in Sielhäuten", technischer Bericht, Institut für Siedlungswasserwirtschaft, Universität Karlsruhe (1995)

[2] **Cadbury, D.**: Attacke auf die Manneskraft; Beitrag in der Sendereihe: "Abenteuer Wissenschaft", BBC Horizon and Süddeutscher Rundfunk, (Deutsche Bearbeitung: C. Wassmann) vom 6.3.1994

[3] **Flemming, H.-C.**: Biofilme und Wassertechnologie, Teil I: Entstehung Aufbau, Zusammensetzung und Eigenschaften von Biofilmen, in: gwf Wasser Abwasser, 132, 197-207 (1991)

[4] **Flemming, H.-C. und Ruck, W.**: Lokalisierung von Schadstoffeinleitern vom Kanalnetz aus, in: Wasser-Kalender 1990, 115-133, Erich-Schmidt-Verlag Berlin (1989)

[5] **Freitag, D. et al.**: Environmental Hazard Profile of Organic Chemicals, in: Chemosphere 14, Nr. 10, 1589-1616 Pergamon Press Ltd. Great Britain (1985)

[6] **Gutekunst, B., H. H. Hahn:** Sielhautuntersuchungen zur Einkreisung schwermetallhaltiger Einleitungen, in: ISWW Schriftenreihe, Bd. 49, Universität Karlsruhe (1988)

[7] **Goverment of Canada:** Canadian Environmental Protection Art, Effluents from Pulp Mills Using Bleaching in: Priority Substances List Acessment Report No. 2 (1991)

[8] **Paris, D. F., Lewis, D. L., Barnett, J. T.**: Bioconcentration of Toxaphene by microorganisms, in: Bull. Environ. Contam. Toxicol. 17, 564-572 (1977)

[9] **Rieger, R. u. Ballschmiter, K.**: Lokalisierung von Polychlordibenzo-p-dioxinen und Dibenzofuranen in das Abwassersystem der Stadt Ulm/Neu Ulm, Posterpräsentation der Abt. Analytische Chemie und Umweltchemie der Universität Ulm zur ANAKON '93, Poster T14

[10] **Urey, J. C., Kircher, J. C., Boylan, J. M.**: Bioconcentration of four pure PCB isomers by Chlorella pyrenoidosa, in: Bull. Environ. Contam. Toxicol. 16, 81-85 (1976)

[11] **Weisbrodt, W:** AOX - Ein Abbild für gefährliche Stoffe ? in: ATV-Berichte Nr. 44, ATV Bundestagung (1994)

[12] **DEV S 18, DEV H 14:** Deutsche Einheitsverfahren zur Wasser-, Abwasser- und Schlammuntersuchung

# FIELD ASSESSMENT SCREENING TEAM (FAST) TECHNOLOGY PROCESS AND ECONOMICS

MARK D. NICKELSON, CHIEF SCIENTIST
DELMAR D. LONG, Ph.D.
Hazardous Waste Remedial Actions Program
Environmental Management and Enrichment Facilities
Martin Marietta Energy Systems, Inc.*
Oak Ridge, Tennessee 37831-7606

The Field Assessment Screening Team (FAST) is a concept of site characterization innovative technologies that have been integrated (1) to expedite the characterization process for hazardous and/or radioactive waste sites, (2) to optimize characterization quality, (3) to reduce field time, and (4) to effect characterization cost reductions in the 60 to 80% range. The FAST system has the following advantages:
- reduces environmental site characterization field time,
- reduces site characterization sampling costs,
- reduces site characterization analytical costs,
- reduces waste generation during field activities,
- improves field worker safety during environmental site characterizations, and
- allows for rapid technology transfer.

## SUMMARY DESCRIPTION

The FAST technology is an integrated system of state-of-the-art components that collectively produce a process that ensures a lower cost, high-quality,

*This manuscript is authored by a contractor of the U.S. Government under contract DE-AC05-84OR21400 with the U.S. Department of Energy. Accordingly, the U.S. Government retains a paid-up, nonexclusive, irrevocable, worldwide license to publish or reproduce the published form of this contribution, prepare derivative works, distribute copies to the public, and perform publicly and display publicly, or allow others to do so, for U.S. Government purposes.

real-time decision-making environmental field screening. The components of the integrated system are as follows:

- Intrusive sampling based on (but not limited to) "push" technology for surface, subsurface, and groundwater media sampling.
- A field mobile laboratory equipped to complement expected site contaminants.
- Computer-assisted design/geographic information system (CAD/GIS) data management with interactive three-dimensional graphics presentation capability.
- A global positioning system to determine sample coordinates.
- A telecommunication linkup for data transmittal and receiving.

Using the integrated components, the technology is designed to determine the horizontal and vertical extent of site soil and/or groundwater contamination with one mobilization of the field investigative team. The system allows the user to make informed field decisions concerning site investigation plans on a real-time basis, thus assuring minimum field time, minimum cost, and optimum data collection.

# DETAILED DESCRIPTION: PROCESS AND ECONOMICS

Consider the following hypothetical example. While the comparative economics that will be presented are for a hypothetical site characterization, the basis for the comparison is taken from actual case histories. The task is to determine the horizontal and vertical extent and the chemical characterization of a groundwater plume. The interface is 35 ft below ground surface, the aquifer depth is 80 ft below ground surface, and groundwater flow direction as well as hydraulic gradient are known.

The conventional method uses five separate functions to determine the answers. First, the intrusive sampling is generally accomplished using some type of flighted auger that drills a hole into the earth for well placement. Using the standard "grid" approach, one determines a need for 28 wells placed at an estimated 3 separate depths. Once in place, a well is developed, a specified number of well volumes of water are removed, and the well is sampled.

The second function is analysis. The samples are placed in shipping containers and shipped to a fixed-base laboratory for analysis. The constituents of concern are volatile organic compounds, and the analytical method is EPA 8240. This procedure, which requires the sample be preserved in 0.008% $Na_2S_2O^3$ and kept at 4°C, has a holding time of 14 days. The cost per sample is $185. Under normal turnaround times, ample results will become available 4 to 6 weeks after the last sample is submitted.

The third function is surveying the wells to determine the vertical and horizontal position of the well head. This information is vital if a visualization of the projected contamination is to be made. Further, this information will be coordinated with the analytical information generated by the field laboratory.

665

The fourth function is visually plotting the test results based on a static condition. This function can only be conducted after receipt of analytical and survey data.

The fifth function, the projection of the next phase of site characterization (Phase II), is accomplished by interpreting available data and implementing a model projecting plume migration (i.e., adjusting the conceptual model of the site) and usually results in a subsequent field investigation report. In this instance, the static condition seen in the fourth function is projected based on a homogenous lithology to determine risk and project status.

So much for background. Consider next the key economics for a single-phase site characterization. This comparison will be made on the basis of per diem costs, system capacities, and a subsequent overlay by the FAST iteration process.

The per diem cost for a conventional site characterization system is estimated at $4000, which includes the cost of a drill rig, labor for its operation, sampling costs, management costs, and cost from a fixed-base laboratory for sample analysis. For this discussion, we will assume this system is capable of placing two wells per day and generating four samples per day for fixed-base laboratory analysis for a full suite of volatiles. In the example cited, at this capacity, the conventional site characterization system would have a single-phase field effort of 14 days (exclusive of mobilization and demobilization). Based on the per diem cost of $4000, this investigation would cost $56,000. Obviously there are other associated costs, but for this comparison this will suffice.

The per diem cost for the FAST system site characterization is also estimated at $4000, which should not be surprising because the fundamentals of both systems are similar. Field experience, however, tells us that FAST is capable of placing five wells per day and generating ten samples per day for field laboratory analysis. Again, using the cited example, at this capacity the FAST process has a single-phase field effort of 6 days. Based on a per diem cost of $4000, this investigation would cost $24,000—a cost savings of $32,000 over the conventional site characterization system—this is a 57% cost savings. But, we are not through! The FAST system has a closed loop iterative system that allows the field manager to **KNOW** the results of previous well placement before beginning subsequent efforts. This iterative process significantly increases the efficiency of the field effort such that the number of wells required for its characterization is significantly reduced. Again, in the example used for this discussion, the number of wells was reduced from 28 to 15—a 46% reduction in the number of wells. Because only 15 wells were then needed to characterize the site, the field time (at five wells per day) dropped from 6 to 3 days—a total characterization cost of $12,000 as compared to the conventional system at $56,000. This produces a net savings of $44,000—or 78.6%.

# EQUIPMENT DESCRIPTION

The following is a description of the four basic components used in the FAST technology.

1. **The Intrusive Sampling System.** The Geoprobe with the following basic characteristics is the unit of choice.

- The unit is a self-contained, motorized, hydraulically powered probe unit operated from a hydraulic system driven from the vehicle motor or an auxiliary engine.
- The probe unit folds for transport and is capable of being set up in seconds. It uses static force (weight of vehicle) and percussion to advance the probing tool.
- The unit drives small-diameter (1- to 1.6-in.-OD) probing tools to depth.

2. **The Field Laboratory.** The contents of this unit are determined by the analytical requirement at the investigation site. Typically, this unit contains a gas chromatograph (sometimes coupled with a mass spectrograph), a metals-detecting system such as an X-ray fluorescence system or inductively coupled plasma, and other analytical systems as needed. Each system downloads to an on-board computer that records, stores, and transmits the developed analytical information.

3. **The Survey System.** The unit of choice is a Ground Positioning System manufactured by Ashtech. This system electronically surveys the surface position of the sample point on the X, Y, and Z axes. The survey information is also downloaded into the field laboratory computer.

4. **The Data Management System.** The stored analytical and survey information is downloaded, either by cellular phone or hard-wire telephone, to a CAD/GIS three-dimensional imaging system on an as-needed basis. This system visually presents the downloaded data and subsequently projects, if requested, the next probable sample location. This information is forwarded to the field laboratory and used as a guide in placing the ensuing sample locations.

# FAST AND THE DATA QUALITY OBJECTIVE PROCESS

The data quality objective (DQO) process acknowledges that site managers must specify acceptable data quality goals by establishing acceptable limits on decision errors. By definition, decision errors occur when variability or bias in data mislead the decision maker into choosing an incorrect course of action. By using the DQO process, the site manager provides the criteria for determining when data are sufficient for site decisions. This provides a stopping rule—a way for site managers to determine when they have collected enough data.

The FAST system is uniquely applicable to the new DQO process. By having real-time data, the site manager maintains full knowledge of site conditions and can invoke the stopping rule when appropriate. Further, by having a full vision of the site characteristics, variability or bias data are minimized, thus ensuring the decision maker optimum information for real-time decisions.

## FAST AND PROGRAMMATIC ISSUES

The FAST system offers benefits to a broad range of programmatic applications.

- **Prioritization of Funding**. FAST allows our customer a relatively low-cost, quatifiable determination of environmental problems, thus allowing optimization of funding individual program or project elements. Past efforts at prioritization using only qualitative data do not ensure that funds are applied to the greatest areas of need.
- **Base Realignment and Closure**. The system allows for a rapid determination of the presence or absence of environmental damage, thus providing a means of expeditiously closing out suspected sites where no contamination can be found.
- **Remediation Closure**. At the completion of a cleanup, it is necessary to verify or demonstrate that the implemented remedial action has met the cleanup levels. The FAST system allows for a rapid, low-cost, high-quality determination of the remedial effort.

The system allows environmental site owners to achieve lower-cost alternatives without sacrificing quality. Transfer of this technology to the private sector will have a significant economic impact on the public's long-term environmental costs. The Environmental Protection Agency needs a system that provides (1) a maximum number of data points at a minimum level of cost and effort without data quality reduction and (2) optimum information for environmental decisions.

## CONCLUSIONS

Using known technology, the worldwide environmental market exceeds one trillion dollars. Of that amount, 40% is attributable to site characterization, the balance to remediation costs. From an historic point, site characterization represents 70% of the timeline for site cleanup. Further, the efficiency of the conventional site characterization process is about 30%. The combination of these figures should cause considerable concern within the ranks of the taxpayer, the potentially responsible party, or any property owner.

The FAST system directly attacks the 40% for site characterization by significantly reducing both cost and timeline without sacrificing characterization quality. The FAST system was born in the field and has been proven in the field. FAST is ready!

# PROGRAM DEVELOPMENT TO IDENTIFY AND CHARACTERIZE POTENTIAL EMERGENCY SITUATIONS AT A PETROLEUM REFINERY AND DETERMINATION OF INDUSTRIAL HYGIENE EMERGENCY RESPONSES

J. J. ORANSKY, S. N. DELP, AND E. A. DEPPEN
Spotts, Stevens and McCoy, Inc.
Reading, Pennsylvania

D. BARRETT
BP Oil Company
Marcus Hook, Pennsylvania

## INTRODUCTION

In the modern world the field of industrial hygiene continues to grow beyond the traditional definition of the profession. Industrial hygienists are constantly demanded to expand their horizons and become involved in the health risk assessment, emergency preparedness, and incident responder roles as required by the Process Safety Management (29 CFR 1910.119) and Hazardous Waste Operation and Emergency Response (29 CFR 1910.120) standards of the Occupational Safety and Health Administration (OSHA). This case study documents the problem solving approach used to identify potential exposures and evaluate industrial hygiene preparedness to handle emergencies due to fire or major spill at a complex multi-process petroleum refinery.

In the recent past an environmental engineer and industrial hygiene consulting firm was retained by a mature, multi-process petroleum refinery to assist in the program development to identify and characterize potential emergency situations due to a fire, major release, or spill. This study would assist the refinery in compliance with the process safety and emergency response standards and to protect refinery operations and fire fighting personnel by minimizing potential exposures and risk when responding to such a major incident.

## KEY WORDS

Emergency Preparedness, Risk Assessment and Communication

## PROBLEM SOLVING APPROACH

In order to develop a practical Industrial Hygiene Emergency Response Program, the problem solving approach addressed the following elements:

A.   Identification and characterization of potential emergency situations.

B.   Documentation of air contaminants, decomposition products, and other stressors from each potential emergency situation.

C.   Recommendation of various types of equipment and collection media needed for conducting instantaneous, short term, and full shift monitoring.

D.   Documentation of detailed sampling and calibration procedures to be used during emergency events.

E.   Specification development for qualified industrial hygiene technicians who will respond to refinery emergency.

A chronological progression of this problem solving technique with conditions unique to this case study follows:

## IDENTIFICATION AND CHARACTERIZATION OF POTENTIAL EMERGENCY SITUATIONS

In order to complete this task, a thorough evaluation of each refinery process, waste treatment operation, and transportation and storage area was completed. Discussions were conducted with representatives from refinery operations, process engineering, industrial hygiene, and safety. In addition, all available process schematics, flow diagrams, and material safety data sheets were thoroughly reviewed. This process was completed for each of the following refinery operations:

- Platformer
- Isocracker
- Naphtha Unit
- Vacuum Gas Oil Desulfurization Unit
- Sulfur Recovery Unit
- Sulfur Recovery Tail Gas Unit
- Boiler House
- Amine Unit
- Crude Stills
- Vacuum Stills
- Kerotreater Unit
- Diesel Treater

670

- Low Line Compressor
- Dry Gas Area
- Fluid Catalytic Cracker
- Alkylation Unit
- Wastewater Treatment Plant
- Oil Movement and Storage
- Railroad Area

For the purpose of this study it was decided to address all possible physical and chemical stressors as well as any potential safety hazards that refinery operations and emergency response personnel may be exposed to in the event of a refinery fire or major spill event.

## DOCUMENTATION OF AIR CONTAMINANTS AND OTHER STRESSORS FROM EACH POTENTIAL EMERGENCY SITUATION

Upon review of all pertinent process information and dialogue with refinery operations, engineering, and safety and health personnel, a detailed list of potential exposures was completed for each operating area. Specific stressors were identified from each of the following sources:

- Petroleum Feed Stock
- Feed Stock Impurity
- Process Stream
- Process Reagent
- Process Additive
- Catalyst
- Catalyst Substrate
- Catalyst Deposit
- Catalyst Regenerating Agent
- Combustion Product
- Incomplete Combustion Product
- Decomposition Product
- Materials from Process Instrumentation
- Materials from Switches and Dispensers
- Structural Material
- Insulation
- Paint

A thorough evaluation of each processing area yielded the following specific stressors:

- Physical Stressors
    - Heat                                   Noise
    - Radioactive Cesium 137

- Organic Chemicals

| | |
|---|---|
| 1,3,-Butadiene | Flammable Hydrocarbons |
| Diethanolamine | Hydrazine |
| Diisopropylamine | Methanol |
| Ethylene Dichloride | Propane |
| Ethylene Oxide | Propanol |

- Aromatics

| | |
|---|---|
| Benzene | Toluene |
| High Boiling Aromatics | Toluene Diisocyanate |
| Phenol | Xylene |
| Polynuclear Aromatics | |

- Poisons

| | |
|---|---|
| Carbon Monoxide | Nitric Oxide |
| Hydrogen Cyanide | Nitrogen Dioxide |
| Hydrogen Sulfide | Phosgene |
| Nickel Carbonyl | |

- Inorganic Gases and Vapors

| | |
|---|---|
| Ammonia | Hydrogen |
| Bromine | Sulfur Dioxide |
| Chlorine | |

- Acids

| | |
|---|---|
| Hydrogen Chloride | Phosphoric Acid |
| Hydrogen Fluoride | Sulfuric Acid |

- Alkaline Materials

| | |
|---|---|
| Calcium Hydroxide | Sodium Hydroxide |
| Potassium Hydroxide | |

- Particulate

| | |
|---|---|
| Alumina | Crystalline Silica or Quartz |
| Asbestos | Magnesium Oxide |
| Coke dust | |

- Metals

| | |
|---|---|
| Chromium | Molybdenum |
| Cobalt | Nickel |
| Iron | Platinum |
| Lead | Tin |
| Magnesium | Tungsten |
| Mercury | Vanadium |

# RECOMMENDATION OF INSTANTANEOUS AND FULL SHIFT SAMPLING EQUIPMENT

Knowing the physical and chemical stressors that may present a problem during a refinery fire, spill, or major release, a detailed evaluation was performed as to the monitoring equipment readily available for use. Research was performed for each stressor under evaluation. Instantaneous exposures could be accurately measured using these types of instruments:

- Colorimetric Detector Tube
- Flammable Gas Meter
- Gamma Ray Survey Meter
- Heat Stress Monitor (WIBGET)
- Hydrogen Fluoride Meter
- Mercury Monitor (Jerome)
- Real Time Aerosol Monitor
- Sound Pressure Level Meter
- Toxic Gas Monitor

Similarly, full shift and OSHA short-term exposure monitoring may be performed in accordance with standard NIOSH recognized procedures using personal sampling pumps and the appropriate collection media:

- Charcoal Tube
- Coated Charcoal Tube
- Hopcalite Tube
- Hydrogen Sulfide Monitor
- Midget Impinger with Appropriate Collection Liquid
- Mixed Cellulose Ester Filter
- Noise Dosimeter
- Orbo 53 Tube
- Polyvinyl Chloride Filter
- Silica Gel Tube
- Teflon Filter
- Thermoluminescent Dosimeter
- Treated Glass Fiber Filter
- Treated Silica Gel Tube
- Triethanol Impregnated Molecular Sieve Tube
- XAD-2 Tube
- XAD-7 Tube

# DOCUMENTATION OF DETAILED SAMPLING AND ANALYTICAL PROCEDURES TO BE USED DURING EMERGENCY EVENTS

Expanding upon the compiled information, rigorous sampling and calibration procedures were prepared for each chemical and physical stressor. Emphasis was placed on the required limits of detection, equipment reliability, ease of operation, generally recognized measurement techniques including NIOSH procedures, OSHA field instruction procedures, and equipment manufacturers' recommended practices. This resulted in an industrial hygiene sampling and analytical summary for each constituent that included the following information:

- Types of Direct Reading Apparatus
- Manufacturers
- Range of Measurement
- Safety and Accuracy
- Potential Interferences
- Calibration
- Miscellaneous Supplies Required
- Collection Media
- Air Flow Rates and Sample Volumes
- Collection Apparatus (Pump)
- Documented Analytical Method (NIOSH and/or OSHA)

## SPECIFICATION DEVELOPMENT FOR QUALIFIED INDUSTRIAL HYGIENE TECHNICIANS WHO WILL RESPOND TO REFINERY EMERGENCIES

Working as a team with refinery health and safety personnel, a brief specification was prepared for any industrial hygiene technician who will respond to refinery emergencies such as a fire or major spill. In summary, the industrial hygiene technician must have a 4 year college degree in a pure or applied science and at least 2 years of practical hands-on field experience. These individuals will work under the direct supervision of an experienced industrial hygienist and be required to keep current in the profession and work towards becoming certified. Additional training in OSHA Hazardous Waste (40 hour course) and EPA asbestos topics are also necessary.

## RESULTS AND DISCUSSION

Incorporating all of the technical principles discussed in the previous section resulted in the preparation of an industrial hygiene response manual for refinery emergency events. This manual contains a section on process descriptions for each operation which includes a general narrative of the unit's purpose, process streams, operating conditions, and resulting products. Each process area also includes a Table of Stressors and specific gravity, vapor pressure, boiling point, source, relative hazard rating, hazard form (vapor, gas, particulate, or physical hazard), OSHA permissible exposure limit, ceiling, short-term, or action levels, refinery reference number, and types of emergency event (fire, spill, or release) for each stressor. Table 1 is a typical process description from the manual for the sulfur recovery unit.

As previously discussed, the manual also includes a section on industrial hygiene sampling procedures for each stressor of interest. Table 2 is a typical entry and documents direct reading measurement techniques for evaluating hydrogen fluoride which may be released from the alkylation unit. Similarly, Table 3 documents full shift time weighted average and short term exposure measurement techniques for hydrogen fluoride using adsorbent tubes and approved NIOSH and OSHA protocols.

Finally, the manual has been prepared and formatted in such a way so it is "user friendly" and a quick reference source for efficient use by refinery industrial hygienists for emergency response in the event of a fire , spill, or major release.

CONCLUSIONS

The industrial hygiene emergency response program discussed in detail has assisted refinery industrial hygiene and safety personnel become better prepared to respond in the event of a major incident such as a fire, spill, or release. The program allows quicker and more efficient response and addresses key issues required by OSHA's Process Safety (29 CFR 1910.119) and Hazardous Waste Emergency Response (29 CFR 1910.120) standards. In summary, the case study shows that the refinery can prepare for future emergency response situations using a practical, methodical, problem solving approach. In the future, industrial hygienists and safety professionals will continue to be called upon to grow and expand their horizons beyond traditional duties to address health risk assessments, emergency preparedness, and incident response.

TABLE 1

PROCESS DESCRIPTION - SULFUR RECOVERY UNIT

During various refinery operations, hydrogen sulfide is produced and subsequently removed and recovered. The hydrogen sulfide is converted into elemental sulfur in the two class units which are part of the Sulfur Recovery Unit.

Feed (hydrogen sulfide) to the Sulfur Recovery Unit (SRU) goes to the Acid Gas Knockout Drum. This IO Drum contains 50-60% hydrogen sulfide. The first step of the two step conversion process begins with combustion in the sulfur incinerator

$$2H_2S + 2O_2 \rightarrow SO_2 + S + 2H_2O \quad \text{(125 Pound Steam)}$$

The remainder of the H2S is mixed with the combustion products and passed over an aluminum catalyst. The H2S reacts with the sulfur dioxide to form sulfur,

$$2H_2S + SO_2 \rightarrow 3S + H_2O$$

This process converts 90-93% of the H2S to Sulfur which is stored in a heated pit until shipped out by truck or rail.

The following table lists environmental stressors of industrial hygiene concern that may exist during an emergency event at the Sulfur Recovery Unit.

| STRESSOR | S.G. | V.P. (mm Hg) | B.P. (F) | SOURCE | REL. HAZ. | HAZ. FORM | OSHA PEL | OSHA CEILING STEL OR ACTION | BP OIL MSDS NO | S & A REF. NO | EMER. EVENT |
|---|---|---|---|---|---|---|---|---|---|---|---|
| Alumina (Aluminum Oxide) | 4.00 | 0 | -- | Reactor - Catalyst Substrate | 3 | P | 10 mg/m³ | -- | A-005 | 1A & 1B | F/S |
| Ammonia | 0.77 | 760 | -28 | Process Stream | 3 | G | -- | 35 ppm | S-011 | 2A & 2B | F/S |
| Carbon Monoxide | 1.25 | >760 | -313 | Incomplete Petroleum Combustion | 1 | G | 35 ppm | C 200 ppm | S-038 | 3A & 3B | F |
| Cobalt | 8.92 | 0 | -- | Reactor - Catalyst | 3 | P | 0.05 mg/m³ | -- | | 11A & 11B | F/S |
| Crystalline Quartz. Silica | 2.60 | 0 | -- | Solid Substrates | 2 | P | 0.1 mg/m³ | -- | | 13A & 13B | F/S |
| Flammable Hydrocarbons | 0.66 | 75 | 248 | Petroleum Feedstock | 2 | V | 400 ppm | -- | S-170 | 18A & 18B | F/S |
| Hydrogen Cyanide | 0.69 | 620 | 79 | Decomposition of Urethane Paint | 3 | G | -- | 5.7 ppm | | 26A & 24B | F |
| Hydrogen Sulfide | 1.54 | 15200 | -76 | Petroleum Feedstock | 1 | G | 10 ppm | 15 ppm | S-173 | 26A & 26B | F/S |
| Mercury | 13.33 | 0.0012 | 675 | Switches and Dispensers | 3 | V | 0.05 mg/m³ | -- | M-183 | 31A & 31B | F/S |
| Molybdenum | 10.20 | 0 | -- | Reactor - Catalyst | 3 | P | 10 mg/m³ | -- | | 33A & 33B | F/S |
| Nickel | 8.90 | 0 | -- | Reactor - Catalyst | 3 | P | 1 mg/m³ | -- | | 34A & 34B | F/S |
| Nickel Carbonyl | 1.32 | 321 | 109 | Incomplete Combustion w/Nickel Present | 1 | G | 0.001 ppm | -- | | 35A & 35B | F |
| Nitric Oxide | 1.34 | -- | -240 | Incomplete Combustion | 2 | G | 25 ppm | -- | S-215 | 36A & 36B | F |
| Nitrogen Dioxide | 1.49 | 720 | 70 | Incomplete combustion | 2 | G | -- | 1 ppm | S-216 | 37A & 37B | F |
| Sulfur Dioxide | 1.43 | >760 | 14 | Combustion of Hydrogen Sulfide | 2 | G | 2 ppm | 5 ppm | S-773 | 48A & 48B | F |
| Toluene Diisocyanate | 1.22 | 0.04 | 484 | Decomposition of Urethane Paint | 3 | P | 0.005 ppm | 0.02 ppm | | 52A & 52B | F |
| Heat | N/A | N/A | N/A | As Encountered | 3 | PH | -- | -- | N/A | 20A & 20B | F |
| Noise | N/A | N/A | N/A | As Encountered | 3 | PH | 90 dB(A) | 85 dB(A) | N/A | 38A & 38B | F/S |

1 - Acutely Toxic or Extreme Safety Hazard
2 - Strong Irritant or Very Toxic
3 - Chronic Toxicity or Not Likely to be present at hazardous concentrations.
While the presence of these materials in hazardous concentrations is remote, the hazard should not be discounted.

| V - Vapor | F - Fire |
|---|---|
| G - Gas | S - Spill or Release |
| P - Particulate | C - Ceiling Concentration |
| PH - Physical | |

# TABLE 2

## INDUSTRIAL HYGIENE DIRECT READING PROCEDURES FOR HYDROGEN FLUORIDE

APPARATUS (OPTION 1): Hydrogen Fluoride Meter (Model 4700)

MANUFACTURER: Gastech

COMMENTS: 0-20 ppm measuring range

Intrinsically safe and RF resistant

No yearly factory calibration required. Sensor cells last for approximately 5 years. Provides peak and average readings.

CALIBRATION: Use Sulfur Dioxide Span Gas 5 ppm (Cat. No. 81-0170/Supplier-Gastech) * Note: HF meter detects Sulfur Dioxide and Hydrogen Fluoride at a 1:1 ratio

MISC. SUPPLIES: Additional accessories are available


APPARATUS (OPTION 2): Gas Detector Pump (Cat. No. 67-26-124)

MANUFACTURER: Drager

COMMENTS: 1.5-15 ppm measuring range

Intrinsically safe

Check manufacturer's literature for cross-sensitivity and/or possible interferences.

CALIBRATION: No calibration required. Perform routine pump for checks according to manufacturer's instructions.

MISC. SUPPLIES: Drager/SKC 800-30301 1.5-15 ppm

## TABLE 3

## INDUSTRIAL HYGIENE TIME WEIGHTED AVERAGE AND SHORT TERM EXPOSURE SAMPLING PROCEDURES FOR HYDROGEN FLUORIDE

| | |
|---|---|
| COLLECTION MEDIA: | Supelco ORBO-53 Tube (Cat. No. 2-0265/ Supplier-Supelco) |
| FLOW RATES: | Max. V: 96 Liters Max. F: 0.2 LPM (TWA) Max. V: 7.5 Liters Max. F: 0.5 LPM (STEL) |
| COLLECTION APPARATUS: | Gilian Sampling Pump (any HFS-513A Model with low flow capabilities) equipped with an adjustable flow holder (Cat. No. 224-26-01/Supplier-SKC) |
| ANALYTICAL METHODS: | NIOSH Method: 7903 Ion Chromatography; IC OSHA Method: ID-165 Ion Chromatography; IC |
| MISC. SUPPLIES: | Gilian Gilibrator Flow cell range: 20 cc - 6 - LPM |

# Groundwater: Modeling and Treatment

# REVIEW OF LIMITATIONS OF PUMP-AND-TREAT SIMULATION MODELS FOR GROUNDWATER REMEDIATION

GERARD P. LENNON
Department of Civil and Environmental Engineering
13 East Packer Ave
Lehigh University
Bethlehem, Pa 18015

## INTRODUCTION

Simulation models have proven to be useful tools in understanding and predicting groundwater flow and contaminant transport, including sites undergoing pump-and-treat (P&T) remediation. Powerful flow codes can handle complex flow behavior such as flow in fractured rock, heterogeneity, complex boundary conditions and density variations. Inverse algorithms provide efficient methods to evaluate aquifer parameters. Contaminant transport codes can simulate phenomena such as desorption from soil using a wide range of algorithms tailored to site-specific conditions. Coupled codes have been developed that simultaneously solve for flow, contaminant transport, and geochemical reactions.

Problems can stem from misuse when applying models, including overkill, improper conceptualization, improper model selection, improper boundary conditions, application of a generic model to a specific site, inappropriate prediction, misinterpretation, misuse of the numerical approximation, and undetected coding error [1].

Although there are many facets to modeling, this paper focuses on the simulation of key factors that affect the design of remediation systems, with a special emphasis on predicting clean-up times. These key factors include presence of pure non-aqueous phase liquids (NAPLs), slow desorption from soil, slow diffusion from low permeability zones, and areas of near stagnant flow. Recent evaluation of P&T at numerous sites indicates that tailing, the long term persistence of contamination levels, is not accurately predicted by simulation models unless these key factors are properly represented as shown in Figure 1 [2]. Although models are available that properly represent such processes, application can be difficult due to proper characteri-

681

zation of the appropriate process [2,3]. Extensive field data may be required to properly document the responsible process, e.g. detection and qualification of NAPL rather than attributing tailing to slow desorption at low aqueous concentrations [3]. A stochastic procedure may be required to represent heterogeneity, especially if the scale of change is smaller than the grid size of the model.

## CATEGORIES OF REMEDIAL ACTIONS

Remedial technologies for groundwater can be broken into two broad categories: *groundwater quality restoration* and *containment of contaminated groundwater* [2]. Although P&T can be designed as a containment system, this paper focuses on restoration, where the goal is to improve groundwater quality by removing groundwater containing contaminants and treating it ex situ. Such systems can include recovery wells, trenches, and/or drains, sometimes with enhanced methods such as horizontal wells or blasting to increase permeability [4].

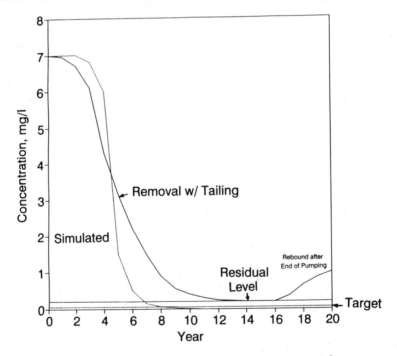

Figure 1. Hypothetical concentration versus time showing simulated clean-up time sooner than response with tailing; rebound occurs after pumping is discontinued (modified from [2]).

Techniques not addressed here include passive containment systems such as slurry walls, which are often used in combination with P&T, and in situ remediation, achieved by chemical or biological neutralization or natural attenuation. Also, enhanced P&T include vapor extraction, injection of solvents, steam flooding, and sparging with air, oxygen, or nitrogen.

## FACTORS HINDERING REMEDIATION PERFORMANCE

Although expectations for P&T are typically met for *containment*, they generally have not been met for *remediation* [2,4,5]. EPA collected information on, and evaluated the performance of, P&T at 112 sites, selecting 19 of these sites for detailed study [4]. The level of contamination measured at monitoring wells was typically reduced dramatically in a moderate period of time, but even when the source was removed, persistence of low levels of contamination occurred which is called tailing [5].

Tailing may require the remediation to be continued for centuries (or indefinitely), or a cause a premature cessation of the remediation and site closure. After closure, many sites experienced rebound, e.g. the contaminant concentrations increased significantly after pumping was discontinued (see Figure 1).

Factors that generate tailing include [2,4,5]:

1. Presence of non-aqueous phase liquids (NAPLs)
2. Contaminant desorption contaminants from sediment
3. Contaminant diffusion from low permeability zones
4. Hydrodynamic isolation (stagnant or dead spots)

At sites involving NAPLs the time for restoration is extremely hard to predict. Interfacial tension forces can cause very slow movement of these liquids, and even highly soluble components may become trapped in the finer pore structure [2,5,9]. To reach the MCL, typically at least 99% of a contaminant such as TCE must be recovered; even big oil companies using enhanced recovery techniques typically recover only 30% to 50% of the available oil.

Groundwater contaminants partition between the water and soil. As groundwater is pumped, the chemicals are held back (retarded) by their adherence to the soil particles. The choice of desorption algorithm can significantly affect predicted clean-up times [2,3,8]. For example, based on soil leaching tests at the Whitmoyer Labs CERCLA site, the nonlinear Freundlich isotherm sorption model was chosen using the EPA's Organic Leachate Model [8]. An estimated clean-up time of 50,000 years was obtained for conventional P&T, whereas a much shorter time was estimated using the linear desorption isotherm, which assumes a constant ratio of soil to aqueous concentration.

Once contaminants have entered a low permeability zone, diffusion back out can be extremely slow, and enhanced methods such as cyclic pumping do not provide a complete solution. Tens or hundreds of pore volume exchanges may need to occur in the higher permeability zones before adequate flushing of the lower permeability zones can occur. Some case studies indicate cyclic pumping yielded an increase in mass removal per unit of water, but a decrease in mass removal per unit time compared to continuous pumping, and no effective reduction in plume concentrations.

A more detailed coverage of related topics including physical processes controlling transport in aqueous and nonaqueous phases, effects of heterogeneties, chemical processes, adsorption, speciation, solubility, dissolution, biorestoration, and limitations of modeling subsurface contaminant fate and transport is available from EPA [6].

## MODELING OF CONTAMINANT MIGRATION

If simulation models do not accurately account for factors causing tailing, a much more optimistic clean-up time may be predicted as shown in Figure 1 [2,5,6]. One problem modelers may face is accounting for heterogeneity when the scale of the heterogeneity is smaller than the scale of the model grid size, called upscaling. Incorporation of upscaling in a simulation model is difficult, and is a current topic of research. For example, if a model grid cell is 100 ft by 100 ft by 20 foot thick, a single value of hydraulic conductivity (coefficient of permeability) will be assigned to this cell. Within this cell, small lenses of low permeability sediments contributing to tailing will not be represented by the model.

Because models are a representation of reality, some uncertainty in prediction always exists. Although the need for a clean-up time guide is recognized complete with the role of simulation models, a final guide receiving widespread acceptance is not available at this time. Existing guides to clean-up times include analytical expressions of mass balance based on flow rates and concentrations [2]. Although ASTM has not developed a guide for clean-up time prediction, their recently developed standard guide for groundwater flow models to a site-specific problems, ASTM provides the recommended steps in applying a flow model, including guidance on calibration and sensitivity analyses using residuals (differences between observed and simulated variables) to provide a quantitative evaluation of the quality of match [11]. Unfortunately, even for the simpler flow models, ASTM still emphasizes that "this standard is not intended to be all inclusive...[and] does not purport to address all...problems" [11].

Even with such guides, a high degree of uncertainty can still exist, and two knowledgeable modelers can produce significantly different predictions [10]. The difficult question is really how much uncertainty is acceptable, and can the uncertainty be evaluated [10]. Currently there is much debate concerning the proper application of models to simulate flow and contaminant transport in the groundwater system [10,12,13]. A common view held by many respected modelers is that ground-water models cannot be validated, only invalidated, and that standard use of *validation* implies the ability to make reliable ground-water flow and transport predictions without qualifications [12]. Alternative terms like *model testing*, *model evaluation*, and *history matching* have been proposed as a substitute for *verification* and *validation* [10,11,13]. However, it is generally agreed that a model should be tested against historical field data, and that it should not invalidate those data. General rules of thumb such as "predictions into the future should not exceed the duration of history match" are general rules which may or may not apply to a particular site [10].

## EXAMPLE OF P&T EFFECTIVENESS

The contaminant concentration in extracted ground-water typically declines rapidly at first, slowly approaching a residual level due to tailing. The causes of tailing were discussed previously; the consequences of tailing are best understood using typical field conditions.

Figure 1 shows a typical concentration decrease in extracted groundwater. More insight is obtained by using the flow and concentration data to perform a mass balance. Table I shows a hypothetical concentration decrease over 20 years, from 7 mg/l to a residual level of 0.20 mg/l, the latter being four times higher than a reasonable target based on the maximum contaminant level (MCL) for drinking water for a contaminant such as arsenic. An extraction rate of 25 gallons of water per minute (gpm) translates into 0.036 million gallons per day (MGD) or 110 million pounds of water per year. For the initial contaminant level of 7 mg/l, the extracted water during the first 3 years contains over 1 ton of contaminants (see Table I). Although another ton will be extracted after about 7 more years of pumping, as concentrations approach the residual value in Year 15, a very small but persistent amount of contaminant in the extracted groundwater (about .05 pounds per day or 22 pounds per year) would be enough to maintain the concentration over the target level. Thus about two tons would be extracted during the first 10 years, and the residual amount, which could easily be on the order of a ton, could provide the small mass removed (22 pounds per year) for decades or even centuries.

## CONCLUSIONS

Can simulation models be used to accurately predict clean-up times for restoring groundwater quality? This begs the question whether it is technically possible to restore groundwater to an environmentally sound condition (target concentration levels) almost regardless of cost. Many investigators believe that it is not possible to restore groundwater quality based on EPA's findings, stating that we should just admit that remediation is unachievable [9]. Rebuttals of others cite at least partial aquifer remediation is achievable at some sites, and admitting it is unachievable would invite little or no action. However, it is generally agreed that residual contamination will persists at some sites for centuries or more, requiring an approach that balances risk and restoration [2,9].

Some researchers feel plume containment and mass reduction should be primary remediation goals, and recognize that restoring aquifers to a pristine condition is not always necessary. For these goals, existing simulation models can be excellent tools when applied properly, even if some degree of uncertainty exists in the prediction of the tailing effect.

To provide accurate clean-up time estimates, simulation models must properly take into account processes that cause *tailing*, e.g. persistence of residual levels of contamination, often above the target level. Tailing can severely limit the effectiveness of P&T, being caused by factors including the presence of NAPL acting as a source, slow desorption at low near-residual contaminant concentrations, slow diffusion from low permeability zones, and hydrodynamic isolation [2,4,5,6].

## REFERENCES

1. Mercer, J., 1991. "Common Mistakes in Model Applications," ASCE Proc. of Symposium on Groundwater, ed. G. Lennon & S. Rouhani, pp. 1-6.

2. Cohen, R., A. Vincent, J. Mercer, C. Faust, and C. Spalding, 1994. "Methods for Monitoring Pump-and-Treat Performance, EPA/600/R-94/123.

3. Hinz, C., A. Gaston, and H. Selim, 1994. "Effect of "Sorption Isotherm Type on Predictions of Solute Mobility in Soil," *Water Resources Research,* 30(11), pp. 3013-3021.

4. Mercer, J. Skipp, D., and Giffin, D., 1990. "Basics of Pump-And-Treat Ground-Water Remediation Technology," EPA/600/8-90/003.

5. Keely, J., 1989. "Performance of Pump-And-Treat Remediations," EPA/540/4-89/005.

6. EPA, 1989. "Transport and Fate of Contaminants in the Subsurface", EPA 625/4-89/019.

7. Hutzler, N., B. Murphy, and J. Gierke, 1989. "State of Technology Review, Soil Vapor Extraction Systems," EPA/600/2-89/024.

8. Stephanatos, B., K. Walter, A. Funk, and A. MacGregor, 1991. "Pitfalls Associated With The Assumption of a Constant Partition Coefficient in Modeling Sorbing Solute Transport through The Sub-Surface," ASCE, *Proc. of Symposium on Groundwater,* ed. G. Lennon and S. Rouhani, pp. 13-20.

9. Travis, C. and C. Doty, 1990. "Can Contaminated Aquifers at Superfund Sites be Remediated?" *Environ. Sci. & Tech.,* 24(10), pp. 1464-1466, 1990.

10. Bair, E. S., 1994. "Model (In)Validation - A View From the Courtroom," *Groundwater,* 32(4), 530-531.

11. American Society for Testing and Materials (ASTM), 1993. "Standard Guide for Application of a Ground-Water Flow Model to a Site-Specific Problem," D5447-93, pp. 1-6.

12. Konkow, L., and Bredehoeft, L., 1992. "Ground-Water Models Cannot be Validated," *Advances in Water Resources,* 15, 75-83.

13. Tsang, C., 1991. "The Modeling Process and Model Validation," *Groundwater,* 29(6), pp. 825-831.

TABLE I. HYPOTHETICAL REMOVAL RATES FOR TYPICAL GRADUAL DECREASE IN CONCENTRATION FOR A EXTRACTION RATE OF 25 GPM, showing rebound after pumping is stopped in Year 15.

| Hypothetical Year Number | Assumed Concentration mg/l | Approx. Pounds Removed in Year | Cumulative Tons Removed |
|---|---|---|---|
| 0 | 7.0 | | |
| 1 | 6.95 | 765 | 0.38 |
| 2 | 6.7 | 749 | 0.76 |
| 3 | 6.1 | 702 | 1.11 |
| 4 | 4.3 | 570 | 1.39 |
| 5 | 3.15 | 409 | 1.60 |
| 6 | 2.20 | 293 | 1.74 |
| 7 | 1.5 | 203 | 1.85 |
| 8 | 0.9 | 132 | 1.91 |
| 9 | 0.55 | 80 | 1.95 |
| 10 | 0.4 | 52 | 1.98 |
| 11 | 0.3 | 38 | 2.00 |
| 12 | 0.23 | 29 | 2.01 |
| 13 | 0.21 | 24 | 2.02 |
| 14 | 0.20 | 23 | 2.03 |
| 15 | 0.20 | 22 | 2.05 |
| 16 | 0.20 | | |
| 17 | 0.35 | | |
| 18 | 0.65 | | |
| 19 | 0.85 | | |
| 20 | 1.00 | | |

# CONFINING UNITS AS BARRIERS TO REGIONAL GROUND-WATER CONTAMINATION: HYDROGEOLOGIC MAPS AS PLANNING TOOLS

AMLETO A. PUCCI, Jr.
Department of Civil and Environmental Engineering
Lafayette College, Easton, PA   18042   puccia@lafvax.lafayette.edu

## INTRODUCTION

In a time when applicability seems increasingly important for publicly funded projects, hydrogeologists and hydrogeochemists may demonstrate potential uses of their products for other disciplines. Full benefits of these efforts are rarely realized because hydrologic intuitions are not obvious in other professions, such as those concerned with land-use planning.

Hydrogeologic maps are typical products of ground-water investigations. The features on these maps can be used by planning commissions to optimize land use. Planners could use confining-unit outcrop maps for siting landfills and hazardous material handling facilities [1]. The implication being that placing landfills over a confining-unit outcrop is a natural protection measure from ground-water contamination.

This paper is partially motivated by an evaluation done in the mid-1980's for locating a hazardous material facility in southwestern Middlesex County, New Jersey [2]. That evaluation considered a confining-unit outcrop map completed for the New Jersey Bond Study [3]. The paper illustrates potential benefits of including hydrogeochemistry in such an evaluation.

This paper examines ground-water chemistry from 53 wells, field measurements, hydrogeologic conditions from a quasi-3-D flow model for predevelopment (before 1900), and 1984 flow conditions [4], and evaluates relationships between them. Several recent reports have examined water quality in the area [5,6,7,8]. The wells for this paper were screened in the Potomac-Raritan-Magothy aquifer system (PRMA) in the northern Coastal Plain of New Jersey in a 184 square mile area which is undergoing rapid growth (Fig. 1). Hydrogeologic conditions considered include aquifer sampled, well location relative to flow-path distance from the outcrop, confining-unit thickness, and confining-unit vertical hydraulic conductivity (Kv). Visual, graphical and principal component analyses were used to evaluate the relationships.

*Only as far as the masters of the world have called in nature to their aid, can they reach the height of magnificence.*   Emerson

## BACKGROUND

Accurate prediction of water quality in regional aquifers requires knowledge of spatial media properties (i.e. hydraulic conductivity, leakance). Because thorough compilation of media properties from point data is beyond the resources of most investigations, alternate and indirect approaches are often attempted. These include geophysical approaches, and conceptualizing aquifer architecture for estimating characteristic hydrogeologic properties, such as interconnectedness [9,10].

For evaluating their effectiveness as chemical barriers, confining-unit Kv's are a principle consideration. Where discontinuities are a factor, confining-unit Kv's in large flow systems can be estimated from regional flow models more accurately than from point measurements [11,12]. In at least one case, water contamination by upward saline migration corroborated large magnitude Kv's caused by confining-unit openings [13].

Many investigators have examined the effects of confining-unit hydrologic properties on water chemistry [14,15,16,17]. It may follow that water chemistry could also be useful for evaluating regional hydrogeologic properties. This evaluation must account for important statistical aspects of water chemistry (including random sources and processes) and regionalized properties.

Barton and others [5], investigated well-water quality for wells within the outcrops and within 1 mile of confinement for the PRMA in the northern Coastal Plain of New Jersey. They reported there were no significant relations between hydrogeologic factors and water-characteristics, major ions, and nutrients, except for beryllium and cadmium. Their analysis included the area of this study but did not examine flow-path and confining-unit leakage. Other studies which have included data from deeper parts of the same flow system reported water chemistry trends for dissolved gases and anthropogenic contaminants corresponded to areas of leaky confining units [7,8].

## HYDROGEOLOGIC SETTING

The PRMA principally is comprised of Cretaceous-age sediments in the subject area which form two highly productive aquifers, the Upper and Middle aquifers, and their confining units (Table I). Both aquifers are composed of fine to coarse-grained sands, and localized thin clay beds [3]. All major units strike northeast (Fig. 1), and dip and thicken to the southeast.

Table I. Geologic and Hydrogeologic Units of the Potomac-Raritan-Magothy aquifer system in the northern part of the New Jersey Coastal Plain (Modified from [3], table 2)

| SYSTEM | GEOLOGIC UNIT | HYDROGEOLOGIC UNIT |  |
|---|---|---|---|
| Cretaceous | Magothy Formation | Potomac-Raritan-Magothy aquifer system | confining Unit |
|  |  |  | Upper Aquifer |
|  | Raritan Formation |  | confining unit |
|  |  |  | Middle Aquifer |
|  |  |  | confining unit |
| Pre-Cretaceous | Bedrock |  |  |

690

## GEOHYDROLOGY

The confining-unit overlying the Upper Aquifer is tight and massive throughout most of the area. The confining unit overlying the Middle aquifer is composed mostly of clays and silts, but its thickness contours indicate it thins in areas (Fig. 1). Lithologic logs [3], hydraulic measurements, and calibrated zones of Kv's from a regional ground-water flow model indicate a good hydraulic connection exists between the Upper and Middle Aquifers through the confining unit overlying the Middle aquifer in the southwest part of Middlesex County [4]. Kv's for the confining unit overlying the Middle Aquifer were much more variable (15 zones, ranging from 4.5 x $10^{-5}$ to 4.5 x $10^{-2}$) (Fig. 2), than for the confining unit overlying the Upper Aquifer (9 zones, ranging from 1.4 x$10^{-4}$ to 2.0 x $10^{-3}$)(not shown). The underlying bedrock confining unit is considered impermeable.

Simulated ground-water flow for predevelopment and 1984 conditions in the PRMA in the subject area for both aquifers showed that lateral regional ground-water movement generally was from topographically high aquifer outcrop areas in southwestern Middlesex County towards low lying discharge areas to the northeast, or downdip to the east and southeast [4]. Calibrated predevelopment leakage through the confining-unit overlying the Middle Aquifer in southwestern Middlesex County was downward, ranging from 1 to 5 in/yr. For the 1984 system, simulated vertical flow through this confining unit in the area was downward, ranging from 5 to 15 in/yr.

## METHODS

Well identifiers, aquifer designations, and select natural and anthropogenic water constituents for 53 well-water samples are given in Table II. Corresponding well locations are shown in Fig. 1. Water samples were collected using U.S. Geological Survey procedures between 1984 and 1986 [6]. Because of the need to avoid spatial autocorrelation of this data, these wells were selected from about two hundred wells by placing a 1-kilometer grid onto a map of the area and choosing no more than one well within each grid cell for this analysis. Of these wells, well-water samples which had an ionic imbalance exceeding approximately 30 percent were eliminated as insufficiently accurate data.

Exploratory analysis examined variations of water quality along three flow paths from the main recharge area into the confined areas of both aquifers. Well distances along flow paths were estimated from predevelopment potentiometric contours by manual reverse particle tracking [4]. The three flow paths generally flowed; a) towards southeast, towards the deepest part of the flow system in Monmouth County, b) northeast, following a shallower path towards Raritan Bay, and c) north-northeast, the shallowest path towards regional drains (Fig. 1). Visual exploration was used to find correlations of major-ion chemistry data plotted on Piper diagrams to hydrogeologic properties at the sampled well locations. Wells which have either lead or cadmium contamination were also plotted on the Piper diagrams.

Principal components (PC) analysis was used to reduce the dimensionality of the water quality and hydrogeologic data, and help express interrelationships among them. It was hypothesized that hydrogeologic factors which effect water quality in these wells included a) flow path distance from the aquifer outcrop, and b) the log value of the model calibrated Kv for the confining unit overlying the well location. With the exception of silica and Eh, the chemical data values were converted to log values which reduced data skewness. Where Eh was not measured, it was computed from the

691

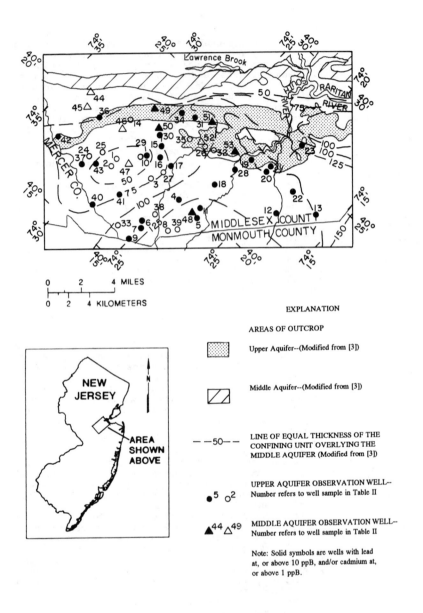

Figure 1. Location of aquifer outcrops, thickness of the confining unit overlying the Middle Aquifer, well locations, and study area location in New Jersey.

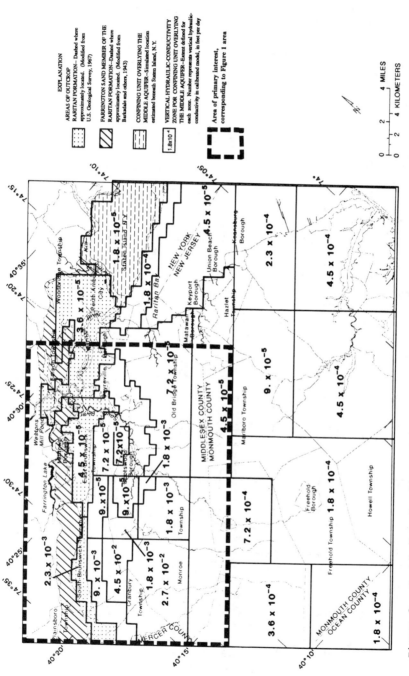

Figure 2. Vertical hydraulic conductivities for confining unit overlaying Middle Aquifer [4].

693

Table II.--<u>Water-quality analysis of select well-water samples from the Upper and Middle aquifers, 1984-6.</u> (Modified from [6], table 4. Chemical concentrations are reported as milligrams per liter, except for lead which is micrograms per liter. Constituents reported as less than detection limit or not detected in a sample are reported as the value of the detection limit for that analysis, i.e. preceded by "<"; Eh in volts. Kv is vertical hydraulic conductivity in ft/d.)

| Map No. | Well ID t | Outcrop distance (miles) | Log Kv | Eh | Dissolved Oxygen | pH | HCO3 | Ca | Mg | Na | K | Cl | SO4 | SiO2 | Fe | Cd | Pb | NO3-N | Dissolved Organic Carbon |
|---|---|---|---|---|---|---|---|---|---|---|---|---|---|---|---|---|---|---|---|
| | | | | | | *UPPER* | | *AQUIFER* | | | *WELLS* | | | | | | | | |
| 1 | 030 | 1.7 | -3.0 | -1.06 | -- | 6.9 | 27 | 6.8 | 3.0 | 13 | 3.5 | 23 | 0.3 | 7.8 | 0.29 | -- | -- | -- | -- |
| 2 | 020 | 2.3 | -3.0 | 4.32 | 7.7 | | 1.5 | 12 | 5.5 | 17.0 | 2.7 | 55 | 3.0 | 9.4 | 0.38 | <1 | <10 | 5.3 | 1.2 |
| 3 | 219 | 3.3 | -3.0 | 5.22 | 3.1 | 4.6 | 1.5 | 5.0 | 2.1 | 5.0 | 1.5 | 13 | 6.5 | 8.6 | 0.38 | <1 | <10 | 2.4 | -- |
| 4 | 218 | 4.3 | -3.0 | -9.09 | 0.2 | 6.5 | 55 | 10 | 3.0 | 2.4 | 2.6 | 3.1 | 42 | 12 | 16.00 | 3 | 20 | <.1 | 9.5 |
| 5 | 773 | 5.2 | -3.7 | -6.72 | 0.1 | 6.0 | 10 | 0.6 | 0.4 | 2.5 | 0.9 | 3.4 | 21 | 5.9 | 11.00 | <1 | 20 | <.1 | 0.7 |
| 6 | 769 | 5.8 | -3.9 | -6.41 | 0.1 | 6.4 | 22 | 3.0 | 0.6 | 2.4 | 0.9 | 2.6 | 5.2 | 8.0 | 5.70 | <1 | 20 | <.1 | 0.7 |
| 7 | 771 | 6.1 | -3.9 | -2.99 | 0.1 | 6.1 | 12 | 2.9 | 0.8 | 1.8 | 0.9 | 2.2 | 5.2 | 7.6 | 1.90 | <1 | 10 | <.1 | -- |
| 8 | 761 | 6.3 | -3.7 | -3.62 | 0.1 | 6.1 | 24 | 5.1 | 1.0 | 1.8 | 1.1 | 2.6 | 8.7 | 7.8 | 2.50 | <1 | <10 | .1 | 0.9 |
| 9 | 770 | 6.9 | -3.7 | -8.52 | 0.1 | 7.3 | 95 | 31 | 3.9 | 2.5 | 2.9 | 2.3 | 39 | 9.0 | 4.40 | 1 | <10 | <.1 | 1.1 |
| 10 | 565 | 2.5 | -3.0 | 9.86 | 6.9 | 5.0 | 9 | 9.6 | 4.8 | 12 | 2.5 | 32 | 0.2 | 11 | 0.03 | <1 | <10 | 5.8 | -- |
| 11 | 225 | 8.0 | -3.7 | -10.68 | 0.1 | 6.9 | 44 | 0.6 | 0.4 | 2.0 | 0.6 | 2.6 | 11 | 7.4 | 19.00 | 1 | <10 | <.1 | 1 |
| 12 | 781 | 12.4 | -3.7 | -8.04 | 0.1 | 6.2 | 31 | 4.1 | 2.3 | 1.6 | 1.1 | 4.7 | 26 | 6.9 | 15.00 | 2 | 20 | <.1 | 2.7 |
| 13 | 783 | 15.0 | -3.7 | -8.76 | 0.2 | 6.3 | 63 | 8.9 | 3.9 | 1.5 | 2.2 | 3.1 | 20 | 9.3 | 18.00 | 3 | 20 | <.1 | 1 |
| 14 | 292 | 0.8 | -1.0 | -7.33 | 0.4 | 6.4 | 60 | 4.3 | 2.8 | 5.3 | 1.3 | 2.5 | 0.4 | 12 | 8.50 | <1 | <10 | <.1 | .9 |
| 15 | 497 | 3.1 | -3.1 | 4.41 | 8.5 | 5.0 | 4 | 2.6 | 1.6 | 5.3 | 1.6 | 15 | 1.5 | 8.4 | 0.32 | <1 | 20 | .8 | 1.3 |
| 16 | 772 | 3.7 | -3.0 | 0.49 | 0.1 | 4.7 | 1 | 2.4 | 2.7 | 11 | 2.2 | 23 | 19 | 9.0 | 2.60 | <1 | 30 | <.1 | 0.8 |
| 17 | 757 | 4.7 | -3.0 | 11.08 | 2.8 | 4.2 | 0 | 7.2 | 3.4 | 11 | 2.3 | 23 | 28 | 8.7 | 0.05 | <1 | 10 | 4.3 | 1.0 |
| 18 | 244 | 8.0 | -3.0 | -7.82 | 0.1 | 6.3 | 41 | 3.6 | 2.9 | 2.1 | 1.4 | 3.0 | 21 | 8.9 | 12.00 | 2 | 10 | -- | 0.7 |
| 19 | 131 | 8.4 | -3.1 | -4.06 | 0.3 | 4.7 | 0 | 4.4 | 4.0 | 5.6 | 2.0 | 7.3 | 77 | 10 | 18.80 | <1 | 20 | .13 | 1.6 |
| 20 | 145 | 10.6 | -2.9 | -4.69 | 0.3 | 5.5 | 15 | 2.3 | 1.8 | 2.5 | 1.0 | 5.9 | 19 | 6.8 | 8.70 | 2 | 30 | .13 | 1.3 |
| 21 | 156 | 10.9 | -1.0 | -8.13 | 0.4 | 5.7 | 15 | 3.1 | 2.6 | 6.1 | 1.5 | 14 | 51 | 6.4 | 30.00 | 2 | 20 | <.1 | .9 |
| 22 | 135 | 11.6 | -2.9 | -2.88 | 0.2 | 4.8 | 0 | 1.8 | 2.2 | 2.9 | 0.1 | 5.6 | 27 | 7.1 | 9.90 | <1 | 20 | <.1 | -- |
| 23 | 192 | 12.8 | -1.0 | -6.58 | 0.2 | 7.0 | 85 | 33 | 6.5 | 11 | 0.1 | 24 | 42 | 4.0 | 2.80 | <1 | 20 | <.1 | -- |
| 24 | 15 | 1.1 | -3.0 | -5.00 | 7.5 | 5.3 | 6 | 15 | 9.2 | 7.5 | 4.7 | 14 | 23 | 8.9 | 1.30 | <1 | <10 | 13.0 | 1 |
| 25 | 24 | 1.3 | -3.0 | 1.38 | 5.3 | 4.6 | 1.5 | 6 | 4.8 | 7.3 | 2.8 | 22 | 7.3 | 9.3 | 0.11 | <1 | <10 | 5.3 | 0.7 |
| 26 | 96 | 0.0 | -1.0 | 1.08 | 7 | 4.9 | 4 | 8.3 | 6.7 | 12 | 2.6 | 25 | 23 | 8.5 | 0.11 | <1 | <10 | 5.2 | 2 |
| 27 | 100 | 2.8 | -3.0 | 1.28 | 2.8 | 4.5 | 5 | 2 | 0.8 | 5.2 | 1.2 | 8.1 | 3 | 8.2 | 0.12 | <1 | <10 | 2.1 | 0.3 |
| 28 | 108 | 0.3 | -3.1 | -10.91 | 0.4 | 5.3 | 9 | 4.1 | 3.5 | 6.5 | 1.7 | 9.4 | 48 | 7.7 | 17.00 | 1 | <10 | 0.0 | 1.2 |
| 29 | 228 | 2.5 | -3.0 | 11.89 | 0.1 | 4.9 | 1 | 7.7 | 2.4 | 8.7 | 2.3 | 25 | 17 | 9.7 | 0.00 | <1 | <10 | 0.3 | 0.5 |
| 30 | 299 | 0.8 | -3.1 | -6.08 | 0.1 | 5.9 | 18 | 0.4 | 0.4 | 32 | 0.4 | 36 | 20 | 4 | 1.60 | <1 | 10 | 0.0 | 1.5 |
| 31 | 328 | 0.2 | -1.0 | -4.92 | 7.8 | 5.6 | 8 | 16 | 7.6 | 13 | 2.3 | 23 | 39 | 6.8 | 1.10 | <1 | 10 | 6.7 | 1 |
| 32 | 442 | 0.0 | -1.0 | -5.00 | 0.2 | 4.2 | 0 | 2.3 | 1.8 | 3.4 | 1.4 | 5.5 | 28 | 8.3 | 2.10 | <1 | <10 | 0.0 | 1.5 |
| 33 | 584 | 6.6 | -3.9 | -8.92 | 0.1 | 7.3 | 103 | 28 | 3.4 | 2.6 | 3.1 | 2.1 | 14 | 9.9 | 3.00 | <1 | <10 | 0.0 | 0.9 |
| 34 | 740 | 0.0 | -1.0 | -8.80 | 0.3 | 4.7 | 1 | 9.5 | 3.8 | 6 | 2 | 10 | 56 | 9.5 | 8.80 | <1 | 30 | 0.1 | 1.5 |
| 35 | 741 | 0.0 | -1.0 | -4.06 | 6.8 | 5.5 | 7 | 7.1 | 6.1 | 11 | 2.2 | 26 | 5.1 | 9.5 | 0.79 | <1 | <10 | 4.8 | 0.9 |
| 36 | 742 | 0.0 | -1.0 | -9.78 | 0.5 | 6.6 | 59 | 5.5 | 3.8 | 5.3 | 1.5 | 3.6 | 11 | 29 | 5.90 | <1 | 10 | 0.0 | 0.8 |
| 37 | 754 | 1.9 | -3.0 | -6.71 | 0.1 | 4.1 | 0 | 1.3 | 3.7 | 7.9 | 1.5 | 13 | 50 | 8.1 | 4.60 | <1 | 20 | 0.0 | 1 |
| 38 | 756 | 6.6 | -3.7 | -11.77 | 0.2 | 6.6 | 69 | 9.6 | 2.6 | 2.7 | 1.9 | 3 | 15 | 9.6 | 14.00 | <1 | <10 | 0.1 | 0.9 |
| 39 | 759 | 8.8 | -3.7 | -10.65 | 0.1 | 6.5 | 41 | 5.5 | 1.1 | 1.8 | 1.1 | 2.5 | 15 | 8.7 | 9.00 | <1 | <10 | 0.0 | 1.3 |
| 40 | 774 | 5.0 | -3.0 | -10.75 | 0.1 | 6.7 | 122 | 24 | 7 | 5.7 | 3.4 | 2.1 | 26 | 15 | 8.60 | 1 | 10 | 0.0 | -- |
| 41 | 775 | 4.7 | -3.0 | -11.41 | 0.1 | 6.6 | 59 | 11 | 2.4 | 2.3 | 1.7 | 2.2 | 18 | 11 | 12.00 | 1 | 20 | 0.0 | 0.8 |
| 42 | 777 | 1.3 | -3.0 | -0.11 | 5.4 | 4.2 | 0 | 10 | 8.4 | 8.6 | 3.8 | 36 | 15 | 8.6 | 0.25 | <1 | 10 | 8.7 | 0.8 |
| 43 | 779 | 1.8 | -3.0 | -5.90 | 0.3 | 4.2 | 0 | 4.5 | 2.8 | 6.7 | 2.3 | 23 | 32 | 7.9 | 3.10 | <1 | 10 | 0.0 | 0.5 |
| | | | | | | *MIDDLE* | | *AQUIFER* | | | *WELLS* | | | | | | | | |
| 44 | 776 | 0.0 | -1.0 | 18.58 | 8.6 | 6.1 | 15 | 2.5 | 0.9 | 4.6 | 1 | 2.4 | 0.5 | 25 | 0.03 | <1 | <10 | 1.1 | 1.1 |
| 45 | 283 | 0.6 | -2.6 | 12.32 | 10.2 | 5.1 | 4 | 14 | 10 | 4.0 | 5.9 | 27 | 1.3 | 9.2 | 0.01 | <1 | <10 | 14.0 | 2.0 |
| 46 | 552 | 2.2 | -2.0 | 5.98 | 6.3 | 5.9 | 16 | 2.2 | 1.0 | 2.8 | 1.2 | 2.8 | 0.7 | 13 | 0.05 | <1 | <10 | 1.0 | 0.2 |
| 47 | 517 | 4.1 | -1.6 | 0.67 | -- | 5.3 | 12 | 9.1 | 4.3 | 8.3 | 3.1 | 27 | 0.8 | 10 | 1.10 | <1 | <10 | 4.9 | 0.6 |
| 48 | 503 | 6.5 | -3.7 | -0.82 | 0.3 | 5.3 | 6.0 | 1.0 | 0.4 | 1.7 | 0.5 | 3.2 | 8.2 | 8.0 | 2.10 | <1 | <10 | <.1 | 1.1 |
| 49 | 319 | 3.5 | -2.6 | 9.56 | 8.2 | 5.1 | 5 | 3.7 | 2.9 | 16 | 1.9 | 16 | 17 | 7.3 | 0.03 | <1 | 20 | 4.3 | 1.2 |
| 50 | 305 | 4.4 | -1.3 | 8.56 | 8.6 | 4.9 | 4 | 6.9 | 4.7 | 6.4 | 2.1 | 18 | 0.1 | 7.8 | 0.06 | <1 | 60 | 5.3 | -- |
| 51 | 506 | 7.7 | -4.0 | -3.45 | 0.3 | 5.3 | 7 | 2.8 | 0.9 | 4.2 | 1.2 | 8.1 | 22 | 9.5 | 6.60 | <1 | 20 | -- | 2.6 |
| 52 | 094 | 7.9 | -4.1 | 2.99 | 0.3 | 4.4 | 1 | 1.3 | 0.5 | 2.2 | 0.6 | 5.0 | 11 | 7.7 | 1.30 | <1 | <10 | <.1 | -- |
| 53 | 499 | 9.6 | -4.1 | -0.04 | 0.3 | 5.3 | 5 | 2.6 | 0.7 | 1.7 | 0.6 | 2.6 | 11 | 7.9 | 1.50 | <1 | 10 | <.1 | .9 |

t - Well Identifiers follow U. S. Geological Survey System, but without county code prefix "23-"

694

Sato algorithm [18]. Lead concentrations less than detection limit
of 10 ppB, or with zero reported values, were assigned values of
0.001 ppB. To simplify the interpretations, cadmium, nitrate-as-
nitrogen, dissolved organic compounds, and dissolved oxygen were
not included in this analysis. These constituents had
proportionately higher numbers of non-detections, or less than
detection-limit values, or non-measurements, or were
autocorrelated.

For the Middle aquifer, where two confining units are overlaying
the sampled well, the lower Kv value of either overlying confining
unit was used. The calibrated values and zones for the confining
unit overlying the Upper Aquifer are reported elsewhere [4]. For
unconfined wells, the Kv is assigned a value of 1 ft/d.

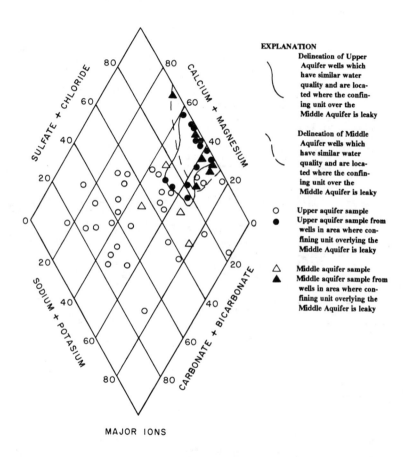

Figure 3.--Piper diagram showing composition of well-water
samples.(Ground-water samples are for the Upper and Middle
aquifers. Locations for the sampled wells is shown in Fig. 1)

## RESULTS AND DISCUSSION

Visual correlation of ground-water chemistry with distance along the two shallower flow paths in both the Upper and Middle aquifers was not tractable like previous flow-path analyses in the area [8]. These less successful explorations probably were impeded by subtle, yet distinct hydrogeochemical factors in the region, such as a "Pinelands" outlier region [5], as well as more complex factors in the shallow flow system than in the deeper system [4].

The Piper diagram for both aquifers generally is scattered (Fig. 3). However, a noticeable similarity exists in water chemistry of the wells within the same aquifer, and for wells in different aquifers if they are located in southwestern Middlesex County. This is indicated by an overlapping delineation of the corresponding wells shown in Figure 3.

The delineated well locations for both aquifers correspond to the area where the Middle Aquifer is thinnest (Fig. 1). Coincidentally, the calibrated Kv's for the confining unit overlying the Middle Aquifer are highest in the area (Fig. 3).

Although some of these delineated wells are in the Middle Aquifer and several miles from outcrop, they contain significant amounts of dissolved oxygen (well nos. 46 and 48), which probably – derives from transport from the overlying aquifer. Similarly, some relatively deep confined Middle Aquifer wells have contamination – typically associated with surface activity such as road salting (well no. 47), or leaky sewers or over fertilization (well nos. 46, 47, 49, and 50).

Several combinations of factors were analyzed using the PC analysis. Initial results were that flow-path distance and lead contamination were highly correlated factors, and Kv was not a significant factor. This result is not intuitive so flow path was explicitly eliminated. Flow-path distance possibly was not an

---

Table III.--Principle component analysis correlation coefficients chemical properties of well water species, dissolved lead, and overlying confining unit vertical hydraulic conductivity. [Correlations with absolute values less than 0.30 have been eliminated for clarity; chemical constituent log values were used and for lead concentrations below detection limit (10 ug/L), a concentration of 1ng/L is assigned; Kv represents the assigned vertical hydraulic conductivity of the confining unit overlying the sampled aquifer.]

|              | Components |       |       |               |
|--------------|-----------|-------|-------|---------------|
|              | 1         | 2     | 3     | communalities |
| Calcium      |           | 0.70  | 0.39  | .79           |
| Magnesium    | 0.51      | 0.76  |       | .84           |
| Sodium       | 0.76      |       |       | .81           |
| Iron (total) | -0.83     | 0.37  |       | .89           |
| Sulfate      | -0.40     | 0.36  | -0.49 | .55           |
| Chloride     | 0.83      |       | -0.45 | .89           |
| Bicarbonate  | -0.50     |       | 0.52  | .68           |
| Silica       |           |       | 0.57  | .48           |
| Lead         |           |       | -0.34 | .27           |
| Eh           | 0.75      | -0.53 |       | .89           |
| Kv           | 0.44      |       |       | .42           |
| PCA Eigenvalue | 3.48    | 1.88  | 1.50  |               |

Variance explained by each component

|            |      |      |      |
|------------|------|------|------|
| Proportion | 46.4 | 25.1 | 20.0 |
| Cumulative | 46.4 | 71.5 | 91.5 |

696

appropriate hydrogeologic factor because for the 1984 flow system there were many wells located in the shallow system and these have more control in the local flow system that predevelopment flow-paths which were used.

Finally, three PC's, which together account for 91.6 percent of the variation and had eigenvalues larger than 1, were considered significant (Table III). About 46.4 percent of the variation is explained by Component one; about 25.1 percent of the variation is explained by Component two; and about 20.0 percent of the variation is explained by Component three. Component one includes positive loadings for sodium, magnesium, chloride and calibrated Kv and negative loadings for sulfate, bicarbonate, iron (total), and lead. The positive high loadings indicate major ion sources are high in sodium, chloride, and magnesium and low in concentrations of sulfate, bicarbonate, and iron. The inverse loadings of Eh and Kv with iron (total) indicate that oxidative waters associated with "leaky" confining units are found where dissolved iron concentrations are low. Where the confining units are leaky, dissolved oxygen is likely to enter the aquifer more easily.

Component two includes positive loadings of calcium, magnesium, sulfate, and iron, and negative loading of Eh. The positive loadings indicate a source of these major ions are associated with of low Eh values which generally are associated with reducing conditions found in the deeper confined area of the aquifers. Release of these constituents from the confining units in the northern Coastal Plain has been reported [7,8].

Component three includes positive loadings of calcium, silica, and bicarbonate, and negative loadings of sulfate, chloride, and chloride. Although trend maps of components were not done, coupled chemical reactions which consume sulfate anions (negative loadings) which leak from confining units and generate bicarbonate anions (positive loadings) along flow paths in these aquifers are documented [7,8]. The component suggests that, in association with the flow path, there are sources of the major ions with positive loadings found in conjunction with sinks of the major ions with negative loadings. Flow path-length controls for this component are also suggested by positive loadings of silica (which should increase with flow contact time) and negative loadings of chloride and lead (which should decrease with distance from the outcrop).

## CONCLUSIONS

The results of water chemistry samples from 53 wells screened in the Upper and Middle Aquifers of a 184 square mile area of the northern Coastal Plain of New Jersey indicate that occurrence of natural and contaminant water quality is partially controlled by hydrogeologic factors. It was demonstrated that evaluating water chemistry in these aquifers is especially benefitted by including the hydrologic properties of confining units. Graphical water-chemistry analysis methods showed a correspondence in water-chemistry types and area. Water chemistry corresponded in areas where confining unit separating the subject aquifers was leaky.

The water chemistry is also greatly influenced by recharge caused by confining-unit leakage. Based on principle components analysis, approximately 46.4 percent of the variation in water chemistry is attributed to this affect. Two other significant principle components were explainable as caused by the effect of tight confinement, which makes reducing conditions, or leakage of constituents from confining units.

The example discussed also reinforces that hydrogeochemical interpretations of water quality data can be used as a tool in the analysis of hydrogeologic contour maps, in order to evaluate the extent that confining units may serve as contamination barriers.

697

**ACKNOWLEDGEMENT**

Helpful comments by Dr. Mary Roth were appreciated.

**REFERENCES**

1. Bernknoph, R. (USGS), February 2-3, 1990. Oral Commun. at Conference on *Geology and Hydrology of South-Central Florida: Planning for future land use.*, Archibald Biological Ctr, FL.

2. Sugarman, P. (NJGS), 1989. Oral commun.

3. Gronberg, J.M., Pucci, A.A., Jr., and Birkelo, B.B., 1991. "Hydrogeologic framework of the Potomac-Raritan-Magothy aquifer system, northern Coastal Plain of New Jersey" US Geol. Surv. Water-Resour. Invest. Rep. 90-4016, 37pp., 8 pl.

4. Pucci, A. A., Jr., Pope, D. A., and Gronberg, J. M., 1994. "Hydrogeology, simulation of ground-water flow, and saltwater intrusion, Potomac-Raritan-Magothy aquifer system, northern Coastal Plain of New Jersey," N.J. Geol. Surv. Geol. Rep. 36, 209 p, 2 pl.

5. Barton, C., Vowinkel, E.F., and Nawyn, J.P., 1987. "PreliminaryAssessment of water quality and its relation to hydrogeology and land use: Potomac-Raritan-Magothy aquifer system, New Jersey," U.S.G.S. Inv. Rep. 87-4023, pp. 79.

6. Harriman, D.A., Pope, D.A., and Gordon, A.D., 1989. "Water-quality data for the Potomac-Raritan-Magothy aquifer system in the Northern Coastal Plain of New Jersey, 1923-86," NJGS Geol. Surv. Rept. 19, 94 p.

7. Pucci, A.A., Jr., T.A. Ehlke, and J.P. Owens, 1992. "Confining unit effects on water quality in the New Jersey Coastal Plain," *Ground Water*, (30)3:415-427.

8. Pucci, A.A., Jr. and Owens, J.P., 1994. "Paleoenvironmental, lithologic, and hydrogeologic controls on the hydrogeochemistry of aquifer systems in the New Jersey Coastal Plain," *The Bulletin, J. N.J. Acad. Sci.*, 39(1):7-16.

9. Fogg, G.E., 1986. "Groundwater flow and sand body interconnectedness in a thick, multiple-aquifer system," *Water Reour. Res.*, 22(5):679-694.

10. Webb, E.K., and Anderson, M.P., 1990. "Tracing contaminant pathways in sandy heterogeneous glaciofluvial sediments using a sedimentary depositional model," in Proceedings of the International Conference and Workshop on Transport and Mass Exchange Processes in Sand and Gravel Aquifers Part 1 (of 2), held in Ottowa, Canada, pp 342-354.

11. Neuzil, C.E., 1986."Ground-Water Flow in Low-Permeability Environments,"*W. Resour.Res.*,22(8):1163-1195.

12. Belitz, K. and Bredehoeft, J.D., 1989. "Role of confining layers in controlling large scale regional ground-water flow" in *Hydrogeology of Low Permeability Environments*, S.P. Neuman and I. Neretnieks, eds., I.A.H. 28th Int. Geol. Cong., Wash., D.C., July 9-19, 1989, pp. 7-17.

13. Maslia, M. (USGS), 1988. Oral comm. at U.S. Geolog. Survey Workshop on the Geology and Geohydrology of the Atlantic Coastal Plain, Reston, VA., 9/28-29/88.

14. Champ, D.R., Gulens, J., and Jackson, R.E., 1979. "Oxidation-reduction sequences in ground water flow systems," *Can. J. Earth Sciences*, 16(1):12-23.

15. Morgan-Jones, M. and Eggboro, M.D., 1981. "The hydrogeochemistry of the Jurassic limestones in Gloucestershire, England," *Quar. J. Eng. Geo.*, 14:25-39.

16. Wicks, C.M., and Herman, J.S., 1994. "The effect of a confining unit on the geochemical evolution of ground water in the Upper Floridan aquifer system," *J. Hydrol.*, 153:139-155.

17. Chapelle, F.H., and Knobel, L.L., 1983, "Aqueous geochemistry and the exchangeable cation composition of glauconite in the Aquia aquifer, Marland," *Ground Water*, 21(3):343-352.

18. Sato, M., 1960. "Oxidation of sulfide ore bodies, 1. Geochemical environments in terms of Eh and pH," *Economic Geology*, 55:928-961.

19. Willatts, E.C., 1970. ""Maps and maidens,", *Cartographic Journal*, 7(1) p.50, <u>cited in</u> D. J. Varnes, 1974, *The logic of geological maps, with reference to their interpretation and use for engineering purposes*, U.S.Geological Survay Professional Paper 837 48 p.

*Maps and Maidens--*

*They must be well proportioned and not too plain;*
*Colour must be applied carefully and discreetly;*
*They are more attractive if well dressed but not over dressed;*
*They are very expensive things to dress up properly;*
*Even when they look good they can mislead the innocent;*
*And unless they are very well bred they can be aweful liars!*

Willatts [19]

# INVESTIGATION AND MONITORING OF A PETROLEUM HYDROCARBON PLUME IN THE BLUE RIDGE PHYSIOGRAPHIC PROVINCE OF SOUTHWESTERN VIRGINIA

ANGELIA C. RISNER
    Delta Environmental Consultants, Inc.
    6701 Carmel Road
    Charlotte, NC 28226
    (704) 541-9890
    (704) 543-4035 fax

## ABSTRACT

Petroleum product was observed in a ditch between two bulk petroleum facilities in southwestern Virginia. Employees of an adjacent furniture factory reported to the local Hazardous Materials Team that petroleum fumes were present in factory buildings. Based on this information, the Commonwealth of Virginia Department of Environmental Quality (DEQ) required a joint investigation of the release by both petroleum distributors. Media coverage as well as questions of responsibility between the two parties resulted in a politically sensitive arena for the investigation to proceed.

A total of thirty seven (37) soil borings and twenty four (24) monitoring wells were installed and sampled. Field measurements indicated that liquid phase petroleum hydrocarbons were not present in the monitoring wells. Laboratory results from soil samples indicated total petroleum hydrocarbons as gasoline were present at concentrations as high as 1,000,000 micrograms per kilogram. Laboratory results of ground water sampling indicated concentrations of benzene as high as 3,100 micrograms per liter.

Due to the urban location of the site, a secure public water source was available. Based on the availability of public water and the absence of liquid phase petroleum at the site, it was determined that the risk to human health from exposure to petroleum compounds was minimal. A Site Characterization Report was prepared to summarize site conditions and the information gathered during the site investigation. The report recommended ground water monitoring based on the lack of complete exposure pathways. The DEQ agreed with the recommendation for monitoring.

## INTRODUCTION

The purpose of the investigation was to gather hydrogeological information concerning the aquifer, define the extent of vapor, soil, free product and dissolved ground water contamination, assess site risk to humans and the environment and develop remedial strategies for corrective action as required by the state of Virginia for VR 680-13-02 Sections 6.3 and 6.4 Initial Abatement and Site Characterization Reports.

The two bulk petroleum facilities are located in the Blue Ridge Physiographic Province of Carroll County, Virginia. A regional topographic map illustrating site location, surface water bodies, and drainage patterns is shown in Figure 1.


## SITE ENVIRONMENTAL HISTORY

Both properties (Distributor #1 and Distributor #2) consist of product storage tanks, distribution lines, and dispensers and have operated as bulk petroleum storage facilities for at least 50 years. The loading areas on the Distributor #1 property were for leaded and unleaded gasoline, kerosene, and #2 diesel fuel oil. Petroleum products from above ground storage tanks were transferred to loading areas by above and below ground distribution lines. The fuel pump island, or loading area, on the Distributor #2 property has underground fuel lines that lead to the loading pad. The fuel lines leading from the pump house to the gasoline and kerosene tanks are above ground.


## EMERGENCY RESPONSE

In 1989 the police department was notified by employees of an adjacent furniture factory, northeast of the site, that petroleum product was flowing under their property via a storm sewer and creating fumes in their building. The storm sewer was traced to the lot owned by Distributor #1, east of the two bulk facilities. Prior to purchase in 1988, this empty lot was part of a railroad. Puddles of black petroleum were observed in the ditch between the two bulk facilities and draining into the storm sewer pipe behind the Distributor #2 property. The fire department was then notified. Oil absorbent pads and hay were placed along the ditch to soak up the petroleum. A foam agent was applied to the surface of the water to hold down the vapors and the incident was then considered resolved.

One week later the hazardous materials team was alerted when employees of the furniture company again reported petroleum fumes entering their building from the storm sewer. The storm sewer discharge was traced back to the storm sewer input pipe behind the Distributor #2 site facility. The discharge was also further traced to the drainage ditch between a storage building on the first site property and the pump house owned by Distributor #2 site. A trench was dug on the empty lot owned by Distributor #1 to prevent runoff from entering the storm sewer. This ditch became filled with water and liquid phase hydrocarbons (LPH) which seeped in through the sides of the trench. Officials from an adjacent county regional hazardous materials team, Virginia Emergency Services and the Virginia State Water Control Board (VSWCB) were summoned to help assess the emergency. The product and contaminated materials were contained and removed to eliminate imminent public hazard.

Initially two narrow trenches were excavated on the empty lot owned by Distributor #1. The excavation was backfilled with gravel and two vertical 18 inch slotted galvanized pipes were set, Sumps 1 and 2 (Figure 2). Approximately 380 cubic yards of soil were excavated from both properties and stored at an adjacent lot. Each pile of soil was lined with plastic, covered with plastic, and surrounded by hay bales.

## DISCUSSION

### Water Supply
Water for commercial and residential facilities supplied by the city is obtained from Chestnut Creek. The city filtration plant, which is located on Chestnut Creek, is approximately 2700 feet upstream from both sites.

### Geology
The sites lie within the Blue Ridge Physiographic Province of southwest Virginia. This part of the central Blue Ridge Anticlinorium is characterized by metamorphic and crystalline rocks. The area is underlain by mica schist and phylite of the Lynchburg Group of the Proterozoic Ashe Formation. Surficially, the sites are underlain by alluvial deposits from Chestnut Creek. As observed in the excavation area, these deposits consist of alternating layers of peat, silty clay, sand, and gravel varying in thickness from a few inches to two feet.

A total of thirty seven (37) soil borings have been advanced on both sites. Soil samples were collected and submitted to the laboratory for analysis. The area is covered by 1-2 feet of gravel fill underlain by alternating layers of micaceous silt, clay, sand, gravel, and peat. Thicknesses of each layer vary from two inches to three feet. As soil boring SB-14 was advanced, bedrock was encountered at approximately 18 feet. The bedrock was described as a metamorphic rock resembling micaceous phylite.

### Hydrogeology
Based on the soil conditions during the installation of the monitoring wells, the aquifer is expected to be anisotropic with a horizontal component of flow greater than the vertical component. Overall horizontal ground water flow is directed to the northeast in the direction of Chestnut Creek.

Significant ground water mounding occurs around the Distributor #1 bulk plant facility to the north. This mounding can be attributed to gravel fill on the Distributor #1 property and around the boundary of the Distributor #2 property. Average horizontal hydraulic gradient in a direction parallel to ground water flow across the site is approximately 0.005 ft/ft.

Based on slug test data collected from monitoring wells MW-1, MW-3, and MW-7, an average discharge velocity through the surficial sediments is expected to be 0.0081 ft/day. Assuming the effective porosity for the sediments is 30%, the average horizontal ground water flow velocity is estimated to be 0.027 ft/day or 6.09 ft/year.

## Soil Contamination

Soil sample field screening and analytical results indicate petroleum hydrocarbons are present in soils throughout each site. Maximum concentrations were measured at the locations of soil borings SB-23 and SB-24, as well as the area surrounding the loading pad on the Distributor #2 property. Analytical results indicate that soil contamination is concentrated near the northern property boundary of the Distributor #2 facility. Results obtained from soil borings and soil excavation indicate the source of contamination is likely from small overfills and spills around the loading pad areas and fill areas during normal facility operations. Another possible source of contamination is from previous railroad activities years ago.

Soil samples collected in from SB-14 through SB-24 were prepared according to EPA SW-846 3550 and 5030 and analyzed for benzene, toluene, ethylbenzene total xylenes (BTEX) and methyl-tert-butyl ether (MTBE) according to Method 602 and total petroleum hydrocarbons (TPH), as well as Method 239.2 for total lead. Concentrations of TPH have been divided into low boiling point TPH, including chemical structures similar to gasoline, and high boiling point TPH, including chemical structures similar to kerosene, diesel fuel, and motor oil.

The maximum concentration for BTEX/MTBE and high boiling point TPH was measured in the soil sample collected from SB-23 (111,900 ug/kg and 968,000 ug/kg respectively). The maximum concentration for low boiling point TPH was measured in the soil sample collected from SB-24 (1,000,000 ug/kg). The maximum concentration for lead in soil is SB-17 (35 mg/kg). The estimated extent of soil displaying levels of high TPH boiling point concentrations in plan view is illustrated in Figure 3.

The vertical extent of contamination is estimated from two feet to eight feet based on the sample depths analyzed. Most of the contamination is localized to the soil near the surface. Soil sample SB-16 was collected at eight feet below grade and contained levels above detection limits for BTEX/MTBE (56,700 ug/kg), high boiling point TPH (465,000 ug/kg), low boiling point TPH (400,000 ug/kg) and lead (18 mg/kg).

## Ground Water Contamination

Analytical results from ground water samples collected April 1994 indicate three distinct zones of ground water contamination. The first zone of ground water contamination is centered around the gasoline loading pad on the Distributor #1 property, the second zone resides around the southeast area of the empty lot owned by Distributor #1, and the third zone was detected on most of the Distributor #2 property. The first and third zones of contamination can be attributed to normal daily operations resulting in small spills and overfills at the facilities. The second zone of contamination may be from one continuous plume from the west. This can be deduced from the analytical results from soil boring SB-16 and the questionable integrity of monitoring well MW-6.

All ground water samples were analyzed according to methods outlined by EPA SW-846 Method 602 for BTEX/MTBE and low boiling point TPH. Method 239.2 for total lead and Method 610 was used for semi-volatiles including naphthalene, 1-methylnaphthalene, 2-methylnaphthalene and high boiling point TPH.

The highest concentration of BTEX/MTBE, low boiling point TPH and semi-volatiles was measured in the ground water sample collected from monitoring well G-5 (11,130 ug/L, 25,000 ug/L, and 1,680 ug/L respectively). The estimated horizontal extent of BTEX/MTBE in plan view is illustrated in Figure 4. The highest concentration of high boiling point TPH was measured in the sample collected from monitoring well G-4 (26,200 ug/L). The highest concentration of lead was measured in the sample collected from drive point DP-1 (1300 ug/L).

The ground water sample collected from monitoring well DW-1 in April 1994 was analyzed for BTEX/MTBE, TPH, lead, and semi-volatiles. The estimated vertical extent of BTEX/MTBE and TPH in ground water is approximately seven feet below grade. The estimated vertical extent of BTEX/MTBE in cross-section is illustrated in Figure 5.

**Other Contaminated Phases**

Free product, or LPH, was observed in monitoring well G-5 in 1989, however no LPH has been observed since 1990. The only evidence of elevated vapors is from soil screening below the ground surface.

**CONCLUSIONS**

- The depth to water ranges from one to four feet below grade.

- Vapor contamination is not present at the either site.

- The horizontal extent of petroleum hydrocarbons in soil have been defined and elevated levels of petroleum hydrocarbons are located throughout both properties.

- LPH were observed in monitoring well G-5 in 1989, however no LPH has been observed since 1990.

- Contamination has not migrated significantly since 1989 as evidenced by analytical results from the downgradient monitoring wells.

- The horizontal extent of dissolved petroleum hydrocarbons has been defined. The mainstay of dissolved constituents are primarily restricted

704

to each property with levels less than 100 ug/l measured immediately downgradient of the property boundary.

- The vertical extent of dissolved petroleum hydrocarbons has been defined. The mainstay of contamination is restricted to the water table fluctuation zone.

## RISK ASSESSMENT CONCLUSIONS

A baseline health risk assessment was conducted at both facilities. The current population identified at or near the sites consist of employees on-site or at nearby businesses. The nearest residential units downgradient of the site are located approximately 1/2 mile southeast of the sites. There are no residential units downgradient of the site between the sites and Chestnut Creek, the discharge area for ground water migrating off the site. Five contaminants of potential concern were identified in ground water at the site; benzene, ethylbenzene, methyl tertiary-butyl ether, toluene, and xylene. Contaminants of potential concern are those contaminants that are considered to be present as a result of past site activities. During the exposure assessment, the exposure setting was characterized, as was the potentially exposed populations. An exposure pathway describes the course by which a contaminant migrates in the environment from a source to a potential receptor. An exposure pathway consists of four elements:

1. A source and mechanism of contaminant release;

2. A transport medium (or media);

3. A point of potential human contact with the contaminated medium (referred to as the exposure point); and

4. An exposure route at the exposure point.

All four elements are necessary to describe a complete exposure pathway. (1) Two probable sources of contamination include the gasoline loading pad on the Distributor #1 property and on most of the Distributor #2 property including the property boundary. (2) The identified transport medium at the site is ground water. (3) No current exposure points were identified. There are no potable water wells on-site or off-site within the estimated boundaries of the contaminant plume. Chestnut Creek was evaluated as a potential exposure point. Contaminated ground water may discharge to the creek in the future. However, it was determined that Chestnut Creek was not an exposure point because the five contaminants of potential concern have high vapor pressures and Henry's Law Constants and are expected to quickly volatilize into the atmosphere after discharging into the creek. (4) No current exposure pathways were identified. Exposure routes are defined as the mechanism by which a contaminant enters the body. These can include ingestion, inhalation, or dermal absorption.

The baseline health risk assessment indicates that no unacceptable risk is present due to exposure to site-specific contaminants resulting from on-site exposure or future potential off-site exposure in Chestnut Creek. It is important to note that this is a future potential risk, and the risk assessment was designed to overestimate, rather than underestimate, potential risk.

## REMEDIATION ASSESSMENT

The following technologies were considered for feasibility in regard to recovery of petroleum hydrocarbons as gasoline at the sites.
  a)  Air Sparging
  b)  In-Situ Bioremediation
  c)  Passive Remediation
  d)  Pump and Treat
  e)  Vapor Extraction
  f)  Excavation

## RECOMMENDATIONS

Remediation strategies for both sites include passive remediation. Based on the risk assessment, ground water monitoring will continue until sufficient data is collected to show that continued monitoring of benzene, toluene, ethylbenzene, and xylene concentrations is not cost effective. The contaminant plume is expected to be adsorbed and degraded to non-detectable concentrations over time.

## STATE OF VIRGINIA'S RESPONSE

The state of Virginia agreed with the passive remediation strategy and monitoring of the sites is currently being implemented. Recently the state of Virginia has modified their environmental regulations towards closing more sites that are no threat to public health or the environment. They concurred that these sites are no threat to either.

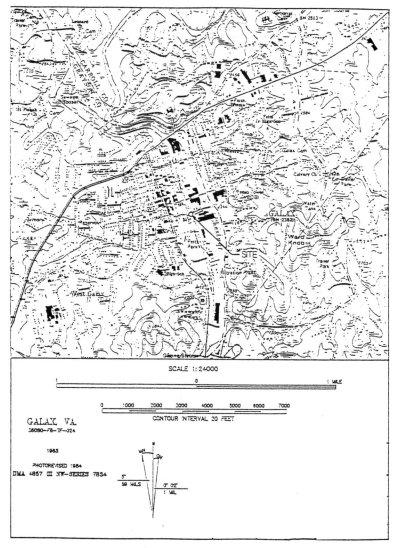

SCALE 1:24000

CONTOUR INTERVAL 20 FEET

GALAX VA.
38080-F8-TF-024

1965

PHOTOREVISED 1984

DMA 4857 II NW—SERIES 7834

Figure 1

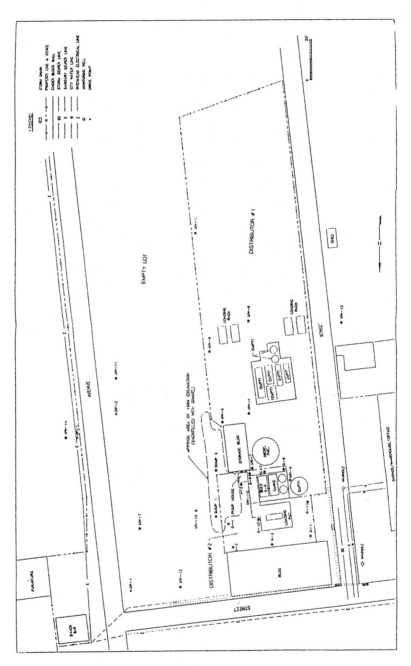

Figure 2

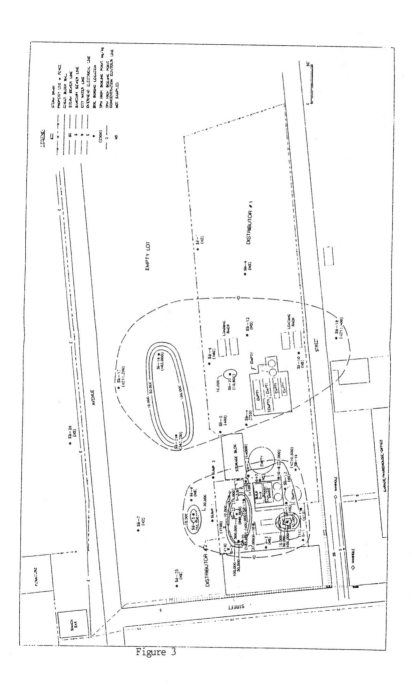

Figure 3

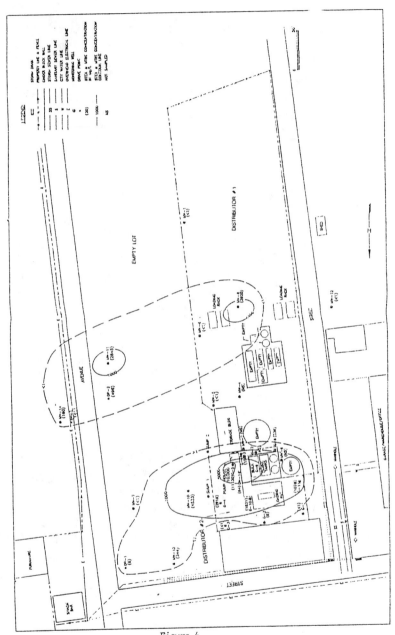

Figure 4

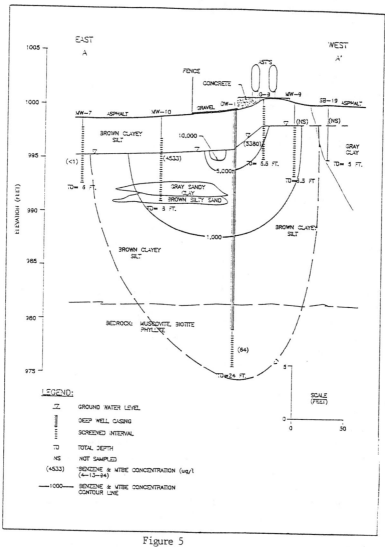

Figure 5

# ENHANCED SLURRY WALLS AS TREATMENT ZONES FOR INORGANIC CONTAMINANTS

JEFFREY C. EVANS
TROY L. ADAMS
KARRIE A. DUDIAK
Bucknell University
Lewisburg, PA 17837

## INTRODUCTION

At present, slurry walls are widely used as passive vertical barriers to control the horizontal flow of contaminated ground water. The most commonly employed slurry wall is known as a soil-bentonite cutoff wall and is composed of a backfill mixture of soil, bentonite, and water. To date, slurry walls have not been used as a medium to treat contaminated ground water which passes through the wall. This paper describes the modification of the present mix design through the addition of alternative clay minerals such as those which act as "molecular sieves" and have superior adsorption capabilities. In this way the enhanced slurry wall will serve to both impede the rate of ground water flow and remove contaminants from the ground water as it passes through the wall.

Geochemical attenuation of lead was studied in an effort to investigate the use of enhanced soil-bentonite slurry walls as active treatment zones for heavy metal ions in solution in the ground water. Distribution coefficients were determined from adsorption isotherms for conventional sand-bentonite and enhanced sand-bentonite-chabazite backfills. These results were 23 ml/g and 144 ml/g, respectively. Using these laboratory data, theoretical breakthrough curves were generated to predict and compare breakthrough time between the conventional barrier and the enhanced barrier. Adsorption test and analytical modeling results demonstrate that adding chabazite to slurry wall backfill does enhance the conventional passive barrier performance.

Based on the results of these studies, including laboratory testing and mathematical modeling, it was found that the enhanced slurry wall mixture considerably increased the time for contaminant breakthrough over conventional slurry wall mixtures.

# BACKGROUND

## Soil Bentonite Slurry Trench Cutoff Walls

A conventional soil-bentonite slurry trench cutoff wall consists of a mixture of soil, bentonite, and water mixture placed vertically into the subsurface of the earth. This technique has been used in various applications for over 40 years. The term slurry trench is derived from the method of construction where a bentonite-water slurry is used in the trench to maintain trench stability. The soil-bentonite backfill material is then placed in the vertical trench, displacing the slurry used for construction stability. Slurry trench technology has historically been used for seepage control and dewatering purposes. More recently slurry trench technology has been used to construct vertical barriers for subsurface control of migration of sewage, acid mine wastes, chemical wastes, and sanitary landfill leachate.

## Transport Processes

Advection, diffusion, and dispersion control the fate and transport of ground water contaminants. It is through modification of the soil parameters that affect these processes that an enhanced barrier can be developed. Although the following equations are available in a number of source, the authors relied heavily upon Shackelford [1] in their preparation of this paper.

Advection is the movement of a solute(s) with the ground water (the solvent) due to a hydraulic gradient. The rate of solvent transport through a porous medium is known as the seepage velocity, or

$$v_s = \frac{v}{n} \tag{1}$$

where $v_s$ is the average linear velocity of the solvent, $n$ is the porosity of the medium, and $v$ is the flux of the solvent. The flux originates from Darcy's law which is written as:

$$v = \frac{Q}{A} = -K\frac{\partial h}{\partial x} = -Ki \tag{2}$$

where $Q$ is the flow rate of the solvent; $A$ is the total cross sectional area perpendicular to flow; $K$ is the hydraulic conductivity; $h$ is the hydraulic head; $x$ is the direction of flow, and $i$ is the hydraulic gradient. The advective mass flux of a specific contaminant can be found by:

$$J_A = vc = Kic = nv_s c \tag{3}$$

where $J_A$ is the advective mass flux, and c is the mass of solute per unit volume of the solvent.

Diffusion is the movement of contaminants driven by a concentration gradient. Fick's first law describes diffusion in a saturated porous medium as:

$$J_D = -D^* n \frac{\partial c}{\partial x} \qquad (4)$$

where $J_D$ is the diffusive mass flux and $D^*$ is the effective diffusion coefficient.

Mechanical dispersion is the spreading of a contaminant due to the encountering of solid material and tortuous path of voids in the medium. The dispersivity flux is expressed as:

$$J_M = -D_m n \frac{\partial c}{\partial x} \qquad (5)$$

where $J_M$ is the dispersive mass flux and $D_m$ is the mechanical dispersion coefficient.

The advective, diffusive, and dispersive fluxes are combined into one term called total flux, $J$, which is written as:

$$J = J_A + J_D + J_M \qquad (6)$$

or,

$$J = n v_s c - D^* n \frac{\partial c}{\partial x} - D_m n \frac{\partial c}{\partial x} \qquad (7)$$

or,

$$J = n v_s c - D_h n \frac{\partial c}{\partial x} \qquad (8)$$

where $D_h$ is the hydrodynamic dispersion coefficient defined as follows:

$$D_h = D^* + D_m = D^* + \alpha_L v_s \qquad (9)$$

where $\alpha_L$ is the longitudinal dispersivity coefficient.

## Adsorption and Geochemical Attenuation

As an contaminant flows through a porous medium, solute ions may adhere to the surface of the solids in a process known as adsorption. Attenuation is the ability a material has to immobilize a constituent in solution. Geochemical attenuation can be quantified by linear adsorption isotherms (Figure 1). The model isotherm shown in Figure 1 illustrates how the contaminant is partitioned between the liquid and solid phase when in contact with the clay minerals. Batch extraction tests and column tests are laboratory procedures that determine the mass of solute adsorbed per mass of solids versus the equilibrium concentration of solute in solution. The slope of the linear isotherm is the distribution coefficient, $K_d$, and has units of ml/g.

714

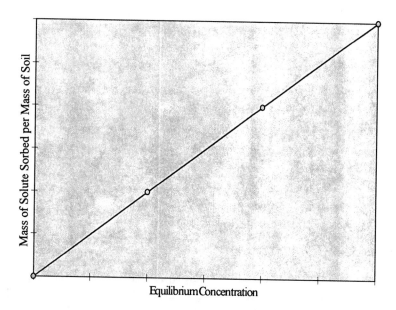

**Figure 1  Linear Adsorption Isotherm**

## Modeling Advection-Dispersion Equation

The advection-dispersion equation is written as:

$$\frac{\partial c}{\partial t} = \frac{D_h \partial^2 c}{R_d \partial x^2} - \frac{v_s \partial c}{R_d \partial x}. \tag{10}$$

Computer programs such as SUTRA are available to solve the advection-dispersion equation and generate theoretical breakthrough curves given $D_h$, $v_s$, and $R_d$, the retardation factor. Retardation is expressed as:

$$R_d = \frac{v_s}{v_r} \tag{11}$$

where $R_d$ is the number of pore volumes it takes for the contaminant to be detected in the effluent; $v_s$ is the seepage velocity; and $v_r$ is the transport rate of the center of mass of the solute [1]. From laboratory testing, the retardation factor is calculated by the equation:

$$R_d = 1 + \frac{\rho_b}{n} k_d \tag{12}$$

where $\rho_b$ is the bulk dry density of the solids. If $R_d > 1$ the solute is being adsorbed; if $R_d < 1$ desorption is occurring.

715

The solution to Equation 10 is [1]:

$$\frac{c(x,t)}{c_o} = \frac{1}{2}\left[ erfc\left( \frac{R_d x - v_s t}{2\sqrt{R_d D_h t}} \right) + \exp\left( \frac{v_s x}{D_h} \right) erfc\left( \frac{R_d x + v_s t}{2\sqrt{R_d D_h t}} \right) \right] \tag{13}$$

where:

$c(x,t)/c_o$ = reaction of the solute concentration at time t and distance x to the initial solute concentration $c_o$,

erfc = complementary error function,

$v_s$ = average linear velocity (cm/s),

$D_h$ = vertical dispersion coefficient (cm$^2$/sec)

t = actual time (sec).

## LABORATORY TEST RESULTS

Laboratory testing initially consisted of fixed wall column tests on a calcareous sand-bentonite backfill using chromium (Cr) as a contaminant. The sample in the fixed wall permeameter was 1.0 inch in height and 4.0 inches in diameter. This test was conducted to evaluate the breakthrough time of the contaminant and to study the effect of chromium on the permeability of the backfill.

The influent concentration of the chromium was 4.0 mg/l. The fixed wall permeameter was connected to a control panel and the influent and effluent volumes of influent were recorded daily. Influent and effluent samples were also extracted daily and chromium concentrations measured using an atomic absorption spectrometer (AA).

The hydraulic conductivity of the backfill increased about two orders of magnitude as a result of permeation with chromium contaminated water (Figures 2 and 3). The Cr contaminated water must have had an adverse affect on the diffuse ion layer of the bentonite particles. The strongly electropositive ions adsorbed into the soil structure reduced the diffuse ion layer thickness and resulted in an increase in hydraulic conductivity. This response is consistent with that expected based on previous studies[2]. The Guoy-Chapman model indicates that as the valence increases, the diffuse ion layer of the bentonite particles increases, therefore; the hydraulic conductivity of the backfill increases.

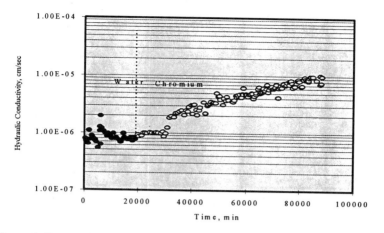

**Figure 2  Hydraulic Conductivity versus Time**

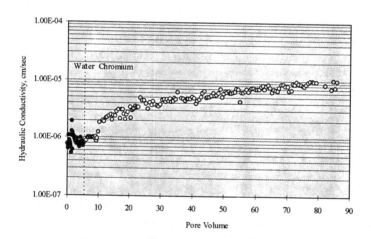

**Figure 3  Hydraulic Conductivity versus Pore Volume Displacement**

The breakthrough curve (Figure 4) illustrates the chromium being trapped in the porous matrix. After 60 days of testing, the concentration of the chromium detected in the effluent was only 2.0 ppm compared with 4.0 ppm in the influent.

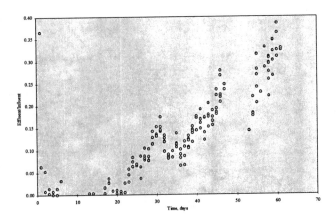

**Figure 4  Breakthrough Curve for Chromium**

Although the permeability test was a valid technique, data was generated slowly. Due to time constraints, it was decided to analyze the adsorption capacity of the calcareous sand-bentonite backfill by adsorption isotherms. Several iterations in the research protocol were required to develop an acceptable test procedure. Research began by placing 31 grams of sand-bentonite backfill into a 100 ml of a 4.5 mg/l chromium solution. The samples were shaken on an orbital shaker at 185 rpm. Sub-samples were extracted periodically for determination of chromium concentration in solution. The concentration in solution was found to be above the AA limit of 15 ppm. Further research into smectitic clays indicate that they are rich in chromium. Considering this interference, it was decided to change the contaminant to lead.

Initial tests using lead ($Pb^{+2}$) also failed to yield reproducible results. Initially, the samples were not centrifuged. As a result, colloidal soil particles scatter the light in the AA and give rise to erroneous readings. Subsequently, all samples were centrifuged at 50,000g for 15 minutes.

The soil:liquid ratio was identified as an important parameter in adsorption testing. A 1:20 ratio was identified from the literature and used for the remainder of the testing. The correct mass of soil (6.25 grams) was added to 125 ml of Pb solution in a 125 ml Erlenmeyer flask.

The equilibrium time, the time when the Pb ions stopped adsorbing on to the soil, was established by shaking a sample up to 24 hours and periodically testing the supernatant after centrifugation. The equilibrium time was found to be 24 hours.

It is known that pH has an affect on adsorption. To simulate natural ground water (having a pH of 5-8) the pH of the samples were adjusted to 6 using 0.05N NaOH and 10N HCl. As the pH of the Pb solution rose above 4, $Pb(OH)_2$ precipitated out of solution. Precipitation complicates the data analysis. When Pb(OH)$_2$ precipitates out, it is neither in solution nor adsorbed on to the soil particles. To avoid precipitation, the pH of the samples was not adjusted, but the pH was recorded before and after shaking.

Since the reproducibility was still limited after the above changes to the test procedure, it was decided to change the calcareous sand-bentonite backfill to silica sand-bentonite backfill. A trial was run, and the results obtained were reproducible. The calcareous sand contained CaCO$_3$, causing pH changes and interfering with the adsorption process. Testing continued using silica sand-bentonite backfills, both unmodified and enhanced.

The partitioning coefficient, $K_d$, of silica sand-bentonite backfill and silica sand-bentonite-chabazite is 23 ml/g and 144 ml/g, respectively (see Figure 5). The greater the $K_d$ coefficient, the greater the adsorptive capacity of the backfill. The adsorptive capacity of the silica sand-bentonite-chabazite is about 6 times that of the silica sand-bentonite backfill. Chabazite is a zeolitic clay. "Zeolite is an aluminosilicate with a skeletal structure containing voids occupied by ions and molecules of water having a considerable freedom of movement that leads to ion-exchange and reversible dehydration."[3] The most commonly used zeolites are analcite, chabazite, clinoptilolite, erionite, and mordenite. Zeolites have the capability to allow water and other cations to move into its structure. The ions can exchange with others while the water can be removed or replaced continuously[3]. Zeolite is referred to as the "molecular sieve" because it allows water to pass through its framework and retain selected molecules within its intracrystalline pores.

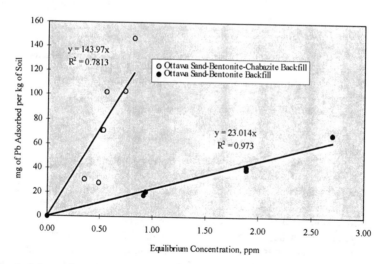

**Figure 5 Adsorption Isotherms: Lead**

## MODELING RESULTS

Using the $K_d$ values of the silica sand-bentonite and silica sand-bentonite-chabazite batch extraction tests, the advection-dispersion equation was solved for "typical" slurry wall conditions. For this modeling, a gradient of one for a one meter thick barrier having a hydraulic conductivity of 1x10-7 was assumed. The results are presented on Figure 5. The modeling show that the contaminants are arriving at the downstream face of the wall at concentrations of 50% of the starting concentration after about 50 years for the soil-bentonite backfill. For the enhanced barrier this breakthrough time is increased to about 300 years.

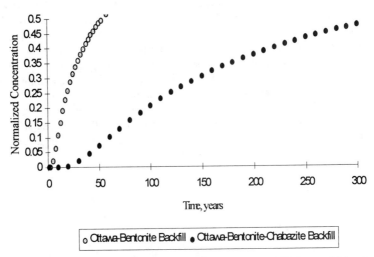

**Figure 6  Breakthrough Curves for Conventional and Enhanced Barriers**

## CONCLUSIONS

The $K_d$ values of the silica sand-bentonite and silica sand-bentonite-chabazite backfills demonstrate that a minerally-enhanced slurry wall will adsorb heavy metal ions in solution. Theoretical breakthrough curves for model conditions show that the breakthrough time is increased from 50 years for a conventional slurry wall backfill to 300 years for an enhanced slurry wall backfill. It is believed that these findings may have a significant effect upon our approach to barrier design. The conventional slurry trench cutoff wall is transformed from functioning merely as a passive ground water flow barrier to a active treatment component "cleaning" the water as it passes through the barrier. This extends the service life and improves the ability of the barrier to reduce contaminant transport. Further, if this improved performance is considered in remedial design, it may reduce pumping and treatment costs typically associated with the conventional soil-bentonite slurry wall applications in site remediation.

## ACKNOWLEDGMENTS

The chabazite was provided by GSA Resources, Inc. of Tucson, Az. The authors appreciate the assistance of Drs. Kaestner, Keithan, Kirby and Prince. Special thanks to Meghan Beck of Bucknell University for her assistance with the data gathering.

## REFERENCES

1. Shackelford, Charles D., *Contaminant Transport*, Geotechnical Practice for Waste Disposal, David E. Daniel, Ed., New York: Chapman and Hall, 1993, pp. 40-41.
2. Alther, G. R., Evans, J. C., Witmer, K. A., and Fang, H. Y., "Inorganic Permeant Effects Upon Bentonite," *Hydraulic Barriers in Soil and Rock*, ASTM STP 874, 1985, pp. 64-74.
3. Tsitsishvili, G.V., T.G. Andronikashvili, and B.N. Kirov L.D. Filizova. Natural Zeolites. New York: Ellis Horwood Limited, 1992, pg. 1.

# USE OF FIELD SCREENING TO DELINEATE A LOW-LEVEL GROUNDWATER PLUME OF ETHYLENE DIBROMIDE

## MICHAEL J. GUNDERSON
Hazardous Waste Remedial Actions Program[*]
Environmental Restoration and Enrichment Facilities
Martin Marietta Energy Systems, Inc.
Oak Ridge, Tennessee 37831

## NELSON M. BRETON
ABB Environmental Services, Inc.
Portland, Maine

## EDWARD L. PESCE
National Guard Bureau
Otis Air National Guard Base, Massachusetts

## R. WAYNE HAMMONS
Solutions To Environmental Problems, Inc.
Oak Ridge, Tennessee

## ABSTRACT

During routine groundwater sampling of monitoring wells downgradient of a groundwater extraction system under construction at the Massachusetts Military Reservation (MMR), a trace level (i.e., <0.05 $\mu$g/L) of ethylene dibromide (EDB) was discovered in one monitoring well. The groundwater extraction system was designed to contain a chlorinated solvent plume. The occurrence of EDB was not anticipated and had only recently been added as a contaminant of concern for ongoing MMR investigations. Because of its low state regulatory limit in groundwater (0.02 $\mu$g/L) and potential impact on the location of the extraction system, a two-phased field screening approach was undertaken to confirm the presence/absence and extent of EDB. The presence

—————————————
[*]Managed by Martin Marietta Energy Systems, Inc., for the Department of Energy under contract DE-AC05-84OR21400

722

of EDB was confirmed; however, a potential source area could not be determined. With regulatory concurrence, a field screening investigation to determine the extent of the EDB contamination in the vicinity of the extraction system and potential impact on design was undertaken. Field screening results confirmed that EDB and volatile organic compounds (VOCs) were downgradient of the extraction system. Based on the screening data of compounds detected, the relative concentrations of the VOCs, and the intervals of detection, it was concluded that the EDB and VOCs represented a separate contaminant plume deeper in the aquifer than the solvent plume. Based on field screening data and with regulatory concurrence, it was determined that the extraction system design would not require modification. The use of screening data vs Contract Laboratory Program data saved approximately 60 workdays.

## INTRODUCTION

The Massachusetts Military Reservation (MMR) is located on Cape Cod, Massachusetts (Figure 1). Historical spills, leaking underground storage tanks and pipelines, landfills, and waste disposal have resulted in soil and groundwater contamination. Seventy-nine individual sites have been identified on the 22,000-acre facility. Both fuel and chlorinated solvent plumes are present; however, because of the recalcitrant nature of the solvents, these plumes have the largest areal extent and present the most significant risk to downgradient receptors. A contaminant of concern for the MMR plumes is EDB, a carcinogen with a regulatory limit in groundwater of 0.02 $\mu$g/L. This paper will focus on a relatively small EDB plume associated with a larger chlorinated solvent plume.

## HYDROGEOLOGY

The geology of western Cape Cod is dominated by Wisconsinian-age glacial sediments deposited from 7,000 to 85,000 years ago. Sediment thickness ranges from 300 ft in the northern portion to less than 150 ft in the southern portion. Glaciofluvial outwash sediments, known as the Mashpee Pitted Plain (MPP), comprise much of western Cape Cod. The MPP deposits consist of poorly sorted, fine-to-coarse sand. Occasionally, finer grained glaciofluvial and glaciolacustrine and till deposits are present below the MPP. The MPP is bounded to the west and north by the Buzzards Bay Moraine (BBM) and Sandwich Moraine (SM), respectively. Both the BBM and SM consist of poorly sorted sand to cobble size material with occasional boulders. Oldale and O'Hara (1984) interpreted the BBM and SM to be composed of glacio-tectonically deformed glacial deposits [1].

A single unconfined aquifer underlies western Cape Cod. Recharge for the aquifer occurs primarily from infiltration of precipitation. MMR is the major

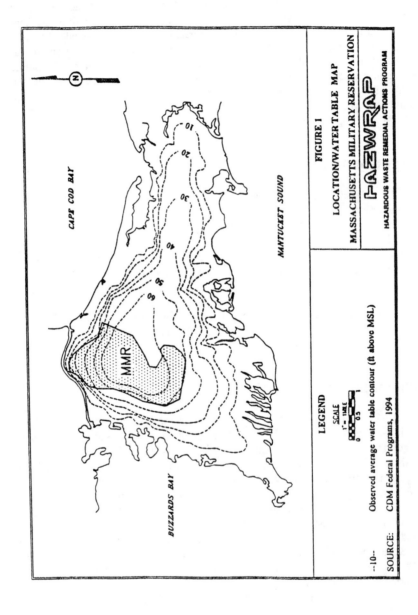

Figure. 1. Location/water table map - Massacchusetts Military Reservation.

recharge area for the aquifer, with groundwater flowing radially from the facility (Figure 2). The surrounding towns of Bourne, Sandwich, Falmouth, and Mashpee use groundwater for their water supply. The aquifer was designated a sole-source aquifer by the U.S. Environmental Protection Agency (EPA) because it is the principal source of drinking water for Cape Cod. Water table gradients on the MMR are approximately 0.002 ft/ft. Groundwater flow is generally horizontal except where kettle hole ponds intercept the water table. Recent remedial investigations and groundwater modeling have shown strong vertical gradients associated with these ponds; however, kettle hole ponds do not exist in the study area.

The CS-4 site was a military vehicle maintenance area operated by the U.S. Army from 1940 to 1946 and by the U.S. Air Force from 1955 to 1973. Potential releases of oils, solvents, antifreeze, battery electrolytes, paint, and waste fuels were identified in a records search at the base. Subsequent investigations identified surface soil contamination and numerous dry wells and maintenance pits where wastes were disposed. In addition, groundwater contamination was determined at the source area. During characterization of CS-4, a zone of groundwater contamination was identified (Figure 3). The contaminants of concern were primarily chlorinated VOCs, principally tetrachloroethylene (PCE), trichloroethylene (TCE), trichloroethane (TCA), and dichloroethylene (DCE) with maximum total concentration exceeding 80 $\mu$g/L. The contaminant plume was estimated to be approximately 800 ft wide, 11,000 ft long, and 40 ft thick. The plume has migrated downward approximately 60 ft into the aquifer with distance downgradient because of rainfall accretion. A Feasibility Study was completed for the groundwater plume [2]. A conventional multiple well groundwater containment system was selected as an interim remedial measure. Construction of the containment system began in December 1992 and was completed in October 1993.


## STUDY AREA HISTORY

Operation began in November 1993.

During routine groundwater sampling of monitoring wells downgradient of a groundwater extraction system under construction, a low-level concentration (0.02 $\mu$g/L) of EDB was discovered in one monitoring well. The occurrence of EDB was not anticipated and had only recently been added as a contaminant of concern for ongoing MMR investigations. Supplemental analyses by two different laboratories of samples from this well confirmed the presence of EDB. The maximum concentration of EDB was 0.06 $\mu$g/L, which exceeds state and federal drinking water standards (0.02 and 0.05 $\mu$g/L, respectively).

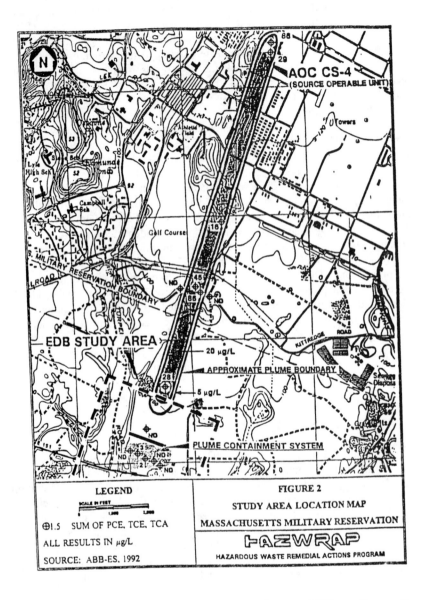

Figure 2. Study area location map - Massachusetts Military Reservation.

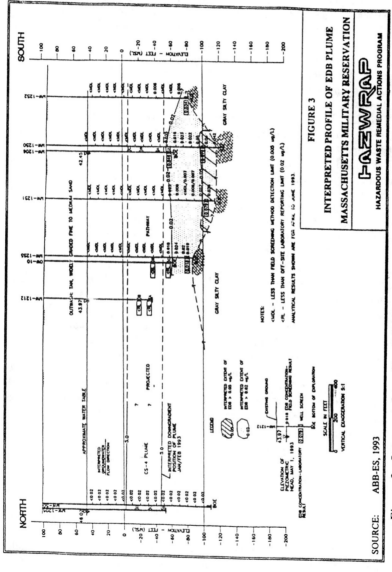

Figure 3. Interpreted profile of EDB plume - Massachusetts Military Reservation.

727

# EDB GROUNDWATER INVESTIGATION

Based on the initial sampling and because of EDB's low state regulatory limit in groundwater and potential impact on the location of the extraction system, a two-phased field screening approach was undertaken to confirm the presence/absence and extent of EDB [3]. The initial field effort consisted of sampling and analysis of existing wells to confirm the detection and determine if an upgradient source could be identified. Approximately 30 samples were collected from potential source areas and areas adjacent to the extraction system. All samples were sent to a local laboratory for analysis of EDB and VOCs. The sampling confirmed the EDB previously detected downgradient of the extraction wells; however, other EDB analytical results were negative at a detection limit of 0.01 μg/L. With regulatory concurrence, an investigation to determine the extent of the EDB contamination in the vicinity of the extraction system and potential impact on design was undertaken.

The investigation relied on screened hollow-stem auger sampling of groundwater at discrete intervals. Groundwater samples were collected by lowering a 2.5-in.-diam. stainless steel submersible pump to the screened interval. An inflatable packer system was used to limit the amount of purge water generated. Each sampling interval was purged of approximately 20 gallons of water before sampling. All purge water was treated on-site using granular activated carbon and discharged to the ground surface. Groundwater samples were collected at a low flow rate directly from the discharge tubing. Samples were analyzed on-site for EDB and VOCs by modified EPA Method 504 and 8010/8020. The detection limit for EDB was 0.005 μg/L. Modifications to EPA Method 504 necessary to achieve the lower detection limits included a larger sample size (25 mL) and hexane extraction [4]. Confirmation samples (10% of total screening samples) were collected and submitted to a fixed-base laboratory (EPA Methods 504 and 8010/8020) to confirm the field screening data. The field screening confirmation samples were collected from both screened auger borings and existing and newly installed monitoring wells. Construction of the extraction system was delayed during the investigation. Eight deep (i.e., approximately 200 ft) screened auger borings were sampled at 10-ft intervals to delineate the extent of the EDB and VOC contamination and its relationship to the chlorinated solvent plume targeted for extraction. Field screening data were also used to select final screened interval depths for monitoring wells.

# RESULTS

Confirmation samples collected showed a good correlation with the field screening data (Table I). The variation in analytical results, although relatively small, likely were because of different sampling methods for screened augers vs wells, temporal affects of field screening and well sampling, loss of volatiles during shipping, and slightly different sampling intervals.

**Table I. Comparison of EDB field screening samples vs fixed-base laboratory analysis**

| Sampling location | Screened interval (ft MSL) | Laboratory analytical results | |
|---|---|---|---|
| | | Field screening | Fixed-base |
| MW-1212A | -35 to -40 | <0.005 | <0.02 |
| MW-1212B | -20 to -25 | <0.005 | <0.02 |
| OW-10B | -20 to -25 | <0.005 | <0.02 |
| OW-10C | -35 to -40 | <0.005 | <0.02 |
| MW-1255 | -53 to -58 | 0.018 | <0.02 |
| MW-1255 | -78 to -83 | 0.019 | 0.018 |
| MW-1251 | -100 to -105 | 0.038 | 0.079 |
| MW-1250 | -150 to -155 | 0.035 | 0.046 |
| MW-1250 | -190 to -195 | 0.040 | 0.072 |
| MW-1252 | -77 to -82 | 0.005 | 0.034 |
| MW-1254 | -115 to -120 | 0.012 | 0.037 |
| MW-1254 | -115 to -120 | 0.048 | 0.036 |

Field screening results indicated an EDB and VOC plume immediately below and downgradient of the CS-4 extraction system (Figure 3). The CS-4 chlorinated solvent plume is projected to be located from 60 ft to 100 ft below the water table at the extraction wells. Field screening results indicate the EDB plume and associated VOCs are located from 100 ft to 130 ft below the water table at the extraction wells. Plume constituents consist of EDB (up to 0.08 $\mu g/L$), PCE (up to 5.7 $\mu g/L$), TCE (up to 3.4 $\mu g/L$), and 1,1,2,2-TCA (up to 6.4 $\mu g/L$). The ratio of PCE to TCA was also different for the CS-4 chlorinated solvent plume (approximately 2:1) and the EDB plume (approximately 1:1). Analytical data indicate the lateral extent of the plume to be similar to the CS-4 chlorinated solvent plume. The upgradient and downgradient extent of the plume was not defined; however, subsequent RI efforts indicate the upgradient extent to be less than 2000 ft [5].

## CONCLUSIONS

Field screening results confirmed that EDB and VOCs were downgradient of the extraction system. Based on the screening data of compounds detected, the relative concentrations of the VOCs, and the intervals of detection, it was concluded that the EDB and VOCs represented a separate contaminant plume deeper in the aquifer than the solvent plume. These conclusions were later confirmed with Contract Laboratory Program (CLP) samples collected from new and existing wells. Based on field screening data and with regulatory concurrence, it was determined that the extraction system design would not require modification. The delay in the extraction system construction totalled 66 days with an additional cost of $10,800. However, the use of screening data vs CLP data saved approximately 60 workdays. Further delay would have severely impacted the construction schedule and resulted in significant additional costs. Additional investigations were recommended to determine the source and extent of the EDB contamination.

The submitted manuscript has been authored by a contractor of the U.S. government under contract DE-AC05-84OR21400. Accordingly, the U.S. government retains paid-up, nonexclusive, irrevocable license to publish or reproduce the published form of the contribution, prepare derivative works, distribute copies to the public and perform publicly and display publicly, or allow others to do so, for U.S. Government purposes.

## REFERENCES

1. Oldale, R. N., and O'Hara, C. J. "Glaciotectonic Origin of the Massachusetts Coastal End Moraines and a Fluctuation Late Wisconsinian Ice Margin." *Geological Society of America Bulletin*, 1988. <u>95</u>, 61-64.

2. ABB Environmental Services, Inc. *Groundwater Focused Feasibility Study West Truck Road Motor Pool (AOC CS-4), Massachusetts Military Reservation*; Portland, Maine; February 1992.

3. Turner, S. A., Twomey, D. M. Jr., Sneed, S. B., Francoeur, T. L., and Baker, L. A. "A Field Analytical Technique for the Rapid Analysis of Trace Levels of Ethylene Dibromide in Groundwater." *Proceedings of the American Defense Preparedness Association, San Antonio, TX*, 1994.

4. ABB Environmental Services, Inc. *Technical Memorandum Monitoring Well MW-1206Z Groundwater EDB Investigation, Massachusetts Military Reservation*; Portland, Maine; October 1993.

5. CDM Federal Programs, Corporation. *Draft Remedial Investigation UTES/BOMARC Area Fuel Spill Groundwater Operable Unit: CS-10D and Hydrogeologic Region II Study, Massachusetts Military Reservation*; Boston, Massachusetts; November 1993.

CATEGORY XI:

# Soil Treatment and Soil Characterization

# ULTRASOUND ENHANCED SOIL WASHING

JAY MEEGODA, W. HO, M. BHATTACHARJEE, C. F. WEI, D. M. COHEN, AND R. S. MAGEE
New Jersey Institute of Technology
Newark, NJ 07102

RAYMOND M. FREDERICK
US Environmental Protection Agency
Edison, New Jersey 08817

## INTRODUCTION

Soil washing has been practiced as a means for removing contaminants from soil since the early 1970s when the EPA funded and operated a crude soil washer, the "Beach Cleaner." Today, it is a commercially available method for treating excavated soil and dredged sediments that are contaminated with toxic or other hazardous pollutants. It involves the application of a set of established engineering principles, unit processes, and equipment that have been used for years in the mining, mineral processing and ore benefaction, and wastewater treatment industries.

Most commercially available soil washing systems utilize mechanical screening devices to remove oversize material, separation systems to generate coarse- and fine-grained fractions, treatment units for washing, systems for scrubbing the separated fractions, equipment for rinsing and de-watering, and water treatment and recycling systems for water management. The specifications for each system are largely driven by individual site characterization.

Higher the percentage of sand and gravel in the soil or sediment, the more effective is soil washing. Soil washing systems usually consist of six distinct process units: pretreatment; separation; coarse-grain (>200 mesh or >74 micron) treatment; fine-grained (<200 mesh or <74 micron) treatment; process water treatment; and residuals management. Contaminated fines and sludges resulting from the process require additional treatment, e.g., incineration; low-temperature thermal desorption; bioremediation; or must be disposed of in a regulated landfill.

Surficial contamination that is attached to sand and gravel fractions through forces of adhesion and compaction is removed from the coarse fraction by abrasive scouring or scrubbing action. This washing step is sometimes enhanced by adding to the wash-water a basic leaching agent, surfactant, pH, adjustment, or chelating agent (such as ethylene diamine tetra-acetic acid or EDTA) to help remove organics or heavy metals. After washing, the coarse soil fraction may be re-washed to further remove residual contaminants and wash-water additives. The spent wash-water and rinse-water are treated to remove the contaminants prior to recycling back to the treatment unit.

An essential part of a soil washing process is the extraction step, in which the soil is brought in contact with the cleaning solution by some form of mechanical agitation. The shearing action achieved by vigorous agitation is needed to dislodge the contamination from the soil particles, and to enhance its solubilization in the cleaning solution.

Ultrasound causes high-energy chemistry. It does so through the process of acoustic cavitation: the formation, growth and implosive collapse of bubbles in a liquid. During cavitational collapse, intense heating of the bubble occur. These localized hot spots have temperatures of roughly 5000°C, pressures of 500 atmospheres, and a lifetime of a few microseconds [1]. Shock waves from cavitation in liquid-solid slurries produce high- velocity interparticle collisions, the impact of which is sufficient to melt most metals. Applications to chemical reactions exist in both liquids (homogeneous) and in liquid-solid (heterogeneous) systems. Of special synthetic use is the ability of ultrasound to create clean, highly reactive surfaces on metals. Ultrasound has also found important uses for initiation or enhancement of catalytic reactions, in both homogeneous and heterogeneous systems.

The velocity of sound in liquids is typically 1500 m/s; ultrasound spans the frequency of roughly 15 kHz to 1 Mhz, associated with acoustic wavelengths of 10.0-0.15 cm. These are not molecular dimensions. Therefore, no direct coupling of the acoustic field with chemical species on molecular level can account for the chemical reactions. Hence chemical effects of ultrasound is derived from several different physical mechanisms, depending on the nature of the system.

Ultrasonic cleaning is an industrial method for removal contaminants such as oil, grease, fingerprints, ink, etc., from solid surfaces. Ultrasonic energy is applied to cleaning of manufactured parts in the metals and electronics industries, among others. Ultrasonic tanks range in size from laboratory size to several thousand gallons. It is also possible to install submersible ultrasonic transducers into ordinary tanks or vats, thereby converting them into ultrasonic baths. In common use, the parts to be cleaned are placed in a bath filled with the cleaning solution and subjected to ultrasonic vibration for 10 - 20 minutes. The liquid is often a detergent solution composed of water, surfactant and other additives. The design and operating variables in ultrasonic cleaning are frequency, power density, temperature, and composition of the cleaning liquid.

All contaminants may be classified as soluble or insoluble according to the nature of their interaction with the detergent. In the removal of insoluble

734

contaminants in chemically neutral solutions the cleaning rate depends on the ultrasonic cavitation intensity. In the removal of soluble contaminants, acoustic streaming, particularly vortex micro-streaming generated in the boundary layer, where it promotes the rapid admission of fresh batches of solvent directly to the surface of the solid, plays a vital role. The reduction in thickness of the boundary layer at the boundary with the solid is the principal factor distinguishing mixing of the liquid in a sound field from any methods of mechanical agitation. This distinction accounts for the effective removal of soluble contaminants at high frequencies, when the sound intensity can prove to be below the threshold value and cavitation is not observed in the liquid.

The great majority of industrial cleaners operate in the frequency range from 18 to 44 kHz. This is the optimum range in the sense of technological efficiency, economy of the process, and safety considerations. The lower frequency range (18 to 22 kHz) is used for the removal of contaminants having a high adhesion to the surface (scale, pickup, polymer films); the higher range (40 to 44 kHz) is used for cleaning in the case of contaminants that are weakly bound to the surface (grease and machining contaminants).

The development of an ultrasonic enhanced soil-washing process requires a comprehensive, well-designed experimental program, with the results carefully analyzed on the basis of known ultrasonic cleaning mechanisms. There has been no systematic work carried out to develop information on the important variables that can affect the efficacy of ultrasonic enhancement of contaminant removal from soil.

The goal of this study is to examine the potential of ultrasonic energy to enhance soil washing and to optimize conditions. Ultrasonic energy potentially can be used in enhancing contaminant removal from the entire soil mix, or it can be used as a polishing operation on the fines portion of the soil mixture after traditional soil washing operations. The research study was designed to demonstrate that ultrasonic energy can:

Improve process performance, e.g., remove contaminants to lower residual concentrations.

Improve process economics, e.g., shorter treatment (residence) times, less surfactant use.

## EXPERIMENTAL PROCEDURE

A soil contaminated with of Polycyclic Aromatic Hydrocarbons (PAHs), was obtained from a contaminated site (Superfund site). It contains a substantial (20-50%) proportion of silt and clay (fines). A single batch of contaminated soil was homogenized and used for the study. The study reported in this paper consisted of five tasks. A detailed description of each task is given after the next section.

Ultrasound Source: Ultrasound energy will be applied to the system using either a) a 1500 Watts probe (Sonics & Materials Inc., Model VC1500, 220 Volts, Power 1500 Watts, Frequency 20 KHz) or b) 500Wx2 bath (Model 4G-500-6,

120 Volts, Power 500x2 Watts, Frequency 40-90 KHz). The actual power received by soil slurry was measured by a power intensity meter. This measurement was not include in the analysis as power intensity levels for the probe type were different at different points within the container.

Soil: Soil samples were Coal Tar contaminated soil fraction finer than US #4 sieve but retained on US #200 (0.075 mm) sieve. The soil was a well-graded sand with silt (16% finer than US # 200 sieve). It had a moisture content of 2%. The total organic content of the soil was 16%.

Surfactant: In a recent study of solubilization of PAHs from soil-water suspensions with several nonionic and anionic surfactants, the most effective surfactants were non-ionic octyl- and nonyl- phenyl-etheoxylates with 9 to 12 ethoxylate units [2]. At soil-water mass ratios of 1:7 to 1:2 greater than 0.1% by volume surfactant dose was required in order to initiate solubilization, with dose of 1% by volume resulting in 70- 90% solubilization. Furthermore, solubalization of PAHs in soil-water systems occurs at surfactant doses much greater than the clean water critical micelle concentration (CMC). In another study involving 22 surfactants on cleaning of diesel contaminated soil, two anionic and one nonionic surfactants were found to be effective [3]. Of the three nonionic surfactants was more effective. Therefore, the surfactant, Octyl-phenyl-ethoxylate, a non-ionic surfactant with a CMC of 2- $3.3 \times 10^{-4}$ moles, was diluted to desired concentrations by adding distilled water.

The glass container with soil slurry was kept inside the wooden cabinet that houses the ultrasonic probe at the top. The aerosolizing action produced by the sonicator probe dipping 4 cm into the soil/surfactant solution kept the soil in suspension. The ultrasound treated soil slurry was allowed to set for 30 minutes and roughly 300 ml of solution was removed to avoid re-deposition of the PAHs. The remaining portion of treated soil slurry was placed in the Buchner funnel to remove the liquid portion and soil was rinsed with additional 500 ml of distilled water. The soil was then placed inside a vacuum oven for 2 hours to dry the soil for analysis. A 10.0g of sodium sulfate was added to 10.0g of dry soil and the mixture was placed inside the extraction thimble. A total of five soil specimens plus untreated soil sample was placed in Soxhlet Extraction for 16 hours followed These samples plus untreated soil sample will be placed in Soxhlet Extraction for 16 hours followed by Kuderna-Danish (K-D) concentrate to 10 ml. All samples were analyzed by Varian Saturn II Ion-Trap GC/MS using SIM (selective ion monitoring) technique to quantify PAHs concentrations. The EPA Method 3540A was employed to extract PAHs from soil sample. To isolating PAHs from liquid portion samples EPA Method 3510A was utilized.

Mobilization: Mobilization consisted of conducting a thorough literature search, including contacting manufacturers of ultrasonic equipment both for scientific and commercial use; installation of the process equipment; installation of the analytical equipment; arrange for the delivery of soil from a contaminated site; preparation of the soil batch for the study, e.g., air drying, screening to remove oversize material, homogenization, particle size analysis, contaminant analysis by particle size distribution; and cold storage in sealed containers.

Analytical Procedures: The analytical procedures included analysis of the key contaminants (e.g., PAHs) in soil and in the extract. The soil analysis was based on EPA Manual SW-846 [4], consisting of soxhlet extraction followed by GC. Analytical procedures were maintained as close to SW-846 as possible using appropriate QA/QC as specified in the methods. For the mass balance calculations wash water had to be analyed. The liquid/liquid extraction and silylation were used to separate analytes from the surfactant. The above was followed by GC on the analytes fraction. The use of GC-MS was required for the identification of the key contaminants in the sample. The method was tested for reproducibility (expressed in terms of standard deviation) and accuracy.

Process Procedures: The process consisted of an extraction step followed by a water rinse. The extraction was carried out with a surfactant solution in distilled water ("solvent"). The rinse will also be done with distilled water. The extraction and the rinse will be carried out in an ultrasonic cell equipped with a heater and a mechanical stirrer.

Baseline Operation: More than fifty trial runs under various conditions were made to establish the baseline set of operating conditions. Each of the runs was paired with a control run, which was operated identically except without ultrasonic energy. The baseline conditions were selected to maximize the ultrasonic effect. Following were the parameters that were supposed to contribute to the enhancement of Soil Washing due to ultrasound.
- Ultrasonic frequency
- Ultrasonic power density
- Dwell (extractor residence time)
- Temperature
- pH
- Surfactant concentration
- Solvent ratio (liquid/soil w/w ratio)

The following were observed from the trial runs. There was an increase in the sample temperature due to the application of ultrasound energy. The rise in temperature was proportional to dwell time. The probe type ultrasound source was much better than the tank type (by comparing two tests involving to ultrasound sources set at 750W power level) in transmitting the ultrasound energy to the container with contaminants. Application of external pressure to the soil/solvent slurry with the probe drastically increased the reflection. This observation eliminated the possibility of applying the ultrasound energy to a container filled with fluid/soil suspension. The above test with external pressure was conducted to determine if the aerosolizing action and cavitation (and not the resonance) were the main reasons for enhancement in removal efficiency. One disadvantage of selecting the probe type source was that the system temperature can not be controlled during the experiment. The other difficulty was the variation of frequency. Since this probe is a 20 kHz at 1500W source, there are no commercially available ultrasound sources with the same power rating but higher frequency. Therefore, the temperature and frequency were eliminated as factors from the statistical model. Initial results showed that the solution pH (if between 2-11) does not contribute to the removal efficiency, i.e., there is no influence of pH on the removal efficiency of the ultrasound enhanced soil washing process (see Figure 1). Therefore, it was decided to keep the pH between 6-7 while varying the other parameters.

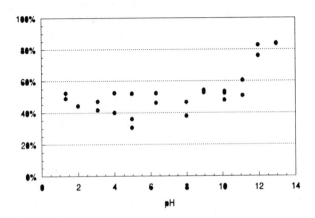

**FIGURE 1- REMOVAL EFFICIENCY OF ANTHRANCENE WITH pH**

Experimental Design: The baseline operation identified four variables (factors) that appeared to have significant contributions to the removal efficiency. They are:

A- Solvent ratio
B- Dwell time
C- Surfactant concentration
D- Ultrasonic power density

For the experimental design each factor was considered in a factorial design at three levels for A, B, C and no power, low power and high power for D; all coded (-1, 0,+1) as follows, namely the following with three levels (low, medium and high except for ultrasound power density where it).

A - Solvent ratio (10, 25 and 50 liquid/soil w/w ratio)
B - Dwell (5, 15 and 30 minute extractor residence time)
C - Surfactant concentration (0.01%, 0.1% and 1%)
D - Ultrasonic power density (0, 50% and 80% power)

Four factors at three levels produce 81 independent experiments. Three surfactant solutions (0.1 g, 1.0 g and 10.0 g of surfactant in 1000 ml of distilled water) were made. Soil specimens weighing 50.0 g, 20.0 g, and 10.0 g were placed in 1000 ml glass container and added 500 ml of the surfactant solution directly into the beaker to yield the desired liquid:soil ratios (i.e., 10:1, 25:1, and

50:1). The probe type Sonicator was set at the 0% (with mechanical agitation), 50% and 80% power output. The glass container with soil slurry will be kept inside the wooden cabinet that houses the ultrasonic probe at the top. The aerosolizing action produced by the sonicator probe dipping 4 cm into the soil/surfactant solution will keep the soil in suspension.

The order of 81 experiments was randomized to eliminate possible biases. Once all 81 experiments were completed, the twelve PAH (Naphthalene, Acenaphthylene, Acenaphthene, Fluorene, Anthracene, Fluoranthene, Benzo(k)fluoranthene, Benzo(ghi)perylene, Chrysene, Pyrene, Indeno(1,2,3-cd)Pyrene and Benzo(a)pyrene) concentrations were determined. The removal efficiency for each compound for each test was calculated. The removal efficiency is computed from the differences in each PAH concentration before and after treatment. Then the 27 experiments with no power were compared with 27 experiments with low power to determine the enhancement due to ultrasound. The enhancement was qualitatively determined by comparing the difference in PAH removal efficiency corresponding no power and low power experiments.

## EXPERIMENTAL RESULTS

Since there are 81 experiments with removal efficiencies for 12 PAHs, following procedure was adopted to analyze the data. For each variable (say power density) three plots were made: 1) with lowest values of other three variables (e.g., 5 minutes, 0.01% surfactant concentration and a solvent ratio of 10), 2) middle values of other three variables and 3) highest values of the other three variables. This resulted in 12 plots. Figure 2 is an example of such plot where removal efficiencies for PAHs with various dwell times are plotted for 1200 Watts ultrasound energy with 1% surfactant concentration at a solvent ratio of 50. Analysis of all 12 plots shows that the 750 Watts power with 30 minute dwell time, and 1% surfactant concentration produced the best removal efficiency. Figure 3 shows a plot of removal efficiency for PAHs with various solvent ratios while keeping power, dwell time and surfactant concentration at the optimum levels. Figure 4 shows that ultrasound energy supplied by a 1500 Watts probe operating at 50% power rating applied for 30 minutes to a container carrying 20 grams of coal tar contaminated soil and 1% surfactant in 500 ml can enhance the soil washing process by more than 100%. Figure 3 indicated that the most economical removal efficiency is obtained at a solvent ratio of 25 with 750 Watts power, 30 minute dwell time, and 1% surfactant concentration. Meegoda and Ratnaweera [5] showed that a surfactant works best for oils when the surfactant weight is more than 50% the weight of the contaminants or surfactant to contaminant ratio of 0.5. For the above optimum combination, soil had 3.2 grams of contaminants and has added 5.0 grams of surfactants or surfactant to contaminant ratio of 1.6. The next lowest surfactant to contaminant ratio was 0.625, for a solvent ratio of 10 with 750 Watts power, 30 minute dwell time (see Figure 3), and 1% surfactant concentration or 0.16, for a solvent ratio of 50 with

750 Watts power, 30 minute dwell time, and 1% surfactant concentration (see Figure 5). It also appears that for a heavily coal tar contaminated soil with ultrasound energy needs surfactant to contaminant ratio of more than 0.625 and a solvent ratio of more than 10 to obtain near perfect removal efficiency.

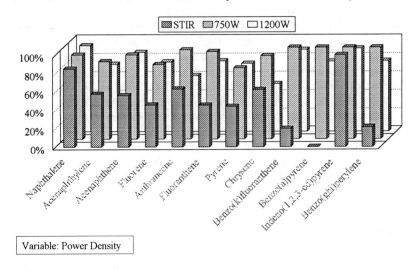

Variable: Power Density

**FIGURE 2-REMOVAL EFFICIENCY OF PAHS FOR DIFFERENT POWER DENSITIES**
(DWELL TIME 30 MINUTES, 1% SURFACTANT CONCENTRATION AND A SOLVENT RATIO OF 50)

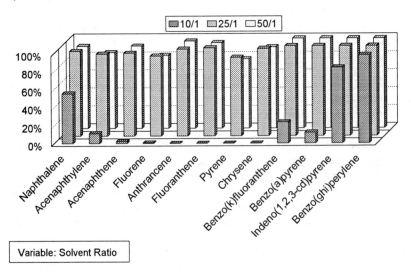

Variable: Solvent Ratio

**FIGURE 3-REMOVAL EFFICIENCY OF PAHS FOR DIFFERENT SOLVENT RATIOS**
(750 WATTS POWER, 1% SURFACTANT CONCENTRATION AND 30 MINUTES DWELL TIME)

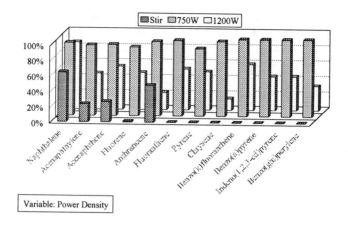

Variable: Power Density

**FIGURE 4-REMOVAL EFFICIENCY OF PAHS FOR DIFFERENT POWER DENSITIES**
(1% SURFACTANT CONCENTRATION, 30 MINUTES DWELL TIME AND A SOLVENT RATIO OF 25)

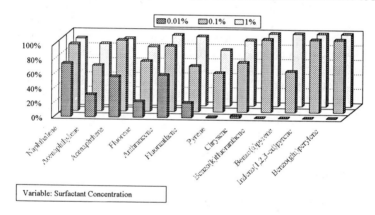

Variable: Surfactant Concentration

**FIGURE 5-REMOVAL EFFICIENCY OF PAHS FOR DIFFERENT SURFACTANT CONCENTRATIONS**
(750 WATTS POWER, 30 MINUTES DWELL TIME AND A SOLVENT RATIO OF 50)

## SUMMARY AND CONCLUSIONS

There was an increase in the sample temperature due to the application of ultrasound energy. The probe type ultrasound source was much better than the tank type in transmitting the ultrasound energy to the container with contaminants. Initial results showed that the solution pH (if between 2-11) does not contribute to the removal efficiency. An experimental design was developed with four variables (power, dwell time, solvent ratio and surfactant concentration) where each variable had three levels (low, medium and high except for ultrasound power density where it is no power, low power and high power) resulting 81 experiments. Test results showed that ultrasound energy supplied by a 1500 Watts probe operating at 50% power rating applied for 30 minutes to a container carrying 20 grams of coal tar contaminated soil with 1% surfactant in 500 ml can enhance the soil washing process by more than 100%. The experimental design suggested that the optimum operation condition was at a solvent ratio of 25 with 750 Watts power, 30 minute dwell time, and 1% surfactant concentration. It also appears that for a heavily coal tar contaminated soil with ultrasound energy needs surfactant to contaminant ratio of more than 0.625 and a solvent ratio greater than 10 to obtain near perfect removal efficiency. This paper describes a research-in-progress. Results stated herein are of preliminary nature and may not accurately reflect the final results published at a later date.

## ACKNOWLEDGMENTS

This research was sponsored by a grant from USEPA through the Northeast Hazardous Substance Research Center at NJIT. Authors wish to acknowledge this support. Authors also wish to acknowledge the efforts of 1)Mr. Mike Borst at USEPA, 2) Mr. Gerard McKenna, NJIT who served as the QA/QC officer and 3) Dr. Itzhak Gotlieb Ghea Associates who was a consultant for this project. Although the research described in this paper has been funded by USEPA, it has not been subjected to the Agency's review process for technical content, quality assurance/quality control (QA/QC) or administrative review. Therefore, it does not necessarily reflect the views of the agency. Mention of trade names or commercial products does not constitute endorsement or recommendation for use.

## REFERENCES

1. Suslick, K. S. 1990. "Sonochemistry" *Science*, 257:1439- 1445

2. Liu, Z., S. Laha, and R. G. Luthy. 1991. "Surfactant Solubilization of Polycyclic Aromatic Hydrocarbon Compounds in Soil-Water Suspensions", *Water Science & Technology*, 23:475-485.

3. Peters, R. W., C. D. Montemagno, L. Shem, and B. A. Lewis. 1992. "Surfactant Screening of Diesel-Contaminated Soil", *Hazardous Waste & Hazardous Materials*, 9(2):113-136.

4. EPA. March 1990. "Toxicity Leaching Characteristics Procedure", Federal register, Vol. No. 261, Method 1311, EPA 530/SW-846, Cincinnati, Ohio

5. Meegoda, N. J. and P. Ratnaweera. 1995. "Treatment of Oil Contaminated Soils for Identification and Classification", *ASTM Geotechnical Testing Journal*, 18(1):41-49.

# EFFECT OF THERMAL GRADIENT ON SOIL-WATER SYSTEM

MESUT PERVIZPOUR
Graduate Research Assistant
Department of Civil and Environmental Engineering
Lehigh University
Bethlehem, PA 18015

SIBEL PAMUKCU
Associate Professor
Department of Civil and Environmental Engineering
Lehigh University
Bethlehem, PA 18015

## INTRODUCTION

The transient behaviour of the excess pore water pressure was used to obtain conductivity and compressibility related parameters of saturated soils. The spatial excess pore pressure measurements due to an applied thermal gradient were made in a one-dimensional non-destructive test setup. The coefficient of consolidation, $C_v$, of a soil sample was determined based on the rate of excess pore pressure migration generated under an applied thermal gradient.

The measured pore pressure propagation data was used in finite difference solution of one-dimensional diffusion equation. The coefficient of consolidation predicted in this way was compared to that obtained from a standard step-load consolidation test.

The purpose of this experiment was to devise a non-destructive testing approach that would maintain the constancy of the original structure of the soil during testing. The application of thermal gradient accomplished this goal by minimizing the load on the soil skeleton and preserving the original structure and volume.

The necessary pore pressure migration data was obtained by loading the liquid phase without the indirect effects of coupled process on the soil skeleton. The test was performed by application of a thermal potential through the ends of a cylindrical shaped saturated soil specimen enclosed in an acrylic cell. The rate of migration of the thermally induced pore water pressure was measured along the soil while a constant gradient is maintained under undrained conditions.

# THEORETICAL BACKGROUND

Pore pressure dissipation due to insertion of different types of probes and cones have been used to obtain an approximation of the lateral coefficient of consolidation, $C_h$ [1], [5], [13]. Some drawbacks of these methods are: there is coupled effect of high strain levels on the pressure measurements; the analyses often require empirical parameters for probe shape and also pore pressure parameters; and remolding and distortion of the adjacent soil may lead to underestimation of the coefficient in comparison to that of the undisturbed soil. The pore-water pressure generation and dissipation based on strain field [1], [4], [8] and velocity field propagation analysis and the cavity expansion models [7], have resulted in better values of $C_v$ for overconsolidated soils, but not necessarily for soft, saturated soils.

Triaxial CIU tests of saturated clay to evaluate the effects of coupled flow under multiple potentials (hydraulic, thermal, chemical), and support of linear superposition showed the potential for an approximate approach to evaluate $C_v$ using the time lag in the pore pressure equalization at either end of a specimen [14].

Thermal potential induces porosity reduction in soils when tested under drained conditions. Consolidation under combined effect of loading and thermal gradients was studied and the effect of heat was observed to be dominant during the secondary compression stage [11]. The pore pressure development in a saturated media under increased temperature is partially due to the thermal expansion of the loosely adsorbed water, especially in overconsolidated soils, which renders the build up to be more pronounced in low-porosity clays [3]. The nonlinearity of the effective thermal expansion coefficient as a function of distance from the mineral surface was found to be significant in overconsolidated clays owing to their less compressible skeleton. The increase in pore pressure during undrained temperature elevation is affected by the differences in thermal expansion characteristics of soil constituents, as well as the physico-chemical effects [9]. The change in pore water pressure with temperature depends strongly on the compressibility and stiffness characteristics of the soil. The stiffer the soil the lower the compressibility and the greater the decrease in effective stress for a given temperature increase under undrained conditions [9]. The pore pressure increase due to elevated temperature is described by Campanella and Mitchell [6] as a function of expansion coefficients and volume compressibility of the components and the physico-chemical soil structure volume change. Baldi et al.[3] suggested the use of effective thermal expansion of adsorbed water to account for the electrochemical or electro-microstructural interactions caused by temperature and pressure effects in the double layer, especially in low-porosity clays.

The relative magnitudes of compressibility, porosity and thermal expansion properties for fluid and soil skeleton determine the amount of excess pore pressure generated in a soil mass for a finite change in temperature.

The non-destructive approach discussed in this paper is based on rate of change of pressure and temperature at a given point inside the soil mass for which net water flow is zero. The approach makes it possible to observe the effects of parameters such as the hydraulic conductivity of the specimen and the rate of temperature increase on the rate of pore water pressure generation and migration in saturated, low porosity soils.

## EXPERIMENTAL RESULTS

The specimens are prepared first by mixing 15% Georgia kaolinite (LL=42%, PL=30%) with 40% silt, 25% sand and 20% coarse sand by weight at a water content of 20%. The mixture was consolidated one dimensionally into an acrylic sample mold of 21 cm in length and 3.5 cm in ID as shown in Figure 1. The final dry density of these mixtures were on the order of 18-20 kN/m$^3$ with measured specific gravity of 2.72.

Backpressure was applied in steps up to 5 psi at the end of consolidation period to assure saturation of the specimen. At the completion of consolidation the mold is detached from the consolidometer base and mounted on the thermal test apparatus without extracting the soil sample. The test equipment consists of three major components. There are two water chambers connected to the ends of the sample mold. The mold carries five pore water pressure and five temperature measurement stations by means of pressure transducers and thermocouples, respectively (Figure 1). The water chambers are connected to volumetric burettes to measure the hydraulic conductivity of the specimen before and after a thermal testing, to monitor sample skeleton constancy (i.e., changes in porosity, water content).

The sample mix formula and consolidation pressure were adjusted to obtain samples with hydraulic conductivity values in the range of $10^{-4}$ - $10^{-6}$ cm/s. This target range is sufficiently low to allow for pore pressure build up and initiate pressure migration with time lag along the sample. The magnitude of the generated pressures are also large enough for accurate measurements with the transducers.

The non-destructive thermal testing is initiated by increasing the temperature of the water in the inlet chamber -thermocouple 1 site- at one end of the specimen by means of an electrical cartridge heater. Variable voltage application provides the control on rate of temperature increase, thereby the rate of excess pore water pressure generation. Thermal gradient was adjusted to maintain a 21°C of difference along the 21 cm long soil sample.

The heater is located away from the soil column, inside the water chamber to delay heat propagation and minimize coupling effects. The water chambers are adjacent to the soil chamber but separated by porous stones. The expansion of water in the inlet chamber at one end of the soil generates pressure which then propagates towards the other end. As the constant temperature difference is maintained between the two ends of the soil, the pressure migration is monitored.

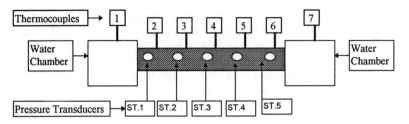

Figure 1. Schematic of the experimental set-up, [12]

A number of assumptions and calibrations were made prior to testing. First, the coupled effect of the migrating pore pressures and the locally generated pore pressures with the heat propagation was assumed to be negligible due to the following reasons:

1) The heater was placed in a water chamber and was not in direct contact with soil;
2) Testing was conducted in an un-insulated set-up, thereby maximizing the heat loss on the periphery of the soil cylinder and maintaining a constant thermal gradient for the duration of testing;
3) Test periods were short not to allow for uniform temperature distribution throughout the soil.

These factors altogether minimized the effect of heat propagation on local pressure development and therefore the measured pore pressures were assumed to be free of the coupling effects. Furthermore, the effect of expansion of the soil and water chambers on the measured values of pore pressure were assumed to be negligible. The expansion of the water chamber due to the temperature and pressure increase had no effect on the computation of $C_v$ since the initial calibrations were made taking those effects into account.

Figure 2 shows typically the measured temperatures along the soil at steady state temperature distribution. Figure 3 shows the distribution of pore water pressure in soil normalized with the pressure generated at the inlet chamber. Pressure distributions at three separate times during the same test are plotted in this figure.

The typical variation of temperature and pore water pressure along the sample at five stations and the effect of the on-off cycle of the heater in the inlet chamber are shown in Figures 4 and 5 respectively. The rate of heat propagation can be enhanced or reduced by adjusting the level of insulation and the voltage input of the test setup. A more uniform temperature distribution is expected for better insulated samples.

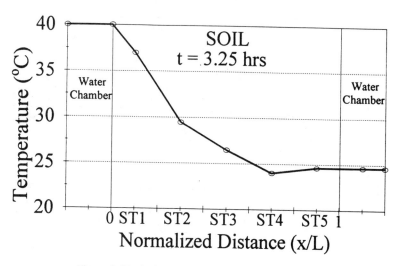

Figure 2. Typical temperature distribution at steady state

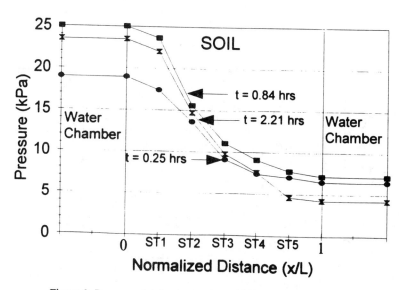

Figure 3. Pressure distribution at three different time segments [12]

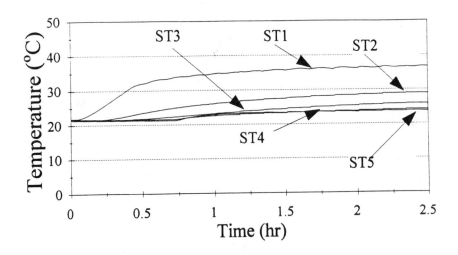

Figure 4. Temperature with time at five stations along the sample

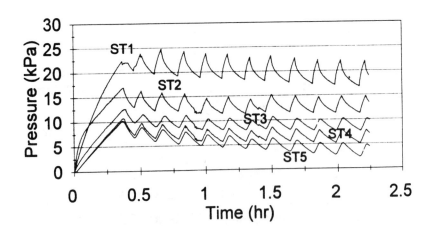

Figure 5. Pressure variation at five stations along the sample [12]

Terzaghi's one dimensional consolidation equation (Equation [1]) with constant coefficient of consolidation is used to reduce the pore pressure data in Cartesian coordinates. A coarse finite difference discretization which matched the actual measurement intervals on the test setup is used.

$$\frac{\partial u}{\partial t} \quad c_v \frac{\partial^2 u}{\partial z^2} \tag{1}$$

where; $u$, $z$, $t$, $c_v$ are the excess pore pressure, longitudinal direction, time and the coefficient of consolidation in longitudinal direction.

The coefficient of consolidation, $C_v$ is derived from Eq. [1] by reducing the pore pressure migration data on a five node finite difference frame. Figure 6a shows the variation of computed $C_v$ with time at station one [ST1] of a soil specimen which was consolidated to 535 kPa. Figure 6b is a plot of the first hour of variation of $C_v$. The perturbation of the computed $C_v$ values about the x-axis, thus the negative sign is due to reversal of propagation during the off cycle of the heater. The negative $C_v$ values are treated by taking their absolute values over the corresponding interval.

The computed values of $C_v$ in these figures can be examined in two parts: a) the warm-up period, b) the on-off cycle. The warm-up period is the initial part of the test up until 0.4 hours during which the temperature in the inlet water chamber rises steadily to 50°C. The following part represents the periodic portion due to the on-off cycle of the temperature controller unit to maintain the temperature within the preset range. During the warm-up period, $C_v$ approaches an asymptotic value of similar order of magnitude in each station. During the on-off cycle period, similar type of exponential decay of the $C_v$ value is observed where it reaches a similar asymptotic value as in the warm-up period.

Average $C_v$ values obtained for laboratory prepared soil specimens of two different gradation and hydraulic conductivity values are presented in Table I. As observed from Table I the $C_v$ obtained for the high conductivity soil is an order of magnitude larger than the low conductivity soil. This is attributed to the longer time period involved in the initial warm-up process which allows the system to reach a steady state pressure distribution before heater is shut off.

The repeatability of the derivation of $C_v$ values from pore pressure migration data was tested four times using a soil specimen consolidated at 535 kPa. These $C_v$ values were on the same order of magnitude as the $C_v$ value obtained from a standard oedometer test of the same soil at the same consolidation pressure. The derived and the measured values are given in Table II. As observed, the average value of $C_v$ obtained during the on-off cycle is closer to the one measured in an oedometer test at the same preconsolidation pressure.

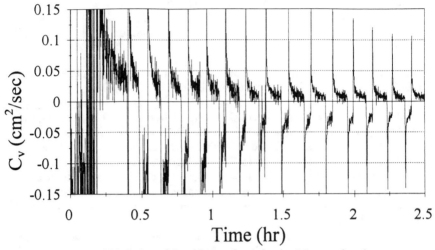

a) Variation of $C_v$ with time for the entire 3 hours of testing

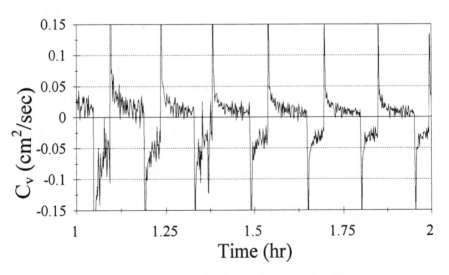

b) Variation of $C_v$ for the first hour of testing
Figure 6. $C_v$ with time at ST1 (Next to heated chamber)

Table-I Average $C_V$ values of 2 soil samples of same preconsolidation pressure (535 kPa), [12]

| Time Interval(seconds) | Coefficient of Consolidation ($C_v$), ($cm^2/sec$) | |
|---|---|---|
| | SAMPLE-A | SAMPLE-B |
| | k = 4E-04 cm/sec | k = 8.7E-06 cm/sec |
| 1000-1700 (Warm-up) | 0.0043 | 0.00051 |
| 2400-2800(On-off cycle) | 0.0052 | 0.00054 |
| 8000-9000 (Cool-down) | N/A | 0.00051 |

Table-II Average $C_v$ from replicate non-destructive thermal tests, [12]

| Test No. | Coefficient of Consolidation ($C_v$, $cm^2/sec$) Non-destructive thermal gradient tests | | Oedometer test |
|---|---|---|---|
| | Warm-up Period | On-off Cycle | |
| 1 | 0.0025 | 0.0039 | |
| 2 | 0.0019 | 0.0034 | 0.0044[*] |
| 3 | 0.0025 | 0.0041 | |
| 4 | 0.0031 | 0.0048 | |
| Average | 0.0025 | 0.0041 | |

[*]At 270 kPa load interval matching the consolidation pressure of the non-destructive test

## DISCUSSION

The undrained nature of the test prohibited fluid flow thereby potentially minimized changes to the soil skeleton. This was checked by measuring the coefficient of hydraulic conductivity, k, before and after the non-destructive thermal testing as given in Table III. The pressure was generated and increased at the boundary of the soil in the water phase gradually. This procedure reduced potential disturbance on the
soil skeleton compared to sudden application of a load step.

Table III presents a summary of evaluated parameters from the non-destructive thermal tests of five different soil samples. The modified set-up with finer port spacing gave lower $C_v$ values than the standard one dimensional consolidation tests of identical samples. It is believed from the magnitude and repeatability of the low values that the results are indicative of low strain levels supported by the diffusion equation (Eq. [1]).

Table III. Summary of k and $C_v$ values for five different soil samples

| Sample No. | k(cm/sec) | | $C_v$ (cm²/sec) | | Comments |
|---|---|---|---|---|---|
| | Before | After | NDT | Cons. press. | Coarse spacing of pore water pressure (3 ports) |
| 1 | 4.0E-04 | 4.0E-04 | 0.0047 | 276 kPa | |
| 2 | 8.3E-06 | 8.7E-06 | 0.0005 | 552 kPa | |
| 3 | 3.0E-06 | 3.0E-06 | 0.0022 | 552 kPa | Finer spacing of pore water pressure (5 ports) |
| 4 | 0.8E-06 | 0.8E-06 | 0.0021 | 276 kPa | |
| 5 | 0.8E-06 | 1.0E-06 | 0.0030 | 276 kPa | |

## CONCLUSIONS

An apparatus developed for non-destructive thermal testing purposes provided the necessary information to evaluate consolidation related parameters. The pore pressure migration data obtained along soil samples was used to calculate $C_v$ values for the soil specimen. The values obtained in this manner showed agreement with those obtained in standard oedometer tests.

The repeatability of non-destructive thermal test was checked for the same sample using different time interval readings and hydraulic conductivity tests after each thermal gradient application. The thermal loading applied on the liquid phase minimized the magnitude of the strain on the soil skeleton.

## ACKNOWLEDGMENTS

This research was made possible through a grant from the National Science Foundation (NSF), Grant No. MSS-9112512.

## REFERENCES

1. Acar, Y.B, Tumay, M.T. (1986). "Strain field around cones in steady penetration", *Journal of Geotechnical Engineering*, Vol. 112, No. 2, February, pp. 207-213.

2. Allen, N.F., Richart, F.E. Jr., Woods, R.D. (1980) "Fluid wave propagation in saturated and nearly saturated sands", *Journal of the Geotechnical Engineering Division*, Vol.106,NoGT3,March,pp.235-254.

3. Baldi, G., Hueckel, T., Pellegrini, R. (1988) "Thermal volume changes of the mineral-water system in low-porosity clay soils", *Canadian Geotechnical Journal*, Vol. 25, June, pp. 807-825.

4. Baligh, M.M. (1985) "Strain path method", *Journal of Geotechnical Engineering*, Vol. 111, No. 9, September, pp. 1108-1136.

5. Baligh, M.M., Leavadoux, J.N. (1986) "Consolidation after undrained piezocone penetration II: Interpretation", *Journal of Geotechnical Engineering*, Vol. 122, No. 7, July, pp. 727-745.

6. Campanella, R.G., and Mitchell, J.K. (1968) "Influence of temperature variations on soil behavior," *Journal of the Soil Mechanics and Foundations Division*, ASCE, May, Vol. 94, No. SM3, pp. 709-734.

7. Carter, J.P., Randolph, M.F., Wroth, C.P. (1979) "Stress and pore pressure changes in clay during and after the expansion of a cylindrical cavity", *International Journal for Numerical and Analytical Methods in Geomechanics*, Vol. 3, pp. 305-322.

8. Gibson, R.E. (1963) "An analysis of system flexibility and its effect on time-lag in pore-water pressure measurements", *Geotechnique*, 13, 1-9.

9. Houston, S.L., Houston, W.N., Williams, N.D. (1985) "Thermo-mechanical behavior of seafloor sediments", *Journal of Geotechnical Engineering*, Vol. 111, no. 11, November, pp. 1246-1263.

10. Leavadoux, J.N., Baligh, M.M. (1986) "Consolidation after undrained piezocone penetration. I: Prediction", *Journal of Geotechnical Engineering*, ASCE, Vol. 112, No. 7, July, pp. 707-725.

11. Paaswell, R.E. (1967) "Temperature effects on clay soil consolidation", *Journal of Soil Mechanics and Foundations Division*, Proceedings of the American Society of Civil Engineers, Vol.93, No. SM3, May,pp9-22.

12. Pervizpour, M. Pamukcu, S. (1995) "Coefficient of consolidation under thermal potential", Proceedings of GEOENVIRONMENT 2000 ASCE Specialty Conference, Feb. 24-26, New Orleans, Louisiana

13. Premchitt, J., Brand, E.W. (1981)"Pore pressure equalization of piezometers in compressible soils", *Geotechnique*, Vol. 31, No 1, pp 105-123.

14. Tuncan, M. (1991), "Coupled flow of water in saturated kaolinite clay under multiple potentials", PhD Dissertation, Department of Civil Engineering, Lehigh University.

# COMPARISON OF EXTRACTION METHODOLOGIES FOR DESORPTION OF PYRENE

D. Raghavan
   Polymer Division
   Department of Chemistry
   Howard University

James H. Johnson, Jr.,
   Department of Civil Engineering
   Howard University

## INTRODUCTION

Polycyclic aromatic hydrocarbons (PAHs) present a significant environmental problem and a serious health concern due to their carcinogenicity and mutagenicity. PAHs occur as natural combustion products of fossil fuels and are introduced into the atmosphere through industrial emissions, exhaust gases of combustion engines, heating plants, and coal tar and creosote wood treatment facilities. Assessment of the efficacy of remediation technologies at PAH contaminated sites necessitates the development of analytical techniques for the complete extraction and analysis of the contaminants. Analysis of environmental samples is often challenging because contaminants are measured at the parts-per-million or lower level in complex matrices such as soil. Accurate quantification of PAHs at these trace levels requires efficient extraction, clean-up, recovery, and detection methods to minimize the interferences contributed by the chemistry of the matrices or the loss of the contaminant. This article compares Soxhlet Extraction, Super Critical Fluid Extraction and Thermal Extraction for the recovery of contaminants from complex matrices.

## ANALYTICAL TECHNIQUES

To test the efficiency of an analytical technique, the soil/sludge/sediment is often spiked with a known amount of contaminant. The percentage recovered by the method is expressed as

$$\% \text{ Recovery} \quad = \quad \frac{\text{mass of PAH recovered}}{\text{mass of PAH spiked}} \quad \text{x } 100\%$$

The history of spiked samples is quite different from that of real samples and information derived from recovery of contaminants from spiked samples cannot be directly applied to real samples without modifications. In spiked samples, the contaminant is solvent deposited which result in some of the matrix sites being occupied by the solvent molecules unlike that of the real sample. The solvent deposition may affect the chemical structure of the sample (1). The solvent used for spiking contaminant, if not properly and completely evaporated may act as a modifier and promote the elution of the contaminant. However, spiked samples are reliable standards for testing the validity of the procedure and the precision of the apparatus.

In real samples there is an aging effect (2,3). Aging has a role to play in the fixation of the contaminant in and on the soil (4). Therefore, for the successful recovery of contaminant from real sample, an understanding of the aging effect, solute-matrix interaction and the manipulation of the extraction methodology is needed (3). Often, the methodology developed for spiked samples, when directly applied to real samples underestimates the amount of contaminant present in the sample. For example, Burford et al. (1) studied the extraction rate of native PAH and calculated it to be approximately four times slower than that of the spiked sample.

## SOXHLET EXTRACTION

The Soxhlet method is generally used to extract nonvolatile and semivolatile contaminants (hydrophobic) from complex matrices. PAH recovery with EPAs' Method 3540 is a two-step system. In the first stage, the contaminant matrix is placed in a thimble and is rinse-extracted by boiling solvent for 16-24 hours. In the second stage, the solvent is evaporated in the Kuderna/Danish apparatus and concentrated. There have been modifications proposed to the Method 3540. For example, Method 3541 is a three step system, with an initial step of lowering of the thimble containing the specimen into a boiling solvent. This stage provides a rapid contact between matrix and solvent as well as the rapid extraction of the contaminant.

## SUPER CRITICAL FLUID EXTRACTION

Supercritical fluid extraction (SCF) is rapidly replacing traditional extraction methods for analyzing the concentration of contaminants on solids. By adjusting the solvation strength of the extraction fluid, preferential dissolution of the contaminant can be achieved. The super critical fluid analysis of the contaminant from the matrix involves four phases (i) extraction, (ii) trapping, (iii) separation and (iv) detection of the contaminant of interest.

The major considerations to be given in developing an efficient supercritical fluid extraction process are the chemistry and type of soil as well as contaminant, contaminant solubility in the supercritical fluid, and contaminant transport to the bulk fluid. SCF has been used for recoveries of high and low molecular weight PAHs from spiked sediment. High molecular weight PAH compounds were less efficiently extracted than the low molecular weight PAHs (5). High molecular weight PAH compounds (4,5 and 6 rings) have a stronger tendency to adsorb into soil than more soluble low molecular weight PAHs. Depending on the soil chemistry, Weber et al. (6) found that the extent of sorption of contaminants to soil may vary from a weak to a strong interaction depending on the carbon content of the soil. To desorb the tightly bound contaminant from the soil, the energy barrier of desorption should be lowered to maximize super critical fluid extraction of the contaminant (7). When super critical fluid extractions were performed for 40 minutes with pure $CO_2$ as the fluid, raising the temperature from 50 to 200°C yielded about a 2 to 6 fold increase in PAH recovery (8). In addition, extraction time, moisture content of sample, sample size, and pressure of the fluid are some of the other factors that influence the extraction efficiency (9).

## THERMAL EXTRACTION

The operating principle of thermal extraction is to volatilize the semi-volatile contaminant without oxidizing or degrading the components of the soil matrix. During thermal extraction, the contaminants are extracted from the soil/sludge by heating the sample in a sealed chamber (10,11). Because the system is closed and at a working temperature of 350°C, the identity of the contaminant is retained. Organic contaminants are vaporized and collected in a nitrogen/helium gas atmosphere, and a soil free of contaminants is obtained. Desorption experiments using thermal analysis have shown that stripping of adsorbed PAH from soil can be achieved (12).

Currently, little is known about the desorption parameters, and much conceptualization of the phenomena remains speculative (13). A well utilized technique for evaluating the parameters of desorption including energy, is thermal extraction/thermal desorption analysis (14). Desorption energy is defined as the energy needed to desorb the irreversibly bound contaminant from the soil. Adsorption studies of PAH contaminants on autoclaved yeast cells have yielded heat of adsorption of 5.2 kcal/mole (15). The stronger the interaction between the soil and contaminant, the higher will be the desorption energy. For strongly adsorbed species, the presence of solvent that will interact with the contaminant and the soil matrix is believed to lower the energy barrier and facilitate the thermal extraction of the contaminant (16,17). Thermal desorption efficiency is dependent on the concentration of contaminant in the soil, the soil type, the residence time of the contaminant in the soil(15-17).

In this article, we will present an on-going research to determine the extraction process, conditions and efficiency of soxhlet extraction, super critical fluid extraction and thermal extraction of contaminant from spiked model soil. Figure 1 schematically compares the three extraction methods

FIGURE 1    SCHEMATIC REPRESENTATION OF SOXHLET
EXTRACTION THERMAL EXTRACTION AND
SUPER CRITICAL FLUID EXTRACTION
METHODOLOGIES

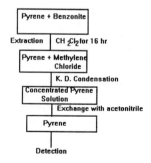

## MATERIALS AND METHOD

### PYRENE DEPOSITION

The benzonite sample was procured from Fisher Scientific. The soil sample was sterilized in an autoclave for 20 minutes at 15 psi and 121°C. 0.5 grams of soil was mixed with 20ml of 50 ppm pyrene in acetonitrile in a round bottom flask. The mixture was allowed to equilibrate for 1 hour. Distillation of the soil-pyrene mixture in a water bath was performed. The distillate was collected and pyrene content was determined by HPLC. Distillation was extended until the soil was dry. The clean unspiked soil was prepared by equilibrating the benzonite with acetonitrile solvent for 1 hour and by performing distillation of the mixture to recover the acetonitrile.

### WATER WASHING

The pyrene deposited sample and 20 ml of deionized water was placed in a amber colored 40 ml bottles. The tubes were sealed with teflon-lined caps and tumbled end-over-end for 3 days to recover water washed pyrene. The soil sample was centrifuged at 8000 rpm for 8 hours to achieve solid-liquid separation. The aqueous phase was analyzed for pyrene content by HPLC.

### DESORPTION STUDY

Soxhlet extraction of 0.5 g pyrene deposited samples was performed with a Soxhlet apparatus with 500 ml of methylene chloride. The intial Solvent cycle time was approximately 16 h. The cycle time was subsequently extended to 24 h from the prescribed 16 h to extract pyrene unrecovered during the 16 h soxhlet extraction. The pyrene along with methylene chloride was concentrated in a Kuderna/Danish apparatus and solvent exchanged with acetonitrile. The pyrene content in the acetonitrile mixture was determined by HPLC. Soxhlet extraction with clean soil samples was also conducted.

### THERMAL EXTRACTION

For performing thermal extraction, a Ruska Laboratories Thermal Extraction Inlet system was used. The Thermal Extraction unit was interfaced to a HP 5890 GC unit. 20 mg of a pyrene deposited soil sample was placed in a pyro cell crucible. The crucible was placed in a loader arm. The crucible was heated to about 250°C. Helium gas was used as the carrier gas, while compressed air serves as the coolant. The thermally desorbed analyte was introduced through the transfer line into the GC capillary column. Thermal extraction experiments were also conducted on clean soil samples to obtain a working base line for comparisons.

# RESULTS AND DISCUSSION

## PYRENE DEPOSITION

Distillation experiment was conducted to spike pyrene onto clean benzonite soil samples. A small fraction of the pyrene was detected in the distillate along with acetonitrile. The difference between the amount of pyrene initially placed and the amount of pyrene collected in the distillate is the amount of pyrene deposited on the soil. By performing a pyrene mass balance, it was concluded that 1760 ppm of pyrene was adsorbed to the soil sample. The results presented are the average of about eight samples.

## PYRENE DESORPTION

It was found that water washings do not remove any pyrene from soil. The physisorbed amount is negligible due to the fact that pyrene deposition was not at ambient temperature but at relatively high temperature (90°C). Trial runs of several samples showed the results were reproducible.

Table 1     EXPERIMENTAL DATA OF SOXHLET EXTRACTION EFFICIENCY OF PYRENE DEPOSITED BENZONITE

| Sample | Efficiency |
|--------|------------|
| #1 | 15% |
| #3 | 9.2% |
| #4 | 26% |
| #5 | (blank) |
| #6 | 20% |
| #7 | 12% |
| #9 | (blank) |

Table 1 shows the soxhlet extraction efficiency of eight samples (6 pyrene deposited and 2 blank samples). Soxhlet extraction of pyrene

759

deposited sample indicated that about 25% of the pyrene was recoverable. However, one of the sample did show about 65% recovery. Further studies are on-going to test the reproducibility and the efficiency of the method.

The Soxhlet extraction method was extended from the prescribed 16 hours to 24 hours to provide time for enhanced desorption of pyrene. No significant amount of pyrene was detected.

Figure 1 shows a composite picture of on-going extraction studies on pyrene deposited benzonite samples. The figure compares and contrasts the three different techniques. The four basic steps involved are extraction, trapping, concentration/separation and detection. In thermal extraction, the contaminant is desorbed at 300°C, while in thermal extraction the contaminant is desorbed at 50°C with methylene chloride. Findings of the research will be presented.

## REFERENCES

1.   Burford, M. D., S. B. Hawthrone, and D. J. Miller. 1993. "Extraction Rates of Spiked Versus Native PAHs From Heterogeneous Environmental Samples Using Supercritical Fluid Extraction and Sonication in Methylene Chloride, " *Analytical Chemistry*, 65 : 1505-1512.

2.   Langenfeld, J. J., S. B. Hawthrone, D. J. Miller, and J. Pawllszyn. 1994. " Role of Modifiers for Analytical-Scale Supercritical Fluid Extraction of Environmental Samples," *Analytical Chemistry*, 66 : 909-916.

3    Alexandrou, N., M. J. Lawrence, and J. Pawliszyn. 1992. " Cleanup of Complex Organic Mixtures Using Supercritical Fluids and Selective Adsorbents, " *Analytical Chemistry*, 64 : 301-311.

4.   Hatzinger, P. B., and M. Alexander. 1995. " Effect of Aging of Chemicals in Soil on Their Biodegradability and Extractability," *Environmental Science and Technology*, 29 : 537-545.

5.   Hawthrone, S. H., and D. J. Miller. 1987. " Extraction and Recovery of Polycyclic Aromatic Hydrocarbons From Environmental Solids Using Supercritical Fluids, "*Analytical Chemistry*, 59 : 1705-1708.

6.   Webber, W. J., P. M. McGinley, and L. E. Katz. 1992. " A Distributed Reactivity Model for Sorption By Soil and Sediments. 1. Conceptual Basis and Equilibrium Assessments," *Environmental Science and Technology*, 26 : 1955-1962.

7.   Langenfeld J. J., S. B. Hawthrone, D. J. Miller and J. Pawliszyn. 1993. " Effects of Temperature and Pressure on Supercritical Fluid Extraction Efficiencies of Polycyclic Aromatic Hydrocarbons and Polychlorinated Biphenyls," *Analytical Chemistry*, 65 : 338-344.

8. Hawthrone, S. B., D. J. Miller, M. D. Burford, J. J. Langenfeld, S. Eckert-Tilotta, and P. K. Louie. 1993. "Factors Controlling Quantitative Supercritical Fluid Extraction of Environmental Samples," *Journal of Chromatography*, 642 : 301-317.

9. Lee, H-B., and T. A. Peart. 1994. "Optimization of Supercritical Fluid Carbon Dioxide Extraction for Polychlorinated Biphenyls and Chlorinated Benzenes From Sediment, " *Journal of Chromatography A*, 663 : 87-95.

10. Shanks, R., and A. T. Trentini. 1994. "Thermal Desorption of SVOC, VOC, and Pesticide Contaminated Soil at the Pristine Facility Trust Superfund Site Reading, Ohio," in *Abstract Proceedings of the Fifth Forum on Innovative Hazardous Waste Treatment Technologies : Domestic and International*, Chicago, IL.

11. Shieh Y-S., 1994. "Thermal Desorption/Base Catalyzed Decomposition (BCD) a Non-oxidative Method for Chemical Dechlorination of Organic Compounds," in *Abstract Proceedings of the Fifth Forum on Innovative Hazardous Waste Treatment Technologies : Domestic and International*, Chicago, IL.

12. Snelling, R., D. King and B. Belair. 1993. "Analysis of PAHs in Soils and Sludges Using Thermal Extraction-GC-MS," *Gas Chromatography*, 1: 1-4

13. Pavlostathis, S. G., and G. N. Mathavan. 1992. " Desorption Kinetics of Selected Volatile Organic Compounds From Field Contaminated Soils, " *Environmental Science and Technology*, 26 : 532-538.

14 Somorjai, G. A., and B. E. Bent. 1995." The Structure of Adsorbed Monolayers. The Surface Chemical Bond ," Progress in Colloid and Polymer Science, 70 : 38-44.

15. Herbes, S. E. 1977. Partitioning of Polycyclic Aromatic Hydrocarbons Between Dissolved and Particulate Phases in Natural Water, *Water Research*, 11 : 493-496.

16. Robbat, A., T-Y. Liu, B. Abraham, and C-J. Liu. 1991. "Thermal Desorption Gas Chromatography-Mass Spectrometry Field Methods for the Detection of Organic Compounds, " in *Second International Symposium on Field Screening Methods for Hazardous Wastes and Toxic Chemicals*, Washington, D.C.

17. Brown R. S., K. Pettit, D. Price, and P. W. Jones. 1991, Thermal Desorption Gas Chromatography : A Quick Screening Technique for Polychlorinated Biphenyls, " *Chemosphere*, 23(8-10) : 1145-1150.

## ACKNOWLEDGMENTS

The authors would like to acknowledge the Great Lakes Mid-Atlantic Center and the Department of Chemistry, Howard University for partially funding the research program. Technical assistance provided by Mrs. L. Wan and Ms. R. Jackson is greatly appreciated.

# VOLATILIZATION AND BIODEGRADATION OF HAZARDOUS WASTE UTILIZING SOIL AGITATION IN A COVERED TREATMENT UNIT

JERRY W. EPLIN, P.E.
Delta Environmental Consultants, Inc.
6701 Carmel Road, Suite 200
Charlotte, NC  28226

SCOTT A. RECKER, P.G.
Delta Environmental Consultants, Inc.
6701 Carmel Road, Suite 200
Charlotte, NC  28226

MARK K. MYRICK, P.G.
Delta Environmental Consultants, Inc.
6701 Carmel Road, Suite 200
Charlotte, NC  28226

## ABSTRACT

During recent years the high cost of hazardous waste treatment and disposal has placed a burden on responsible parties at RCRA sites where correction action is necessary.   Classical methods of disposal often involve off-site transportation to a hazardous waste landfill or to an incineration unit resulting in extremely high cradle-to-grave costs and increased risk to the responsible party. Recent changes in the regulatory perspective have provided opportunities to explore alternative options for management of hazardous waste at such sites.

Such an innovative technology has been utilized at a RCRA site in North Carolina.   The management method involves the on-site treatment of soil containing spent solvents and petroleum hydrocarbons.   Releases of 1,1,1-trichloroethane resulted in contamination of soil.   The soil was subsequently excavated and placed in a temporary on-site treatment unit.   The unit consisted of a ventilated greenhouse placed on an asphalt surface.   Soil was placed in the unit for treatment utilizing volatilization and biodegradation.   The soil was agitated on a regular basis to maximize exposure of the soil to air and promote volatilization  and  biodegradation  of  the  spent  solvents  and  petroleum hydrocarbons. Following approval of a Solid Waste Management Unit (SWMU)

Closure Plan, complete treatment of approximately 144 cubic yards of soil was completed within 18 months.

## INTRODUCTION

Since the initiation of the Resource Conservation and Recovery Act (RCRA) handling of contaminated media at sites with releases to the environment has been difficult to address. The regulations apply specifically to wastes generated as part of an industrial process but do not directly address contaminated media (soil and ground water) that may be generated as part of an environmental cleanup. Historically, these contaminated media fall under the same regulatory statutes as industrial wastes, but must, by their nature and greater volume, be handled differently. In many instances it is necessary to remove a large volume of material to properly manage risk to human health and welfare or to prevent or minimize continued migration of contamination in the environment. This often results in large volumes of contaminated media, classified as hazardous waste, to be managed and disposed.

The cost associated with disposal of soil classified as hazardous with spent solvents continues to be high. During the 1980s, cost for disposal could be as high as $800 per 55 gallon drum of contaminated material. The incineration cost for large volumes of material is often in the range of $800 to $1100 per ton if the material is "land banned" or contains concentrations higher than the RCRA land banned standard for the compounds contained in the soil. Hazardous waste landfill disposal costs may be as high as $500 per ton. Neither of these disposal costs take into account increased risk to the responsible party through off-site transportation of the waste or the continued risk of placing waste in a landfill that will be monitored for a period of no less than 30 years.

During the early 1990s, the EPA realized that in some situations the cost, risk, and application of industrial management statutes may be avoided for some contaminated media and wastes could be treated on-site in temporary treatment units. The "Corrective Action Management Unit" or CAMU was designated for use on hazardous waste sites where RCRA corrective action is taking place. Delta Environmental Consultants, Inc. (Delta) has successfully utilized an alternative temporary treatment unit at a site in North Carolina. The approach resulted in a lower waste management cost and decreased risk to the responsible party where corrective action was taking place.

## BACKGROUND INFORMATION

From 1965 to the early 1980s, a small brick lined disposal basin was used to dispose of unknown quantities of machine cutting oils and 1,1,1-trichloroethane (TCA). The TCA was used to clean parts and as a degreasing agent for cold metal surfaces prior to painting and therefore is designated as an F001 hazardous waste according to 40 CFR 261.31. The drains to the disposal basin became permanently clogged in the late 1970s and the basin was capped with concrete in 1985. TCA storage was transferred to a 1500 gallon above ground storage tank. During 1987, facility personnel identified a leak in the storage tank discharge. Subsequent soil investigations indicated that continued use of the disposal basin prior to capping, and leaks within the system, resulted in soil and ground water contamination on the property. Initial concentrations in soil in the area of the disposal unit were as high as 770,000 ug/kg TCA. The disposal basin was designated as a Solid Waste Management Unit (SWMU) and subsequently managed by the North Carolina Hazardous Waste Section (HWS).

The HWS required that a subsurface investigation ensue to define the extent of soil and ground water contamination. As part of the investigation the responsible party also opted to apply an interim remedial measure fashioned to remove TCA associated with the disposal basin. A soil vapor extraction (SVE) system was designed and installed in the vicinity of the disposal basin. Operation of the SVE system resulted in the removal of approximately 556 gallons of TCA over a period of 785 days. However, it is believed that the high levels of petroleum hydrocarbons in the soil reduced permeability and increased the adsorption of the TCA onto the soil and therefore inhibited the effectiveness of the SVE system. Concentrations of TCA ranging from 2.3 to 67,000 ug/kg remained in the soils. Because the SVE system was not adequate to remediate the soils within an agreed upon time frame, an alternative treatment technology was required to accelerate remediation of the site. The HWS required that a Closure Plan be prepared according to 40 CFR 265 Subpart G and an Administrative Order on Consent was negotiated between the responsible party and the HWS. The Closure Plan called for the excavation of the disposal basin materials followed by installation of a concrete cap over the SWMU. Soils excavated from the SWMU would have to be handled in a manner acceptable to the HWS and pertaining to the rules in 40 CFR 261.

## TREATMENT OPTIONS

Several treatment options for soil removed from the SWMU were explored. Through sampling of material in and around the former disposal basin it was determined that land disposal was not an option because concentrations of TCA were greater than land banned values. Off-site transportation and incineration at an authorized facility appeared to be the only option available. The cost to incinerate with transportation and management, however, was in the range of $120,000. This would have placed a major strain on the financial stability of the responsible party, therefore another treatment option was considered.

The chosen option was to utilize a Corrective Action Management Unit (CAMU). By doing so, soil would not have to be transported but could be treated passively on-site resulting in extensive cost savings for the responsible party. The selected technology included excavation of the disposal basin soils for land treatment by volatilization and passive bioremediation to remove the hazardous constituents from the soil. The primary objectives were to excavate the contaminated soils within the unsaturated zone and to spread the soils within a treatment unit. The soils would be turned to achieve aeration to enhance volatilization and bioactivity. After chemical analysis determined that the hazardous constituent levels were below practical quantitation levels, the material would no longer be classified as a hazardous waste. It was anticipated that residual levels of petroleum hydrocarbons would necessitate that the material be disposed of off-site by thermal treatment at a non-hazardous handling facility. The expected cradle to grave operating cost of the unit was $50,000 verses the estimated cost of disposal of the land banned material at $120,000.

The treatment unit consisted of a greenhouse constructed on a new asphalt liner which was placed over an existing asphalt surface (Figure 1). A concrete block berm was constructed around the outside of the asphalt liner to keep material inside the unit during treatment and to help prevent surface runoff from contacting the soil. The greenhouse was constructed of fence grade prefabricated metal tubing. The frame anchors were cemented into the concrete blocks used for the berm which also added treatment unit strength. The greenhouse was covered by a 6 mil white nursery film guaranteed for one year. To ensure integrity of the treatment unit cover, plastic clips were used to secure the plastic cover to the frame. A blower was placed within the treatment unit to ventilate the open space of the greenhouse. Circulation of fresh air through the enclosed unit greatly reduced condensate build up, supplied oxygen to promote aerobic biological activity to facilitate degradation of petroleum hydrocarbons, maximized volatilization of the hydrocarbons and provided adequate air supply for workers.

**IMPLEMENTATION**

Prior to excavating the soils, the bottom of the treatment unit was lined with plastic to prevent any soil moisture from contacting the underlying asphalt liner. The soil was excavated in August 1992 and was placed directly into the treatment unit. The soil was spread to an approximate depth of 16 inches. The greenhouse was then constructed over the treatment unit and an 8000 cfm fan was installed in on end of the greenhouse with vents in the other end.

Upon placement of soil in the treatment unit, three discrete soils samples, one for each fifty cubic yards of soil in the treatment unit, were collected and analyzed by EPA Method 8240 (volatile organic compounds), EPA Method 907 (Total Petroleum Hydrocarbons (TPH)), and, TCLP metals. The analytical results are summarized in Table I. The TCA concentrations ranged from 30,000 ug/kg to 52,000 ug/kg. TPH levels ranged from 15,200 ug/kg to 38,700 ug/kg.

To provide soil agitation and aeration, a dedicated bobcat loader was placed inside the greenhouse to turn the soil. The soil was turned on a weekly basis using the loader.

Final rules for CAMU were issued by EPA in February 1993. The rule stated that CAMUs could only be applied in situations where the treater, disposer or storer of hazardous waste had completed a RCRA Part B permit application and had obtained a Part B permit to treat, dispose and store hazardous waste. Since the facility had not completed a Part B permit application the HWS asked that the CAMU be designated as a "treatment unit" and CAMU was not used as a reference during the remainder of the soil remediation.

## RESULTS

The treatment unit was operated for a period of 15 months between August 1992 and November 1993. The progress of soil remediation was monitored by collection of soils samples in May 1993, August 1993, and November 1993. The analytical results are presented in Table I. As shown in Figure 2, the average TCA concentration decreased from an initial level of over 27,000 ug/kg to the desired remedial level of 5 ug/kg over the time period. The average perchloroethylene level decreased from over 3000 ug/kg to less than the practical quantitation level of 5 ug/kg, likewise, xylenes were reduced from an average of 2000 ug/kg to less than 5 ug/kg. Reduction of these F001 listed constituents to practical quantitation levels rendered the soil non-hazardous according to 40 CFR 261.

In addition to the remediation of the volatile compounds, the TPH levels found in the soils were greatly reduced during the remediation process. Initial TPH levels from the three sampling locations ranged from 11,400 ug/kg to 18,700 ug/kg. These levels were reduced to between 2,370 ug/kg and 6,630 ug/kg over the operational period. Figure 2 displays the results for the average concentration values and Figures 3, 4, and 5 display the individual results for each sampling location.

Throughout the operational period of the greenhouse treatment unit, no attempt was made to determine what percentage of the remediation was occurring as a result of bioremediation and what percentage was the result of volatilization. In this case, the method by which remediation was occurring was not important to the responsible party. As a consequence, costs were kept low and no unnecessary sampling was performed. Because significant remediation of both the volatile solvents and the non-volatile petroleum hydrocarbons was achieved, it is felt that both factors were occurring.

When the soil was rendered non-hazardous it could be treated as a petroleum hydrocarbon contaminated soil. The material was removed from the treatment unit and disposed of in a brick kiln by incorporating the soil into the brick manufacturing process and removing residual petroleum hydrocarbons by incineration. The greenhouse was disassembled and the liner and soil beneath the liner were sampled for hazardous constituents and found to contain none. The treatment system construction material was disposed of as construction debris.

## TABLE I: SOIL SAMPLE ANALYTICAL DATA

| Compound | Sample Location 1 (ug/kg) | | | Sample Location 2 (ug/kg) | | | | Sample Location 3 (ug/kg) | | | |
| --- | --- | --- | --- | --- | --- | --- | --- | --- | --- | --- | --- |
| | 08/13/92 | 05/07/93 | 08/12/93 | 11/09/93 | 08/13/92 | 05/07/93 | 08/12/93 | 11/09/93 | 08/13/92 | 05/07/93 | 08/12/93 | 11/09/93 |
| 1,1,1-TCA | 30000 | 100 | 7 | 5 | 5200 | 90 | 9 | 5 | 46000 | 20 | 20 | 5 |
| PCE | 2700 | 200 | <5 | <5 | 3200 | 300 | <5 | <5 | 3200 | 40 | 9 | <5 |
| Xylenes | 1200 | <6 | <5 | <5 | 2400 | <10 | <5 | <5 | 2400 | <6 | <6 | <5 |
| TPH | 18700 | 12800 | 9540 | 6120 | 15200 | 11400 | 8240 | 6630 | 38700 | 14800 | 11200 | 2370 |

ALL TCLP METALS WERE BELOW THEIR RESPECTIVE REGULATORY LEVEL

768

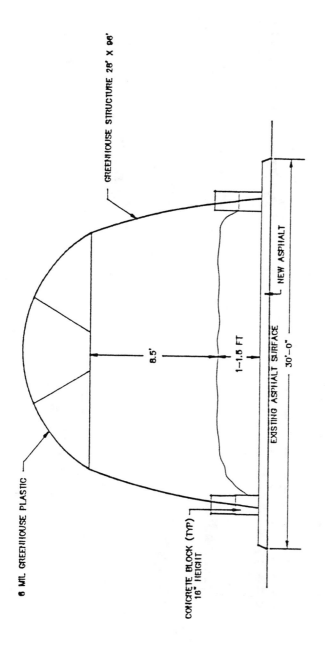

FIGURE 1: TREATMENT UNIT CONSTRUCTION

769

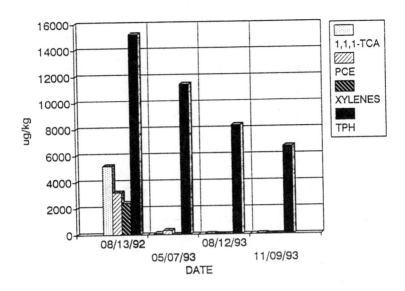

FIGURE 2: AVERAGE SOIL SAMPLE ANALYTICAL RESULTS

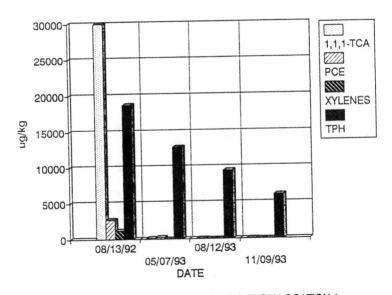

FIGURE 3: SOIL ANALYTICAL RESULTS FROM LOCATION 1

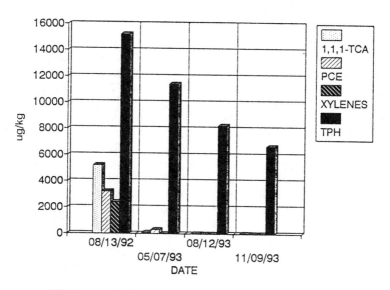

FIGURE 4: SOIL ANALYTICAL RESULTS FROM LOCATION 2

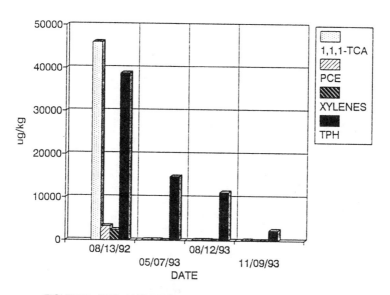

FIGURE 5: SOIL ANALYTICAL RESULTS FROM LOCATION 3

## DISCUSSION

Problems with a former disposal basin resulted in soil being contaminated with spent solvents and petroleum hydrocarbons. In-situ remediation was initially attempted using soil vapor extraction. The SVE system performed well, removing approximately 556 gallons of TCA over 785 days of operation. However, the petroleum hydrocarbons apparently interfered with the volatilization rate resulting in the need to use an alternative technology to enhance the rate of remediation. Excavation and subsequent incineration of the hazardous soils appeared to the most appropriate technology, however, the associated cost were prohibitive. Alternatively, the soil was remediated in an on-site treatment unit which utilized volatilization and biodegradation. The treatment unit consisted of a greenhouse constructed on an asphalt liner and a block base. The soil was placed in the treatment unit and was turned weekly to achieve aeration. The soil treatment unit was operated for less than 18 months. Chlorinated solvents had been remediated to at, or below, the 5 ug/kg practical quantitation limit rendering the soil non-hazardous. Petroleum hydrocarbon levels in the soil were also reduced to acceptable levels for treatment using incineration in the brick manufacturing process.

The remediation process utilizing soil agitation to achieve volatilization and biodegradation in a greenhouse type structure proved to be a simple and low cost alternative to incineration of land banned hazardous waste soil.

# OPERATION OF A SOIL VAPOR EXTRACTION AND AIR SPARGING SYSTEM AT A FORMER GASOLINE SERVICE STATION

GREGORY J. GROMICKO
ROBERT C. KLINGENSMITH
    Groundwater Technology, Inc.
    600 Clubhouse Drive, Suite 200
    Moon Township, PA 15108

DUANE K. SIMPSON
    Quaker State Corporation
    P.O. Box 989
    Oil City, PA 16301

## INTRODUCTION

Closure activities for three underground storage tanks (USTs) were conducted at Quaker State Corporation's (QSC) former service station in Conneaut, Pennsylvania as part of a property transfer during July, 1991. The facility, formerly owned by QSC, was operated from construction (early 1960's) through the sale of the property (early 1980's). Subsequent to sale of the facility, the property has been re-sold and the building reconfigured several times.

The facility is located on a corner lot located along state highway Route 322 in the business district of Conneaut Lake as shown in Figure 1. Across the highway to the north, is Conneaut lake. The site is bordered by a residential property to the south and commercial properties on the east and west. A Pennsylvania State Game Commission Game Lands, is located approximately 150 feet southeast of the property.

Quaker State again became involved with the property in 1991 when the current owner attempted to sell the property and the lender for the prospective purchaser identified the presence of USTs. Subsequent to the confirmation of the USTs, UST closure activities were initiated. Subsurface investigations were conducted to delineate the extent of potential petroleum impacts and corrective actions were initiated which are on-going today.

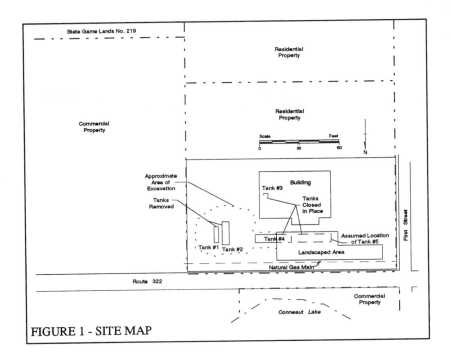

FIGURE 1 - SITE MAP

## PREVIOUS INVESTIGATIONS

The scope of work associated with the UST closure activities included the closure by removal of UST #1 and UST #2, the closure in place of UST #3, and the determination of the presence of UST #4 and UST #5.

UST's #1, #2, and #3 were known to exist at the site based upon a site inspection completed by the QSC. The presence of UST #4 and UST #5 were reported, but a records search was inconclusive based upon the dated nature of the initial property sale. Information provided by the current owner of the property indicated that UST #4 and UST #5 were located under the landscaped area as illustrated in Figure 1.

In accordance with American Petroleum Institute Recommended Practices 1604 and 2015, UST #1 and UST #2 were permanently closed by removal, and UST #3 was permanently closed in place [1,2]. The presence and location of UST #4 and UST #5 was confirmed through the enlargement of the area excavated for closure of UST #1 and UST #2. During the UST closure activities, a series of four test pits were excavated on the western side of the property to determine if any other USTs were present at the site. All test pits were excavated to a depth of five feet and no evidence of UST systems or petroleum impacts were noted. The location of all USTs determined to be at the site are illustrated in Figure 2.

Visual and photoionization inspections of the excavations, the resulting excavated soils and the groundwater that entered the excavations indicated that petroleum hydrocarbons had been released to the surrounding soils and groundwater.

Following UST closure activities, soil samples were collected and analyzed in accordance with the Pennsylvania Department of Environmental Resources (PADER) site characterization guidelines as outlined in the UST Closure Phase I Investigation. The soil samples were analyzed for total petroleum hydrocarbons (TPH) by EPA

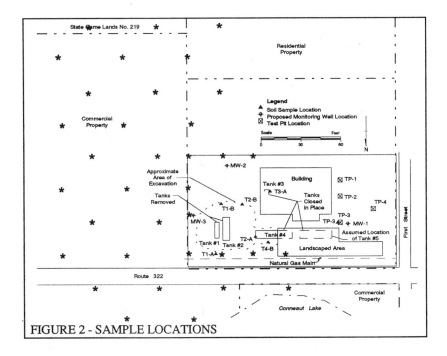

FIGURE 2 - SAMPLE LOCATIONS

Method 418.1 and benzene, toluene, ethylbenzene, and xylene (BTEX) by EPA Method 8020. Locations of soil sampling locations are illustrated in Figure 2.

Based upon observations noted during the UST closure and the analytical results of the soil samples collected subsequent to the UST closure activities, the presence of petroleum hydrocarbons in the surrounding soils was confirmed [3].

Following completion of the UST Closure Phase I Investigation, a Phase II Investigative Work Plan was developed to determine the extent and nature of the petroleum impacts to the subsurface soils and first water bearing zone. This Investigative Work Plan was issued to the PADER for concurrence. Upon review, the PADER gave approval to initiate the plan.

The Phase II Investigation consisted of a soil and groundwater sampling and analysis program that was completed by advancing three soil borings and constructing monitoring wells (MW-1, MW-2, and MW-3). The soil boring and monitoring well locations are shown in Figure 2.

Soil samples were collected during soil boring operations to characterize soil conditions outside of the area of the closed USTs. Two groundwater sampling and analysis events were completed to characterize groundwater quality. The analytical results from the soil boring and groundwater sampling and analysis events indicated that the soils and first water bearing zone about the closed USTs was impacted with petroleum products [4]. The field data also indicated that the groundwater flow direction was in a south to southeastern direction. Following discussions and gaining concurrence with the PADER, a soil gas survey was conducted to determine the aerial extent of petroleum impacted groundwater.

The soil gas survey was used as a preliminary screening tool to determine the extent of migration of the impacts. By measuring the gaseous phase of the volatile organic compounds migrating to the surface above and impacted area, the source area and plume extent may be delineated [5]. The advantages of this technique as a screening tool instead of a soil boring and well installation program are that the soil gas sampling method is relatively non-disruptive and less costly, and can provide a thorough evaluation due to the close spacing of the sampling points.

The soil gas survey was completed over a 50 by 50 foot grid pattern as shown in Figure 2. This grid pattern was surveyed into existing datum points on the site to correlate known site soil and groundwater conditions with the soil gas survey. Of primary concern was evaluating the potential impact of the State Game Land located approximately 150 feet beyond the southeast corner of the site. A total of 22 soil gas and 12 groundwater samples were collected. The analytical results from this survey indicated that the impacted area was confined to the site [5].

Review of the soil boring logs and groundwater elevation data for the monitoring wells (MW-1, MW-2, MW-3) indicated that migration of the BTEX compounds were limited to the area in the vicinity of the former USTs. This conclusion was developed based upon the low permeability of the fine clayey silt and sandy silt soils, and the low hydraulic gradient within the uppermost water bearing unit.

## CORRECTIVE ACTION

Following the investigative activities, corrective action alternatives were evaluated. Based upon site and constituent characteristics, soil vapor extraction (SVE) integrated with air sparging (AS) technologies were selected as the recommended corrective action alternative.

Soil vapor extraction is the process of removing volatile constituents from soil pore spaces by causing air to flow through the subsurface [6]. In SVE, air flow enters through the surface adjacent to the treatment zone and is drawn through the treatment zone, mixing with the soil vapors as it moves. The SVE vapors are withdrawn using a vacuum source through extraction wells or trenches placed within the treatment area. Flow patterns for the air are dependent upon the soil permeability, soil type, soil conditions, and placement of the extraction points. Upon removal, SVE vapors may be treated using an above-ground treatment process, such as activated carbon, prior to discharge of the gas.

Air sparging is the treatment of saturated soil and groundwater by providing air to the subsurface to volatilize organic constituents within the treatment zone [7]. Air sparging, using deep well points within the saturated zone, is intended to deliver air to the subsurface in order to transfer the remaining constituents to the vapor phase. Within the unsaturated zone, AS through the shallow well points is intended to supply oxygen for biological activity, and induce air flow to the SVE system. Nutrients may also be added to the AS system to stimulate growth of the indigenous microbial population and promote biodegradation of these constituents [8]. Placement of the injection points are dependent upon site-specific factors such as soil types, permeability, and depth of the water table.

The SVE/AS system installed at this site is an integrated remedial system that has been designed to remove the volatile BTEX constituents from the treatment area. A layout of the SVE/AS system is shown in Figure 3.

The goal of the system is to balance the amount of air that is introduced through AS and removed through SVE with the rate of volatilization and natural biodegradation. Vapor recovery trenches were installed around the perimeter of the site as shown in the Figure 3. The southern trench is approximately 100 feet in length and the northern trench extends approximately 170 feet. Construction of these trenches was completed by excavating to a depth of 10 feet below ground surface and backfilling the trench with permeable select fill (2 inch limestone gravel) to a depth of 4 feet below ground surface. On top of the gravel, perforated 4 inch, PVC piping

776

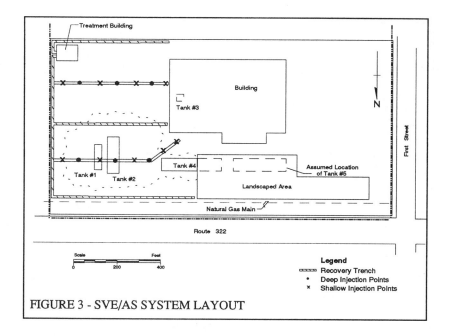

FIGURE 3 - SVE/AS SYSTEM LAYOUT

was placed, and the trenches were backfilled to grade with less permeable native material. An additional recovery trench was installed in the area of the former UST #1 and UST #2. This third trench, extending approximately 50 feet east to west midway within the treatment zone, was installed to address the petroleum impacts in the shallow soil immediately in the former UST area. This trench was constructed in a similar manner, but only excavated to a depth of 5 feet below ground surface.

Figure 4 shows a schematic of the above-ground components of the system. A positive displacement blower was used to create a vacuum in the recovery trenches. To prevent any moisture collected from the recovery trenches from damaging the blower, a moisture knockout tank was installed in front of the blower. The SVE system is capable of recovering a maximum of 100 SCFM at a vacuum of up to 4 inches of mercury. Instrumentation and controls for monitoring the SVE system include vacuum gauges, regulators, flowmeters, and control valves. The extracted vapors are treated prior to discharge to the atmosphere using an air-phase, granular activated carbon (GAC) system. The GAC system utilizes two carbon canisters in series to treat the 100 SCFM flow. The system is equipped with sampling ports to monitor the vapors prior to discharge to the atmosphere.

The AS system consists of a positive displacement blower capable of a maximum flow rate of 75 SCFM at a pressure of 12 psi. The air is cooled through a heat exchanger prior to movement to the shallow and deep injection points. Instrumentation and controls consist of pressure regulators, pressure gauges, temperature indicators and control valves.

The injection trenches and injection points are shown in Figure 3. Injection trench #1 is 50 feet in length and injection trench #2 is 60 feet in length. The trenches were excavated to 3 feet below ground surface to permit installation of the injection points and header piping below-grade. Two-inch, schedule 40, PVC wellpoints were utilized with a 2 1/2 foot screened interval. The shallow points were installed to a

777

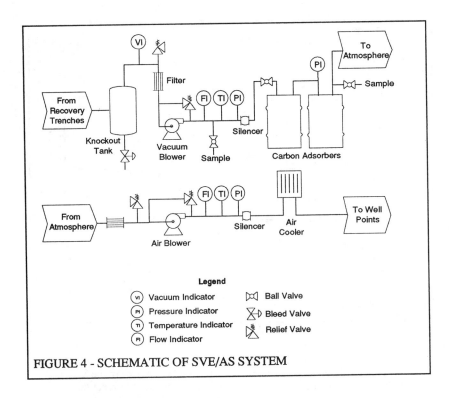

FIGURE 4 - SCHEMATIC OF SVE/AS SYSTEM

depth of approximately 6 feet and the deep injection points were installed to a depth of approximately 18 feet. Sand was placed around the screened intervals and the remaining riser pipes were grouted to the surface using bentonite and concrete. The injection points are connected to header piping leading to the injection blower.

All trenches were then backfilled and the entire site surface was paved with asphalt to reduce infiltration of surface run-on and to induce air flow through the treatment area. A small treatment building was erected on the southeast corner of the property to house all above-ground equipment. Separate utilities were installed to allow operation of the system, independent of site operations.

## OPERATION

Operation of the SVE/AS system was initiated in August 1993. Start-up consisted of setting flowrates and collecting samples of the recovered vapors. The system operates continuously and is monitored monthly. Vacuum and pressure readings are recorded and air flowrates are monitored. Mechanical operation of the equipment is also inspected during the monthly monitoring. System maintenance includes changing the lubricating oil from the blowers, greasing bearings, and cleaning the air intake filter.

Air sampling of the SVE vapors is also performed monthly. Samples are collected prior to the carbon adsorbers to represent recovered vapors and following the carbon adsorbers as a discharge sample. Samples are collected by inserting a syringe directly into the SVE line via a sampling port. The sample is then extracted and injected into an evacuated vial and immediately shipped to the laboratory for analyses for BTEX using modified EPA Methods 3810 and 8000.

Groundwater samples are collected quarterly from MW-1, MW-2, and MW-3 and analyzed for BTEX, using EPA Method 8020, to determine the progress of the remediation. The first sampling event occurred prior to start-up in August 1993 and serves as a baseline.

Baseline soil samples were not collected during start-up of the SVE/AS system. However, results from soil sampling conducted in October 1991 have been used as baseline data. Soil samples were collected following one year of operation to determine the effect of the SVE/AS system in addressing the vadose zone soil. Samples were collected near the three initial sampling locations at MW-1, MW-2, and MW-3 and analyzed for BTEX using EPA Method 8020.

RESULTS AND CONCLUSIONS

The results from the SVE air sampling are presented in Table 1. Data show a general reduction in the BTEX recovered in the soil gas. Initial gas samples contained high levels of BTEX and the concentrations decreased rapidly during the initial months of operation. Additionally, after the decrease, the concentrations generally remained non-detectable. This trend indicates that the easily volatilized constituents were removed initially. As expected, BTEX concentrations in the recovered SVE vapors increased during the summer months because of increased volatilization of the compounds at the higher ambient temperatures.

TABLE 1 - SVE AIR SAMPLING RESULTS (Results in ppmv)

| Date | Benzene | Toluene | Ethylbenzene | m&p Xylenes | o Xylenes |
|------|---------|---------|--------------|-------------|-----------|
| 9/93  | 0.45   | 3.82   | 1.07   | 13.95  | 6.36   |
| 9/93  | 0.38   | 3.14   | 0.93   | 11.98  | 5.45   |
| 9/93  | <0.07  | 0.73   | 0.22   | 3.26   | 2.01   |
| 10/93 | 0.11   | 0.21   | <0.07  | 0.23   | 0.16   |
| 10/93 | <0.07  | 0.23   | <0.07  | 0.34   | 0.25   |
| 1/94  | 0.13   | 0.2    | <0.07  | <0.07  | <0.07  |
| 2/94  | <0.07  | <0.07  | <0.07  | <0.07  | <0.07  |
| 3/94  | <0.07  | <0.07  | <0.07  | <0.07  | <0.07  |
| 4/94  | <0.07  | <0.07  | <0.07  | <0.07  | <0.07  |
| 5/94  | <0.07  | <0.07  | <0.07  | <0.07  | <0.07  |
| 6/94  | 0.7    | 0.3    | <0.07  | 0.2    | <0.07  |
| 7/94  | 0.16   | 0.13   | <0.07  | <0.07  | <0.07  |
| 8/94  | 0.37   | 0.33   | <0.07  | 0.13   | 0.08   |
| 9/94  | <0.07  | <0.07  | <0.07  | <0.07  | <0.07  |
| 10/94 | <0.07  | <0.07  | <0.07  | <0.07  | <0.07  |
| 11/94 | <0.07  | <0.07  | <0.07  | <0.07  | <0.07  |
| 1/95  | <0.07  | <0.07  | <0.07  | <0.07  | <0.07  |
| 2/95  | <0.07  | <0.07  | <0.07  | <0.07  | <0.07  |

## TABLE 2 - GROUNDWATER ANALYTICAL RESULTS (Results in mg/L)

|              | 10/91  | 7/93   | 10/93 | 2/94 | 5/94 | 8/94 | 11/94 | 2/95 |
|--------------|--------|--------|-------|------|------|------|-------|------|
| **MW-1**     |        |        |       |      |      |      |       |      |
| Benzene      | 0.227  | <0.2   | **50**| **17**| 1.3 | 0.4  | **8.2**| <0.2 |
| Toluene      | <0.2   | <0.2   | 6     | 1.1  | <0.2 | <0.2 | <0.2  | <0.2 |
| Ethylbenzene | <0.2   | <0.2   | 3     | 1.4  | <0.2 | <0.2 | 1.4   | <0.2 |
| Xylene       | <0.3   | <0.3   | 5     | 6.9  | <0.3 | <0.3 | 1.2   | <0.2 |
| **MW-2**     |        |        |       |      |      |      |       |      |
| Benzene      | **6,510** | **6,080** | **45** | <20 | **8.9** | **39** | 0.8 | **12** |
| Toluene      | **4,070** | 360    | 15    | <20  | 0.45 | <2   | <0.2  | <0.2 |
| Ethylbenzene | **2,620** | **3,840** | 161 | 24  | 18   | 81   | 4.1   | 2.3  |
| Xylene       | **17,800** | **24,800** | 2,100 | 155 | 78 | 140 | 6.4 | 3.8 |
| **MW-3**     |        |        |       |      |      |      |       |      |
| Benzene      | **2,950** | **13,500** | **25** | **19** | 1.5 | 3 | <0.2 | <0.2 |
| Toluene      | **2,870** | <200   | <220  | 91   | 0.52 | 0.5  | <0.2  | <0.2 |
| Ethylbenzene | **1,530** | **6,770** | 35  | 103  | 9.6  | 11   | 1.3   | <0.2 |
| Xylene       | **13,400** | **41,200** | 229 | 769 | 78 | 35 | 2.1 | <0.2 |

## TABLE 3 - SOIL ANALYTICAL RESULTS (Results in mg/Kg)

|                    | 10/91       | 8/94      |
|--------------------|-------------|-----------|
| **SB-1 (5-7 ft bgs)** |          |           |
| Benzene            | 37          | 35        |
| Toluene            | 40.5        | <20       |
| Ethylbenzene       | BDL         | <20       |
| Xylene             | 118         | <30       |
| **SB-1 (9-11 ft bgs)** |         |           |
| Benzene            | BDL         | <20       |
| Toluene            | BDL         | <20       |
| Ethylbenzene       | BDL         | <20       |
| Xylene             | BDL         | <30       |
| **SB-2 (5-7 ft bgs)** |          |           |
| Benzene            | BDL         | 62        |
| Toluene            | **11,600**  | 56        |
| Ethylbenzene       | **37,100**  | 54        |
| Xylene             | **297,000** | 300       |
| **SB-2 (9-11 ft bgs)** |         |           |
| Benzene            | **3,570**   | **4900**  |
| Toluene            | **32,900**  | **1000**  |
| Ethylbenzene       | **67,100**  | **1500**  |
| Xylene             | **473,000** | **8000**  |
| **SB-3 (5-7 ft bgs)** |          |           |
| Benzene            | 106         | 46        |
| Toluene            | 60.3        | <20       |
| Ethylbenzene       | 249         | 42        |
| Xylene             | **2000**    | 250       |
| **SB-3 (7-9 ft bgs)** |          |           |
| Benzene            | BDL         | <1        |
| Toluene            | 462         | **6500**  |
| Ethylbenzene       | **2,620**   | **9.800** |
| Xylene             | **20,100**  | **160,000** |

The results of the groundwater analyses are presented in Table 2. The highlighted values are those that exceed the current PADER clean-up criteria. As shown, a general reduction in groundwater BTEX concentration has occurred over time. MW-1 and MW-3 have met the clean-up criteria. The total BTEX removal percentages in these two wells is 100%. The total BTEX removal percentage for MW-2 is >99.9%. The BTEX removal percentage in MW-2 has not yet reached the PADER clean-up levels. The system will be operated until groundwater clean-up criteria have been achieved at all wells.

The results of the soil sampling are presented in Table 3. Also shown in the table are the results of the soil sampling performed during the Phase II investigation, which serves as a baseline. As can be seen, the BTEX concentrations in the soil at depths of 5-7 feet and 9-11 feet, decreased during operation of the SVE/AS system. The only increase was observed at location SB-3 at the 7-9 foot depth. This is attributed to the potential presence of a "smear zone" that may contain residual BTEX and is affected by the groundwater table. Following continued operation of the system, and once groundwater BTEX concentrations are below clean-up criteria, the soil will again be sampled to confirm that treatment levels have been achieved.

Following treatment, a closure report will be prepared and submitted to the PADER for approval. The site may then be considered "clean" and the site may once again be used commercially.

## REFERENCES

1. American Petroleum Institute. 1987. "Removal and Disposal of Used Underground Storage Tanks," Recommended Practice 1604 Second Ediition, API.

2. American Petroleum Institute. 1987. "Safe Entry and Cleaning of Petroleum Storage Tanks," Recommended Practice 2015 API.

3. Keystone Environmental Resources, Inc. August 1991. "Phase I Soils Sampling Analytical Results and Proposed Phase II Subsurface investigation Work Plan in The Area or Former Underground Storage Tank Systems," Quaker State Corporation Former Conneaut Lake Facility.

4. Keystone Environmental Resources, Inc. May 1992. "Soil Gas Survey," Quaker State Corporation Former Conneaut Lake Facility.

5. Environmental Protection Agency. June 1987. "Soil-Gas Measurement for Detection of Subsurface Organic Contamination," EPA/600/S2-87/027.

6. Johnson, P.C., Stanley, C.C., Kemblowski, M.W., Byers, D.L., Colthart, J.D. 1990. "A Practical Approach to the Design, Operation, and Monitoring of In Situ Soil Venting Systems," GWMR.

7. Griffin, C.J., Armstrong, J.M., Douglass, R.H. 1991. "Engineering Design Aspects of an In Situ Soil Vapor Remediation System (Sparging)," in In Situ Bioreclamation Applications and Investigations for Hydrocarbon and contaminated Site Remediation. Butterworth-Heinemann.

8. Environmental Protection Agency. March 1992. "A Citizen's Guide To Bioventing," EPA/542/F-92/008.

9. Pennsylvania Department of Environmental Resources. December 1993. "Closure Requirement for Underground Storage Tank Systems".

# THE USE OF DIGITAL SIGNAL PROCESSING AND NEURAL NETWORKS FOR CHARACTERIZATION OF COMPOSITES OF RESIDUAL MATERIALS

SALOME ROMERO
Graduate Research Assistant
Department of Civil and Environmental Engineering
Lehigh University
Bethlehem, PA 18015

SIBEL PAMUKCU
Associate Professor
Department of Civil and Environmental Engineering
Lehigh University
Bethlehem, PA 18015

## INTRODUCTION

Due to variations in source production and handling, residual materials are difficult to identify or categorize when they are used in constructed facilities. Standard tests do not often produce unique and repeatable results. However, proper characterization and quality control is needed if residual materials are to be used to construct geocomposites. One promising new technique for the characterization, identification as well as quality control of such composites is nondestructive testing using wave propagation analysis.

Previous applications of wave propagation techniques include determination of location of flaws in concrete structures [1,2], location of cracks in pavement structures [3], and crack detection in soil [4]. Such nondestructive techniques employ low-strain wave propagation testing which preserve the constancy of the medium, yet can accurately detect the presence and location of defects. Integrity testing is achieved by producing an impulse generated by a hammer which causes waves to propagate through a medium. Accelerometers or geophones are used to receive reflected waves. By recording the time interval between reflections, wave velocity and elastic modulus can be calculated. Analysis can be performed in either the time or frequency domain.

In the work presented here, a resonant column and piezoelectric elements are used to find unique signature patterns for a given geocomposite. Both apparatus are used to generate low-strain distortion or compression waves which propagate through a sample. Signature patterns are established by

782

analysis of the waveform obtained at the resonant frequency. Signature pattern identification may be difficult since slight changes are not readily discernible. Due to the complexity of matching signature patterns, an expert system capable of identifying similar patterns is required.

Expert systems require explicit mathematical relationships between inputs and outputs to be defined. The complexity between sample parameters and corresponding signature patterns makes it difficult to construct this type of relationship. Artificial neural networks (ANN), which are a form of an expert system, have the advantage of establishing these relationships through "training". Training teaches the neural network to identify trends between inputs and outputs. With proper training, the neural network will distinguish between different parameters and categorize a material within a stated tolerance.

To determine the applicability of neural networks in recognizing and characterizing similar materials, tests were run with Ottawa sand at various densities. Since density is one of the major physical properties which affects wave velocities and dynamic properties in materials, a neural network trained to recognize characteristic wave velocities should be able to determine existing differences in sand specimens of various densities.

## SIGNATURE PATTERN DETERMINATION

Signature patterns were determined using a nondestructive wave propagation technique. Two apparatus were used to determine signature patterns. Both resonant column apparatus and piezoelectric chips (referred to as bender elements here) use the concept of wave propagation through media to determine dynamic properties at low strains. If comparative results are obtained from both apparatus, then it can be concluded that signature patterns are unique to the material and insensitive to the measuring apparatus, thereby characterization is "universal".

### Resonant Column Tests

A Drnevich Long-Tor Resonant Column (Figure 1) was used to test various geocomposites. This apparatus is used to excite a cylindrical sample either longitudinally or torsionally by means of a function generator. The function generator applies an oscillatory force to magnets located at one end of the sample. The generated waves travel along the length of the sample and are reflected at the bottom. The input and reflected waves are recorded on an oscilloscope. The frequency of the input wave is adjusted until a resonant condition exists. Resonance occurs when the transmitted and reflected waves are in phase.

The resonant column can apply both torsional and longitudinal excitation to the sample. Torsional motion is produced by four torsional coils and magnets. This movement results in shear wave propagation. Longitudinal motion is produced by a longitudinal coil and magnet producing

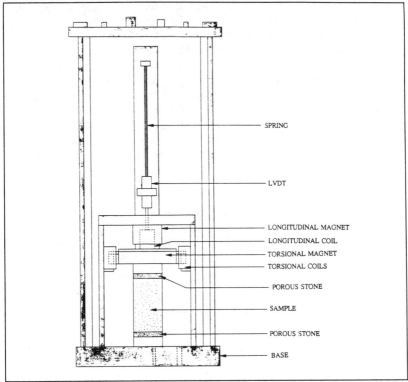

**Figure 1:** Drnevich Long-Tor Resonant Column

compressional waves. Using a resonant column, low strain as well as strain dependent dynamic properties such as Young's and shear modulus and damping ratio can be calculated from wave velocity and energy dissipation measurements.

### Bender Element Tests

Bender elements (Figure 2) are small piezoelectric metal chips embedded in porous stones and are cantilevered into the top and bottom of a cylindrical sample. Each element is composed of two piezoelectric pieces which are connected together. A voltage is applied using a function generator (Figure 3) which causes one plate to lengthen and the other to shorten. Since the two plates are attached, bending occurs which causes shear waves to propagate through the sample. The receiving bender element picks up the wave motion. Both the input and output waves are recorded on an oscilloscope.

The shear wave velocity ($v_s$) is calculated by dividing the length of the sample between the bender elements by the time of arrival of the wave. The shear modulus can then be calculated by equation (1)

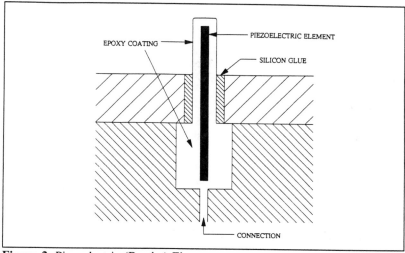

**Figure 2**: Piezoelectric (Bender) Element

$$G = \rho (v_s)^2 \tag{1}$$

where $\rho$ is the mass density of the sample. Although the bender elements described above cannot produce compressional waves, Poisson's Ratio ($\mu$) can be used to approximate compressional wave velocity by equation (2)

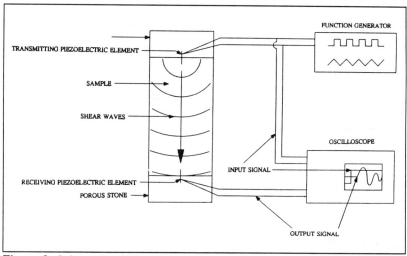

**Figure 3**: Schematic of Piezoelectric Element Set-up

$$v_p = [\frac{2(1 - \mu)}{(1 - 2\mu)}]^{1/2} v_s \qquad (2)$$

Ottawa sand samples were tested using both the resonant column and bender elements.

**Signal Processing**

Signals obtained from both bender elements and resonant column tests are recorded in the time domain. Figure 4 shows the time domain signal from a sand specimen at resonance using the resonant column and bender elements. Time domain signals often suffer from noise and indiscernible patterns. Signal processing alleviates this problem. The time domain signals are converted to the frequency domain using the Fast Fourier Transform (FFT)

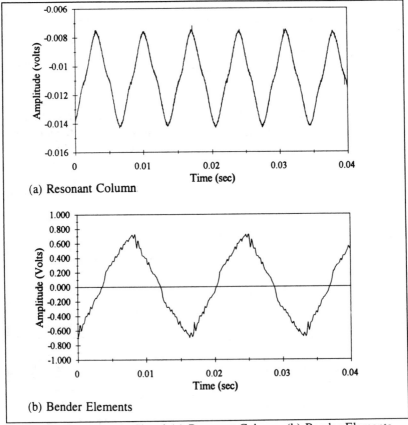

**Figure 4:** Time Domain Signal (a) Resonant Column; (b) Bender Elements

algorithm given in equation (3)

$$X(k+1) = \sum_{n=0}^{N-1} x(n+1) W_N^{kn} \qquad\qquad W_N = e^{-j(2\pi/N)} \qquad (3)$$

where x is the signal in the time domain, X is the signal in the frequency domain, N is the length of the time domain signal (x), k corresponds to the length of the frequency domain signal (X), and j represents the imaginary part of the function. The frequency domain displays peak frequencies or resonant frequencies more clearly than the time domain. The frequency domain spectra is called the power spectral density. The time domain signals are processed to attain the power spectral density in Figure 5, which clearly shows the dominant frequency of resonance.

Figure 5 shows a sand sample tested using both a resonant column and bender elements. The shift in dominant frequencies can be attributed to different densities of the two samples. The resonant column sample had a density of 1692 kg/m$^3$ and was tested at a pressure of 136 kPa (20 psi). The bender element sample had a density of 1684 kg/m$^3$ and was tested at a pressure of 68 kPA (10 psi). Although the differences may appear small, the sensitivity of the equipment is such that minor changes are readily noticeable. More work must be done to determine the sensitivity of the equipment and tolerable ranges for characterization.

The signal processing using power spectral density analysis was performed on all the data using Matrix Laboratory (MATLAB). MATLAB also handles neural networks, which made it convenient to identify and characterize the tested materials in one system.

## APPLICATION OF ARTIFICIAL NEURAL NETWORKS

The concept and architecture of artificial neural networks are based on biological neural networks in the human brain. They are part of the artificial intelligence field and have been used increasingly in recent years for engineering applications. Neural networks are well suited to engineering applications because of their ability to handle complex problems efficiently and without the need for explicit relationships to be given. Neural networks have been applied to pattern, speech, and image recognition [5, 6]. They are used to locate and characterize cracks and voids in concrete pavements [7, 8], identify instabilities in structural systems [9], predict flow in rivers [10], and many other applications in all disciplines of engineering. Neural networks are well suited to such problems because of their ability to learn. Hence, no relationship between inputs and outputs is needed. Hajek and Hurdal [11] discuss the differences between rule based expert systems and neural networks which develop associative relationships. Neural networks process information and determine trends between inputs and outputs. After sufficient training,

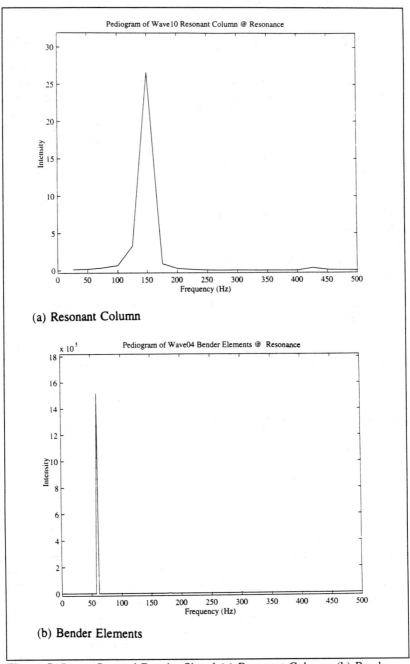

(a) Resonant Column

(b) Bender Elements

**Figure 5:** Power Spectral Density Signal (a) Resonant Column; (b) Bender Elements

the network will identify patterns between inputs and outputs based on the training information given.

A typical architecture of the neural network used to characterize soils is shown in Figure 6. The system assigns a weight (W) to each input, sums the weighted values, and adds a bias. The result is passed through a transfer function which then assigns a numerical value to the number. This result may again be weighted, summed, biased, and passed through another transfer function. Transfer functions include linear functions, unit-step or hard limit functions, and log-sigmoid functions. Each transfer function is a neuron. Each combination of weights, biases, summations, and transfer functions is termed a "layer". The layer which produces the final output is the output layer and all other layers are hidden layers. The number of layers and neurons are set by the user. Increasing the number of neurons and layers may increase the accuracy, but may lead to long and unacceptable processing times.

There are various techniques available to train the network. One of the most valuable is backpropagation which is also the technique used here. Unlike the perceptron learning rule which can only classify linearly separable inputs, or the Widrow-Hoff clearning rule which produces linear solutions, backpropagation uses multiple layers to develop nonlinear solutions to problems. Therefore, backpropagation is able to solve more complex problems than either the perceptron or Widrow-Hoff learning rules.

Backpropapation trains the network by comparing the computed output with the true or target output. It then adjusts the weights and biases by working backwards through the network. Once all weights and biases are adjusted, a new output is obtained which is compared to the target output. Training is complete when the difference between the computed and target outputs are within a predetermined tolerance. The training process is separated into four phases: presentation, check, backpropagation, and learning. Each complete pass through the training process is an epoch.

The neural network system developed for the study at hand was used to categorize Ottawa sand samples at different densities. Two methods of categorization can be applied. The first approach uses sample characteristics such as wave velocity, density, resonant frequency, modulus, and damping ratio to identify similar samples. This method does not utilize signal processing since all inputs are measured during testing or calculated from results attained. One hidden layer is used since increased hidden layers show little improvement in characterization [8]. The log-sigmoid function is used in the hidden layer, and a linear transfer function was used in the output layer. Five neurons are included in the hidden layer, and three neurons, corresponding to the three outputs, are in the output layer.

The second method of characterization involves the signal processing methods previously discussed. Unlike the first method which simply takes sample data, this approach replicates the wave pattern attained from the power spectral density plot. The signature pattern is represented by combinations of sine and cosine functions. The Fourier Series representation for a function is

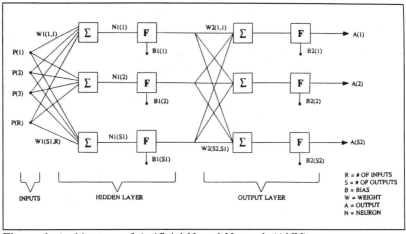

**Figure 6:** Architecture of Artificial Neural Network (ANN)

given in equation (4)

$$f(x) \approx \frac{a_0}{2} + \sum_{n=1}^{\infty} [a_n \cos nx + b_n \sin nx] \qquad (4)$$

where $a_n$ and $b_n$ are the Fourier coefficients, and f(x) is the function to be approximated.  Using this representation, the Fourier coefficients are the inputs to the neural network.  The corresponding output classifies the sample according to density.

Although both approaches to sample characterization are to be implemented, currently the first method is being applied.  This is a simpler approach to sample classification, however is sufficient to show the applicability of neural networks to this process.  With additional signal processing, the second approach can be introduced to validate the results from the first method, and also classify with higher precision.  At the time this document was written, the neural network system was undergoing training.  Therefore, conclusive results cannot be presented here.  However, the general approach is new and it is described with sufficient detail to warrant its use and feasibility.


## CONCLUSIONS

The application of digital signal processing, particularly Fast Fourier Transforms, to characterization of residual materials is a promising technique. With proper identification, such materials will be readily classified.  Artificial neural networks are used to further simply classification through associative relationships between sample parameters.

# REFERENCES

1. Pratt, Donald, and Mary Sansalone. 1988. "Impact-Echo Signal Interpretation Using Artificial Intelligence." *ACI Materials Journal*, 89 (2): 178-187.
2. Olson, Larry D., Dennis A. Sack, Kenneth H. Stokoe II, Kenneth W. Buchinski. 1994. "Stress-Wave Nondestructive Testing of Tunnels and Shafts." *Transportation Research Record 1415*, Transportation Research Board, Washington, D.C. 88-94.
3. Rosenblad, Brent L., Chine Chung Chiang, Kenneth H. Stokoe II, and Jose M. Roesset. 1995. "Suitability of Stress Waves to Detect Flaws in Rigid Pavements While Moving." The University of Texas at Austin, Presented at Transportation Research Board 74th Annual Meeting, Jan. 22-28, 1995, Preprint 950645.
4. Imran, Irsal, Soheil Nazarian, and M. Picornell. 1995. "Crack Detection Using Time Domain Propagation Technique." *Journal of Geotechnical Engineering*, 121 (2), 198-207.
5. Mah, R.S.H., and V. Chakravarthy. 1992. "Pattern Recognition Using Artificial Neural Networks." *Computers in Chemical Engineering*, 16 (4), 371-377.
6. Udpa, L., and S.S. Udpa. 1991. "Neural Networks for Classification of Nondestructive Evaluation Signals." *IEE Proceedings-F*, 138 (1), 41-45.
7. Attoh-Okine, Nii O. 1995. "Analysis of Learning Rate and Momentum Term in Backpropagation Neural Network Algorithm Trained to Predict Pavement Performance." Florida International University, Presented at Transportation Research Board 74th Annual Meeting, Jan. 22-28, 1995, Session 176A, Preprint 950553.
8. Alexander, A. Michel, and Richard Haskins. 1995. "Application of Artificial Neural Networks to Ultrasonic Pulse Echo System for Detecting Microcracks in Concrete." Technical Report REMR-CS, Waterways Experiment Station, Corps of Engineers, Department of the Army.
9. Elkordy, M.F., K.C. Chang, and G.C. Lee. 1994. "Application of Neural Networks in Vibrational Signature Analysis." *Journal of Engineering Mechanics*, 120 (2), 250-265.
10. Garrett, James H. Jr., Michael P. Case, James W. Hall, Sudhakar Yerramareddy, Allen Herman, Ruofei Sun, S. Ranjithan, and James Westervelt. 1993. "Engineering Applications of Neural Networks." *Journal of Intelligent Manufacturing*, (4), 1-21.
11. Hajek, Jerry J., and Brian Hurdal. 1993. "Comparison of Rule-Based and Neural Network Solutions for a Structured Selection Problem." *Transportation Research Record 1399*, Transportation Research Board, Washington, D.C, 1-7.

# CATEGORY XII:
# *Electrokinetic Processes*

# ELECTROKINETIC DECONTAMINATION OF MILLPOND SLUDGE

LUTFUL I. KHAN
Assistant Professor
Cleveland State University
Cleveland, Ohio-44115

M. RAHMAN
Graduate Assistant
Cleveland State University
Cleveland, Ohio-44115

## INTRODUCTION

Electrokinetic decontamination of high clay containing soils is a developing technology. EPA has recently designated electrokinetic method as a viable insitu process and interested parties are attempting to apply this method at contaminated sites which have inherently low permeability soils and otherwise difficult to remediate. Electrokinetic process induces a high water flow rate in clayey soils by the mechanism known as electroosmosis and is primarily suitable for heavy metal removal. However, chemical reactions and sorption of metal ions within the soil matrix may adversely effect the decontamination process. Presence of a significant amount of heavy molecular weight organic matter (humus substances) within the soil pores may reduce the mobility of the heavy metals due to the formation of insoluble organometallic compounds. There has been some research in removing heavy metals and low concentration organic matter from soil by the electrokinetic method [1- 3]. The effects of soil organic matter on heavy metal removal by electrokinetic method has not been adequately investigated and several unanswered questions remain about the efficiency of the process under such circumstances. This paper is a partial result of an electrokinetic decontamination investigation for Zn, Pb and Mn removal from a soil with high organic matter content.

A foundry millpond sludge from North East Ohio was chosen as a high organic matter containing matrix. It is a fine-grained material with low permeability ($\approx 10^{-7}$ cm/sec) which satisfies the basic conditions for electrokinetic treatment. Upon burning at $400^{\circ}$ C, it turns into a reddish brown silty material. Table I list the various characteristics of the millpond sludge.

## TABLE I: CHARACTERISTICS OF THE MILLPOND SLUDGE

| | |
|---|---|
| Zn | 28,000 mg/kg |
| Pb | 6,000 mg/kg |
| Mn | 8,000 mg/kg |
| Organic content | 25% |
| Plasticity | Low, primarily silty |
| Water content | 72 % avg. |
| Appearance | Black, fine grained |
| Initial pH | 6.5 |

## ELECTROOSMOSIS IN SOIL

The electroosmotic flow rate $q_{eo}$ in a porous medium of length $L$, porosity $n$, area $A$ and degree of saturation $S$, may be represented by the following equation:

$$q_{eo} = \frac{\Psi_d \varepsilon_0 D}{\eta} I_s (\frac{R_s}{L})(nAS) \tag{1}$$

Where $\Psi_d$ is the potential at the slipping plane, $\varepsilon_0$ is the permeability of free space, $D$ is the dielectric constant of the pore fluid, $\eta$ is the pore water viscosity, $I_s$ is the current carried by surface conductance and $R_s$ is the surface resistance of the porous medium (soil) [4].

Unlike water flow under pressure, electroosmosis depends on the electric current through the soil and originates at the electric double layer of the soil pores. The electroosmotic water flow in soil is directly proportional to the surface charge. In case of most soils, the surface charge is negative and the electroosmotic flow occurs from the positive electrode (anode) towards the negative electrode (cathode).

A low pH front moves into the soil during electrokinetic decontamination and lowers the system pH. Since most heavy metals are soluble in acidic environment, this lowering of pH assists the remediation effort by dissolution and desorption of the metals. If, however, organic matter are present within the soil pores, the metals may not be easily extracted by a low pH solution due to metal-organic matter bonding.

## SOIL-ORGANIC MATTER INTERACTION

The organic phase of the soils may be humus or nonhumus. The high molecular weight humus organic substances have a high affinity for metals and are form water insoluble metal complexes. Nonhumus substances of low molecular weight, such as organic acid and bases, are relatively soluble when complexed with metals.

Metal ions are bound to organic molecules by complexation and chelation. A complex is formed when an electron rich atom in an organic molecule shares a pair of electrons with a metal ion having an empty outer shell resulting in a coordination compound. Chelation occurs when two or more coordination positions are occupied by two or more donor groups from the same organic molecule. The resulting organometallic ring gives the complex a high degree of stability. The presence of humus substances in soils would therefore greatly reduce the mobility of metals. The metals, so bonded, cannot be easily ionized by an acidic environment. This would be a disadvantage for the electrokinetic decontamination process.

The metal-organic matter bond within the soil pores may be broken by the action of a sequestering ligand. Ethylenediaminetetraacetate (EDTA) is such a ligand and used in this investigation for Zn, Pb and Mn extraction. The action of metal and ligand may be expressed by the following equation:

$$Me + L \Leftrightarrow MeL, \quad k_1 = [MeL]/[Me][L] \qquad (2)$$

where $Me$ represents a metal cation, $L$ represents a ligand anion and $k_1$ is the formation constant. A stepwise formation takes place with the addition of ligand.

$$MeL + L \Leftrightarrow Me(L)_2, \quad k_2 = [Me(L)_2]/[MeL][L] \qquad (3)$$
$$\cdots\cdots$$
$$\cdots\cdots$$
$$Me(L)_{n-1} + L \Leftrightarrow Me(L)_n, \quad k_n = [Me(L)_n]/[Me(L)_{n-1}][L] \qquad (4)$$

Depending on the type of ligand, the metal-ligand complex may be soluble or insoluble in water. For a soluble metal-ligand complex, the charge of the ligand dictates if the resluting comples will be +ve, -ve or neutral.

## EXPERIMENT SETUP

The millpond-sludge was compacted in a consolidation apparatus under a 20 psi pressure for 24 hours. The final dimension of the samples were 3.5 cm in diameter and 7.5 cm in length. Electrokinetic decontamination was conducted under a 2.5 volts/cm gradient in an apparatus where the electrode was separated from the solid phase by a liquid (water) medium. The water filled chamber containing the anode electrode is referred to as the anode chamber and the water filled chamber containing the cathode electrode is referred to as the cathode chamber. During electrokinetic decontamination, water from anode chamber flows through the sample due to electroosmosis and evolve at the cathode chamber. This results in reduction in water volume at the anode chamber and increase in water volume in the cathode chamber. A schematic diagram of the apparatus used is shown in Figure 1. Readers are referred to Khan,1991[5] for a detail description of the Electrokinetic Apparatus.

797

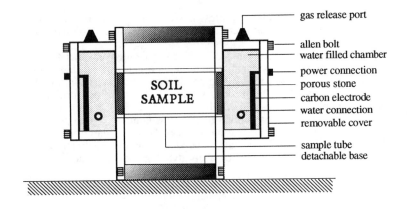

gas release port

allen bolt
water filled chamber

power connection
porous stone
carbon electrode
water connection
removable cover

sample tube
detachable base

SOIL
SAMPLE

Figure 1: Electrokinetic Apparatus

## ANALYSIS

After the completion of experiments, the millpond sludge samples were divided into five equal segments and the pH, water content and metal concentrations in each of these segments were measured.

## RESULTS AND DISCUSSIONS

In the first phase of the investigation decontamination of the millpond sludge samples were carried out for 5, 10 and 20 days. The cumulative electroosmotic water flow through the millpond sludge samples during these durations are shown in Figure 2. It is seen that the three samples had consistent flow with time, each maintaining almost a constant slope. The total cumulative flows for 5,10 and 20 days were about 32, 52 and 84 cc respectively. Unlike electroosmosis in many natural soils, the flow rate did not significantly reduce after 20 days of experiment which indicates minimal adverse electrochemical reactions within the system.

Metal removal from the different segments of the samples were analyzed by an Atomic Absorption Spectrophotometer. The results of these analyses are shown in Figures 3 and 4 where C/Co is the ratio of the metal concentration in the segment to the initial metal concentration and L/Lo represents the normalized distance from the anode end of the sample. After the first 5 days of experiment only about 10% zinc was removed (Figure 3) from the sample showing slightly higher removal toward the anode end. This is not unusual since water moved from anode to cathode direction during the decontamination process. After 10 days of decontamination, an average of about 31% zinc was removed from the sample. More zinc was removed from the center regions (36%). After 20 days of decontamination, concentration at the anode end decreased significantly (47% removal) but the concentrations towards the cathode end increased compared to

798

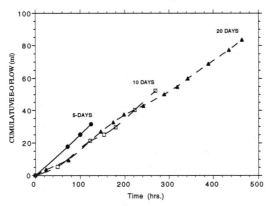

Figure 2 : Cumulative electroosmotic flow in millpond slude with time.

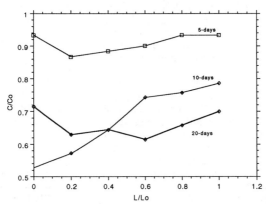

Figure 3: Zinc removal from millpond sludge at various durations.

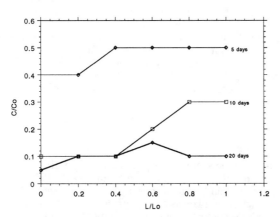

Figure 4 : Lead removal from millpond sludge at various durations.

10-day test. This indicates accumulation of the zinc at the cathode end. But this accumulation is significantly less than other reported results [Pamukcu 199 , Khan, 199 14]. It not unreasonable to think that a longer duration would increase the zinc removal.

Lead removal from the millpond-sludge was more successful. After five days about 50% lead was removed. The removal increased significantly after 10 days showing a removal gradient of 90% at the anode end to about 70% at the cathode end. After 20 days, the average lead removal was about 90% (Figure 4). Comparison of Zn and Pb removals at the end of 20 days experiment is shown in Figure 5. It is notable that Pb removal was more than Zn removal. This is peculiar in the sense that Pb is less mobile and usually more difficult to remove than Zn. Negligible amount of Mn removal was noted.

Figure 6 is a plot of C/Co ratios for zinc and lead at different durations. It is apparent that both lead and zinc removal slowed down significantly after 10 days. The reason for low removal of Zn was attributed to the Zn-organic matter bonding. A follow up experiment was conducted with EDTA solution in to determine if the immobile Zn could be extracted by the action of a ligand.

A 0.02 M sodium-EDTA solution was added the to the anode cell water with the notion that the EDTA would flow within the soil pores by electroosmosis and form soluble complexes with the immobile heavy metals. EDTA is a polydentete sequestering chelating agent which is environmentally benign upto a large concentration. It was expected that the heavy metal which did not flush out under electroosmosis with tap water would show some mobility.

The immediate effect of EDTA addition in the anode chamber was that the electroosmotic flow increased drastically. This effect is shown in Figure 7. The cumulative flow at the end of 10 day period amounted four times greater than that without EDTA addition. This phenomena may be explained by the fact that Na-EDTA dissociates into hexadentete negatively charged molecules which could be sorbed by the organic matter of the millpond sludge; such sorption would increase the negative surface charge. Since the electroosmotic flow is directly proportional to the surface charge, the consequence will be an increase in electroosmotic water flow. From the previous data, it appears that the increased flow would increase the metal removal but analysis of the sample at the end of the experiment revealed otherwise. The metal, complexed with EDTA was found to be moving towards the anode chamber, a conclusive evidence that the complexes were -ve ly charged. Such migration is opposite to the electroosmotic water flow direction. Therefore stagnation of the metals within the soil is not unlikely. The distribution of Zn within a sample is shown in Figure 8. The cathode end of the sample shows good removal but the metal seemed to have accumulated close to the anode region where the C/Co ratio had increased to 2. This accumulation is opposite to that observed in pure kaolinite clay without EDTA. The redistribution of Zn within the sample suggests that EDTA had effectively combined with Zn.

No significant difference in Pb or Mn removal was observed with the EDTA solution.

**Variation of pH**

The pH gradient in the samples after approximately 5,10 and 20 days duration are shown in Figure 9. The initial pH was about 6.5; It is seen that the pH reduced in the anode end and increased at the cathode end of the sample. However the reduction of pH was not significant except for the 20 days duration. Most heavy metals do not ionize under the pH environment that existed within the samples. The high pH may be another probable cause for the immobility of Zn

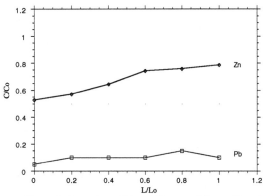

Figure 5: Comparison of Zn and Pb removal from millpond
sludge after 20 days.

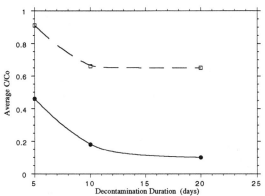

Figure 6: Zn and Pb removal from millpond sludge with
experiment duration.

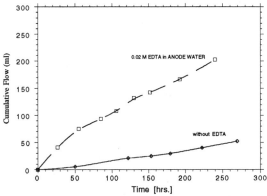

Figure 7: Comparison of 10 day electroosmotic flow in
millpond sludge with and without EDTA.

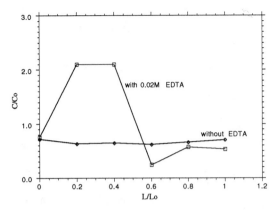

Figure 8: Zn removal from the millpond sludge after 10 days.

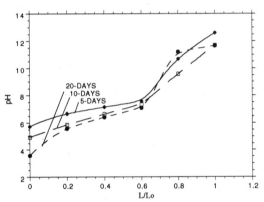

Figure 9: Variation of pH within the millpond sludge sample after 5,10 and 20 days.

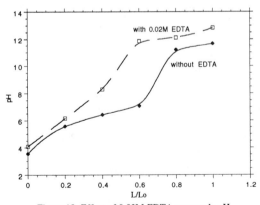

Figure 10: Effect of 0.02M EDTA on sample pH.

and Mn. The effect of 0.02 M EDTA solution on system pH is shown in Figure 10. It seems that EDTA had increased the pH at the center region of the sample.

## CONCLUSIONS

The organic matter present in the millpond sludge prevents electrokinetic process from extracting heavy metals from the soil pores. The millpond sludge also exhibits pH buffer capacity and prevents the lowering of pH within the samples. About 90% Pb and only 35% Zn can be extracted form the millpond sludge in 20 days by electrokinetic process with tap water as influent. When 0.02 M EDTA solution is used an influent, the electroosmotic flow increases about four fold but increased metal extraction is not achieved even though the metals may redistribute within the sample and show higher accumulation towards the cathode end. Mn was not removed or redistributed within the sample.

## REFERENCES

1. Pamukcu, S., Khan, L.I. and Fang, H. Y. 1990 'Zinc Detoxification of Soils by Electroosmosis', TRB No. 1288, pp 41-46

2. Shapiro, A. P. and Probstein, R. F. 1993 'Removal of Contaminants from Saturated Clay by Electroosmosis', Environmental Science and Technology, Vol 27, pp 283 - 291

3. Acar, Y. B., Li, H. and Gale, R.J. 1992' Phenol Removal from Kaolinite by Electrokinetics', ASCE J. of Geotechnical Engineering, Vol. 118, Nov. 1992, pp 1837-1851

4. Khan, L. I. 1991, Ph.D. Dissertation, Lehigh University, Bethlehem, PA

5. Khan, L. I., Pamukcu, S. and Fang, H. Y. 1991 "Apparatus for Testing Electrokinetic Decontamination" , Proc. ASTM Conf. on Agric. Chemical Properties of Soil, ASTM, Philadelphia, Pa.

# ELECTROKINETIC (EK) REMEDIATION OF A FINE SANDY LOAM: THE EFFECT OF VOLTAGE AND RESERVOIR CONDITIONING

JOEL RAMSEY
West Virginia University
Department of Civil / Environmental Engineering
Engineering Sciences Building
Morgantown, WV 26505

Dr. Brian Reed
West Virginia University
Department of Civil / Environmental Engineering
Engineering Sciences Building
Morgantown, WV 26505

## INTRODUCTION

Recently, attention has focused on developing cost effective in situ techniques to remove contaminants from soils. Examples of these technologies are soil flushing and *in-situ* biological treatment. *In-situ* technologies have had success predominately at sites with high permeability, sandy soils, but have not shown much promise with fine-grained, low permeability clayey soils. The primary difficulty associated with these low permeability soils lies in transporting groundwater, chemical additives, and contaminants through the subsurface by hydraulic means. Unfortunately, soils with low hydraulic conductivities usually contain clays and silts that sequester large quantities of inorganic and organic contaminants. Thus, soils having low permeabilities are generally efficient in holding pollutants but are resistant to *in-situ* remediation techniques because of the difficulty encountered in promoting groundwater movement.

There is a need for cost effective and more efficient technologies that can be used to remediate low permeability soils *in-situ*. A candidate technology for this type of remedial measure is electrokinetic (EK) soil flushing. Electrokinetics was first used for dewatering and consolidation of soils for slope stability, not for remediation purposes. During the period following the first application of electrokinetics for consolidation in 1947, there have been numerous geotechnical studies pertaining to the its mechanics in the dewatering and consolidation area. Recently, attention has focused on electrokinetics as a remediation technology. However, there is a need for experimentation on methods that may potentially enhance the effects and minimize the cost of EK soil flushing on the remediation of contaminated soil systems.

## OBJECTIVES

Based on the fact that lead (Pb) is a predominate constituent found in many Superfund sites across the United States, this research focused on applying the EK soil flushing process to removing Pb from an artificially contaminated soil system. The objectives of this research were to determine the effect of voltage and reservoir conditioning on energy expenditure and time required for lead removal.

## BACKGROUND

A schematic of the EK soil flushing process is presented n Figure 1. A DC current is applied to the soil to promote water and contaminant movement. Cations are attracted to the negatively charged cathode and anions are attracted to the positively charged anode. Because most soils have a negative surface charge, water flows from the anode to the cathode. This movement of water during electrokinetic treatment (electroosmotic [EO] flow) is not a function of pore size as is hydraulically induced flow. Therefore, there can be significant movement of water in soils of low permeability.

Phenomena that occur because of the applied potential are electroosmosis, ion migration, electrophoresis, electrolysis of water, and plating at the cathode. Each of these phenomena will be discussed in the following subsections

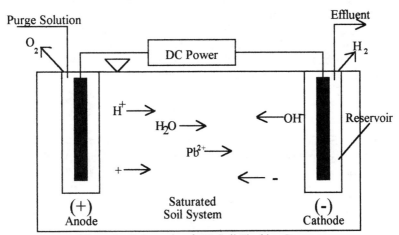

Figure 1. Schematic of EK Soil Flushing Process

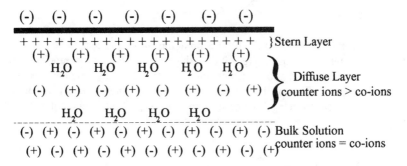

Figure 2. Schematic of the Electric Double Layer

## Electroosmotic (EO) Flow

Electroosmosis (EO) is the movement of a liquid under an applied potential. Net flow results when momentum transfer between migrating ions of one sign and the surrounding water molecules exceeds that of the ions of the opposite charge. In Figure 2, a schematic of the electric double layer (EDL) for a negatively charged particle (*e.g.*, clay particle) is presented. Counter-ions (cations is this case) collect near the particle surface in the stern layer where they are relatively immobile. Both counter-ions and co-ions exist in the diffuse layer, but the counter-ions outnumber the co-ions. Water molecules also reside in the diffuse layer. In the bulk solution, the total charge of the cations is equal to the anions. As an applied potential is applied to the system, cations move toward the cathode and anions move toward the anode. Water surrounding the ions are dragged along via frictional forces. In the bulk solution, the net flow is zero since the charge of the anions is equal to the charge of the cations. However, inside the EDL the number of cations is greater than that of the anions and flow results in the direction of the cathode. Note, all parameters that affect the EDL will also affect EO flow (*e.g.*, pH, ionic strength, surface charge, *etc.*). A Darcy's Law type approach is used to model EO flow [1]:

$$q_e = k_e i_e A = k_i I \tag{1}$$

where $q_e$ = EO flow rate (mL/s), $k_e$ = coefficient of electroosmotic permeability (cm$^2$/V-s), $i_e$ = potential gradient (V/cm), A = cross sectional area (cm$^2$), $k_i$ = coefficient of water transport efficiency (mL/amp-s), and I = current (amps).

## Ion Migration

Cations move toward the negatively charged cathode and anions move toward the positively charged anode. The effective ion mobility is a measure of how fast ions migrate toward the oppositely charged electrode. Acar [2] has reported that the effective ion mobility can be represented by the following equation:

$$u^* = D\tau nzF/RT \tag{2}$$

where $D$ = diffusion coefficient of ion in free solution at infinite dilution ($cm^2/s$), $\tau$ = tortuosity factor, $n$ = porosity, $z$ = valence of the ion, $F$ = Faraday Constant, $R$ = Universal Gas Constant, and $T$ = absolute temperature.

## Electrophoresis

Electrophoresis is the movement of a charged particle or colloid under an applied potential. For a compacted system, such as a soil system, electrophoresis is relatively unimportant. However, electrophoresis is important when EK technology is used for dewatering slurries, during the transfer of colloidal particles such as surfactants, and if EK is used during bioremediation.

## Electrolysis of Water

The water electrolysis reactions are defined by Equations 3 and 4:

$$2H_2O - 4e^- \rightarrow O_2 + 4H^+ \quad E_0 = -1.23 \text{ V} \quad \text{(Anode)} \tag{3}$$

$$2H_2O + 2e^- \rightarrow H_2 + 2OH^- \quad E_0 = -0.83 \text{ V} \quad \text{(Cathode)} \tag{4}$$

These reactions are the primary cause of the changes in soil chemistry during EK. $H^+$ produced at the anode is swept toward the cathode via ion migration, EO advection, and diffusion forming an acid front. In the absence of reservoir conditioning, $OH^-$ produced at the cathode is transported into the soil against the flow of water via ion migration and diffusion resulting in a base front near the cathode. The pH changes that occur during EK flushing affect the sorption/speciation/solubility of the metal and thus, the efficiency of the EK process. Specifically, the low pH front generated at the anode causes desorption, dissolution, and ionization of most cationic heavy metals. If there is no reservoir conditioning, precipitation or re-adsorption of the migrating heavy metals occurs in the high pH zone near the cathode, preventing their movement into the cathode reservoir liquid (where they could easily be removed and treated *ex-situ*).

## Plating Reactions

Cationic metals that transported into the cathode reservoir can receive electrons and plate on the cathode. for example:

$$Pb^{2+} + 2e^- \Rightarrow Pb^0 \quad E_o = -0.126\,V \qquad (5)$$

Metal plating on the electrode can be minimized by proper selection of the electrode material. However, plating of the heavy metal contaminant may be beneficial in that aqueous water treatment could be minimized (*i.e.*, aqueous metal → plated metal and electrode removed and disposed.

## EXPERIMENTAL DESIGN

The soil used in this study was classified as a galen fine sandy loam. The soil was obtained in Erie County, NY. The soil was contaminated with Pb at a concentration of 1000 mg/kg of dry soil. There were a total of six different testing conditions with three different reservoir conditioning schemes at two different voltages (30 and 60 volts). The three reservoir conditioning schemes were: (1) 500 microsiemens (μs) sodium nitrate ($NaNO_3$) in the anode and cathode, (2) 500 μs $NaNO_3$ in the anode and 1.0 M acetic acid (HAc) in the cathode, and (3) 0.1 M hydrochloric acid (HCl) in the anode and 1.0 M HAc in the cathode. These conditioning schemes are abbreviated as 500/500, 500/HAc, and HCl/HAc respectively. Within each testing condition, six experiments were performed . The duration of these experiments were 1, 5, 10, 21, 30, and 40 days. The testing conditions are presented in Table I.

Table I. Summary of Experimental Conditions Used for EK Soil Flushing

| Exp. No. | Initial Soil Pb (mg/kg) | Anode Reservoir | Cathode Reservoir | Voltage (volts) |
|---|---|---|---|---|
| 1D | 1,000 | 500 μs NaNO3 | 500 μs/cm | 60 |
| 2D | 1,000 | 500 μs NaNO3 | 500 μs/cm | 30 |
| 3D | 1,000 | 500 μs NaNO3 | 1.0 M HAc | 60 |
| 4D | 1,000 | 500 μs NaNO3 | 1.0 M HAc | 30 |
| 5D | 1,000 | 0.1 M HCl | 1.0 M HAc | 60 |
| 6D | 1,000 | 0.1 M HCl | 1.0 M HAc | 30 |

## MATERIALS AND METHODS

The soil was pulverized and mixed with a PbNO$_3$ stock solution to Pb concentration of 1000 mg/kg. The soil/Pb slurry was then transferred to a pneumatic consolidator where pressure was applied in 10 p.s.i. increments for 24 hours. Then, the soil specimen (7.5 cm long by 3.5 cm diameter) was transferred to the EK reactor. A constant voltage of 30 or 60 volts was maintained throughout the duration of each test. The current and EO flow were measured daily and the reservoir liquid pH and conductivity were measured at three day intervals. At the end of each test, the soil specimen was segmented into eight equal slices and analyzed for conductivity, pH, water content, and Pb concentration. The reservoir solutions, EO flow effluent, and electrodes were also analyzed for Pb content in order to complete the mass balance.

The soil pH and conductivity were determined using a 5:1 water:soil ratio. The soil was digested for 48 hours using a 5N HNO$_3$ solution at an 8:1 liquid to soil ratio. The electrodes were serially acid washed with 8N HNO$_3$. Lead concentrations were determined using atomic absorption spectrophotometry.

## RESULTS AND DISCUSSION

This section will be divided into two main sections: (1) The effect of voltage on time required for lead removal and energy expenditure and (2) the effect of reservoir conditioning on time required for lead removal and energy expenditure.

### Effect of Voltage on EK Remediation

In Table II, the difference, between 30 and 60 volts, in time required to remove specific amounts of lead for each of the testing conditions is presented. In all three of the reservoir conditioning schemes, the 60 volt tests were more than 100 percent faster at removing 250 mg/kg of the initial 1000 mg/kg. As the amount of lead removed increased, the difference in time between the 30 and 60 volt tests to achieve the specified removal decreased. The 60 volt tests were significantly less than 100 percent faster at removing 750 mg/kg than the 30 volt tests. This is because the metal/soil binding energy increases as the metal concentration decreases.

In Table III, the difference, between 30 and 60 volts, in energy required to remove specific amounts of lead for each of the testing conditions is presented. In an ideal circuit, the energy expended is directly proportional to the applied voltage. Therefore, it would be expected that the energy expended would increase by 100 percent as the voltage was doubled from 30 to 60 volts. The experiments simulated an ideal circuit with modest variation. The increase in energy expenditure from 30 to 60 volts ranged from 63 percent at 500 mg/kg removal in the 500/500 tests to 150 percent at 750 mg/kg removal in the

Table II. Effect of Voltage on Time of Remediation

| Tests (conditions) | %Removal (mass of Pb removed) | Voltage (volts) | Time (days) | % Increase From Lowest |
|---|---|---|---|---|
| 1D & 2D (500/500) | 25 | 30 | 6.7 | 139 |
| | (250 mg/kg) | 60 | 2.8 | |
| | 50 | 30 | 24.1 | 141 |
| | (500 mg/kg) | 60 | 10.0 | |
| | 75 | 30 | 39.3 | 34.6 |
| | (750 mg/kg) | 60 | 29.2 | |
| 3D & 4D (500/HAc) | 25 | 30 | 5.4 | 135 |
| | (250 mg/kg) | 60 | 2.3 | |
| | 50 | 30 | 8.0 | 100 |
| | (500 mg/kg) | 60 | 4.0 | |
| | 75 | 30 | 17.0 | 22.3 |
| | (750 mg/kg) | 60 | 13.9 | |
| 5D & 6D (HCl/HAc) | 25 | 30 | 1.4 | 367 |
| | (250 mg/kg) | 60 | 0.3 | |
| | 50 | 30 | 3.3 | 175 |
| | (500 mg/kg) | 60 | 1.2 | |
| | 75 | 30 | 5.0 | 67 |
| | (750 mg/kg) | 60 | 3.0 | |

HCl/HAc tests. In an ideal circuit, the resistance remains constant as the voltage increases. Electrolysis of water at the anode and cathode combined with ion migration and EO advection change the soil chemistry and, thus, the soil's electrical resistance with time. This accounts for the variations from the expected 100 percent increase.

**Effect of Reservoir Conditioning on EK Remediation**

Percent Pb removed versus time, for each of the tests, is presented in Figure 3. The rate and extent of removal increased from 500/500 to 500/HAc to HCl/HAc. For both 30 and 60 volts, the HCl/HAc tests were, on average, six times faster

Table III. Effect of Voltage on Energy Expended

| Tests (conditions) | %Removal (mass of Pb removed) | Voltage (volts) | Energy Expended (kW-hr/ton) | % Increase From Lowest |
|---|---|---|---|---|
| 1D & 2D (500/500) | 25 | 30 | 70 | 120 |
| | (250 mg/kg) | 60 | 154 | |
| | 50 | 30 | 140 | 62.9 |
| | (500 mg/kg) | 60 | 228 | |
| | 75 | 30 | 174 | 93.7 |
| | (750 mg/kg) | 60 | 337 | |
| 3D & 4D (500/HAc) | 25 | 30 | 65 | 100 |
| | (250 mg/kg) | 60 | 130 | |
| | 50 | 30 | 115 | 47.8 |
| | (500 mg/kg) | 60 | 170 | |
| | 75 | 30 | 185 | 146 |
| | (750 mg/kg) | 60 | 455 | |
| 5D & 6D (HCl/HAc) | 25 | 30 | 60 | 91.6 |
| | (250 mg/kg) | 60 | 115 | |
| | 50 | 30 | 100 | 65 |
| | (500 mg/kg) | 60 | 165 | |
| | 75 | 30 | 140 | 150 |
| | (750 mg/kg) | 60 | 350 | |

than the 500/500 test and three times faster than the 500/HAc test at removing the same amount of lead. The addition of HAc inhibited formation of a base front, thereby hindering Pb precipitation and enhancing Pb transport through the soil sections near the cathode. The addition of HCl to the anode increased the strength (desorbing capability) of the acid front transported through the soil by EO flow. The maximum removals were 76, 90, and 96 percent for the 500/500, 500/HAc, and HCl/HAc 60 volt tests respectively.

In Figure 4, energy expended versus percent lead removed is presented. As stated in the previous subsection, the 30 volt tests consumed less energy than the 60 volt tests. Although adding acid to the reservoir solutions changed the extent and rate of removal, the acid additions did not considerably affect the energy expended to remove specific amounts of lead. None of the reservoir conditioning

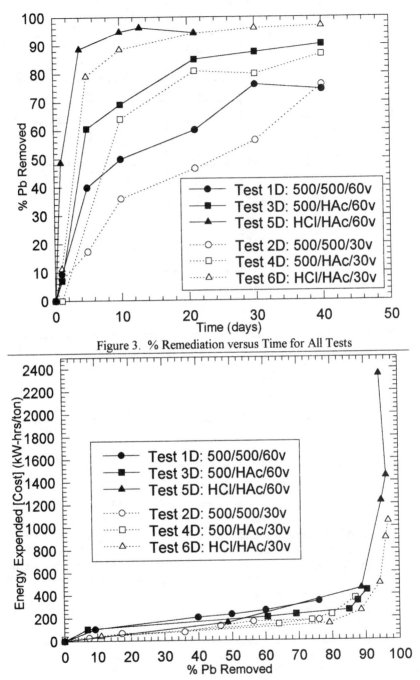

Figure 3. % Remediation versus Time for All Tests

Figure 4. Energy Expended versus % Remediation for All Tests

schemes, compared separately at 30 and 60 volts, were consistently higher or lower than the others in terms of energy expended.

## SUMMARY

The objectives of this research were to determine the effect of voltage and reservoir conditioning on an EK remediation system by conducting experiments at 30 and 60 volts using the following reservoir conditioning schemes: (1) 500 microsiemens ($\mu s$) sodium nitrate ($NaNO_3$) in the anode and cathode, (2) 500 $\mu s$ $NaNO_3$ in the anode and 1.0 M acetic acid (HAc) in the cathode, and (3) 0.1 M hydrochloric acid (HCl) in the anode and 1.0 M HAc in the cathode. Specifically the effect of voltage and reservoir conditioning on time required for lead removal and energy expended were examined. In the early stages of treatment (through about 25 percent lead removal) the lead removal rate was nearly proportional to the voltage. However, as the amount of lead removed increased, the difference in time, between 30 and 60 volts, required to removed specified amounts of lead decreased. Concerning energy expenditure, the 60 volts test generally used twice as much energy as the 30 volt tests.

The addition of acid to the reservoirs had a significant effect on the extent and rate of Pb removal. The maximum removals were 76, 90, and 96 percent for the 500/500, 500/HAc, and HCl/HAc 60 volt tests respectively. For both 30 and 60 volts, the HCl/HAc tests were, on average, six times faster than the 500/500 test and three times faster than the 500/HAc test at removing the same amount of lead.

## REFERENCES

1. Mitchell, J.K. (1976). *Fundamentals of Soil Behavior.* John Wiley and Sons, New York, N.Y.

2. Acar, Y.B. and Alshawabkeh, A.N. (1993). "Principles of Electrokinetic Remediation," **Environmental Science and Technology,** December 1993.

# TRANSPORT OF HEXAVALENT CHROMIUM IN POROUS MEDIA: EFFECT OF AN APPLIED ELECTRICAL FIELD

K.R. MCINTOSH
Conestoga-Rovers & Associates
2055 Niagara Falls Blvd., Suite #3
Niagara Falls, NY 14304

Department of Civil Engineering
University of Delaware
Newark, Delaware 19716


C.P. HUANG
Department of Civil Engineering
University of Delaware
Newark, Delaware 19716

## INTRODUCTION

Contamination of soil and groundwater by chromium has occurred in many industrial areas[1]. Sources of chromium waste include metal-plating, steel fabrication, paint and pigment production, wood treatment, leather tanning, and chromium mining and milling[1,2]. In the past, a common method for disposal of chromium containing liquids was through the use of seepage basins[1]. This practice often resulted in chromium contamination in both the saturated and unsaturated zones. Use of chromium contaminated fill material, such as occurred in Hudson County, New Jersey[3], has also caused soil and groundwater contamination.

In the natural aqueous environment chromium exists at significant levels in two oxidation states: trivalent (Cr[III]) and hexavalent (Cr[VI]). Trivalent chromium compounds are of low toxicity and are generally less soluble and mobile in water than the hexavalent compounds. Hexavalent chromium compounds are systemic toxicants and suspected human carcinogens[2]. Furthermore, Cr(VI) occurs as the oxyanions $HCrO_4^-$ and $CrO_4^{2-}$

which are highly soluble and mobile in groundwater. Therefore, with respect to potential human exposures, the presence of Cr(VI) in soil and groundwater is generally the primary concern at hazardous waste sites with chromium contamination. The investigations described herein focused on the potential removal of Cr(VI) from soil using an electrokinetic process.

## BACKGROUND

Application of DC voltage to a soil/water system changes the pH of the soil solution. At the anode, hydrolysis produces oxygen gas and hydrogen ions, lowering the solution pH in the vicinity. At the cathode, hydrogen gas and hydroxides are produced, increasing the pH in the vicinity. Anions such as Cr(VI) have little potential for adsorption to kaolinite and other clay minerals at typical environmental pH ranges. However, at low pH, anion adsorption can be a significant factor influencing transport in soils. Anions have a greater tendency to be adsorbed onto soil particles under the low pH conditions which develop near the anode. For kaolinite, the particle charge is highly dependent on the solution pH. At pH values lower than the zero point of charge (ZPC), which is approximately 6.5 for kaolinite[16], anions will be adsorbed onto the crystal edge groups. At pH higher than 6.5, the net particle charge is negative and adsorption is generally negligible due to electrostatic repulsion[17].

The electrokinetic process involves application of a direct current (DC) voltage to the soil/water system. The electrical field induces the migration of mobile anions and cations in the soil pore water. The net movement of soil pore water occurs in the direction of the cathode -- a phenomenon termed electro-osmosis. Therefore, the soluble uncharged contaminants and cations tend to be induced to migrate toward the cathode. Conversely, anions migrate toward the anode. Electro-osmosis has been employed for several decades by geotechnical engineers to consolidate clays and fine-grained soils[4,5]. More recently, the application of electrokinetic principles for removal of contaminants from soil has received significant attention. Lab-scale studies have demonstrated successful removal of both organic and inorganic contaminants[6-10]. The principles and theoretical development of electrokinetic remediation have been reported[11-14]. Field applications of electrokinetic remediation techniques have shown the potential feasibility of in situ remediation of metal-contaminated soils[15]. Much of the previous work has focused on the removal of cationic species from saturated soils. The current study has investigated the transport of Cr(VI) anions in both saturated and unsaturated soil under an applied electrical field.

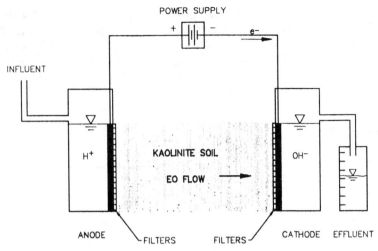

Figure 1 - Schematic of Electrokinetic Test Cell.

**EXPERIMENTAL METHODS**

Experiments were conducted using kaolinite clay as the soil. Sodium chromate ($Na_2CrO_4$) was used as the contaminant. Cr(VI) was analyzed by the 1.5-diphenylcarbohydrazide chromate complex at 540 nm[18]. Soil pH was measured using USEPA SW-846 Method 9045A[19].

The electrokinetic test cell was made using a 8.9 cm inner diameter of acrylic extruded cylinder. The test cell consisted of three parts: anode reservoir, soil specimen chamber (10 cm in length), and cathode reservoir (Figure 1). The volume of each reservoir was 650 mL. To separate the soil from the water solution, a set of two nylon meshes (140 μm polypropylene filter, Spectrum, Houston, TX) with a paper filter in between (Whatman #1 filter, Maidstone, England) was used both in the anode and cathode reservoirs. Two sets of graphite rod electrodes, one set at the anode and one at the cathode, were installed on each side of the soil specimen.

The current study included batch equilibrium sorption experiments, electro-osmosis experiments, and Cr(VI) transport experiments as described below.

*Batch Sorption Experiments*

To assess the partitioning behavior of Cr(VI) between the aqueous and sorbed phases, batch equilibrium sorption experiments were conducted. Kaolinite and water suspensions (5 g/L) were allowed to equilibrate (on a reciprocating shaker) with a range of Cr(VI) concentrations in 0.5 M NaNO3 solutions with pHs ranging from 1 to 13.

*Electro-Osmosis Experiments*

Three experiments were conducted to investigate the rates of electro-osmosis pore water flow at different moisture contents: 56.5 percent, 40 percent, and 30 percent. The soil pore water used was a solution of 0.5 M NaNO3, in distilled deionized water. The constant applied DC voltage was approximately 12 volts (1.2 V/cm). Electro-osmosis flow, electric current, and pH in the anode and cathode reservoirs were monitored during the tests. No pH control was used in these tests. Additions of 0.5 M NaNO3 solution to the anode reservoir were made as necessary.

*Chromium Transport Experiments*

Two chromium transport experiments were conducted by introducing $Na_2CrO_4$ to the cathode reservoir after at least one month of electro-osmosis testing as described above. The initial Cr(VI) concentration was $10^{-2}$ M (520 mg/L). Cr(VI) concentration was measured over time in the anode reservoir and the flux through the kaolinite was calculated.

## RESULTS AND DISCUSSION

Figure 2 presents the results of the batch adsorption experiments. While these experiments using dilute suspensions did not simulate

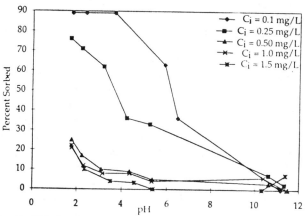

Figure 2 - Cr(VI) Sorption as a Function of Solution pH.

the actual conditions in the test cell, where the proportion of solids was much higher, the results can be used to show the relative adsorption of Cr(VI) throughout a range of initial concentrations and solution pHs. For Cr(VI) concentrations less than 0.25 mg/L, significant adsorption occurs at pH values less than 6.5. This will tend to retard Cr(VI) migration near the anode. However, at higher initial aqueous concentrations, the percent adsorbed is less than 30 percent at pHs below 2.0, reflecting an upper limit to the quantity of Cr(VI) anions which can be adsorbed onto the kaolinite crystal edges where the pH dependent charge occurs.

As indicated on Figure 3, the lower limit of soil moisture content for significant electro-osmotic flow in kaolinite under the test conditions is between 30 and 40 percent. At 30 percent moisture content, air space in the soil causes sufficient resistance such that measurable electro-osmotic flow did not occur at the applied voltage gradient of 1 volt/cm. However, although electro-osmotic flow was not induced at 30 percent moisture content, a low (0.1 mA) current was sustained. This suggests that some induced migration of anions is possible at low moisture content.

Figure 4 shows the pH and moisture content profiles across the kaolinite specimen (with an initial moisture content of 40 percent) after 37 days of treatment. The moisture content profile shows a gradual increase in moisture content in the direction of the electro-osmotic flow. Soil near the anode became less saturated (approximately 36 percent moisture content) and soil near the cathode became more saturated (approximately 44 percent moisture content). The pH profile indicates that acidity produced at the anode has migrated nearly throughout the soil core. Elevated pH occurs only in the immediate vicinity of the cathode. This indicates a greater migration rate of hydrogen ions compared to hydroxide ions and a tendency for an "acid front" to move through the soil during electrokinetic treatment.

Figures 5 and 6 present concentrations of Cr(VI) measured in the anode reservoir during the transport experiments conducted at moisture contents of 56.5 percent and 30 percent, respectively. Figure 7 shows the depletion of Cr(VI) in the cathode reservoir.

The initial breakthrough of Cr(VI) occurred at similar times for the two moisture contents: 14 days at 56.5 percent and 12 days at 30 percent. This suggests that the net velocity of migration of Cr(VI) anions is similar at the two moisture contents but that the flux is lower. The lower flux may be attributed to the higher air space at low moisture content since there is less cross-sectional area of interconnected water-filled pore space through which ion migration can occur. This is analogous to the relationship between hydraulic conductivity and moisture content.

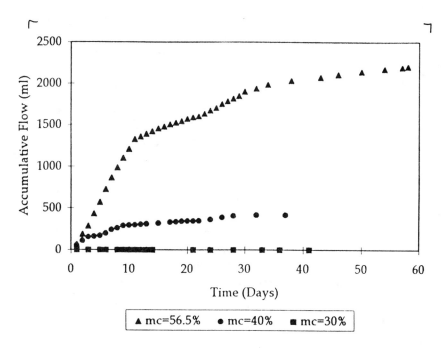

Figure 3 - Accumulative Flow of Effluent Over Time.

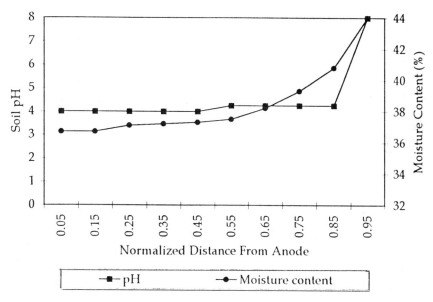

Figure 4 - pH and Moisture Content Access the Kaolinite Specimen
after 37 Days (Initial Moisture Content - 40 Percent).

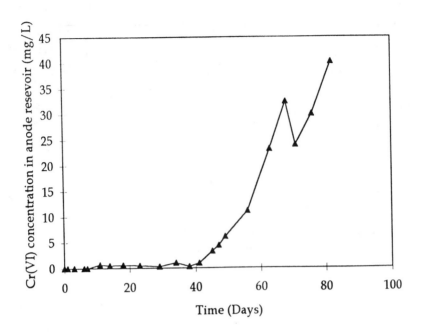

Figure 5 - Cr(VI) Concentration in Effluent vs. Time
(Moisture Content = 56.5 Percent).

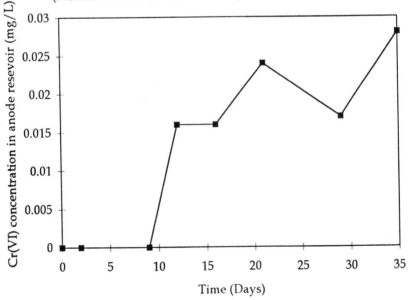

Figure 6 - Cr(VI) Concentration in Effluent vs. Time.
(Moisture Content = 30 Percent).

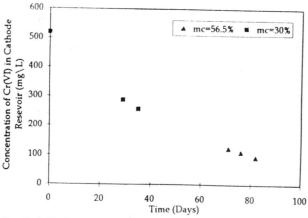

Figure 7 - Cr(VI) Concentration in the Cathode Reservoir vs. Time.

The potential contribution of molecular diffusion to the measured concentrations of Cr(VI) in the anode reservoirs was assessed using the one-dimensional solution to Fick's Second Law[20]. Using an apparent diffusion coefficient of $5\times10^{-11}$ $m^2/s$, the estimated molecular diffusion would not contribute significantly to the Cr(VI) concentration in the anode reservoir.

Empirical evidence from these experiments suggest that the Cr(VI) flux is dependent on the electric current and moisture content. The flux of Cr(VI) into the anode reservoir during the first 20 days following the initial breakthrough was $1 \times 10^{-2}$ mg/day at 56.5 percent compared to $8 \times 10^{-4}$ mg/day at 30 percent moisture content -- a factor of 12.5 times higher. This is approximately equal to the product of the ratios of the electrical currents (7:1) and the moisture contents (1.9:1).

## CONCLUSIONS AND FUTURE STUDY

The results of this study indicate that electrokinetic processes may be useful in removing Cr(VI) anions from partially saturated soils. As shown on Figures 5, 6, and 7, the transport tests did not reach steady state conditions after 82 days for the 56.5 percent moisture content test and 35 days for the 30 percent moisture content test. It is planned that both transport tests will be continued until steady state conditions are reached.

The possibility of removal of anions using electrokinetic processes is also intriguing because contaminant removal from groundwater may be feasible without implementing a physical or hydraulic containment. Furthermore, if electrically induced anion migration occurs at a high rate compared to advective transport in groundwater, it may be practical to construct barriers to anion migration by installing appropriately designed arrays of anodes and cathodes without creating hydraulic barriers by groundwater pumping and/or electro-osmosis. The feasibility of such remediation depends on the rate of anion migration which can be induced electrically compared to the rate of advective migration in groundwater flow.

An experimental apparatus has been designed to measure anion migration at different groundwater flow rates. This unit is currently being tested by the authors.

## REFERENCES

1. Nriagu, J.O. and E. Nieboer, 1988. *Chromium in the Natural and Human Environment*, John Wiley and Sons, New York.

2. Casarett, L.J. and J. Doull, 1980. *Toxicology, the Basic Science of Poisons*, Macmillan Publishing Company, New York, pp. 441-442.

3. Montclair Environmental Management Team (MEMT), 1990. "Management plan for chromium-contaminated soil." *Intern, J. Environ. Studies.* 35:263-275.

4. Casagrande, L. 1949. "Electro-osmosis in Soils." *Geotechnique*, 1(1):1959-1977.

5. Casagrande, L., 1983. "Stabilization of soils by means of electro-osmosis-state-of-the-art." *Journal of Boston Society of Civil Engineering*, 699(2):255-302.

6. Ray, C. and R.F. Probstein, 1993. "Removal of heavy metals in wastewater in a clay soil matrix using electro-osmosis." *Environmental Progress*, 6(30:145-149.

7. Hamed, J., Y.B. Acar and R.J. Gale, 1991. "Pb(11) removal from kaolinite by electrokinetics." *Journal of Geotechnical Engineering*, 117(2):241-271.

8. Bruell, C.J., B.A. Segall and M.T. Walsh, 1992. "Electro-osmotic removal of Gasoline Hydrocarbons and TCE from Clay." *Journal of Environmental Engineering*, 118(1):68-83.

9. Shapiro, A.P. and R.F. Probstein, 1993. "Removal of Contaminants from Saturated Clay by Electroosmosis." *Environ. Sci. Technol.*, 27(2):283-291.

10. Pamukcu, S., and J. Wittle, 1992. "Electrokinetic Removal of Selected Heavy Metals from Soil." *Environmental Progress,* 11(3):241-250.

11. Acar, Y.B. and A.N. Alshawabkeh, 1993. "Principles of electrokinetic remediation." *Environ. Sci. Technol.,* 27(13):2639-2647.

12. Alshawabkeh, A.N. and Y.B. Acar, 1992. "Removal of contaminants from soils by electrokinetics: A theoretical treatise." *Journal of Environ. Sci. Health,* A27(7):1835-1861.

13. Segall, B.A. and C.J. Bruell, 1992. "Electro-osmotic contaminant-removal processes." *Journal of Environmental Engineering,* 118(1):68-83.

14. Acar, Y.B., J.G. Robert, G.A. Putanam, Hamed, J., and R.L. Wong, 1990. "Electrochemical processing of soils: Theory of pH gradient development by diffusion, migration, and linear convection." *Journal of Environ. Sci. Health,* A25(6):687-714.

15. Lageman, R., 1993. "Electroreclamation: Applications in the Netherlands." *Environmental Science and Technology,* 27(13):2648-2650.

16. Schofield, R.K. and H.R. Samson, 1953. "The Deflocculation of Kaolinite Suspensions and the Accompanying Change-over from Positive to Negative Chloride Adsorption." *Clay Miner. Bull.* 2:45-51.

17. Bohn, Hinrich, McNeal, Brian, and George O'Connor, May 1985. *"Soil Chemistry."* Second Edition, John Wiley & Sons, New York, pp124-127.

18. American Society of Testing Materials (ASTM), 1990. "Standard Test Methods for Chromium in Water." *American Society of Testing Materials,* ed. pp. D1987-86.

19. USEPA Office of Solid Waste, November 1986. "Test Methods for Evaluating Solid Waste." *SW-846, Third Edition,* Method 9045, Soil pH.

20. Cherry, J.A., and R.A. Freeze, 1979. *Groundwater,* Prentice-Hall, Inc. Englewood Cliffs, New Jersey, p. 393

# UTILIZATION OF SOLUBILIZING AND STABILIZING AGENTS IN ELECTROKINETIC PROCESSING OF SOILS

**ANTOINETTE WEEKS**
Graduate Research Fellow
Department of Civil and Environmental Engineering
Lehigh University,
Bethlehem, Pa 18015

**SIBEL PAMUKCU**
Associate Professor
Department of Civil and Environmental Engineering
Lehigh University,
Bethlehem, Pa 18015

## INTRODUCTION

Throughout the globe there are numerous inactive waste sites that have resulted from a combination of engineered and unplanned releases at research and production facilities. Metals are found to be among the most prevalent contaminants throughout these sites. Common contaminants include lead, chromium, arsenic, zinc, copper, cadmium, and nickel. Many of these sites also have other waste materials such as organic acids, solvents and complexing agents. These contaminants exist in the vadose zone and in some cases are found to have migrated towards the ground water.

The use of electrokinetics in soil and groundwater decontamination promises to be an effective method for both cost savings and reduction in health and environmental risks aspects. In field application of the process, the soil being remediated need not to be disturbed to cause release of the contaminants. The process controls the direction of contaminant transport which may be captured in collection wells or other installations, such as exchange resins, carbon filters and membranes. Past research shows that the key component in success of electro-kinetics in soil decontamination is the site specific engineering design of the total system, which includes:

1. Site characterization and contaminant identification,
2. Electrodes and their lay-out,
3. Enhancement and control of the transport,
4. Collection or in-situ stabilization of the transported/targeted contaminants.

Only when the system is designed relevant to the site being remediated based on an informed and systematic approach, the end product will be a better engineered electrokinetics process for soil decontamination in field, whether it is used alone or in conjunction with another technology for enhancement.

This paper looks into the feasibility of two applications to enhance electrokinetic treatment of heavy metal contaminated soils, as given in items 3 and 4 above. The first application deals with the enhancement of the transport and the second one deals with in-situ stabilization of targeted substances using electrokinetic and electro-chemical methods.

## OVERVIEW

Electrokinetics have been used for dewatering of soils and sludges since the first recorded use in the field by Casagrande (1949). Work and subsequent research in electrokinetic decontamination of soils has accelerated in recent years, stimulated by a report of the detection of high concentrations of metals and organic compounds in electroosmotically drained water of a dredged sludge in the field by Segall and co-workers (1980). Since than, successful applications of the electrokinetic decontamination technique was demonstrated on pure soil-contaminant-mixtures in the laboratory by several researchers (Hamed, et al., 1991; Pamukcu and Wittle, 1992; Shapiro and Probstein, 1993; Pamukcu, 1994).

The electrokinetic treatment technique uses *electroosmosis-electrophoresis*, and *ion migration* to enhance the transport and subsequently removal or stabilization of contaminants in soils. The two primary mechanisms of mobilizing contaminants are: (1) movement of the charged species by electromigration and electropheresis, and (2) transport of contaminants in the pore space by the advection of water or electroosmotic flow. These processes are dependent upon the type of soil, nature of contamination and moisture content of the soil.

Extraction of contaminants by electrokinetic method is based on the assumption that the contaminant is in the liquid phase in the soil pores. Electroosmotic advection is able to transport non-ionic and non-polar as well as ionic species through soil toward the cathode. This is best achieved when the state of the material (dissolved, suspended, emulsified, etc.) is suitable for the flowing water to carry it through the tight pores of soil without causing an immovable plug of concentrated material to accumulate at some distance from an electrode. Electroosmotic advection is probably most useful to transport contaminants in clays and very low permeability soils, since the electroosmotic permeability of such material is often several orders of magnitude higher than their hydraulic permeability.

Speciation and precipitation are major factors in mobilization and transport of heavy metal constituents by ion-migration component of electrokinetics. The speciation is dependent upon a number of fairly well understood parameters including pH, redox potential, and concentration. These same factors influence the equilibrium conditions relating to the soil and the contaminants. The migrating ions may be made to remain in solution and not be available for adsorption or precipitation by arresting them with complexing agents or ligands (Wittle and Pamukcu 1992; Pamukcu and Wittle, 1994).

One of the important aspects of electrokinetics in soil-water systems is the transient migration of an acid front from the anode site to the cathode site during treatment (Acar, et al., 1990; Shapiro, et al., 1989). When electrolysis of water takes place, hydrogen is produced at the cathode and oxygen is produced at the anode. Subsequently the hydronium ions produced at the anode migrate toward the cathode, whereas the hydroxide ions produced at the cathode migrate toward the anode. The ionic mobility of the hydronium ion is nearly twice as high as that of the hydroxide ion and its migration is further enhanced by the electroosmotic flow of water toward the cathode. One benefit of increased hydrogen ion concentration (lower pH) is extraction of metals, since hydrogen ion will tend to exchange with metal ions held on clay surfaces and low pH condition is favorable for the dissolution of most metal precipitates. Natural soils with high buffering capacity would tend to neutralize the acid front and maintain a higher pH environment.

Past research projects looked at the feasibility of electrokinetic transport and removal of mixed contaminants from pure clay media, synthetic reference matrix soils, and retrieved soil specimens typically found at contaminated sites. A discussion of the electrokinetic treatment of those materials found at typical sites is covered in a report to Department of Energy (DOE/ CH-9206 ;Wittle and Pamukcu, 1992). In that particular study, extensive laboratory work on the electrokinetic treatment of a number of combinations of soil, pore fluid and contaminant types was conducted. The inorganic contaminants included the classes of metals (Cd, Hg, Pb, Ni, Zn) , surrogate radionuclides (Co, Ce, Sr U), and anions ($HAsO_4$, $Cr_2O_7$). Five soil types were studied: kaolinite clay, Na-montmorillonite clay, sand with 10% Na-montmorillonite, kaolinite clay with simulated groundwater, and kaolinite clay with humic substances solution. Table 1 is an abbreviated table showing the results of heavy metals removal from the anode end side third (cathode end for the anions) of the soil samples tested..

## INVESTIGATION

Past experience with electrokinetic treatment (E-K) technology shows that the process is most effective when the transported substances are ionic, surface charged or in the form of small micelles with little drag resistance. Therefore, utilization of solubilizing agents that break apart molecules into their ionic constituents, or arrest existing ions so they are no longer available for precipitation and soil surface adsorption; or transform complex molecules into micro-micelles which are readily transportable with water advection, enhances greatly the success of electrokinetic treatment of contaminated soils. As observed from Table 1, the removal rates of a number of metals tested in the synthetic soil samples are fairly good. This is mostly owed to the low pHs attained (2 to 3) during the

826

**Table 1.** Percent Removal of Heavy Metals from Clays and Clay Mixtures by Electrokinetic Treatment (from report DOE/CH-9206; Wittle & Pamukcu, 1992))

| Metal | Percent Metal Removal | | | | |
|---|---|---|---|---|---|
| | SOIL TYPE* | | | | |
| | KS | KG | KH | MS | SS |
| As | 54.7 | 56.8 | 27.2 | 64.3 | 54.7 |
| Cd | 94.6 | 98.2 | 92.7 | 86.6 | 98.0 |
| Co | 92.2 | 93.9 | 95.9 | 89.4 | 97.5 |
| Cr | 93.1 | 94.8 | 97.6 | 93.5 | 96.8 |
| Cs | 71.9 | 80.1 | 74.7 | 54.7 | 90.5 |
| Hg | 26.5 | 13.1 | 42.5 | - | 78.3 |
| Ni | 88.4 | 95.4 | 93.9 | 93.6 | 95.9 |
| Pb | 69.0 | 75.2 | 66.9 | - | 83.-0 |
| Sr | 97.8 | 99.5 | 96.0 | 92.3 | 99.0 |
| U | 79.3 | 84.3 | 67.4 | 39.8 | 33.0 |
| Zn | 54.6 | 43.3 | 36.3 | 64.4 | 54.5 |

\*    KS: Kaolinite; KG: Kaolinite & simulated groundwater; KH: Kaolinite & humic substances;   MS: Montmorillonite; SS: Clayey sand

electrokinetic processing of these synthetic matrices which help to keep the metals in solution and thus migratory or readily transportable. In natural soils, pH often remains neutral or basic which inhibits the solubilization and thus transportation of most metals to a collection well where the products may be pumped and treated in closed systems using treatment processes appropriate to the contamination by use of ion exchange resins, carbon filters, chemical treatment, etc. Complete removal of those metals that possess complex aqueous chemistry and tendency for speciation and form hydroxide complexes is particularly difficult under variable pH and redox conditions.

There are a number of solubilizing agents, such as inorganic acids or organic compounds such as EDTA or EDA available to enhance E-K treatment. However, utilization of such products requires that they too be taken out of the ground since they may be harmful to the ecological make-up of the sub-surface. Poly Lactic Acid, PLA is an agricultural by-product currently utilized in polymerization technologies (Bonsignore et al., 1990). The fact that it is a readily biodegradable substance in its pure form makes it a desirable substance to put into the ground for remediation purposes since it would not be required to extract from the ground.

The alternative to removal of the contamination could be the electrokinetic delivery of oxidizing or reducing reagents to convert the targeted substance(s) into a *non-toxic form*, or *stabilize* by precipitation and/or adjust the pore fluid conditions to enhance *adsorption and incorporation* into the soil structure. An example of this treatment in place would be the reduction of chromate with iron two to produce chrome oxide which is a more stable species.

### Feasibility of PLA as an Solubilizing Agent

The feasibility of using poly (lactic acid ) as a solubilizing or complexing agent was investigated through batch tests using kaolinite and standard reference soil and five selected metals. Each salt solution was prepared at **5000 ppm** aqueous concentration of the metal. Table 2 presents the salts used and the initial pH values of the stock solutions.

**Table 2.** Selected metal solutions for PLA extraction batch tests

| METAL | SALT FORMULA | SOLUTION pH |
|---|---|---|
| Lead | $Pb(NO_3)_2$ | 4.49 |
| Nickel | $Ni(NO_3)_2 . 6H_2O$ | 4.51 |
| Zinc | $ZnCl_2$ | 5.57 |
| Mercury | $Hg(NO_3)_2 . H_2O$ | 2.57 |
| Cesium | $CsNO_3$ | 5.87 |

Two sets of ten batch samples were prepared. One set of samples contained kaolinite clay only, and the other standard reference soil (SRS), which was synthetically mixed to represent natural soil. The SRS contained the following ingredients based on EPA standard of SRM (Synthetic Reference Matrix): 30% - Clay (Kaolinite); 25% - Silt (ML); 20% - Sand; 20% - Top Soil (Commercial potting soil); 5% - Gravel (% retained between the no. 4 and 10 sieves). Each batch sample was prepared by mixing 5.0 grams of soil with 100 ml of the stock solutions. After mixing, all samples were placed in a mechanical rotary extractor and agitated continuously for 24 hours. pH readings were taken at regularly spaced intervals during the batching operation. After the batching operation, all samples were centrifuged for one hour at 15,000 revolutions per minute (rpm) to separate the liquid and solid portions of each sample. The ten samples (five kaolinite and five SRS) without PLA were centrifuged immediately. A predetermined volume of PLA (molecular weight 180) was added to the other half of the batch mixtures at a ratio of 1:5 moles of metal to PLA. These samples were allowed to sit at 4°C for approximately 24 hours prior to being centrifuged. pH readings of these samples were also obtained before and after centrifuging. After centrifuging, the liquid portion from each batch sample was decanted for analysis using the AA.

Figures 1 and 2 show the pH variations of SRS and kaolinite mixtures with duration of agitation and addition of PLA using a metal to acid molar ratio of 1:5, respectively. As observed from these figures, the mixture pH does not change appreciably with agitation for both of the soils. The pH of the mixture increases significantly with addition of SRS in the case of cesium solution. This is due to the affinity of the organic substances to cesium in the SRS matrix. There is slight increase in pH with the addition of SRS in case of Nickel also. The others appear to be unaffected both with SRS and kaolinite addition. The PLA causes the pH to decrease to a median value of 2.62 for all batch mixtures.

Table 3 presents the mass fraction of three metals extracted from soil matrices simply by mixing PLA. The data indicates that PLA improved the extraction of all three metals, zinc being the most and consistently affected regardless the soil matrix. Electrokinetic injection of PLA into lead, zinc and nickel contaminated soil samples were conducted. At the time this paper was prepared the results were not available, therefore are not included here.

**Table 3.** Mass Fraction of PLA Extracted Metals from Kaolinite and SRS

| Metal | Mass of Metal in Decant Liquid - SRS | | Mass of Metal in Decant Liquid - Kaolinite | | Percent Extracted by PLA (%) | |
|-------|--------|------|--------|------|------|-----------|
|       | BLANK  | PLA  | BLANK  | PLA  | SRS  | Kaolinite |
| Pb    | -      | -    | 431.31 | 447  | -    | 3         |
| Ni    | 165    | 200  | 262.5  | 342.5| 7    | 16        |
| Zn    | 300    | 375  | 297.5  | 380  | 15   | 16.5      |

## Feasibility of Stabilizing Cr(VI) in Kaolinite Using Fe(II)

Chromium exists in two possible oxidation states in soils: the trivalent Cr(III) and the hexavalent Cr(VI) chromium. The hexavalent chromium is more toxic than trivalent chromium. At low pH conditions (2 to 6.5) the predominant form of the hexavalent chromium is chromate or dichromate ion. Due to their negative charge they are not readily adsorbed or exchanged at clay surfaces, therefore remain in soil pore water and be readily transported. This was observed by successful removal of the chromate ion with electrokinetic migration in the opposite direction of water flow, as shown in Table 1. However at sufficiently low pH, the soil surface sites become positively charged and tend to retain anions, such as chromate. Therefore complete removal may not be achieved unless precise control of pH is maintained during an electrokinetic process.

Hexavalent chromium can be reduced to Cr(III) under normal soil and pH conditions, for which soil organic matter has been identified as the electron donor (Rai et al., 1987; Vitale et al., 1995). Bartlett (1991) reported that in natural soils, this reduction may be extremely slow, requiring years. In subsurface soils where there is less organic matter, the Fe(II) containing minerals reduce Cr(VI) at pH less than 5 (Eary and Rai, 1991). Electrokinetically injected Fe(II) into a matrix of soil containing hexavalent chromium should facilitate the reduction of Cr(VI) since electrokinetic process produces low pH conditions. This theory was tested in the laboratory where Fe(II) and Cr(IV) were caused to migrate into soil from the anode and the cathode electrode water chambers, respectively. At the end of 122 hrs (approximately 0.5 pore volume of water flow) very little chromium was detected at the anode water chamber as well as at the anode end of the soil.

## Electrokinetic (EK) Testing

Fe(II) and Cr(VI) were introduced to the kaolinite sample as tracer solutions. They were prepared at 20 °C by mixing iron (pentahydrate) sulfate ($FeSO_4.7H_2O$) and potassium dichromate ($K_2Cr_2O_7$) with distilled water, respectively. Each solution was prepared at concentrations below its respective solubility level (CRC Handbook of Chemistry and Physics, 1988). The pH of the prepared solutions $K_2Cr_2O_7$ and $FeSO_4$ were 4.05 and 2.95, respectively. The iron solution was injected into the anode chamber anticipating the migration of Fe(II) into the soil towards cathode. The chromate solution was injected into the cathode chamber, expecting $Cr_2O_7^{2-}$ it to migrate into the soil towards anode.

A saturated, homogeneous soil sample was prepared by mixing 253.70 g of kaolinite with 154.00 mL of distilled water and consolidated for 41.0 hours. After consolidating, approximately 103.1 g of the kaolinite sample was placed into the E-K cell. The water content of the soil sample was 41.7%. Approximately 139.0 mL of $K_2Cr_2O_7$ (220 ppm) and 186.00 mL of $FeSO_4$ (140 ppm) solutions were added to the cathode and anode chambers, respectively. Four iron nails, weighing 4.0 g total, were added to the anode chamber to increase the concentration of iron introduced to the soil.

The E-K test was performed for an elapsed period of approximately 122 hours. During the E-K test, inflow, outflow, resistivity, current, and pH data were obtained. The cathode chamber served as the inflow source to the E-K sample, while the anode chamber was the outflow. The variation of flow and current with duration of treatment are shown Figure 3.

Upon completion of the E-K test, the soil sample was divided into four equal parts. Each portion weighed approximately 27.75 g. A representative sample, weighing 10.0 g, was obtained from these sections. Each 10.0 g sample was divided into two 5.0 g samples for acid and water extractions. Water content samples and pH readings were obtained for each test sample. Water and acid extraction procedures were performed to obtain the soluble and insoluble portions of iron and chromium ions, respectively. After the extraction procedures, the samples were analyzed using the AA. The mass fraction of chromium and iron (with respect to total mass of each injected into the electrode chambers) are plotted in Figure 4. Superimposed on this graph is the distribution of pH throughout the soil at the sampling locations and the water chambers.

**Chemical and Equilibrium Analysis for $K_2Cr_2O_7$ and $FeSO_4$:**

The half reactions used to define $K_2Cr_2O_7$ and $FeSO_4$ are shown in equations 1 and 2 as $Cr_2O_7^{2-}$ and $Fe^{2+}$, respectively:

$$Cr_2O_7^{2-} + 14H^+ + 6e^- \rightarrow 2Cr^{3+} + 7H_2O \tag{1}$$

$$Fe^{2+} \rightarrow Fe^{3+} + e^- \tag{2}$$

The net reaction for the combination of $Cr_2O_7^{2-}$ and $Fe^{2+}$ is shown by equation 3:

$$Cr_2O_7^{2-} + 6Fe^{2+} + 14H^+ \rightarrow 2Cr^{3+} 6Fe^{3+} + 7H_2O \tag{3}$$

The standard electromotive force (emf) or cell reaction ($E^\circ$) of the half reactions for $Cr_2O_7^{2-}$ and $Fe^{2+}$ defined by equations 1 and 2 are 1.33 V and - 0.770 V, respectively (Bard, 1980). The net emf ($E^\circ_{Rxn}$) as defined by the net reaction in equation 3, is 0.560 V. A net, positive emf implies that the overall net reaction of $Cr_2O_7^{2-}$ and $Fe^{2+}$ yielding the products to $Cr^{3+}$ and $Fe^{3+}$ is spontaneous. In addition, a net, positive emf value suggests that Gibbs free energy ($\Delta G^\circ$), used to define the direction of the reaction, is negative; as such, the overall reaction is spontaneous and occurring in the direction shown in equation 3. $\Delta G^\circ_{(Rxn)}$ for the net reaction defined in equation 4 is determined as -324,240.00 J/moles. The equilibrium constant ($K_{eq}$) for the net reaction can be determined as a function of $\Delta G^\circ_{Rxn}$ as shown by equation 5.

$$\Delta G^\circ_{Rxn} = nFE^\circ_{Rxn} \tag{4}$$

$$\Delta G^\circ_{Rxn} = -nFE^\circ_{Rxn} = -RT(lnK_{eq}) \tag{5}$$

$K_{eq}$ determined for the net reaction is approximately $6.86 \times 10^{-56}$. A positive value of $K_{eq}$, as determined by $\Delta G^\circ_{Rxn}$, further denotes the spontaneity of the reaction. For $K_{eq} \gg 1$, the reaction is considered to occur at a very fast rate. As such, the net reaction for $Cr_2O_7^{2-}$ and $Fe^{2+}$ is expected to occur quickly; with Cr(VI) being reduced to Cr(III) and Fe(II) oxidized to Fe(III) from $Cr_2O_7^{2-}$ and $Fe^{2+}$, respectively. However, at such a fast reaction rate, all the Cr(VI) may not be reduced to Cr(III); subsequently, all Fe(II) may not be oxidized to Fe(III).

Within the scope of the analysis relative to the final pH of the soil samples tested, its is believed that the reduction of Cr(VI) to Cr(III) by Fe(II) did occur. Cr(III) forms hydroxy complexes in presence of (OH⁻). These species precipitate at pH 4.5 and complete precipitation occurs at 5.5. Trivalent chromium ion which occur at lower pH is readily adsorbed by soils (Baes, 1976 and Rai et al., 1987). Observing Figure 4, it can be seen that there is a sharp decrease of the soluble form of Cr while an increase of acid extractable form of Cr at the anode side of the soil . Most of the Cr appears to have accumulated in the center, and little has migrated towards the anode, which suggests that the emerging iron front may have been met at that location. With the final ranges of pH distribution for the sample tested (between 2.15 and 2.5) it is expected that most of the chromium is either in hydroxy complexes or adsorbed by the soil in its Cr(III) ionic form.

## REFERENCES

Acar, Y.B., Li, H., Gale, R.J., Putman, G., Hamed J. and Wong, R., Electro Chemical Processing of Soils: Theory of pH Gradient Development by Diffusion and Linear Convection, J. of Envir. Sci. and Health, Part (A); Envir. Science and Engineering, Vol. 25, No. 6, pp. 687-714, 1990.

Bard, A. J. and Faulkner, L. R., Electrochemical Method, John Wiley, 1980.

Bartlett, R.J., Chromium Recycling in Soils and Water: Links, Gaps, and Methods, Environmental Health Perspective, Vol. 92, pp. 17-24, 1991.

Bonsignore, P. V., Coleman, R. D., and Mudde, J. P., "Potato Peels to Degradable Plastics," 22nd Congress of Nat. Agric. Plastics Assoc., Canada, May, 1990.

Casagrande, L., Electro-Osmosis in Soils, Geotechnique, V. 1, No. 3, p.159, 1949.

Eary, L.E. and Rai, D., Chromate Reduction by Surface Soils Under Acidic Conditions, J. American Soil Science Society, V. 55, pp. 676-683, 1991.

Hamed, J., Acar, Y.B., and Gale, R.J., Pb(II) Removal from Kaolinite by Electrokinetics, J. Geotech. Engr., ASCE, V. 117, No. 2, 1991.

Pamukcu, S. and Whittle, J.K., Electrokinetic Removal of Selected Heavy Metals from Soils, Environmental Progress, V. 11, No. 3, 1992.

Pamukcu, S. and Whittle, J.K., Electrokinetic Treatment of Contaminated Soils, Sludges, and Lagoons, DOE/CH-9206, Contract No. 02112406, 1994

Rai, D., Sass, B.M., and Moore, Chromium (III) Hydrolysis Constants and Solubility of Chromium (III) Hydroxide, J. Chem., V. 26, pp. 345-349, 1987.

Segall, B.A., O'Bannon, C.E., and Matthias, J.A., Electro-Osmosis Chemistry and Water Quality, J. Geotech. Eng., ASCE, V. 106, No. GT10, 1980.

Shapiro, A.P. and Probstein, R.F., Removal of Contaminants from Saturated Clay by Electroosmosis, Env. Science & Technology, V. 27, pp.283-291, 1993.

Vitale, R.J., Mussoline, R.G., Petura, J.C., and James,, B.R., Hexavalent Chromium Quantification in Soils: An Effective and Reliable Procedure, American Environmental Laboratory, 1994.

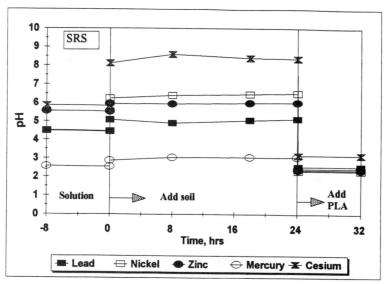

Figure 1.    pH Variations for SRS with Duration of Agitation and Addition
of PLA at a Metal to Molar Acid Ratio of 1:5.

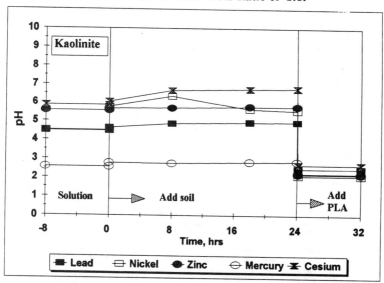

Figure 2.    pH Variations for Kaolinite with Duration of Agitation and
Addition of PLA at a Metal to Molar Acid Ratio of 1:5.

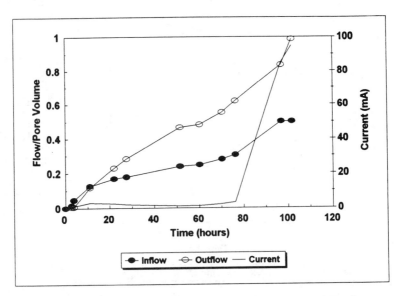

**Figure 3.    Variation of Flow and Current with Duration of Testing.**

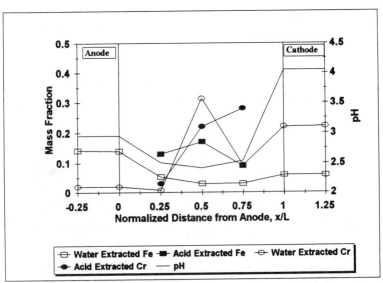

**Figure 4.    Mass Fractions of Chromium (Cr) and Iron (Fe); Superimposed pH Distributions Along Length of Sample and Water Chambers.**

# IN-SITU REMOVAL OF PHENOLS FROM CONTAMINATED SOIL BY ELECTRO-OSMOSIS PROCESS

LUÍS R. TAKIYAMA AND C. P. HUANG
Department of Civil Engineering, University of Delaware
Newark, DE 19716

## INTRODUCTION

Since the first successful attempts to utilize electroosmosis in foundation engineering by Casagrande in 1941 [1], the electrokinetic phenomenon has found increasing application for the drainage and treatment of problem soils. Many researchers in the 1950's and 1960's have directed special attention to particular topics such as physico-chemical description of electroosmosis; prediction of soil dewatering by electroosmosis; consolidation by electroosmosis and special applications of the technique for foundation soil treatment. Despite the mechanics of electroosmosis stabilization are not completely understood, many fortunate field applications have been demonstrated. Bjerrum et al. [2] described an electroosmotic treatment carried out at the site of an excavation for a sewage treatment plant in quick clay; Chappell and Burton [3] reported an embankment stabilization to permit a construction of a dry dock, Mitchell and Wan [4] showed a stabilization of fine grained soils and many others have succeeded practice field electroosmosis uses.

Although the utilization of electroosmosis has been focused mainly on foundation engineering for many years, some investigations have discovered other applications such as dewatering of foams, sludges and dredgings [5]; ground-water lowering and barrier systems [6]; chemical grout injection [7]; increasing of petroleum production [8]; and separation and filtration of materials in soils and solutions [9]. More recently, special attention has been given to the potential of the technique for waste remediation. This fact resulted in the initiation of several studies in the environmental field [10-14].

In the present work, the feasibility of the technique to remove some organic compounds such as phenol, 2-chlorophenol, 3-chlorophenol and 4-chlorophenol was investigated as well as the influence of some physical properties of the soil core over the removal process.

## EXPERIMENTAL PROCEDURES

Electro-osmosis cell

Figure 1 shows the apparatus employed to perform the experiments. The electro-osmosis cell consisted in an acrylic unit with a central cylinder of 11.5 cm in length and 8.9 cm in internal diameter, where the soil sample was located and a cathode and anode compartments.

Graphite rods (type "F" grade 014144-08 U7/SPK, Ultra Carbon Co., Bay City, MI) were utilized as electrodes and a series of 8 rods were held at each compartment near the central cylinder. To separate the soil from the electrolyte solutions at the cathode and anode reservoir, a set of two nylon meshes with a filter paper in between was employed.

Soil sample preparation

The soil used was a combination of Ottawa sand (U.S. Silica Co., Ottawa, IL) and Georgia kaolinite (Georgia Kaolin Co, Elizabeth, NJ) at ratio 1:1 (w:w). A solution of the phenolic compound (in NaCl $10^{-3}$ M) were added, mixed well and let stand for about 24 hours to reach an equilibrium and consenquently to have a uniform distribution of the contaminant in the soil. The mixture were then carefully placed in the acrylic cylinder to avoid the formation of large air spaces and all the apparatus were set up for the testing.

Testing

The electrodes were connected to a 12 volts power supply (model E-12/158, Power/Mate Corporation, Hackensack, NJ) and the anode section was kept filled with NaCl $10^{-3}$ M electrolyte solution and daily water samples were taken at the cathode side. Parameters such as amount of water flow, current, effluent contaminant concentration, pH of catholyte and anolyte were monitored as a function of time.

After the conclusion of test, the soil sample was removed from the cell and sliced into 10 sections. Each one was analyzed for water content, pH and contaminant concentration.

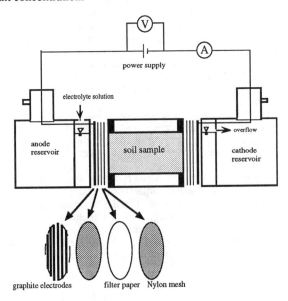

Figure 1. Electro-osmosis apparatus

Analytical methods

The contaminants in the soil were extracted by adding 2 mL HNO$_3$ (1M) and 3 mL of hexane and to approximately 1 g to 2 g of soil sample and shaking at room temperature for 48 hours.

The analysis of the phenolic compounds in solution (hexane phase) was done by gas chromatography method (model 5890 II, Hewlett-Packard, Wilmington, DE) equiped with electron capture detector (GC-ECD).

## RESULTS AND DISCUSSION

Flow and coefficient of permeability (k$_e$)

Figure 2a shows the amount of electro-osmotic flow produced as a function of time. In general, the flow reaches a maximum value and then decreases gradually possibly due to changes in the electrical properties of the packed soil cores originated from the electrochemistry associated with the electro-osmosis process. By applying a potential, water will be decomposed to H$^+$ and O$_2$ at the anode and these hydrogen ions will flush across the cell modifying the original conditions of the pore fluid.

Noticeable differences among the experiments were found related to the water flow. Figure 2b presents the diagram of accumulative flow versus time. The test with 2-chlorophenol produced the largest flow while the blank test showed the lowest water flow. Refering to Table I, it is noticed that the water flow is related to the pore volume of the soil core. Larger the pore volume, larger is the electro-osmotic water flow since more water is available to be transported. Obviously there is a limitation. If the pore volume is too large (see blank test) in which the sample behaves as a free electrolyte solution and less electro-osmotic flow are recorded [15]. Figure 3 exhibits the total average flow of each experiment versus pore volume; increasing the pore volume there is an increase in the average flow until a point that an abrupt decrease is noticed.

Figure 4 shows the results obtained for the coefficient of electro-osmotic permeability. As expected, the profile has the same pattern as for the flow since the flow and electro-osmosis permeability are closely related:

$$Q_e = k_e \, i_e \, A \qquad (1)$$

Equation (1) was used to estimate the k$_e$ values [16]. In the present case the potential gradient i$_e$ and the cross section area A were constant and the k$_e$ values were found to be consistent with those reported in the literature [16] of about 10$^{-5}$ cm$^2$/V.s.

Influent-effluent pH changes

The influent and effluent pH variation during the experiments are presented in Figures 5 and 6, respectively. For all experiments the pH at the anode (influent) decreased from values around 5 to approximately 2.5-3.0 while the effluent pH (cathode) rose to values up to 12-13 and then decreased gradually due to the acid front generated at the anode. As mentioned, hydrogen ions were produced from the oxidation of water causing the drop in pH at the anode and hydroxyl ions from the reduction of H$_2$O were responsible for the pH increase at the effluent.

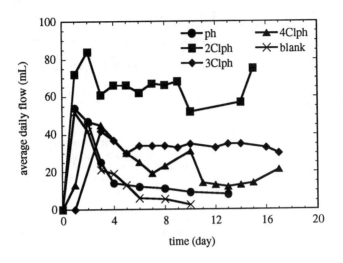

Figure 2a. Average daily flow as a function of time.

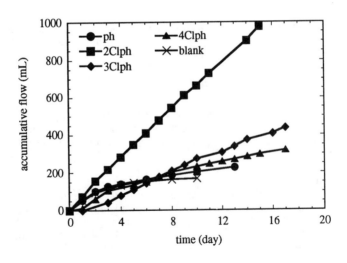

Figure 2b. Accumulative flow as a function of time.

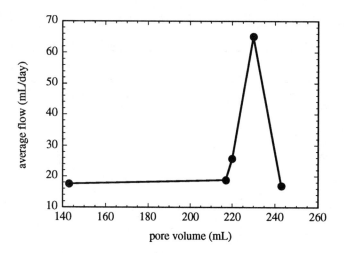

Figure 3. Average of total flow versus pore volume.

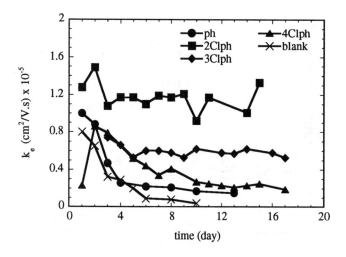

Figure 4. Coefficient of electro-osmotic permeability as a function of time.

Table I. Physical properties of the specimen and experimental conditions.

| Test # | Blank | Test 1 | Test 2 | Test 3 | Test 4 |
|---|---|---|---|---|---|
| compound | | (phenol) | (2-Clph) | (3-Clph) | (4-Clph) |
| Weight of dry soil (g) | 983 | 1245 | 1253 | 1287 | 1294 |
| Total weight (g) | 1147 | 1452 | 1441 | 1471 | 1479 |
| Volume ($cm^3$) | 622 | 622 | 715 | 715 | 715 |
| Bulk density ($g/cm^3$) | 1.58 | 2.00 | 1.75 | 1.80 | 1.81 |
| Particle density ($g/cm^3$) | 2.60 | 2.60 | 2.60 | 2.60 | 2.60 |
| Porosity | 0.39 | 0.23 | 0.33 | 0.31 | 0.30 |
| Pore volume ($cm^3$) | 244.1 | 143.3 | 233.5 | 220.4 | 217.7 |
| Water content (%) | 14.3 | 14.3 | 13.0 | 12.5 | 12.5 |
| Contaminant concentration ($\mu g/g$ dry soil) | --- | 166 | 143 | 143 | 143 |
| Test time (h) | 240 | 312 | 360 | 408 | 408 |
| Potential gradient (V/cm) | 1.21 | 1.21 | 1.05 | 1.05 | 1.05 |

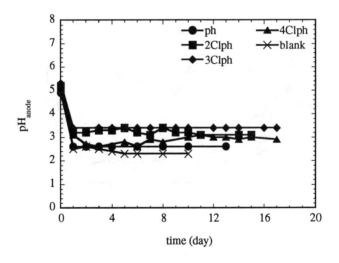

Figure 5. Influent (anode) pH as a function of time.

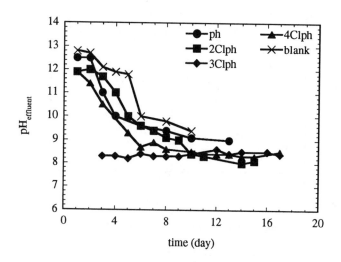

Figure 6. Effluent (cathode) pH as a function of time.

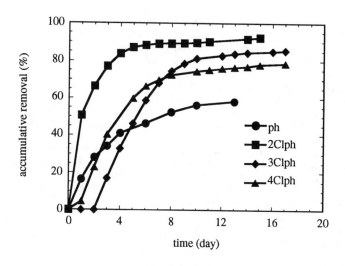

Figure 7. Accumulative percentage removal as a function of time.

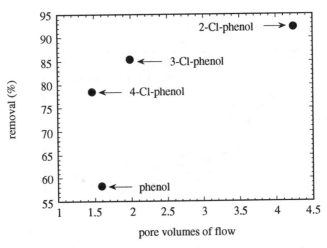

Figure 8. Total removal versus total flow.

Contaminant removal

Figures 7 and 8 present the percent removal as a function of time and the removal related to the total water volumes flushed through the soil cores, respectively. The results demonstrate that the removal efficiency was proportional to the amount of water passed through the soil samples. The 2-chlorophenol was almost completely removed from the soil while only 58% of the phenol was carried out by the electro-osmosis process. The removal of 3-chlorophenol and 4-chlorophenol were 85% and 79%, respectively.

pH profile, water content and contaminant distribution

After the completion of the tests, the samples were sliced into 10 sections and analyzed for pH, water content and contaminant concentration. The resulted pH profiles shown in the Figure 9 were originated exclusively from the redox reactions of water at the electrodes and demonstrated to have the same pattern as the profiles determined by Hamed and Acar [11]. The acidic electrolyte solution generated at the anode reservoir flows across the soil sample lowering the pH to values around 3-4 until near to the cathode the pH rises to values about 8-9 because of the basic conditions produced by the reduction of water.

Water content was measured in order to determine the concentrations of the phenolic contaminants per gram of dry soil and it was found to have a uniform distribution across the cell with an average value of 15% as presented in the Figure 10. This result was expected since the water was supplied continually at the anode section.

Figure 11 shows the distribution in relative concentration of the phenolic compounds remained in the soil. Small amounts of 4-chlorophenol and high concentration of phenol were retained while no 2-chlorophenol and 3 chlorophenol were found in the soil. An accumulation of phenol above the initial concentration (represented by the solid straight line) was detected near the cathode, meanning that the contaminant does not move uniformily across the soil core.

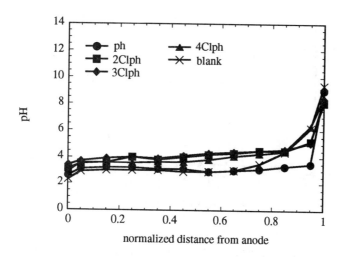

Figure 9. pH profile across the soil samples as a function of distance from anode.

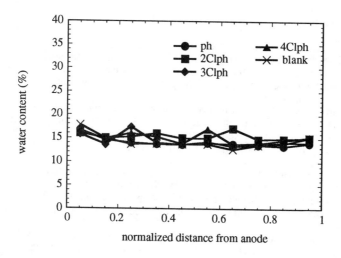

Figure 10. Water content distribution as a function of distance from anode.

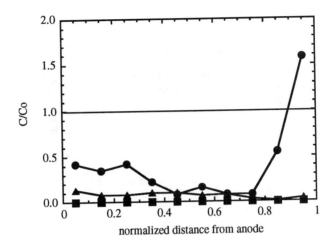

Figure 11. Relative concentration of the contaminants in the soil as a function of distance from anode.

The mass balance for the phenol and mono-chlorophenols are summarized in the Table II. The differences can be attributed to the diffusion of the compounds to the anode reservoir and to the efficiency of the extraction method used to determine the contaminant concentrations in the soil.

**CONCLUSIONS**

Electrochemistry plays an important role during an electro-osmosis process. The generation of pH gradients and possible effects such as ion exchange and changing in the soil composition caused by the acid front originated at the anode ensue a diminishing of the water flow compromising the efficiency of the process.

As an inovative contaminant removal method, the technique demonstrated to be effective depending upon some conditions related to the physical-chemical properties of the packed soil core. Among these conditions, the pore volume was proved to be important and proportional to the removal efficiency.

The phenol distribution obtained in the soil core demonstrated that the contaminant movement is not uniform throughout the medium and accumulates near to the cathode. This could be used as a tool to concentrate the pollutants in a specific area between the electrodes for further remediation treatment.

A 93% removal was achieved for the 2-chlorophenol test showing that electroosmosis can be an effective way to remediate contaminated soils.

Table II. Mass balance of the phenols.

|  | Test 1 | Test 2 | Test 3 | Test 4 |
|---|---|---|---|---|
|  | (phenol) | (2-Clph) | (3-Clph) | (4-Clph) |
| removed from the soil (mg) | 120.6 | 165.5 | 157.1 | 144.2 |
| remained in the soil (mg) | 81.3 | N.D.[*] | N.D.[*] | 10.4 |
| total (mg) | 201.9 | 165.5 | 157.1 | 154.6 |
| initial mass (mg) | 207.0 | 179.2 | 183.9 | 185.0 |
| difference (mg) | 5.1 | 13.7 | 26.8 | 30.4 |

[*] Non detected

## REFERENCES

1. Casagrande, L., "The drainage of Fine Soils", *Strasse*, No. 19-20, 1941

2. Bjerrum, L., Moum, J., and Eddie, O., *Geotechnique*, Vol. 17, No. 3, pp. 314-235, 1967.

3. Chappell, B. A., and Burton, P. L., *J. Geotech. Eng. Div.*, Vol. 101, No. 8, pp. 733-740, 1975.

4. Mitchell, J. K., and Wan, T-Y, *Proc. IX ICSMFE*, Tokyo, Japan, No. 1, pp. 219-224, 1977.

5. O'Bannon, C. E., Segall, B. A., and Mathias, J. S., *U.S. Corps Of Engineers Report*, Vicksburg, Miss, 1976.

6. Gray, D. H., *Nature*, Vol. 223, No. 5204, pp. 371-374, 1969.

7. Esrig, M. I., *J. Soil Mech. Fdn. Div.*, Am. Soc. Civ. Engs., Vol. 93, No. 6, pp. 109-128, 1967.

8. Amba, S. A., Chilingerian, G. V., and Beeson, C. M., *J. Canadian Petroleum Tech.*, Vol. 3, No.1, pp. 8-14, 1964.

9. Yukawa, H., Chigira, H., Hoshiro, T., and Iwata, M., *Journal of Chemical Engineering of Japan*, Nagoya, Japan, No. 4, pp. 370-376, 1971.

10. Mitchell, J. K., and Yeung, T-C, *Transp. Res. Records*, No. 1289, 1991.

11. Acar, Y. B., Hamed, J., and Gale, R.J., *J. Geotech. Eng. Div.*, Vol. 117, No. 2, pp. 241-271, 1991.

12. Ray, C., and Ramsey, R. H., *Environmental Progress*, Vol. 6, NO. 3, pp. 145-149, 1987.

13. Khan, L. I., Pamukcu, S., and Kugelman, I., *2nd Int. Symp. on Envir. Geotech.*, Shangai, China, Envo Publishing, Bethlehen, PA, 1, pp. 39-47, 1989.

14. Coolbroth, F. P., "The Sewage Osmosis Concept for on-site Disposal Systems - Clay Soils," in N.J. McClelland, Ed., *Proceedings of Third National Conference, National Sanitary Foundation*, pp. 131-137, Ann Arbor Science, Ann Arbor, Mich. (1976).

15. Kezdi, A. "Handbook of Soil Mechanics," Vol. 1, pp. 178-181. Elsevier Scientific Publishing Co., Amsterdam, 1974.

16. Casagrande, L., *Journal of the Boston Society of Civil Engineers*, Vol. 39, pp. 51-83, 1952.

17. Esrig, M. I., and Majtenyi, S., *Highway Research Record*, No. 111, pp. 31-41, 1966.

18. Schmid, G., *Series of papers in Z. Electrochem.*, Vol. 54, p. 424, 1950; Vol. 55, pp. 229, 684, 1951; Vol. 56, pp. 35, 181, 1952.

19. Gray, D. H., and Mitchell, J. K., *J. Soil Mech. Fdn. Div.*, Am. Soc. Civ. Engs,.Vol. 93, No. 6, pp. 209-236 1967.

# CATEGORY XIII:
# In-Situ *Treatment*

# IN SITU BIOREMEDIATION OF CHLORINATED SOLVENTS

W.A.SACK, P.E.CARRIERE, C.S.WHITEMAN, M.P.DAVIS
    S.RAMAN and J.E. CUDDEBACK
Civil and Environmental Engineering Department
West Virginia University
Morgantown, WV  26506-6103

A.K.SHIEMKE
Department of Biochemistry
West Virginia University
Morgantown, WV, 26506

## INTRODUCTION

Chlorinated aliphatic hydrocarbons (CAHs), such as tetrachloroethylene (PCE) and trichloroethylene (TCE), are among the most common contaminants of groundwater and soils. TCE is sufficiently soluble in water so as to be easily transported and broadly disseminated. Because of TCE's wide dispersion at relatively low concentrations, in situ treatment technologies are sorely needed. Currently available remediation methods for subsurface environments include air sparging of the groundwater, vacuum extraction of contaminants from the vadose zone, and extraction of contaminated water for air-stripping. These techniques do not destroy contaminants, but rather transfer them to another medium or concentrate them for later handling and often pose potential health hazards.

In situ bioremediation of chlorinated organic solvents is receiving growing support and widespread testing in the field. It is an attractive alternative with the potential to destroy contaminants almost completely. The research seeks to exploit the natural symbiotic relationship between methanogenic and methanotrophic microorganisms. The methanogens are able to carry out anaerobic reductive dehalogenation of highly chlorinated solvents while producing methane. The methanotrophs in turn utilize the end products of the methanogens, including the methane, to aerobically degrade the residual CAH compounds to environmentally acceptable end products. Both groups of organisms degrade the CAH compounds cometabolically and require a primary substrate.

The purpose of the research is to evaluate and optimize the ability of methanotrophic, methanogenic, and other selected bacteria for cost-effective biotransformation of TCE and other volatile organic compounds (VOC's). Both anaerobic and aerobic microcosm studies have been performed to evaluate TCE mineralization as reported [1] earlier. This paper describes initial studies using separate anaerobic and aerobic columns. As soon as the initial column studies are complete, the anaerobic and aerobic columns will be combined in both sequential and simultaneous modes to evaluate complete CAH destruction.

## BACKGROUND

The methanogen's ability to dehalogenate more oxidized (more halogenated) pollutants, generating vinyl chloride, is complemented by the ability of the methanotrophs to mineralize vinyl chloride and other small mono-chlorinated hydrocarbons. The rate limiting step of mineralization under anaerobic conditions is the dehalogenation of vinyl chloride. Thus, it might be advantageous to induce methanotrophic growth at the point where all (or most) of the more chlorinated compounds have been dehalogenated to vinyl chloride. The number of chlorines dramatically effects the rate of anaerobic degradation: the more chlorines the faster the rate of degradation [2].

Although methanogenic bacteria will not grow in the presence of oxygen, the reductive dehalogenation reaction is somewhat oxygen tolerant. The rate of TCE degradation is reduced under micro-aerophilic conditions, but not completely blocked [3]. Under these conditions a suitable source of reducing equivalents must be provided (eg. methanol, hydrogen, acetate, and formate). It is interesting to note in this regard that Kastner [4] reported that an aerobic enrichment culture's ability to dechlorinate cis-1,2-dichloroethylene was shown to be dependent on a cyclic transition from aerobic to anaerobic conditions and limited oxygen supply. It may also be possible to increase biodegradation rates by alternating methanogenic and methanotrophic growth conditions, thus limiting the accumulation of potentially toxic byproducts, as well as optimizing the alternating production of methane rich and oxygen rich conditions.

Methanotrophic bacteria produce methane monoxygenase

850

(MMO) to begin catabolism of methane (the primary substrate) for assimilation into biomass or mineralization. Fortunately, the MMO is relatively non-specific, and the enzyme also is active (cometabolically) in transforming TCE to TCE-epoxide. The epoxide is unstable and undergoes chemical abiotic transformation to TCE-diol and eventually biological mineralization to $CO_2$ and $H_2O$. It is the nature of cometabolic processes that the presence of the primary substrate *both* increases cometabolic transformation by increasing the available biomass and decreases cometabolic transformation by competing for the active site of the enzyme. The competitive inhibition of methane has been demonstrated both in the laboratory and the field [5]. It is clear that careful application of methane, either in a pulse feed mode or at low concentrations will be necessary to avoid problems with competitive inhibition.

While the MMO enzyme is responsible for the initial epoxidation of methane to methanol, other enzymes are responsible for further metabolism for biomass assimilation or energy production in the form of reduced nicotinamide adenine dinucleotide ($NADH_2$). Further oxidation of methane oxidation products (methanol, formaldehyde and formate) yields reducing equivalents in the form of NADH. Methanol has been investigated as both an electron acceptor and a carbon source. Broholm et al. [6] reported field investigations showing TCE transformation could be supported by methanol addition in the absence of methanol for a period of 60 hours. Thus, the addition of methanol as an alternative primary substrate to methanol as a source of electrons and as a growth substrate along with pulse feeding of methane seems an attractive alternative to maintain a viable population while avoiding problems with competitive inhibition of methane.

## MATERIALS AND METHODS

**Microorganisms:** The organisms used in the anaerobic column were originally obtained from the Morgantown Wastewater Treatment Plant anaerobic digester. The aerobic column was inoculated with three methanotrophs: *Methylococcus capsulatus, Methylosinus trichosporium* and *Methylocystis parves.*

**Anaerobic Basal Salts Medium:** The following constituents were fed to the anaerobic column in the simulated groundwater: $NH_4Cl$ (55.0 mg/L), $K_2HPO_4$ (2.92 mg/L), $KH_2PO_4$ (2.11 mg/L), $MgCl_2*6H_2O$ (5.76 mg/L), $MnCl_2$ $*4H_2O$ (1.92 mg/L), $CoCl_2*6H_2O$ (3.84 mg/L), $ZnCl_2$ (1.92

mg/L), CaCl$_2$*2H$_2$O (6.34 mg/L), HBO$_3$ (0.580 mg/L), NiCl$_2$6H$_2$O (1.92 mg/L), Na$_2$MoO$_4$*2H$_2$O (0.768 mg/L), L-Cysteine C$_3$H$_7$NO$_2$S$^-$HCl H$_2$O (17.28 mg/L), NaHCO$_3$ (76.8 mg/L), Yeast Extract (1.92 mg/L), FeCl$_2$*4H$_2$O (1.15 mg/L), Resazurin (0.0384 ug/L)

**Aerobic Basal Salts Medium:** The following constituents were fed to the aerobic column in the simulated groundwater: NaNo$_3$ (85 mg/L), KH$_2$PO$_4$ (53 mg/L), Na$_2$HPO$_4$ (86 mg/L), K$_2$SO$_4$ (17 mg/L), MgSO$_4$*7H$_2$O (3.7 mg/L), CaCl$_2$*2H$_2$O (0.7 mg/L), FeSO$_4$*7H$_2$O (.05 mg/L), ZnSO$_4$*7H$_2$O (.04 mg/L), MnCl$_2$*4H$_2$O (.002 mg/L), CoCl$_2$*6H$_2$O (.005 mg/L), NiCl$_2$.6H$_2$O (.001 mg/L), H$_3$BO$_3$ (.0015 mg/L), EDTA (.0025 mg/L), CuSO$_4$*5H$_2$O (.0125 mg/L), KHPO$_4$ (2.6 mg/L), Na$_2$HPO$_4$ (3.3 mg/L), Biotin (.2 ug/L), Folic Acid (.2 ug/L), Thiamine HCl (.5 ug/L), Calcium Pantothenate (.5 ug/L), B12 (.1 ug/L), Riboflavin (.5 ug/L), Nicotamide (.5 ug/L)

**Column and Media Description:** Both the anaerobic and aerobic columns are 50 X 300 mm glass chromatography columns fitted with glass threaded adapter tubes for sampling at the one-third and two-thirds the height of the column in addition to the column effluent. Each sample port is connected to a miniature inert value by a Teflon threaded adapter. The columns were filled with standard Ottawa sand with a porosity of approximately 0.40. The sand has an effective size of 0.58 mm and a uniformity coefficient of 1.29.

**Sample Analysis:** CAH concentrations were determined using a Hewlett Packard 5890, Series II gas chromatograph equipped with an OI Corporation Model 4420 Electrolytic Conductivity Detector and a Tekmar, LSC 2000 Purge and Trap Concentrator. EPA Method 601 is used for the determination.

**General Conditions and Procedures:** The temperature during the research reported in this paper varied from 24 to 29 oC. Samples were generally taken daily for pH, dissolved oxygen (aerobic column) and CAH levels during runs.


## RESULTS AND DISCUSSION

**Anaerobic Column Studies:** The anaerobic column was operated in a continuous upflow mode under a variety of conditions. The influence of two different concentrations of influent TCE and two different methanol concentrations (primary substrate) were investigated. In addition, a shock load of TCE was fed to the column for three days which provided valuable data on the ability of the system to handle significantly higher loads.

Methanol and acetic acid were used as the primary substrates to accomplish RD of the TCE. Unless otherwise noted, the concentrations of methanol and acetic acid fed were 751 and 183 mg/L respectively. The composition of a basal salts media utilized as a simulated groundwater was presented earlier. The TCE and primary substrates were fed using a syringe pump while the simulated groundwater was fed from a piston pump. The two flows joined in a manifold at the base of the column. The column was operated at a hydraulic detention time of 48 hours with a feed flow rate of approximately 5 ml/hour. The pH varied from 6.7 to 7.5.

*Influence of TCE concentration.* Two different (1.6 and 5.0 mg/L) feed TCE concentrations were utilized to evaluate effect on performance. Figure 1 presents the TCE mass flow rate (concentration X time) in microgram per hour (ug/h) of TCE in the feed and at the bottom, middle and top ports at a feed TCE concentration of 1.6 mg/L. The 1.6 mg/L TCE concentration corresponded to a volumetric loading rate of 0.326 mg/d L. An average mass removal rate of about 5.5 ug/h TCE was obtained resulting in an average of about 58% overall removal. It is interesting to note (see Figure 1) that most of the TCE removal occurred at the bottom port (the first 100 mm of the column). Only about 1 to 2% additional removal occurred from the bottom to the top port of the column. The mass flow rate of the daughter products leaving the column, 1-DCE and 1,2-DCE, were respectively 0.08 and 0.24 ug/h.

The TCE feed concentration was increased to 5.0 mg/L corresponding to a loading of 1.02 mg/d L. Mass flowrate data are shown in Figure 2 (days 0 to 7). The rate of TCE removal averaged about 14.5 ug/h while overall removal averaged 67%. The mass flow of the daughter produce, 1,1-DCE, fell to essentially zero for a short period after the TCE concentration was increased and then rose to about 0.11 ug/h. The formation rate of 1,2-DCE was essentially unchanged when the TCE concentration was increased averaging about 0.23 ug/h.

Comparing the results of the two TCE feed levels, the mass removal rate and percentage removal of TCE were significantly improved at the higher concentration. The rate of formation of daughter products (DP's) at the two feed levels (ug/h) was approximately the same although the rate of formation 1,1-DCE decreased somewhat at the higher TCE concentration. In addition, the actual concentration of DP's was approximately the same at the two TCE concentrations. For example, the concentration of 1,1-DCE in the column averaged 16.1 and 20.1 ug/L at the TCE feed concentrations of 1.6 and 5.0 mg/L respectively. The concentration of 1,2-DCE averaged 44.4 and 40.7 ug/L at the feed concentrations of 1.6 and 5.0 mg/L respectively.

*Effect of Methanol Concentration.* The effect of the concentration of the primary substrate (methanol) on TCE degradation was investigated at a feed TCE of 5.0 mg/L. During the run shown in Figure 2, the methanol

853

concentration was decreased from 751 to 185 mg/L at the end of day 7. In Figure 2, it may be noted that as the methanol concentration was decreased, the average mass rate of TCE leaving the column rose significantly. The mass rate of TCE removal fell from 14.5 to 11.6 ug/h and the percent removal fell from 67 to 49 %. As would be expected, the concentration of primary substrate has a significant effect on TCE degradation. Even though the lower primary substrate level significantly lowered overall removal of TCE, the mass flowrate of daughter products leaving the column was essentially unchanged.

*Effect of TCE Slug.* A short-term (3 day) shock load (approximately 19.0 mg/L) was allowed to enter the system. The shock translates into a loading rate of 3.813 mg/d L. The excursion lasted for 72 hours. After 3 days, the influent concentration was dropped to 1.6 mg/L as it had been before the shock. In Figure 3, the TCE concentration before and after the shock is presented showing that the column TCE rose to a maximum of around 5.0 mg/L at the bottom port and around 4 mg/L at the top port one day later. The DP's rose and fell roughly in proportion to the TCE.

A rough mass balance was performed in order to estimate removal for the three days the slug was fed to the system and the next 9 days until the system once again produced a constant effluent. It was estimated that the system achieved an overall TCE removal of around 49% in handling the shock load. It appears that the methanogens were able to handle a short term shock which was an order of magnitude higher than they had been receiving with no undue stress on the system. TCE removal returned to normal within a relatively short period (about 7 days) after cessation of the slug load. The system will be subjected to a planned series of shock loads in future runs so as to determine response to much higher CAH concentrations.

**Aerobic Column Studies:** Aerobic mineralization of TCE using methanotrophs and other heterotrophs is being evaluated in an upflow column. Parameters evaluated to date include the concentration of primary substrate (methanol) and an alternative electron acceptor (hydrogen peroxide). The pH varied from 6.8 to 7.2. TCE concentration fed to the column was held at 5.0 mg/L.

Methane and Methanol are being used as the primary substrates to accomplish oxidation of TCE. The TCE and methanol are continuously fed using a syringe pump. The basal salts media (simulated groundwater) was continually fed using a peristaltic pump from two separate reservoirs at the same time. The groundwater (GW) in one reservoir contained the electron acceptor (oxygen or hydrogen peroxide) while the GW in the other reservoir was sparged with methane three times per week to maintain methane in the feed. The GW flow and the flow from the syringe pump merged in a manifold before entering the column. Total flow into the aerobic column was approximately 10.2 mL/h to give a detention 24 hours.

Column operation began using oxygen in the GW as the electron acceptor. Since this source was limited by the solubility of oxygen, the methanol concentration initially fed to the column was limited to 5.0 mg/L so as not to produce anaerobic conditions in the column. In order to improve TCE removal, it was desired to increase the level of methanol added to the column. Limitations with respect to DO level may be overcome by adding an alternative hydrogen acceptor such as hydrogen peroxide

Hydrogen peroxide reacts with methanol as shown in the following equation:

$$3H_2O_2 + CH_3OH \rightarrow 5H_2O + CO_2$$

Since it was desired to increase the methanol concentration to around 30 mg/l, approximately 96 ppm of $H_2O_2$ would be required according to the above noted equation (assuming no synthesis). However, to avoid a possible toxic shock to the microorganisms due to the 96 mg/L of $H_2O_2$, the concentration of peroxide is being raised in 25 mg/L increments per week. To date, the $H_2O_2$ concentration has been raised to 50 mg/L while the methanol concentration in the feed has been increased to 16 mg/L. TCE removal under these conditions has reached about 9 ug/h with overall removal of 20 %. It is anticipated that removals will improve as the concentrations of methanol and peroxide are increased. As expected, the addition of the peroxide increased dissolved oxygen levels leaving the column and the effluent DO is now about 7 mg/L.

As noted earlier, methane (primary substrate) may actually interfere with removal of TCE by competitive inhibition. It was therefore decided to begin pulse feeding of methane by alternating the use of methane sparged and unsparged groundwater every other day.

## CONCLUSIONS

The anaerobic column was operated under a variety of conditions. Parameters investigated included TCE concentration, methanol feed level and response to a short-term TCE shock. TCE feed levels of 1.6 and 5.0 mg/L were evaluated with respect to column performance. It was found that the removal of TCE improved from 58 to 67 % as the feed concentration was increased from 1.6 to 5.0 mg/L. The mass removal rate of TCE increased from 5.5 ug/h at 1.6 mg/L to 14.5 ug/h at the 5.0 mg/L TCE concentration. However, the concentration of daughter products was about the same at both feed TCE levels.

As expected, the feed concentration of the primary substrate (methanol) had a significant effect on TCE removal. As methanol was increased from 185 to 751 mg/L, TCE removal improved from 49 to 67 % respectively. A slug of TCE (about 19 mg/L) was imposed on the system for about three days. It was found that the methanogens were able to handle the shock load and still achieve an overall removal of about 49 % with no apparent toxic effects.

Parameters investigated for the aerobic column included the concentration of primary substrate (methanol) and an alternative electron acceptor (hydrogen peroxide). It was found that the overall removal rate of TCE and percentage removal at the top port reached 9 ug/h and 20% respectively at a fed methanol concentration of 16 mg/L. It is anticipated that TCE removal will continue to improve as the methanol concentration is increased.

## ACKNOWLEDGEMENT

Funds for this research were provided by the U.S. Department of Energy, Office of Fossil Energy, Morgantown Energy Technology Center, Under Contract No. DE-FC21-92M29467. Views expressed in this paper do not necessarily reflect the official position of the U.S. Department of Energy.

## REFERENCES

1. Cuddeback, J.E., W.A.Sack, P.E.Carriere, and C.S. Whiteman. 1994 "In-Situ Bioremediation of Chlorinated Aliphatic Hydrocarbons in Soil and Groundwater," *Proc. of the Twenty-Sixth Mid-Atlantic Industrial Waste Conf.*, Technomic Publ. Co. Inc., Lancaster PA, pp 131-140.

2. Sims, J.L., R.C. Sims, and J.E. Mathews 1990 "Approach to Bioremediation of Contaminated Soil", *Hazardous Waste and Hazardous Materials*, V. 7, pp 117-149.

3. Freedman, D.L. and J.M. Gossett. 1989 " Biological Reductive Dechlorination of Tetrachlorethylene and Trichlorethylene to Ethylene Under Methanogenic Conditions", *Appl. and Env. Microbiol.*, pp 2144-2151.

4. Kaster, M. 1991 "Reductive Dechlorination of Tri- and Tetrachloroethylenes Depends on Transititon from Aerobic to Anaerobic Conditions", *Appl. and Environ. Microbiology*, Vol 57, No. 7, pp. 2039-46.

5. Semprini,L., G.D. Hopkins, and P.L McCarty. 1994. "A field and modeling comparision of in situ transormation of TCE by methane utilizers and phenol utilizers," in *Bioremediation of Chlorinated and Polycyclic Aromatic Hydrocarbon Compounds*, Lewis Publishers, Ann Arbor, MI, pp. 248-254.

6. Broholm, K., T.H. Christensen, and B.K.Jensen. 1993. "Different Abilities of Eight Mixed Cultures of Methane-Oxidizing Bacteria to Degrade Trichloroethylene." *Water Resources*, Vol. 27, p. 215-224.

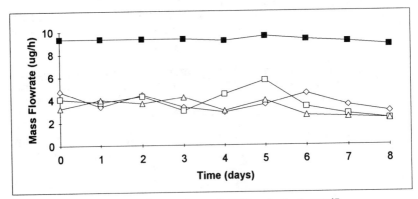

Figure 1.  Mass Flowrate of TCE at 1.6 mg/L

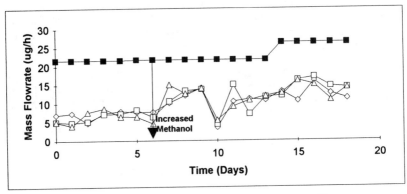

Figure 2. Mass Flowrate of TCE at 5.0 mg/L

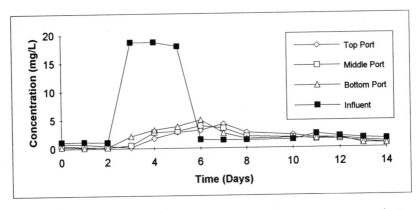

Figure 3.  Concentration of TCE During Shock Period

# INDUSTRIAL SOIL VAPOR EXTRACTION ENHANCED BY HOT AIR INJECTION

WILLIAM H. MAC NAIR, JR., PE
Air Products and Chemicals, Inc.
7201 Hamilton Blvd.
Allentown, PA 18195-1501

KENT V. LITTLEFIELD, PG
R. E. Wright Environmental, Inc.
3240 Schoolhouse Road
Middletown, PA 17057-3595

## INTRODUCTION

Soil vapor extraction (SVE) is a soil remediation technique that takes advantage of the physical properties of volatile organic compounds (VOCs) by vacuuming them out of the ground in the vapor phase. In many applications, SVE is more flexible, rapid, and cost-effective than competing technologies.

At an industrial site in New Jersey, SVE enhanced with hot air injection was used to clean up xylene contamination in soil at an active tank farm. The remediation was complicated by construction over the area and by gasoline contamination in a non-aqueous phase liquid (NAPL), originating from a neighboring site, underlying the area. Air injection facilitated control over the size of the SVE treatment area, and the heating of the injected air resulted in a rapid cleanup. Recycling the SVE vacuum pump exhaust as injection air eliminated the need for an air emissions permit for the system and provided the air heating.

The project was required because the xylene residual was considered a Solid Waste Management Unit (SWMU) under RCRA and because the site was needed for construction of a new hazardous waste storage tank. Before the new tank could be constructed, the residual from the spill had to be evaluated and cleaned up to New Jersey Department of Environmental Protection (NJDEP) specifications. The work phases included:

1. Investigating the depth of xylene infiltration into the soil.
2. Conducting a vapor extraction pilot test and designing a Vapor Extraction System (VES).

3. Constructing the VES.
4. Operating and monitoring the VES from initial start-up through attaining clean-up standards.

## SVE TECHNOLOGY

SVE takes advantage of the partitioning properties of volatile organic compounds (VOCs). All VOCs partition to some extent when in contact with air. The amount of the compound that exists in the vapor phase depends on its volatility, which depends on the chemical's vapor pressure. Contaminants with a higher vapor pressure will exist in the vapor phase to a greater degree. Since vapor pressure increases with temperature, heating of the vapor space will increase the VOC in the vapor phase. Other factors that determine the amount of a chemical that exists in the vapor phase include its Henry's Law constant and the soil permeability, moisture content, and organic carbon content.

Soil vapor extraction removes the vapor phase in the soil by promoting air flow through the soil by generating a vacuum. Vacuum is applied by installing extraction wells in the contaminated area and connecting these wells through a header system to a vacuum pump. The exhausted air must typically be treated to remove the organic contaminants. The predominant treatment processes available are vapor combustion, catalytic combustion and carbon adsorption.

## SITE CHARACTERIZATION

The key information that must be obtained in any site characterization is the volume and chemical analysis of the contaminated soil. At this site, an additional issue was the delineation between the xylene spill and the xylene in the gasoline NAPL underlying the site. The shallow xylene contamination was the result of a release from an aboveground storage tank that occurred in the mid 1980's. The water table with the gasoline NAPL is about 20 feet below ground surface. The NAPL is being contained by an area-wide pumping system.

Since gasoline contamination includes xylene, it was difficult to distinguish between the two contamination sources. The "xylene" product that was spilled contained 87% xylene, 12% ethylbenzene and 1% toluene. The floating NAPL contained 17% toluene, 8% benzene, 23% ethylbenzene and 52% xylene. Considering this comparison, toluene was identified as the key compound delineating the gasoline contaminated soil from the soil contaminated by the surface xylene release.

Results from a typical soil boring are shown in Figure 1 [1]. Soil samples taken at less than four feet below grade level were relatively clean because this soil had been replaced when the failed above ground storage tank was removed. Xylene and ethylbenzene were the predominant chemicals found immediately below this level  There was a zone of no BTEX between 10.5 and 13.5 feet in this soil boring. Toluene was only detected at depths greater than about 14 feet below grade surface. The soil contamination below this level was attributed to

860

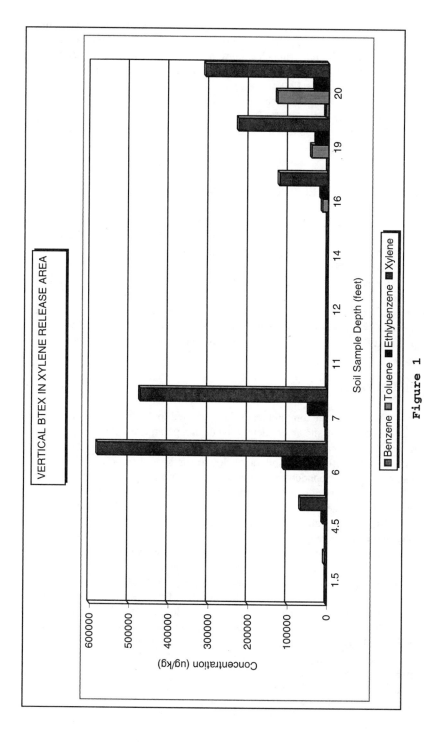

Figure 1

the NAPL. Cleanup responsibility for the spill was limited to soils above this level. The contaminated soil was determined to be in a 24 by 40 foot area. The total volume of soil requiring cleanup was about 7000 cubic feet (260 cubic yards). The maximum xylene concentration found in any sample in the contaminated area was 22,000 mg/kg.

Since xylene and ethylbenzene have high vapor pressures, SVE was considered prime candidate for corrective action. Pilot testing and design of a VES were conducted so that it could be compared to soil excavation and off-site disposal.

## REGULATORY REQUIREMENTS

The corrective action for this spill site was handled under a Memorandum of Agreement (MOA) with NJDEP. All project plans and reports were reviewed by NJDEP under this MOA. After the site characterization, the data was reviewed with NJDEP to determine clean-up requirements. NJDEP required that xylene levels be reduced to less than 10 mg/kg in the soil. NJDEP also agreed that the use of toluene as an indicator of the petroleum NAPL was appropriate. Therefore, the clean-up was limited to the zone from the surface down to 14 feet below grade level. NJDEP required an air emission source permit if SVE was utilized and the exhaust was vented to atmosphere.

## PILOT TESTING

A vapor extraction pilot test was conducted to determine whether SVE was feasible at the site and to determine design parameters. The test was used to determine soil permeability, the concentration of BTEX in the extracted vapor, and the radial influence of soil vapor flow from the extraction vent.
The pilot test was conducted by constructing four vacuum monitoring wells and extraction vents (wells) at two of the soil sampling locations. During the test, vacuums were measured in inches of water with a magnehelic gauge at all of the wells. In addition, air velocity was measured at the extraction wells. Measurements were generally taken at 5 to 10 minute intervals. The concentration of organic vapors in the blower exhaust was measured using a flame ionization detector. Testing of system response at vacuums ranging from 20 to 98 inches of water column was completed. A plot of vacuum pressure versus distance from extraction well is shown in Figure 2. This plot shows a logarithmic relationship between the vacuum and the distance from the extraction well. A vacuum of 0.1 inch or more indicates that an extraction well has an influence at that point [2]. This analysis indicated that SVE was certainly feasible at this location and that a single well could probably yield adequate extraction over the entire remediation area. However, since controlling air flow patterns was critical to cleaning up only the spill area xylene, a more detailed flow analysis was done during the design.

862

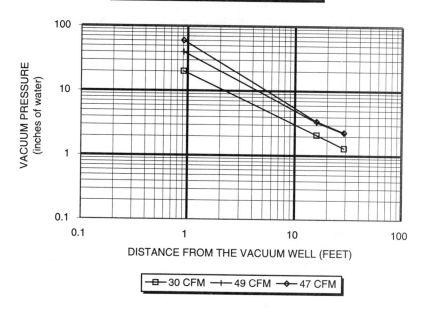

**Figure 2**

The data collected from the pilot tests was used to determine the soil permeability. The equation used for determining soil permeability was:

$$P = \frac{Q}{4\pi m(k/\mu)}\left[-0.572 - \ln\left(\frac{r^2\varepsilon\mu}{4kP_{Atm}}\right) + \ln(t)\right] \qquad (1)$$

Where:

| | |
|---|---|
| Q | = volumetric vapor flow rate from extraction well |
| m | = stratum thickness |
| P | = "gauge" pressure measured at distance r and time t |
| k | = soil permeability to air flow |
| $\mu$ | = viscosity of air = $1.8 \times 10^{-4}$ g/cm-s |
| $\varepsilon$ | = air-filled soil void fraction |
| t | = time |
| $P_{atm}$ | = ambient atmospheric pressure = 1.0 atm = $1.013 \times 10^6$ [3]. |

Equation 1 predicts a plot of P Vs ln(t) should be a straight line with a slope A equal to:

$$A = \frac{Q}{4\pi m(k / \mu)} \qquad (2)$$

The y-intercept is not needed to calculate k if Q and m are known.

Using this method to evaluate the pilot test data, the soil permeability was determined to range from 3.9 to 10.7 Darcy (0.0038 to 0.0098 cm/sec). These values were used in a steady state radial flow model during design

BTEX concentrations were measured in the extracted vapors at several hours into each pilot run. These concentrations ranged up to 1470 mg/l.


## SYSTEM DESIGN

The vapor extraction system design requires the evaluation of vent layout, system vacuum, air flow rates, and exhaust vapor treatment. The worst case estimate for the xylene spill was 500 gallons (about 3800 pounds). The dimensions of the contaminated soil volume were 24' by 40' by 14' depth.

In order to maintain a nearly horizontal gas flow, which would minimize the vaporization of constituents in the NAPL, a system of 4 extraction wells and 6 injection wells was designed using the Airflow model [4]. A schematic of the process is shown in Figure 3. The injection wells were placed on site periphery while the extraction wells were placed at the center of the contaminated area. Two of the extraction wells were screened from 5-8 feet and two were screened from 8-14 feet. The air injection wells were screened from 8-14 feet. The well casing was 2-inch diameter Schedule 40 galvanized steel. The wells were piped to a 7.5 horsepower blower (vacuum source). A moisture separator ("knock-out pot") and a granular activated carbon (GAC) adsorber were located between the wells and the blower. The GAC system contained two 1000 pound canisters of carbon.

To eliminate the need for a NJDEP air permit, the air from the blower was recirculated to the injection wells. In addition to simplifying regulatory constraints, the recirculated air became heated as it was pumped through the system. This heated air increased the vapor pressure of the xylene and improved its removal efficiency.


## SYSTEM OPERATION

System start-up occurred on June 10, 1993. The system was monitored continuously for the first four days, biweekly for the following two weeks, and weekly for the next three weeks of operation. Monitoring consisted of obtaining airflow, temperature, and pressure readings, as well as collecting air quality samples for analysis. For the first two weeks of operation, flow was relatively constant at 140 cfm and vacuum was 50 inches of water. The initial injection air

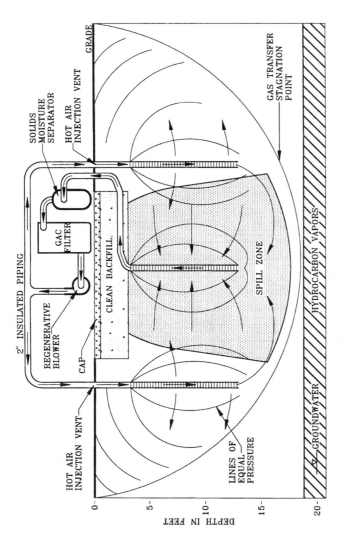

SCHEMATIC OF SOIL VAPOR VAPOR EXTRACTION
AND HOT AIR INJECTION PROCESS

Figure 3

temperature was 225°F. Fluctuation in influent and effluent temperature during this period was due primarily to the atmospheric temperature change. After this initial two week operation, the two deeper interior extraction wells were shut off because of an increase in gasoline-like hydrocarbons. This resulted in an increase in the vacuum to 73 inches of water and a decrease in air flow to 103 cfm. The injection temperature increased to between 270 and 300 degrees Fahrenheit.

Air quality samples were taken from the activated carbon system influent and effluent. During the initial week of operation, air samples were analyzed on-site using a portable GC. Xylene concentrations decreased sharply from 5,050 ppm after 11 hours to 0.6 ppm after 33 hours. Thereafter, the decrease in influent xylene concentrations became more gradual, and concentrations declined to 0.205 ppm after 14 days of operation. These influent air samples indicate that 217 pounds of xylene were removed during the first month of operation. The activated carbon effluent air samples indicated that the carbon was spent after 5 days of operation. The 1000 pound primary carbon canister was replaced at this time. The rapid decrease of xylene concentration and loading rate indicated that the xylene was remediated from the soil by the vapor extraction system. Soil samples showed that all BTEX compounds were below the analytical detection limit (2 ppb). The clean-up was complete. The system was shut off on July 28, 1993.

## SYSTEM PERFORMANCE -- PREDICTED VS ACTUAL

The system performance was predicted using "Venting" software [5]. The program uses vapor flow equations described in Johnson et al [3]. The chemical mass balance over time is defined by solving the following equation:

$$\frac{dM_i}{dt} = -\eta Q C_i^{eq} \tag{3}$$

Where:
$M_i$     = the total number of moles of component i in the soil [mole]
$Q$     = the total gas flow rate through the contaminated zone [$L^3 T^{-1}$ ]
$C_i^{eq}$     = the equilibrium molar gas phase concentration of species i [mole $L^{-3}$]
$\eta$     = an efficiency factor to account for nonequilibrium effects.
The solution for this equation includes vapor pressure, temperature and soil:organic partitioning factors.

A comparison of actual and predicted treatment times, with and without hot air injection, is shown in Figure 4. The venting software predicted a clean-up time of about 90 days when the spill quantity was estimated to be 500 gallons of xylene. Since the actual quantity removed was 217 pounds, the clean-up time was re-estimated using that amount. The actual time was faster than the

866

predicted time, probably due to re-injection effects that were not modeled. Many factors must be estimated when using this model.

## COSTS

The cost of the soil vapor extraction system including site investigation, pilot testing, design, equipment installation and operation was approximately $200,000. The cost for transportation and disposal at a hazardous waste landfill would be $200,000 based on unit costs of $3.50 per loaded mile for transportation and $250 per ton for disposal. This price does not include any waste stabilization or treatment prior to landfill. The close proximity of other structures would have made soil removal extremely difficult. SVE proved to be a cost effective solution for the remediation of xylene contamination at this site.

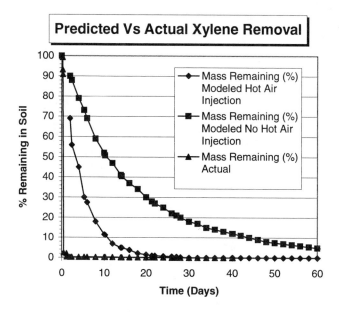

**Figure 4**

**REFERENCES**

1. R. E. Wright Associates, Inc. 1993. "Progress Report at the Air Products and Chemicals, Inc., Chemical Group Paulsboro, New Jersey Facility." 3240 Schoolhouse Road, Middletown, PA 17057-3595.

2. Keech, D. A., 1989. Surface Venting Research and Venting Manual by the American Petroleum Research Institute. Presented at the Workshop on Soil Vacuum Extraction, R. S. Kerr Environmental Research Laboratory. Ada, OK. pp. 27-28.

3. Johnson, P. C., Stanley, C. C., Kemblowski, M. W., Byers, D. L., Colthart, J. D. 1990. "A Practical Approach to the Design, Operation, and Monitoring of In Situ Soil-Venting Systems," Spring 1990 Ground Water Monitor Review. pp. 159-178.

4. Franz, T, Guiguer, N. Airflow, Steady-State Radial-Symmetric Vapor Flow Simulation Model, Version 2. Waterloo Hydrogeologic Software. 200 Candlewood Crescent, Waterloo, Ontario, Canada N2L 5Y9.

5. Environmental Systems & Technologies, Inc. 1991. Venting, A Program for Estimating Hydrocarbon Recovery from Soil Vacuum Extraction Systems, USER'S GUIDE, Version 2.0. P. O. Box 10457, Blacksburg VA 24062-0457.

# THE DESIGN OF AN IN-SITU SPARGING TRENCH

Michael C. Marley
Stewart H. Abrams, P.E.
Edward X. Droste, P.E.
   Envirogen, Inc.
   Princeton Research Center
   4100 Quakerbridge Road
   Lawrenceville, NJ 08648

## INTRODUCTION

In a high percentage of remediation programs, it is necessary to prevent off site migration of dissolved phase volatile organic compounds (VOCs). Traditionally, this has been achieved through hydraulic control (pump and treat). This type of control can be costly with regards to treatment of the pumped contaminated water prior to discharge, and system operation and maintenance.

An alternative approach to the prevention of off-site contaminant migration is the use of in-situ interception/treatment techniques. One such technique is the use of zero valence metal interception trenches. Another relatively new technique is the use of an interception sparging trench. Dependent on the contaminants of concern, the sparging trench may be designed to mitigate off-site migration through stripping of the target contaminant from the ground water or through the destruction of the target contaminant by in-situ biodegradation. Since, in either case, ground water is not extracted, no costly ground water treatment equipment, discharge permits or maintenance costs are required. If the mechanism of contaminant control is stripping the sparged air may be captured via a soil vapor extraction and treatment system or may be discharged directly to the atmosphere.

The purpose of this paper is to present the general design of a sparging trench system. The main considerations in the design of a sparging trench are: ensure ground water will flow through, not around the trench; ensure target VOCs can be removed to predetermined clean-up levels; and ensure that the trench manifold is designed to provide a relatively uniform air flow over the trench length. Figure 1 shows a conceptual layout of an interception sparging trench.

In order to achieve the above stated design criterion, the physical trench system parameters (trench dimensions and fill material properties) and operating parameters (air injection flow rates, sparging manifold radius, port size and manifold pipe lengths) must be established. In many cases, a passive or active trench SVE system design may be required. However, the design of a sparging trench SVE system is not included in this paper.

The design methodology presented is based on target contaminant removal via volatilization/stripping.


## Trench Design

The trench design is based on the principles of multi-fluid phase flow and the principles of mass transfer. The problems domain is defined as the water saturated soil zone within the trench. On the onset of sparging, the degree of water saturation in the trench material is reduced as a result of the presence of air. The van Genuchten (1980) formulae are then applied to determine the effective hydraulic conductivity of the trench material based on the change in the degree of water saturation.[1] The effective hydraulic conductivity of the trench material is less than the saturated hydraulic conductivity and will be dependent on the type of material and the air flow rate and distribution in the trench. To ensure that ground water will flow through and not around the trench, the net effective hydraulic conductivity of the trench material must be greater than the saturated hydraulic conductivity of the native soils.

The air flow rate into the trench must not only be designed low enough to ensure that the net effective conductivity of the trench material is adequate, but the flow rate must be sufficient to strip the target contaminant from the ground water during its passage through the trench.

Two general methods were used in the evaluation of the stripping of the target contaminants from the dissolved phase (dissolved phase contamination only is assumed since we are dealing with the plumes leading edge). The first method considers mass transfer across an air water interface (bubble transfer) and the second method assumes instantaneous vapor phase - dissolved phase equilibrium (Henry's Law) is valid.

Consider a typical trench as shown in Figure 1. The trenches effective depth (from the water table to the bottom of trench) is h and the width is L. If the theoretical rate of removal of VOCs by sparging in the trench is greater than the rate of VOC's entering the trench, VOC removal will be achieved. The condition for this mass balance (on a unit trench length basis) is:

$$V \ (LhC_o - LhC_t) \geq hLC_oV \tag{1}$$

Where

| | | |
|---|---|---|
| L | = | Width of trench |
| h | = | Depth of trench |
| $C_o$ | = | Initial or inlet concentration |
| $C_t$ | = | Concentration at time t (or outlet concentration |
| V | = | Water flow velocity into the trench |
| t | = | Hydraulic residence time in trench |

According to Sellers and Schreiber (1992),[2] the dissolved VOC concentrations within a sparging zone at any given time can be defined as:

$$C_t = C_oe^{-Bt} \tag{2}$$

Where

| | | |
|---|---|---|
| B | = | Empirical parameter as defined by Sellers and Schreiber (1992) [2] |

B is a function of many factors which are difficult to measure (e.g., diffusion coefficient of the VOCs in water, diffusive distance around an air bubble, bubble terminal rise velocity and average effective surface to volume ration of a bubble). For the calculations presented herein an empirical B value was used based on existing literature values.

Although it is well understood that sparging in native soils results in a random distribution of air channels, bubble theory may be more appropriate in the design of a sparging trench.[3] This is due to the fact that the trench backfill material will generally be coarse grained (greater than 2 mm particle diameter) which can result in bubble formation and movement in soils [4].

Combining equations (1) and (2) with the van Genuchten formulae (1980) [1], the equation that defines the conductivity of the trench material and the required removal efficiency is as follows

$$K_1 \leq K_2 \leq ((-LBK_{max}h)/(q \ ln(C_C/C_O))) \ (1 - (C_C/C_O)) \tag{3}$$

Where

| | | |
|---|---|---|
| $K_1$ | = | Hydraulic conductivity of native soils |
| $K_2$ | = | Effective Hydraulic conductivity of trench backfill materials |

871

$K_{max}$ = Maximum effective hydraulic conductivity of the trench material as determined by the van Genuchten formulae

$q$ = Ground water flow rate through the trench

$C_C$ = Target VOC concentration level in the trench discharge

When the design satisfies the conditions of equation (3), the system will achieve the required VOC removal rates and will allow ground water to pass through the trench. From equation (3) we can derive the following observations

1) The trench material's effective hydraulic conductivity K2 is in direct proportion to the trench dimensions.
2) K2 is inversely proportional to the injected air flow rate and the aquifer's ground water flow rate.
3) K2 has a non-linear relationship with VOC concentrations.

Figure 2 illustrates a typical relationship between the air injection rate and the trench materials effective hydraulic conductivity as described in equation (3). Table 1 lists the input parameters.

Table 1

| | |
|---|---|
| Sparging parameter B (from literature) | 0.02 |
| Saturated trench depth (ft) | 4.92 |
| Trench material porosity | 0.25 |
| Irreducible saturation value of trench material | 0.12 |
| van Genuchten empirical soil parameter (from literature) | 1.1 |
| Allowable effluent VOC concentration | 5 |
| Influent VOC concentration | 7000 |
| Hydraulic gradient through trench | 0.004 |

As illustrated in Figure 2, as the air injection flow rate increases, the trench material's effective hydraulic conductivity decreases in a non-linear fashion. For example, if the trench width is 1.6 ft (0.5 m), and the trench materials saturated hydraulic conductivity is 0.0114 ft/min (5 meters per day), the air injection flow rate will need to be less than 75 cfm (0.035 cubic meter per second). If the trench's effective hydraulic conductivity is less than that of the native soils, ground water may be diverted around the trench.

For given trench dimensions (i.e., saturated height and width) the trench backfill material can be designed. Figure 3 illustrates the effective and saturated hydraulic conductivity of a trench material for varying air injection rates. Assuming that the native soils hydraulic conductivity is 0.0114 ft/min and the hydraulic gradient is 0.004, Figure 3 shows that the trench material's saturated hydraulic conductivity ($K_S$) would have to be greater than 0.934 ft/min (410 m/d) when the air injection flow rate is greater or equal to 160 cfm (0.08 cubic meter per second) to ensure ground water will flow through the trench.

As the $K_S$ required for a given trench design varies with the trench operating conditions, trench dimensions and air injection flow rates can be varied to achieve the most cost effective design.

When the design $K_S$ is selected, the trench material particle size distribution can be determined through the relationship between $K_S$ and $D_{10}$ [5].

$$K_S = 100\ D_{10}^2 \tag{4}$$

Where $D_{10}$ is the soil particle diamter for which 10% of the soil weight is finer.

The second method of evaluating the VOC removal rate from a sparging trench is based on equilibrium partitioning [6]. VOC removal rates are based on Henry's Law relationships. The governing equations are as follows.

$$E = S/(1 + S) \tag{5}$$

Where

| | | |
|---|---|---|
| E | = | Theoretical fractional efficiency of the sparging trench |
| S | - | Dimensionless sparge number |

$$S = HR_g/(RTyzv) \tag{6}$$

Where

| | | |
|---|---|---|
| H | = | Henry's Law constant for the target VOC |
| $R_g$ | = | Air injection rate |
| R | = | Universal gas constant |
| T | = | Temperature degrees Kelvin |
| yz | = | Cross sectional area of trench perpendicular to ground water flow |
| v | = | Darcy velocity of ground water through trench |

## Air Distribution System Design

One of the assumptions used in the trench design is that the injected air is distributed relatively uniformly throughout the trench. To achieve uniform air flow throughout the trench manifold, an analysis procedure based on the energy equation for fluid flow through a pipe is used. The main assumptions for this analysis are: 1) Air discharge ports on the sparging manifold are circular and oriented at 90 degrees to the pipe axis; 2) Manifold pipe friction losses are negligible; and 3) manifold cross sectional area is constant. The fluid velocity in any port along the manifold pipe can be described as [7]:

$$V_{Li} = V_{L1}(B_1/B_i)^{0.5} \tag{7}$$

and

$$V_{Li} = (Q_a/A_{Li}(B_1)^{0.5}) \\ (\text{summation } i=1 \text{ through } n \ (1/B_i)^{0.5})^{-1} \tag{8}$$

Where

| | | |
|---|---|---|
| $V_{Li}$ | = | The velocity in port i |
| $A_{Li}$ | = | The area of port i |
| $B_1$ | = | Head loss component for port 1 |
| $B_i$ | = | Head loss component for port i |
| $Q_a$ | = | The total manifold air flow rate |
| $n$ | = | Number of ports in the manifold |

Equations (7) and (8) were incorporated into a numerical program which iterates through the equation's using sets of input parameters to produce the optimal air flow distribution.

Figures 4 and 5 illustrate simulation results for a typical sparging manifold. The simulation assumes a 50 ft long, 2 inch diameter manifold pipe (RM) with 1000 uniformly distributed ports (approximately 40 ports per linear foot). The sparging manifold is installed in the trench 3 feet below the ground water table. An air entry pressure of 0.2 ft of water into the trench material is assumed. Assuming an air injection flow rate of 24 cfm (0.4cts), Figure 4 shows that the radius of the porthole has to be less than 0.036 inches to allow air flow throughout the manifold pipe length. If the porthole radius is greater than 0.036 inches, the air injection pressure will be too small to overcome the entry and static water pressure throughout the manifold pipe length. Figure 5 demonstrates the air velocity distribution along the sparging manifold pipe under the simulation conditions. Increasing the air flow will increase the ability to provide flow throughout the length of the manifold pipe. However, increasing the air flow will also decrease the effective hydraulic conductivity of the trench backfill material. In designing the total system, the equations provided can be used to design an optimal system solution.

## SUMMARY

The design of air sparging trenches in conceptually simple. However, without consideration of the key design parameters, trenches can and have been designed that will not achieve their remedial objectives. A general methodology for in-situ sparging trench design is presented. The main considerations for the design of a sparging trench are that ground water should flow through, not around the trench; target VOCs are required to be removed to predetermined clean-up levels; and the air flow needs to be distributed uniformly along the length of the trench.

Mathematical equations have been developed and utilized to demonstrate the relationships between the trench dimensions, properties of the trench materials and the air injection rates. This design process can be utilized to cost effectively evaluate the most efficient system design to meet site specific remedial objectives.

# REFERENCES

1 van Genuchten, M. Th.; *A Closed-form Equation for Predicting The Hydraulic Conductivity of Unsaturated Soils*; Soil Science Society of America Journal, Vol. 44, No. 5, September-October 1980.

2 Seller, K.L. and Schreiber, R.P.: *Air Sparging Model for Predicting Groundwater Cleanup Rate*; Conference Proceedings of the 1992 Petroleum Hydrocarbons and Organic Chemicals in Groundwater; Prevention, Detection, and Restoration, The Weston Galleria, Houston, Texas November 4-6, 1992.

3 Marley, M.C., Droste, E.X., and F.J. Cody; *Mechanisms that Govern the Successful Application of Sparging Technologies*; Proceeding of the Air & Waste Management Association's 87th Annual Meeting and Exposition, June 1994.

4 Ji, W., Dahmani, A., Ahlfeld, D., Lin, J. and Hill, E., Jr.; *Laboratory Study of Air Sparging; Air Flow Visualization*; Ground Water Monitoring and Remediation, Vol. 13, No. 4, 1993.

5 Lambe T.W. and Whitman, R.V.; *Soil Mechanics*; John Wiley & Sons, Copyright 1969, Page 290.

6 Pankow, J.F., Johnson, F.L. and Cherry, J.A.; *Air Sparging in Gate Wells in Cutoff Walls and Trenches for Control of Plumes of Volatile Organic Compounds (VOCs)*; Ground Water, Vol 31, No. 4, July-August 1993.

7 Hudson, H.E., Jr., Uhler R. B. and Bailey, R.W.; *Dividing-Flow Manifolds With Square-Edged Laterals*; Journal of the Environmental Engineering Division, ASCE, Vol. 105, No. EE4, August, 1979, Page 745-755.

8 Chaudhry, F.H. and Reis, L.F.; *Calculating Flow in Manifold and Orifice Systems*; Journal of the Environmental Engineering Division, ASCE, Vol. 118, No.4, August, 1992, Page 585-596.

Key Words

Air Sparging
Ground Water Remediation
Remedial Design

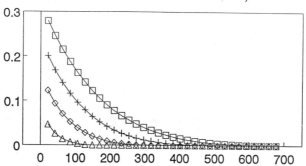

FIGURE 2

Relationship Between Qa and Ke
(Materials Saturated K=0.0114 ft/min)

Air Injection Flow Rate(cfm)

□ Trench Width=6.5'  + Trench Width=4.9'
◇ Trench Width=3.3'  △ Trench Width=1.6'

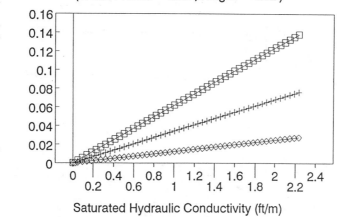

FIGURE 3

Relationship Between Ks and Ke
(Trench Width = 3.28'; Height = 4.92')

Saturated Hydraulic Conductivity (ft/m)

□ Qa=60 cfm  + Qa=100 cfm  ◇ Qa=160 cfm

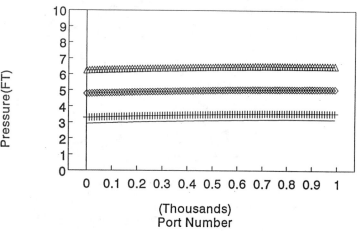

FIGURE 4
Pressure Distribution Along Pipe
(Qa=0.4 cfs, Rm=2", 1000 portholes)

—Rp=0.036"  + Rp=0.035"  ◇ Rp=0.032"  △ Rp=0.03"

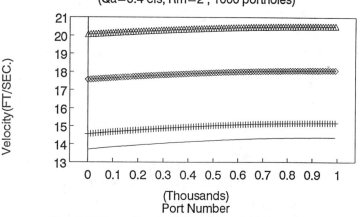

FIGURE 5
Velocity Distribution Along Pipe
(Qa=0.4 cfs, Rm=2", 1000 portholes)

—Rp=0.036"  + Rp=0.035"  ◇ Rp=0.032"  △ Rp=0.03"

g:\tech\papers\ngwa94f4.wk1

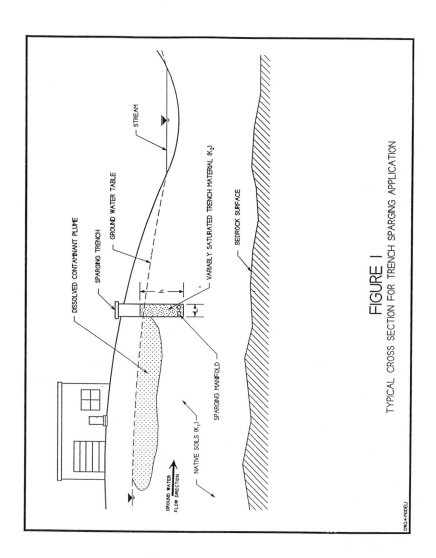

## FIGURE I

### TYPICAL CROSS SECTION FOR TRENCH SPARGING APPLICATION

DWG.—MODEL1

# INSITU VACUUM EXTRACTION/BIOVENTING OF A HAZARDOUS WASTE LANDFILL

DEBORAH M. HEUCKEROTH
Dames & Moore
2325 Maryland Rd.
Willow Grove, PA 19090

MICHAEL F. EBERLE AND MICHAEL J. RYKACZEWSKI
Dames & Moore
2325 Maryland Rd.
Willow Grove, PA 19090

## ABSTRACT

Insitu vacuum extraction (VE)/bioventing of volatile and semivolatile organic compounds (VOCs and SVOCs) was performed at an inactive hazardous waste landfill in the northeast United States. A VE system was used to extract VOCs from a horizontal well (trench), thereby delivering oxygen to the subsurface to biodegrade VOCs and SVOCs insitu. In areas where oxygen concentration increases were observed in the subsurface, microbial populations increased by orders of magnitude, and temperatures increased from 18°C to as high as 27°C. In these areas, up to 99% reductions were observed in specific VOC and SVOC concentrations after 3 months of operation.

Based on system monitoring, the remediation of soil contaminants was approximately 90% attributable to insitu biodegradation and 10% attributable to volatilization.

## INTRODUCTION

The use of VE and bioventing was approved by the state regulatory agency to remediate an inactive landfill, comprised of a complex mixture of pharmaceutical intermediates, finished pharmaceutical products, research wastes, filter cakes, still bottoms, and oils. A VE system was used to extract vapors from a trench and to deliver oxygen to the subsurface. The constituents of concern with the highest concentrations relative to the site-specific regulatory cleanup goals were benzene (0.756 mg/kg goal), phenol (0.33 mg/kg goal), and 4-methyl phenol (1.8 mg/kg goal). Other constituents of concern included chlorinated VOCs and polycyclic aromatic hydrocarbons. The total organic carbon (TOC) concentration in the soils ranged from 2 to 5%. In soils at greater than 1.8 m below ground surface (bgs), where the process was monitored, 75 to 98% of the TOC was solvent extractable, which indicates that the source of TOC for the microbes was primarily landfill contamination. Both landfill materials (fill) and the underlying silt soils were addressed by the process; however, the fill is the primary focus of this paper. The silt is submerged by seasonally high groundwater for approximately 3 months per year. This paper addresses findings from pilot study operations from September 1993 through March 1994. VE and bioventing were selected to remediate VOCs and SVOCs in the landfill. The findings presented herein provided the design basis for a full-scale VE/bioventing system which is currently operational at the site.

## PROCEDURES

A rotary lobe blower was operated at approximately 250 cfm (7 $m^3$/min), 8 to 12 h per day typically, extracting from a trench to evaluate the use of VE in the fill and silt soils, and to deliver oxygen to the subsurface to evaluate the use of bioventing. Operations began September 22, 1993, and after 2.5 months, system operation was discontinued for one month during December/January due to high groundwater conditions and mounding of the groundwater in the vicinity of the extraction trench. Operations restarted in the latter half of January 1994. Figure 1 illustrates the process and the monitoring points.

Two parallel trenches were similarly constructed and located approximately 18 m apart. The trenches are approximately 15 m long and extend into the silt, which underlies the more permeable fill. Liner and geotextile layers were installed to provide vapor seals to prevent air from migrating from the surface through the trench itself, i.e., short-circuiting. One trench was used as an extraction trench and the other as an air inlet trench.

Vapor monitoring probes were installed at 3 depths (using separate boreholes for each depth) and at 11 locations to monitor induced vacuum, oxygen concentrations, oxygen uptake rate, and carbon dioxide production rate. Induced vacuum was monitored at all of the probes on 10 different days during system operations to evaluate the area of influence of the VE system. Oxygen concentrations were monitored at all of the probes approximately weekly to evaluate the distribution of oxygen in the subsurface as a result of bioventing operations.

Oxygen uptake and carbon dioxide production measurements were performed on November 4, 1993 at each vapor monitoring probe. After the VE unit operated for an extended period of time to saturate the subsurface with oxygen, the unit was turned off, and oxygen and carbon dioxide concentrations were

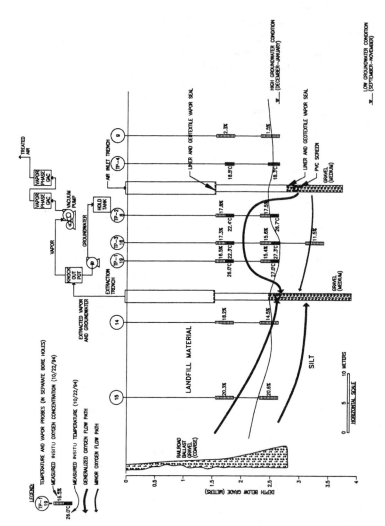

Figure 1. Cross section of extraction trench, air inlet locations, and system monitoring probes. Representative oxygen concentrations and temperatures in the subsurface are indicated.

monitored periodically during a 4.5-hour (or greater) time period. The oxygen concentrations for each probe were plotted versus time to determine the rate of change in oxygen during the monitoring period. The oxygen uptake rate was calculated for each probe location by determining the slope of the line for oxygen concentrations greater than 5%. This rate is considered indicative of the overall rate of microbial activity, since microbial activity generally decreases significantly or ceases below 5% oxygen concentration [1].

Temperature probes were installed at 2 depths (using separate boreholes for each depth) and at 10 locations to monitor subsurface temperatures. Temperatures were measured 2 to 3 times per week during system operations, and was used as an indicator of microbial activity.

Soil samples were collected before system startup, and 5 times thereafter. The samples were collected from 2 locations between the trenches to evaluate the area which was provided the most oxygen. At each location, samples were collected from the fill at 1.8 m depth bgs and from the silt at 3 m depth bgs. Samples were analyzed for VOCs by EPA Method SW846, SVOCs by EPA Method 8270, microbes, nutrients, and pH. Total heterotrophic microbes and VOC- and phenol-degrading microbes were analyzed using standard methods for microbial enumerations [2].

VE/bioventing system flowrates, pressures, temperatures, and vapor concentrations were monitored 2 to 3 times per week. These data were used to estimate mass flowrates, and to determine the need for carbon changeouts. Benzene concentrations in the vapor were measured on site using a gas chromatograph.

## PROCESS MONITORING RESULTS

The system extracted approximately 900 kg of VOCs during 1,072 hours of extraction over 6 months, with 90% of the VOCs extracted during the first week of operation. The VOCs consisted primarily of benzene; toluene, 3-methyl thiophene, ethyl ether, methane, and mercury were also detected in the extracted vapor. In addition, the system extracted approximately 30,000 kg of carbon dioxide over 6 months at a relatively consistent rate during system operation.

The VE area of influence of the trench was evaluated by measuring induced vacuums at the probes at an applied vacuum of 84 mm Hg. The area of influence was extrapolated to be approximately 3,500 m$^2$, based on a minimum induced vacuum of 13 mm of water at the probes. Oxygen concentrations at the probes indicated that the effective oxygen delivery area of influence was approximately 1,500 square meters, based on a minimum oxygen concentration of 5% at the probes. Subsurface oxygen concentrations were highest between the extraction trench and the air inlet locations, as Figure 1 shows.

Temperatures in the oxygen area of influence increased significantly during operations, which is indicative of microbial activity [3]. Figure 2 illustrates temperatures measured over time at a probe between the trenches and at a probe outside the oxygen area of influence. The outlying temperature probe

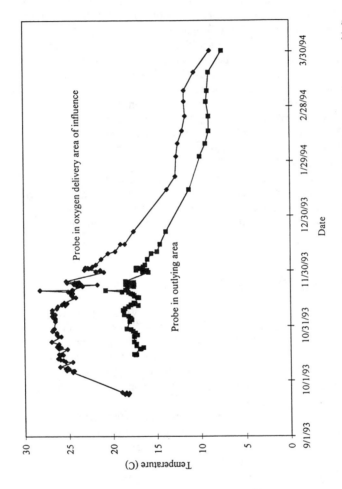

Figure 2. Temperatures in the subsurface versus time at probes within and outside the oxygen delivery area of influence.

was installed in an area that had low oxygen concentrations, but was within the VE area of influence of the extraction trench. Subsurface temperatures measured using this probe remained relatively constant at approximately 18°C. Conversely, the probes within the oxygen delivery area of influence exhibited increases in temperature, up to approximately 27°C within the first month of operation, as shown in Figures 1 and 2. Temperatures remained elevated until system operation was interrupted due to high groundwater. The decreases in soil temperatures are partially attributable to decreases in ambient temperatures, but microbial activity may have begun to decrease due to decreased operating times (reduced oxygen delivery).

## SOIL SAMPLE RESULTS AND MICROBIAL POPULATIONS

Overall, VOC and SVOC concentrations decreased significantly in the study area; however, fluctuations in the results due to high detection limits, matrix interferences, and sampling variability hindered the use of soil sampling as a means of process monitoring. At one fill sample location between the trenches after 6 months, concentrations were reduced by 99.9% for benzene (from 140 to 0.081 mg/kg), 89% for phenol (from 7.3 to 0.77 mg/kg), and 90% for 4-methyl phenol (from 75 to 7.5 mg/kg), achieving the cleanup goal for benzene, and approaching the cleanup goals for phenol and 4-methyl phenol. At the other fill sample location between the trenches, however, the benzene concentration remained constant at approximately 80 mg/kg, and phenol and 4-methyl phenol reductions could not be evaluated since their initial concentrations were not detected (<3.7 mg/kg).

Indigenous microbial populations within the oxygen area of influence increased significantly and consistently. As shown in Figure 3, from September 1993 to April 1994, phenol-degrader and VOC-degrader populations increased by approximately 3 orders of magnitude at a sample location between the trenches. Microbial populations at both sample locations between the trenches increased similarly. The concentration of biphenyl, which is considered to be a site-specific representative compound, decreased during this time period as shown on Figure 3. Although it is not a constituent of concern, biphenyl is prevalent in the landfill and is believed to be representative of the biodegradable constituents in the landfill, as it is in the mid-range of the molecular weights and sizes of the constituents.

Soil samples in each sampling round were analyzed for pH, nitrogen, and phosphorus. The pH ranged from 6 to 8, which is acceptable for bioremediation. Nitrate and nitrite concentrations were less than 1 mg/kg in all soil samples. Ammonia nitrogen concentrations ranged from 9.8 to 113 mg/kg. Total kjeldahl nitrogen (TKN) concentrations ranged from 344 to 4,800 mg/kg. Phosphate concentrations ranged from 23 to 7,700 mg/kg. The nutrient concentrations varied depending on the sample location, but were considered sufficient for biodegradation to occur using the indigenous microbes [4].

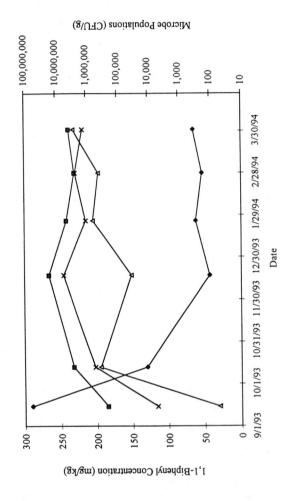

Figure 3. Microbe populations and biphenyl concentrations versus time at a sample location between the trenches.

## OXYGEN UPTAKE AND BIODEGRADATION RATE CONSTANTS

Oxygen uptake measurements were used to estimate oxygen uptake rates and biodegradation rate constants [1]. The relationship between oxygen concentration and time was modeled linearly to estimate oxygen uptake rates. Biodegradation rate constants were calculated as a function of these rates and the mass ratio of biphenyl to oxygen. Oxygen uptake rates at the probes in the oxygen area of influence ranged from 1.0 to 3.5%/h, and biodegradation rate constants in this area ranged from 22 to 79 mg/kg·day of organics as biphenyl. At a probe outside the oxygen delivery area of influence, but within the VE area of influence, the oxygen uptake rate was 0.28%/h, and the biodegradation rate constant was 6.4 mg/kg·day.

## CONTAMINANT BIODEGRADATION VERSUS EXTRACTION

During system operation, contaminated, oxygen-depleted soil vapor was removed from the subsurface through the extraction trench, and oxygen-rich air was drawn into the subsurface through the air inlet trench. In the oxygen-depleted, extracted soil vapor, carbon dioxide concentrations were elevated compared to air, typically at concentrations of 3 to 4%, due to microbial respiration. To evaluate the mass of organics degraded insitu, the total carbon dioxide produced was assumed to be due to the complete mineralization of biphenyl:

$$C_{12}H_{10} + 14.5\ O_2 \rightarrow 12\ CO_2 + 5\ H_2O \tag{1}$$

Based on this stoichiometry, the 30,000 kg of carbon dioxide recovered in the extracted vapor corresponds to 8,750 kg of organics as biphenyl biodegraded insitu. An additional 900 kg of organics were extracted, thus the remediation of the total of 9,650 kg of landfill constituents was accomplished 90% by biodegradation and 10% by volatilization.

Using these results to estimate a biodegradation rate constant, 8,750 kg of organics degraded from approximately 6.8 million kg of soil in 135 days (September 22, 1994 through March 18, 1995, omitting system downtime due to high groundwater) corresponds to 9.5 mg/kg day. This is significantly lower than the rates calculated based on oxygen uptake rates at the probes. This lower rate is expected for carbon dioxide-based estimates based on previous studies [5,6,7]. In addition, some of the carbon which resulted from the biodegradation of contaminants would have been immobilized in the subsurface as microbial biomass [7].

Many assumptions were used in calculating these values, and there are inherent uncertainties in soil sample results. This analysis, however, combined with the microbial population increases, solvent-extractable TOC results, relatively high oxygen uptake rates, carbon dioxide increases, and temperature increases in the subsurface, indicates that biodegradation is occurring and is a prime contributor to contaminant reduction in the fill. Reductions in VOCs in the soil can be attributed to removal by VE and to insitu biodegradation by bioventing. Reductions in SVOCs, however, are almost solely attributable to insitu biodegradation because of their lower volatility.

## BIODEGRADATION OF CONTAMINANTS IN THE SILT

Benzene in the silt was reduced by over 99% (from 3,000 to 20 mg/kg) at the two sample locations between trenches. The 4-methyl phenol concentration in the silt was reduced by approximately 73% (from 37 to <20 mg/kg) at one location between the trenches. Other reductions in phenol or 4-methyl phenol concentrations could not be quantified due to high detection limits (<38 mg/kg in the initial sampling event to <20 mg/kg in the final sampling event). The biphenyl concentrations in the silt decreased by approximately 93% at both sample locations between the trenches (from approximately 7,000 to 500 mg/kg). Initially, microbial populations in the silt were approximately 3 orders of magnitude less than those in the fill (100 colony forming units/gram [CFU/g] in the silt compared to 200,000 CFU/g in the fill). However, by March 1994, microbial populations comparable to those observed in the fill were achieved in the silt between the trenches (approximately 2,000,000 CFU/g in the fill and the silt). At a fill probe between the trenches, the oxygen uptake rate was 1%/h, compared to that of an adjacent probe in the silt, which was significantly lower at 0.33%/h. The biodegradation rate constant at this location in the silt was 7.4 mg/kg·day based on the oxygen uptake rate. These lower rates indicate that less microbial activity is occurring in the silt than in the fill soils. Based on a comparison of the fill and soil types, the permeability of the silt is approximately 3 orders of magnitude lower than the fill. Because of its lower permeability, less oxygen is delivered to the silt, and it is expected that a longer time period will be required to achieve the same contaminant reductions as in the fill.

## CONCLUSIONS

The findings presented herein provided the design basis for full-scale VE/bioventing of the landfill soils and the underlying silt. The full-scale system is currently operational. Oxygen, carbon dioxide, and temperatures in the subsurface soils are used to monitor microbial activity in the subsurface. Use of these indicator parameters minimizes analytical costs, and provides realtime process monitoring data which can be used routinely to control system operations. Therefore, trends in the concentrations of contaminants and indigenous microbial populations are used less frequently to monitor the progress of the remedy. By constructing and operating the system as a bioventing system, the extracted VOC concentrations are significantly decreased, since the VOCs are degraded in situ. This significantly reduces the costs associated with vapor treatment.

Constructed air inlet locations are required to effectively deliver oxygen to the subsurface. To implement bioventing using a VE system to deliver oxygen to the subsurface, the permeability and stratigraphy of the soil must be evaluated to ensure that sufficient oxygen will be provide to the subsurface at depth. The success of the insitu bioventing process is dependent on its ability to deliver oxygen to the subsurface. Air flow must be induced by creating a vacuum in the subsurface and by providing constructed air inlet locations to direct the air to the contaminated soils in the subsurface. If air inlet locations at depth are not available, air flow will enter the soil from the surface, and the oxygen in the air may be consumed by microbes in the shallow soils. In this situation, the air flow to deeper soils may still provide a medium for extraction of volatiles from the soils, but will not deliver oxygen to the deeper soils to enhance biodegradation of contaminants using the indigenous microbial population.

## REFERENCES

1. Kittel, J.A., R.E. Hinchee, R. Miller, C. Vogel, and R. Hoeppel. 1993. "In-Situ Respiration Testing: A Field Treatability Test for Bioventing." In *Proceedings of the 1993 Petroleum Hydrocarbons and Organic Chemicals in Groundwater: Prevention, Detection, and Restoration*, pp. 351-366. Houston, Texas.

2. Remediation Technologies Incorporated (ReTec). 1991. "Standard Method for Microbial Enumerations." January, pp. 1-11. Chapel Hill.

3. U.S. Environmental Protection Agency. 1979. "Composting." *Process Design Manual Sludge Treatment and Disposal*, p. 13. Municipal Environmental Research Laboratory, Cincinnati.

4. Morgan, P., and R.J. Watkinson. 1992. "Factors Limiting the Supply and Efficiency of Nutrient and Oxygen Supplements for the In-Situ Biotreatment of Contaminated Soil and Groundwater." *Water Resources* 26(1): 73-78.

5. Hinchee, R.E., and S.K. Ong. 1992. "A Rapid In-Situ Respiration Test for Measuring Aerobic Biodegradation Rates of Hydrocarbons in Soil." *Journal of Air and Waste Management Association* 42:1305-1312.

6. Newman, W.A., and M.M. Martinson. 1992. "Let Biodegradation Promote In-Situ Soil Venting." *Remediation* Summer, pp. 277-291.

7. Rasiah, V., R.P. Voroney, and R.G. Kachanoski. 1992. "Biodegradation of an Oily Waste as Influenced by Nitrogen Forms and Sources." *Water, Air, and Soil Pollution* 65:143-151.

CATEGORY XIV:
# Case Studies and Economic Analysis

# CASE STUDY: BENEFICIAL USE OF A MINERAL WASTE

Jeffrey C. Carlton, P.E.
  Minerals Technologies Inc.
  640 N. 13th Street
  Easton, PA 18042

Thomas W. Christopher, P.E.
  Minerals Technologies Inc.
  640 N. 13th Street
  Easton, PA 18042

## INTRODUCTION

The reuse and recycle concept is a foundation of
responsible environmental management and sustainable
development.  Reducing solid waste volumes by
eliminating waste streams and beneficially using what
cannot be eliminated are strategies promoted by many
different organizations of various shades of green and
gray.  However, beneficially using a waste stream is
not the simple task which one would expect.  This
paper will describe one company's experiences in
identifying and implementing  beneficial use
alternatives, and offer suggestions for others
interested in exploring waste reuse options.

## PROCESS AND WASTE

Specialty Minerals Inc (SMI) is a wholly-owned
subsidiary of Minerals Technologies Inc., a producer
and marketer of specialized inorganic mineral
products.  SMI has developed a proprietary process to
manufacture Precipitated Calcium Carbonate (PCC), an
inorganic chemical used in a number of industries,
especially the paper industry.  SMI produces PCC at 37
small facilities located at paper mills in North
America and Europe.
Chemically, PCC is similar to high purity
limestone.  PCC is produced by combining water,

calcium oxide, and carbon dioxide in a series of carefully controlled operations. Calcium oxide (lime) and water are combined to produce calcium hydroxide, then carbon dioxide is added to produce calcium carbonate. SMI has developed technology which utilizes stack emissions from combustion sources at a paper mill as the source of carbon dioxide. The final product, which is a suspension of PCC in water, is pumped directly to the paper mill.

The only process by-product generated by the PCC process is oversized screenings from the calcium hydroxide and PCC slurries. These screenings consist primarily of the following materials:

- Calcium Carbonate
- Calcium Hydroxide
- Calcium Oxide
- Silica
- Iron

The chief influence upon the characteristics (physical and chemical) of the screenings is the lime which is used as a raw material in the PCC process. The appearance of the screenings is similar to damp gray sand. The screenings contain no organic or toxic materials and are odorless. Due to the presence of the calcium oxide and calcium hydroxide, the screenings are an alkaline material. The metals content of the screenings is very low; toxicity characteristic leachate procedure tests have never identified metals above trace levels.


## DISPOSAL

The primary means of disposing of the screenings is to use a non-hazardous waste landfill. SMI sends screenings to landfills owned by the paper companies as well as to municipal and private landfills. However, there are many costs and concerns associated with the use of landfills.

Easily identified costs are the hauling charges and the landfill tipping fees. SMI's experience is that these costs can vary significantly, depending on the plant location. Hauling and landfill disposal costs range from $20 to $100 per ton for the U.S. PCC plants. Additional costs include expenses for testing, landfill approval fees, solid waste permit fees and solid waste taxes.

In addition to the direct costs associated with disposing of the screenings, there are other less quantifiable costs. Chief among these are the potential future liabilities associated with all disposal sites. The past decade is rife with examples of companies which have been designated Potentially Responsible Parties for cleanup actions at sites where those companies properly disposed of non-hazardous wastes.

Finally, disposing of the screenings in a landfill is simply a poor alternative to beneficially using the screenings in another application. The screenings use up landfill space which should be reserved for non-reusable wastes.

## REUSE PROGRAM

All of these are excellent reasons for SMI (or any company) to seek alternatives to waste disposal via landfills. Unfortunately, there are a number of hurdles which must be overcome before potential uses can become actual uses. Implementing a by-product reuse program is a multi-step endeavor.

1.   The first step is to examine the by-product material and the process which generates it in detail. Often the by-product material represents lost product or raw material. The best method for handling a by-product is to avoid making it in the first place.

The screenings from SMI's PCC process represent materials which cannot be returned to the process without impacting the performance of PCC in paper applications. SMI cannot avoid producing the screenings, nor can the volume of screenings be reduced.

2.   The second step is to investigate the characteristics of the by-product. What are the physical and chemical properties of the material? Is the material a consistent by-product or does it vary? Does the by-product have any characteristics that would make it acceptable for use in other applications?

SMI is fortunate in that the screenings from all PCC plants are similar. The nature of the screenings and the requirements of the PCC process also ensure that the quality and quantity of the screenings are consistent. These facts made the evaluation of the screenings straightforward. The task would be more difficult for a by-product whose qualities varied over time or between sites.

3.   Given this information, the next step is to identify possible alternatives for reusing the by-product. Identifying and evaluating reuse options

requires a marketing rather than environmental attitude. The by-product should be viewed as a material that must meet the criteria (quality, consistency, quantity, reliability, appearance, odor, etc.) which the market requires. Additionally, the producer should consider the specific quirks of the market, such as location, seasonality, and the price of virgin raw materials.

There are many sources which can be used in developing a list of potential reuse applications. These include:

- Internal Marketing Department
- Waste Exchanges
- Trade Organizations
- Industrial Groups
- Universities
- Consultants
- Regulatory Agencies
- Publications

SMI's experience with calcium mineral products had already proven the viability of using the PCC screenings as an agricultural liming agent. In addition, the alkalinity of the screenings opens the potential for using them to neutralize acidic wastes.

Further investigation by SMI identified other reuse options. Three of the companies which supply lime to SMI suggested using the screenings to neutralize acid mine drainage from abandoned coal mines. Contacts with the National Lime Association, an industry organization, provided information about the use of alkaline materials in specific grades of asphalt. The screenings could also be used as a calcium source material for cement or lime kilns. Additional sources indicated that the screenings could be used as an alkaline additive in composting municipal sewage sludge and in stabilizing certain soils for construction purposes.

4. A potential user for the material must be identified. Supply and demand considerations are critical. The demand for the by-product should parallel the supply, that is, the potential user must be able to successfully accept the material at the rate the generator produces the by-product. The user should also be able to handle any production changes which could affect the supply of the by-products. Both the generator and user should have options available in case the by-products cannot be produced or used for a period of time. The user must also be able to accept the quality of the by-product as well as be able to tolerate the range of characteristics

896

which the by-product may exhibit. Finally, the by-product must have sufficient benefits (cost, availability, quality, etc.) to overcome any reluctance the user may have about replacing a raw material with a by-product.

The reuse process must be carefully managed. The producer should work closely with the user to ensure that the material meets the users' requirements. The producer should visit the user periodically to ensure that the user has no concerns about the by-product. Visits should also verify that the by-product is being stored and used properly. The generator is always responsible for the by-product, even if the user is the one who handles or disposes of the material improperly.

SMI's process generates a steady stream of screenings with little change in the quality or volume of the material. Unfortunately, the volume of screenings produced by a typical PCC plant (less than 5000 dry tons per year) is not sufficient to interest lime or cement manufacturers whose volume is normally much larger. Similarly, the amount of screenings is too great for a typical asphalt plant which rarely requires an alkaline additive. The amount of material can easily be used by the agricultural market which surrounds the typical PCC plant. However, the screenings must be stockpiled at the PCC plant or at an off-site location during winter or rainy periods.

SMI facilities have approval to send the screenings to a non-hazardous landfill if the alternative users cannot accept the screenings.

5.   Another issue that must be addressed is cost. Typically, by-products have a very low market value. After including costs such as freight (if the generator pays this), handling costs, and any preparation/processing costs, the reuse option will probably continue to be an expense rather than a profitable activity. However, this expense should be compared to the full cost of disposal, including potential future liability. It is unlikely that the price will be the same as the price of the raw material being replaced.

SMI's intent is to at least break even in beneficial use arrangements. Local business situations have a major impact on the results.

6.   Finally, another hurdle that must be dealt with is the regulatory requirements for reusing by-products. In almost every instance, state agencies

endorse the beneficial reuse of by-products.  However, due to legitimate concerns about sham recycling and speculative accumulation, the agencies have developed regulations that can delay and discourage development of legitimate reuse programs.  There are federal requirements for reusing hazardous wastes, used oil, and municipal treatment plant sludge.  Many states also have requirements for reusing non-hazardous materials.

SMI has only had experience with state requirements.  SMI has found that face-to-face meetings with the regulators, either at the facility or at the agency headquarters, is very important in obtaining the approvals.

**EXPERIENCES**

SMI is now providing screenings to neutralize acid mine drainage and for agricultural soil additives. Use of screenings in cement kilns and asphalt mix additives remains a potential in the right situation.
Because beneficial use of industrial by-products is new and specialized, people working within State agencies do not always know all of the State's requirements.  In one case, SMI was told by a local solid waste inspector that agricultural reuse was the responsibility of the State Department of Agriculture. Six months were spent working with the Department of Agriculture in an effort to have the screenings classified as a Registered Liming Agent.  This included obtaining a registered trademark for the screenings.  Later it was learned that the screenings could not be a registered liming agent, as they did not meet the State's size requirements.  Faced with the added cost of milling to reduce the size, the State environmental agency was again consulted. During this second meeting SMI learned that the screenings would have to be managed as a Beneficial Reuse of Waste, even if the Department of Agriculture's standards were met.  However, the screenings could be approved for agricultural use at specific sites by the environmental agency without any size requirements.  SMI was required to obtain an environmental permit (a two month procedure) but was able to avoid the cost of milling the screenings. Several months of effort were invested before the correct path was determined.
Enhanced local presence may help speed approvals. SMI, in another case, arranged with a local aggregates firm to handle storage and marketing of screenings to agriculture.  This allowed SMI management to focus on PCC business, not agriculture and gives the screenings

effort a local presence.  The State had a procedure in
its regulations to approve reuse of waste, but it was
specifically written for sewage sludge.  After a
meeting with SMI and the local firm, the State
environmental agency concluded that the regulation was
not appropriate for SMI waste and that approval of the
plan would be handled on an ad hoc basis.  Information
on the material, handling plans, and storage site was
provided to the State environmental agency.  The same
information plus agriculture test results from other
States was provided to the State agriculture agency.
When the agriculture agency approved use of the
screenings, both SMI and the local aggregates firm
used this information to keep the environmental agency
approval process moving.


## CONCLUSION

Reuse of by-products and waste materials results in
significant benefits to the producer, the user, and to
society.  However, patience and perseverance are
required to address the numerous issues that the
producer will face before a by-product may be reused.
The necessary parts of a successful beneficial use
effort are:

- A Useable Material          - a by-product that can
                                meet a market need.

- A Marketing Orientation  - the material should be
                             thought of as a
                             product, not a waste.

- A Local Presence            - using a broker or
                                extension agent if
                                there is no long term
                                company presence.

- A Local Focus               - working with local
                                officials and
                                investigating reuse
                                options in the
                                immediate area is
                                usually most effective.

- Persistence                 - essential in the
                                effort to get
                                approvals, to find a
                                market, and to
                                penetrate it.

# A UNIQUE APPROACH TO ENVIRONMENTAL EVALUATION AND CLEANUP ACTIVITIES UNDER ECRA

H. JAMES ARCHER, P.E.
Environmental Resources Management Group
855 Springdale Drive, Exton, PA 19341

JEFF CASE
Environmental Resources Management Group
855 Springdale Drive, Exton, PA 19341

## INTRODUCTION

The effectiveness of an overall strategy and Remedial Action Program for any site under the Environmental Cleanup Responsibility Act (ECRA) is directly proportional to the information and data input to evaluate the site. Careful consideration must be given to the location of the site, it's history, and the contaminants that would be suspected based on past operations and those ultimately found. The approach taken for this site was different in that emphasis was placed on defining the past operations and history of the site and adjacent properties. This Case History for example, highlights a familiar situation many companies have found themselves in when trying to sell a facility or operation in New Jersey. The New Jersey Department of Environmental Resources (NJDEP) wanted to perform an extensive remedial action program for the site including excavation of a major portion of the site, and incineration or off-site disposal of the excavated material.

ERM became involved in this project in 1990 after the client had performed a preliminary site investigation and submitted an ECRA Cleanup Plan that was rejected by the NJDEP.

## SITE HISTORY

ECRA was developed in New Jersey to ensure responsible property transfer during the sale of a property in any commercial merger or acquisition transaction. The original trigger of ECRA requirements for this site was the sale of company ownership and operations in 1986. The ECRA reporting forms (GIS and SES) were submitted and approved by NJDEP in 1987. The former owner then proceeded to perform an "at risk" soil and ground water sampling program. An ECRA Cleanup Plan, which summarized the sampling results and provided recommendations for site remediation, was then submitted. The NJDEP rejected this plan and requested additional site characterization. At this time, ERM became involved and, together with the client, established an investigation program that included development of a detailed site history.

The section of Newark where this site is located (Figure 1) has been a heavily industrialized area since the land along the Passaic River was filled around the turn of the century. Based on the heavy industrialization in this area, the detection of soil and ground water contamination caused by historical operations, fill material, and/or neighboring industrial operations would be expected on most properties.

The effect of regional industrial operations on ground water quality in the Newark area along the Passaic River is well documented by NJDEP. Consequently, the ground water in this area is not regarded as suitable for potable purposes.

A review of the facility and regional history had shown that the site was originally marshland and was filled sometime during the period between 1873 and 1890. Approximately 8 to 12 feet of mixed fill material, including coal, slag, and ash from coal-gasification facilities and other coal-fired operations in the vicinity of the site, were deposited at the site.

The identification of the fill material composition for the site was critical for characterization of site concerns. The primary constituents of concern detected in on-site soils were polycyclic aromatic hydrocarbons (PAH), both carcinogenic and non-carcinogenic, and trace heavy metals (Table I and Table II). A major source of these compounds is the non-combusted waste (slag, ash) from various coal-fired operations (boilers, heaters, coal-gasification units). Consequently, the site strategy formulated was to show that the PAH and trace metals contamination found in on-site soils should be considered background in the area around the site.

Coal-fired boilers and heaters were commonly used during the late 1800's and early 1900's in all areas of the country. During that time it was common practice to dump the residual onto the ground or take them to a common dumping ground. Due to the absence of large construction equipment (bulldozers, loaders) during the operating periods of these facilities, movement and disposal of waste materials was a very labor intensive process.

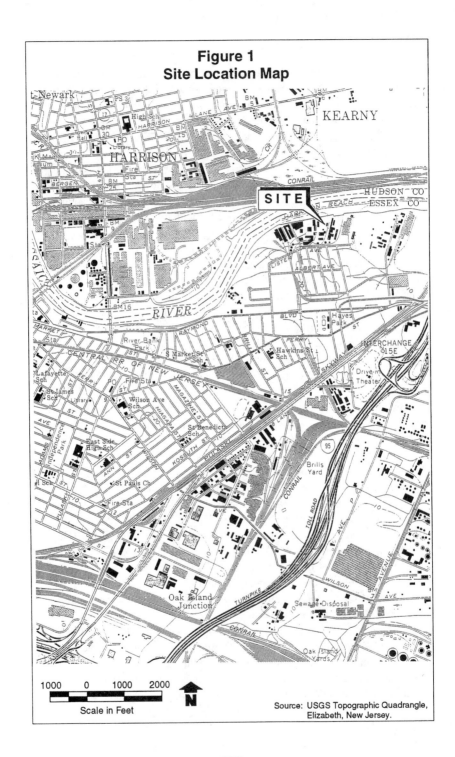

# Figure 1
# Site Location Map

Scale in Feet

1000    0    1000    2000

N

Source: USGS Topographic Quadrangle,
Elizabeth, New Jersey.

902

**TABLE I — COMPARISON OF PAH LEVELS REPORTED IN COAL ASH VERSUS ONSITE FILL MATERIAL, NEWARK, NEW JERSEY**

PAHs
Sample Location and Sample Type

| Constituent | Coal Tar | Project Site Soils (0 to 2.5 ft) | 2 Nearby Sites Soils (0 to 2 ft) | Soils (0 to 2 ft) |
|---|---|---|---|---|
| acenaphthene | 10000 | 0.35 - 31 | 0.24 - 4.6 | 0.26 - 2.4 |
| anthracene | 9000 | 0.46 - 71 | 0.63 - 3 | 0.16 - 3.9 |
| benzo(a)anthracene | 0.007 | 0.46 - 409 | 0.51 - 47 | 0.16 - 16 |
| benzo(a)pyrene | 30 | 1.1 - 344 | 0.56 -44 | 0.27 - 16 |
| benzo(b)fluoranthene | 3000 | 0.98 - 317 | 0.94 - 71 | 0.37 - 4.1 |
| benzo(k)fluoranthene | NS | 0.65 - 119 | NS | 4.1 - 5.4 |
| benzo(ghi)perylene | NS | 0.80 - 322 | 3.3 - 32 | 0.13 - 3.8 |
| chrysene | 4000 | 0.60 - 404 | 1.4 - 120 | 0.19 -12 |
| dibenz(a,h)anthracene | NS | ND - 92 | NS | 0.17 - 1.5 |
| fluoranthene | 6000 | 0.64 - 201 | 0.33 - 64 | 0.16 - 15 |
| fluorene | 10000 | 0.46 - 43 | 0.25 - 0.32 | 0.36 - 2.5 |
| indeno(1,2,3-cd)pyrene | NS | 0.63 - 237 | 0.48 - 21 | 0.3 - 4 |
| naphthalene | 90000 | 0.41 - 67 | 0.2 - 11 | 0.16 -1.8 |
| phenanthrene | 9000 | 0.613 - 242 | 0.25 - 61 | 0.23 - 15 |
| pyrene | 3000 | 0.94 - 434 | 0.28 - 78 | 0.23 - 39 |

Note: All concentrations are in mg/kg

NS: not Sampled
ND: not detected

903

## TABLE II — COMPARISON OF METALS LEVELS REPORTED IN COAL ASH VERSUS ONSITE FILL MATERIAL, NEWARK, NEW JERSEY

### METALS
Sample Location and Sample Type

| Constituent | Fly Ash | Project Site Soils (0 to 2.5 ft) |
|---|---|---|
| arsenic | 6 - 1200 | ND - 42 |
| barium | 100 -1074 | NR |
| cadmium | 0.29 - 51 | NR |
| chromium | 15 - 900 | 3 - 2500 |
| copper | 16 - 400 | 16 - 430 |
| iron | 49 - 235 | NR |
| lead | 11 - 800 | 13 - 510 |
| manganese | 100 - 1000 | NR |
| selenium | 6.9 - 760 | NR |
| silver | 3 | NR |
| zinc | 50 - 9000 | 22 - 480 |

NR: not reported
ND: not detected

Note: All concentrations are in mg/kg

904

Subsequently, local disposal options for the slag, ash, and waste coal were important in order to control operating costs. It stands to reason that these operations, knowing that the area along the Passaic River was being filled, would have taken full advantage of the opportunity to dispose of large quantities of waste material at these local filling operations.

According to the deed records, no evidence could be found regarding industrial operations at this site during the late 1800s. At various times in its history the site was owned by the an agricultural chemical company involved in the manufacturing of inorganic fertilizers, several small oil companies who used the site as a loading/unloading facility for ships, and by a manufacturer of pigments and dyes for automotive, textile, and cosmetic industries.

## PREPARATION OF A CLEANUP PLAN

All information collected during the investigations at the site was used to formulate a Cleanup Plan. The plan focused on the following main issues:
- The site had been contaminated by constituents present in the material used to backfill the site (i.e., slag, coal, etc.).
- A plan for remediation of site-related contamination was proposed that leaves background fill material (PAHs) in place.
- An alternate cleanup standard petition had been presented.

An environmental and human health based risk assessment was performed for the site. The risk assessment used an industrial exposure scenario, and successfully demonstrated that there is no appreciable threat to human health or the environment from contaminants found on the site. This risk assessment generated cleanup levels for contaminants detected in on-site soil and ground water. The goals of the risk assessment were to calculate cleanup levels that would be protective of human health from chemical exposures via direct soil contact and to protect the designated uses of the Passaic River from the discharge of ground water to the river.

The plan also presented a petition for no action for the fill material at the site and incorporated elements of NJDEP's proposed alternate cleanup standard process. The petition demonstrated that the contamination resulting from fill material was not from a facility-related discharge and that the alternate cleanup standard proposed would accomplish the same degree of protection of human health and the environment as meeting the standards. Protection of human health and the environment was provided in this plan through the use of engineering controls to eliminate the same direct contact exposure pathway. The risk assessment and information presented pursuant to N.J.A.C. 7:26D-(c)(2) support the petition for no further action regarding the fill material.

This cleanup plan contained provisions to perform limited soil remediation in areas that could be directly related to current site activities, various housekeeping activities to prevent potential future contaminant releases, and imposed deed restrictions on the site (Figure 2). To further restrict direct human contact with on-site soil, land use deed restrictions were proposed that would require all land areas to be covered in the future, and would regulate all subsurface construction activities.

Only hexachlorobenzene detected in samples from one area of the site exceeded the risk-based cleanup levels calculated in the risk assessment. Concentrations of contaminants detected in ground water on site did not exceed any of the ground water cleanup levels. Cleanup levels were not generated for the PAH compounds in soil, primarily due to the ubiquitous nature of PAHs to the site and surrounding area and the proposal to limit exposure potentials by capping the site and imposing deed restrictions. The capping program will provide asphalt paving on all currently unpaved areas of the site. The program also outlines the repair and improvement of existing paved areas and implementation of a long-term maintenance program for the cap.

The ground water investigation showed that although the majority of the fill material was within the aquifer (less than three feet from ground surface), any contaminants that were found to be present in the soil were not leaching to the water. The only apparent elevated concentrations of contamination that were of concern were found in ground water at two separate areas of the site. Each of these areas is located adjacent to a property boundary where ground water is documented to flow onto the subject property. In one of the areas, the contaminants detected were detected on the adjacent property at higher concentrations and the ground water was documented to flow radially from that property.

The second area contained high levels of an organic solvent that is believed to be a major raw material utilized in operations at the adjacent property. This area is upgradient of the subject site. An added note in this area, is that the excavation activities completed to remediate hexachlorobenzene was in the immediate area of the well where the solvent was detected. No elevated levels of the solvent or potential source area was detected during the excavation, providing further indication that the contamination was most likely from an off-site source.

Since the ground water cleanup levels were not exceeded by existing contaminant concentrations, a five-year ground water monitoring program was proposed to track contaminant concentrations and to demonstrate that the cleanup levels were not exceeded. Further, the deed and land use restrictions will also include restrictions of the use of ground water at the site for potable purposes.

The plan was ultimately approved by NJDEP in February of 1994, and remedial activities commenced in April 1994. Currently all remedial activities are complete and the ground water monitoring program is continuing.

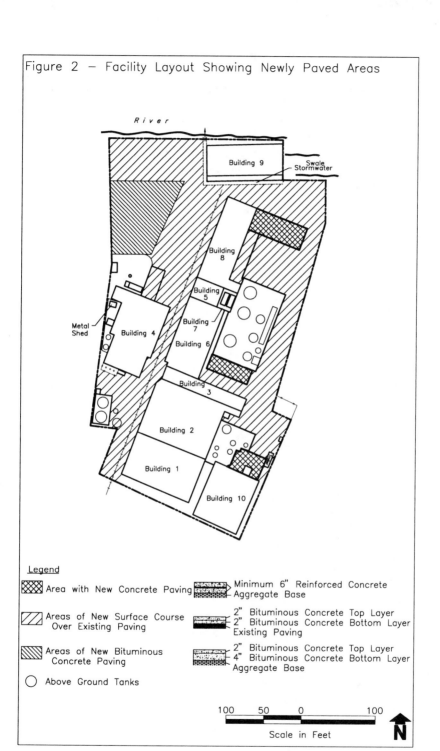

Figure 2 — Facility Layout Showing Newly Paved Areas

River

Building 9

Swale
Stormwater

Building 8

Building 5

Metal Shed

Building 4

Building 7

Building 6

Building 3

Building 2

Building 1

Building 10

Legend

Area with New Concrete Paving

Areas of New Surface Course Over Existing Paving

Areas of New Bituminous Concrete Paving

Above Ground Tanks

Minimum 6" Reinforced Concrete
Aggregate Base

2" Bituminous Concrete Top Layer
2" Bituminous Concrete Bottom Layer
Existing Paving

2" Bituminous Concrete Top Layer
4" Bituminous Concrete Bottom Layer
Aggregate Base

100    50    0    100

Scale in Feet

N

## SUMMATION OF THE PROJECT

A substantial amount of time was spent by ERM and the client to investigate the history of the site and adjacent properties. Through this search, information was found that detailed the source of the major contamination found on the site. This information was used successfully to convince the NJDEP that the site was consistent with the entire area along the Passaic River at this junction of the waterway and that remediation of the historical related operations would not provide any additional level of protection of human health or the environment.

# APPLICATION OF A PASSIVE SOIL VAPOR SURVEY AT A FORMER MANUFACTURED GAS PLANT

MARK J. WRIGLEY, P.G.
W. L. Gore & Associates, Inc.
P.O. Box 1100
Elkton, MD 21922-1100

## INTRODUCTION

Manufactured gas plants (MGPs) operated from the mid-nineteenth century to the 1950's, producing wastes characterized by tars and oils comprised of compounds including volatile organic compounds (VOCs) and polynuclear aromatic hydrocarbons (PAHs) [1]. Active soil gas methods have been used at MGP sites to help locate these wastes in soil and ground water as part of initial site investigations. However, active methodology is generally limited to detection of only the volatiles present, while passive methods have shown limited success at MGP sites [2].

A new passive soil vapor survey has been developed and validated over the past two years which helps to address these limitations. By using a time-integrated approach and a unique soil vapor collector the GORE-SORBER® Screening Survey has proven to successfully detect PAH compounds as well as volatiles. This technology features a patented, passive sorbent collector constructed from GORE-TEX® expanded polytetrafluoroethylene (ePTFE), which allows for efficient vapor transfer from the subsurface to the adsorbent, but prevents liquid water and particulate from impacting sample integrity and sensitivity.

This discussion describes the unique qualities of this passive soil gas technology and presents the results of a recent application at a former MGP site in the Eastern United States with comparison to the results from previous soil and ground water investigations conducted at the site.

## SCREENING TECHNOLOGIES AT MGP SITES

Screening level methods employed at MGP sites range from immunoassay test kits to active and passive soil gas techniques. While immunoassays can provide

909

quick, on-site answers for a preselected range of hydrocarbons, they are limited to discretely collected soil samples only. One advantage of some soil gas methods as compared to immunoassays is the ability to indirectly detect hydrocarbons located at greater depths than the site of actual sample collection - in some cases detecting dissolved-phase contaminants in the saturated zone 27 feet below the sampling depth [3].

Active soil gas methods can provide relatively rapid results but have traditionally been limited to volatile organic compounds and to sites with permeable soils. These factors significantly limit their application to MGP site investigations where low vapor pressure compounds are of primary interest, and where low permeability soils may further limit their application.

Due to their time-integrated sampling approach, passive soil gas techniques have shown some increased utility on MGP sites by detecting low vapor pressure, high molecular weight organics and overcoming some of the limitations inherent with low permeability soils. However, collector design and analytical techniques have limited more widespread successful applications. Recent advances address some of the limitations of early passive technologies.

## NOVEL PASSIVE COLLECTOR DESIGN

A typical GORE-SORBER® Module consists of three (3) separate GORE-SORBER® Passive Sorbent Collection Devices (sorbers). A typical sorber is 30 millimeters (mm) long, with a 3 mm inside diameter (ID), and contains 40 milligrams (mg) of a suitable granular adsorbent material depending on the specific compounds to be detected. Typically, Tenax-TA® resin is used for it's affinity for a broad range of VOCs and SVOCs. The three sorbers are sheathed in the bottom of a four (4) foot length of vapor-permeable insertion and retrieval cord. This construction is termed a GORE-SORBER Screening Module (Figure 1). Both the retrieval cord and sorbent container are constructed solely of inert, hydrophobic, microporous GORE-TEX® ePTFE (similar to Teflon®).

A unique feature of ePTFE membranes are that they are hydrophobic and exclude liquid water, yet they do not retard vapor transfer, thus allowing VOC and SVOC vapors to freely penetrate the module and collect on the adsorbent material. This ability to protect the sorbent media from contact with ground water and soil pore water without retarding soil vapor diffusion facilitates the application of GORE's soil vapor screening methods to saturated and very low permeability, poorly drained soils.

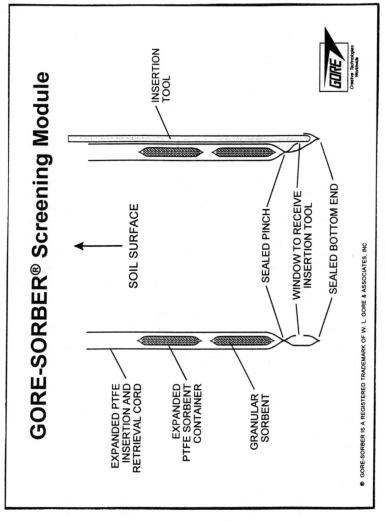

FIGURE 1

® GORE-SORBER IS A REGISTERED TRADEMARK OF W. L. GORE & ASSOCIATES, INC.

911

# SITE APPLICATION OF GORE-SORBER SCREENING SURVEY

## OBJECTIVE

The objective of this survey was to demonstrate the applicability of GORE-SORBER® Screening Surveys (GSSS) in delineating the relative subsurface distribution of low volatility, high molecular weight, organic chemical contaminants (including polynuclear aromatic hydrocarbons (PAHs)), often associated with town gas manufacturing facilities. The study area was selected because of the availability of some soil and ground water data showing the nature and location of subsurface impact by PAHs/SVOCs related to historical MGP operations.

## SITE INFORMATION & SURVEY DESIGN

The site is located in the eastern U.S. and is characterized by the presence of artificial fill/soils consisting of silty to clayey sand with intervals of wood, sawdust, natural organic debris, coal fines, brick and concrete rubble extending to a depth of 4 to 15 feet below ground surface (bgs). Immediately underlying the fill is a laterally extensive clay layer which extends to a depth of 40 to 55 feet bgs, and serves as a confining bed throughout most of the site. The average hydraulic conductivity (K) of the fill and clay units are $6.1 \times 10^{-3}$ and $3.4 \times 10^{-7}$ cm/sec, respectively. Depth to ground water at the site generally averages 2-4 feet.

Forty five (45) GORE-SORBER® Modules were deployed throughout a selected portion of the site covering an approximate 85,000 sq. ft. (1.95 acre) area. Module installation depth was approximately 2.5 to 3 ft. and exposure time was 25 to 26 days. All of the modules were noted to be in ground water at the time of retrieval. They were subsequently returned to Gore's lab and analyzed using thermal desorption, gas chromatography and mass selective detection (TD/GC/MSD). Replicate sorbers from each module were also analyzed using solvent extraction (SOLVEX) and GC/MS.

## RESULTS

Five light PAHs (naphthalene, 2-methyl naphthalene, fluorene, acenaphthene and acenaphthylene) quantitated using TD/GC/MS and four heavier PAHs (phenanthrene, anthracene, fluoranthene and pyrene) quantitated using SOLVEX/GC/MS were used to compare the results of the GSSS to available data from wells and borings located within the study area. A sample chromatogram identifying the PAHs detected using solvent extraction is shown in Figure 2. The results of the GSSS are summarized in Table I and indicate that target PAHs were detected in greatest quantities in the areas northwest, south and west of the transformer house, and to the north of the former rail spur. The relative distributions of combined masses of the five light PAHs and four heavier PAHs are respectively depicted in Figures 3 and 4. For comparison purposes, respective ground water concentrations (in μg/l) of the mapped analytes from four shallow ground water monitoring wells located in the study

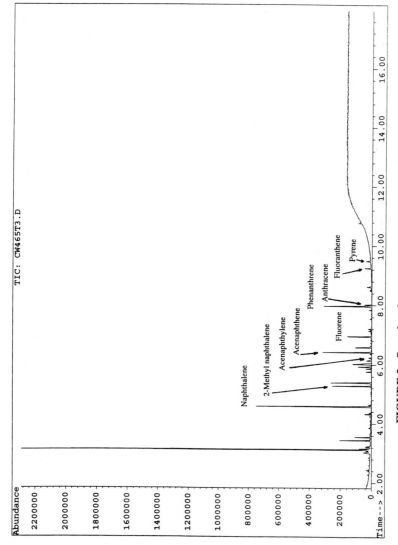

**FIGURE 2: Sample solvent extraction gas chromatogram**

| | | C11, C13, | | 5 PAHs, ug | 4 PAHs, ug |
|---|---|---|---|---|---|
| Module # | BTEX, ug | &C15, ug | TMBs, ug | TD | SOLVEX |
| 104462 | 0.47 | 0.06 | 0.08 | 1.21 | 0.00 |
| 104463 | 0.00 | 0.02 | 0.00 | 0.01 | 0.00 |
| 104464 | 0.09 | 0.03 | 0.01 | 0.02 | 0.00 |
| 104465 | 6.20 | 0.84 | 7.15 | 295.40 | 96.10 |
| 104466 | 0.15 | 0.04 | 0.02 | 0.07 | 0.00 |
| 104467 | 0.01 | 0.01 | 0.00 | 0.00 | 0.00 |
| 104468 | 0.01 | 0.01 | 0.00 | 0.01 | 0.00 |
| 104469 | 0.08 | 0.03 | 0.14 | 0.28 | 0.00 |
| 104474 | 0.13 | 0.05 | 0.02 | 0.04 | 0.00 |
| 104475 | 0.01 | 0.01 | 0.00 | 0.00 | 0.00 |
| 104476 | 0.02 | 0.01 | 0.01 | 0.00 | 0.00 |
| 104477 | 0.00 | 0.02 | 0.00 | 0.00 | 0.00 |
| 104478 | 0.60 | 0.08 | 0.05 | 2.69 | 1.74 |
| 104479 | 2.00 | 0.21 | 0.10 | 3.30 | 1.91 |
| 104530 | 0.14 | 1.00 | 0.27 | 52.22 | 26.52 |
| 104531 | 0.04 | 0.13 | 0.12 | 1.69 | 0.00 |
| 104532 | 0.07 | 0.05 | 0.07 | 29.54 | 17.72 |
| 104533 | 0.21 | 0.02 | 0.06 | 0.08 | 0.00 |
| 104534 | 8.32 | 1.09 | 12.82 | 128.05 | 29.98 |
| 104535 | 5.46 | 5.16 | 10.52 | 90.19 | 3.68 |
| 104536 | 0.08 | 12.39 | 0.20 | 12.63 | 7.18 |
| 104537 | 0.09 | 0.17 | 0.16 | 31.91 | 9.10 |
| 104538 | 7.03 | 0.78 | 3.94 | 92.50 | 28.54 |
| 104539 | 0.68 | 0.27 | 0.51 | 2.61 | 0.00 |
| 104541 | 0.04 | 0.12 | 0.09 | 18.74 | 13.22 |
| 104544 | 0.07 | 1.60 | 0.04 | 11.81 | 4.37 |
| 104545 | 0.01 | 0.14 | 0.01 | 13.02 | 5.07 |
| 104546 | 0.67 | 0.01 | 0.06 | 0.17 | 0.00 |
| 104547 | 6.99 | 0.05 | 2.64 | 22.79 | 0.00 |
| 104548 | 0.06 | 0.01 | 0.03 | 0.25 | 0.00 |
| 104551 | 0.11 | 0.01 | 0.00 | 0.00 | 0.00 |
| 104552 | 0.03 | 0.01 | 0.01 | 0.01 | 0.00 |
| 104553 | 1.07 | 1.08 | 1.27 | 71.04 | 15.26 |
| 104554 | 0.04 | 3.87 | 0.01 | 7.37 | 1.12 |
| 104971 | 0.08 | 0.03 | 0.02 | 0.63 | 1.12 |
| 104972 | 0.42 | 0.02 | 0.05 | 0.21 | 0.45 |
| 104973 | 0.04 | 0.03 | 0.02 | 0.01 | 0.00 |
| 104975 | 0.01 | 0.01 | 0.01 | 0.01 | 0.00 |
| 104976 | 0.12 | 0.12 | 0.16 | 0.23 | 0.00 |
| 104977 | 0.30 | 0.05 | 0.10 | 2.31 | 0.00 |
| 104978 | 0.64 | 3.25 | 0.08 | 3.01 | 1.12 |
| 104979 | 0.15 | 1.36 | 0.13 | 22.91 | 3.98 |

Notes:

1) C11,C13&C15 = diesel range alkanes (undecane, tridecane and pentadecane

2) TMBs = combined 1,2,4- and 1,3,5-trimethylbenzenes

3) 5 PAHs = naphthalene, 2-methylnaphthalene, acenaphthylene, acenaphthene, and fluorene

4) 4 PAHs = phenanthrene, anthracene, fluoranthene, and pyrene

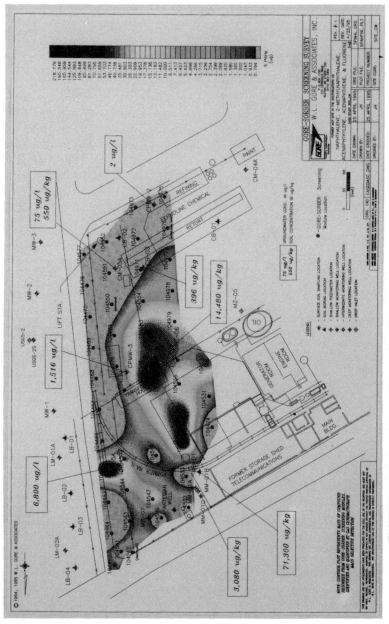

**FIGURE 3: Contour of 5 PAH mass distribution with soil and groundwater data shown**

915

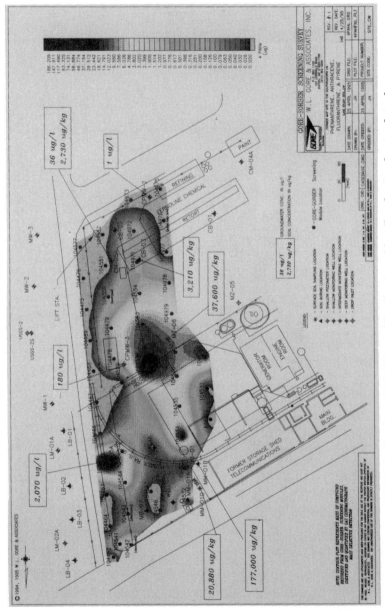

FIGURE 4: Contour of 4 PAH mass distribution with soil and groundwater data shown

area have been posted on Figures 3 and 4. Although the number of shallow monitoring wells is limited, there is good agreement between the results of the GSSS and these available water quality data.

Previously, soil samples had been collected from five locations in or adjacent to the study area. The highest concentrations of target PAHs in subsurface soils were detected south west of the transformer house (MZ-06,), and east of the former storage shed (MB-01, MM-01D). Low levels of PAHs were detected in soil samples collected in the southern portion of the study area at CB-03 and CM-05A. These data also are generally consistent with the results of the GSSS. Shallow site-specific soil quality data are posted on Figures 1 and 2.

## CONCLUSION

The results of the GSSS demonstrate the capabilities of this unique soil gas sensor in successfully screening for low volatility organic compounds even in saturated soils.

## REFERENCES

1. Srivastava, Vipul J. 1993. "Manufactured Gas Plant Sites: Characterization of Wastes and IGT's Innovative Remediation Alternatives", *Paper Presented at the Symposium for Hazardous and Environmentally Sensitive Waste Management in the Gas Industry, Albuquerque, NM, January, 1993*
2. Gas Research Institute (GRI). 1992. "Evaluation of Three Soil Gas Techniques at an MGP Site", Topical Report, September 1990-April, 1992
3. Stutman, Mark B. 1993. "A Novel Passive Sorptive Method for Site Screening of VOCs and SVOCs in Soil and Ground Water", *Proceedings of the 8th Annual Conference on Contaminated Soils, Amherst, MA, September, 1993.*

# INVESTIGATION AND REMEDIATION OF A CHLOROBENZENE SPILL - A CASE STUDY

EUGENE J. DONOVAN, JR., PE
HydroQual, Inc.
1 Lethbridge Plaza
Mahwah, New Jersey 07430

LESLIE A SPARROW
HydroQual, Inc.
1 Lethbridge Plaza
Mahwah, New Jersey 07430

## INTRODUCTION

During the remedial investigation of an industrial facility, an area containing elevated concentrations of chlorobenzene in the soil and groundwater was identified. No record of chlorobenzene use or reference to a spill could be found at the site, that had recently been shut down after more than 80 years of operation. The lack of historical information suggested that the spill occurred at least 20 years prior. The area, located in New Jersey, was adjacent to railroad tracks frequented by chemical tanker cars. The site has been under an Administrative Consent Order (ACO) with the New Jersey Department of Environmental Protection (NJDEP) since 1988 to investigate and remediate contamination at the site. Except for the chlorobenzene, only minor amounts of volatile organics had been found elsewhere on the site. Figure 1 shows the location of the spill area and groundwater plume.

## SOILS INVESTIGATION

The extent of chlorobenzene in the soils was investigated through an extensive series of soil borings in the spill area. The site stratigraphy consisted of approximately 35 feet of horizontally discontinuous layers of glacial till and fine to coarse sands overlaying weathered diabase bedrock. Results of

918

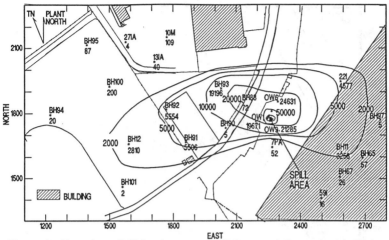

Figure 1.  Location of Chlorobenzene Spill Area and Groundwater Plume

chlorobenzene was concentrated in a zone 10 to 20 feet below ground, and essentially all of the chlorobenzene was no more than 25 feet below ground. The tight clayey soils and the DNAPL (dense non-aqueous phase liquid) properties of chlorobenzene had limited the migration of the chlorobenzene in the soils to a relatively small area, for many years after the spill. Concentrations of chlorobenzene in soil samples varied considerably by depth and location. The highest concentration found in a sample was 9840 mg/kg chlorobenzene.

Data analyses were performed using the computer based gridding and contouring program SURFER$^{TM}$.  The volume of soils in the chlorobenzene spill zone was estimated to be 25,600 cubic yards.  It was estimated that the soils contained approximately 3,300 pounds of chlorobenzene, or the equivalent of about 440 gallons of 100 percent chlorobenzene.

Early on in the site investigation, excavation was retained as a possible first-step for remediation of this area.   Site redevelopment with construction of warehousing, to provide for an overall site remediation cap, was proceeding faster than anticipated. It became necessary to pursue a short-term remediation approach as well as evaluating other possibly less costly but more time consuming options.   Due to the limited migration of the DNAPL, analysis showed that excavation of the higher concentration soils would greatly reduce the amount of soil removed for treatment or disposal while still removing a high percentage of the chlorobenzene.   Figure 2 illustrates the spill area and shows various cutoff concentrations and the amount of chlorobenzene that would be removed for each one. As shown in Figure 2, excavation of 7,600 cubic yards of soil containing greater than 100 mg/kg chlorobenzene would remove about 3,160 pounds, or 95.5% of the chlorobenzene in the soils.   Further data examination showed that by selective excavation of only the high concentration soils by removing the low concentration overburden, the total volume excavated for treatment and disposal could be reduced further to about 5000 cubic yards.

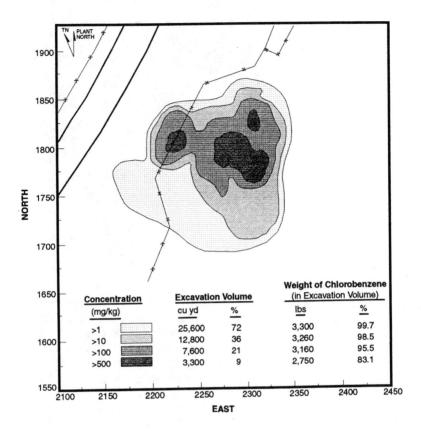

| Concentration (mg/kg) | Excavation Volume | | Weight of Chlorobenzene (in Excavation Volume) | |
|---|---|---|---|---|
| | cu yd | % | lbs | % |
| >1 | 25,600 | 72 | 3,300 | 99.7 |
| >10 | 12,800 | 36 | 3,260 | 98.5 |
| >100 | 7,600 | 21 | 3,160 | 95.5 |
| >500 | 3,300 | 9 | 2,750 | 83.1 |

Figure 2. Chlorobenzene Concentration in Soil in Spill Area

## GROUNDWATER MONITORING

Monitoring wells were installed to define the chlorobenzene plume in the groundwater around the spill area. Water level elevations from wells in and around the site showed a groundwater divide across the spill area. The plume had moved out in two directions away from the spill area, with highest concentrations off-site to the west.

A stratigraphic cross section through the plume delineation area and centered on the spill area is shown on Figure 3. The upper and lower sand strata to the west separated by a till layer can be observed. To the east the upper sand lens sloped toward the river. Because of the dense nature of the chlorobenzene, the higher concentration of chlorobenzene in the groundwater was found in the lower sand zone to the west. Lower concentrations of chlorobenzene were observed in the upper sand zone both to the west and east of the spill zone.

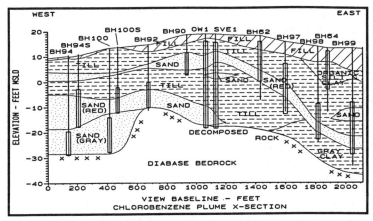

Figure 3. Stratigraphic Cross Section through Plume Delineation Area

The wells and area encompassed by the groundwater chlorobenzene plume are shown on Figure 1. Groundwater chlorobenzene concentrations decreased to low levels over a short distance beyond the periphery of the 2,000 $\mu$g/l contour. Based on samples collected in June 1994, the average concentration of chlorobenzene in the groundwater within the plume delineation area was 5,600 $\mu$g/l. There is an estimated 7 million gallons of groundwater within the lower sand zone of the plume delineation area. This represented an estimated 330 pounds of chlorobenzene in the delineation area groundwater, or about 10% of the amount estimated to be in the soil in the spill zone.

## REMEDIAL ALTERNATIVE ANALYSIS

Several remedial alternatives for the chlorobenzene area were considered for evaluation as part of the overall site feasibility study. Laboratory studies to evaluate biodegradation of chlorobenzene by indigenous microorganisms were conducted. It was determined that in situ biological treatment would not be feasible due to slow degradation rates and slow movement of groundwater in the poorly permeable till soils.

Soil vapor extraction (SVE) for soils in the spill area was recommended by the NJDEP and was evaluated. SVE is a demonstrated and widely used technology for in situ extraction of volatile organic compounds from soils in the vadose or unsaturated zone. The water table in the spill area, however, was approximately 4 feet below ground, and the target soils were predominantly saturated. A combined pumping and SVE remedial alternative was therefore considered.

A pump test to determine soil and groundwater characteristics was conducted. An initial step pump test in the chlorobenzene area indicated a maximum yield from a 35 ft screened zone in the till-sand of about 3 gpm and an hydraulic

921

conductivity of 0.7 ft/day. A 72 hour pump test yielded variable horizontal hydraulic conductivity values in the saturated site soils at the pumping well and 8 surrounding monitoring wells of from 0.55 to 18.1 ft/day averaging 4.4 ft/day. An average vertical conductivity of 0.7 ft/day was also determined. These values are characteristic of values reported for glacial till consisting of variable mixtures of sands, silts, and clay. A maximum area of influence of about 100 to 120 feet around the pumping well was observed, although most of the drawdown occurred 25 to 50 feet from the pumping well.

Following the pump test, a combined pump and treat and SVE pilot plant was designed, constructed and operated for several months. The average removal rate from the test vapor extraction well was 1.1 pounds of chlorobenzene per day, with a maximum of 6.4 pounds per day at startup. Figure 4 shows the cumulative amount of chlorobenzene removed during the pilot study. SVE performance was limited by the clayey soils and high water table. Dewatering of the area was hampered by periods of heavy rainfall during the study.

Groundwater pumping from the vapor extraction well and a second pumping well began one month prior to the start up of the vapor extraction system, and continued throughout the testing program. Over the 4-month period, approximately 263,000 gallons of groundwater containing about 600 pounds of chlorobenzene were extracted from the two pumping wells and treated by activated carbon adsorption. The flow rate was 3.5 gpm from the dual pumping/SVE well, and 0.5 gpm from the second pumping well. Figure 5 illustrates the influent and effluent concentrations of the carbon treatment system and the cumulative volume of groundwater treated. Groundwater chlorobenzene concentrations gradually increased for several weeks, peaking about the theoretical saturation value of chlorobenzene in water (about 450 mg/l) indicating the possible presence of colloidal pure product. The concentration

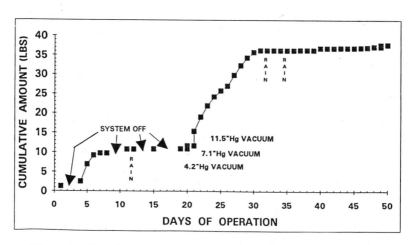

Figure 4. Cumulative Amount of Chlorobenzene Removed by SVE

then gradually decreased during the later stages of the testing program. It was concluded that most of the chlorobenzene in the well vicinity was removed; however, a gradual tailing of the groundwater concentration typical for pump and treat systems was expected. The peak chlorobenzene effluent concentrations observed were the result of the actual carbon canister replacement sequence, versus required replacement time that often occurred over weekend periods, and the time to obtain analytical results. Figure 6 illustrates chlorobenzene removal by activated carbon. During the study it was found that the activated carbon loaded to exhaustion could remove up to 0.28 pounds of chlorobenzene per pound of carbon at the groundwater concentrations treated.

A full scale SVE system layout was designed. It was estimated that it would require 20 or more wells to fully dewater the area, and to draw air through the soils. A program of intermittent operation of the wells was anticipated. Remediation of the soils using a combined SVE system was estimated to take from 1 to 2 years or more to reduce chlorobenzene in the soil to an acceptable level. In addition, due to the nature of the chlorobenzene and soils characteristics, it was concluded that it would be difficult to determine and actually accomplish adequate remediation by the combined soil vapor extraction, pump and treat method.

Site specific factors such as clayey soils and a high water table, combined with time constraints imposed by impending new construction, focused selection of remediation on methods of short duration. The remedial action determined to best meet remediation requirements and site constraints was soil excavation. The NJDEP cleanup level of chlorobenzene in industrial non-residential soils was 600 mg/kg. Cleanup requirements are much more stringent in cases where a groundwater aquifer was affected. In this case, the groundwater was not a useable source nor did it effect any useable aquifer. The NJDEP agreed on using 100 mg/kg as a basis for design of the excavation. The excavated soils would be remediated by on-site low temperature thermal desorption and the cleaned soils reused as fill on the site. Groundwater treatment of the plume area by a pump and treat system would proceed following removal of the source chlorobenzene soils.

EXCAVATION AND TREATMENT

Selective excavation of soils containing greater than 100 mg/kg was carried out over a 3-day period. Soils near the ground surface that contained less than 100 mg/kg were removed to a clean area. Layers of soils containing chlorobenzene were then excavated down to 15 to 25 feet depth depending upon the location of chlorobenzene during the drilling program. Observation of the soils and field screening measurements were made of soils as they were excavated to direct the actual excavation areas. The predesigned area of excavation was extended in a southerly direction to capture soils with higher readings than had been indicated by soil borings. Results from sampling of soils on the excavation perimeter indicated that soils with chlorobenzene concentrations greater than about 1 mg/kg had been removed, ensuring the goal

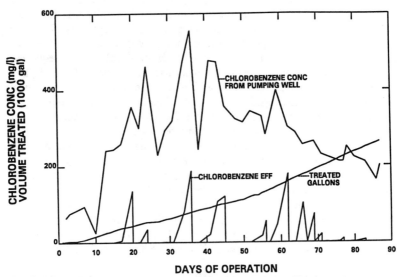

Figure 5. Groundwater Chlorobenzene Treatment

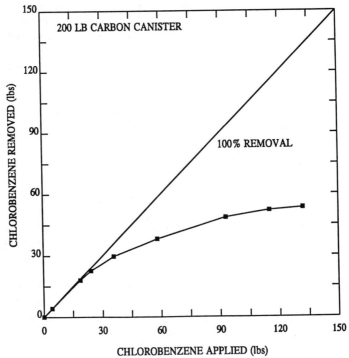

Figure 6. Chlorobenzene Removal by Activated Carbon

of 100 mg/kg was achieved. The soils were transported to an on-site storage and treatment area away from the new construction. Approximately 5,000 cubic yards of soil were excavated and stored for subsequent treatment.

A low temperature, thermal desorption system (IRV100) was provided by a subcontractor, McLaren/Hart of Albany, New York. This system was selected after investigating several other proposed systems, and the construction of a non-proprietary system from available used equipment. The IRV 100 low temperature thermal desorption unit (LTDU) utilized on the site consisted of four 9-foot by 18-foot by 2-foot steel ovens capable of holding up to eight cubic yards of soil each. Well screens on the oven bottoms were packed in pea gravel and protected by steel tracker. Soil was loaded into the unit to a depth of 12 to 18 inches using a front-end loader. The top of the LTDU contained 16 propane-fired infrared heaters, and was rolled into place after loading. The soil was heated for about 1.5 to 2.5 hours as chlorobenzene was extracted by hot air drawn through the soil into the well screens by the vacuum system. The air containing chlorobenzene was cooled and passed through a proprietary resin adsorption system manufactured by "PURUS". The extracted air was discharged to the atmosphere. Chlorobenzene extracted was concentrated into a liquid, stored on site and subsequently disposed off site by the contractor.

Soil from each treatment batch was stockpiled and sampled. The specification required that the treated soil measure less than 10 mg/kg chlorobenzene. Several batches, particularly during initial operations, did not meet this level and were reprocessed after adjustments to the thermal system operating parameters were made. Typically, concentrations measured in acceptable batches were from trace to 2 mg/kg.

The thermal treatment treated 5,080 cubic yards of soil in 46 days. Based on pre-treatment soil samples, and results from treated samples collected each shift, approximately 3,950 pounds of chlorobenzene were removed from the soils. This was somewhat higher than the original estimate of 3,300 pounds of chlorobenzene. Measurement variability, the variable concentrations of the DNAPL in the soil and somewhat greater volume then originally anticipated probably account for the higher removal.

CHLOROBENZENE PLUME REMEDIATION

Following removal of the source area of the chlorobenzene, a work plan was developed and approved by the NJDEP for treatment of the groundwater containing chlorobenzene. A groundwater pump and treat system has been designed, with the startup scheduled for mid 1995. Pumping will focus initially at the excavation, which was backfilled with coarse gravel. Two pumping wells will be installed within the backfilled excavation. Four additional pumping wells will be installed in the lower sand zone in the plume area containing greater than 20,000 $\mu$g/l chlorobenzene. Pumping will continue in these wells with regular monitoring of the chlorobenzene concentration.

Figure 7 illustrates the location of the pumping wells and treatment system. The estimated area of influence over the initial 90 days of pumping is illustrated

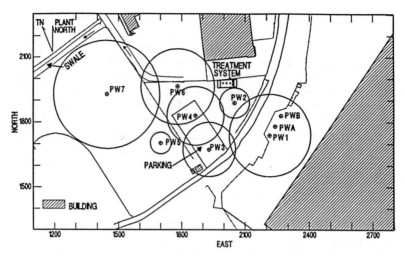

Figure 7. Location of Pumping Wells and Treatment System

by the circles around each well. The time and water trace estimated by a modeling program "QuickFlow" and shown on Figure 8 illustrates the expected coverage of the area by the wells. Input data included soil hydraulic conductivity and sand zone thickness in the area of the pumping wells. Actual data will be obtained during operation and analyzed.

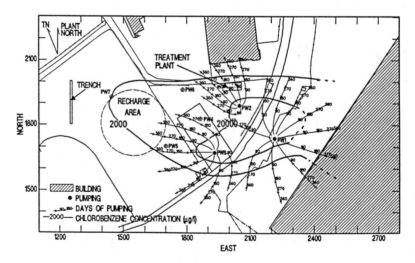

Figure 8. Estimated Area of Influence of Pumping Wells

926

Groundwater pumped from the wells will be piped to a treatment system utilizing activated carbon to adsorb chlorobenzene. Three 1000 pound units in series will be used. Treated effluent will be discharged to the groundwater system through a surface recharge system and a discharge trench to help maintain a supply of groundwater moving through the system. It is planned to monitor the results of the pump and treat program and to modify the pumping program based on results. Pumping will be continued until there is no longer a reasonable improvement in the groundwater quality commensurate with the site characteristics. There is little potential for environmental effect due to a low residual chlorobenzene concentration in this groundwater area as previously indicated. The possibility of enhancing natural degradation over the long term by natural biological means will be considered as the pump and treat program achieves its purpose.

# INDEX

ABRAMS, S. H., 433, 869
Acid Extraction, 551
Activated Carbon, 248, 538, 622
Activated Sludge, 274, 314
Activated Sludge System, 155
ADAMS, T. L., 712
Adsorbent Resins, 909
Adsorption, 503, 518, 600, 610, 644, 712
Adsorption Columns, 538
Aeration, 322
AGARWAL, R. K., 155
Air Emission, 165
Air Sparging, 433, 773, 869
Alum Recovery, 33
Anaerobic Biodegradation, 187
ANTUSCH, E., 655
AOX, 655
Aquifer Evaluation, 689
Aquifer Remediation, 681
ARCHER, H. J., 900
Artificial Neural Networks (ANN), 782
Asphalt Mix, 72

BANDYOPADHYAY, A., 233
BANDYOPADHYAY, N., 175
BARLAZ, M. A., 450
BARRETT, D., 669
BASHEER, I. A., 538
Beneficial Reuse, 450
Beneficial Use, 893
Bentonite, 712
Benzene, Toluene, Ethylbenzene, Xylene (BTEX), 773
BERNSTEIN, C., 62
BHAT, N. N., 371
BHATTACHARJEE, M., 733
BHAUMIK, S., 187
Binding Energy, 754
Bio-Attenuation, 304
Bioavailability, 233

Biodegradable Plastic, 23
Biodegradation, 284, 294, 763
Biofilter, 264, 322
Bioreactor, 241
Bioremediation, 194, 213, 849
Biosorption, 223, 248
Biotransformation, 203
Bioventing, 880
BISHOP, N., 43
BLEAM, R., 194
BOD, 241
BORDEN, R. H., 450
BOWERS, D. L., 333
BRANDL, P. J., 551
BRETON, N. M., 722
BREWSTER, M. D., 632
BROOKS, J., 322
Bulking, 274
Butylchloride, 352

CAAP, 129
Cadmium, 588
CAMU, 763
Carbon, 511
Carbon Adsorption, 918
CARGNEL, D. A., 600
CARLTON, J. C., 893
CARRIERE, P. E., 264, 294, 322, 414, 424, 622, 849
CASE, J., 900
Case Studies, 3
Ceramic, 511
Chelant Extraction, 632
Chelating Exchangers, 561
Chemical Grouts, 450
Chemical Stabilization, 450
Chemostat, 203
CHEN, C.-H., 518
CHEN, C.-W., 361
CHENG, S.-C. J., 610
CHIA, T.-S., 383
Chlorinated Compound, 518, 528

929

Chlorinated Compounds, 144, 213
Chlorinated Solvents, 849
Chlorobenzene, 371, 918
Chlorophenol, 342, 361
Chlorophenols, 655
CHRISTODOULATOS, C., 187, 284
CHRISTOPHER, T. W., 893
Chromatography, 722
Chromium, 223, 588, 814
Chromium Contaminated Soils, 470
Clarification, 314
Clarifier Sludge, 33
Coal Tar Contaminated Soils, 733
COD, 241
COHEN, D. M., 470, 733
Coke Oven Plant, 480
COLE, C. A., 600
Cometabolism, 203
Competitive Adsorption, 528
Complex Cyanide, 480
Complexation Process, 480
Composite Ion Exchange
    Membrane, 33
Composting, 23
Confining Units, 689
Construction, 91
Construction Materials, 470
Contaminant Transport, 712
Contamination, 91
Coolant Waste, 424
Copper, 588, 644
COREE, B. J., 72
Cost, 91
Cost/Benefit, 118
Coupled Flow, 743
Cr(VI) Reduction, 568
Creosote, 294
CUDDEBACK, J. E., 849
Cyanide, 480

DASGUPTA, A., 165
DAVIS, A. P., 62
DAVIS, M. P., 849
DAVIS, W. T., 175
DE ROSE, N., 304
Delays, 91
DELP, S. N., 669
DEPPEN, E. A., 669
Desorption, 233, 294, 754
Dewatering, 395
Dibromoethane, 722
Dichlorophenol, 383
Digital Signal Processing (DSP), 782

Dinitrotoluene, 284
Dissolved Air Flotation, 314, 414
DiSTEFANO, T. D., 213
DONG, C., 361
DONOVAN, E. J., JR., 918
DROSTE, E. X., 433, 869
Drying, 395
DUDIAK, K. A., 712
DUNN, C., 424
DURAL, N. H., 518, 528

EBERLE, M. F., 880
Economy, 100
ECRA, 900
EDTA, 450, 795
Effluent Treatment, 241
EKMEKYAPAR, A. 644
Electro-Osmosis, 814, 835
Electrokinetic, 814
Electrokinetic Decontamination, 795
Electrokinetic Processes, 824
Electrokinetics, 804, 835
Electroless Nickel, 588
Electroplating, 490
Electropolishing, 62
Emergency Preparedness, 669
Emission Inventory, 175
ENDY, D., 255
Environment, 100
Environmental Evaluation, 900
Environmental Issues, 109
Environmental Permit, 135
EPLIN, J. W., 763
Ethylenediaminetetraacetate (EDTA),
    578
EVANS, J. C., 712
Excavation, 918
EZELDIN, S., 470

Ferrous Sulfate, 480, 568
Filtration, 503
FISHER, J. R., 405
Fluid Treatment, 503
Fluoride, 490
Fluoroborate, 490
Fly Ash, 588
FORSYTH, J. V. III, 194
Foundry Wastes, 450
FREDERICK, R. M., 733
Gasification, 43
Geocomposites, 782
GIETKA, P. M., 62
Gore-Tex® ePTFE, 909

Granular Activated Carbon, 223
GROMICKO, G. J., 773
Ground Water Remediation, 433, 869
Groundwater, 304, 681, 722, 814, 849
Groundwater Extraction, 681
Grout, 450
GÜL, R., 644
GUNDERSON, M. J., 722
GUROL, M. D., 352, 371

HAHN, H. H., 655
HAMMONS, R. W., 722
HARTNETT, R., 135
Hazardous Waste, 763
Hazardous Wastes, 849
HEALY, T. C., 551
Heavy Metal, 578
Heavy Metal Decontamination, 795
Heavy Metal Removal, 503
Heavy Metals, 223, 561, 622, 632, 644
    804
Heterogeneous Catalysis, 371
Heterogeneous Oxidation, 352
HEUCKEROTH, D. M., 880
Hexachlorobenzene, 900
HMX, 129
HO, W., 733
HOWELL, J. A., 241
HSIEH, Y.-H., 342, 383
HUANG, C., 551
HUANG, C. P., 814, 835
HUANG, K.-S., 342
Hydraulic Barrier, 11
Hydraulic Conductivity, 443
Hydrocyclone, 53
Hydrogen Peroxide, 352, 361
Hydrogeochemistry, 689

Indirect Discharges, 655
Industrial Wastes, 109
Industrial Wastewater Treatment, 333
In Situ Soil Treatment, 824
Ion Exchange, 33, 578
Iron Ore, 600
Iron Oxide, 352, 371, 600
Isotherms, 712

JABLONSKY, J., 511
JACOBSON, P. R., 91
JAMIL, M., 622
JANCUK, W. A., 405
JOHNSON, J. H., 203, 754

KAMOLPORNWIJIT, W., 470
KARANJKAR, A. M., 248
KERECZ, B. J., 314
KHAN, L. I., 795
KIM, Y., 274
KLINE, J. P., 165
KLINGENSMITH, R. C., 773
KNEY, A. D., 578
KOCH, J. R., 53
KOCHER, W. M., 450
KODALI, S., 284

Laboratory Wastewater, 405
LAIRD, H. S., 91
Landfill, 880
Landfills, 11
LANE, R. D., 53
LARSON, M. A., 23, 155
LAUMAKIS, T. M., 610
LAWTON, L. J., 144
Leaching, 450
Lead, 588, 610, 632, 804
LENNON, G. P., 681
LI, P., 33
LI, W., 632
LIBRIZZI, B., 470
Ligand, 578
LIN, C.-J. J., 568
LIN, S.-S., 352
LITTLEFIELD, K. V., 859
LONG, D. D., 664
LORKOWSKI, T., 414
LOVELL, C. W., 72, 109
LYNN, J. D., 53

MacNAIR, W. H., JR., 859
MADABHUSHI, B. S., 264
MAGEE, R. S., 733
Magnetite, 600
MALONE, J. M., 450
Manufactured Gas Plants, 909
MARLEY, M. C., 433, 869
MARTIN, J. P., 610
MARTINO, L. E., 632
MARTINS, C., 223
Massachusetts, 722
McINTOSH, K. R., 814
McKENNA, G. F., 470
MEEGODA, J., 470, 733
Membrane, 241, 561
MESANIA, F. A., 294
Metal Removal, 588
Metals, 248

Metals Removal, 600
Microfiltration, 511
Microorganisms, 233
MIHALICH, J. P., 144
MILLER, D. R., 314
MILLER, G., 632
MILLER, T. L., 175
Millpond Sludge, 795
Modeling, 175, 284
MOHAGHEGH, S. D., 264
Monitoring, 700
Mono-Chlorophenol, 835
MOO-YOUNG, H. K., JR., 11
MUELLER, R. T., 470
MYRICK, M. K., 118, 763

NAJJAR, Y. M., 538
Naphthalane, 233
NETO, P., 223
Neural Networks, 538
NICKELSON, M. D., 664
Nitroglycerin, 187
Nitrous Oxide, 255
NONAVINAKERE, S., 129, 588
Non-Destructive Testing (NDT), 743,
    782
NOVAL, B. A., 470
NPL, 129

O'BRIEN, N. M., 23
Oil Removal, 322, 414
ORANSKY, J. J., 669
Organic, 622
Organics, 43, 248
Oxidation, 352, 371
Oxygen Uptake, 880
Ozonation, 371
Ozone, 175

Packed-Bed Reactor, 284
PAH, 655
PAL, N., 187, 284
PAMUKCU, S., 610, 743, 782, 824
Paper Sludge, 11
PARK, T., 72
PARTRIDGE, G. P., 165
Passive Soil Vapor Survey, 909
PATTON, T. L., 632
PENG, D., 528
Performance Optimization, 314
Permeability, 11
PERVIZPOUR, M., 743
PESCE, E. L., 722

PETERS, R. W., 632
Petroleum, 304
Petroleum Hydrocarbon, 700
pH Variation, 795
Phanerochaete Chrysosporium, 284
Phenol, 538, 835
Photocatalysis, 342, 383
Photocatalyst, 361
Photodegradation, 342, 383
PIPES, W. O., 274
Pollutant Transport, 165, 700
Pollution, 100
Pollution Prevention, 3, 62
Polynuclear Aromatic Hydrocarbons
    (PAHs), 909
Polyvinyl Alcohol, 23
Precoat, 503
Process Control, 62
PUCCI, A. A., JR., 689
Pump and Treat, 681, 918
PVOH, 23
Pyrene, 754
Pyrolized Carbon Black, 72

RAGHAVAN, D., 754
RAHMAN, M., 795
RAMAN, S., 849
RAMSEY, J., 804
RAPPA, P. III, 129
RDX, 129
Reactive Chemical, 135
RECKER, S. A., 763
Reclamation, 450
Recycle, 893
REDOX Manipulation, 632
Reduced Costs, 664
Reduced Field Time, 664
Reduction, 255
REED, B. E., 322, 414, 424, 588, 622,
    804
Refractories, 53
Regional-Flow Model, 689
Regulation, 100
Remedial Design, 433, 869
Remedial Investigation, 918
Remediation, 304, 804, 814, 859, 900
Remediation Cost, 144
Remediation Vitrification, 470
Removal Action, 129
Reservoir Conditioning, 804
Residual Materials, 610, 782
Respirometric, 294
Reuse, 470

932

Reuse/Recycle, 109
*Rhodocossus,* 203
RICHTER, M. F., 135
RIPP, C., 655
Risk, 118
Risk Assessment, 118, 669
Risk Communication, 118, 669, 763
Risk-Based Cleanup, 144
Risk-Based Cleanup Level, 900
RISNER, A. C., 700
ROBINSON, K. G., 233
ROMANO, F. J., 333
ROMERO, S., 782
RYAN, M. E., 551
RYKACZEWSKI, M. J., 880

SACK, W. A., 849
Safety, 135
SAILING, J. H., 85
SANTORA, S., 470
SAWICKI,, J. E., 333
SCOTT, J. A., 241, 248
Screening, 722
SENGUPTA, A. K., 33, 561, 578
SENGUPTA, S., 561
SERT, A., 644
Settling Tank, 274
Sewer Slime, 655
SHIEMKE, A. K., 849
SIESS, A. A., JR., 100
Signature Patterns, 782
SIMPSON, D. K., 773
Simulation Models, 681
Site Characterization, 118, 664
Site Closure, 118
Slag, 53
Sludge, 395
Sludge Treatment, 333
Sludges, 561
Slurry Walls, 712
SMITH, J. R., 503
SMITH, K. L., 241
Soil, 518, 528, 644, 814
Soil-Bentonite, 443
Soil Colloids, 233
Soil Mixing Equipment, 443
Soil Remediation, 835
Soil Stabilization, 450
Soil Vapor Extraction (SVE), 773, 859, 918
Soil Washing, 450, 551, 733
Solid Phase, 561
Solid Waste, 644, 893

Solid Waste Disposal, 33
Solids Storage, 274
Solubilizing Agents, 824
Solvent Extraction, 450
SPARROW, L. A., 918
Stainless Steel, 62
Steel Mill, 53
Subgrades (Waste Disposal Facility), 610
Surfactant, 294, 551
Surfactants, 733
Sustainable Development, 109
SWMU, 763
Syngas, 43
Synthetic Sorbent, 578

TAKIYAMA, L. R., 835
TAVARES, T., 223
Technology Transfer, 85
Temperature, 880
Tetrachlorophenol, 383
THARAKAN, J. P., 203
Thermal Consolidation, 743
Thermal Desorption, 918
Thermal Extraction, 754
THOMAS, B., 622
Titanium Dioxide, 342, 361, 383
TNT, 129
Transformation Products, 187
Transport Modelling, 165
Treatability, 490
Trichloroethylene, 849
Trichlorophenol, 383
Trinitrotoluene, 203

UAM, 175
Ultrafiltration, 424, 511
Ultrasound, 733
Underground Storage Tanks, 304
USINOWICZ, P. J., 3, 490
UV Light, 361

VACCARI, D. A., 470
Vacuum Extraction, 880
VAN BENSCHOTEN, J. E., 551
VARUNTANYA, C. P., 480
VESILIND, P. A., 395, 568
VOC, 175, 264
VOC Emissions, 155
Volatile Organic Compounds, 264, 538
Volatilization, 763
Voltage, 804

933

WALDEN, L., 470
WANG, K.-H., 342
Waste, 91
Waste Disposal, 11
Waste Disposal Facility, 610
Waste Minimization, 3, 62
Waste Tires, 72
Waste Treatment, 503
Wastes, 100
Wastewater, 490
Wastewater Treatment, 255, 314, 405
Water, 395
Water Quality, 689
Water Treatment, 568
WATKINS, D. M., 490
WEBBER, T., 443
WEEKS, A., 824
WEI, C. F., 733
WELSH, G. E., 203

Wet Oxidation, 333
WHITEMAN, C. S., 849
WILLIAMS, M., 443
Windrow, 23
WRIGLEY, M. J., 909
WRUBEL, N., 194
WU, S.-S., 361

Xylene, 859

YARTAŞI, A., 644

ZABBAN, W., 480
Zeolite, 712
ZHU, X., 414
ZIMMIE, T. F., 11
Zinc, 600
Zinc Removal, 53

# INDICES FOR 22ND, 23RD, 24TH, 25TH & 26TH INDUSTRIAL WASTE CONFERENCES

Abandoned Mine Lands, XXV, 468
Abandoned Wastes, XXV, 305
Abatement Costs, XXVI, 380
ABT, S. R., XXII, 570
ABU-ORF, M., XXIV, 39
Accumulation, XXV, 146
Acetylene, XXII, 144
Acid Generation, XXVI, 715
Acid Mine Drainage, XXIV, 687
Acid Mine Drainage, XXVI, 715
Activated Carbon, XXIII, 148
Activated Carbon, XXIV, 101, 120, 563, 573
Activated Carbon, XXV, 105, 116, 283
Activated Carbon, XXVI,, 250, 259, 267
Activated Sludge, XXIII, 175
Activated Sludge, XXIV, 277, 312
Activated Sludge, XXV, 248
Activated Sludge Foaming, XXV, 159
ADAMS, V. D., XXVI,, 69
Adsorption, XXII, 57, 77, 83, 355, 382, 457, 626
Adsorption, XXIII, 188, 374
Adsorption, XXIV, 101, 120, 563, 573
Adsorption, XXV, 105, 116, 146, 283, 506
Adsorption, XXVI, 241, 275, 285, 311, 733
Advanced Oxidation Processes, XXVI, 178, 186
Aerobic, XXII, 57, 341, 383, 414, 425, 503, 626, 645, 669
Aerobic, XXIII, 198
Aerobic Treatment, XXIV, 593
AHMED, T., XXVI, 69
Aircraft Engines Plant, XXVI, 701
Airports, XXV, 330
Air Emissions, XXV, 42
Air Injection, XXV, 238
Air Pollutants, XXV, 31
Air Quality, XXII, 503, 693, 702, 743

Air Quality, XXIII, 337
Air Sparging, XXIV, 468
Air Stripping, XXII, 41, 269, 317, 445, 669
Air Stripping, XXIII, 24
Air Stripping, XXV, 136, 523
Air-Water Exchange, XXV, 449
AKRAM, M. H., XXVI, 34, 567
AL-GHUSAIN, I. A., XXVI, 61
AL-YOUSFI, A. B., XXIII, 111
ALBERICI, R. M., XXVI, 230
ALFARO, F. DE M., XXVI, 311
Alginate, XXVI, 275
ALHUMOUD, J. M., XXVI, 388
Alkalimetry, XXIV, 602
Alkaline Neutralization, XXV, 506
ALLEN, H. E., XXIV, 359
ALLEN, H. E., XXVI, 742
ALLEN, K. L., XXII, 290
ALMES, W. S., XXII, 251
Aluminum Processing, XXII, 243
AMD, XXVI, 715
Ammonia, XXIV, 79
Ammonia Removal, XXIII, 25
Ammunition Waste, XXVI, 61
Anaerobic, XXV, 209, 419, 551
Anaerobic (Anoxic), XXII, 57, 341, 382, 412
Anaerobic-Aerobic Treatment, XXIII, 175, 267
Anaerobic Digestion, XXIV, 602
Anaerobic Digestion, XXVI, 115
Anaerobic Lagoon, XXVI, 109
Anaerobic Toxicity Assay, XXVI, 742
Anaerobic Treatment, XXIV, 197, 593
Anaerobic Treatment, XXV, 218
Analysis, XXV, 348
Analysis of Variance, XXVI, 221
ANDERSON, P., XXII, 223
Aniline, XXVI, 241
Anionic, XXIII, 3
Anionic Surfactants, XXVI, 742

Antifreeze, XXIV, 238
Appropriate Treatment, XXVI, 109
Aquatic Life (Fish, etc.), XXII, 223, 321, 626, 695
Aquifer Contamination, XXIV, 27
Aquifer Tests, XXIV, 27
Arsenic, XXII, 322
Arsenic, XXV, 83
ARUMUGAM, S. V., XXII, 485
ARUNACHALAM, S., XXIV, 101
Ash, XXII, 155, 171, 243, 251, 470, 613, 743
Ash, XXV, 197
ASH, J. M. XXVI, 629
Assessment, XXIII, 293
ATA, XXVI, 742
Atrazine, XXVI, 194
Audit, XXIII, 306
Automotive Coolant, XXIV, 238
AWAAD, G. S., XXVI, 42
Azo Dyes, XXIII, 359
Azo Dyes, XXIV, 593
Azo Dyes, XXVI, 186


BACKUS, D., XXII, 453
Bacteria, XXVI, 151
BAKER, B., XXIV, 687
BALDWIN, P. N., XXIII, 3
BALDWIN, P. N., XXIV, 341
BANTA, L., XXV, 321
BARCLAY, P., XXIII, 348
BARNHORN, L. K., XXIV, 69
Barrier, XXII, 66, 195, 261, 278, 294
BARTLETT, C. L., XXVI, 87
BARTONE, M., XXII, 759
BARTEK, L. R., XXIV, 27
BASS, D. H., XXIV, 458
Batch Reactors, XXVI, 123
Batteries, XXII, 6
BDAT, XXII, 33
BEETSCHEN, L. J., XXVI, 396
BENSON, C. H., XXII, 195
BENSON, C. P., XXVI, 20
Bentonite, XXII, 77
Bentonite, XXVI, 311, 522
Benzene, XXIV, 583
Benzene, XXVI, 647
BERCHTOLD, S. R., XXIV, 573
BERG, M. T., XXV, 55, 105, 283, 478
BERG, M. T., XXVI, 480, 514, 715
BERNARDO, M., XXVI, 522
BERRY, S. W., XXII, 425

Best Management Practices, XXVI, 575
BHADA, R., XXV, 449
BHADA, R. K., XXIV, 89
BHADRIRAJU, M., XXIV, 312
BHAGAT, S. K., XXII, 485
BHOWMICK, M., XXV, 136, 478
Bicarbonate, XXIV, 602
BIEHL, F. J., XXII, 594
Bioaccumulation, XXII, 224, 626, 695
Bioassay, XXIV, 79
Bioassays, XXII, 358, 695
Biodegradation, XXII, 57, 95, 156, 377, 399, 414, 626, 645
Biodegradation, XXIII, 198, 211
Biodegradation, XXIV, 135, 563, 573, 583, 421
Biodegradation, XXV, 209, 238, 330, 551, 561
Biodegradation, XXVI, 115, 647
Biogas, XXIII, 267
Biological Filter, XXVI, 141
Biological Oxidation, XXIV, 290
Biological Process, XXVI, 396
Biological Remediation, XXVI, 629
Biological Treatment, XXII, 339, 355, 377, 398, 412, 425, 669
Biological Treatment, XXIV, 179
Biological Treatment, XXVI, 141, 241, 522
Biomedia, XXV, 248
Biomonitoring, XXIV, 79
Bioreactor Design, XXIII, 163
Bioremediation, XXII, 41, 57, 425, 626
Bioremediation, XXIII, 211, 259
Bioremediation, XXIV, 157, 197
Bioremediation, XXV, 187, 439
Bioremediation, XXVI, 53, 131, 629
Biosolids, XXV, 409
Bioventing, XXV, 238
Bioventing, XXVI, 488, 528
BIRD, N., XXVI, 557
BISHOP, P. L., XXIII, 385
BLAIR, J. E., XXII, 669
BLATT, R. G., XXIII, 52
BLISS, S. G., XXIII, 267
BMPs, XXVI, 575
BOARDMAN, G. D., XXIV, 593
BOARDMAN, G. D., XXV, 218
BOE, R. W., XXV, 218
BOONE, D. A., XXV, 330
BOONE, S., XXVI, 557

BOTTS, J. A., XXVI, 362
BOUWER, T. J., XXIV, 197
BOWDERS, J. J., XXII, 251
BOWDERS, J. J., XXIV, 687
BOWDERS, J. J., JR., XXV, 197
BOWEN, R. B., XXV, 248
BOWERS, A. R., XXIV, 135
BOYLE, E., XXV, 409
BOZZELLI, J. W., XXVI, 417
BOZZINI, C., XXV, 83
BROSNAN, T. M., XXVI, 657
BROSS, J. M., XXII, 290
BROWN, A., XXIV, 687
BROWN, G. J., XXIII, 267
BROWN, W. A., XXIII, 3
BROWN, W. A., XXIV, 341
BROWNAWELL, B. J., XXV, 531
BROWNING, J. S., III, XXII, 594
BRUGGEMANN, E. E., XXIII, 188
BRUNNER, M. R., XXIV, 505
BRYERS, J. D., XXVI, 151
BUCHANAN, R. J., XXII, 43
BUCHANAN, R. J., JR., XXVI, 3
BUGLASS, R. L., XXV, 541
BULLER, P. G., XXII, 213
BURKS, S. L., XXII, 355
BURROWS, W. D., XXII, 339
BURROWS, W. D., XXIII, 188
BUTLER, C. R., XXV, 197
BYER, H. G., JR., XXII, 43

CADENA, F., XXII, 77, 718
Cadmium, XXIV, 359
Cadmium, XXV, 72
Calibration, XXIV, 368
California List, XXII, 32
CAMPBELL, M. P., XXVI, 141
Capillary Suction Time, XXVI, 742
Capping, XXIII, 127
Capstone Course, XXV, 449
CAPUTI, J. R., XXII, 31
CARBERRY, J. B., XXIII, 198, 259
Carbon, XXIII, 188, 374
Carbon Adsorption, XXII, 269, 311, 355, 399, 445, 522
CARRIERE, P., XXVI, 619
CARRIERE, P. E., XXVI, 131
CARROLL, C. G., XXII, 355
Catalytic Oxidation, XXV, 523
CATHERMAN, D. R., XXV, 348
Cation Exchange, XXII, 80
Cation Exchange, XXV, 93
Cationic, XXIII, 3

Cellular Nitrogen, XXIV, 312
Cement, XXII, 155, 172, 252, 543, 594
Cement, XXIII, 11
Cement, XXV, 3
Cement, XXVI, 567
Centrifuge, XXV, 178
CERCLA, XXIII, 283
CERCLA (Superfund), XXII, 10, 31, 37, 155, 438, 453
CHADDERTON, R. A., XXIV, 677
CHAMBERLAIN, B. J., XXVI, 338
CHANG, C.-F., XXV, 23
CHANG, C.-Y., XXV, 491
CHANG, C.-Y., XXVI, 221
CHANG, F.-Y., XXV, 146
CHANG, K.-T., XXV, 42
CHANG, M.-C., XXVI, 186
CHANG, S.-C., XXV, 42
CHARLTON, T. J., XXVI, 629
CHATTERJEE, S., XXIV, 27
CHAVEZ, R. P., XXIII, 148
CHAWLA, R. C., XXIII, 317
Checklist for Emergency Plan, XXIII, 348
Chelate Agent, XXVI, 595, 611
Chemical Oxidation, XXII, 41, 57, 355, 382
Chemical Oxidation, XXIII, 211
Chemical Oxidation, XXIV, 135
Chemical Precipitation, XXVI, 109
Chemical Vapor Hazards, XXIII, 348
CHEN, B., XXIV, 359
CHEN, J. M., XXV, 541
CHEN, J.-N., XXV, 3, 266, 491
CHEN, J.-N., XXVI, 221
CHEN, N., XXVI, 585
CHEN, T. N., XXIV, 120
CHEN, W.-T., XXV, 168
CHENG, J. M., XXVI, 61
CHERN, H.-T., XXVI, 417
CHIANG, P.-C., XXV, 42
CHITIKELA, S., XXVI, 742
Chitosan, XXVI, 275
Chlorinated Aliphatic Hydrocarbons, XXVI, 131
Chlorinated Gas, Inhalation, XXIII, 348
Chlorinated Hydrocarbons, XXIII, 368
Chlorinated Hydrocarbons, XXV, 523
Chlorinated Organics, XXIII, 211
Chlorine Cylinder Repairs, XXIII, 348
Chlorobenzene, XXIV, 197

Chloroform, XXII, 97
Chloroform, XXV, 551
2-Chlorophenol, XXV, 491
Chlorophenol, XXVI, 472
CHOU, C. P., XXII, 17
Chromium, XXII, 459
Chromium, XXIV, 257
Chromium, XXV, 3, 187, 366
Chromium Contaminated Soil, XXVI, 496
CHU, B. J., XXII, 613
CHU, C.-S., XXVI, 178
CLARKSON, C. C., XXV, 178
CLAY, XXII, 66, 77, 155, 195, 243, 543, 580, 594
CLINE, S. R., XXV, 93
Closed Loop, XXV, 136
CMC, XXVI, 303
COAD, R. M., XXII, 321
Coagulation/Flocculation, XXII, 382, 522
Coal, XXIII, 63, 67, 385
Coal, XXV, 468
Coalescence, XXV, 500
Coal-Tar, XXIII, 148
Coal Waste, XXII, 120, 155, 171, 251, 594
Coastal Aquifers, XXVI, 347
COCCI, A. A., XXIII, 267
COD, XXV, 506
COLE, C. A., XXIII, 52
COLE, C. A., XXIV, 300
COLLINS, A. G., XXIII, 295
COLLINS, M., XXVI, 362
Color Removal, XXV, 218
COLSMAN, M. R., XXVI, 285
COLVIN, R. J., XXIV, 277
Combustion, XXIV, 529
COMEL, XXII, 377
Community, XXII, 4, 9, 377
Compliance, XXII, 26, 144
Complexation, XXV, 168
Compost, XXV, 429, 439
Composting, XXII, 501
Composting, XXIII, 247
Compressive Strength, XXIV, 341
Computer Aided Instruction, XXIII, 295
Conditioning, XXIV, 39
CONNOR, J. M., XXII, 213
Construction Fill, XXV, 381
Construction Materials, XXII, 155, 171, 243, 251, 559

Consumer Price Index, XXII, 130
Containment, XXIV, 687
Containment, XXV, 321
Containment, XXVI, 87
Containment Cover, XXIV, 665
Contaminant Fate and Transport, XXIII, 106, 127
Contaminant Modeling Hazard Assessment, XXIII, 96
Contaminant Transport, XXVI, 96, 724
Contaminant Transport Migration, XXII, 195
Contaminant Transport Model, XXVI, 724
Contaminate Mobility, XXIV, 341
Contaminated Aquifer, XXVI, 629
Contaminated Groundwater, XXV, 523
Contaminated Soil, XXV, 13, 65, 93, 310, 439
Contaminated Soil, XXVI, 480
Contamination, XXVI, 77
COOPER, A. B., XXII, 521
COOPER, W. J., XXII, 95
Copper, XXV, 126, 168
Corrosion, XXIII, 317
Corrosion, XXV, 541
Corrosives, XXII, 33, 146
Cost Analysis, XXIII, 67
Cost Recovery, XXIII, 283
Creosote, XXII, 453
CROUSE, G., XXVI, 454
Crude Oil Contamination, XXIV, 157
CST, XXVI, 742
CUDDEBACK, J. E., XXV, 228
CUDDEBACK, J. E., XXVI, 131
CULLINANE, M. J., XXIV, 333
Cyanide, XXV, 3
Cyanides, XXII, 32

2,4-D, XXV, 266
DAILY, T., XXII, 129
Dairy Wastes, XXII, 66
DAMERA, R., XXVI, 267, 637
Data Quality, XXV, 31
DAVIDSON, J. H., XXII, 655
DAVIES, S., XXIV, 411
DAVIS, A. P., XXIV, 110, 359
DAVIS, A. P., XXV, 275
DAVIS, A. P., XXVI, 170
Debris, XXVI, 454
Delmarva, XXVI, 329

Denitrification, XXIV, 197
Denitrification, XXVI, 123
DENNIS, R. M., XXIV, 515
DENNIS, R. M., XXV, 65
DENTEL, S. K., XXIV, 39
DENTEL, S. K., XXVI, 303, 742
DePINTO, J. V., XXIII, 295
DESF, XXVI, 724
Design, XXIV, 391
Desorption, XXVI, 472
DESTEPHEN, R. A., XXVI, 20
Dewatering, XXII, 454, 470, 485,
    501, 521
Dewatering, XXV, 178
DI TORO, D. M., XXVI, 667
2,4-Dichlorophenol, XXVI, 212, 221
2,4-Dinitrotoluene, XXVI, 61
DICKENSON, K., XXV, 500
Diesel, XXII, 425
DIETRICH, A. M., XXV, 218
Diffusion Dialysis, XXII, 136
DIMARINO, L. S., JR., XXV, 178
1,4-Dioxane, XXVI, 178
Dioxin, XXII, 32, 223
Dioxin, XXV, 478
Direct Photolysis, XXVI, 178
Disaster Management, XXIII, 348
Disaster Training, XXIII, 348
Disposal, XXII, 43, 66, 103, 130, 683
DiVINCENZO, J. P., XXVI, 303
DNT (dinitrotoluene), XXVI, 61
DOLAS, R., XXIII, 293
Dortmund Clarifier, XXIV, 290
DOUGLASS, C. H., XXVI, 53
DOWNEY, W. F., JR., XXII, 559
Drain Enhanced Soil Flushing, XXVI,
    724
Drains, XXII, 300, 321
Drilling, XXV, 178
Drills, XXIII, 348
Drip Fluids, XXV, 257
Drums, XXII, 438
DUDLEY, J., XXII, 223
DUJARDIN, C. L., XXVI, 667
DUNCAN, J., XXII, 269
DUNCAN, S. G., XXIV, 58
Durability, XXII, 167, 172, 542,
    559, 580
Durability, XXV, 23
DURANTE, J., XXIV, 529
DUSING, D. C., XXIII, 385
DWORKIN, D., XXIV, 515
DWYER, T. E., XXVI, 354

DYER, J. A., XXVI, 380
DZOMBAK, D. A., XXIII, 329

E. Coli, XXV, 187
EBCT, XXVI, 619
Economics, XXII, 13, 129, 412
ECRA, XXV, 295
EDTA, XXIII, 175
Education, XXII, 718
Education, XXV, 449
Effluent, XXIII, 111
Effluent, XXVI, 657
EHRLICH, R. S., XXVI, 472
EICHELBERGER, J. G., XXIV, 79
Electric Industry, XXIII, 67
Electrical Resistivity, XXII, 118
Electricity, XXIII, 67
Electro-Fenton, XXVI, 178
Electro-Osmosis, XXVI, 496
Electrochemical Oxidation, XXVI, 585
Electrochemical Process, XXVI, 611
Electrodialysis Processes, XXVI, 585
Electrokinetic, XXVI, 436, 480, 496,
    514
Electrokinetics, XXV, 55
Electron Beam, XXII, 95
Electroosmosis, XXV, 55
Electroplating, XXV, 366
Electroplating & Circuit Board
    Manufacturers, XXVI, 557
Elutriation, XXII, 459
Emergency Response Plan, XXIII, 84,
    348
Emissions, XXIV, 529
Empty Bed Contact Time, XXVI, 619
Emulsions, XXV, 257, 500
End-of-Pipe Treatment Costs, XXVI,
    380
Energy Recovery, XXIII, 52
Enhanced Biodegradation, XXIII, 211
Environmental, XXV, 389
Environmental Design Contest, XXV,
    449
Environmental Equity, XXVI, 338
Environmental Remediation, XXV,
    458
Environmental Site Assessment,
    XXIII, 306
EP Toxicity, XXII, 38, 40, 249, 565
ESSEL, A. A., XXIV, 179
Estuary, XXVI, 675
Ethylene Glycol, XXIV, 238
Ethylene Production, XXVI, 602

ETTINGER, R. A., XXVI, 462
EVANS, J. C., XXII, 542
EVANS, J. C., XXVI, 522
Ex-Situ, XXVI, 629
Expanded Bed, XXV, 209
Expert Systems, XXIII, 295
Explosives, XXII, 103, 339
Explosives, XXV, 338, 439
Exposure, XXVI, 678
Extraction, XXII, 459, 613
Extraction, XXV, 491
Extreme Conditions, XXIII, 111
EYRAUD, P., XXII, 377

FANG, H. J., XXIII, 329
FANG, H. Y., XXII, 702
FANG, H. Y., XXIV, 263
FANG, H. Y., XXV, 381
Fate, XXV, 478
Fate and Transport, XXII, 224, 324,
    691, 702
FAULKNER, M., XXVI, 522
Fe(II)-EDTA Recovery, XXVI, 595, 611
Fe(III) Reduction, XXVI, 595, 611
Feasibility Study, XXIII, 127
FELDMAN, H., XXVI, 371
Fenton's Reagent, XXIII, 211
Fenton's Reagent, XXVI, 178, 194,
    205, 212
FGD Gypsum, XXIII, 11
Fiberglass Waste, XXII, 759
Filtration, XXV, 500
Filtration, XXVI, 34
FINGER, F. J., XXIV, 677
Finite Element Modeling, XXVI, 96
FIORUCCI, L. C., XXIV, 229
FIORUCCI, L. C., XXV, 31
First Aid for Chlorine Exposure,
    XXIII, 348
Fixation, XXIII, 385
Fixation, XXIV, 631
Fixed Bed Bioreactors, XXIII, 163
Fixed Film, XXIII, 163
Fixed Film Reator, XXVI, 141
FLEMING, E. C., XXVI, 285
FLEMMING, E. C., XXIV, 333
Flexible Membrane, XXII, 66
Flotation, XXIV, 290
Flow Modeling, XXVI, 96
Flow System Analysis, XXVI, 354
Flue Gas Cleaning Wastes,
    Desulfurization Sludge, XXIII, 385
Fluidized Bed, XXIV, 529

Fluidized Bed Bioreactor, XXIV, 583
Fly Ash, XXII, 155, 171, 243, 251,
    594, 613
Fly Ash, XXIII, 385
Fly Ash, XXIV, 687
Fly Ash, XXV, 506
Fly Ash, XXVI, 34, 311, 567
FOGG, S. R., XXVI, 123
Food Chain, XXII, 224
FORDHAN, S. M., XXVI, 69
FRANK, M., XXII, 438
FREEMAN, H., XXIV, 3
Frequency, XXIII, 111
FRY, V., XXII, 223
FU, G., XXVI, 585
Fuel Spill Models, XXIII, 106
Fugacity, XXV, 449
Fungal Biomass, XXV, 146
FUSCALDO, A., XXIII, 293

GAAL, M. D., XXV, 561
GABR, M. A., XXIV, 665
GABR, M. A., XXVI, 34, 42, 567, 724
GAC, XXIII, 374
GAC, XXV, 209
GAC, XXVI, 619
GADDIPATI, P., XXIV, 135
GALBRAITH, M., XXIV, 411
GALLAGHER, D. P., XXIV, 391
GALYA, D., XXII, 223
GAME, R. E., XXII, 521
GAO, Y.-M., XXVI, 293
GARG, A. K., XXVI, 53
GARON, K. P., XXVI, 87
GARRETT, K. E., XXVI, 285
Gas Chromatography, XXV, 330
Gas Leaks, XXIII, 348
Gas Phase, XXV, 136
Gas Phase, XXVI, 230
Gasoline, XXIV, 468
GAUDY, A. F., XXIV, 157
GAVASKAR, A. R., XXIV, 238
GC/MS, XXV, 349
Geomembranes, XXIII, 234
Geophysics, XXII, 118
GESALMAN, C. M., XXII, 3
GHASEMI, A., XXIV, 89
GHASSEMI, A., XXV, 305, 449
GIACOMINI, D., XXVI, 13
GLOWACKI, M. L., XXV, 338
GOBER, J. W., XXV, 409
GONG, R., XXIII, 385
GRAFTON, J. A., XXIV, 381

Granular Activated Carbon, XXVI, 619
GRATZ, J. H., XXVI, 575
GREENWALD, B. P., XXII, 425
GRIGGS, W. E., XXII, 144
GROBLER, F., XXV, 458
Groundwater, XXII, 13, 50, 57, 78, 118, 155, 213, 269, 295, 303, 438 626, 683
Groundwater, XXIII, 75, 84
Groundwater, XXIV, 359, 621
Groundwater, XXV, 228
Groundwater, XXVI, 3, 77, 96, 131, 178, 285, 321, 629, 637
Groundwater Contamination, XXIV, 197
Groundwater Recharge, XXVI, 354
Groundwater Remediation, XXVI, 637
Grouting, XXIV, 687
Grouting Jet, XXIV, 631
Guidelines, XXVI, 338
GUPTA, A., XXV, 551
GUPTA, M., XXV, 551
GUROL, M. D., XXII, 57
GUROL, M. D., XXIII, 211
GUSTAFSON, J. B., XXVI, 462
GUYER, P. D., XXII, 501

HALL, J. C., XXIV, 381
HALLORAN, A. R., XXV, 83
HANEWALD, R. H., XXIV, 257
HAO, O. J., XXIV, 110
HAO, O. J., XXV, 159, 541
HAO, O. J., XXVI, 61, 267
HARMON, C. D., XXIII, 259
HARRIS, R. H., XXIV, 69
HARTLEB, J. R., XXV, 257
HARTMANN, B. J., XXVI, 329
HATFIELD, J. H., XXV, 55
HATFIELD, J. H., XXVI, 436
HAYES, C. A., XXII, 269
Hazardous Substances Constituents, XXII, 5, 41, 57, 130, 355, 613, 626
Hazardous Waste, XXII, 32, 95, 213, 269, 398, 522, 683, 702, 718, 743
Hazardous Waste, XXIII, 52, 293
Hazardous Waste, XXIV, 665
Hazardous Waste, XXVI, 141, 388
Hazardous Waste Managers, XXVI, 338
HAZEBROUCK, D. J., XXIV, 468
HAZEN, C., XXIV, 431
HE, H. Y., XXIV, 368

HEAD, W. J., XXVI, 567
Heavy Metal, XXVI, 275, 293
Heavy Metals, XXIV, 515
Heavy Metals, XXV, 65, 72, 105, 126, 187, 283, 310, 366, 429, 508
Heavy Water, XXVI, 69
Herbicide, XXIV, 421
HERZBRUN, P. A., XXIV, 437
HICKEY, D. XXIII, 24
HIJAZI, H., XXII, 155
HILLIARD, G., XXVI, 454
HITCHENS, D., XXIV, 263
HMX, XXIII, 188
HNAT, J. G., XXII, 759
HOC's, XXII, 33
HOLLEY, D. F., XXIV, 245, 401
HOUGH, B. J., XXIV, 277
Household Cleaning Products, XXVI, 742
HOWE, R. A., XXVI, 285
HSIEH, H. N., XXII, 459
HSIEH, H.-N., XXVI, 115
HSIEH, Y.-H., XXVI, 163
HSU, M.-C., XXVI, 611
HSWA, XXII, 31, 43, 129
HUANG, C., XXVI, 275
HUANG, C. P., XXIV, 120, 359
HUANG, C. P., XXV, 266, 506
HUANG, C. P., XXVI, 178, 194, 205, 212, 230, 472, 496, 506, 585, 595, 611
HUANG, C.-R., XXVI, 186
HUANG, H. S., XXVI, 178, 595, 611
HUANG, J., XXV, 159
HUANG, L., XXV, 541
HUANG, S., XXIV, 583
HUANG, Y.-C., XXVI, 178
HUBER, W. A., XXII, 13
Humic Acid, XXV, 168
HUNT, C. D., XXVI, 691
Hybrid Inorganic, XXVI, 293
HYDE, L. M., XXVI, 329
Hydraulic Connector, XXVI, 354
Hydrogen Peroxide, XXII, 58
Hydrogen Peroxide, XXIII, 368
Hydrogen Peroxide, XXIV, 135
Hydrogen Peroxide, XXVI, 647
Hydrogen Precipitation, XXIV, 120
Hydrogeology, XXII, 118, 216, 271, 290, 438

IMAN, J. L., XXV, 523
Immobilization, XXIII, 385

Immobilization, XXVI, 275
In-situ, XXV, 310, 322
In-situ, XXVI, 13, 20, 131, 425, 480,
    496, 514
In-situ Bioremediation, XXIV, 179
In-situ Treatment, XXII, 41, 57, 155,
    425
Incineration, XXII, 41, 453, 533, 613,
    743, 759
Incineration, XXV, 31, 42
Indirect Photolysis, XXIII, 359
Industrial Discharge, XXVI, 362
Industrial Facility, XXII, 31, 43, 103,
    438, 453, 771
Industrial Waste, XXII, 14, 95, 355
    425, 521, 645, 655, 743
Industrial Waste, XXIV, 277
Industrial Wastewater, XXIII, 24, 84
Industrial Wastewater, XXIV, 391
Industrial Wastewater, XXVI, 585
Infectious Waste, XXVI, 53
Infiltration, XXII, 197, 321, 378,
    580, 594
Influent, XXIII, 111
Injection, XXII, 31, 438, 747
Innovative Technology, XXIV, 468
Inorganic Solids, XXV, 248
Intake, XXVI, 678
Investigation Exploration, XXII, 40,
    108, 118, 270, 292, 303, 321,
    425, 470
IRELAND, J. S., XXII, 501
Irradiation, XXII, 95
Irreversibility, XXV, 409
Isotope, XXVI, 69

JACKSON, D., XXII, 453
JAMIL, M., XXV, 283
JAMIL, M., XXVI, 250, 259, 619
JANWADKAR, A., XXV, 561
JARDIM, W. F., XXVI, 230
JENNINGS, A. A., XXIV, 216
JENSEN, J. N., XXIV, 421
JERGER, S. W., XXII, 425
Jet-A, XXV, 330
JOHNSON, J. H., XXIII, 247
JOHNSON, J. H., JR., XXV, 439
JOHNSON, P. C., XXVI, 462
JOHNSON, P. W., XXIII, 337
JOHNSON, T. A., XXVI, 123
JOHNSON, T. M., XXII, 269
JONES, D., XXIV, 135
JONES, E. G., XXVI, 678

JONES, J. A., XXIV, 238
JONES, K. D., XXV, 228
JONES, N. L., XXVI, 347
JOURNICK, F. X., JR., XXV, 295
JUPIN, R. J., XXIII, 96

KAO, M.-M., XXV, 146
KEENER, T., XXIII, 385
KELLER, E., XXIV, 652
KELLY, R., XXII, 3
KHODADOUST, P., XXV, 209
KIM, B.-J., XXVI, 115
KIM, R. P., XXIV, 300
KIM, S., XXV, 458
Kinetics, XXIV, 421
Kinetics, XXV, 168
KLOEKER, J., XXVI, 575
KODUKULA, P. S., XXIV, 245
KONG, W. S., XXIV, 135
KOPCOW, D. R., XXVI, 629
KORNHAUSER, A. A., XXVI, 53
KU, Y.-C., XXVI, 446
KUMAR, S., XXVI, 170
KUNCHAKARRA, P. V., XXVI, 96
KUNTZ, C. S., XXIV, 207
KURUCZ, C. N., XXII, 95
KURUCZ, L., XXV, 500
KYLES, J. H., XXIV, 437

Laboratory, XXV, 348
LACKEMACHER, P., XXIV, 50
LAI, M. S., XXIV, 421
LAIRD, H. S., XXII, 321
Land Application, XXII, 14
Land Disposal, XXII, 31, 130, 321,
    377, 521, 570, 580, 594, 613
Land Disposal, XXIII, 225
Land Disposal, XXVI, 13
Land Treatment, XXII, 425
Land Treatment, XXVI, 541
Landfarming, XXIII, 259
Landfill, XXII, 14, 129, 156, 195, 213,
    377, 412, 470, 559, 580, 594, 613
Landfill, XXIV, 665
Landfill, XXV, 561
Landfill, XXVI, 388
Landfill Cap Cover, XXII, 215, 251,
    571, 580, 594
Landfill Leachate, XXIII, 127
Landfills, XXIII, 75, 293
LANDINE, R. C., XXIII, 267
Land Spreading, XXV, 419
LANG, L. J., XXV, 295

LaPlante, C., XXII, 580
Larsen, B., XXIII, 24
Larson, A. C., XXV, 523
Latex, XXIII, 337
Law, XXV, 295
LDR, XXII, 31
Leachate, XXII, 195, 213, 321, 377, 412, 594, 613
Leachate, XXV, 197, 506, 561
Leachate, XXVI, 141, 163
Leachate Collection, XXII, 215, 321, 559
Leachate Treatment, XXII, 377, 412, 568, 645
Leaching, XXIII, 385
Leaching, XXV, 23
Lead, XXIV, 101, 257, 341, 359
Lead, XXV, 13, 55, 72, 93, 105, 283, 310
Lead, XXVI, 180, 250, 259, 425, 619
Lead Contaminated Particulates, XXVI, 678
Lead Contaminated Soil, XXVI, 514
Lead Removal, XXVI, 506, 619
Leakage Factor, XXIV, 27
Leakance, XXIV, 27
Lee, C.-M., XXVI, 647
Lee, D.-Y., XXV, 168
Lee, J., XXIV, 359
Lee, S. H., XXIII, 198
Lee, S. H., XXVI, 733
Lees, R. E., XXII, 645
Lees, R. E., XXIII, 306
Legal, XXV, 295
Levine, A. D., XXVI, 141
Lewandowski, J., XXIV, 17
Lewis, D. A., XXVI, 691
Liability, XXIII, 306
Liao, C.-H., XXVI, 647
Liaw, C.-T., XXV, 366
LIMB, XXIII, 385
Lime, XXII, 144, 155, 171, 252, 506, 529, 594, 613
Lime, XXVI, 109
Lin, C.-C., XXV, 3
Lin, C. F., XXVI, 61
Lin, C.-F., XXV, 168
Lin, C.-K., XXV, 3
Lin, H.-C. J., XXVI, 347
Lin, W., XXIV, 563
Lindhult, E. C., XXII, 303
Lindhult, E. C., XXIV, 484
Lindner, N. J., XXIV, 58

Liner, XXIV, 665
Liner Systems, XXVI, 42
Liners, XXII, 66, 195, 213, 251, 426, 559
Liners, XXIII, 234
Liou, M.-R., XXVI, 275
Liquid Chlorine Eye Contact, XXIII, 348
Liquid Chlorine Skin Contact, XXIII, 348
Liu, B.-W., XXVI, 163
Liu, C.-B., XXVI, 275
Liu, J. C., XXV, 72
Lo, K. S., XXV, 168
Lowe, W. L., XXIV, 515
Low Temperature Desorption, XXVI, 267
Low Temperature Thermal Treatment, XXV, 83
Loyd, C. K., XXIV, 593
Lu, C.-J., XXVI, 647
Lu, M.-C., XXV, 266
Lucy, J. T., XXIII, 137
Lukas, M. J., XXVI, 13
Lumpkin, P. D., XXVI, 338

Magley, L. M., XXII, 569
Maloney, S. W., XXIV, 110, 573
Mann, T. M., XXVI, 28
Marks, S. D., XXIV, 207
Marley, M. C., XXIV, 468
Martinelli, D., XXV, 321
Masten, S. J., XXIV, 411, 642
Mathavan, G. N., XXIV, 610
Matsik, G. A., XXIII, 163
Matsumoto, M. R., XXV, 93, 310
McFarland, M. A., XXVI, 528
McGhee, T. J., XXII, 66
McGill, W. A., XXIV, 505
McIlvain, J. B., XXIII, 283
McKeown, J. J., XXII, 223
Medical Waste, XXVI, 53
Megahan, H., XXV, 330
Membrane Separation, XXV, 126
Memon, A. A., XXII, 559
Mercaptan, XXVI, 602
Merkle, P. B., XXV, 531
Mercury Fulminate, XXII, 114
Metal Finishing, XXII, 129
Metal Recovery, XXVI, 557
Metal Removal, XXIV, 101
Metal Removal, XXV, 146
Metal Toxicity, XXV, 541

943

Metals, XXIII, 385
Metals, XXV, 419
Metals, XXVI, 657, 667, 675, 691
Metals, Heavy, XXII, 32, 39, 50, 77,
    438, 485, 526, 613, 694
Metals Removal, XXIII, 24, 137
Methanogenic, XXV, 551
MICHELSEN, D. L., XXIV, 593
MICHELSEN, D. L., XXV, 218
Microorganisms, XXII, 57, 359, 379,
    626, 645, 669
Microtox Bioassay, XXV, 266
MIDDLEROOKS, E. J., XXVI, 69
Migration, XXIII, 106
Migration, XXVI, 733
MILLER, D. L., XXIII, 148
MILLER, R. E., XXIV, 391
MILLER, R. L., XXVI, 667
MIODUSKI, K. A., XXVI, 194
MITAL, H. K., XXVI, 637
Mixed Liquor, XXV, 248
Mobile, XXV, 348
Modeling, XXV, 449, 478
Modeling, XXVI, 87
Modelling, XXIII, 127, 329
Modelling, XXIV, 368
Models, XXII, 195, 223, 626, 645, 692
MOHSIN, M. B., XXV, 439
MONADJEMI, P., XXVI, 109
Monitoring, XXII, 14, 118;
    Wells, 14, 50, 218, 290, 303
MONSERRATE, M. L., XXII, 103
MOORE, R., XXVI, 425
MOORE, R. E., XXIV, 207
MOORE, R. E., XXV, 310
MOORIS, R. L., XXIV, 290
MOOSE, R. D., XXII, 303
MORRIS, F. R., XXVI, 96
MORRIS, T. L., XXVI, 362
MORRISON, D., XXIII, 175
MORRISON, M., XXV, 381
Municipal Solid Waste, XXV, 429, 561
Municipal Waste, XXVI, 388
Municipal Wastewater Treatment
    Plant, XXIII, 111
Munitions, XXVI, 28
Mutagenicity, XXV, 42

N-Nitrosodimethylamine, XXVI, 285
NANGUNOORI, R. K., XXIV, 349
NARKIS, N., XXVI, 241
Negotiation, XXVI, 404
NELSON, J. D., XXII, 570

NEUFELD, R. D., XXII, 669
NEUFELD, R. D., XXIII, 96, 111
Neutralization, XXII, 129, 136, 152,
    341, 613
New Jersey Harbor, XXVI, 667, 675,
    691
New Jersey Law, XXV, 295
New York Harbor, XXVI, 657, 667,
    675, 691
NHLOAD, XXVI, 371
NIACKI, S., XXII, 669
Nickel, XXIV, 257
Nickel Hydroxide, XXII, 485
NIMBY, XXII, 3, 745
Nitrification, XXIII, 175
Nitrification, XXVI, 123, 396
Nitrobacter, 396
Nitrocellulose, XXV, 338
Nitrocellulose, XXVI, 115
Nitrotoluene, XXVI, 170
Nocardia, XXV, 159
Non-Aqueous Phase Liquid (NAPL),
    XXIII, 106
Non-Woven Geotextile, XXVI, 34
NONAVINAKERE, S., XXV, 13
NORIEGA, M. D., XXII, 521
NOYES, G., XXV, 429
NPDES, XXII, 13, 144, 340, 355
NPDES, XXIII, 148
NPDES, XXIV, 79
Nuclides, XXVI, 733
Nutrients, XXII, 57, 348, 414


ODE, R. H., XXIV, 290
Odors, XXII, 503
OGDEN, D. A., XXV, 178
OH Radicals, XXIV, 411
Oil, XXV, 500
Oil Contaminated Soil, 446
Oil Pollution, XXIV, 610
Oil-Sorbed Peat, XXIV, 610
OLENBUTTEL, R. F., XXIV, 238
OLIX, W. F., XXII, 569
OLSON, C. L., XXII, 103
On-Site, XXV, 348
ONG, S. K., XXVI, 678
Onsite Treatment, XXII, 41, 413
ONUSKA, J. C., XXIV, 257
Open-Burning, XXV, 338
Operating Parameters, XXVI, 417
Optimization, XXVI, 221
Organic Amines, XXIV, 391

944

Organic Compounds, XXII, 57, 95, 355, 626, 655, 669
Organic Content, Soil, Sediment, XXII, 224, 594, 702, 769
Organics, XXIV, 515
Organics, XXV, 257
Orphan Waste, XXV, 305
ORSHANSKY, F.,241
Oxidation, XXIII, 198
Oxidation, XXV, 275
Oxidation, XXVI, 194, 212, 585
Oxidation/Reduction, XXII, 57, 96, 616, 769
Oxygen, XXIII, 317
Ozonation, XXIV, 642
Ozonation, XXV, 491
Ozone, XXIV, 135, 411, 642
Ozone/Peroxide, XXIV, 411
Ozone/UV, XXIV, 411
Ozonolysis, XXVI, 221

PACT, XXVI, 241
PADAKI, M., XXV, 218
PAH, XXIII, 247
PAH, XXV, 42, 449
PAH's (PNA's), XXII, 453
PAH's, XXIV, 642
Paint Solids, XXIII, 52
Palmerton Zinc Site, XXV, 65
PAMUKCU, S., XXII, 155
Papermill Waste, XXV, 359
PARATE, N. S., XXIII, 67
PARIKH, U. R., XXVI, 575
PARK, S.-W., XXVI, 733
PASIN, B., XXIII, 359
PAVLOSTATHIS, S. G., XXIII, 175
PAWLOWSKI, T., XXIV, 529
Pb-naphthalene, XXVI, 425
PCB, XXV, 449
PCB's, XXII, 33
PCP, XXII, 57
PCP, XXVI, 637
Peat, XXIV, 610
Peat, XXVI, 311
PEDERSEN, T., XXV, 330
PENNINGTON, J. C., XXVI, 285
Pentachlorophenol, XXV, 209
Pentachlorophenol, XXVI, 637
Pentachlorophenol (PCP), XXIII, 211
Performance, XXIII, 111
Permangante, XXIV, 135
Permeability, XXII, 66, 155, 195, 254, 290, 377, 559, 580, 594, 615

Permits, XXII, 13, 41
Pesticide, XXVI, 454
Pesticides, XXV, 83
Pesticides, XXIV, 515
PETERS, R. W., XXII, 77, 718
PETERS, R. W., XXIII, 37
PETERS, R. W., XXIV, 718
PETERSON, W. A., XXVI, 123
Petroleum, XXIII, 259
Petroleum Hydrocarbons, XXIV, 179
Petroleum Refining, XXII, 355, 399, 425, 542, 594
Petroleum Refining, XXVI, 557
Petroleum Spills, XXIII, 106
PFRP, XXV, 409
Pharmaceutical Industry, XXVI, 557
Phenol, XXVI, 241, 259, 311
Phenol, XXIII, 374
Phenol, XXIV, 563
Phenols, XXII, 355, 398, 453
Phenols, XXV, 13, 116, 187
Photocatalysis, XXV, 266, 275
Photocatalysis, XXVI, 163, 170, 230
Photocatalytic Degradation, XXIII, 368
Photodegradation, XXIII, 359
Photographic Waste, XXIII, 175
Photooxidation, XXV, 136
Photooxidation, XXVI, 186
PHULL, K. K., XXIV, 110
Physical-Chemical Treatment, XXII, 521
Pickling Acid, XXII, 136
Pilot Plant, XXII, 7, 357, 388, 400, 448, 470
Pilot Plant Study, XXIII, 267
PINTO, D., XXIII, 368
Plant Survey, XXV, 159
Plastics, XXVI, 557
Plume, XXII, 20, 213, 269
Pollution Control, XXVI, 657
Pollution Prevention, XXV, 389
Pollution Prevention, XXVI, 380, 549, 575
Polymer, XXVI, 303
Polymer Modification, XXV, 23
Polymerization, XXV, 116
Polymers, XXIV, 39
Polystyrene, XXV, 42
POON, C. P. C., XXII, 613
POPOVICS, S., XXII, 243
PORI ST Technology, XXV, 409
Post Closure, XXII, 14, 570, 580, 594

Potato Processing Plant, XXIII, 267
POTW, XXII, 78, 130, 382, 413, 472, 645
POTWs, XXVI, 362, 701
Poultry Processing, XXIV, 79
POURHASHEMI, S. A., XXIII, 317
POWELL, G., XXII, 438
POWELL, G., XXIV, 368
Pozzolan, XXII, 156, 172, 215, 542, 594, 613
Pozzolanic, XXIII, 3
Pozzolanic Chemistry, XXIV, 341
Precipitation, XXII, 129, 382, 456, 487
Precipitation, XXIII, 137
Precipitation, XXV, 366
Predictive Modeling, XXIV, 157, 277
Pretreatment, XXII, 129, 412
PRICE, T., XXII, 425
PRINCE, M., XXVI, 522
Prioritization, XXVI, 404
Priority Pollutants, XXII, 38, 57
Priority Pollutants, XXVI, 178
Probability, XXII, 195, 683
Probability Distribution, XXIII, 111
Process Water, XXII, 129, 145, 216, 340, 669
Project Management, XXV, 458
Propoxur, XXV, 266
PSARIS, P. J., XXII, 521
Public Relations, XXII, 3, 21, 377
Publicly Owned Treatment Works, XXVI, 362, 701
Pulp and Paper, XXII, 5, 225, 425, 645
Pulse Drying, XXIV, 58
Pump Test, XXIV, 27
Pump-and-Treat, XXVI, 20, 77
Pyrene, XXIII, 247
Pyritic Coal, XXVI, 715
Pyrometallurgical, XXIV, 257

Quality Models, XXIII, 96
QUIMBY, J. L., XXIV, 602

RABOSKY, J. G., XXV, 257
RACHELSON, M. L., XXII, 66
Radioactivity, XXIII, 67
Radon, XXII, 702
RAIDER, R. L., XXIV, 381
RANDALL, P. R., XXIV, 238
RANDALL, T. L., XXVI, 602
RANDOLPH, S. A., XXIII, 234

RAO, G., XXV, 458
RAO, M. G., XXIII, 317
RASCHKE, S. A., XXIII, 234
RAVIKUMAR, J. X., XXII, 57
RAVIKUMAR, J. X., XXIII, 211
RCRA, XXII, 31, 42, 43, 129, 213, 580, 745
RCRA, XXIII, 225
RCRA, XXIV, 665
RCRA, XXV, 178
RCRA—Liste "K" and "F" Wastes, XXVI, 541
RDF, XXV, 197
RDX, XXIII, 188
Real Estate Transaction, XXIII, 306
Recalcitrance, XXII, 57, 95, 680
Recharge Trench, XXVI, 354
Recharge Wells, XXVI, 354
Reclamation, XXII, 570
Reclamation, XXV, 468
Recovery, XXII, 303, 438
Recovery, XXV, 366
Recovery XXVI, 28
Recycle Materials, XXVI, 549
Recycled-Based Manufacturing, XXVI, 549
Recycling, XXIV, 238, 257
Recycling, XXVI, 549
Recycling Reuse, XXII, 145, 155, 171, 243, 760
REDDY, V., XXV, 458
REED, B. E., XXIV, 101
REED, B. E., XXV, 13, 55, 93, 105, 283, 310, 478
REED, B. E., XXVI, 250, 259, 425, 436, 480, 514, 619, 715
REED, J. R., XXIV, 79
Refractory Toxicity Assessment, XXVI, 362
Refuse, XXIV, 529
REGAN, R. W., XXIII, 106
Regeneration, XXV, 283
Regeneration, XXVI, 250
Regulations, XXII, 3, 14, 31, 43, 103, 129, 144, 330, 340, 355, 470, 530, 570, 613, 683, 743, 761
Regulatory Interaction, XXVI, 404
REINHOLTZ, C. F., XXVI, 53
Remedial Investigation, XXVI, 96
Remediation, XXIV, 468, 515, 621, 642
Remediation, XXV, 228, 295, 338
Remediation, XXVI, 3, 13, 20, 77, 321, 446, 637

Remediation Cleanup, Corrective Action,
  Mitigation, XXII, 10, 15, 26, 31, 37,
  43, 103, 269, 290, 303, 321, 425,
  438, 453, 683
Remining, XXV, 468
Removal, XXII, 438, 453
Removal Efficiency, XXVI, 417
REN, Y. G., XXVI, 657
Research, XXII, 718
Research, XXV, 458
Respirometry, XXIV, 157, 277
Respirometry, XXVI, 241
Restrictions, XXVI, 13
Resuscitation, XXVI, 151
Retrofit, XXII, 129, 144, 416
Reuse, XXV, 381
Reverse Osmosis, XXII, 142
REZNIK, Y. M., XXII, 118
REZNIK, Y. M., XXIII, 75
RFA, XXII, 43, 46
RFI/CMS, XXII, 43, 49
RICHARDS, D. R., XXVI, 347
RICKABAUGH, J., XXIII, 359, 368
RICKABAUGH, J., XXIV, 652
RICKABAUGH, J., XXV, 429
RI/FS, XXII, 103, 321, 453
Risk, XXII, 4, 16, 195, 223, 683
Risk, XXV, 478
Risk, XXVI, 678
Risk Analysis, XXIII, 127, 306
Risk Assessment XXVI, 13
Risk Levels, XXVI, 3
Rivers, XXII, 223, 340, 448
RIZVI, R., XXII, 77
Road Construction, XXIII, 11
ROAM, G.-D., XXV, 266
ROBERTSON, J. P., XXVI, 250
Robotics, XXV, 324
Rocky Mountain Arsenal, XXVI, 285
ROLL, R. R., XXII, 521
ROMANOW, S., XXV, 83
ROSENBAUM, W. E., XXIV, 621
ROSS, D., XXV, 238
ROSSI, A., XXVI, 611
Rotary Kiln, XXII, 743
Rotating Biological Contractor, XXII, 398
ROTH, M. J. S., XXV, 65
Rotor, XXIV, 50
ROZICH, A. F., XXIV, 157, 277
RTA, XXVI, 362
RUNNER, M., XXIV, 687

s-triazine, XXIV, 421

SACK, W. A., XXIV, 312
SACK, W. A., XXV, 197, 228
SACK, W. A., XXVI, 131
Safety, XXV, 305
SAFFERMAN, S. I., XXV, 209
SALT, XXIV, 312
SALVATORI, L. C., XXIII, 306
SALVITO, D., XXVI, 675
Sand Columns, XXVI, 647
Sand Filtration, XXIII, 148
SANDERS, P., XXIV, 359
SANDERS, T. M., XXIII, 337
SANDS, R. N., XXII, 13
Sandy Loam, XXVI, 425
SANIN, S. L., XXVI, 151
SANTOLERI, XXII, 743
SAPIENZA, V., XXVI, 657
SARA, XXII, 10, 751
SAROFF, S., XXIV, 368
SARTAIN, H. S., XXII, 655
SATHIYAKUMAR, N., XXII, 412
SAYLES, G. D., XXV, 551
Scalped, XXIV, 50
SCHMITT, T., XXVI, 362
SCHULER, V. J., XXIV, 79
SCM, XXVI, 733
SCOULAR, R. J., XXV, 500
Scrap, XXV, 381
Scrubber, XXIII, 11
Scrubber Sludge Grouts, XXVI, 567
Sea Water, XXVI, 347
Sealant, XXII, 68, 294
SEALS, R. K., XXIII, 11
Sedimentation, XXII, 215, 226, 341, 522
Sediments, XXII, 226, 321
Segregations, XXVI, 575
SEMMENS, M. J., XXV, 136
SENGUPTA, A. K., XXV, 126
SENGUPTA, A. K., XXVI, 293
SENGUPTA, S., XXV, 126
Sensitizers, XXIII, 359
SENTURK, C., XXV, 381
Separation, XXV, 257
SERIE, P. J., XXII, 3
SEX, XXIII, 188
SHAFER, D. R., XXIV, 141
Shallow Groundwater, XXVI, 528
Shallow Soil Mixing, XXIV, 631
SHANG, J., XXIV, 547
SHAW, B. A., XXIV, 505
SHEKHER, C., XXVI, 488
SHEN, H., XXV, 187
SHEN, H.-S., XXV, 146

SHEPHERD, T. A., XXII, 570
SHEM, L., XXIV, 718
Shock Loads, XXIV, 312
SHIEMKE, A. K., XXV, 228
SHIEMKE, A. K., XXVI, 131
SHIN, H. M., 506
SHORTELLE, A., XXII, 223
Shortfall Reducation, XXVI, 557
SHRIVASTAVA, S. R., XXII, 412
SHU, G. Y., XXV, 72
SHU, H.-Y., XXVI, 186
Silicon Slag, XXII, 559
Silt Clay Loam, XXVI, 436
SINGH, A., XXIV, 135
SINGHAI, P., XXV, 105
Site Assessment, XXII, 103, 310
Site Remediation, XXIII, 127
Site Remediation, XXV, 65
Site Stabilization, XXVI, 454
Sludge, XXII, 129, 144, 157, 243, 350,
    398, 414, 453, 470, 485, 501, 521,
    542, 594, 743
Sludge, XXIII, 247
Sludge, XXIV, 39
Sludge, XXV, 23, 126, 419
Sludge, XXVI, 13
Sludge Drying, XXIV, 58
Sludge Filtration, XXII, 480, 486, 521
Sludge Recycle, XXIV, 58
Slurry Wall, XXVI, 87
SMITH, D. E., XXIII, 295
SMITH, D. P., XXVI, 141
SMITH, J. L., XXV, 389
SMITH, J. R., XXIII, 148
SMITH, L. M., XXII, 438
Software, XXV, 389
Soil, XXV, 72, 381
Soil, XXVI, 131, 321, 454, 472
Soil and Groundwater Models, XXIII,
    106, 211
Soil Contamination, XXII, 57, 78,
    103, 155, 271, 303, 425, 438, 459,
    683, 702
Soil Flushing,425, 436
Soil Gas Survey, XXII, 280, 291, 304
Soil Remediation, XXVI, 462, 480,
    514
Soil Vapor, XXV, 283, 330
Soil Vapor Extraction, XXVI, 462, 488
Soil Vapor Remediation, XXVI, 528
Soil Vapor Technology, XXVI, 20
Soil Venting, XXVI, 488
Soil Washing, XXV, 65, 93, 310

Soil Washing, XXII, 459
Soil Washing, XXVI, 506
Solidification/Stabilization, XXIII, 385
Solidification/Stabilization, XXV, 3,
    13, 23, 83
Solids Separation, XXII, 129, 661
Solubility, XXII, 613, 632
Solubilization, XXV, 419
Solvent, XXII, 32, 38, 50, 116,
    269, 304
Solvent Extraction, XXII, 355, 459
Solvents, XXV, 228, 275
SOMERVILLE, T., XXV, 330
Sorption, XXV, 126
Sorption, XXVI, 293, 303
Sorption Phenomena, XXIII, 106
Source Reduction, XXIII, 295
Source Reduction, XXV, 389
Source Reduction, XXVI, 575
SPARKS, D. L., XXIV, 359
SPEECE, R. E., XXII, 626
Spent Caustic, XXVI, 602
Spills, XXII, 57, 110, 157, 225
SRIDHAR, K., XXIII, 175
ST. JOHN, J. P., XXVI, 667
Stabilization, XXII, 41, 156, 171,
    215, 243, 251, 506, 542, 571, 594
Stabilization, XXIII, 3, 11
STALLINGS, T. A., XXVI, 549
Starvation, XXVI, 151
State-of-the-Art, XXVI, 347
STEELE, M. D., XXVI, 321
STEIN, R. M., XXV, 359
STEPHANATOS, B. N., XXII, 683
STILLEY, T. E., XXVI, 404
Stochastic Analysis, XXII, 195, 227,
    683
Strategy, XXVI, 404
Streaming Current, XXIV, 39
Streaming Potentials, XXIII, 75
Strength, Mechanical, XXII, 155, 181,
    542, 594
Stripping, XXIII, 317, 329
STUBIN, A. I., XXVI, 657
STUMPF, A., XXV, 458
Subsurface Cavities, XXVI, 42
Subsurface Contaminant Transport,
    XXIII, 106
SUDANO, P., XXV, 238
SUIDAN, M. T., XXIII, 374
SUIDAN, M. T., XXIV, 573
SUIDAN, M. T., XXV, 116, 209, 551
Sulfate-Reducing, XXV, 541, 551

Sulfide, XXV, 541
Sulfide, XXVI, 602
Sulfide Oxidation, XXVI, 585
Sulfide Precipitation, XXIV, 120
Sulfides, XXII, 146
Sulfite, XXIII, 175
Sulfur Recovery, XXVI, 585
Supercritical, XXVI, 472
Superfund, XXIII, 127, 148
Superfund, XXVI, 77
Superfund Site, XXVI, 637
Surface Catalysis, XXV, 116
Surface Complexation Model, XXVI, 733
Surface Impoundment, XXIII, 24
Surface Impoundment Lagoon, Pond, XXII, 32, 68, 122, 195, 470, 594
Surface Water Quality, XXII, 155, 215, 223, 321
Surfactant, XXVI, 303, 742
SUSARLA, M., XXVI, 715
Suspended Solids, XXII, 144, 655
SVE, XXVI, 462, 488
SWINDOLL, C. M., XXVI, 488
SYLVESTER, K. A., XXVI, 354
SYLVIA, T. E., XXIV, 458

Taguchi-Designed Experiment, XXVI, 221
TAI, F.-J., XXVI, 115
Tailings, XXII, 570
TAKIYAMA, L. R., XXVI, 496, 595, 611
TAKIYAMA, M. M. K., XXVI, 178, 230, 595
TALLEY, W. F., XXII, 759
TANG, W. Z., XXVI, 212
TANSEY, W. J., XXVI, 541
TARSAVAGE, J. M., XXIV, 484
TAX, XXIII, 188
TAYLOS, W. C., XXVI, 380
TCDD, XXV, 478
TCE, XXVI, 230, 522
TCLP, XXII, 35, 40, 249, 544, 594, 613
TCLP, XXIII, 385
TEL, XXVI, 506
Tensile Strain, XXVI, 42
TERRANOVA, N., XXVI, 205
Tetraethyl Lead, XXVI, 506
Tetramethyl Ammonium, XXII, 77
Textile Wastewater, XXV, 218
Thermal Desorption, XXVI, 417, 541
Thermal Process, XXVI, 446

Thermal Treatment, XXII, 43, 536, 613, 743, 759
Thermophilic Biodegradation, XXIII, 247
THIELKER, J. L., XXVI, 96
Thiosulfate, XXIII, 175
THM, XXVI, 205
THMF, XXVI, 205
THOMAS, M. A., XXV, 338
THOMPSON, J. C., XXV, 55
THOMPSON, J. C., XXVI, 436
THOMPSON, J. L., XXII, 501
THUOT, J. R., XXIII, 37
TIEN, C.-T., XXVI, 371
TISCHUK, M. D., XXII, 303
Titanium Dioxide, XXVI, 163, 170, 178, 230, 595
TITTLEBAUM, M. E., XXIII, 11
TJHO, S.-K., XXVI, 557
TNT (trinitrotoluene), XXVI, 61
TOBLIN, A. L., XXVI, 96
TOKUZ, R. Y., XXII, 398
Toluene, XXIV, 583
Toluene, XXVI, 267, 647
TORRENTS, A., XXVI, 61
Tower Biology, XXIV, 290
TOXI4, XXIII, 96
Toxic Chemical, XXV, 31, 266, 275, 491
Toxic Release Inventory, XXII, 5
Toxicity, XXII, 356, 399, 613, 626, 645, 683, 702
Toxicity, XXIII, 188
Toxicity, XXV, 218
Toxicity, XXVI, 678
Trace Metal, XXVI, 691
TRACY, M., XXVI, 575
Transient Simulation, XXIV, 368
Treatability, XXII, 377, 399, 445, 474, 626
Treatability, XXV, 257, 523
Treatment, XXIV, 563, 621
Treatment, XXV, 275, 338
Treatment Standards, XXII, 33, 39, 129, 149, 340, 355, 683
Trends, XXVI, 657
1,3,5-Trichlorobenzene, XXIV, 411
Trichloroethylene, XXIII, 198
Trichloroethylene, XXV, 228
Trichloroethylene, XXVI, 230, 522
Trickling Filter, XXII, 339, 386
TRIMBATH, W. D., XXIII, 84
2,4,6-Trinitrotoluene, XXV, 439

Tritium, XXVI, 69
TROY, M. A., XXII, 425
TROY, M. A., XXIV, 179
TRUAX, D. D., XXII, 655
TSAI, C.-H., XXV, 366
TSENG, J. M., XXIV, 120
TSENG, P.-C., XXVI, 163
TUNNEL, D. M., XXV, 65
Two-Phase System, XXV, 491
TYRIAN, G. P., XXV, 359
2,4-Dinitrobenzene, XXIV, 573

UDUJI, C., XXIII, 293
Ultra Violet Light, XXII, 58
Ultrafiltration, XXII, 132
Ultrafiltration, XXVI, 701
Ultrafiltration Treatment Plant,
    XXVI, 701
Ultrasound, XXVI, 178
Ultraviolet, XXVI, 170, 186, 230, 595
Ultraviolet Light, XXIII, 188, 368
Uncertainty, XXV, 458
Underground Storage Tanks, XXII,
    155, 290
Underground Storage Tanks, (UST),
    XXIII, 84, 106
United Kingdom, XXVI, 321
Universities, XXV, 449
Unsaturated Polyester, XXV, 3
Uranium, XXII, 570
USMEN, M. A., XXII, 171
Utilization, XXIII, 11
UTP, XXVI, 701
UV Light, XXIII, 359

Vacuum Extraction, XXII, 269, 304
Vacuum Extraction, XXV, 238
Vadose Zone, XXII, 58, 78, 269
Vadose Zone, XXVI, 528
Vapor, XXII, 269, 290, 501
Vapor Extraction, XXVI, 462
Variability, XXV, 458
VARUNTANYA, C., XXIV, 141
VARUNTANYA, C. P., XXVI, 701
VEIL, J. A., XXV, 468
VERON, XXII, 377
VIDIC, R. D., XXIII, 374
VIDIC, R. D., XXV, 116
VIRARAGHAVAN, T., XXIV, 610
VIRARAGHAVAN, T., XXV, 500
VIRARAGHAVAN, T., XXVI, 311
Vitrification, XXII, 759
VOC, XXIII, 329

VOC, XXV, 31, 136
VOC, XXVI, 629, 637
VOC's, XXII, 37, 50, 269, 291,
    416, 438, 669
VOHRA, M. S., XXV, 275
Volatile Acids, XXIV, 602
Volatile Organic Carbons, XXIV, 468
Volatile Organic Compounds, XXVI,
    629, 637
Volume Reduction, XXV, 409
VORUGANTI, R. S., XXVI, 53

WACHOB, B. G., XXVI, 285
WAGNER, J. A., XXV, 209
WAITE, T. D., XXII, 95
WALKER, A. D., XXIV, 631
WALKER, M. L., XXVI, 575
WALKER, T. E., XXVI, 53
WALSH, M. T., XXIV, 468
WALTER, D. K., XXIV, 731
WAN, L., 247
WAN, L. W., XXV, 439
WANG, J., XXVI, 724
WANG, Y.-T., XXV, 187
WANG, Z., XXV, 366
WANTLAND, G., XXIV, 665
WARNACUT, W., XXIV, 539
WASHBURN, S. T., XXIV, 69
WASP4, XXIII, 96
Waste, XXV, 178
Waste Combustion, XXV, 197
Waste Contaminants, XXIII, 234
Waste Dewatering, XXIV, 58
Waste Fixation, XXIII, 3
Waste Load Allocation, XXVI, 667
Waste Management, XXVI, 380
Waste Minimization, XXII, 129, 340,
    718, 743, 759
Waste Minimization, XXIII, 295
Waste Minimization, XXV, 359, 389
Waste Minimization, XXVI, 380, 541,
    557, 575
Waste Reuse/Recycling, XXII, 144,
    155, 171
Waste Segregation, XXII, 140, 356, 718
Waste Tires, XXV, 381
Waste Water, XXV, 146, 248
Wastewater, XXIII, 188
Wastewater, XXVI, 311, 371, 585
Wastewater Effluents, XXII, 13, 34,
    66, 95, 109, 129, 144, 339, 355,
    358, 413, 453, 459, 521, 645,
    655, 669

Wastewater Plant, XXVI, 329
Wastewater Treatment Plant, XXVI, 742
Water Balance, XXV, 359
Water Plant, XXVI, 329
Water Pollution, XXII, 13, 213, 223, 412, 655
Water Quality, XXV, 468
Water Quality, XXVI, 675
Water Recycling, XXVI, 541
WATSON, R. P., XXV, 506
WATTRAS, R. P., XXIII, 84
WEI, S.-M., XXV, 366
WENG, C. H., XXVI, 496
WENTZ, C. A., XXII, 718
WEST, D. A., XXVI, 691
WESTON, R. F., XXIII, 225
Wet Oxidation, XXVI, 602
Wet Scrubber, XXVI, 595, 611
Wetlands, XXV, 561
Wheat Starch Waste, XXVI, 109
Whey, XXII, 66
WHITEMAN, C. S., XXVI, 131
WIK, J. D., XXIII, 259
WILSON, D. S., XXIII, 127
WILSON, D. S., XXIV, 17
WILSON, E., XXVI, 362
WILSON, K., XXII, 144
WILSON, R., XXV, 13

WOKASIEN, S., XXVI, 724
WOODFORD, J., XXVI, 619
WOODFORD, J. S., XXVI, 259
WOODHULL, P. M., XXII, 425

Xenobiotics, XXVI, 151

YANCHESKI, XXVI, T. B., 528
YANG, G. C. C., XXV, 23
YANG, G. C. C., XXVI, 446
YANG, H.-M., XXV, 366
YANG, J. Y., XXIV, 368
YANG, W.-C., XXV, 366
YAPIJAKIS, C., XXV, 419
Yard Waste, XXV, 429
YEH, G.-T., XXVI, 347
YOU, J.-H., XXV, 42
YUSUF, D. S., XXIII, 247

ZANIKOS, I., XXVI, 13
ZANONI, C., XXV, 419
ZARLINSKI, S. J., XXII, 542
ZHANG, M., XXVI, 77
ZIMMIE, T. F., XXII, 580
Zinc, XXIV, 120
Zinc, XXV, 72
ZITOMER, D., XXII, 626
ZUPKO, A. J., XXIV, 515